Advances in Science, Technology & Innovation

IEREK Interdisciplinary Series for Sustainable Development

Advances in Science, Technology & Innovation (ASTI) is a series of peer-reviewed books based on important emerging research that redefines the current disciplinary boundaries in science, technology and innovation (STI) in order to develop integrated concepts for sustainable development. It not only discusses the progress made towards securing more resources, allocating smarter solutions, and rebalancing the relationship between nature and people, but also provides in-depth insights from comprehensive research that addresses the **17 sustainable development goals (SDGs)** as set out by the UN for 2030.

The series draws on the best research papers from various IEREK and other international conferences to promote the creation and development of viable solutions for a **sustainable future and a positive societal** transformation with the help of integrated and innovative science-based approaches. Including interdisciplinary contributions, it presents innovative approaches and highlights how they can best support both economic and sustainable development, through better use of data, more effective institutions, and global, local and individual action, for the welfare of all societies.

The series particularly features conceptual and empirical contributions from various interrelated fields of science, technology and innovation, with an emphasis on digital transformation, that focus on providing practical solutions to **ensure food, water and energy security to achieve the SDGs.** It also presents new case studies offering concrete examples of how to resolve sustainable urbanization and environmental issues in different regions of the world.

The series is intended for professionals in research and teaching, consultancies and industry, and government and international organizations. Published in collaboration with IEREK, the Springer ASTI series will acquaint readers with essential new studies in STI for sustainable development.

More information about this series at http://www.springer.com/series/15883

Abdelazim M. Negm
Editor

Flash Floods in Egypt

Editor
Abdelazim M. Negm
Faculty of Engineering
Zagazig University Faculty of Engineering
Zagazig, Egypt

ISSN 2522-8714 ISSN 2522-8722 (electronic)
Advances in Science, Technology & Innovation
IEREK Interdisciplinary Series for Sustainable Development
ISBN 978-3-030-29637-7 ISBN 978-3-030-29635-3 (eBook)
https://doi.org/10.1007/978-3-030-29635-3

This Springer imprint is published by the registered company Springer Nature Switzerland AG
The registered company address is: Gewerbestrasse 11, 6330 Cham, Switzerland

Preface

Due to the scarcity of water in Egypt, the national water research plan of Egypt 2017–2037 focuses on using each drop of water including the rainfall water and flash floods water, in addition to the reuse of wastewater to close the gap between the demand and the supply. Springer published several books on the different aspects of water resources in Egypt without sufficient information on flash floods. Therefore, the idea of publishing a book on the flash floods comes to complete the picture of the water resources in Egypt. Why Egypt? Because it is a good example of arid countries located in MENA which are also almost arid regions. Consequently, the lessons learned from this book could be of great benefit for other countries in MENA regions particularly those in north Africa.

This book is divided into 17 chapters, written by more than 25 researchers, scientists, and hydrologist experts from several institutions and more than one country, and have a good experience in the domain of water resources and particularly the topic of flash floods and rainfall. One goal of the book is to assess the flash floods environment, its hazards, risk, harvesting, utilization, and management in Egypt as a typical example of an arid country with water scarcity. The book also highlights some important experiences and successful case studies that reflect the present challenges and their possible solutions.

The 17 chapters are presented in 5 parts in addition to the Introduction and Conclusions chapters. Part II is about Analysis and Design Aspects of the flash floods, while Part III is about the Recent Technologies for Investigating Flash Flood. The Environmental Approaches in Flash Flood are covered in Part IV. Part V discusses the Hazards, Risk, Harvesting, and Utilization of Flash Floods and Part VI is devoted to Prediction and Mitigation of Flash Flood in Egypt. The introduction chapter presents a summary of the technical elements of each chapter, while the conclusions chapter summarizes the most important conclusions and recommendations of the book in addition to presenting an update of the literature on flash floods.

The editor wants to introduce his great thanks to all who contributed in one way or another to make this unique, high-quality book a real source of knowledge and latest findings in the field of flash floods in Egypt. He appreciates too much the contributions of all authors. Without their patience and efforts in writing and revising the different versions of the manuscripts to satisfy the high-quality standards of Springer, it would not have been possible to produce this high-quality book. Many thanks are also owed to the editors of the series and his team for constructive comments, advice, and critical reviews. Acknowledgments must be extended to include all members of the Springer team who have worked long and hard to produce this book and make it a reality for the researchers, graduate students, and scientists in Egypt and worldwide, particularly the countries with similar aridity conditions.

The book editor would be pleased to receive any comments to improve future editions. Comments, feedback, suggestions for improvement, or new chapters for the next editions are welcomed and should be sent directly to the volume editor. The emails of the editor can be found inside the book in several chapters.

Zagazig, Egypt
November 2019

Abdelazim M. Negm

Contents

Hazards, Risk, Harvesting and Utilization of Flash Floods

Mitigation of Flash Flood

Conclusions

Introduction

Introduction to "Flash Floods in Egypt"

Abdelazim M. Negm and El-Sayed E. Omran

Abstract

This chapter highlights the hydrogeological technical elements contained in the 17 chapters presented in the book according to its five themes. Therefore, this chapter contains information on analysis: extreme rainfall analysis and critical design aspects of flash floods, recent technologies for investigating flash flood, environmental approaches in flash flood, hazards, risk and utilization of flash floods, and prediction and mitigation of flash flood.

Keywords

Egypt • Sinai • Aswan • Extreme rainfall • Statistical analysis • Design aspects • Risk analysis • Hazard • Prediction • Harvesting • Early warning system • Monitoring • Environment • Assessment • Mitigation • Management

1 Background

Egypt which is located in the north of Africa is an arid country. Its water demand is now more than 100 billion cubic meters per year and about 50% of this demand is fulfilled form the Nile River waters. With the rapid growth of population, agricultural expansion, among other activities, the water demand increased. Egypt is planning to reuse the agriculture drainage water and the treated wastewater, and to harvest the flash floods from everywhere in Egypt as well. The book could thus greatly assist decision-makers in maximizing storing and using water from the harvest flash floods. From our results, climate change seems to affect patterns of rainfall. Therefore, Egypt received a different rainfall pattern for the last few years. One of these observations is recognized in the Aswan governorate in October 2019. It is documented in this book to help the concerned authorities to harvest, manage, and ultimately utilize the harvested heavy rains for the benefits of the communities in Aswan. The presented materials in the book are useful for different arid countries in the MENA regions as well. This book could be considered a follower to the previously published books by Springer on the water resources and agriculture in Egypt, Negm (2019a, b).

2 Themes of the Book

Therefore, the book intends to address the following main theme.

- Analysis and Design Aspects
- Recent Technologies For Investigating Flash Flood
- Environmental Approaches In Flash Flood
- Hazards, Risk, Harvesting, and Utilization of Flash Floods
- Prediction, Mitigation, and Management of Flash Flood.

3 Chapters' Summary

The next subsections present the main technical elements of each chapter under its relevant theme briefly.

A. M. Negm (✉)
Water and Water Structures Engineering Department, Faculty of Engineering, Zagazig University, Zagazig, 44519, Egypt
e-mail: amnegm85@yahoo.com; amnegm@zu.edu.eg

E.-S. E. Omran
Soil and Water Department, Faculty of Agriculture, Suez Canal University, Ismailia, 41522, Egypt
e-mail: ee.omran@gmail.com

E.-S. E. Omran
Institute of African Research and Studies, Nile Basin Countries. Aswan University, Aswan, Egypt

A. M. Negm (ed.), *Flash Floods in Egypt*, Advances in Science, Technology & Innovation,
https://doi.org/10.1007/978-3-030-29635-3_1

3.1 Analysis and Design Aspects

This theme is covered in three chapters. Chapter one deals with the analysis of extreme rainfall events and the critical design aspects of flash floods with examples from Egypt and Oman.

Chapter 2 is titled "Statistical Behavior of Rainfall in Egypt." In this chapter, the statistical characteristics of extreme rainfall in Egypt have been investigated based on historical daily rainfall records for 30 stations throughout the country. Six types of rainfall data from daily records have been collected: monthly rainfall, annual rainfall, the monthly number of rainy days, the annual number of rainy days, monthly maximum daily rainfall, and annual maximum daily rainfall. Generally, there is a great variation in all different aspects of the rainfall across the country. The rainfall indices have higher values in the north of the country than those in the middle and the south. The results indicate that Egypt's rainy season extends from October to March when a significant amount of precipitation is received by some stations. The dry season, on the other hand, extends from April to September. Furthermore, this research seeks to derive the design rainfalls through the GEV distribution along with the L-moments using annual maximum daily rainfalls in these stations. In the case study, the GEV distribution fits well with the annual maximum daily rainfall data. Although all stations not located on the north have low values of rainfall characteristics, some locations (e.g., Hurghada) have high values of design extreme rainfall. Therefore, further studies are required about the study of extreme rainfall at these sites.

Following this, Chap. 3 is concerned with the "Critical Analysis and Design Aspects of Flash Flood in Arid Regions: Cases from Egypt and Oman." It presents a brief review of flood definitions in general and in particular, flash floods. Also, two main case studies related to flash floods are presented in this chapter. The two main case studies from the Arab Republic of Egypt and Sultanate of Oman are investigated. Extreme events caused by cyclonic and convective storms are covered by the cases discussed. The presented cases caused damages to the properties and fatalities. For each case, detailed hydrological and hydraulic analyses carried out and the results are presented. Hydrological and hydraulic analyses are used to study the considered two cases to evaluate the effect of each case and compare it to the regular flood events (if any). Meanwhile, this chapter casts light on the difference between the effect of thunderstorms and cyclones on the corresponding peak flow and flood volume. It showed that cyclonic events were accompanied by a very high value of rain depth, which may be corresponding to thousands of years on the frequency distribution curves. These events of precipitation have created huge volumes of water which fill all dams and run the spillways to near maximum design flow values. The distribution of storms over 24 h, however, was almost uniform, suggesting moderate strength of rainfall.

3.2 Recent Technologies for Investigating Flash Flood

Chapters 4 and 5 cover this theme. Chapter 4 is titled "Developing an Early Warning System for Flash flood in Egypt: Case Study Sinai Peninsula." It explains how to develop an early warning system for flash floods in Sinai Peninsula. An early warning system is essential as flash floods can occur anywhere. When it comes to arid and mountain catchments where catastrophic events are more prevalent than wet and flat areas, it becomes more critical. Egypt can be considered one of the arid and semi-arid Arab countries in the coastal and Nile Wadi systems facing flash floods. In Egypt, particularly in the Eastern Desert, Red Sea Mountains, and Sinai, flash floods are occurring widely.

The Sinai Peninsula can be considered as the most important Egyptian areas common and subject to synoptic circumstances causing heavy rainfall due to its natural geography and context complex, causing severe damage due to flash flooding. This makes the study and analysis of the flash flood on the Sinai Peninsula a crucial and significant job. Because of the above-mentioned challenges, the latest numerical mesoscale meteorological model named the Weather Research and Forecasting (WRF-ARW) is selected to forecast rainfall and simulate the synoptic situation related to some flash flood events over Egypt. Mainly the US National Centre develops WRF for Atmospheric Research (NCAR) in collaboration with many other research centers and universities. WRF allows forecasting weather in complex terrains such as the one in Sinai and at the same time is considered for its orographic features' characteristics.

Also, flash floods occur along the Red Sea, where a series of mountains are there. Chapter 5 is presented to discuss "Flash Flood Investigations in Northern Red Sea Governorate Case El-Gouna." Understanding the flooding and its effects on human life and ecosystem requires fundamental analysis of the boundary conditions that geography, geology, atmosphere, and hydrology usually provide for the proper modeling of the flooding event for forecasting. In this chapter, all of these aspects were systematically analyzed and findings were introduced in a hydrological flow model. A morphological analysis of slopes in correspondence to the geological setting of the catchment of Wadi Bili is presented and the results of all analyses are discussed.

3.3 Environmental Approaches in Flash Flood

Chapters 6–9 address the environmental solution to flash floods. Chapter 6 is titled "Environmental Flash Flood Management in Egypt." It is well know that flash floods have an adverse impact on human health and ecosystems. This may be extended to cause damage to infrastructure, livestock, and plants. Floodwater can cause a number of significant health effects, including deaths and/or injuries, contamination of drinking water, loss of electricity, disruption of the environment, and displacement among other effects. Therefore, this chapter will present the Egyptian national environmental action plan to face the natural environmental hazards (earthquakes–flash floods–dust and sand storm). Moreover, it will highlight the environmental and disaster risk reduction in Egypt and Arab Countries, and present the tools for implementing disaster risk reduction in environmentally hazardous zones.

The Egyptian experience with the floods has shown that much is remaining to be done to improve the use of water resources and protect our infrastructure that crosses dangerous areas. However, there are major challenges in the region, such as lack of continuous data for both flow and precipitation; lack of real-time data transmission; soil erosion problems in mountain areas; maximization of flood uses rather than flood risk management; the absence of a flash flood early warning system; the absence of a disaster risk management plan; little interest in the users of the flood; and the availability of a model to describe the hydrological conditions of the Arab region.

Due to the environmental problems of the flash flood, harvesting and management are essential. Chapter 7 with the title "Flash flood management and harvesting via groundwater recharging in wadi systems: An integrative approach of remote sensing and direct current resistivity techniques" aims to discuss how to (i) mitigate the flashflood hazards and assess the water resources in wadi systems using an integrative approach of remote sensing (RS), geomorphological, and geophysical data. The chapter addressed two case studies in two Egyptian wadis in which we use an innovative integrative approach to flash flood research and management. In workflow design, the chapter introduces a suggested strategy using the surface geological data and geophysical measurements. To reduce the uncertainty of geophysical data inversion, the conventional and non-conventional inversion algorithms are applied. Based on geophysical data inversion results, this chapter shows that it is possible to recognize sites for successful dam construction and groundwater bearing zones exploration. From the hydrogeological point of view, RS and GIS are used to estimate a hydrograph and runoff modeling for wadi systems. Accordingly, the calculated flashflood total discharges and the storage capacity can be recognized. Finally, the chapter provides some solution for scare water management, via locating potential areas suitable for surface water harvesting to promote the percolation of trapped water into the alluvium aquifers.

Harvesting and management of flash floods, and monitoring of needs: Chap. 8 deals with the "Environmental Monitoring and Evaluation of Flash Floods Risks Using Remote Sensing and GIS Techniques." Therefore, this chapter attempts to synthesize the relevant database in a spatial framework to evolve a flood risk map for some study areas with the application of remote sensing and GIS techniques. The study has also focused on the identification of factors controlling flash floods risks in two study areas in Egypt. Satellite images of the study area have been collected and required field data have been gathered. Depending on the digital elevation model (DEM), this study extracts drainage networks and watersheds using ArcGIS Basins' watersheds, drainage network, stream order, flow direction, and other basin characteristics. The watershed model was created in the GIS of the first study area, Firan catchment in Sinai. Hydrological characteristics of basins have been identified. Also, the runoff volume of flash floods has been calculated for the second study area Wadi Hashim 2 in the Egyptian Northern coast through the identification of land use from GIS, soil texture, and rainfall from field measurements. The location of the required Dam for flash floods has been proposed. Finally, GIS and remote sensing techniques have proved to be a successful tool for flash flood monitoring and assessment.

Following the above chapter, Chap. 9 discusses the "Sustainable Development of Mega Drainage Basins of the Eastern Desert of Egypt: Halaib-Shalatin as a Case Study Area." The authors assess the South Eastern Desert of Egypt for the rainfall water harvesting (RWH) capabilities, with the determination of their optimum methods and techniques. Achieving this aim will assist in poverty alleviation, Bedouin and urban allocation, supporting animal husbandry, accelerating agricultural development, improved agricultural and food production for local inhabitants, combating desertification, resolving unemployment problems, and raising individual incomes. Bedouin and natives as the main end-users will be a major target of the project. Innovative ways to improve the capture, storage, and use of rainwater will have their own-bearing on the sustainable and profitable production of dry season vegetable crops in South Eastern Desert. According to the worldwide trends and techniques in RWH, which is applied aggressively in many neighboring countries, Egypt should enter the era of catching every water droplet for domestic and agricultural development. The findings of the current research work could set a good example to be applied both in other parts of the country and around the world. The chapter , therefore, focuses on using effective tools of monitoring and management of natural resources, based on the integration of modern techniques of remote sensing (RS), geographic information systems (GIS), and watershed modeling systems (WMS) to provide a plan for the RWH. Sustainable water supply is vital for the development of communities in arid regions, such as that of the South Eastern Desert of Egypt. The economic importance of the area is enormous, besides the fact that it has long been a target zone for mineral resources excavation and mining. One of the challenges facing this arid area is the limited water resources needed for agricultural, industrial, mining, or domestic uses. Bedouin depends mainly on rainwater, which constitutes the main source feeding their hand-dug wells and fracture springs. Rainwater harvesting (RWH), as a historical and worldwide trend, could fulfill the gap of water scarcity in arid or semi-arid regions. RWH is the accumulation and storage of rainwater for reuse before it reaches the aquifer system (Groundwater). The RWH could be used also for maximizing the recharge possibilities of groundwater. As a non-conventional water resource, RWH could provide water for gardens, livestock, irrigation, mining, cleaning of bathrooms as in the first flush, etc. The collected water is diverted to a deep pit with percolation in many places with similar climatic conditions to refill the groundwater for later use and protection, particularly in accumulations of structurally regulated groundwater. The harvested water could be used as drinking water if the storage is a tank that can be accessed and cleaned when needed. Additionally, the chapter's recommendations will be a good source for the up-to-date databases, which could be used effectively by the decision-makers, researchers, executive authorities, planners, and related governorates.

A second case study titled "Torrents Risk in Aswan Governorate, Egypt" is reported by Prof. El-Sayed Omran, the Dean of the Institute of African Research and Studies and Nile Basin Countries, Aswan University, Aswan, Egypt. In October 2019, Egypt was hit by a flood, which the country has not seen in terms of rising rates for 50 years. Floods started in mid-August on average and then increased in October, with rainfall on the Ethiopian plateau significantly higher than previous rates. Compared to previous years, the water level in Lake Nasser has reached a high level this year, 2019. Torrents and rain helped to raise Lake Nasser's level. This culminated, in the first time employees, were seen in the High Dam terminal (a river port on Lake Nasser, a gateway for passengers and commercial goods between Egypt and Sudan. The construction of the port began in 1964 after the transformation of the Nile River into the construction of the High Dam) the presence of high water in a pavement-side area that has long been totally dry.

Aswan's torrents arise as a result of precipitation on the government's eastern hills, where water flows into a group of valleys west to the Nile. The most prone areas are Wadi al-Sarraj area, Wadi Ajam area, Umm Habbal area, and Wadi Haymour Allaqi area. The streams that flow into Lake Nasser represent the minimum risk of torrents as there are no urban communities. Eastern Nile basins in the area between Edfu and Aswan cities are very risky, particularly in the area of Kom Ombo and east of Aswan city. In May 1979, the flood flow disrupted the railway lines and affected Edfu, Kom Ombo, and Aswan centers. About 300 families have been displaced and the falling of torrential rocks on some parts of the agricultural road and railway lines. Floods were repeated in 1980 and 1987, 2005, 2010, and 2014 (Saber et al. 2017). In 2010, some villages in Aswan city were severely hit by that torrent. About 500 families were evacuated from their houses that were at risk and lost their livestock and harvest, but they were indemnification by the Government through the donation account they have created for the Egyptian flash floods (Al-Momani and Shawaqfah 2013).

3.4 Hazards, Risk, Harvesting, and Utilization of Flash Floods in Sinai and Egypt

On the one hand, Sinai is one of the most vulnerable parts in Egypt, which is subjected to severe flash floods almost every year. On the other hand, Sinai is progressively suffering from an overwhelming water crisis. Flash flood and runoff water could be an answer to this issue. Therefore, this section which consisted of Chaps. 11–14 is devoted to discuss the flash flood issues in Sinai.

Chapter 11 is titled "Egypt's Sinai Desert Cries: Flash Flood Hazard, Vulnerability, and Mitigation." This chapter has three objectives. The first chapter objective is to determine the flood hazard occurrence in the vulnerable areas of Sinai. Remote sensing (RS) and geographic information system (GIS) are utilized to provide improved spatial considerate of basin response to storm rain events and flood monitoring. It is critically essential to precisely predict the occurrence of flash floods in terms of both timing and magnitude.

The second objective is to draw the vulnerability map for several wadis in southern Sinai. Flash floods in Sinai are an inadequately understood feature due to a lack of accurate environmental and hydrological data, which are challenging and expensive to develop and manage in such a region. It is important to understand that risk is determined not only by the climate and weather events, i.e., the hazards but also by the exposure and vulnerability to hazards, which have been induced by human activity. The produced risk map is useful to know the locations that have a high flood risk in order to avoid loss of life and reduce damages to property.

To mitigate flash flood damages and efficiently harvest the highly needed freshwater, the third chapter's objective is to manage and mitigate the flood hazard and minimize their effect. The main watersheds flowing through Sinai are classified into four categories where 4% of watersheds have very high risk, 10% have high risk, 38% have moderate risk, and 48% have moderate to low risk.

Flood risk assessment and flood mapping will help to show which places are most at risk and in what circumstances. After that, governments can take the correct strategy for flood risk reduction or mitigation. Because of the rapidity of flash flood occurrence and its power, flash flood experts recommend the use of early warning systems for reducing vulnerability. Flood risk assessment helps to create flood vulnerable map, and from the historical rainfall data, we can make an early warning system. The early warning system is very important to protect the city by reducing the losses and victims in the region.

However, the Chap. 12 titled "Egypt's Sinai Desert Cries: Utilization of Flash Flood for a Sustainable Water Management" is presented to discuss the utilization strategies of flash floods in Sinai. This is important for Sinai because its flash floods constitute a potential for non-conventional sources of freshwater. However, most floodwater discharges waste as runoff into the Suez Gulf, and if used effectively, this could satisfy some of the water requirements for a variety of uses. The wise use of floodwater to enable the sustainable management of water resources is a significant challenge in these fields. Therefore, the chapter objective is to put the best ways to mitigate and utilize the floods water as a new supply for water harvesting in Sinai.

Different low-cost storage mechanisms for floodwater harvesting were identified to suit the different technical and socio-economic conditions. Firstly, an underground concrete reservoir is one of the most appropriate water harvesting techniques and easily maintained by the Bedouins themselves. Secondly, a Haraba is a low-cost alternative to capture floodwater, often used by Bedouins. Thirdly, one of the potential technically and highly requested by the stakeholders is a low-cost gabions dam with an underground reservoir as the one constructed in Wadi Ghazala.

Low earthen or stone dykes in the Wadi beds (locally known as Oqum) are recognized. They are usually protected by vegetation remains. Masonry dams for the storage of water are also identified.

The total amount of rainfall and flash floods that could be used annually is estimated at around 1.3 billion cubic meters. This quantity can be increased to 1.5 billion cubic meters.

The establishment of the various dams leads to the presence of communities around the areas of these dams to work with agriculture and grazing, as well as to reduce wastage of water, and reduce the speed of the flooding and thereby protect the soil from water erosion. Disasters such as floods disaster in El-Arish (2010) may not return to the floods alone but as a result of the random nature of the establishment of the buildings in the corridors of the floods and without the work of the previous geological study of the area where the various facilities will be installed on them.

Chapter 13 is titled "Flash Flood Risk Assessment in Egypt." It presents an assessment of flash floods in Egypt. The assessment of the flash flood risk in Egypt is classified in this chapter based on three main perspectives.

1. How to deal with the current situation since all catchments are draining toward urban areas, agricultural land, and other assets?

After assessing the causes of some previous incidents, it was clear that the lack of drainage structures (whether due to poor design or flood plain encroachments), and lack of maintenance of existing ones are the main source of these catastrophic incidents. The 100 year return period was selected for the peak discharge calculations, that is subjected to stakeholder decision based on the allocated budget for flood mitigation measures. Due to the large variance of the catchments peak discharge and runoff volume, the box plot technique was employed to eliminate the ranking outlier values.

A map of peak discharge standardized risk classified into five categories (Very High Risk, High Risk, Moderate Risk, Low to Moderate Risk, and Low Risk) is proposed to highlight the reassessment priorities of the flood mitigation measures to control the catchment outlets affecting the exposed human lives, agricultural lands, and any other assets.

2. How to prioritize the rainfall harvesting projects to support in current water stress problem?

A map of runoff volume standardized risk classification is also provided for the same five categories used to assess the peak discharge. The classification based on the runoff volume can guide the designer accounting for rain harvesting projects that would increase the rate of investment return from both flood mitigation and the reduction of freshwater stress.

3. What to do in future planning for unavoidable urban and agricultural expansion?

A two-dimensional HEC-RAS rainfall-runoff model is conducted for Ras-Gharib City by using 30×30 DEM files. The DEM files could not capture the effect of the levels of Ras-Gharib El-Sheikh-Fadl road on the flow directions. The DEM file has been updated based on the available road topographical survey data. The flood plain, flow depths, and velocities were obtained, and accordingly, the flood intensity was calculated for all streams affecting Ras-Gharib City. The model was verified versus aerial photos for the 2016 incident. In order to assess the effectiveness of the newly constructed culvert (16 vents, 3 m × 3 m box culvert) with attached two dikes, another updated two-dimensional HEC-RAS rainfall-runoff model has been conducted and the results showed significant improvement in flood intensity values in Ras-Gharib City.

In order to harvest and use the water of flash floods, it is necessary to identify the potential location. Consequently, Chap. 14 comes with the Determination of Potential Sites and Methods for Water Harvesting in the Sinai Peninsula by Application of RS, GIS, and WMS Techniques to handle this issue with application to Wadi Dahab. Wadi Dahab has very high importance in a new development in southeastern Sinai, for its touristic position and promising water resources. RS, WMS, and GIS techniques are modern research tools that proved to be highly effective in mapping, investigation, and modeling the runoff processes and optimization of the RWH. In the present work, these tools were used to determine the potential sites suitable for the RWH in W. Dahab. The performed WSPM for determining the potentiality areas for RWH depended on the hydro-morphometric parameters of drainage density, infiltration number, maximum flow distance, overland flow distance, basin slope, basin area, the volume of the annual flood, and basin length. The WSPM model was accomplished through three scenarios: equal criteria weights (scenario I), authors' judgment (scenario II), and weights justified by the sensitivity analysis (scenario III). The obtained WSPM maps for defining the RWH potentiality areas classified W. Dahab basin into five RWH potentiality classes ranging from very low to very high. There are good matches between the three performed WSPMs' scenarios in results for the very high and high RWH potentiality classes, which are very suitable for RWH applications.

There are good matches between the three performed WSPMs' scenarios in results for the very high and high RWH potentiality classes, which are very suitable for RWH applications. These classes are frequently represented generally by El-Ghaaib, Dahab Trunk Channel, Zoghra, Nassab, Saal, and Ganah sub-watersheds, which represent about 62.94%, 56.95%, and 73.83% of the total area of the basin for scenarios I, II and, III, respectively. RWH utility system has been proposed to store runoff water and reduce flash flood risks in identified optimal locations.

3.5 Prediction, Mitigation, and Management of Flash Flood

This theme is covered in Chaps. 15 and 16. Chapter 15 is titled "Prediction and Mitigation of Flood in Egypt." The chapter presented an overview of flash flood including flood definition, flood causes, and comprises types of floods and damages caused by the flood. The chapter also presented the application of prediction and mitigation methods to a real case study in Egypt (Wadi Sudr, Sinai). Egypt has alluvial (wadi) systems, formed during fluvial time of the Tertiary and Quaternary Periods. These wadis suffer from flash flood, consequent to heavy precipitations. Wadi Sudr, Sinai has selected a case study to study the prediction and mitigation of flood in this area. The runoff flow paths are detected across the study area and their flow magnitudes under different rainfall events of 10, 25, 50, and 100-year return periods that have been used for designing the flood mitigation measures. Rational and SCS methods are used to facilitate the simulation process during this study and used as tools to convert rainfall to runoff discharges to determine flood quantity throughout the study area.

Once the flash flood or rainfall pattern is predicted, it could be harvested or collected in different ways. Chapter 16 utilizes Alexandria as a case study to switch in stormwater management from gray to green infrastructure. The chapter is titled "Gray-to-Green Infrastructure for Stormwater Management: an Applicable Approach in Alexandria City, Egypt." The green infrastructure systems have recently found successful applications for stormwater management and flood control. This chapter aims at reviewing the recent applications of the management of stormwater drainage projects. The discussed green infrastructures include bioswales, retention basins, ponds, wetlands, rain gardens, permeable pavements, and urban green spaces. These stormwater infrastructures tend to control runoff volume and timing and promote ecosystem services. In addition, this work would provide a better understanding of the barriers and facilitate factors affecting the management of reclaimed stormwater.

Alexandria is the second-largest city in Egypt that has been suffered from periodic flash floods due to rapid urbanization and various infrastructure-related problems. Stormwater management can be extended to an Alexandria Governorate case study to demonstrate the impact of climate change and urbanization on the performance of a city's drainage system, subject to repeated periods of heavy rain, flash flooding, and strong winds. The combination of traditional drainage systems with green infrastructure could be a viable solution to mitigate the stormwater runoff in Alexandria city, Egypt. The outputs of this work can assist water resource managers, government, professionals, and private and public sectors in maintaining flood risk management, especially in Egypt.

The book ends with the 17th chapter of conclusions and recommendations. Chapter 17 contains an update of recently published research works on flash floods including references from Bruins et al. (2019), Vema et al. (2019), Vemula et al. (2019), Sörensen and Emilsson (2019), Shariat et al. (2019), Alves et al. (2019), Piro et al. (2019), Osti (2018), Abdelkarim et al. (2019), Saber et al. (2017), and Al-Momani and Shawaqfah (2013) among others.

Acknowledgements The writers who wrote this chapter would like to acknowledge the authors of the chapters for their efforts during the different phases of the book including their inputs in this chapter.

References

Abdelkarim, A., Gaber, A. F. D., Youssef, A. M., & Pradhan, B. (2019). Flood Hazard assessment of the Urban area of Tabuk city, Kingdom of Saudi Arabia by integrating spatial-based hydrologic and hydrodynamic modeling. *Sensors, 19,* 1024.

Al-Momani, A. H., & Shawaqfah, M. (2013). Assessment and management of flood risks at the city of Tabuk, Saudi Arabia. *The Holistic Approach to Environment, 3*(1), 15–31.

Alves, A., Gersonius, B., Kapelan, Z., Vojinovic, Z., & Sanchez, A. (2019). Assessing the co-benefits of green-blue-grey infrastructure for sustainable urban flood risk management. *Journal of Environmental Management, 239,* 244–254.

Bruins, J. H., Guedj, H. B., & Svoray, T. (2019). GIS-based hydrological modelling to assess runoff yields in ancient-agricultural terraced wadi fields (central Negev desert). *Journal of Arid Environments*.

Moawad, M. (2013). Analysis of the flash flood occurred on 18 January 2010 in Wadi Al-Arish, Egypt. *Geomatics, Natural Hazards and Risk Journal, 4*(3). https://doi.org/10.1080/19475705.2012.731657.

Negm, A. M. (Ed.). (2019a). *Conventional water resources and agriculture in Egypt*. In HBC series: Springer.

Negm, A. M. (Ed.), (2019b) *Unconventional water resources and agriculture in Egypt* (Vol. 166, pp. 91–107). In HBC series: Springer.

Osti, R. P. (2018). Integrating flood and environmental risk management principles and practices. ADB east Asia working paper series. NO. 15. Rabindra Osti is a senior water resources specialist at the Asian Development Bank.

Piro, P., Turco, M., Palermo, S.A., Principato, F., & Brunetti, G. (2019). A comprehensive approach to Stormwater management problems in the next generation Drainage Networks. In F. Cicirelli, A. Guerrieri, C. Mastroianni, G. Spezzano, & A. Vinci (Eds.), *The internet of things for smart Urban ecosystems. internet of things (Technology, communications and computing)* (pp. 275–304). Cham: Springer.

Saber, M., Kantoush, S., Abdel-Fattah, M., & Sumi, T. (2017). Assessing Flash Floods Prone Regions at Wadi basins in Aswan, Egypt. *DPRI annuals* (No.60, pp. 853–863).

Shariat, R., Roozbahani, A., & Ebrahimian, A. (2019). Risk analysis of urban stormwater infrastructure systems using fuzzy spatial multi-criteria decision making. *Science of the Total Environment, 647,* 1468–1477.

Sörensen, J., & Emilsson, T. (2019). Evaluating flood risk reduction by urban blue-green infrastructure using insurance data. *Journal of Water Resources Planning and Management, 145*(2), 04018099.

Vema, V., Sudheer, K. P., & Chaubey, I. (2019). Fuzzy inference system for site suitability evaluation of water harvesting structures in rainfed regions. *Agricultural Water Management, 218,* 82–93.

Vemula, S., Raju, K., Veena, S., & Kumar, A. (2019). Urban floods in Hyderabad, India, under present and future rainfall scenarios: A case study. *Natural Hazards, 95*(3), 637–655.

Analysis and Design Aspects

Statistical Behavior of Rainfall in Egypt

Tamer A. Gado

Abstract

The extreme rainfall events have critical impacts on hydrologic systems and the society, especially in arid countries. Recently, Egypt has been subject to some flash floods, due to extreme rainfall events, in particular regions (e.g., Sinai, North Coast, and Upper Egypt) that caused severe damages in lives and vital infrastructure and buildings. This chapter investigates, therefore, the statistical characteristics of rainfall over Egypt based on historical daily rainfall records at 30 stations throughout the country. Six types of rainfall data were extracted from daily records: monthly rainfall, annual rainfall, the monthly number of rainy days, the annual number of rainy days, monthly maximum daily rainfall, and annual maximum daily rainfall. Rainfall frequency analysis, based on the generalized extreme value (GEV) distribution along with the L-moments parameter estimation method, was applied to analyze annual maximum rainfall series. Results of the numerical application indicate a great variation over the whole country in all different aspects of rainfall. The rainfall indices have higher values in the north of the country than those in the middle and the south. Also, the annual maximum daily rainfall data in the case study has been well described by the GEV distribution with negative values of the shape parameter for all stations except only one station (Ras Sedr).

Keywords

Precipitation • Rainfall frequency analysis • Rainfall extremes • Generalized extreme value • Egypt

1 Introduction

Extreme rainfall events have severe consequences on human society. Storms are natural hazards that cause a significant proportion of deaths and devastated infrastructure around the world. Consequently, an accurate estimation of extreme rainfall (magnitude, duration, and frequency) is vital. Furthermore, extreme rainfall estimation is crucial for the design, operation, and management of various hydraulic structures. Thus, improving the accuracy of extreme rainfall estimation has been the main objective of many studies. Rainfall frequency analysis (RFA) is usually used in this regard to estimate extreme rainfall at a specific location. In RFA, it is extremely important to find an accurate relationship between an extreme rainfall magnitude P and the corresponding recurrence interval T. The choice of a suitable method for estimating this relationship depends on the availability of the rainfall record at the site of interest. The commonly used extreme rainfall estimation models are based on annual maximum series (AMS). The AMS of rainfall data contains only the maximum peak rainfall in each year; therefore, for an n-year rainfall record, the AMS consists of n annual maximum rainfall values.

In Egypt, the severe damage caused by several flash floods, due to extreme rainfall events, points to the necessity to understand the characteristics of rainfall events to improve means of flood protection. Nevertheless, scanning the literature for statistical analysis of rainfall events in Egypt reveals that little has been done regarding estimation of extreme rainfall events. Thus, in this chapter, a thorough study of the statistical characteristics of rainfall in Egypt will be conducted to develop methods for extreme rainfall estimation in gaged and ungaged sites in Egypt.

T. A. Gado (✉)
Department of Irrigation and Hydraulics Engineering, Faculty of Engineering, Tanta University, Tanta, Egypt
e-mail: tamer.gado@f-eng.tanta.edu.eg; tamergado@hotmail.com

A. M. Negm (ed.), *Flash Floods in Egypt*, Advances in Science, Technology & Innovation,
https://doi.org/10.1007/978-3-030-29635-3_2

2 Literature Review

Characteristics of precipitation were widely investigated in the literature. For instance, storm precipitation in a mountainous watershed was studied in the southwestern British Columbia, Canada (Loukas and Quick 1996). They concluded that the time distribution of precipitation was not affected by the elevation, type of storm, storm duration, or storm precipitation depth. Spatial characteristics of precipitation over Jordan were depicted by calculating basic statistical parameters such as the amplitude of frequencies (Tarawneh and Kadioglu 2003). A database of daily rainfall, compiled from 75 gages in Catalonia covering the period 1950–2000, was used to obtain new statistical patterns of daily rain amounts (Burgueno et al. 2005). They evaluated the annual number of rainy days and its coefficient of variation. It was shown that the average annual number of rainy days has high values in the north of the country, whereas the relative dispersion is higher in the south. In north central Italy, the statistical properties of rainfall extremes were investigated, for storm duration ranging from 15 min to 1 day, and significant relationships between these properties and the mean annual precipitation (MAP) were detected (Baldassarre et al. 2006). Consequently, a regional model for estimating the rainfall depth for a given storm duration and recurrence interval was developed of the study region. In arid regions, short-duration rainfall data was analyzed in order to propose design storm distributions (Awadallah and Younan 2012). It was found that Bell ratios are adequate to represent rainfall patterns in arid regions for rainfall durations less than two hours.

To estimate rainfall intensity of different durations and return periods, Intensity-Duration-Frequency (IDF) relationships were developed by using different distribution models in different neighborhoods, see, e.g., (Nhat et al. 2006; Vivekanandan 2012; Jiang and Tung 2013). Many studies evaluated the performance of various probability models in order to identify the most suitable distribution that could provide accurate extreme rainfall estimates in different regions in the world, see, e.g., (Nguyen and Mayabi 1990; Nguyen et al. 2002; Vivekanandan 2015; Wdowikowski et al. 2016). For example, the design extreme rainfalls were estimated by using annual maximum daily rainfall in 38 stations in Korea (Lee and Maeng 2003). They used the L-moments method to estimate the parameters of three different probability distributions: GEV, GLO, and GPA. In Catalonia, the maximum daily rainfall depths, for several established return periods, were determined with a high spatial resolution by using the Gumbel distribution along with the L-moments (Casas et al. 2007). The maximum daily rainfalls in 12 stations located in Upper and Middle Odra River Basin (Poland) were estimated based on nine probability distributions (Wdowikowski et al. 2017).

An extensive empirical investigation on the extreme value distribution was performed by using a collection of 169 of the longest available rainfall records worldwide (Koutsoyiannis 2004). The results showed that the shape parameter of the GEV distribution was constant for all examined geographical zones with value $\kappa = 0.15$. Furthermore, it was concluded that the shape parameter κ was very hard to estimate on the basis of an individual series, even in series with length 100 years or more. The GEV distribution was fitted to the annual maximum daily rainfall of 15,137 records from all over the world, with lengths varying from 40 to 163 years (Papalexiou and Koutsoyiannis 2013). This analysis revealed that the record length strongly affected the estimate of the GEV shape parameter, and the geographical location might also affect its value. In recent times, a general procedure was presented for assessing the performance of ten commonly used probability distributions in rainfall frequency analyses based on their descriptive and predictive abilities in Ontario, Canada (Nguyen et al. 2017). Their proposed assessment approach was shown to be more efficient and more robust than the traditional selection method. Also, they concluded that the GEV model was the best model for describing the distribution of annual maximum rainfalls in this region. Additionally, an improved procedure was proposed to describe the distribution of extreme rainfalls in consideration of the scale-invariance properties of extreme rainfall processes for different durations by using the GEV distribution (Punlum et al. 2017). The proposed procedure was applied on a network of nine rain gage stations located in the north and northeast region of Thailand during the period 1950–2010.

Recently, few studies have analyzed extreme rainfall events in Egypt. For Wadi Watir in Sinai, an early warning system (EWS) for flash floods was developed (Cools et al. 2012). The study stated that only 20 significant rain events were measured over a period of 30 years (1979–2010); nine of them resulted in flash floods. In Wadi El Arish in Sinai, the flash flood that occurred on January 18, 2010 was analyzed based on the integration of remote sensing and GIS (Moawad 2013). The wadi was subjected to severe thunderstorms on January 17 and 18, 2010 followed by an extreme violent flood that claimed six victims, tens of injured, and devastated vital infrastructure and hundreds of houses (Moawad 2013). This flash flood was also analyzed by using the Weather Research and Forecasting (WRF) model over different parts of Sinai Peninsula (El-Afandi et al. 2013). The results show that the WRF model was able to capture heavy rainfall events in the case study. Furthermore, the performance of WRF model was evaluated in heavy rainfall prediction in Egypt (Ibrahim and El-Afandi 2014).

To estimate rainfall intensity of different durations and return periods, Intensity-Duration-Frequency (IDF) relationships were developed by using different distribution models. IDF curves were constructed for Sinai Peninsula by using rainfall frequency analysis (El-Sayed 2011; Fathy et al. 2014). Regional IDF formula was proposed to estimate rainfall intensity for various return periods and rainfall durations at ungaged sites in Sinai (El-Sayed 2011). A recent study (Salama et al. 2018) assessed the performance of some of the most popular probability distributions to identify the most suitable model that could accurately estimate extreme rainfall events in different regions in Egypt. The results indicated that the Log-Normal and Log-Pearson Type III distributions were the best models for most studied stations in Egypt.

Apart from Sinai, very few studies investigated extreme rainfall events in other parts of Egypt. In Alexandria, two rainfall prediction models were developed (El-Shafie et al. 2011): Artificial Neural Network (ANN) and Multi Linear Regression (MLR). The study concluded that the performance of the ANN model is better than that of the MLR and recommended more detailed studies to quantify the uncertainties inherent in the ANN models. Changes in extreme precipitation over Alexandria region during the period 1979–2011 were discussed (Said et al. 2012). The impact of extreme precipitation events on both water resources quality and water supplies was investigated in Alexandria region and Upper Egypt (Yehia et al. 2017). In Qena, the flash flood that occurred on January 28, 2013 was studied, and the surface runoff was estimated (Moawad et al. 2016). To the best of the author's knowledge, this chapter introduces the first study for the analysis of extreme rainfall events in the whole of Egypt.

3 Statistical Analysis

3.1 Study Area

Egypt is located between latitudes 22° and 32 °N, and longitudes 25° and 35 °E with an area of 1,001,450 km^2 (Fig. 1). The majority of Egypt's landscape is desert except

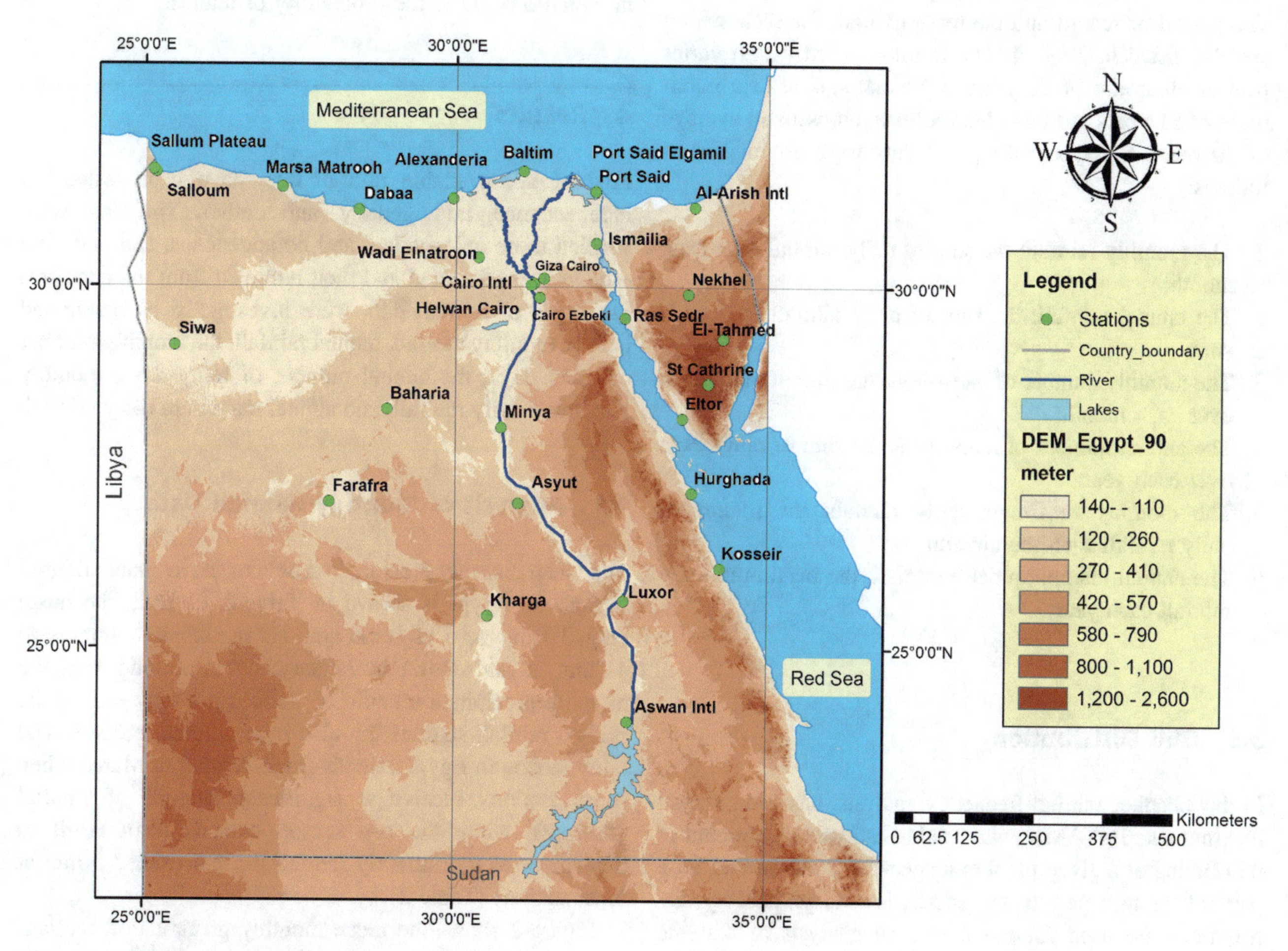

Fig. 1 Map of the studied rainfall stations in Egypt

the Nile Valley and few scattered oases. The climate of Egypt is characterized by four seasons: winter, spring, summer, and autumn. Most of the rainfall occurred in winter when it is cold and moist. The summer, on the other hand, is hot, dry, and rainless. Whereas spring and autumn are transitional seasons. Special characteristics of rainfall over Egypt are generally affected by the closeness to the Mediterranean coast, where rainfall is concentrated in the Egyptian north coast and decreases rapidly inland becoming rare in the south. However, in autumn and winter, extreme rainfall events in some cases cause flash floods over the Red Sea Mountains and the Sinai Peninsula (Ibrahim and El-Afandi 2014). The common characteristics of rainfall events in Egypt are locality, convective, high spatial variability, and short duration.

3.2 Data

In this chapter, the database consists of 30 stations throughout Egypt (Fig. 1). The database was taken from a previous study (Gado 2017). The characteristics of the selected stations used in this chapter are shown in Table 1. The period of record and the record length for each station are also listed in Table 1. The rainfall record length varies from a minimum of 14 years at Nekhel station to a maximum of 81 years at Marsa Matrooh station, with an average of 30 years. Six types of rainfall data were established as follows:

1. The monthly rainfall: the sum of daily rainfall over each month;
2. The annual rainfall: the sum of daily rainfall over each year;
3. The monthly number of rainy days: the sum of rainy days over each month;
4. The annual number of rainy days: the sum of rainy days over each year;
5. The monthly maximum daily rainfall: the maximum daily rainfall each month; and
6. The annual maximum daily rainfall: the maximum daily rainfall each year.

3.3 GEV Distribution

In this chapter, rainfall frequency analysis, based on annual maximum series (AMS), was used to estimate the amount of rain falling at a given point in a specified duration and for a particular return period. In general, several probability distributions are used for the frequency analysis of extreme rainfalls. Here, the Generalized Extreme Value (GEV) distribution along with the L-moments parameter estimation method were used to analyze the AMS as they were found to be an efficient approach for extreme event estimation according to the literature, see, e.g., (Hosking et al. 1985; Gado and Nguyen 2015). The cumulative distribution function (CDF) or the probability of non-exceedance [F(q)] for the GEV distribution is given as (Gado and Nguyen 2016):

$$\begin{aligned} F(q) &= \exp\left[-\left(1-\frac{\kappa(q-\xi)}{\alpha}\right)^{1/\kappa}\right] \quad \kappa \neq 0 \\ &= \exp\left[-\exp\left(-\frac{q-\xi}{\alpha}\right)\right] \quad \kappa = 0 \end{aligned} \tag{1}$$

where q is the extreme observation rainfall; and ξ, α, and κ are, respectively, the location, scale, and shape parameters of the distribution. The quantiles (P_T) can be computed using the following relation:

$$\begin{aligned} P_T &= \xi + \frac{\alpha}{\kappa}\{1-[-ln(p)]^{\kappa}\}\kappa \neq 0 \\ &= \xi - \alpha ln[-ln(p)]\kappa = 0 \end{aligned} \tag{2}$$

in which $p = 1/T$ is the probability of interest.

4 Results

First of all, available rainfall data have been tested for independence, homogeneity, and outlier. The data were verified to be independent and homogeneous, and very few outliers were detected and then removed from the database. Six types of rainfall data were investigated as mentioned before: monthly rainfall, annual rainfall, the monthly number of rainy days, the annual number of rainy days, monthly maximum daily rainfall, and annual maximum daily rainfall.

4.1 Analysis of Monthly Rainfall Data

The mean monthly precipitation data of the available rainfall stations in Egypt is shown in Table 2. Overall, the mean ranged from zero at some stations in different months to 41 mm at Alexandria in January. Great variations of the monthly precipitation could be noticed not only among the studied rainfall stations but also over different months. The rainy season in Egypt extends from October to March when some stations receive a significant amount of rainfall (Table 2), while the dry season extends from April to September as the monthly rainfall did not exceed 5 mm at most stations (Table 2).

Figure 2 shows the mean monthly precipitation for January at all stations in descending order. For January, the

Table 1 Summary of the characteristics of the studied rainfall stations in Egypt (Gado 2017)

Station ID	Station name	Latitude	Longitude	Elevation (m)	Period of record	Missing years	Record length (years)
62337	Al-Arish Intl	31.07	33.84	36.9	1985–2015	–	31
62318	Alexandria Intl	31.18	29.95	−1.8	1957–2015	1967–1972	53
62414	Aswan Intl	23.97	32.78	200	1957–2015	1967–1972	21
62393	Asyut	27.04	31.01	226	1957–2014	1967–1972	19
62420	Baharia	28.33	28.90	130	1957–2015	1967–1982, 1992	24
62325	Baltim	31.55	31.08	2	1994–2015	–	22
62366	Cairo Intl	30.12	31.41	75	1944–2015	1947–1956, 1967–1972	51
147728	Cairo Ezbekiya	30.05	31.25	20	1909–1957	1941–1943	46
62309	Dabaa	31.03	28.44	18	1963–2015	1967–1998	21
	El-Tahmed	29.30	34.30	625	1922–1955	–	31
62459	Eltor	28.21	33.65	35	1960–2015	1967–1993	22
62423	Farafra	27.05	27.98	92	1957–2014	1967–1975, 1977–1992	16
62375	Giza Cairo	30.03	31.21	28	1924–1957	1941–1943	31
147730	Helwan Cairo	29.86	31.34	116	1904–1957	–	54
62463	Hurghada	27.18	33.80	14	1991–2015	2000, 2006, 2008, 2009, 2011, 2012	16
62440	Ismailia	30.59	32.25	13	1987–2013	–	26
62435	Kharga	25.46	30.53	73	1957–2014	1958–1959, 1967–1972, 1975–1981	15
62465	Kosseir	26.14	34.26	11	1960–2014	1967–1972	25
62405	Luxor	25.67	32.71	99	1944–2015	1946–1956, 1967–1972	32
62306	Marsa Matrooh	31.33	27.22	30	1920–2015	1923, 1941–1944, 1967–1973, 1975, 1976	81
62387	Minya	28.08	30.73	37	1957–2015	1967–1972, 1978	30
62452	Nekhel	29.91	33.74	403	2001–2014	–	14
62333	Port Said	31.28	32.24	2	1957–2015	1967–1972, 1974–1978, 1990	22
62332	Port Said Elgamil	31.28	32.24	6	1987–2014	–	27
62455	Ras Sedr	29.58	32.72	16	2000–2015	–	15
62300	Salloum	31.53	25.18	6	1957–1995	1967–1979	26
62305	Sallum Plateau	31.57	25.13	6	1996–2015	2013	19
62417	Siwa	29.20	25.48	–12	1920–2015	1923, 1941–1944, 1967–1978, 1984, 1986, 2014	59
623664	St Catherine Intl	28.69	34.06	1331	1934–2006	1938–1979	31
62357	Wadi Elnatroon	30.40	30.36	1	1996–2015	1999, 2013. 2014	17

mean exceeded 30 mm at four stations located on the North Coast (Fig. 1): Alexandria Intl (41 mm), Baltim (36 mm), Dabaa (36 mm), and Marsa Matrooh (35 mm). On the other hand, the mean was zero at four stations (Baharia, Aswan Intl, Asyut, and Minya) located in the middle and the south of the country (Fig. 1). Figure 3 displays the average monthly rainfall data for Alexandria station estimated from the period 1957–2015. It can be noticed that the highest month in total rainfall in Alexandria was January then December.

Table 2 Average of the monthly rainfall (mm) of the studied stations in Egypt

Month station	1	2	3	4	5	6	7	8	9	10	11	12
Al-Arish Intl	23	15	16	5	4	0	0	1	0	6	5	16
Alexandria Intl	41	27	12	2	1	0	0	0	1	8	25	37
Aswan Intl	0	0	0	1	0	1	0	0	0	0	0	0
Asyut	0	1	1	2	0	1	2	0	0	1	0	2
Baharia	0	0	1	0	0	0	0	0	0	0	0	0
Baltim	36	34	12	4	1	0	0	0	5	9	12	34
Cairo Intl	5	6	6	1	1	0	0	1	1	1	5	5
Cairo Ezbekiya	5	4	4	2	2	0	0	0	0	2	2	5
Dabaa	36	15	10	2	4	0	0	0	1	16	14	19
Eltor	1	0	1	1	0	0	0	0	3	0	2	3
Farafra	1	0	0	3	0	0	1	0	0	1	1	0
Giza Cairo	3	3	4	1	1	0	0	0	0	3	4	4
Helwan Cairo	6	5	4	2	3	0	0	0	0	1	3	6
Hurghada	4	0	10	0	10	0	0	3	5	5	5	7
Ismailia	12	7	7	6	0	0	1	0	4	4	1	4
Kharga	1	0	0	1	1	1	0	0	0	0	0	1
Kosseir	1	0	0	1	1	0	1	1	1	1	0	0
Luxor	1	0	1	1	3	1	1	0	0	0	1	2
Marsa Matrooh	35	21	16	8	7	3	0	0	1	12	20	34
Minya	0	2	0	2	0	0	0	1	2	1	1	0
Nekhel	7	4	3	2	0	0	0	0	2	4	0	2
Port Said	3	3	2	2	0	0	0	0	0	3	3	4
Port Said Elgamil	22	16	7	8	6	0	0	2	4	6	13	8
Ras Sedr	3	2	2	0	1	0	0	0	0	0	1	2
Salloum	16	9	5	1	2	3	4	0	1	5	6	10
Sallum Plateau	12	11	3	2	2	1	0	0	6	5	2	8
Siwa	1	1	1	1	1	0	0	0	0	1	0	2
Wadi Elnatroon	6	11	2	1	1	0	2	0	0	3	2	8

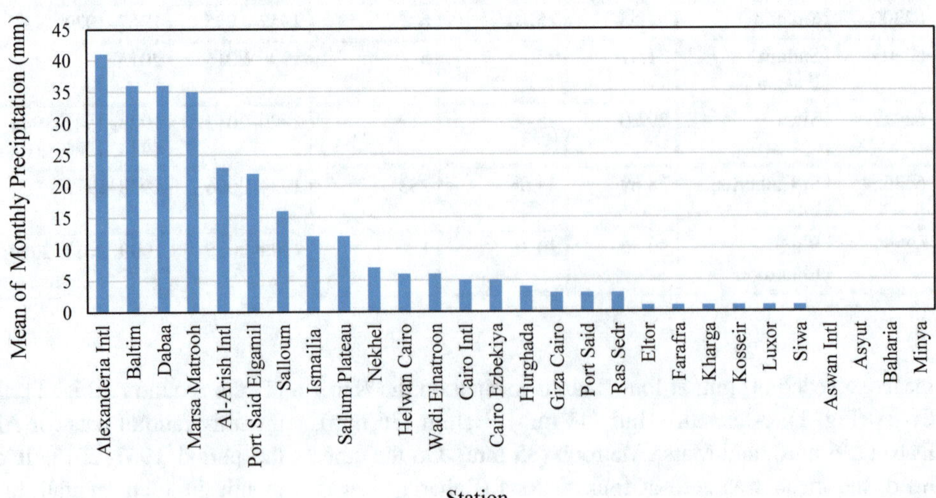

Fig. 2 Mean of monthly precipitation of the studied stations for January

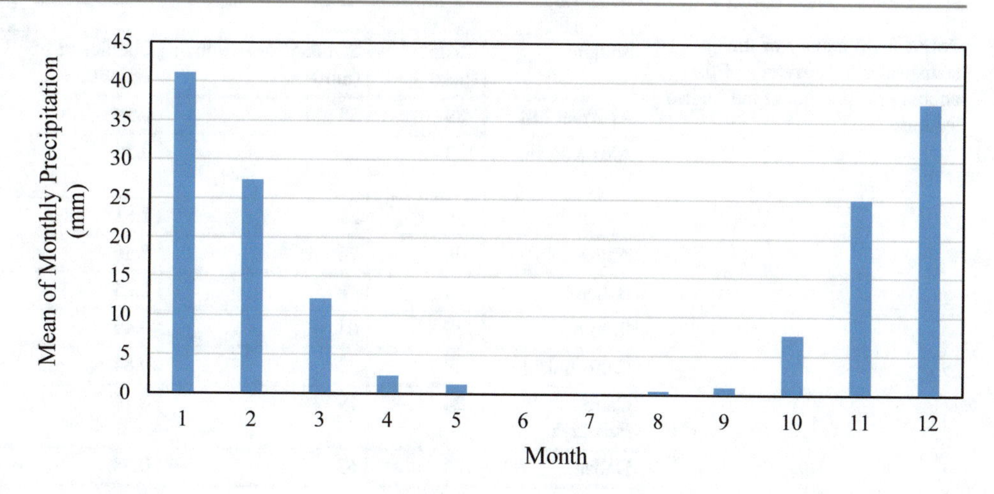

Fig. 3 Average monthly rainfall data for Alexandria station (1957–2015)

4.2 Analysis of Annual Rainfall Data

Basic statistical characteristics such as the mean, the standard deviation, the coefficient of variation, the maximum, and the minimum were derived from the annual precipitation data of the available rainfall stations in Egypt (Table 3). Great variations of the statistical characteristics of the annual precipitation among the studied rainfall stations could be noticed. Overall, the mean ranged from 5 to 153 mm; the standard deviation (6–85 mm); the coefficient of variation (0.46–2.81); the maximum (27–401 mm); and the minimum (0–9 mm). The maximum of the annual precipitation in the case study was 401 mm, and it occurred at Marsa Matrooh in 1994 (Table 3).

The mean annual precipitation of all stations is shown in Fig. 4 in descending order. The mean exceeded 100 mm at four stations located on the North Coast (Fig. 1): Alexandria Intl (153 mm), Baltim (139 mm), Marsa Matrooh (122 mm), and Dabaa (108 mm). On the other hand, the mean annual precipitation did not exceed 10 mm at four stations located on the middle and the south of the country (Fig. 1): Baharia (5 mm), Aswan Intl (9 mm), Siwa (9 mm), and Ras Sedr (10 mm). Figure 5 displays the annual rainfall of Alexandria station for the period 1957–2015. The annual rainfall varied between a minimum amount of 7.4 mm in 1975 and a maximum of 352 mm in 2004 (Table 3).

4.3 Analysis of the Monthly Number of Rainy Days Data

The mean monthly number of rainy days of the studied stations in Egypt is presented in Table 4. Overall, the mean ranged from zero days at some stations in different months to nine days at Baltim in January. Table 4 confirms that the dry season in Egypt includes six months from April to September when the mean of the monthly number of rainy days was equal or less than two days at almost all stations.

Figure 6 shows the mean monthly number of rainy days for January at all stations in descending order. For January, the mean exceeded seven days at four stations located on the North Coast (Fig. 1): Baltim (9 days), Alexandria Intl (8 days), Dabaa (8 days), and Marsa Matrooh (7 days). On the other hand, the mean was one day at nine stations (Baharia, Aswan Intl, Asyut, Farafra, Hurghada, Kharga, Kosseir, Luxor, and Minya) located in the middle and the south of the country (Fig. 1). Figure 7 displays the average monthly number of rainy days at Alexandria station estimated for the period 1957–2015. It can be concluded that Alexandria witnessed at least one rainy day each month on average during the recorded period.

4.4 Analysis of the Annual Number of Rainy Days Data

Basic statistical characteristics of the annual number of rainy days at the studied stations are presented in Table 5. The characteristics include the mean (1–31 days), the standard deviation (0.63–13.31 days), the coefficient of variation (0.36–1.03), the maximum (3–64 days), and the minimum (1–6 days). The maximum of the annual number of rainy days in the case study was 64 days which occurred at Dabaa in 2000.

The mean annual number of rainy days for all stations was shown in Fig. 8 in descending order. The mean exceeded 20 days at six stations located on the North Coast: Baltim (31 days), Alexandria Intl (30 days), Dabaa (30 days), Marsa Matrooh (28 days), Port Said Elgamil (23 days), and Al-Arish Intl (20 days). In contrast, the mean did not exceed 5 days at 12 stations located on the middle and the south of the country (Fig. 1): Farafra, Aswan Intl,

Table 3 Summary of the statistical characteristics of the annual precipitation of the studied stations

Station	Mean (mm)	Standard deviation (mm)	Coefficient of variation	Maximum (mm)	Minimum (mm)
Al-Arish Intl	89	53	0.59	205	9
Alexandria Intl	153	85	0.56	352	7
Aswan Intl	9	25	2.81	117	1
Asyut	30	76	2.56	316	1
Baharia	5	6	1.25	27	0
Baltim	139	64	0.46	236	4
Cairo Intl	38	32	0.84	129	1
Cairo Ezbekiya	24	19	0.76	92	1
Dabaa	108	63	0.58	257	5
Eltor	16	23	1.39	83	1
Farafra	13	18	1.39	70	1
Giza Cairo	22	16	0.74	61	3
Helwan Cairo	30	20	0.68	92	1
Hurghada	42	45	1.05	147	1
Ismailia	45	67	1.49	310	3
Kharga	14	25	1.80	73	1
Kosseir	14	15	1.10	53	1
Luxor	20	31	1.53	124	1
Marsa Matrooh	122	80	0.65	401	1
Minya	15	24	1.54	80	1
Nekhel	22	17	0.76	60	4
Port Said	45	45	0.99	168	1
Port Said Elgamil	89	76	0.86	397	7
Ras Sedr	10	7	0.69	32	1
Salloum	63	52	0.84	189	1
Sallum Plateau	51	41	0.80	173	3
Siwa	9	10	1.09	42	1
Wadi Elnatroon	34	33	0.98	111	2

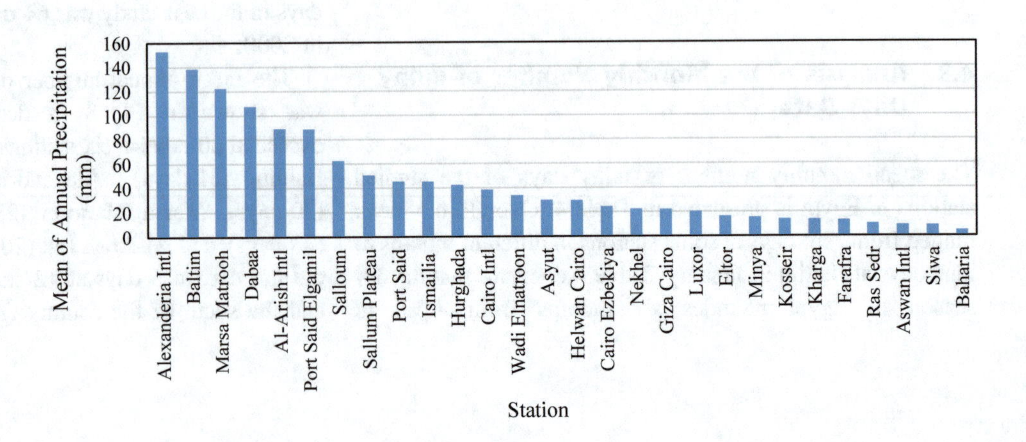

Fig. 4 Mean of annual precipitation of the studied stations in Egypt

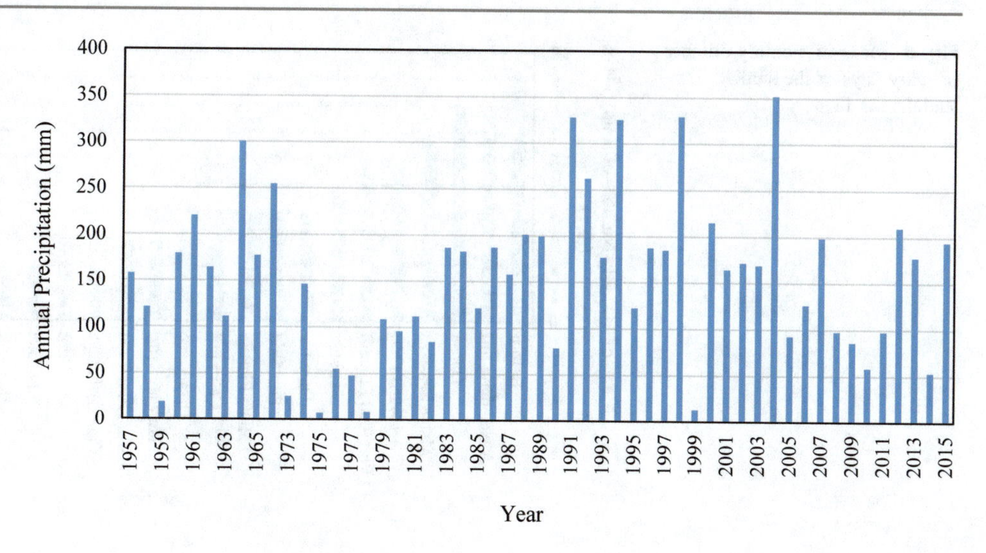

Fig. 5 Annual rainfall of Alexandria station (1957–2015)

Table 4 Average of the monthly rainy days (day) of the studied stations in Egypt

Month station	1	2	3	4	5	6	7	8	9	10	11	12
Al-Arish Intl	6	5	4	2	1	1	2	1	0	2	2	4
Alexandria Intl	8	6	4	2	1	1	2	1	1	3	4	6
Aswan Intl	1	1	1	1	1	1	2	1	2	2	1	1
Asyut	1	1	1	2	1	1	2	1	0	1	1	2
Baharia	1	2	1	1	1	0	2	1	1	2	1	1
Baltim	9	8	4	2	2	0	2	0	1	3	3	7
Cairo Intl	3	2	2	2	1	1	2	1	2	1	2	3
Cairo Ezbekiya	2	2	2	1	1	1	0	0	0	2	1	3
Dabaa	8	6	4	2	2	0	2	0	2	3	4	6
Eltor	2	1	2	1	2	0	2	0	1	1	1	2
Farafra	1	1	1	1	1	0	2	0	0	1	2	2
Giza Cairo	3	2	2	1	2	0	0	1	1	2	2	3
Helwan Cairo	3	3	2	2	2	1	0	0	0	1	2	3
Hurghada	1	0	1	0	1	1	2	1	1	1	2	2
Ismailia	4	3	2	2	2	0	2	0	1	2	2	3
Kharga	1	1	2	1	1	1	2	0	1	0	1	1
Kosseir	1	1	1	1	1	1	1	1	1	1	1	2
Luxor	1	1	2	2	1	1	2	1	1	2	1	1
Marsa Matrooh	7	6	4	2	2	1	2	2	1	3	4	5
Minya	1	1	1	2	1	0	1	1	1	2	1	1
Nekhel	2	2	2	2	2	0	0	0	1	2	1	2
Port Said	4	4	3	2	1	0	0	0	1	3	3	4
Port Said Elgamil	6	5	4	2	2	1	1	1	1	3	3	4
Ras Sedr	2	2	1	1	2	0	1	0	0	1	1	2
Salloum	5	4	3	2	2	1	3	0	1	3	2	4
Sallum Plateau	4	4	2	1	2	1	2	0	2	3	2	3
Siwa	2	1	2	1	2	1	2	0	1	2	1	2
Wadi Elnatroon	3	2	2	1	2	0	1	0	0	1	1	2

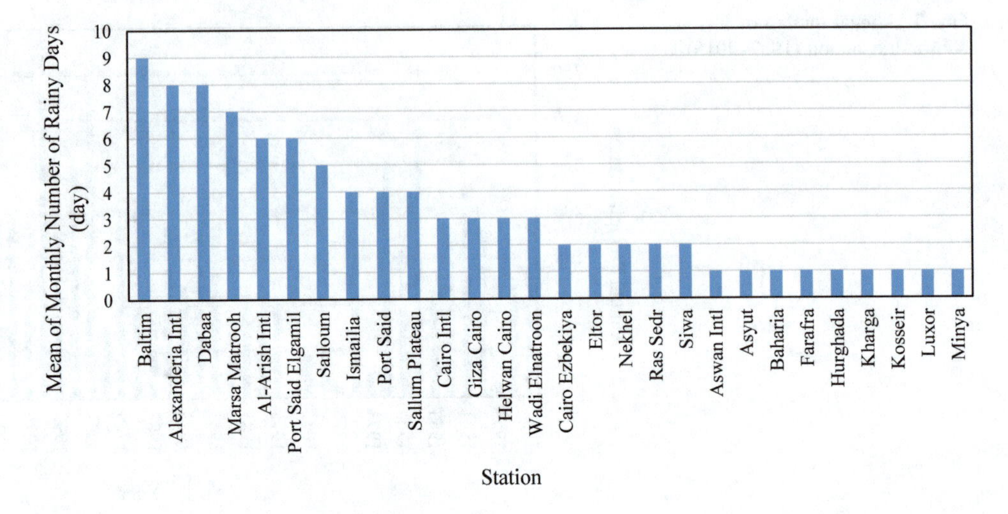

Fig. 6 Mean of monthly number of rainy days of the studied stations for January

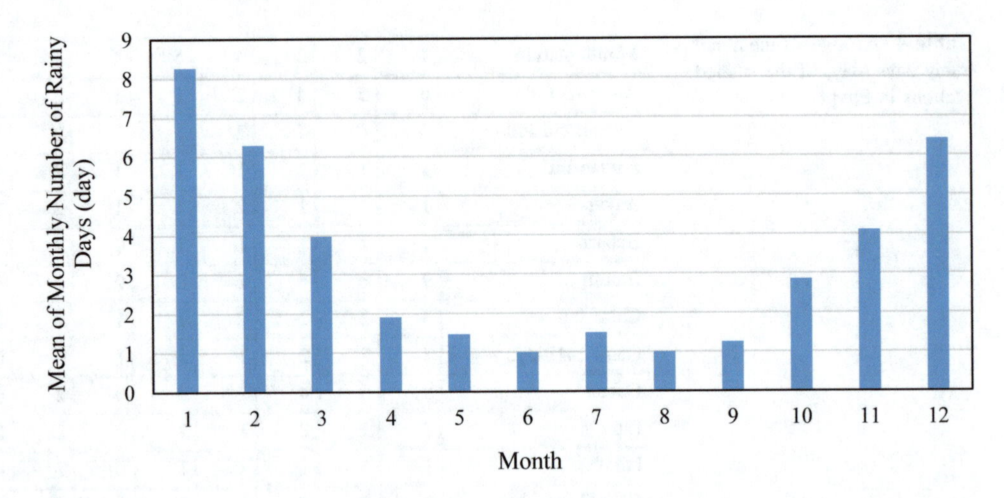

Fig. 7 Average monthly number of rainy days of Alexandria station (1957–2015)

Asyut, Baharia, Hurghada, Kharga, Kosseir, Luxor, Eltor, Minya, Siwa, and Ras Sedr. Figure 9 displays the annual number of rainy days at Alexandria station for the period 1957–2015. The annual number of rainy days varied between a minimum of two days in 1975 and a maximum of 56 days in 1991 (Table 5).

4.5 Analysis of Monthly Maximum Daily Rainfall Data

The mean of the monthly maximum daily rainfall of the studied stations in Egypt is shown in Table 6. Overall, the mean ranged from zero at some stations in different months to 19 mm at Port Said Elgamil in January. Table 6 confirms the great variations among the stations over the different months. Figure 10 shows the mean of the monthly maximum daily precipitation for January at all stations in descending order. For January, the mean exceeded 12 mm at four stations located on the North Coast (Fig. 1): Port Said Elgamil (19 mm), Alexandria Intl (14 mm), Marsa Matrooh (14 mm), and Dabaa (12 mm). Whereas the mean was zero at five stations (Aswan Intl, Asyut, Baharia, Farafra, and Minya) located in the middle and the south of the country (Fig. 1). Figure 11 displays the average monthly maximum daily rainfall at Alexandria station estimated for the period 1957–2015.

4.6 Analysis of Annual Maximum Daily Rainfall Data

Here, the annual maximum daily rainfall series from 30 stations in Egypt have been analyzed to determine the annual maximum daily precipitation for several established return periods in different regions of Egypt.

Basic statistical characteristics such as the mean, the standard deviation, the coefficient of variation, the maximum, and the minimum were derived from the annual maximum daily rainfall data of the 30 available rainfall

Table 5 Summary of the statistical characteristics of the annual number of rainy days of the studied stations

Station	Mean (day)	Standard deviation (day)	Coefficient of variation	Maximum (day)	Minimum (day)
Al-Arish Intl	20	9.12	0.45	37	3
Alexandria Intl	30	13.24	0.45	56	2
Aswan Intl	2	1.03	0.65	4	1
Asyut	2	1.79	0.87	8	1
Baharia	2	1.46	0.71	6	1
Baltim	31	11.56	0.37	48	4
Cairo Intl	10	5.58	0.55	30	1
Cairo Ezbekiya	7	3.74	0.51	15	1
Dabaa	30	13.31	0.45	64	6
Eltor	3	1.93	0.66	8	1
Farafra	1	0.63	0.44	3	1
Giza Cairo	7	4.51	0.67	23	1
Helwan Cairo	9	4.47	0.50	22	1
Hurghada	2	0.80	0.41	3	1
Ismailia	12	5.19	0.44	20	3
Kharga	2	0.74	0.48	3	1
Kosseir	2	1.54	0.67	7	1
Luxor	2	1.91	0.86	10	1
Marsa Matrooh	28	13.31	0.47	54	1
Minya	3	1.46	0.58	6	1
Nekhel	6	2.58	0.45	10	2
Port Said	16	9.92	0.61	35	1
Port Said Elgamil	23	8.52	0.36	47	4
Ras Sedr	5	2.90	0.54	13	1
Salloum	17	10.71	0.61	37	1
Sallum Plateau	14	8.31	0.58	32	3
Siwa	3	3.37	1.03	19	1
Wadi Elnatroon	8	4.27	0.51	15	2

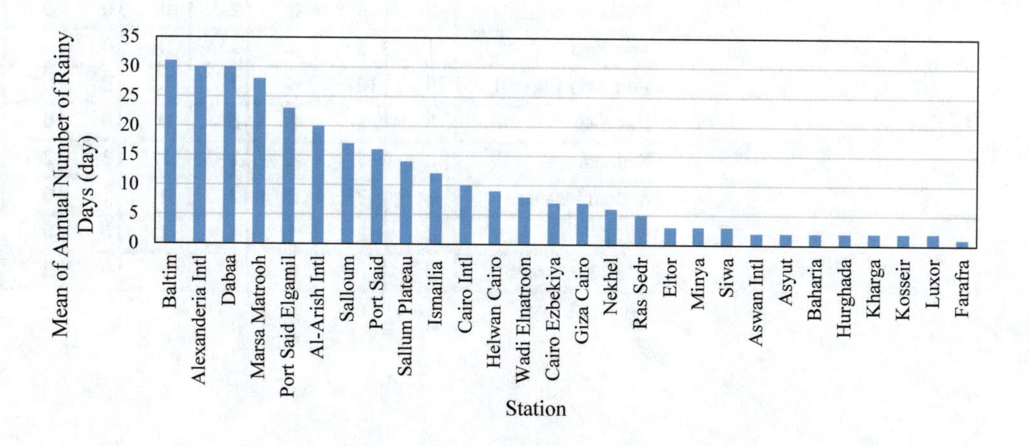

Fig. 8 Mean of annual number of rainy days of the studied stations in Egypt

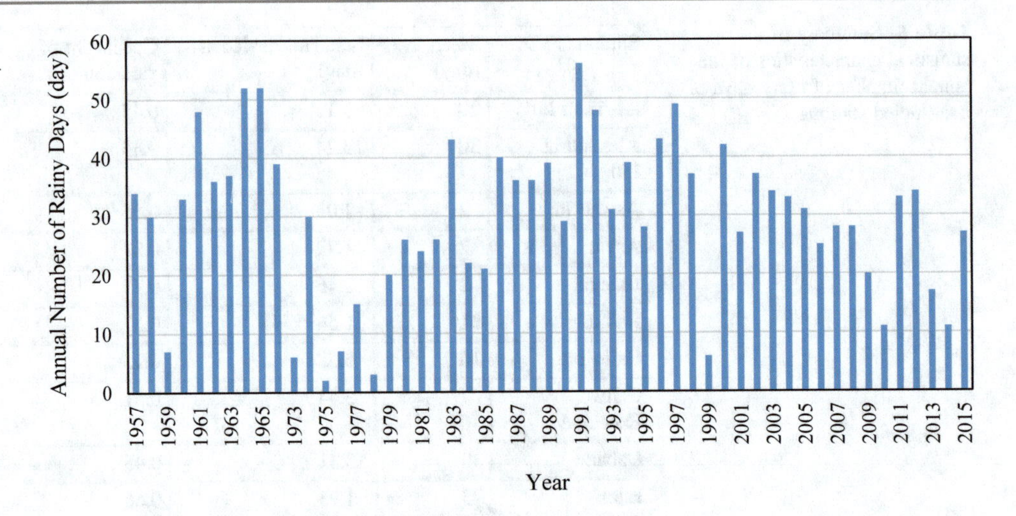

Fig. 9 Annual number of rainy days of Alexandria station (1957–2015)

Table 6 Average of the monthly maximum daily rainfall (mm) of the studied stations in Egypt

Month station	1	2	3	4	5	6	7	8	9	10	11	12
Al-Arish Intl	11	9	11	3	4	0	0	1	0	5	3	10
Alexandria Intl	14	12	7	2	1	0	0	0	1	5	13	16
Aswan Intl	0	0	0	1	0	1	0	0	0	0	0	0
Asyut	0	1	1	1	0	1	1	0	0	1	0	1
Baharia	0	0	1	0	0	0	0	0	4	0	0	0
Baltim	11	13	7	2	1	0	0	0	5	6	6	14
Cairo Intl	3	5	7	3	4	3	3	5	5	5	9	9
Cairo Ezbekiya	4	3	3	1	2	0	0	0	0	1	2	3
Dabaa	12	6	6	1	4	0	0	0	0	12	8	9
Eltor	1	0	1	1	0	0	0	0	3	0	2	2
Farafra	0	0	0	3	0	0	1	0	0	1	0	0
Giza Cairo	2	2	3	1	1	0	0	0	0	3	3	2
Helwan Cairo	4	3	3	2	2	0	0	0	0	1	2	4
Hurghada	4	0	12	0	12	0	0	3	6	6	4	8
Ismailia	10	6	5	4	0	0	1	0	4	3	1	2
Kharga	1	0	0	1	1	1	0	0	0	0	0	1
Kosseir	1	0	0	1	1	0	1	4	1	1	0	0
Luxor	1	0	0	1	3	1	1	0	0	0	1	2
Marsa Matrooh	14	10	9	7	7	2	0	0	1	8	11	15
Minya	0	2	0	2	0	0	0	1	2	1	1	3
Nekhel	5	3	2	2	0	0	0	0	2	3	0	1
Port Said	1	2	1	1	0	0	0	0	0	2	2	3
Port Said Elgamil	19	10	4	7	5	0	0	2	3	4	7	4
Ras Sedr	3	2	1	6	0	0	0	0	0	0	1	2
Salloum	8	5	4	1	1	3	2	0	1	3	4	6
Sallum Plateau	6	7	2	2	1	1	0	0	6	4	2	5
Siwa	1	2	1	1	1	0	0	0	0	1	0	2
Wadi Elnatroon	5	9	2	0	1	0	2	0	0	3	2	8

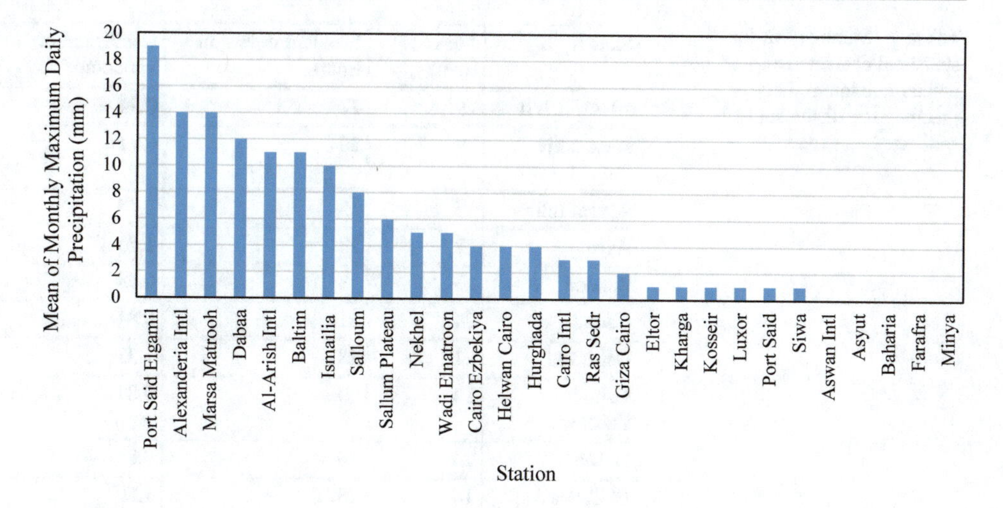

Fig. 10 Average monthly maximum daily precipitation of the studied stations for January

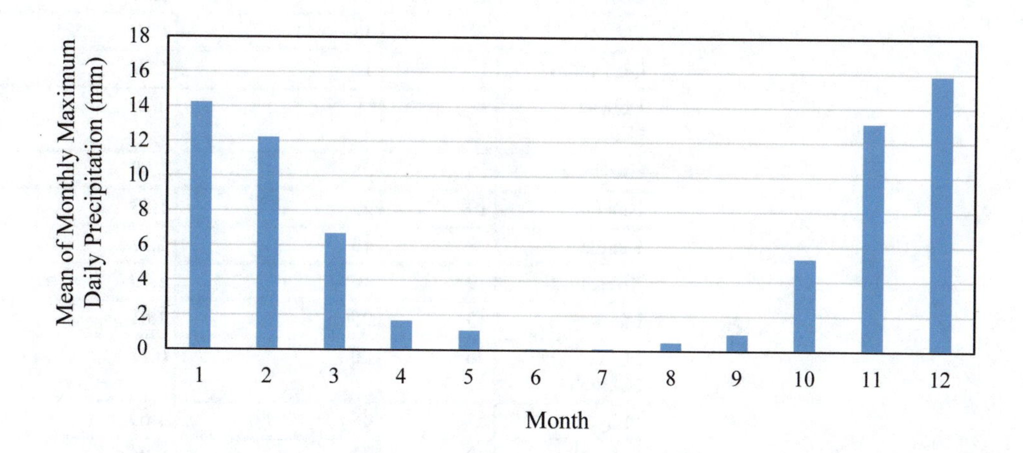

Fig. 11 Average monthly maximum daily rainfall of Alexandria station (1957–2015)

stations (Table 7). Overall, the mean ranged from 3 to 35 mm; the standard deviation (2–38 mm); the coefficient of variation (0.5–2.5); the maximum (8–142 mm); and the minimum (0–6 mm). The maximum of the annual maximum daily rainfall in the case study was 142 mm that occurred at El-Tahmed station in 1925 (Table 7). Figure 12 shows the mean annual maximum daily precipitation of all stations in descending order. It can be observed that the mean exceeded 25 mm at only four stations (Fig. 12): Hurghada (35 mm), Al-Arish Intl (29 mm), Alexandria Intl (29 mm), and Marsa Matrooh (26 mm). On the other hand, the mean did not exceed 5 mm at two stations (Fig. 12): Baharia (3 mm) and Ras Sedr (4 mm). Figure 13 displays the annual maximum daily rainfall data at Alexandria station for the period 1957–2015. The annual maximum daily rainfall varied between a minimum of 6 mm in 1978 and a maximum of 102 mm in 2004 (Table 7).

GEV distributions were fitted to each of the 30 annual maximum series of daily precipitation depths using the L-moments method. The values of the parameters in the GEV functions fitted to each series are shown in Table 8. The averages over all stations and the dispersion characteristics (minimum and maximum values and standard deviations) of the three parameters of the GEV distribution are shown in Table 9. The most important parameter is the shape parameter (κ) because it determines the shape of the distribution and consequently the behavior of its tail. However, the estimation of the shape parameter involves a great deal of uncertainty because it depends on the skewness whose value cannot be specified accurately (Koutsoyiannis 2004). The estimated values of the shape parameter range from -0.75 to 0.11 as shown in Table 8. It can be noticed that the estimated κ is negative for all stations except only one station (Ras Sedr) whose $\kappa = 0.11$ (Table 8) which implies an upper bound. In this case, it may be better to use the GEV with shape parameter equal to zero, i.e., the Gumbel distribution (Papalexiou and Koutsoyiannis 2013).

The maximum daily precipitation for each station corresponding to return periods between 5 and 1000 years have been estimated by using the GEV distribution with L-moments (Table 8). For the return period of 100 years as an example, the daily extreme rainfall estimation ranges from 11 mm at Ras Sedr to 203 mm at Hurghada (Fig. 14). The 100-years daily extreme rainfall (P_{100}) exceeded

Table 7 Summary of the statistical characteristics of the annual maximum daily precipitation of the studied stations

Station	Mean (mm)	Standard deviation (mm)	Coefficient of variation	Maximum (mm)	Minimum (mm)
Al-Arish Intl	29	27	0.94	99	4
Alexandria Intl	29	20	0.71	102	6
Aswan Intl	5	13	2.45	62	1
Asyut	8	13	1.53	52	1
Baharia	3	3	0.98	12	0
Baltim	24	21	0.87	102	2
Cairo Intl	21	28	1.33	106	1
Cairo Ezbekiya	10	9	0.87	43	1
Dabaa	24	20	0.83	99	2
El-Tahmed	18	28	1.50	142	0
Eltor	13	20	1.52	70	1
Farafra	12	18	1.50	70	1
Giza Cairo	10	11	1.07	53	2
Helwan Cairo	12	9	0.75	37	1
Hurghada	35	38	1.08	121	1
Ismailia	19	29	1.53	102	1
Kharga	8	13	1.57	40	1
Kosseir	9	10	1.09	32	1
Luxor	15	25	1.61	100	1
Marsa Matrooh	26	20	0.77	99	1
Minya	11	18	1.69	76	1
Nekhel	12	11	0.95	34	2
Port Said	13	11	0.86	44	1
Port Said Elgamil	24	28	1.19	103	1
Ras Sedr	4	2	0.53	8	1
Salloum	17	16	0.96	70	1
Sallum Plateau	20	20	1.03	92	2
Siwa	5	6	1.10	28	0
St Catherine Intl	12	16	1.29	76	1
Wadi Elnatroon	23	32	1.40	99	1

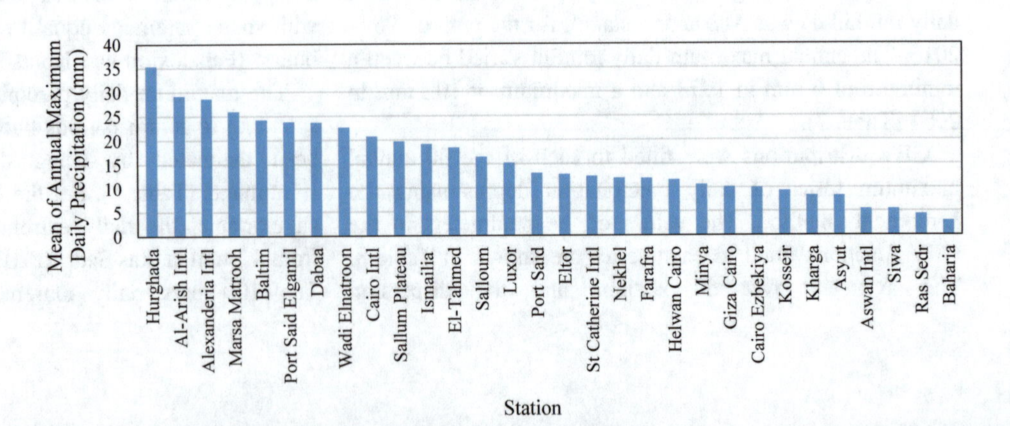

Fig. 12 Mean of annual maximum daily precipitation of the studied stations in Egypt

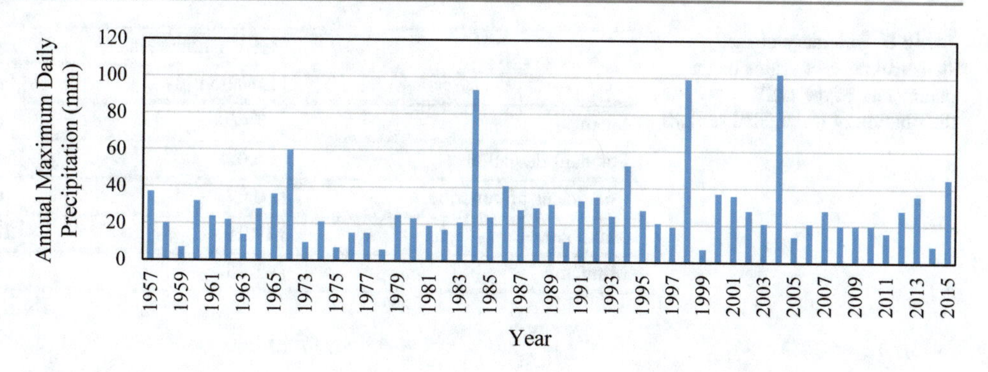

Fig. 13 Annual maximum daily rainfall data for Alexandria station (1957–2015)

Table 8 Values of the parameters of the GEV distribution and daily extreme rainfall estimation of different return periods (P_T) of the studied stations

Station	GEV parameters			P_T (mm)						
	Location (ξ)	Scale (α)	Shape (κ)	P_5	P_{10}	P_{25}	P_{50}	P_{100}	P_{200}	P_{1000}
Al-Arish Intl	15.66	11.86	−0.37	39	57	88	119	159	210	394
Alexandria Intl	19.25	10.37	−0.25	38	51	70	88	109	134	212
Aswan Intl	1.25	1.21	−0.75	5	8	17	30	50	84	282
Asyut	2.51	3.51	−0.52	11	18	32	47	70	103	244
Baharia	1.49	1.44	−0.32	4	6	10	13	17	21	38
Baltim	14.44	8.89	−0.35	32	45	67	89	116	152	275
Cairo Intl	7.93	8.39	−0.50	27	43	74	108	157	225	511
Cairo Ezbekiya	5.80	4.29	−0.30	14	20	29	37	48	61	104
Dabaa	15.21	9.66	−0.24	33	44	61	77	95	117	183
El-Tahmed	7.06	6.14	−0.57	22	35	63	95	144	215	543
Eltor	4.16	5.20	−0.53	16	27	48	72	107	157	376
Farafra	3.57	5.46	−0.50	16	26	47	69	101	147	337
Giza Cairo	5.16	4.14	−0.39	14	20	32	44	59	79	155
Helwan Cairo	7.31	5.41	−0.19	17	22	31	38	47	56	84
Hurghada	15.55	22.64	−0.23	56	83	123	160	203	253	407
Ismailia	5.76	7.31	−0.56	23	39	72	110	167	250	632
Kharga	2.40	3.20	−0.57	10	17	32	49	74	112	284
Kosseir	4.12	5.17	−0.31	14	21	32	43	57	74	129
Luxor	4.01	6.28	−0.56	19	32	60	92	138	207	518
Marsa Matrooh	16.19	11.36	−0.22	36	49	69	87	108	132	204
Minya	2.65	4.03	−0.59	12	22	41	65	100	153	406
Nekhel	6.24	5.15	−0.36	17	24	37	50	67	89	165
Port Said	7.48	5.74	−0.29	18	26	38	49	64	81	137
Port Said Elgamil	10.91	9.08	−0.47	31	47	78	111	158	221	479
Ras Sedr	3.36	2.12	0.11	6	8	9	10	11	12	14
Salloum	8.81	8.16	−0.27	24	34	50	66	84	106	176
Sallum Plateau	10.37	8.01	−0.38	27	39	60	82	110	146	278
Siwa	2.53	2.51	−0.37	8	11	18	25	33	44	84
St Catherine Intl	5.27	5.07	−0.46	16	25	43	61	87	122	263
Wadi Elnatroon	7.25	10.23	−0.49	30	49	86	128	185	266	603

Table 9 Summary of the statistical characteristics of the parameters of the GEV distribution of the studied stations

	GEV parameters		
	Location (ξ)	Scale (α)	Shape (κ)
Mean	7.46	6.73	−0.39
Standard deviation	5.03	4.18	0.17
Coefficient of variation	0.67	0.62	−0.42
Maximum	19.25	22.64	0.11
Minimum	1.25	1.21	−0.75

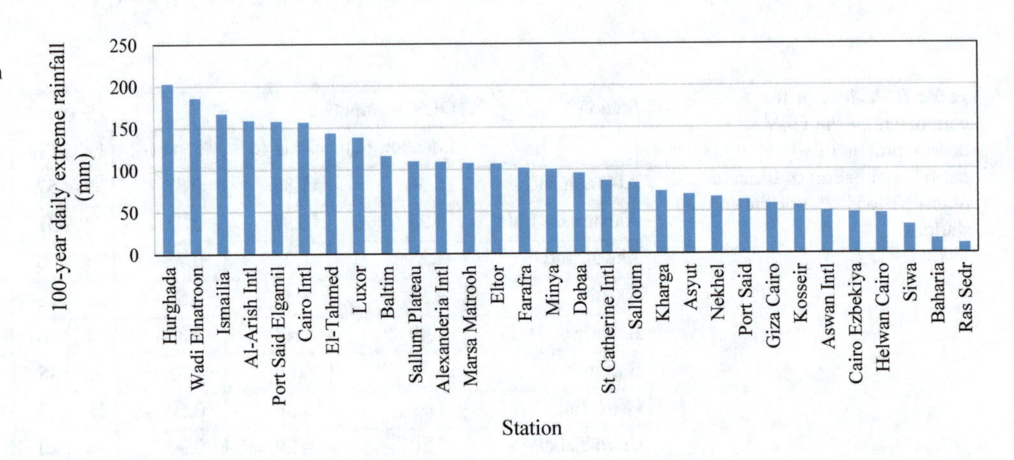

Fig. 14 100-year daily extreme rainfall of the selected stations in Egypt

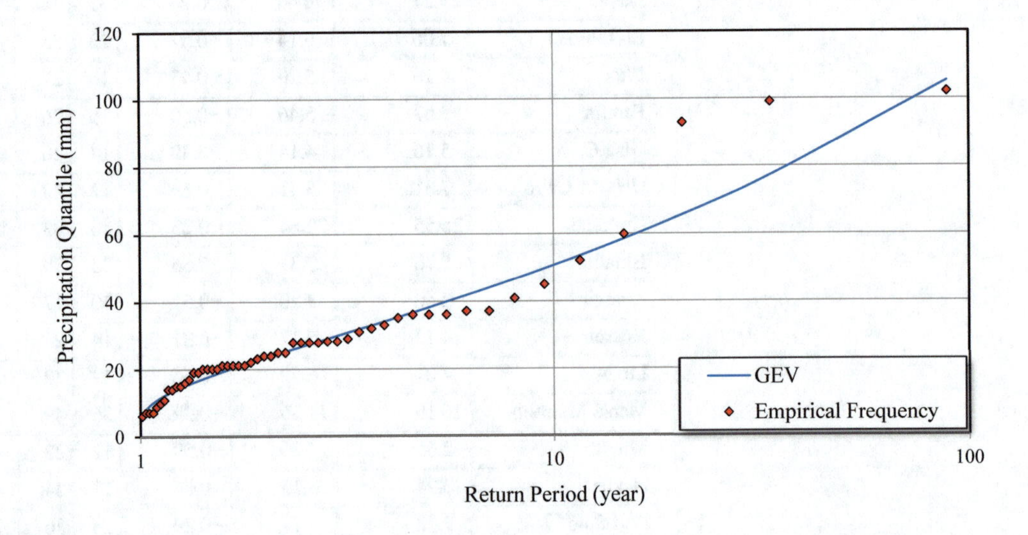

Fig. 15 Daily rainfall quantiles of different return periods for Alexandria station

150 mm at six stations (Fig. 14). These stations include Hurghada (203 mm), Wadi Elnatroon (185 mm), Ismailia (167 mm), Al-Arish Intl (159 mm), Port Said Elgamil (158 mm), and Cairo Intl (157 mm). In contrast, the P_{100} did not exceed 50 mm at six stations (Fig. 14): Ras Sedr (11 mm), Baharia (17 mm), Siwa (33 mm), Helwan (47 mm), Cairo Ezbekiya (48 mm), and Aswan Intl (50 mm). Figure 15 displays the daily rainfall quantiles of different return periods at Alexandria station. It can be concluded that the GEV fits well with the annual maximum daily rainfall data at Alexandria.

5 Conclusions

The statistical characteristics of extreme rainfall in Egypt have been investigated based on historical daily rainfall records for 30 stations throughout the country. Six types of rainfall data were extracted from daily records: monthly rainfall, annual rainfall, the monthly number of rainy days, the annual number of rainy days, monthly maximum daily rainfall, and annual maximum daily rainfall. In general, there is a great variation over the whole country in all different aspects of rainfall. The rainfall indices have higher values in the north of the country than those in the middle and the south. The results conclude that the rainy season in Egypt extends from October to March when some stations receive a significant amount of rainfall. The dry season, on the other hand, extends from April to September.

Furthermore, this research seeks to derive the design rainfalls through the GEV distribution along with the L-moments using annual maximum daily rainfalls in these stations. In the case study, the GEV distribution fits well the annual maximum daily rainfall data with a shape parameter having negative values at all stations except only Ras Sedr which implies an upper bound. Thus, the Gumbel distribution may be better than the GEV in this station. Although all stations not located in the north have low values of rainfall characteristics, some locations (*e.g.*, Hurghada) have high values of design extreme rainfall. Consequently, more investigations are needed in regard to extreme rainfall analysis in these sites.

6 Recommendations

The inferences made in this chapter represent a starting point in regard to the analysis of extreme rainfall events in Egypt. As the case study is based on rainfall data from only 30 stations, a recommendation for future studies is directly related to the use of more stations to have a better understanding of the statistical characteristics of extreme rainfall events in Egypt. Furthermore, different probability distributions along with different parameter estimation methods can be used to identify the best method that could provide the most accurate extreme rainfall estimation in the country. Also, regional rainfall frequency analysis should be used to estimate extreme rainfall quantiles at ungaged sites in Egypt. Lastly, a comprehensive study of the effect of climate change on rainfall characteristics in Egypt should also be conducted to develop robust methods for estimating extreme rainfalls considering climate change.

Acknowledgements A short version of this study was presented at the Twentieth International Water Technology Conference, IWTC20, Hurghada, Egypt, 18–20 May 2017.

References

Awadallah, A. G., & Younan, N. S. (2012). Conservative design rainfall distribution for application in arid regions with sparse data. *Journal of Arid Environments*, *79*, 66–75.

Baldassarre, G. D., Castellarin, A., & Brath, A. (2006). Relationships between statistics of rainfall extremes and mean annual precipitation: An application for design-storm estimation in northern central Italy. *Hydrology and Earth System Sciences*, *10*, 589–601.

Burgueno, A., Martinez, M. D., Lana, X. & Serra, C. (2005). Statistical distribution of the daily rainfall regime in Catalonia (Northeastern Spain) for the years 1950–2000. *International Journal of Climatology*, 25, 1381–1403.

Casas, M. C., Herrero, M., Ninyerola, M., Pons, X., Rodrıguez, R., Rius, A., & Redano, A. (2007). Analysis and objective mapping of extreme daily rainfall in Catalonia. *International Journal of Climatology*, *27*, 399–409.

Cools, J., Vanderkimpen, P., El-Afandi, G., Abdelkhalek, A., Fockedey, S., El-Sammany, M., Abdallah, G., El-Bihery, M., Bauwens, W., & Huygens, M. (2012). An early warning system for flash floods in hyper-arid Egypt. *Natural Hazards and Earth System Sciences*, *12*, 443–457. https://doi.org/10.5194/nhess-12-443-2012.

El-Afandi, G., Morsy, M., & El-Hussieny, F. (2013). Heavy rainfall simulation over Sinai Peninsula using the weather research and forecasting model. *International Journal of Atmospheric Sciences, Hindawi Publishing Corporation*, *2013*, Article ID 241050.

El-Sayed E. A. H. (2011). Generation of rainfall intensity duration frequency curves for ungauged sites. *Nile Basin Water Science & Engineering Journal*, *4*(1).

El-Shafie, A. H., El-Shafie, A., El Mazoghi, H. G., Shehata, A., & Taha, M. R. (2011). Artificial neural network technique for rainfall forecasting applied to Alexandria, Egypt. *International Journal of the Physical Sciences*, *6*(6), 1306–1316. https://doi.org/10.5897/ijps11.143.

Fathy, I., Negm, A. M., El-Fiky, M., Nassar, M., & Al-Sayed, E. (2014). Intensity duration frequency curves for Sinai Peninsula, Egypt. *IMPACT: IJRET*, *2*(6), 105–112, ISSN (E): 2321-8843; ISSN (P): 2347-4599.

Gado, T. A. (2017). Statistical characteristics of extreme rainfall events in Egypt. In *Twentieth International Water Technology Conference, IWTC20* (pp. 18–20). http://iwtc.info/wp-content/uploads/2017/05/44.pdf.

Gado, T. A., & Nguyen, V. T. V. (2015). Comparison of homogeneous region delineation approaches for regional flood frequency analysis at ungauged sites. *Journal of Hydrologic Engineering*, 04015068. https://doi.org/10.1061/(asce)he.1943-5584.0001312.

Gado T. A., & Nguyen V. T. V. (2016). An at-site flood estimation method in the context of nonstationarity. II: Statistical analysis of floods in Quebec. *Journal of Hydrology*, *535*, 722–736.

Hosking, J. R. M., Wallis, J. R., & Wood, E. F. (1985). An appraisal of the regional flood frequency procedure in the U.K. flood studies report. *Hydrological Sciences Journal 30*(1), 85–109.

Ibrahim S., & El-Afandi G. (2014). Short-range rainfall prediction over Egypt using the weather research and forecasting model. *Open Journal of Renewable Energy and Sustainable Development*, *1*(2).

Jiang, P., & Tung, Y.-K. (2013). Establishing rainfall depth–duration–frequency relationships at daily raingauge stations in Hong Kong. *Journal of Hydrology*, *504*, 80–93.

Koutsoyiannis, D. (2004). Statistics of extremes and estimation of extreme rainfall: II. Empirical investigation of long rainfall records. *Hydrological Sciences Journal*, *49*, 591–610.

Lee, S. H., & Maeng, S. J. (2003). Frequency analysis of extreme rainfall using L-moment. *Irrigation and Drainage*, *52*, 219–230.

Loukas, A., & Quick, M. C. (1996). Spatial and temporal distribution of storm precipitation in southwestern British Columbia. *Journal of Hydrology*, *174*, 37–56.

Moawad, M. B. (2013). Analysis of the flash flood occurred on 18 January 2010 in wadi El Arish, Egypt (a case study). *Geomatics, Natural Hazards and Risk*, *4*(3), 254–274. https://doi.org/10.1080/19475705.2012.731657.

Moawad, M. B., Abdelaziz, A. O., & Mamtimin, B. (2016). Flash floods in the Sahara: A case study for the 28 January 2013 flood in Qena, Egypt. *Geomatics, Natural Hazards and Risk*, *7*(1), 215–236. https://doi.org/10.1080/19475705.2014.885467.

Nguyen, V. T. V., & Mayabi, A. (1990). Probabilistic analysis of summer daily rainfall for the Montreal region. *Canadian Water Resources Journal*, *15*(3).

Nguyen, V. T. V., Tao, D., & Bourque, A. (2002). On selection of probability distributions for representing annual extreme rainfall series. *Global Solutions for Urban Drainage*.

Nguyen, T.-H., El Outayek, S., Lim, S. H., & Nguyen, V. T. V. (2017). A systematic approach to selecting the best probability models for annual maximum rainfalls—A case study using data in Ontario (Canada). *Journal of Hydrology*, *553*, 49–58.

Nhat, L. M., Tachikawa, Y., & Takara, K. (2006). Establishment of intensity-duration-frequency curves for precipitation in the monsoon area of vietnam. *Annuals Disaster Prevention Research Institute* of Kyoto University, (49 B).

Papalexiou, S. M., & Koutsoyiannis, D. (2013). Battle of extreme value distributions: A global survey on extreme daily rainfall. *Water Resources Research*, *49*, 187–201.

Punlum, P., Chaleeraktrakoon, C., & Nguyen, V. T. V. (2017). Development of IDF relations for thailand in consideration of the scale-invariance properties of extreme rainfall processes. In *World Environmental and Water Resources Congress*, Sacramento, California. https://doi.org/10.1061/9780784480618.063.

Said, M. A., El-Geziry, T. M., & Radwan, A. A. (2012). Long-term trends of extreme climate events over Alexandria region, Egypt. *INOC-CNRS, International Conference on "Land-Sea Interactions in the Coastal Zone" Jounieh—LEBANON*, 06–08.

Salama, A. M., Gado, T. A., & Zeidan, B. A. (2018). On selection of probability distributions for annual extreme rainfall series in Egypt. In *Twenty-first International Water Technology Conference, IWTC21, Ismailia* (pp. 383–394).

Tarawneh, Q., & Kadioglu, M. (2003). An analysis of precipitation climatology in Jordan. *Theoretical and Applied Climatology*, *74*, 123–136.

Vivekanandan, N. (2012). Probabilistic modelling of hourly rainfall data for development of intensity-duration-frequency relationships. *Bonfring International Journal of Data Mining*, *2*(4).

Vivekanandan, N. (2015). Rainfall frequency analysis using l-moments of probability distributions. *International Journal of Engineering Issues*, *2015*, 65–72.

Wdowikowski, M., Kaźmierczak, B., & Ledvinka, O. (2016). Maximum daily rainfall analysis at selected meteorological stations in the upper Lusatian Neisse River basin. *Meteorology Hydrology and Water Management*, *4*(1).

Wdowikowski, M., Kotowski, A., Dąbek, P. B., & Kaźmierczak, B. (2017). Probabilistic approach of the Upper and Middle Odra basin daily rainfall modeling. In *E3S Web of Conferences* (Vol. 17). https://doi.org/10.1051/e3sconf/20171700096.

Yehia, A. G., Fahmy, K. M., Mehany, M. A. S., & Mohamed, G. G. (2017). Impact of extreme climate events on water supply sustainability in Egypt: Case studies in Alexandria region and Upper Egypt. *Journal of Water and Climate Change*. https://doi.org/10.2166/wcc.2017.111.

Critical Analysis and Design Aspects of Flash Flood in Arid Regions: Cases from Egypt and Oman

Alaa El Zawahry, Ahmed H. Soliman, and Hesham Bekhit

Abstract

This chapter addresses several cases of peculiar flash flood events that took place recently in the arid region. The addressed cases covered extreme events generated from Cyclonic and convective storms. All cases caused damages to the properties and loss of lives in some events. For each case, detailed hydrological and hydraulic analyses are carried out and the results are presented. The hydrological analyses covered the rainfall frequency analysis, the storm pattern as a function of time, the catchment properties and the computed flood (flow and volume). Moreover, the hydraulic study presented the degree of impacts of such events. The results indicated that all cyclonic events were accompanied by a very high value of rain corresponding to thousands of years on the frequency distribution curves. Such rainfall events generated huge volumes of water filling all dams and running the spillways to near maximum design flow values. However, the storm distribution over 24 h was almost uniform, indicating moderate rainfall intensity. The study covered cases in Sultanate of Oman and in Arab Republic of Egypt.

Keywords

Flash floods • Oman cyclones • Egypt floods

A. El Zawahry (✉) · A. H. Soliman · H. Bekhit
Irrigation and Hydraulics Department, Faculty of Engineering, Cairo University, Giza, Egypt
e-mail: alaa.zawahry@gmail.com

A. H. Soliman
e-mail: a.soliman@cu.edu.eg

H. Bekhit
e-mail: heshambm@hotmail.net

A. M. Negm (ed.), *Flash Floods in Egypt*, Advances in Science, Technology & Innovation,
https://doi.org/10.1007/978-3-030-29635-3_3

1 Introduction

The temporary covering of a specific land by water, which is not normally covered by water, is defined as flood. Floods usually occurred in the same areas called floodplains due to well-known flood conditions and reasons. The floodplains are characterized as flat and fertile lands, which attract a lot of people to live on. The amount of flood is extremely affected by urbanization and climate change. So, the vulnerability of communities inhabit in the floodplains is increasing worldwide due to the exposure of flood hazards (El-Hames and Al-Wagdany 2012). Meanwhile, flash floods occurred due to heavy rainfall intensities within 6–12 h (Gaume et al. 2009). Also, flash floods can be defined as the rapid flow of large amount of water into a usually dry area or a rapid increase of water level in a river or stream (National Weather Service glossary 2016). A major part of the world may be categorized as arid or semi-arid. Such a climate is well-known for flash flood development due to convective rainfall and poor soil. Such a combination creates a hazard, noticeably different from other climates. During the last three decades; about 40% of worldwide floods occurred in Asia, 22% in Africa, 21% in America, 14% in Europe and 3% in Oceania (The OFDA/CRED International Disaster Database). Due to the remoteness of such areas and sparsely population, the flash flood information and data required for simulation are lacking.

The mountainous arid areas in Egypt and the gulf (Sultanate of Oman) are suffering from flash flood attacks. The two countries have been selected for detailed investigation due to the availability of enough data to carry out detailed simulations. The available data covers temporal and spatial long rainfall and flow records. Moreover, Sultanate of Oman has experienced severe cyclone over the recent decades.

2 Sultanate of Oman Cyclones

Sultanate of Oman is famous for Cyclone attacks generated from the Arabian Sea and the Gulf of Oman. Table 1 presents a list of cyclones rainfall that attacked Oman in recent decades.

In this chapter, the last three cyclones that attacked Oman (Gonu, Phet and Mukuno) are analyzed, with focus on Gonu as the most severe cyclone. The analyses address three main hydrological design parameters: Rainfall, Flood Volume and Peak Flood.

3 Gonu Cyclone

Gonu as a Super Cyclonic Storm is the strongest tropical cyclone on record in the Arabian Sea and the northern Indian Ocean. Gonu developed from a persistent area of convection in the eastern Arabian Sea, started on 1 June 2007. On 6 June , it attacked the eastern tip of Oman. The cyclone caused about $4.2 billion in damage (2007 USD) and a causality of 50 deaths in Oman. Gonu was considered the nation's worst natural disaster.

3.1 Gonu's Data Availability

The available data was limited to 12 rain gauges surrounding the area of Muscat. The rain records included the rainfall in mm against time in minutes (Hyetograph) for the entire period of the cyclone around one day. Additionally, the daily volume recorded in two wadi gauges, Wadi Aday and Wadi Al Ansab.

3.2 Rainfall Analysis

The available data was limited to 12 rain gauges surrounding the area of Muscat. The rain records included the rainfall in mm against time in minutes (Hyetograph) for the entire period of the cyclone around one day. Additionally, the daily volume recorded in two wadi gauges, Wadi Aday and Wadi Al Ansab.

3.2.1 Daily Rainfall

Table 2 summarizes the rainfall recorded on 6 June 2007. The maximum recorded value reached 613.2 mm at HAYFADH. This value is near the estimated Probable Maximum Precipitation (PMP) at Muscat zone. The estimated PMP was based on AMRAT rain-gauge using all records up to 2006. The recorded rainfall in Muscat rain gauge is near to 50-years return period using records up to 2006. Figure 1 depicts the daily rainfall contour lines for the listed rain gauges.

3.2.2 Daily Rainfall

Figure 2 shows the cumulative rainfall percentage for Yiti rain gauge (total 350 mm on 6 June). It is clear that the rainfall was somehow continuous all over the day. However, the variation of the rainfall intensity along the day was not large. The slope of the cumulative rainfall with respect to time is mild compared with the adopted storm in Oman (Wheater and Bell 1983). The slope of the cumulative rainfall can be simply assumed as 12.5 mm/h. This value is much less than the rainfall intensity adopted by Wheater and Bell (1983), which can be taken as 90 mm/h for 100-years return period. It is important to point out that the total daily precipitation is much higher than the 100 years, but the intensity is much less. The same was applicable for all gauges with some degree of freedom.

Table 1 Oman cyclones rainfall

Period	Max. total precipitation (mm)	Max. 12 h precipitation (mm)	Region
23–26 October 1948	156.8	33.0	Dhofar
23–26 May 1959	117.2	81.8	Dhofar
26–29 May 1963	236.1	134.0	Dhofar
12–13 November 1966	202.3	178.1	Dhofar
14–18 June 1977	122.7	70.3	Dhofar
4 April 1983	NA	127.0	Dhofar
11–13 May 1996	98.0	NA	Dhofar
9–12 May 2002	191.8	145.6	Dhofar
29–30 September 2004	116.2	116.2	Dhofar
5–6 June 2007	903	688	North
2–5 June 2010	485.4	277.4	North
25–26 May 2018	505	NA	Dhofar

Table 2 Recorded rainfall (mm), 6 June 2007

Station		6 June 2007
Code	Name	Rainfall (mm)
FL609065AF	Yiti	286.2
FL672275AF	Buei near Buei	378.4
FL681271CF	Fuwad 2 at Fuwad	373.8
FL771702AF	Taba at Taba	514.8
FL788017AF	Hayfadh at Hayfadh	613.2
FL950429CF	Mazara 3 at Mazara	322.0
FL974085AF	Qurayat near Qurayat	187.8
FM104840AF	Al Khaud	118.8
FM415095AF	Al Khuwair	168.8
FM517016AF	Ruwi at Ruwi	223.8
FM593824AF	Amrat	438.4
FM612285BF	Muscat near Muscat	139.4

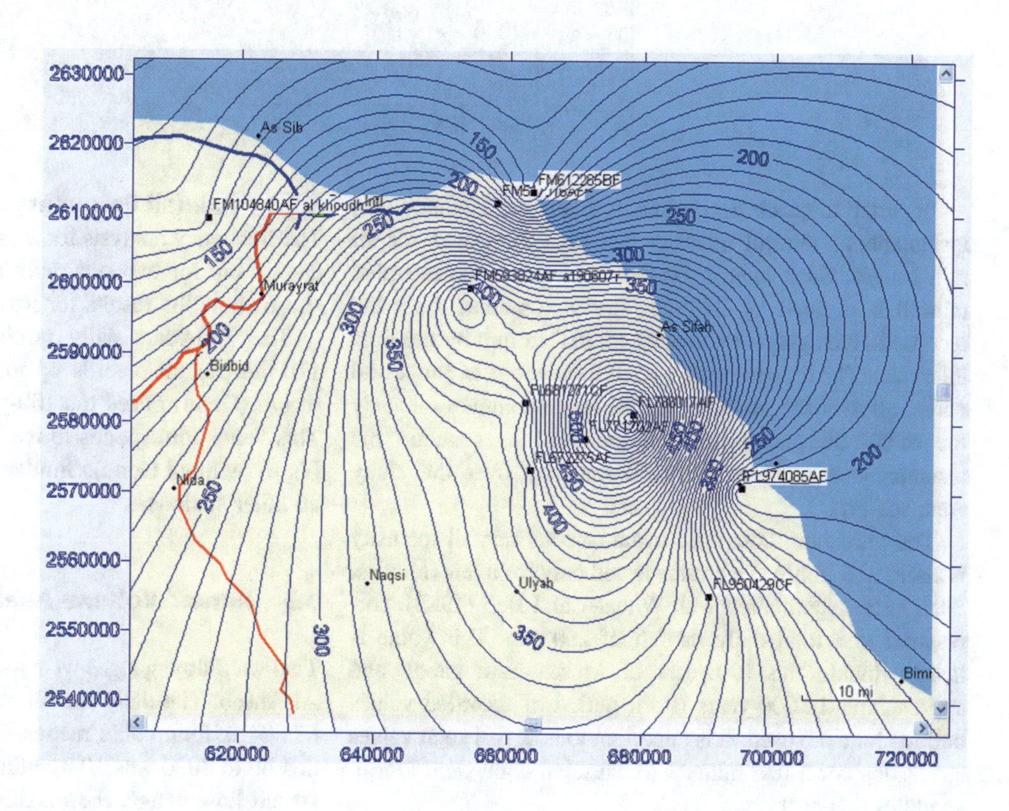

Fig. 1 Daily rainfall distribution

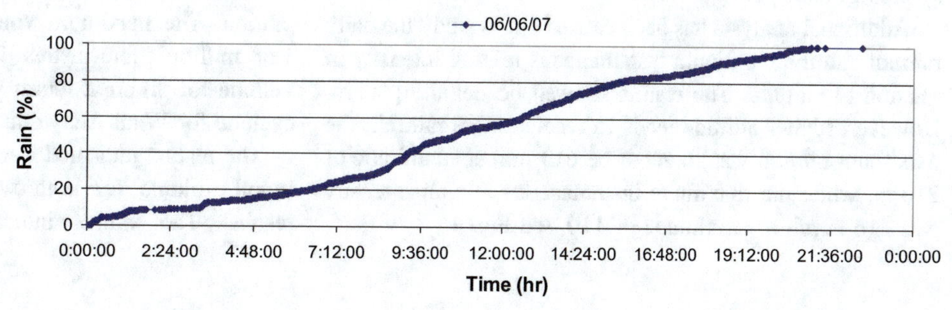

Fig. 2 Rainfall% of the total for Yiti rain gauge (6 June 2007)

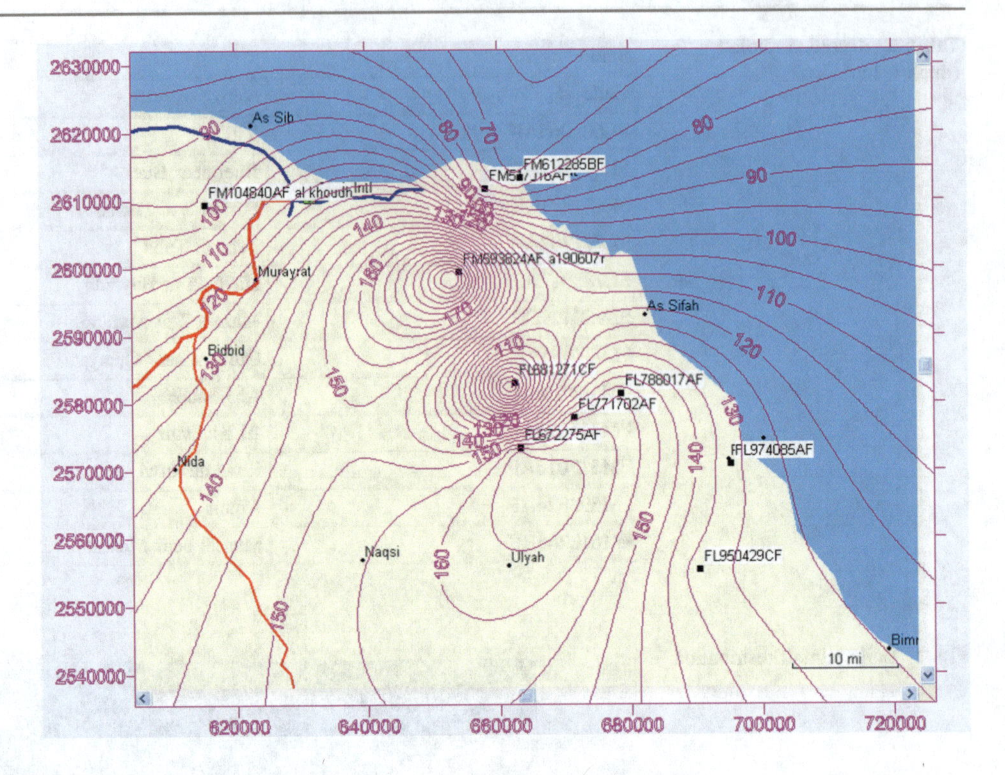

Fig. 3 Maximum instantaneous rainfall intensity

In order to reach a conclusion about the intensity, the instantaneous rainfall intensity has been computed for all rain gauges. Such intensity is defined here, as the recorded rainfall is divided by its time interval (obtained from the recorded Hyetograph by divided rainfall in mm by the time intervals). The time intervals varied from gauge to gauge and varied all over the period of record. The maximum intensity for each rain gauge has been adopted to generate the instantaneous maximum rainfall intensity over the study area, see Fig. 3.

The maximum computed instantaneous rainfall intensity is about 216 mm/h. This value lasted only for a few minutes. Such values, compared with Wheater and Bell (1983), correspond to a total daily rainfall of 260 mm. This value is nearly double the 100 years of Muscat rain gauge and approaching 10,000-years return period, if recorded values before Gonu are used. It is important to note that such values are lasting for a few minutes to cause an equivalent (corresponding) damage.

Additional analysis has been carried out to study the daily rainfall and the maximum instantaneous rainfall intensity in relation to altitude. The results showed no definite pattern. However, higher altitude tends to receive more rainfall; the maximum rainfall was found to be 613 mm at an altitude of 210 m, while the maximum instantaneous rainfall intensity was 216 mm/h at an altitude of 110, see Fig. 4.

3.2.3 Rainfall Frequency Analysis

The frequency analysis for maximum daily rainfall has been carried out for two sets, with and without Gonu. Figures 5 and 6 show the results for Ruwi rain gauge.

The 100-years daily rainfall at Ruwi station reached 110 mm for all records up to 2006. The addition of 2007 record (Gonu) raised the 100-years daily rainfall up to 200. This value corresponds to the 10,000-years return period in Fig. 6 (without Gonu). Similar results have been observed in all other stations.

3.3 Runoff Volume Analysis

Two wadi flow gauges were available, Wadi Aday and Wadi Al Ansab. The daily runoff volume for wadi Aday reached 67.824 million cubic metres. The maximum volume recorded up to 2006 was 10.8 million cubic metres. For Wadi Al Ansab flow gauge, the maximum recorded was about 13.5 million. The maximum volume recorded up to 2006 was 1.68 million cubic metres. Figures 7 and 8 show the daily volume for different return periods with and without Gonu cyclone for Wadi Aday and Wadi Al Ansab.

The results indicated a tremendous increase in the daily runoff volume for both wadis. Table 3 summarizes the results. The values indicate that the runoff volumes

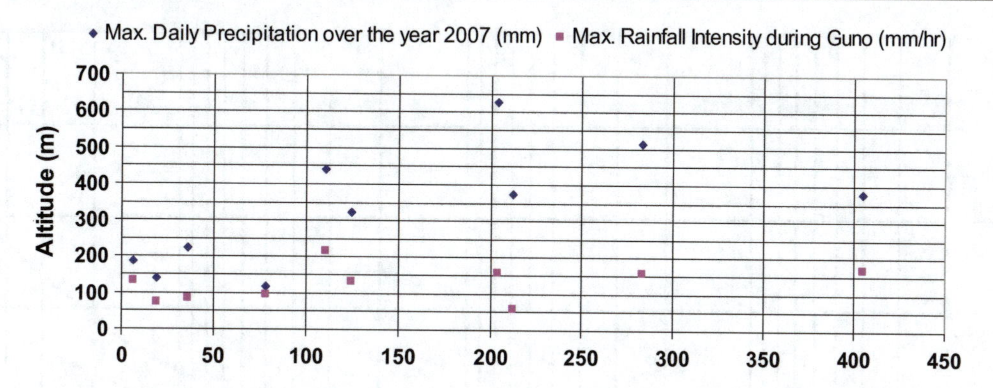

Fig. 4 Daily rainfall and maximum instantaneous rainfall intensity in relation to altitude

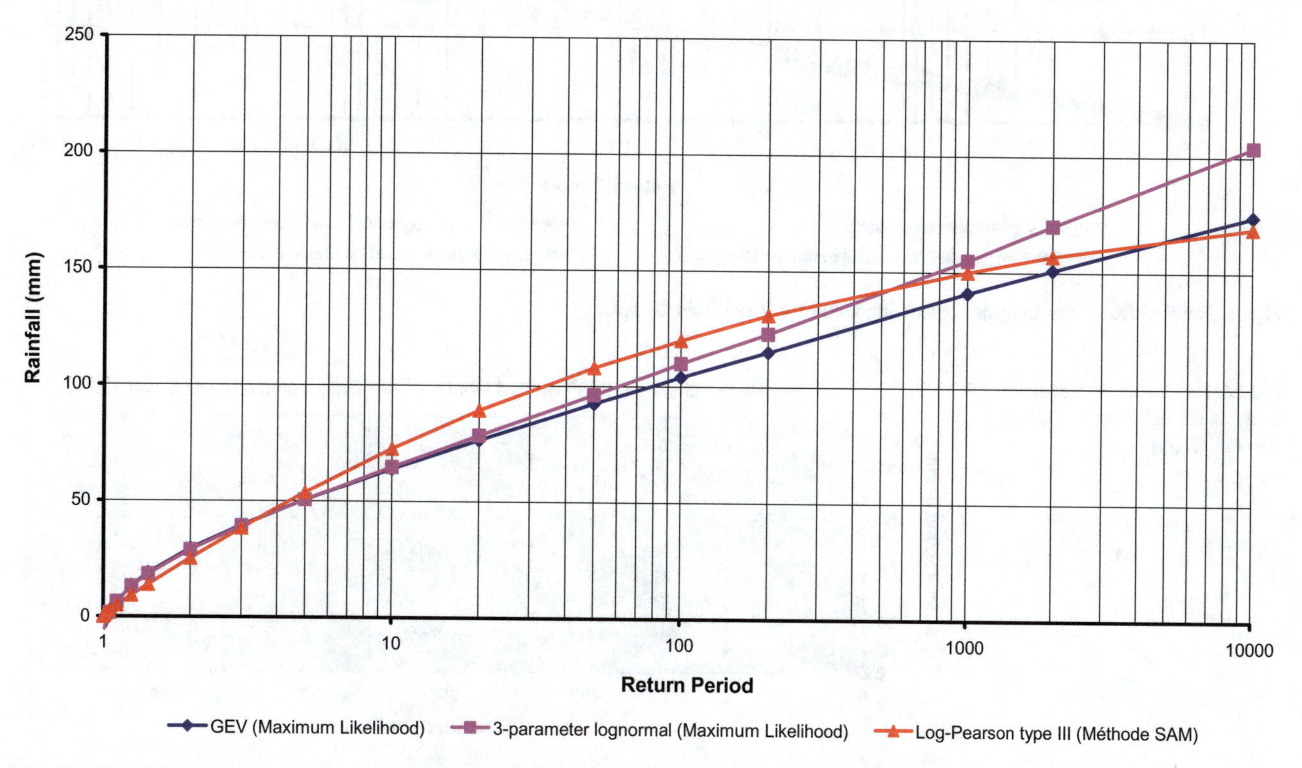

Fig. 5 Rainfall frequency analysis without considering Gonu for Ruwi Rain Gauge

generated from Gonu for both wadis were never experienced over the available period of records. The 100-years daily runoff volume including Gonu exceeds four times the values previously adopted for the design. The volume recorded for Gonu corresponds to 2000 and 10,000 years for Wadi Aday and Al Ansab, respectively. The recorded volumes during and after Gonu are shown in Fig. 9.

3.4 Peak Flow Analysis

The obtained data did not include any peak flow records. In order to obtain feelings of the peak floods associated with Gonu, the following assumptions have been adopted:

1. The SCS unit hydrograph approach is used.
2. The maximum recorded rainfall of Hayfadh at Hayfadh (613 mm) is adopted and assumed over the entire catchment. Figure 10 shows the hyetograph.
3. The actual recorded design storm of HAYFADH AT HAYFADH is used instead of Wheater and Bell (1983) storm distribution. Figure 11 shows the two design storms.
4. The rain is covering the entire catchment (area reduction factor = 1).
5. The runoff coefficient equals 0.8.

The results of applying the above assumptions on Wadi Aday are given in Fig. 12. Based on the maximum recorded

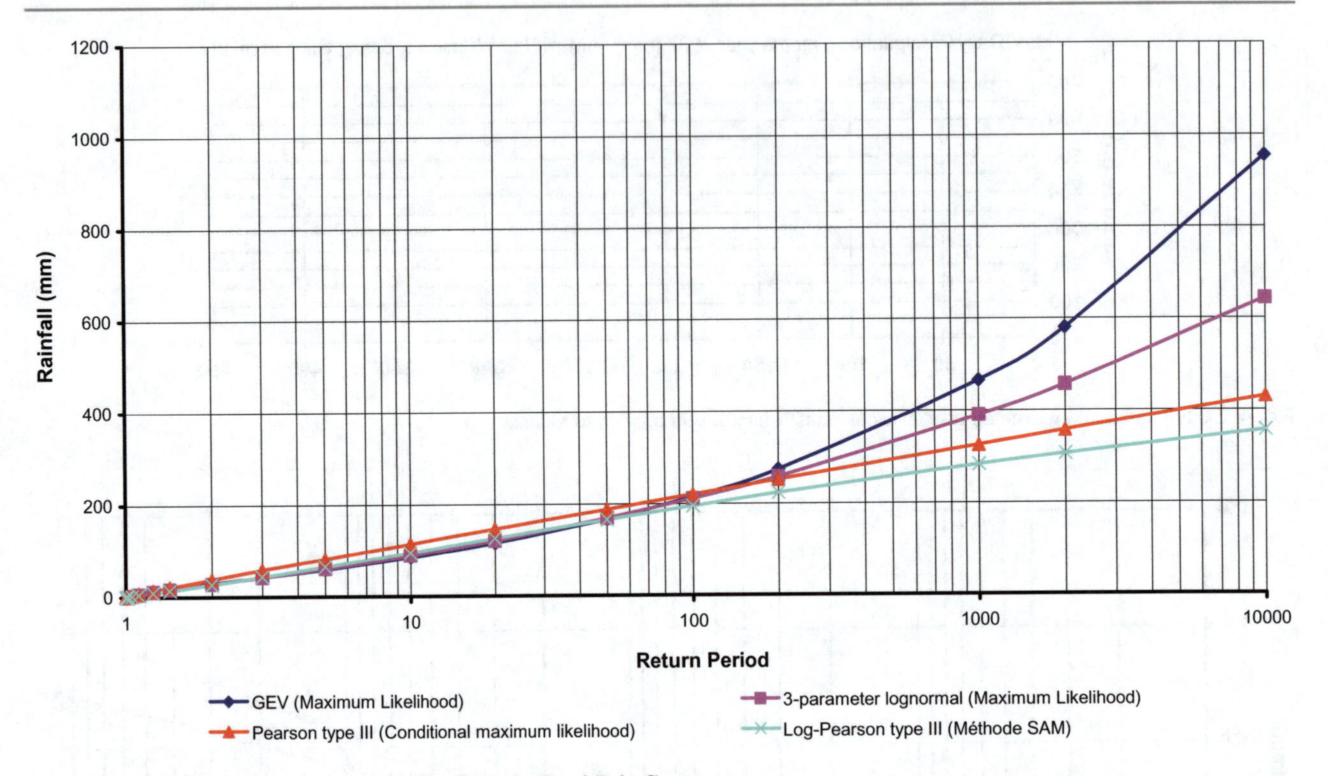

Fig. 6 Rainfall frequency analysis considering Gonu for Ruwi Rain Gauge

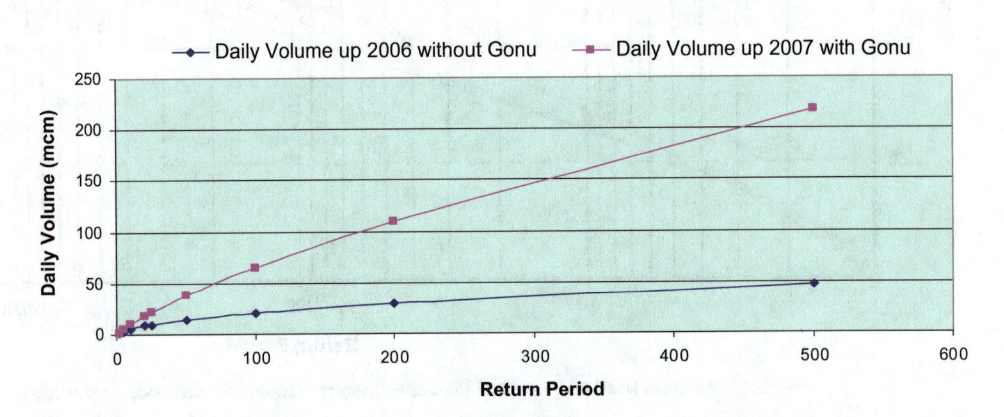

Fig. 7 Daily volume frequency analysis for Wadi Aday (with and without Gonu)

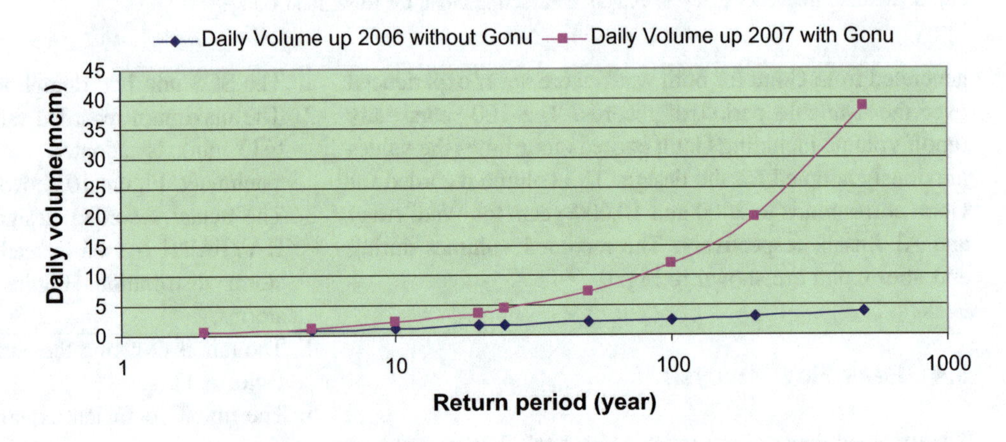

Fig. 8 Daily volume frequency analysis for Wadi Al Ansab (with and without Gonu)

Table 3 Wadi Aday daily runoff frequency analysis (with and without Gonu)

	Wadi Aday		Wadi Al Ansab	
Return period (yr)	Daily volume up 2006 without Gonu	Daily volume up 2007 with Gonu	Daily volume up 2006 without Gonu	Daily volume up 2007 with Gonu
2	1.06	1.37	0.46	0.47
5	3.16	5.15	0.87	1.20
10	5.35	10.10	1.20	2.14
20	8.34	18.29	1.56	3.67
25	9.53	21.95	1.68	4.34
50	14.13	38.03	2.10	7.26
100	20.55	64.90	2.58	12.04
200	29.51	109.82	3.13	19.88
500	47.06	218.61	3.97	38.44

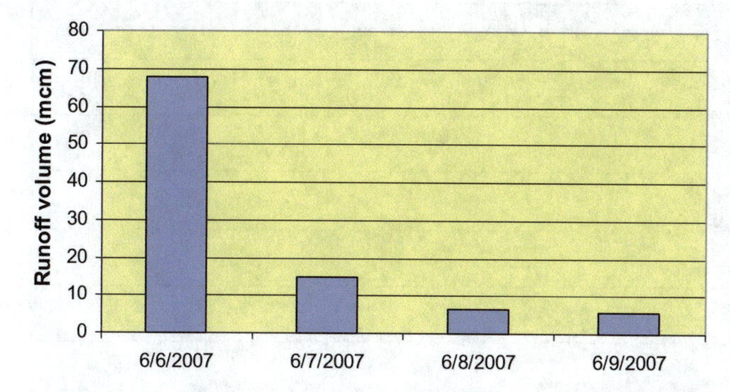

Fig. 9 Recorded daily runoff volume (mcm)

rainfall, the peak flood reached 3600 m^3/s, and the runoff volume 155 million cubic metre. The recorded runoff volume during and after 6 June 2007 are given in Fig. 9. The recorded volume over the hydrograph period (30 h) can be assumed as 75 m cm. This indicates that the generated volume from the hydrograph (Fig. 12) is double the recorded. The only assumption, out of the above, needs to be changed is the effective rainfall storm. The numbers indicate that the daily rainfall over the entire catchment has to be reduced to 300 mm instead of 613 mm, keeping all other parameters the same. In such a case, the peak flood may be estimated as 1600 m^3/s. The rainfall has a return period of 10,000-years compared with the used values before Gonu, while the peak flood is slightly higher than the 100-years return period.

3.5 Conclusions from Gonu

The above analysis has been based on the available data, additional data of peak floods and more rain gauges and flood gauges that will enhance the analysis and reach more

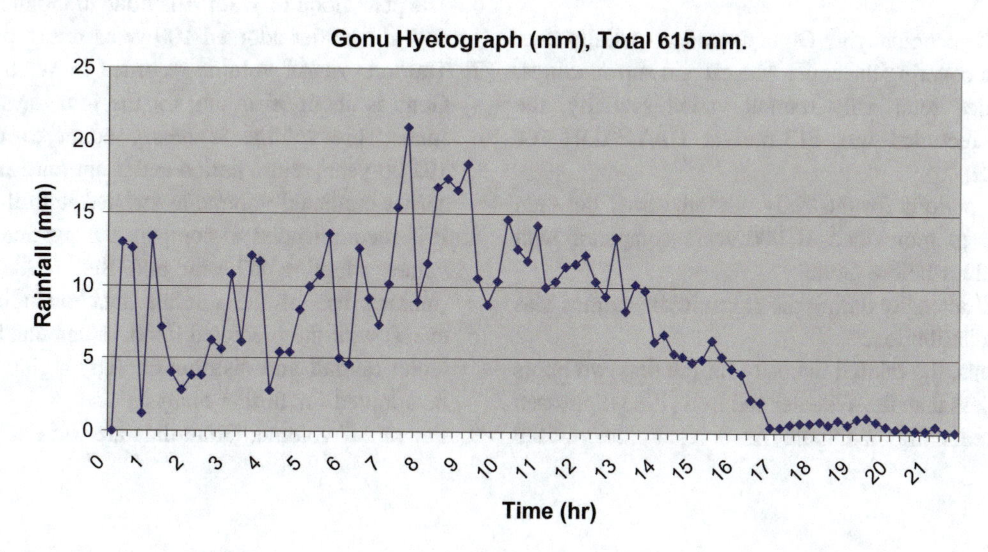

Fig. 10 Gonu hyetograph (max recorded, FL788017AF)

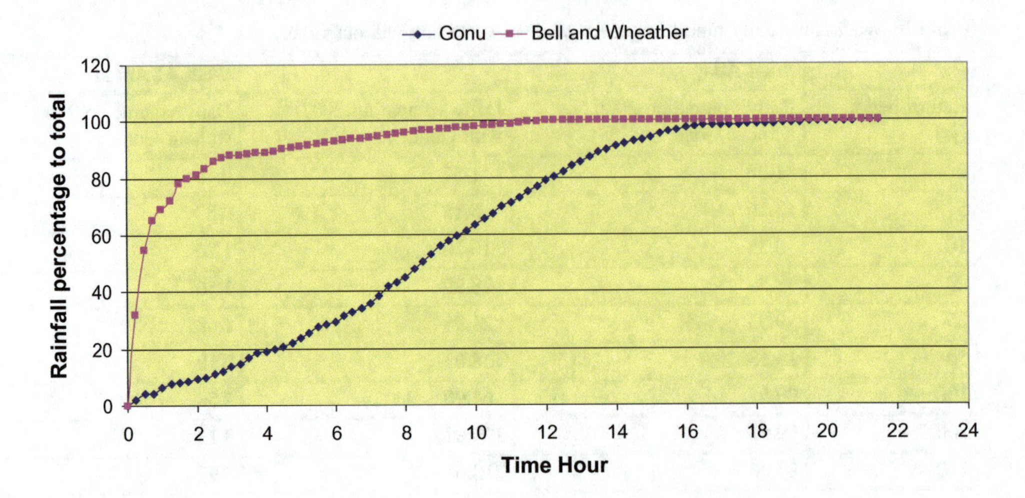

Fig. 11 Gonu design storm against Wheater and Bell (1983)

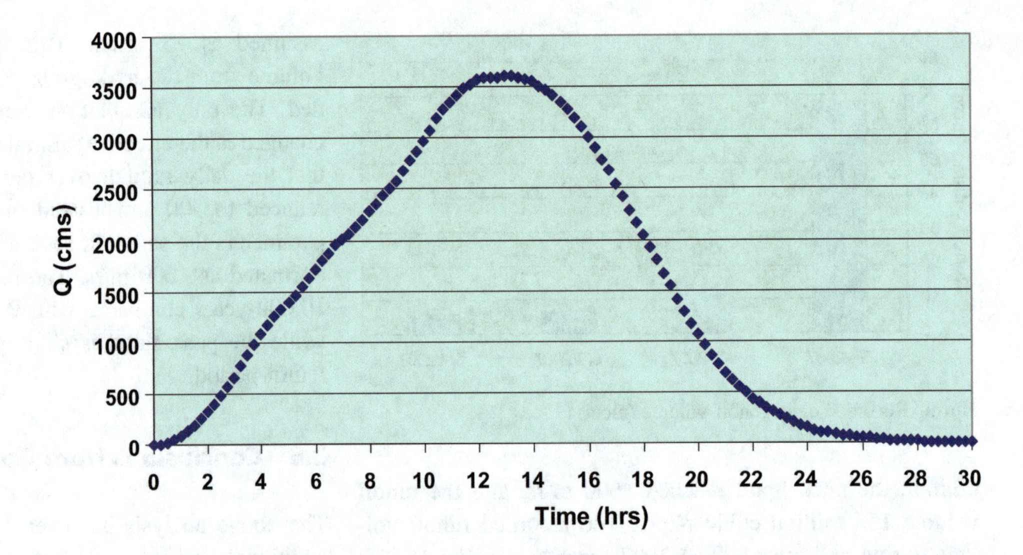

Fig. 12 Wadi Aday hydrograph using maximum daily rainfall of 613 mm

solid conclusions. The following assumptions may be drawn at this stage:

1. The rainfall accompanying Gonu extended spatially over a huge area covering the entire Muscat and surroundings.
2. The recorded total daily rainfall varied spatially, the maximum recorded was 613 mm at HAYFADH AT HAYFADH.
3. The return periods for the daily rainfall varied between 50 and up to more than 10,000 years compared with adopted values before Gonu.
4. The rainfall intensity during the storm in all stations was similar in distribution.
5. In all stations, the rainfall intensity for the first two hours was much less than the Wheater and Bell (1983) approach adopted frequently in Oman. As a result, the revised predicted peak floods did not change dramatically from the earlier calculations; although the total daily rainfall does.
6. The peak flood in Wadi Aday due to Gonu is comparable with the earlier adopted 100-years return period.
7. The total runoff volume recorded in Wadi Aday due to Gonu is about 96 m cm, for the four days starting 6 of June. This volume is nearly equal to the estimated 10,000-years return period and is approaching 75% of the earlier predicted volume associated with the PMF.
8. It is recommended to compare the predicted peak flood values adopting Wheater and Bell (1983) distribution (making use of the rainfall data excluding the Gonu event) with the predicted flood values making use of the Gonu rainfall and distribution. The higher values are to be adopted for further analysis.
9. For runoff volume, Gonu data are to be used.

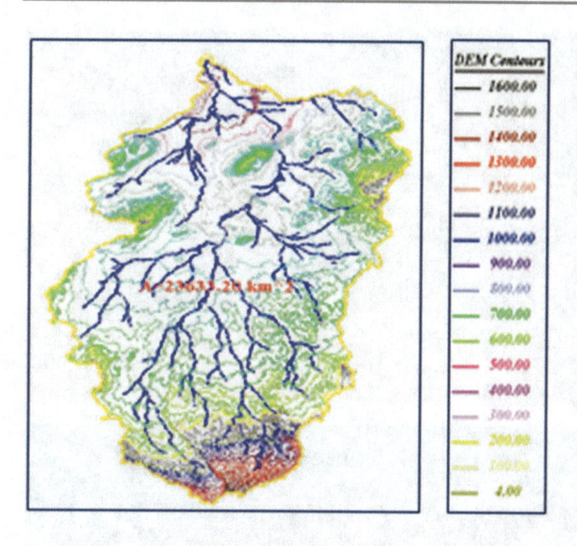

Fig. 13 Wadi El Arish delineated catchment

3.6 Oman Highway Design Standards

The Experience of Gonu impacts on Oman infrastructures and properties, and necessitates Engineering actions towards the adopted design standards before Gonu. All available hydrological records including Gonu have been used in a detailed study to obtain a new Highway Design standard dealing with such severe storms and Cyclones.

4 Egypt Case

Several sites in Egypt are frequently attacked by flash floods. Even though the peak floods are not high compared with Oman, the impacts are still severe. This part of the chapter is allocated for analysis of one of the most severe flash floods that attacked Wadi El Arish area.

4.1 Wadi El Arish

Wadi El Arish is located at North Sinai and drains to the Mediterranean Sea. The wadi has a total catchment area of about 24000 km^2 and a maximum stream length of about 324 km (Fig. 13). In 2010 El Arish City was attacked by a severe flood that removed all the major roads across the city including the international shoreline road, as shown in Fig. 14.

Several rainfall stations are available around Wadi El Arish Catchment as presented in Fig. 15.

Five of the available rainfall stations recorded 17–18 January 2010 rainfall event as presented in Table 4. The average rainfall duration of the recorded storms for all stations is about 12.7 h with an arithmetic mean of the same stations as 29.76 mm.

The rainfall mass curve of El-Tahmed rainfall station is presented in Fig. 16.

The maximum daily rainfall recorded at each one of the same five rainfall stations before January 2010 rainfall event is presented in Table 5. The average rainfall duration of the recorded storms for all stations is about 8.5 h while the arithmetic mean of the same stations is 16.02 mm. The comparison between January 2010 event and the historical rainfall events shows that the rainfall depth of January 2010 event is about twice the historical records and extended along about one and half times the duration of historical records. This quick analysis explains the reason for considering January 2010 event as a severe rainfall event.

5 Conclusions

This chapter presented several extreme events that took place in the arid regions. By the four cyclonic cases in Sultanate of Oman the following may be concluded, for the major cyclone Gonue took place in June 2007:

Fig. 14 Roads and properties damage due to Wadi El Arish Flash Flood in 2010

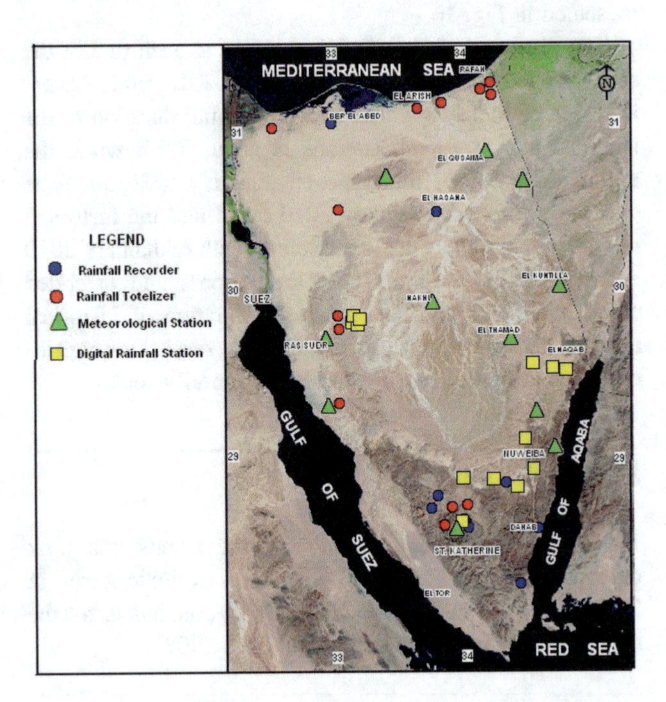

Fig. 15 Available rainfall stations around Wadi El Arish Catchment

- The maximum rainfall was 904 mm/d, recorded on the mountainous rain gauge. Such value corresponded to 10,000-years return period. Such storm generated huge volume of water.
- All dams were full, and the spillways reached the maximum design values, the risk of dam failure was very near.
- The damage took place was major.
- The storm distribution as function of time was nearly uniform over 24 h, and the generated flows were comparable with generated by Wheater and Bell (1983) design storm of Oman.
- Based on the above, Oman developed a new Highway Design Standards (HDS) (2010) to accommodate such events.
- The HDS has been updated in 2017 (Highway Design Standards (HDS) 2017).
- It was observed in all study cases that the Right of Way (ROW) of all wadis is not respected, with massive constructions by the local.
- Some of the protection works constructed over the last few decades were not adequately designed/constructed.

Table 4 Rainfall records of 17–18 January 2010 event

Station	Rainfall depth (mm)	Duration (h)
ST. Katherine	25.1	11.0
Rawafaa	19.1	14.0
El-Nekhael	50.4	15.0
El-Tahmed	21.2	8.5
Al-Godairat	33.0	15.0

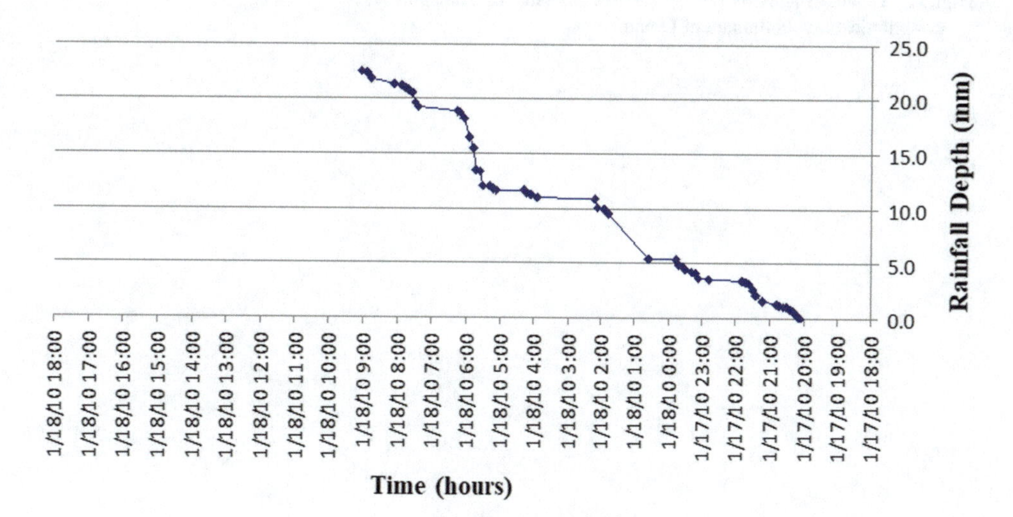

Fig. 16 Rainfall mass curve of El-Tahmed rainfall event (17–18 January 2010)

Table 5 Rainfall records of 17–18 January 2010 event

Station	Rainfall depth (mm)	Duration (h)
ST. Katherine	20.6	8.0
Rawafaa	25.0	10.5
El-Nekhael	9.8	1.0
El-Tahmed	9.0	13.0
Al-Godairat	15.7	10.0

- There is a clear lack of maintenance in the hydraulic and drainage structures, which caused blocking of the drainage area causing overtopping, flooding and failure of the structures.

6 Recommendations

The main recommendations are as follows:

- The floodplains and its buffer zones corresponding to 100 years return period should be defined for all wadis.
- The building inside the buffer zones of all wadis should be prohibited.
- Laws related to waterways and its floodplains should be strengthened and strongly applied.
- The rainfall distribution and duration should be considered carefully especially in cyclone cases.
- The floodplains and buffer zones of all wadis should be reviewed and updated periodically to check the effect of new rainfall events of the rainfall depths corresponding to different return periods.
- The maintenance of all wadis, streams, waterways and hydraulic structures should be conducted periodically to attain all of them working correctly during flood events.
- All design standards related to flood events should be updated periodically to cope with the new flood events.

References

Al-Hames, A., & Al-Wagdany, A. (2012). Reconstruction of flood characteristics in urbanized arid regions: Case study of the flood of 25 November 2009 in Jeddah, Saudi Arabia. *Hydrological Sciences Journal, 57*(3). https://doi.org/10.1080/02626667.2012.665995.

Gaume, E., Bain, V., Bernardara, P., Newinger, O., Barbuc, M., Bateman, A., et al. (2009). A compilation of data on European flash floods. *Journal of Hydrology, 367*(1–2), 70–78.

Highway Design Standards (HDS). (2010). Ministry of Transport and communications, Sultanate of Oman.

Highway Design Standards (HDS). (2017). Ministry of Transport and communications, Sultanate of Oman.

National Weather Service glossary (2016). http://w1.weather.gov/glossary/index.php.

The OFDA/CRED International Disaster Database, Emergency Events Database (EMDAT). https://www.emdat.be/.

Wheater, H. S., & Bell, N. C. (1983). North Oman flood study. *Proceedings of the Institution, Civil Engineers Part,* 2(75), 453–473.

Recent Technologies for Investigating Flash Flood

Developing an Early Warning System for Flash Flood in Egypt: Case Study Sinai Peninsula

Gamal El Afandi and Mostafa Morsy

Abstract

Heavy rainfall is one of the major severe weather conditions over the Sinai Peninsula, and causes many flash floods over the region. Good rainfall forecasting is very much necessary for providing an early warning to avoid or minimize flash flood disasters. In the present study, heavy rainfall events that occurred over the Sinai Peninsula, on 24 October 2008 caused a flash flood, and was investigated using the Weather Research and Forecasting (WRF) Model. This flash flood has predicted and analyzed over different parts of the Sinai Peninsula. The predicted rainfall in four dimensions (space and time) has been compared with the recorded measurements from rain gauges. "The results show that, the WRF model was able to capture the heavy rainfall events over different regions of Sinai." The results reveal the capability of the WRF model in capturing extreme rainfall along different parts of the Sinai Peninsula. It observed that the WRF model was able to predict rainfall in a significant consistency with real measurements. At the same time, WRF has succeeded to represent small or mesoscale hazards such as severe thunderstorms, squall lines and heavy rainfall that caused a flash flood. Therefore, WRF is a reliable short-term forecasting tool for severe and heavy rainfall events over the Sinai Peninsula. One may conclude that the WRF model is an important and good tool in developing early warning systems of flash flood over the Sinai Peninsula. These systems will help and support the related authorities and decision-makers, to avoid many disasters accompanied by the flash flood. In addition, this research could apply and spread over the whole of Egypt as it gives warnings several days in advance.

Keywords

Early warning • WRF model • Heavy rainfall • Flash floods • Sinai Peninsula • Egypt • Synoptic situation • Statistical analysis

1 Introduction

Flash floods are a worldwide problem and represent one of the most dangerous weather phenomena and flood kinds. In the arid regions, flash floods are among the devastating natural hazards in terms of human life losses, economical damages (Abdel-Fattah et al. 2015), costliest and deadliest natural hazards (Eliwa et al. 2015). It represents a risk for vital sectors like agriculture and industry (Baldi et al. 2017), and adverse effects to society and the ecosystem (Papagiannaki et al. 2015), because they combine the flood destructive power and unpredictability with little or no warning. Slow-moving or multiple thunderstorms cause most of flash floods over the same area. On the other hand, heavy rain or flash flood water can be the potential source for non-conventional fresh and groundwater, especially in arid and semi-arid zones (Baldi et al. 2017), where mostly the water recourses in arid and semi-arid areas are limited to groundwater and rainfall (Hassan 2011). The flash floods destructive power, their size and their time thresholds which vary in different regions from minutes to several hours depend on several factors including intensity and duration of rainfall, topography, soil conditions, ground cover, geomorphological and hydro-climatological characteristics of the region. Therefore, flash flood is not only a meteorological phenomenon, but also a hydro-meteorological one, because it combines their characteristics.

G. El Afandi (✉)
College of Agriculture, Environment and Nutrition Sciences, Tuskegee University, Tuskegee, AL 36088, USA
e-mail: gelafandi@tuskegee.edu

G. El Afandi · M. Morsy
Astronomy and Meteorology Department, Faculty of Science, Al Azhar University, Cairo, 11884, Egypt
e-mail: mostafa_morsy@azhar.edu.eg

A. M. Negm (ed.), *Flash Floods in Egypt*, Advances in Science, Technology & Innovation,
https://doi.org/10.1007/978-3-030-29635-3_4

2 Flash Flood Main Events at Egypt

Flash floods may happen anywhere, but the catastrophic events are more common in arid and mountain catchments than humid and flat regions. Egypt can be regarded as one of the arid and semi-arid Arabian countries that face flash flood in the coastal and Nile Wadi systems (Abdel-Fattah et al. 2015). Flash floods extensively occur in Egypt, especially in the Eastern Desert, Red Sea Mountains and Sinai Peninsula. Abdel-Fattah et al. 2015 showed that during the last decades between 1972 and 2015 more than 15 destructive flash flood events have occurred, as shown in Table 1. These ones cause a severe damage to the ecosystem, infrastructure, loss of properties and the most important loss of human lives.

3 Case Study: Sinai Peninsula

The Sinai Peninsula can be considered as the most important Egyptian regions that frequently and exposed yearly, for few times, to synoptic situations that cause heavy rainfall that may lead to flash flooding causing severe damage, because of its natural complex of geography and context. This makes studying and analyzing the flash flood over Sinai Peninsula becomes a vital and important task to help decision-maker to avoid or at least mitigate the flash flood hazards and harvest floodwater to be used in sustainable development in this arid Egyptian area.

The Sinai Peninsula is a triangular peninsula in Egypt about 60,000 km^2 in area. As shown in Fig. 1, it is situated between the Mediterranean Sea to the north, and southwest and southeast shores of Gulf of Suez and Gulf of Aqaba of the Red Sea. It is considered as one of the coldest provinces in Egypt because of its mountainous topography with two famous mountains, Mount Sinai that rises 2285 m and Mount Catherine whose summit reaches 2637 m. El Afandi et al. 2013 showed that Sinai Peninsula is characterized by the Mediterranean climate in the north, and close to the desert and semi-desert climate to the south. Much of the Sinai is warm, or very warm, with greater inland temperatures, but close to the north shore and above the hills, there is a more temperate region. The average daily maximum is 28–37 °C in the north, 31–42 °C near the south coast and 35–41 °C inland during the period May/June to September/October. Average minimum temperatures in the summer range from 20 to 25 °C. By day-highs in the mid-teens and possible 20s, the winter season is a little less harsh, and evenings often fall to about 6–10 °C, and may drop below 0 °C. The climate of Sinai that is generally hot and dry receives an average rainfall less than 100 mm per year. The Sinai Peninsula is classified as arid land (desert) since the surface water loss due to the evaporation exceeds the surface water gain from precipitation. The peninsula of Sinai extends 32.25° E and 34.8° E; 27.8° N and 31.2° N latitude. Geologically it is possible to divide Sinai approximately into three fields. The northern region is made up of

Table 1 Flash flood main events at Egypt (Abdel-Fattah et al. 2015)

Date	Affected area	Recorded damages
Feb 2015	Sinai, Red Sea region	Road damages
Mar. and May. 2014	Taba, Sohag, Aswan, Kom Ombo	Dam failure at sohag, road damages
2013	South Sinai	2 death, road damage
2012	Wadi Dahab, Catherine area	Dam failure, destroyed houses
Jan. 2010	Along the Red Sea Coast, Aswan, Sinai	12 death, damage houses and roads
Oct. 2004	Wadi Watir	Road damage
May. 1997	Safaga, El-Qusier	200 death, destroy roads, demolished houses, damaged vehicles
Nov. 1996	Hurghada, Marsa Alam	
Sep., Nov. 1994	Dahab, Sohag, Qena, Safaga, El-Qusier	
Mar., Aug. 1991	Marsa Alam, Wadi Awag	Destroyed houses
Oct. 1990	Wadi El-Gemal, Marsa Alam	
Jan. 1988	Wadi Sudr	5 death
Oct. 1987	South Sinai	1 death, road damage
May, Oct. 1979	Aswan, Kom Ombo, Idfu, Assiut, Marsa Alam, ElQusier	23 death, demolished houses
Feb. 1975	Wadi El-Arish	20 death, road damage
1972	Giza	Destroyed houses, roads, and farms

Fig. 1 Study area and rain gauges location

sand dunes and fossil lakes created during glacial periods two million years ago by the evolving levels of the Mediterranean Sea. The landscape is flat and uniform, interrupted by a few large sand and calcareous mountains only (Elmoustafa and Mohamed 2013).

Figure 2 displays yearly average precipitation falling over Egypt. One should note that the highest amount falls along a narrow strip along the Mediterranean with values that exceed 200 mm/year. Precipitation rates drop rapidly as one moves away from the coast, with values around 20 mm/year in middle and 2 mm/year in Upper Egypt. The figure also indicates similar distribution over the Sinai Peninsula, but the drop rate towards the south is less. So Sinai obtains yearly on larger amount of rain than other Egyptian regions. In Egypt, it is known that winter is the wettest season, although heavy rain events occur generally during the spring and autumn.

Along the Mediterranean Coast, 60% of the rain occurs in the winter, while 40% falls during the transitional seasons.

During autumn and winter seasons, rainfall is ranging from medium to heavy, particularly in some high topography areas, while it generally not occurs in the summer season (El Afandi et al. 2013).

The main causes of heavy rainfall events in arid areas, such as the Sinai Peninsula in Egypt, are the availability of the squall line and convective cloud mechanisms as well as low-intensity frontal rain, causing storm floods (El Afandi 2010). The weather is mostly influenced in a variety of ways by the complexity of the terrain. Under atmospheric stability conditions, the terrain produces internal gravity waves that have the responsibility of momentum distribution over wider areas. The occurrence of such processes may be associated with strong winds and turbulence that influence air traffic.

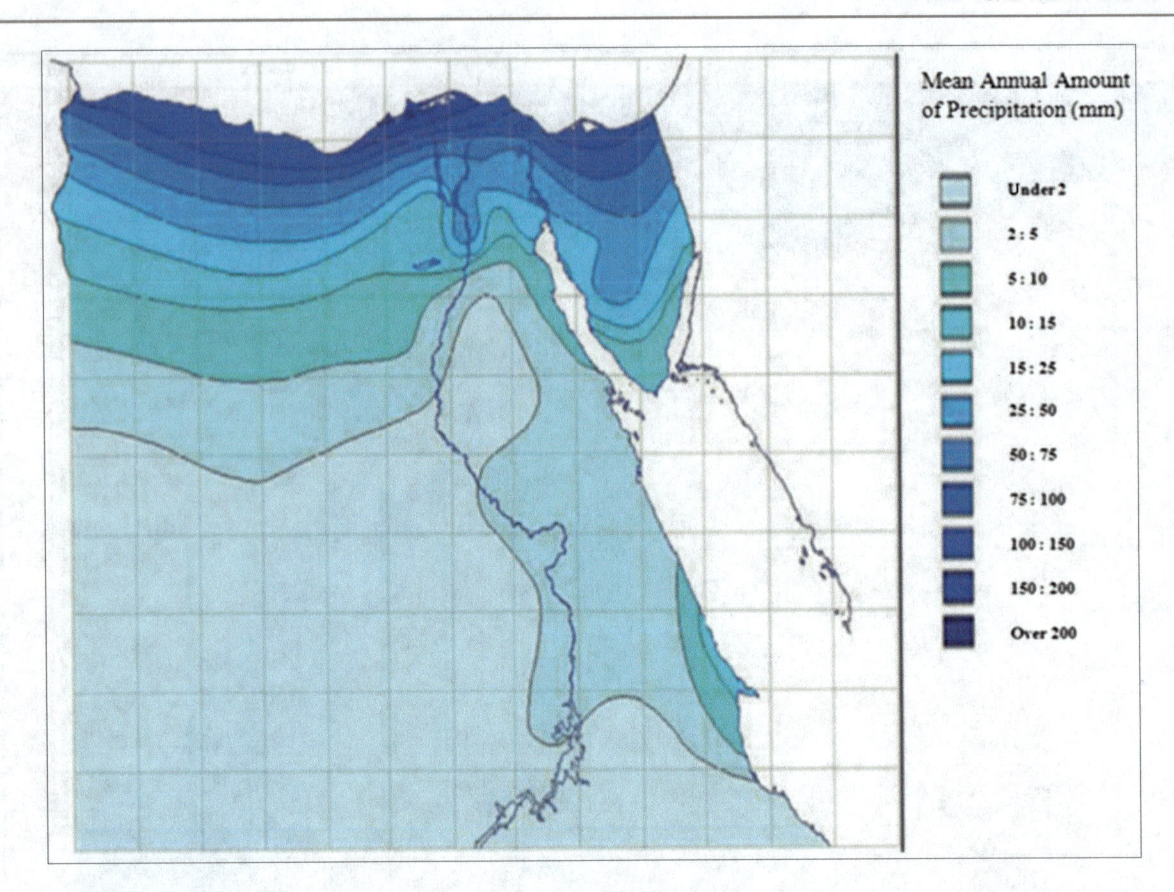

Fig. 2 Average annual precipitation in Egypt (mm/year). *Source* EEAA (2010)

On the other side, under unstable atmospheric conditions, the atmospheric instability leads to the occurrence of convective clouds, thunder activities and finally precipitation over complex terrain, which can grow into severe thunderstorms. All aforementioned processes occur in spatial length scales smaller than 100 km, usually even smaller than 10 km (El Afandi et al. 2013). Therefore, forecasting severe convection and flash flooding can be a considerable challenge because flash floods are associated with different storm types and, additionally, are a function of storm location and movement within the hydrological basin.

Because of the above-mentioned challenges, the latest numerical mesoscale meteorological model named the Weather Research and Forecasting (WRF-ARW) is selected to forecast rainfall and simulate the synoptic situation related to some flash flood events over Egypt. Mainly the US National Centre develops WRF for Atmospheric Research (NCAR) in collaboration with many other research centres and universities. WRF allows forecasting weather in complex terrains such as the one in Sinai and at the same time considers the orographic characteristics. It is appropriate for a wide range of applications along different scales ranging from metres to thousands of kilometres. Furthermore, in 2009, Egypt in collaboration with Belgium has installed the first early warning system called Flash Flood Manager (FlaFloM, which was co-funded by the European Commission under the LIFE Third Countries Fund) in Wadi Watir catchment, southern Sinai. The FlaFloM System consists of models for rainfall forecasting, rainfall-runoff modelling, hydraulic modelling and a warning system, each one sending its output to the next model (Vanderkimpen et al. 2010; Cools et al. 2012) as shown in Fig. 3.

There were other recorded flash floods over Egypt not mentioned in Table 1, like the occurred one on 24 October 2008 event. It caused flash flood over Wadi Watir in Sinai which destroyed the roads, as shown in Fig. 4.

This case study will be the main core of this research, to evaluate the performance of the WRF model in predicting heavy rainfall events that caused flash flood over the area.

Main objectives of the current study are as follows:

1. To analyze the synoptic situation associated with the flash flood occurrence over Sinai, Egypt.
2. To investigate the capability of WRF, and to simulate the rainfall quantity during flash flood event.

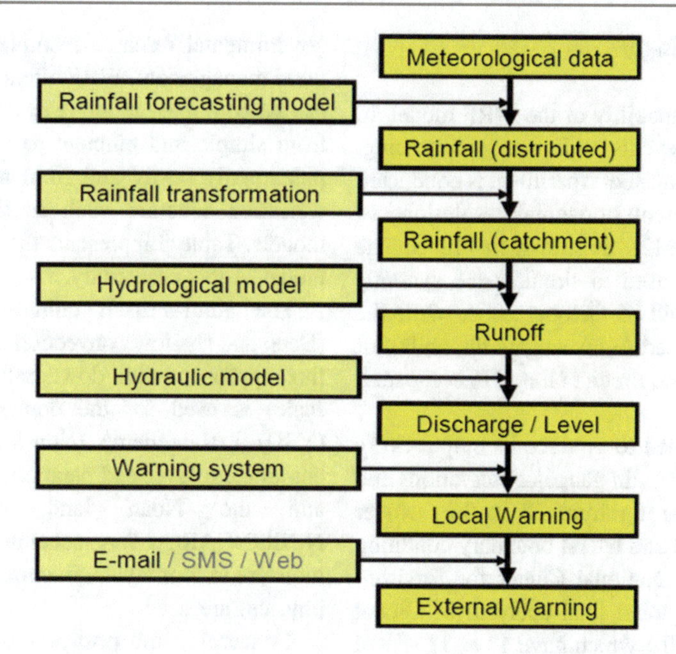

Fig. 3 FlaFloM early warning system components (Vanderkimpen et al. 2010; Cools et al. 2012)

Fig. 4 Some of the damages caused by flash flood in 2008, Wadi Watir, Sinai

4 Data and Methodology

This research used nesting capability of the WRF model, to simulate the heavy rainfall events with fine grid spacing. During this study, a two-way nested experiment is conducted for two domains with different horizontal resolutions of 18 km (DM1) and 6 km (DM2), as shown in Fig. 5. The first domain (DM1) will be used to simulate the synoptic situation over Egypt on 24 and 25 October 2008, while the last domain (DM2) will be used to investigate the ability of the WRF model to simulate and predict rainfall to compare it with rain gauge at Sinai.

The model output is adopted to produce its output every hour to be compatible with the rain gauges observations and two-way nesting domain starting from 23 to 26 October 2008. The WRF model initial and lateral boundary condition has been obtained from the National Center for Environmental Prediction (NCEP), global final analyses of Global Forecasting System (GFS/FNL) which have $1° \times 1°$ spatial resolution and is updated every six hours with 27 vertical levels and the model top of 50 hPa.

The hourly observational rainfall data was obtained from the rain gauges stations, Table 2, installed by the Water Resources Research Institute (WRRI); WRRI is the Egyptian governmental research institute with the mandate for flash flood management. WRF offers multiple physics options that can be combined in any way. These options typically range from simple and efficient to sophisticated and more computationally costly and from newly developed schemes to well-tried schemes such as those in current operational models. Table 3 represents the selected physics of the WRF model in this case study.

The Kain–Fritsch cumulus parameterization scheme (Deep and shallow convection sub-grid scheme using a mass flux approach with downdrafts and CAPE removal time scale) is used for the domains. The Yonsei University (YSU) PBL scheme (Non-local-K scheme with explicit entrainment layer and parabolic K profile in unstable mixed) and the Noah land surface scheme (Unified NCEP/NCAR/AFWA scheme with soil temperature and moisture in four layers, fractional snow cover and frozen soil physics) are used.

Concerning the post-processing system, the ARW post program is used to read the WRF model outputs and convert it for other readily available graphics packages such as Grid Analysis and Display System (GRADS): to display and analysis the model outputs and simulate the synoptic situations for troposphere pressure layers.

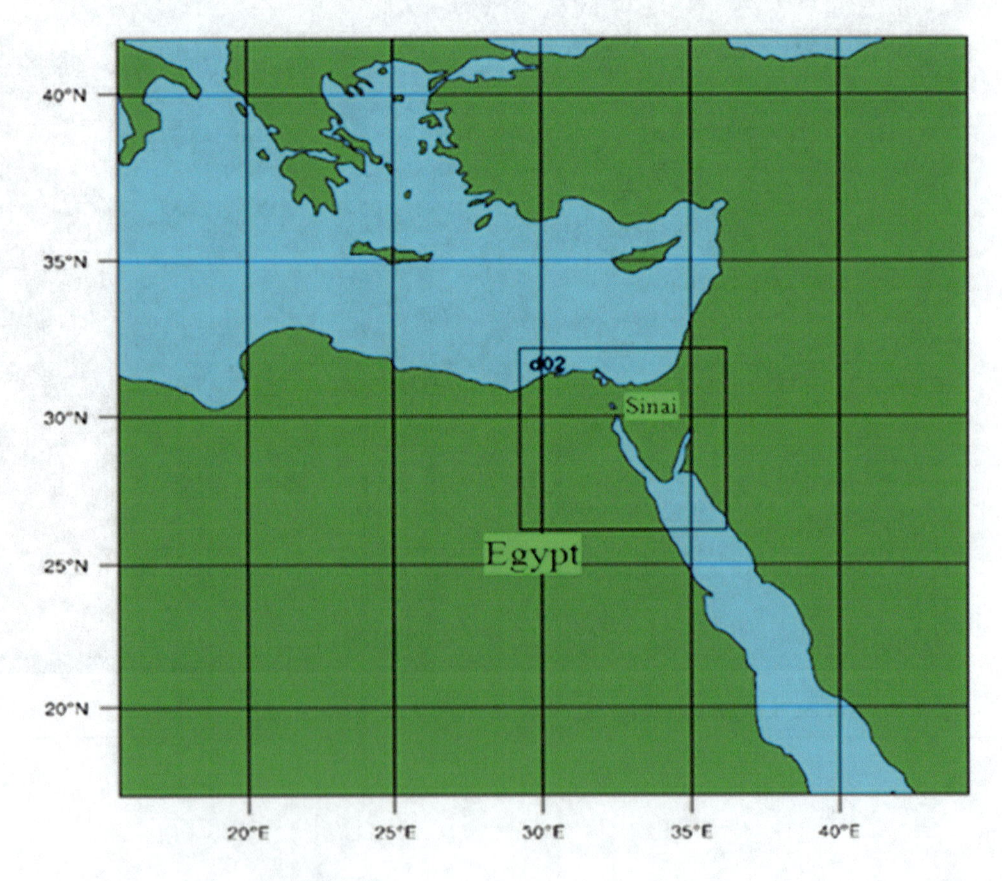

Fig. 5 Model domains of 18 km (DM1) and 6 km (DM2) horizontal resolution

Table 2 Rain gauge name and location that used in the comparison

Station	Lon.	Lat.	Elevation (m)
Gof El-Dalal	33° 56′ 51″	28° 49′ 52″	1236
Sorah	34° 10′ 10″	28° 48′ 53″	988

Table 3 Selected physics of the WRF model options

Physics scheme	Type of the scheme that used in WRF run
Microphysics	Single-moment 3-classWSM3 scheme
Longwave radiation	Rapid Radiative Transfer Model (RRTM) scheme
Shortwave radiation	Dudhia scheme
Land surface Noah	Land Surface Model scheme
Surface and boundary layers	Yonsei University scheme
Cumulus parameterization	Kain-Fritsch scheme

5 Results and Discussion

This study will mainly deal with two different procedures: the first one is the synoptic situation over the area of study and the second one will be concentrated on the statistical analysis during 24 October 2008 over Sinai Peninsula, Egypt.

5.1 Synoptic Situation

Developments of intensive weather events that invade Egypt during 24 October 2008 were characterized as "exceptional and extremely heavy rainfall," which affected the Northern coast of Egypt and the Sinai Peninsula. This heavy rainfall event is accompanied by strong winds and thunderstorms. Because these severe weather conditions were not predicted in advance, these disasters caused significant infrastructural damages, population displacement and loss of some Bedouin lives.

The WRF model was used to simulate the synoptic situations that lead to development of this severe weather and thunderstorm event that causes flash flood occurrences over Egypt and particularly the Sinai Peninsula.

Generally, the autumn season in the Middle East is known to be associated with the development of low-pressure systems (Cyprus lows) over the eastern Mediterranean Sea (Babu et al. 2011). Such system has been observed at upper levels, particularly at 500 hPa. The Mediterranean region considered one of the most cyclogenesis areas in the world, usually favours the development of weak low-pressure systems. Occasionally these systems develop into deep cyclogenesis that cause a series of severe weather events as they cross the Mediterranean (El Afandi et al. 2013).

Kahana et al. (2004) found that Mediterranean Sea cyclones deepen, in general, in association with upper-level troughs. Upper-level troughs are considered as the main factors in atmospheric activity in mid-latitude areas, such as the Mediterranean cyclonic system, especially in their front and sides, where the positive vorticity advection and enhanced convection take place (Holton 1992). However, the polar continental air from central Asia moving towards the eastern Mediterranean (Ferraris et al. 2002), and interacting with the relatively warm Sea Surface Temperature (SST), produces enhanced lower level instability. Figure 6 (A–D) shows the Mean Sea Level Pressure (MSLP) for different times every six hours for 24 and 25 October 2008. It is noticed that the active Red Sea Trough (RST) (inverted V-shape) which is an extension of Sudan monsoon low affects the weather of Egypt. The Sudan monsoon low is mostly concentrated over Eastern desert, Red Sea Mountains and Sinai Peninsula. This situation brings thunderstorms and heavy rainfall over the mentioned area mostly causing flash flood. Therefore, these situations coincide with Elfandy (1948) who has showed that the so-called small oscillation accompanies the passage of trough of low pressure or secondaries associated with depressions travelling further north and east over the Eastern Mediterranean, Eastern Europe and Eastern part of Arabian Peninsula.

The Red Sea trough extension and the meridional orientation of the upper trough permit favourable conditions for the formation and development of cyclone, heavy cloud and sudden and intense rainfall leading to flash floods occurrences. The occurrence of severe instability, heavy rainfall and flash floods during autumn generally resulted from the existence of this barometric RST at the Earth's surface. It transports moist air (warm and wet) from tropics and subtropics with relatively high temperature to North Africa. This gives an enhancement to the wind and becomes more active, and leads to increase of air temperature near the earth's

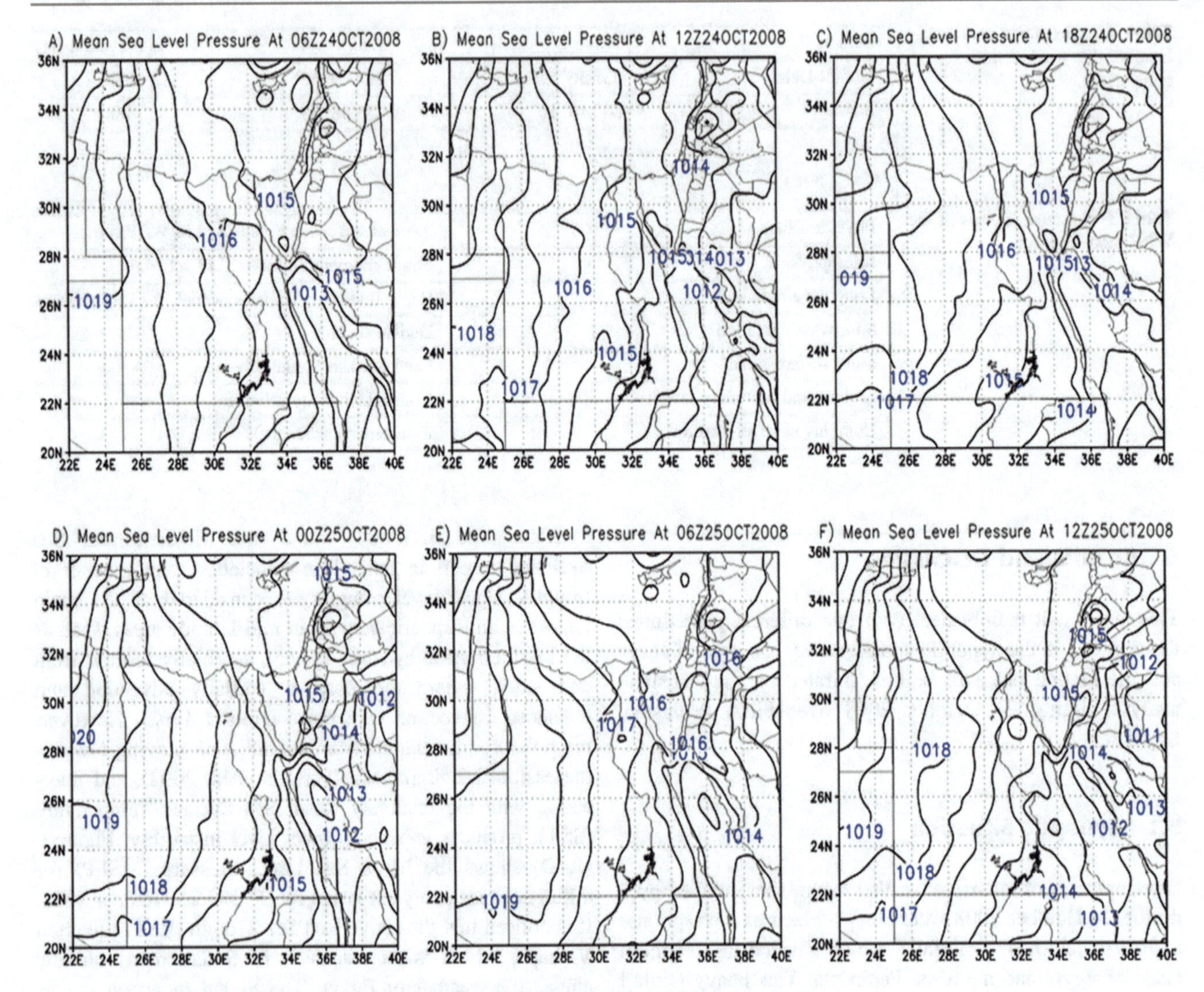

Fig. 6 Mean Sea Level pressure (inverted V-shape) every six hours on 24 and 25 October 2008

surface. Warm air is rising (upward motion) and coincides with a pronounced quasi-stationary upper-level trough that develops over Egypt and brings a cold air mass from central and southeast Europe to the Mediterranean Sea in the upper atmosphere layers. This leads to development of severe convective storms. One of the most important levels and variables to analyze and better understand the synoptic situations are the mid-troposphere (500 hPa) level and geopotential height. The complete situation of 500 hPa geopotential height and wind speed fields is shown in Fig. 7.

The axis of the trough line of the low pressure located over east Europe shifts southeast. This trough line orientation indicates that the affected area has extended and engulfed the whole east Mediterranean area. This development agrees with Krichak et al. (1997a, b); Krichak and Alpert (1998), where they showed that the meridional orientation of the upper air trough line represents one of the major synoptic processes which helps in the East Mediterranean cyclones development.

The relative vorticity is a significant parameter that helps to steering and intensifying the trough, where the positive relative vorticity advection leads to move the upper atmospheric trough from west to east and vice versa (steering of trough). As shown in Fig. 8, the positive relative vorticity (shaded) occurred and lead to move the trough (contours) from west to east and reached to Sinai Peninsula that helped to intensify the thunderstorm and heavy rainfall and flash flood occurrences over the study area.

Not only the relative vorticity advection can steer the trough and helps to intensify the thunderstorms, but also the temperature advection can deepen the trough and helps to intensify the thunderstorms. Where the warm region or advection leads to deepening and intensifying the trough and is associated with thunder activities and vice versa are shown

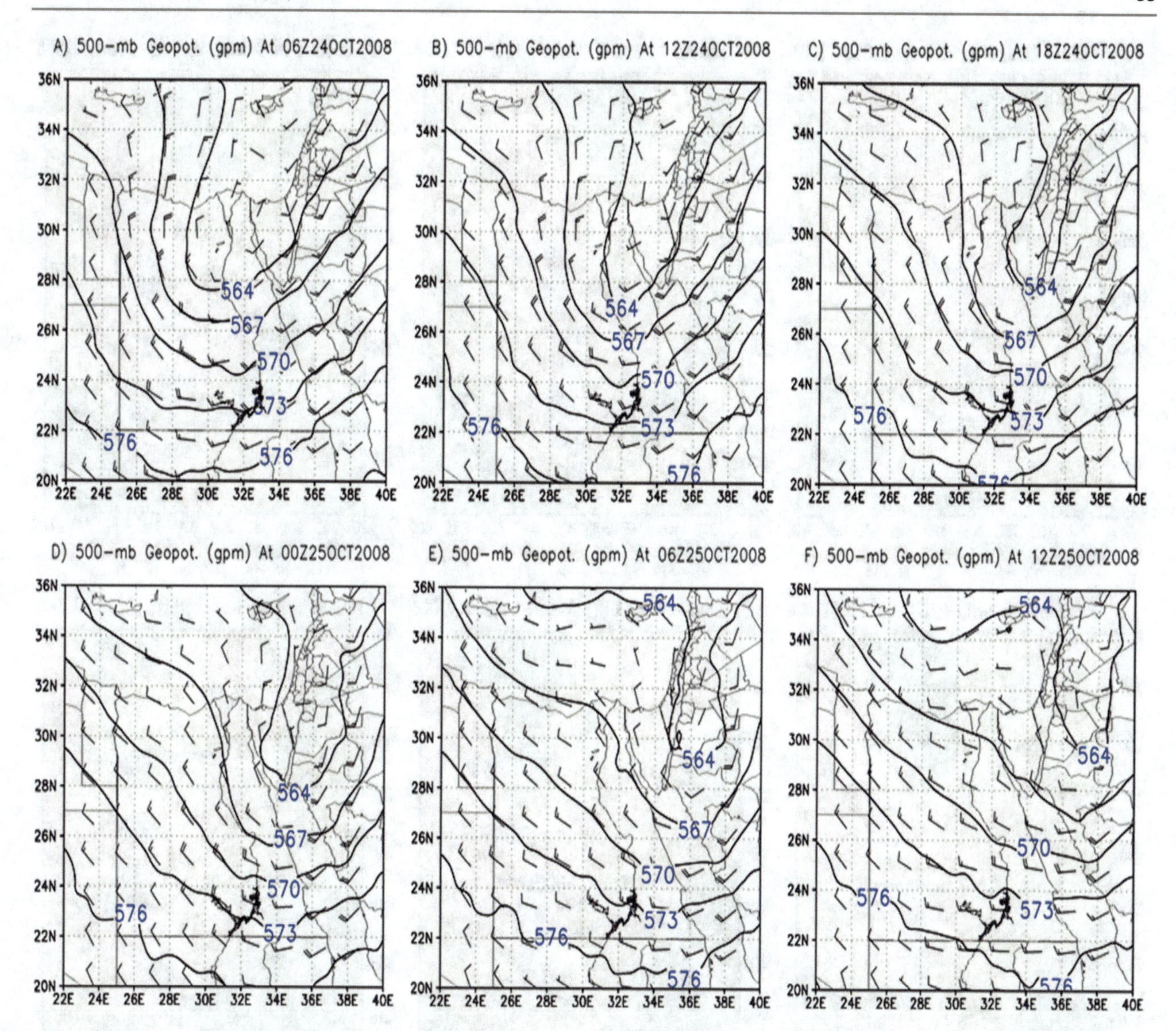

Fig. 7 Geopotential height at 500 hpa every six hours on 24 and 25 October 2008

in Fig. 9. Since, the southward warm advection in the eastern part of the domain is corresponding to the east branch of the subtropical jet, the eastward cold advection in the northeastern part of the domain is associated with the west branch. Where the growth of mid-latitude cyclone is mostly accompanied by meridional temperature gradient. One can expect that the cold advection of severe cold cyclone is responsible for the temperature decreasing over Egypt in that period (Awad 2009). Both positive vorticity and warm advections are increasing with height above the surface low pressure. It leads to upward motion, upper-level divergence and lower level convergence.

All of these factors lead to intensifying the instability in the atmosphere cause of thunderstorms and convective clouds as shown in Fig. 10.

All of these factors lead to the occurrence of heavy rainfall that develops flash floods especially in complex terrain and Wadis catchments area in Sinai Peninsula as shown in Fig. 11. The rainfall amount increases as the RST moves northward and the upper air low-pressure system deepens, where the precipitation area extended eastward following the occurred system.

Not only, these factors cause this flash flood event but also the existence of the Subtropical Jet stream (STJ) over

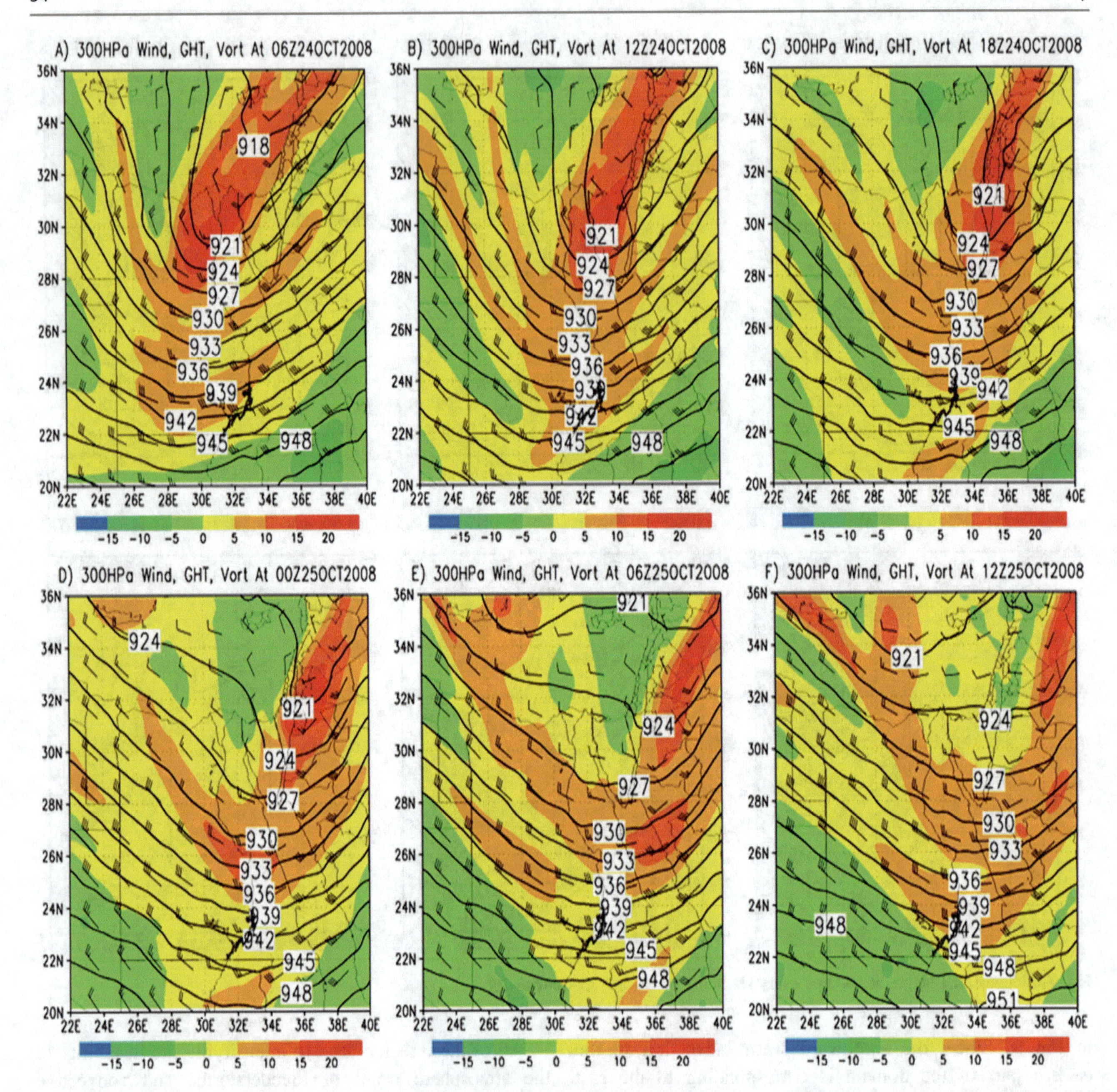

Fig. 8 Relative vorticity advections (shaded) with geopotential height (contours) and wind (barbs) at 300 hpa every six hours during 24 and 25 October 2008

northeastern Africa combined with a ridge over Syria, as shown in Fig. 12. It illustrates the movement and oscillation of STJ during the autumn season. With increasing of time, the subtropical jet has weakened, shifted to the south and divided into two branches: a south-westerly branch to the east and a north-westerly branch to the west. The east branch of the jet was stronger than the west branch. A marked STJ with wind speed exceeding 80 m s^{-1} and reaching to 95 m s^{-1} at the 250 hPa level (30% above normal) was observed during these two extremely widespread major flash floods. The existence of the Inter Tropical Conversions Zone (ITCZ) to the north of its usual position over the tropical part of the eastern Africa in the Northern Hemisphere could be the reason for excessive pumping of the tropical air moisture into the mid-latitudes by this STJ which caused an increase of static instability.

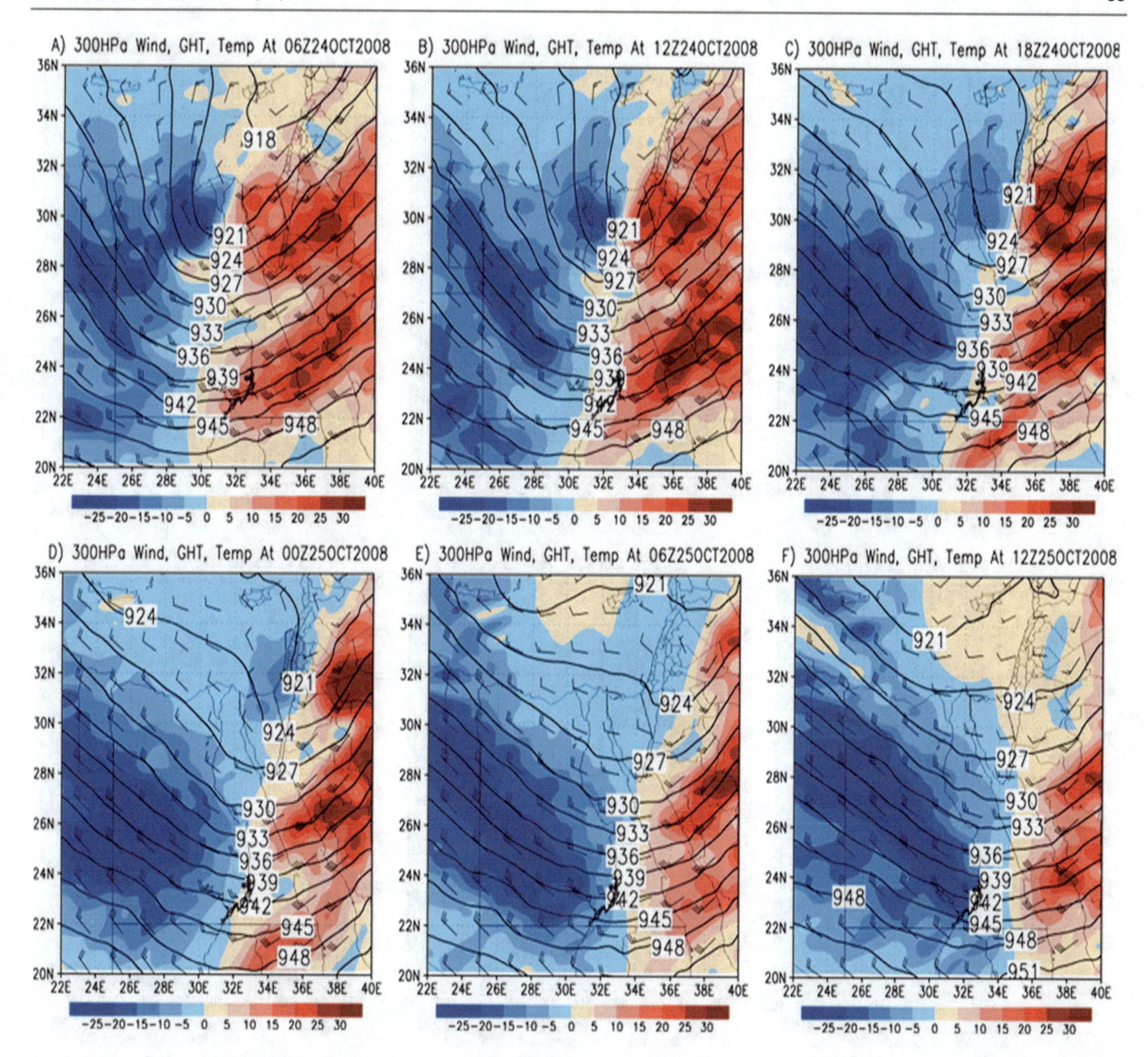

Fig. 9 Thermal advection (shaded) with geopotential height (contours) and wind (barbs) at 300 hpa every six hours on 24 and 25 October 2008

5.2 Statistical Analysis

The evaluation of the performance of WRF model to predict heavy rainfall and thunderstorms that lead to flash flood event over Sinai Peninsula will be done by comparison of the predicted rainfall from WRF with the measured one from rain gauges.

Root Mean Square Error (RMSE) can be considered as one of the very important statistical tools which give a good fit overall measure of model performance. The best model fit is corresponding to RMSE of zero and vice versa. Also, the Mean Bias (MB) is also used to evaluate the model performance, which refers to the degree of correspondence between the mean predicted and observed values. Lower numbers around zero are the best and values less than zero indicate that the model is underestimated and vice versa. The equations for RMSE, MB and its percentages are given by

$$\mathrm{RMSE} = \left(\left(\sum_{i=1}^{n} (X_p - X_o) \right)^2 \bigg/ n \right)^{0.5} \tag{1}$$

$$\mathrm{RMSE\%} = (\mathrm{RMSE}/\bar{X}_o) * 100 \tag{2}$$

$$\mathrm{MB} = \sum_{i=1}^{n} X_p - X_o \bigg/ n \tag{3}$$

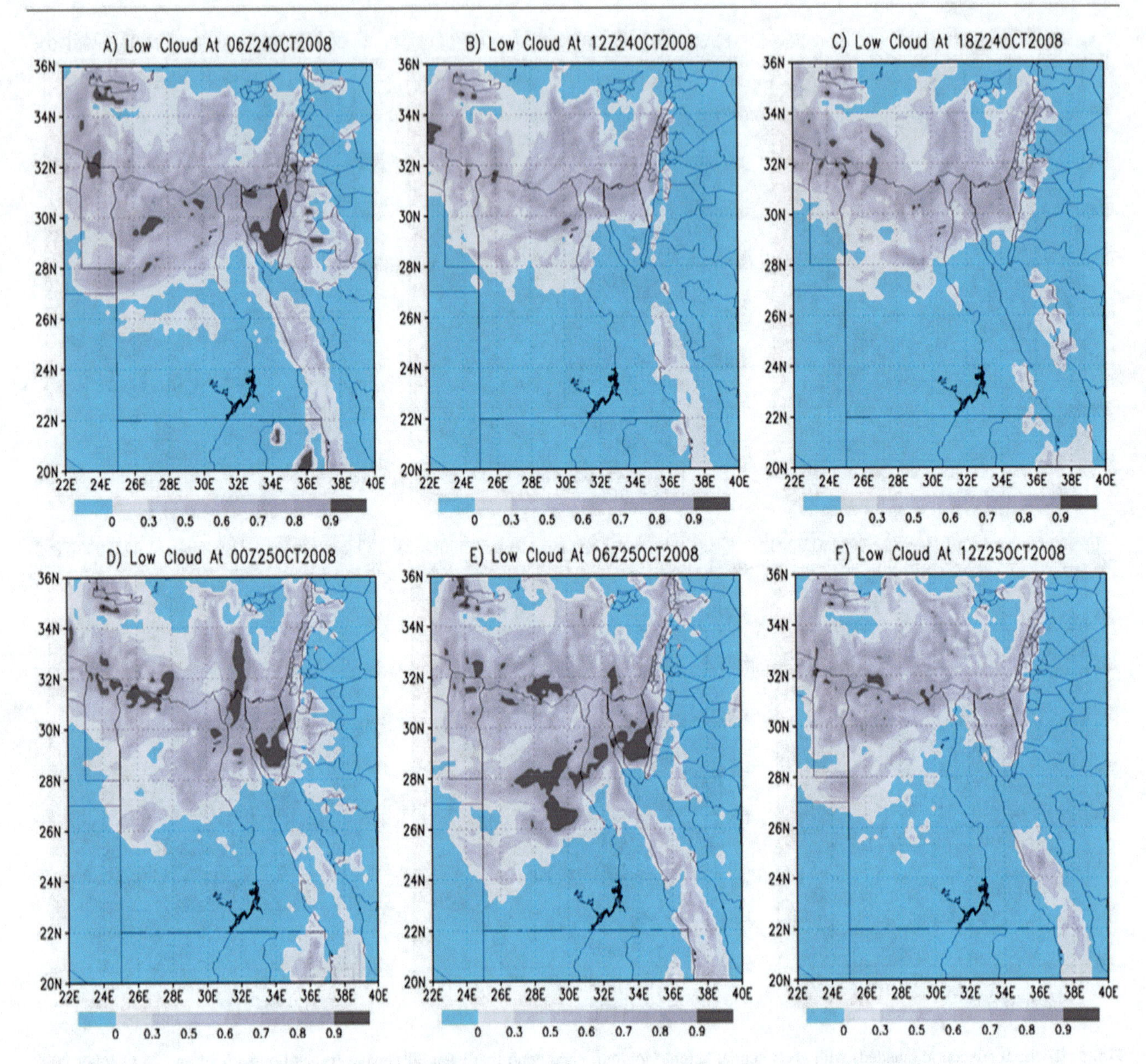

Fig. 10 Low clouds every six hours on 24 and 25 October 2008

$$MB\% = \left({}^{MB}\!/_{\bar{X}} \right) * 100 \quad (4)$$

where n is the number of observations, X_p and X_o are the predicted and observed values, respectively, and $\bar{X}_o$ is the average of observed values. These statistical measures are used extensively for evaluation of model forecasts of precipitation (e.g. Mesinger et al. 1990; Mesinger 1996; Lagouvardos et al. 2003).

Figure 13 shows the comparison between hourly predicted rainfall from WRF Model and measured one from available rain gauges in different regions over Sinai Peninsula during 24 October 2008. It can be noticed from this figure that the predicted rainfall by WRF model is in a good consistency and harmony with observed rainfall for all rain gauge stations. In Gof El-Dalal and Sorah stations, rainfall prediction by WRF is very close to the measured ones. Table 4 shows the percentages of root mean square error and mean bias for predicting rainfall and the measured ones at different stations. From this table, it can be found that the maximum values of RMSE (8%) and mean bias (above measured by 5%) are for Sorah station, while the minimum values of root mean square error (3%) and mean bias (1%)

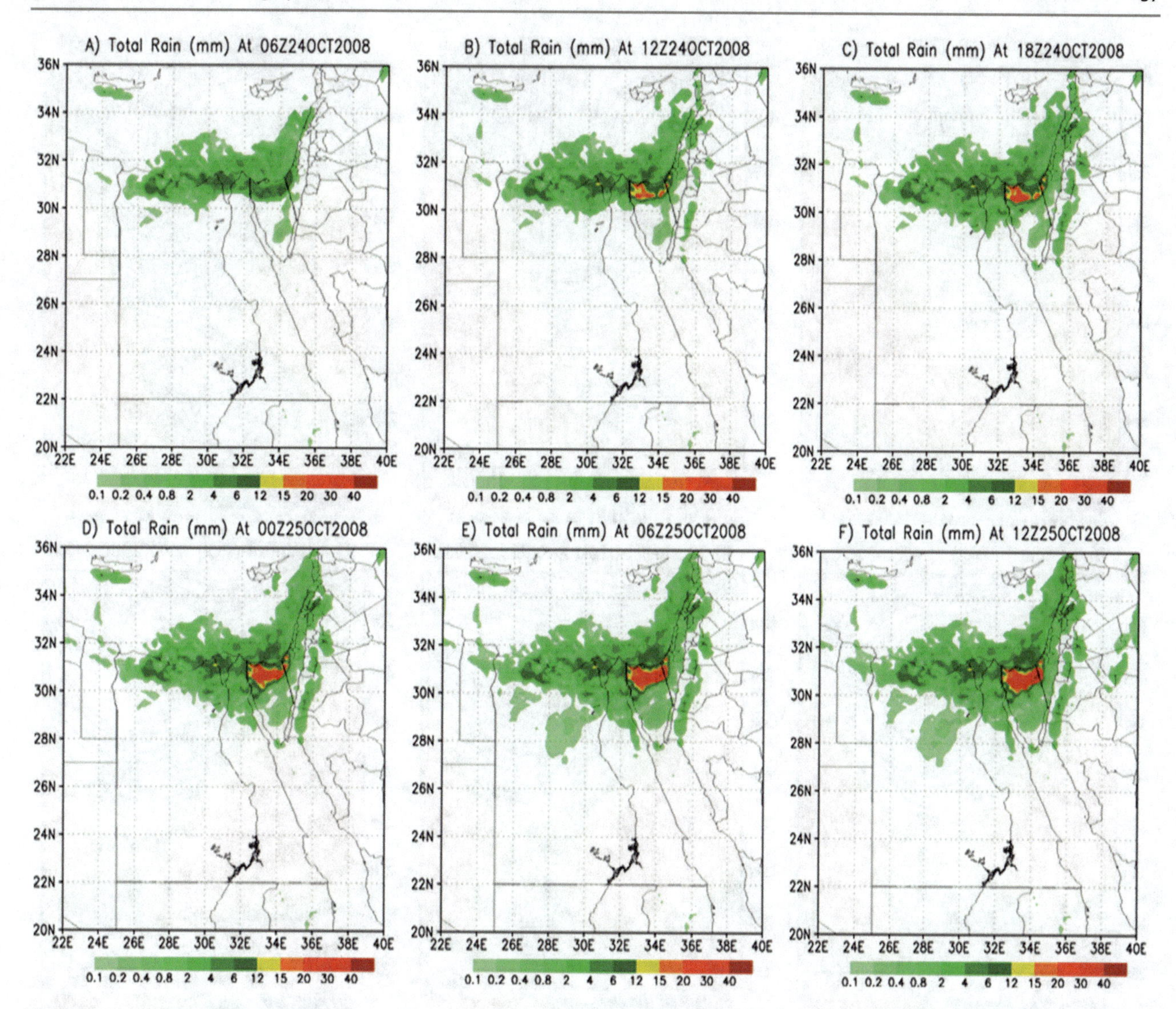

Fig. 11 Accumulated rainfall every six hours with the maximum volume of rainfall over Sinai Peninsula on 24 and 25 October 2008

are for Gof El-Dalal station. This means the WRF model has high-performance rainfall prediction for the two stations. Also, the patterns and curves of predicted rainfall of the WRF model are in similar accordance and in harmony with observed ones at all rain gauges. Thus, the WRF model was able to simulate the synoptic situation which invaded Egypt and is accompanied by heavy rains starting at the north coast, the Red Sea and the Sinai Peninsula during 24 October 2008. Also, the WRF simulation for the synoptic pattern was in a good agreement with the previous studies. Finally, We may conclude that the WRF model approximately gave a good and reasonable accuracy in rainfall prediction compared to observations at the selected two stations over the Sinai Peninsula.

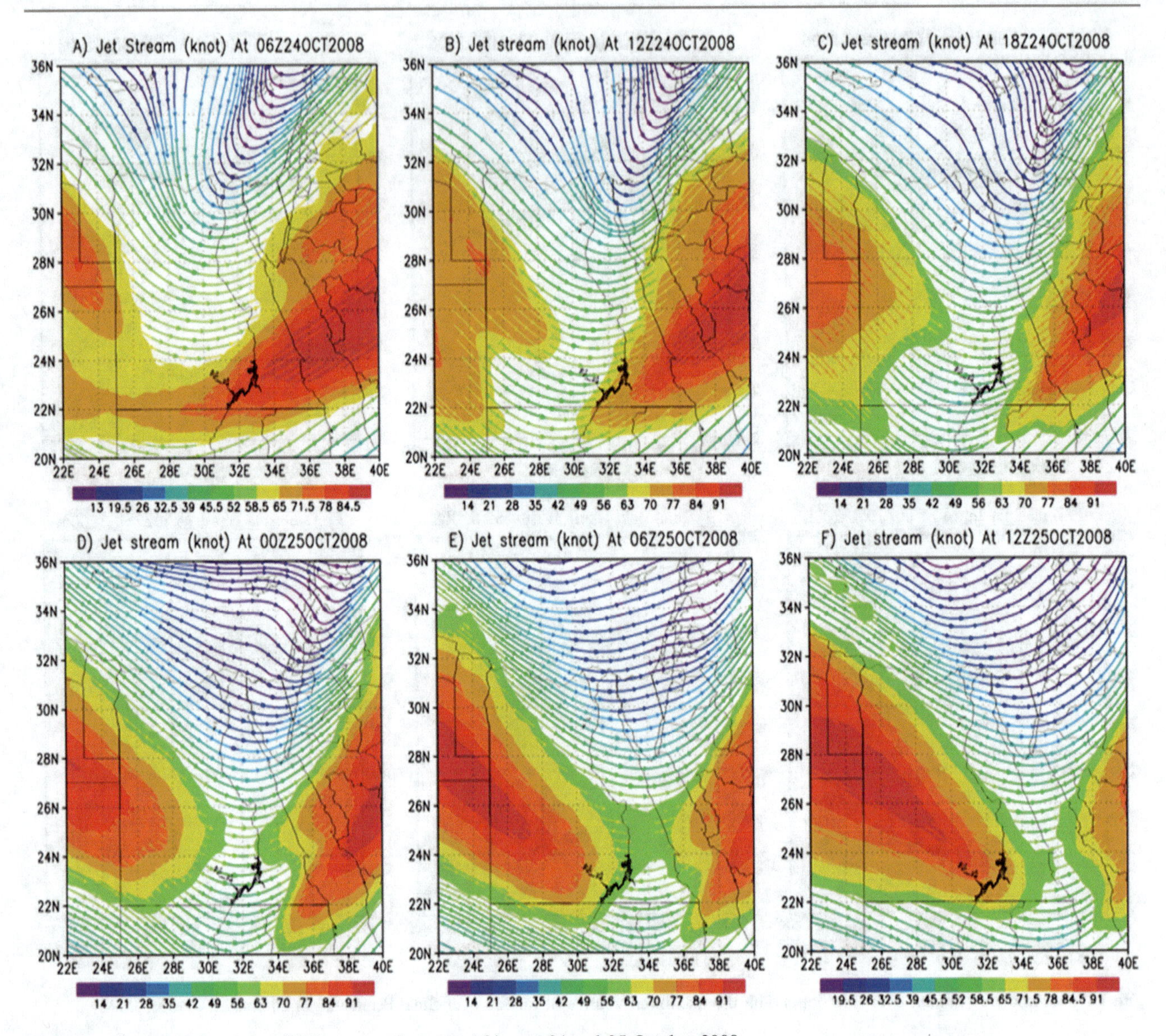

Fig. 12 Subtropical Jet stream (STJ) over northeastern Africa on 24 and 25 October 2008

6 Conclusion

Evaluation of the performance of WRF model to predict heavy rainfall, thunderstorms and flash floods in different study cases that occurred over Sinai Peninsula is investigated in this study via comparing its predictions with ground stations' measurements. The results revealed that WRF model can be used to predict the synoptic situations that cause thunderstorms and heavy rainfall. So, it may be concluded that WRF model can be used as a pre-warning system for flash floods forecasting.

The predicted total rainfall by WRF model is in a similar accordance and in harmony with observed rainfall in all rain gauges, which show very significant consistency between WRF and measurements; the WRF is presumably reliable in short-term forecasting. WRF has success to represent small or mesoscale hazards such as severe thunderstorms, squall lines and flash floods so it can be used as a good tool in assessment of the threat and a rapid dissemination of warnings to the population and appropriate authorities. Useful forecasts of the behaviour of the larger scale weather systems such as tropical storms and cyclones, intense

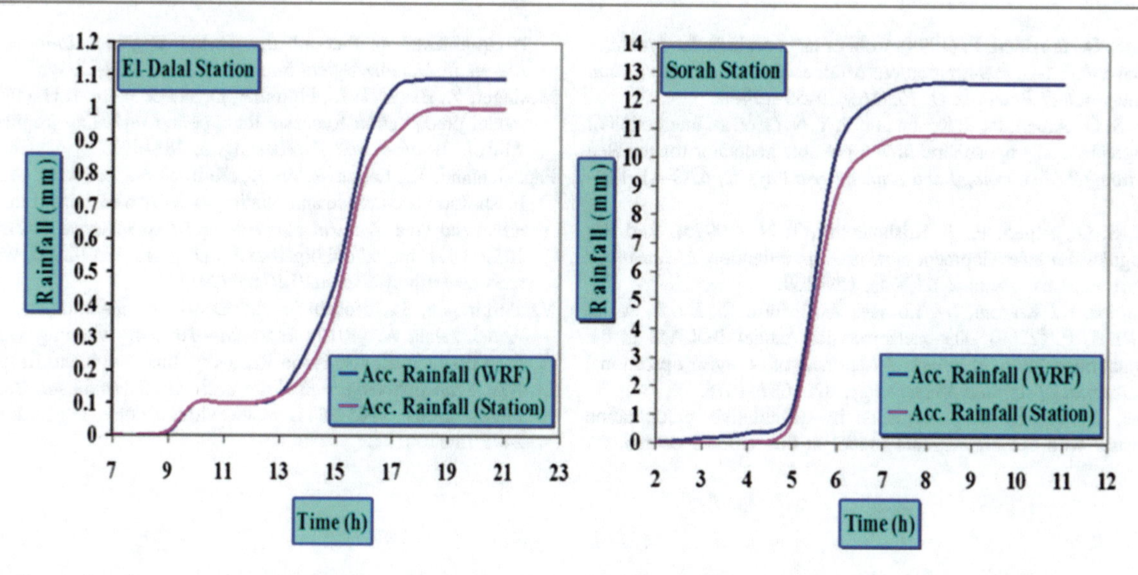

Fig. 13 Comparison between accumulated observed and predicted precipitation (mm) for Gof El-Dalal and Sorah stations during 24 October 2008

Table 4 The percentages of root mean square error and mean Bias for rainfall from WRF model and the measured ones at different stations

Station	RMSE (%)	MB (%)
El-Dalal	3	1
Sorah	8	5

depressions and large flash floods can be prepared several days in advance using WRF model.

7 Recommendations

The authors are strongly recommending further investigations and validations of the WRF model for different extreme rainfall events which caused flash flood over the Sinai Peninsula.

References

Abdel-fattah, M., Kantoush, S., & Sumi, T. (2015). Integrated management of flash flood in Wadi system of Egypt: Disaster preventing and water harvesting, *Annuals of Disaster Prevention Institute* (Kyoto University) No. 58 B.

Awad, A. M. (2009) Numerical simulation of a severe cold and rainy event over the east Mediterranean. *Meteorology, Environment and Arid Land Agriculture Sciences, 20*(2), 95–109.

Babu, C. A., Samah, A. A., & Varikoden, H. (2011). Rainfall climatology over middle East region and its variability. *International Journal of Water Resources & Arid Environments, 1*(3), 180–192.

Baldi, M., Amin, D., Al Zayed, I. S., & Dalu, G. A. (2017). Extreme rainfall events in the Sinai Peninsula. In *Geophysical Research Abstracts, EGU General Assembly* (Vol. 19), EGU2017-13971.

Cools, J., Vanderkimpen, P., El Afandi, G., Abdelkhalek, A., Fockedey, S., El Sammany, M., Abdallah, G., El Bihery, M., Bauwens, W., & Huygens, M. (2012) An early warning system for flash floods in hyper-arid Egypt. *Natural Hazards and Earth System Sciences, 12*, 443–457. https://doi.org/10.5194/nhess-12-443-2012, www.nat-hazards-earth-syst-sci.net/12/443/2012/.

EEAA. (2010). *Egypt second national communication under the United Nations framework convention on climate change*. Cairo: Egypt Environmental Affairs Agency.

El Afandi, G. (2010). Developing flash-flood guidance in Egypt's Sinai Peninsula with the weather research forecast (WRF) model. *International Journal of Meteorology, 35*(347), 7582.

El Afandi, G., Morsy, M., & El Hussieny, F. (2013). Heavy rainfall simulation over sinai peninsula using the weather research and forecasting model, Hindawi Publishing Corporation. *International Journal of Atmospheric Sciences, 2013*, Article ID 241050, 11. http://dx.doi.org/10.1155/2013/241050.

Elfandy, M. G. (1948). The effect of the Sudan monsoon low on the development of thundery conditions in Egypt, Palestine and Syria. *Quarterly Journal of the Royal Meteorological Society, 74*, 31–38.

Eliwa, H. A., Murata, M., Ozawa, H., Kozai, T., Adachi, N., & Nishimura H. (2015). Post Aswan High Dam flash floods in Egypt: Causes, consequences and mitigation strategies. In *Bulletin of Center for Collaboration in Community, Naruto University of Education* (No. 29).

Elmoustafa, A. M., & Mohamed, M. M. (2013). Flash flood risk assessment using morphological parameters in sinai peninsula. *Open Journal of Modern Hydrology, 3*, 122–129. https://doi.org/10.4236/ojmh.2013.33016.

Ferraris, L., Rudari, R., & Siccardi, F. (2002). The uncertainty in the prediction of flash floods in the northern Mediterranean environment. *Journal of Hydrometeorology, 3*, 714–727.

Hassan, S. S. (2011). Impact of flash floods on the hydrogeological aquifer system at delta wadi El-Arish (North Sinai). In *Fifteenth International Water Technology Conference, IWTC-15*, Alexandria, Egypt.

Holton, J. (1992). *An introduction to dynamic meteorology* (3rd ed.). New York, NY, USA: Academic Press.

Kahana, R., Ziv, B., Dayan, U., & Enzel, Y. (2004). Atmospheric predictors for major floods in the Negev Desert, Israel. *International Journal of Climatology, 24*(9), 1137–1147.

Krichak, S. O., & Alpert, P. (1998). Role of large scale moist dynamics in November 1–5, 1994, hazardous Mediterranean weather. *Journal of Geophysical Research D, 103*(16), 19453–19468.

Krichak, S. O., Alpert, P., & Krishnamurti, T. N. (1997a). Interaction of topography and tropospheric flow a possible generator for the Red Sea trough? *Meteorology and Atmospheric Physics, 63*(3–4), 149–158.

Krichak, S. O., Alpert, P., & Krishnamurti, T. N. (1997b). Red Sea Trough/cyclone development numerical investigation. *Meteorology and Atmospheric Physics, 63*(3–4), 159–169.

Lagouvardos, K., Kotroni, V., Koussis, A., Feidas, C., Buzzi, A., & Malguzzi, P. (2003). The meteorological model BOLAM at the national observatory of athens: Assessment of two-year operational use. *Journal of Applied Meteorology, 42,* 1667–1678.

Mesinger, F. (1996). Improvements in quantitative precipitation forecasts with the eta regional model at the national centers for environmental prediction: the 48-km upgrade. *Bulletin of the American Meteorological Society, 77*(11), 2637–2649.

Mesinger, F., Black, T. L., Plummer, D. W., & Ward, J. H. (1990). Eta model precipitation forecasts for a period including tropical storm Allison. *Weather and Forecasting, 5,* 483–493.

Papagiannaki, K., Lagouvardos, K., Kotroni, V., & Bezes, A. (2015). Flash flood occurrence and relation to the rainfall hazard in a highly urbanized area. *Natural Hazards and Earth System Sciences, 15,* 1859–1871. https://doi.org/10.5194/nhess-15-1859-2015, www.nat-hazards-earth-syst-sci.net/15/1859/2015/.

Vanderkimpen, P., Rocabado, I., Cools, J., El-Sammany, M., & Abdelkhalek, A. (2010). FlaFloM—An early warning system for flash floods in Egypt, Flood Recovery, Innovation and Response II. *WIT Transactions on Ecology and the Environment, 133* (WIT Press). ISSN 1743-3541, www.witpress.com, https://doi.org/10.2495/friar100171.

Flash Flood Investigations in El Gouna, Northern Red Sea Governorate

F. Bauer, A. Hadidi, F. Tügel, and R. Hinkelmann

Abstract

Flash floods along the Red Sea coast are getting more and more into consideration as with rising population and tourists, and the damages caused by flash floods are stronger impacting the life in this region. Investigations in El Gouna close to Hurghada concentrate on a basic description of the basin, geological features, and climate conditions and in the second place on modeling rainfall-runoff during storm events close to the city. The Wadi Bili catchment, which is connected to the coastal plain of El Gouna, is morphologically and geologically structured into the high mountains of the Red Sea Hills, the Abu Shar Plateau between the high mountains and the steep slope of Esh El Mellaha which separates the coastal plain from the upper parts of the catchment. 3D geological modeling of the plain on the basis of stratigraphic correlation separated three different aquifers where the uppermost is of interest for abstraction of groundwater for desalination. Porosities are about 0.2 and provide good conditions for artificial recharge, even the drainable porosity is much lower. A hydrodynamic flash flood model for the region of El Gouna is presented to simulate the spreading of water and flooding areas generated by heavy rainfall inside the domain as well as by the runoff from the large catchment of Wadi Bili. The model supports the understanding of flow processes during and after heavy rainfall events. It can be applied to investigate structural mitigation measures such as retention basins and dams to create risk maps raising the awareness regarding flood-prone areas and is capable to become part of an early warning system.

1 Introduction

Flash floods along the Red Sea coastline are causing harmful events which are more and more reported by local and international press. The rising number of inhabitants along the Red Sea also increases the impact of destructive flash floods. Recent flash flood events caused several casualties (Table 1). In January 2010, a flash flood event on Sinai Peninsula (Vanderkimpen et al. 2010) resulted in five deadly accidents. In October 2016, a severe flooding took place along the Red Sea coast, South Sinai, Sohag, and Assiut (IFRC 2017) (All locations are plotted in the subfigure in Fig. 1). Especially the city of Ras Gharib was strongly affected by about 120 million cubic meters of rainfall and the resulting flooding. During this event 26 people died. The strongest impact of flash floods on human lives in the last decades in Egypt was recorded in November 1994 where 600 people lost their lives during a 2–4 days flooding event (Krichak and Alpert 1998; De Vries et al. 2013).

Prediction of flash flood events is first of all a challenge in terms of meteorology. Long-term knowledge of changing weather conditions in the future is almost impossible and therefore, reliable forecasts are only available for a few days. But besides the amount of precipitation as the source of flooding, the morphology and composition of the hydrologic basin affects the impact of a flash flood as well.

Rainfall events which cause major flash floods are very rare in the Eastern Desert and their occurrence is infrequent (Hadidi 2016). The periodicity of heavy thunderstorms in this area is about a decade, a medium rainfall event returns every three to four years (Saleem 1990). Besides the

F. Bauer (✉) · F. Tügel · R. Hinkelmann
Department of Water Engineering, Technische Universität Berlin, Campus El Gouna, Hurghada, Egypt
e-mail: bauerflorian@hotmail.com

A. Hadidi
Department of Applied Geology, GUTech, German University of Technology, Halban, Oman

A. M. Negm (ed.), *Flash Floods in Egypt*, Advances in Science, Technology & Innovation,
https://doi.org/10.1007/978-3-030-29635-3_5

Table 1 Reported flash floods in the Eastern Desert of Egypt (modified after De Vries et al. (2013))

Date	Precipitation	Location	Casualties and Damages	Reference
Dec. 1923	No information	Wadi Baroud	Severe destructions	Saleem (1990)
Nov. 1934	34 mm	Quseir	No information	Saleem (1990)
Dec. 1954	28 mm	Daedalus	No information	Saleem (1990)
Oct. 1979	48 mm	No information	50 casualties	Wagdy et al. (2003)
Oct. 1987	No information	No information	30 casualties	De Vries et al. (2013)
Nov. 1994	60 mm	No information	500 people to lose their lives	Krichak et al. (2000)
Oct. 1996	80 mm	No information	No information	Wagdy et al. (2003)
Oct. 1997	40 mm	No information	6 casualties	Dayan et al. (2001)
Mar. 2014	34 mm	El Gouna	Some destructions	Hadidi (2016)
Nov. 2016	No information	Ras Gharib	26 casualties and destructions	IFRC (2017)

casualties, destruction of infrastructure is the major negative effect to human life. Reports of destruction of mines, telephone lines, houses, and roads are very common after a flash flood event (Saleem 1990; IFRC 2017). A composition of heavy flash flood incidents is given in Table 1.

Very impressive to understand the power of those floods is the observation of a ten tons weighted boulder which was moved by the water in 1923 on Sinai Peninsula (Saleem 1990). According to the World Meteorological Organization (WMO 2012), a flash flood event is "*A flood of short duration with a relatively high peak discharge.*" For the Eastern Desert, precipitation rates causing heavy flash flood events with high peak discharge are recorded in a range of about 30 mm/event to about 80 mm/event. The time of rainfall usually is about a few hours. In El Quseir, 34 mm were measured in 1954, in Daedalus close to Marsa Alam 28 mm (Saleem 1990). In 1979, 1994, 1996, and 1997 (Wagdy et al. 2003) recorded 48, 60, 80, and 40 mm during severe flash flood events. The case study of Hadidi (2016) which is described closely in this chapter took place in March 2014 with a total amount of rainfall of about 34 mm over a period of four hours, recorded close to the Red Sea shoreline.

These differences in reporting flood occurrences may be explained because of two main reasons: first, rainfall events are usually of small-scale convective systems (5–10 km) (Hadidi 2016); second, the Eastern Desert is a remote area, and there is no official observation for all wadis. According to the Emergency Center in the Red Sea Governorate, there are floods almost every year in the Red Sea Governorate in Egypt, but in different wadis with a wide variety of intensities.

The following work describes the specific situation of flash floods around El Gouna, Red Sea, and is focused on general morphological, geological, and hydrological settings of the catchment and detailed estimation of flooding effects close to the city using numerical modeling. The strongest event took place in March 2014 where several roads were destroyed, and water was covering the cities infrastructure over days. This event is intensively observed, and the data is used for the numerical model.

2 Hydrology and Morphology

2.1 Location

The city of El Gouna is located about 20 km north of the Red Sea capital Hurghada (Fig. 1). It is a touristic place which is famous for kite boarding and diving. During holidays, up to 20,000 inhabitants and tourists are living in the city in small villas and hotels, most of them are connected to the Red Sea by artificial lagoons.

To the west, a plain of about 10 km spreads in the vicinity of the city until it reaches a steep mountain ridge (Esh El Mellaha) which extends parallel to the coastline in a north–south direction (Fig. 1). In front of the mountain, ridge gardens are present which provide land for agriculture. Between the city and the agricultural lands, the landscape is dominated by desert and the highway from Safaga to Cairo.

The mountain ridge is crossed by a steep valley named Wadi Bili from west to east. It is the connection of the El Gouna plain to the catchment area of the Wadi Bili catchment in the west. It covers the high desert plain of Abu Shar and the western mountain ridge which separates the Red Sea catchments from the Nile river catchments. To the north, the catchment is limited by the catchment of the Wadi Umm Masaad, to the South by the Wadi Umm Diheis.

The water sources in the El Gouna area are wells of different depths. Close to the mountain ridge deep wells of up to 100 m provide fresh and brackish water for irrigation to the agriculture lands. In the city itself, brackish water is pumped by wells of the shallow aquifer to provide the local desalination plants with water. In the last years, destructions

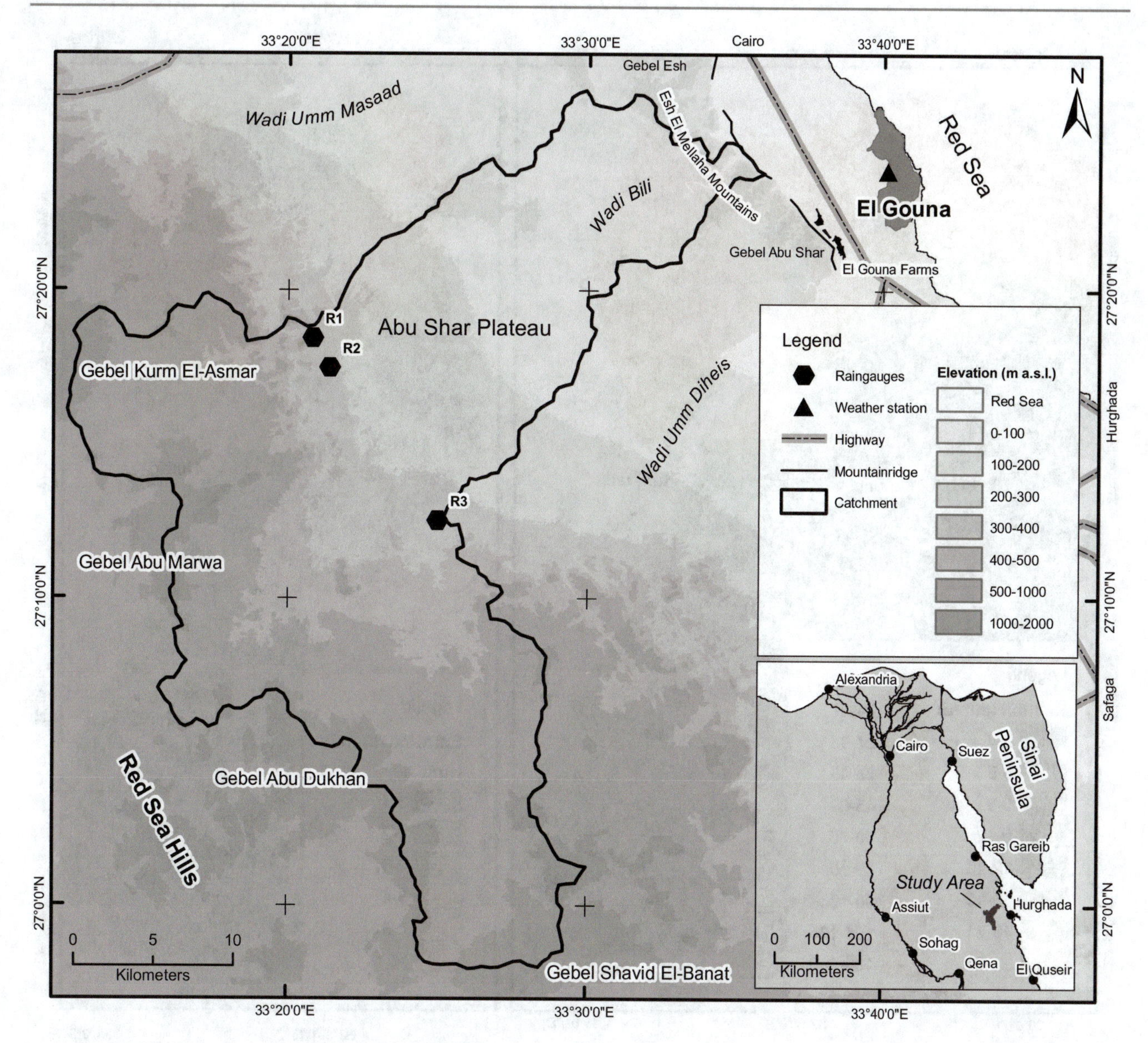

Fig. 1 Location and morphology of catchment Wadi Bili

caused by flash floods of the infrastructure, especially roads were observed.

2.2 Climate Conditions

Besides literature review, the basis of the description of the climate conditions is measurements at an automatic meteorological station close to the main building of the Technische Universität Berlin in El Gouna. Since 2013, records of rainfall, temperature, humidity, air pressure, wind speed, and wind direction for every ten minutes are available. Before the flash flood event in March 2014 three rainfall gauges (R1–R3 in Fig. 1) were installed in the distal areas of the catchment and data about the amount of precipitation were manually collected (Hadidi 2016). At the Red Sea, only a few meteorological stations are present, and they are located close to the shoreline. Meteorological direct measurements of rainfall in the mountainous area are not available. Long-term data of rainfall is present only at a few meteorological stations in Egypt. Additionally to the measurements of the meteorological station in El Gouna satellite data is used to understand the broad distribution of rainfall and intensity of heavy rainfall events. The Tropical Rainfall Measurement Mission (TRMM) (Huffman et al. 2007) datasets were chosen over a period of ten years (2005–2015). Long-term assessment of the average rainfall amounts in the Eastern Desert was taken from the calibrated data in Milewski et al. (2009) (Fig. 2). The measurements are available in time periods of three hours each and therefore can show

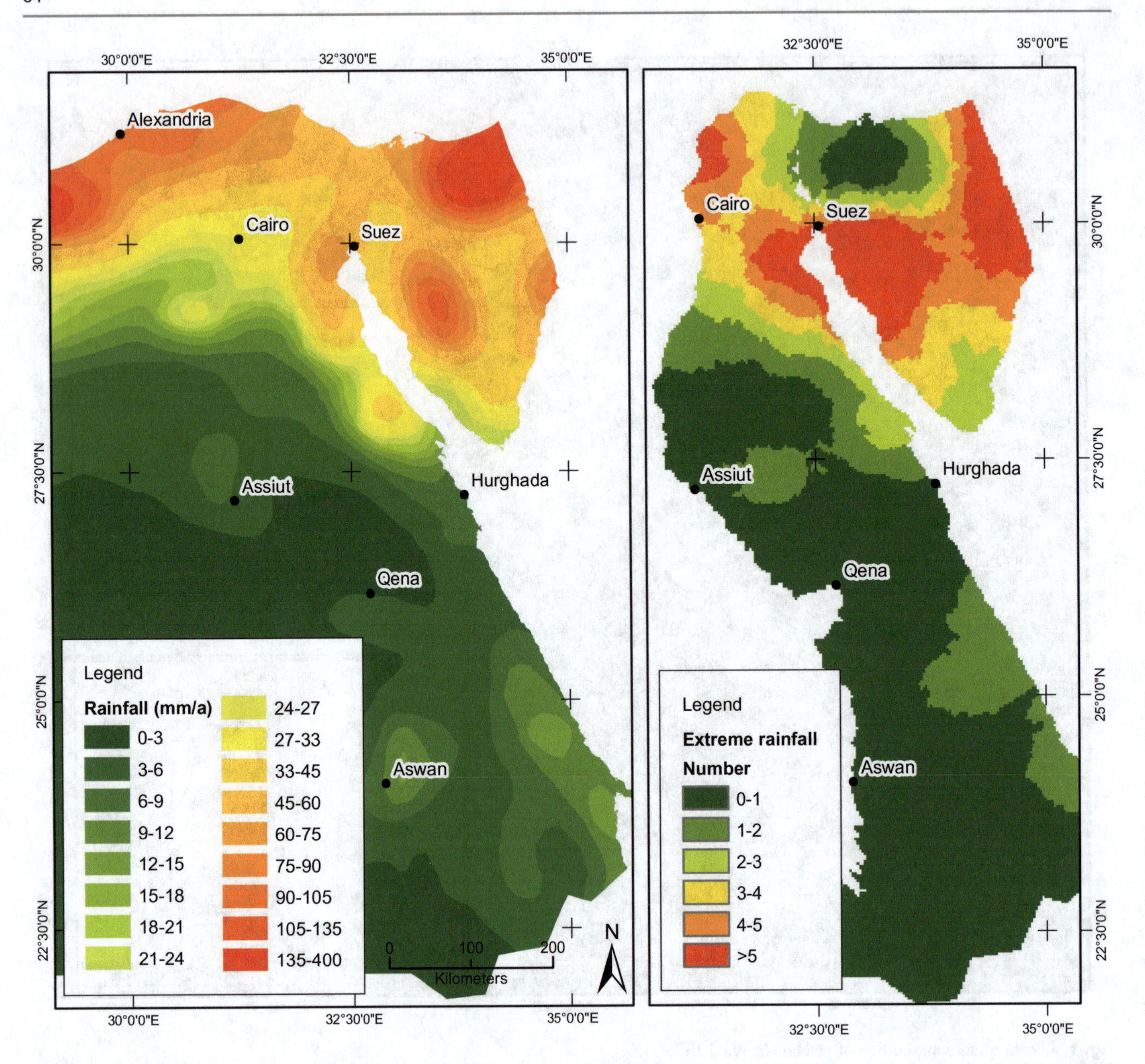

Fig. 2 Distribution of average annual rainfall (1997–2008) (left) modified after Milewski et al. (2009) and a number of extreme rainfall events of a time period of 10 years (2005–2015) (right)

short-term high rainfall events which are needed to produce flash floods in the desert. The total amount of rainfall in the Eastern Desert is a sum of all data over the ten years as an average annual value. The occurrence of heavy rainfall is a count of 3 h events which exceeded the value of about 8 mm. Calibration of the satellite data was done by referencing the TRMM measurement to the direct rainfall measurements at the rain gauges in the catchment of Wadi Bili on March 2014 and the meteorological weather station in El Gouna.

Driving force of the climatic conditions in the northern Red Sea area is the Active Red Sea Trough (ARST) (De Vries et al. 2013). The Red Sea Trough (RST) is a region of low pressure with its maximum at the eastern Mediterranean Sea north of the Sinai Peninsula. It is extended from south to north and follows the Red Sea which is surrounded by high mountains and therefore forms a morphological depression (Ashbel 1938; Krichak et al. 2012). This morphological feature is generating the trough as a primary factor (Krichak et al. 1997a, b). In the summertime, the conditions of moving air are relatively static. Beginning of autumn warm and moist air originated at the Arabian Sea is moving from south to the Red Sea region and the Mediterranean. While meeting cold air in the north, convective movement of the air is induced and causing rainfall up to the Mediterranean Sea. This phenomenon is known as the ARST. The periodicity of those rainfall events is very hard to predict as according to the lack of data, a detailed understanding of the physical forces is not understood until now (Krichak et al. 2012).

The distribution of rainfall is separated into three main regions of different rainfall intensities (Fig. 2). In the north at the Mediterranean coast high amount of rainfall is observed especially in the northern Sinai and Israel. The average rainfall can reach up to 400 mm/A. It decreases significantly to the center of the Sinai and the Gulf of Suez where it does not exceed 40 mm/A. Toward west in vicinity of the city Alexandria the precipitation increases close to the coastline with more than 200 mm/A. South of the Egyptian capital Cairo the desert climate causes very little rainfall amounts around 10 mm/A decreasing to almost no rainfall in the Sahara desert (Western Desert) southwest of Assiut. Along the northern Red Sea, coast rainfall amounts are higher and following a north–south trend up to the city of Ras Gharib. The trend continues to Hurghada close to the investigation site and has its minimum in the El Quseir region south of Hurghada. Heading to the south, rainfall rates increase again up to 40 mm/A at the Sudanese border.

The amount of heavy rainfall events in the Eastern Desert (Fig. 2) generally follows the same distribution as the average total rainfall amount. The highest amounts of events were detected in western Sinai at the Red Sea coast. To the south, the number decreases to about 3 very strong events close to the city of Ras Gharib which was heavily affected by a flash flood in 2016. The Hurghada area is less affected by storm events, the Wadi Bili catchment shows 2 events over the period of 10 years (Fig. 2).

Rainfall in Hurghada area at the Red Sea coastline is characterized by the absence of rain in the summertime in a 30 year average. The rainy period starts in October and ends in April and creates an average annual rainfall of about 5 mm/A. Calculations from Masoud (2007) show an average annual rainfall in the period from 1945 to 2000 of about 2.2 mm/A which is close to the calculations of Saleem (1990) of about 3.3 mm/A. Shahin (2007) estimated an annual rainfall of about 10 mm/A for the Hurghada region. The World Meteorological Organization (WMO 2013) mentions the highest amount of rain in November with an average of 2 mm. In the months October, December, and April the average precipitation is between 0.5 and 1 mm; January, February, and March it is less than 0.5 mm (Fig. 3). During the hydrological year 2014/2015 measurements of the weather station in El Gouna document an annual rainfall of about 38 mm/A. 34 mm rained during the March event. The rain gauge R1 (Fig. 1) recorded 32 mm; higher values were measured at R2 (42 mm) and R3 (47 mm). This experience shows the significance of time dealing with rainfall records and considers long-term average values as not representative for the occurrence and intensity of flash flood events.

Egypt and its Eastern Desert are meant to be one of the regions in the world with the highest amount of sunshine hours. In a long-term average 3530 h of sunshine are recorded (NOAA 2015). In the hottest months of the year, in July and August, average maximum temperatures exceed 35 °C (Fig. 3). At night, the air temperature decreases to about 25 °C. In wintertime colder air maximum daily temperatures around 20 °C and at night down to 10 °C can be observed for January and February. Between both temperature extremes in summer and winter, a smooth continuous transition of temperatures takes place. The average annual humidity of about 47% varies from 53% in the winter (October and November) to 41% in summer. It corresponds to the rainfall distribution but does not vary in a big range between the seasons (Saleem 1990). The Red Sea coast is

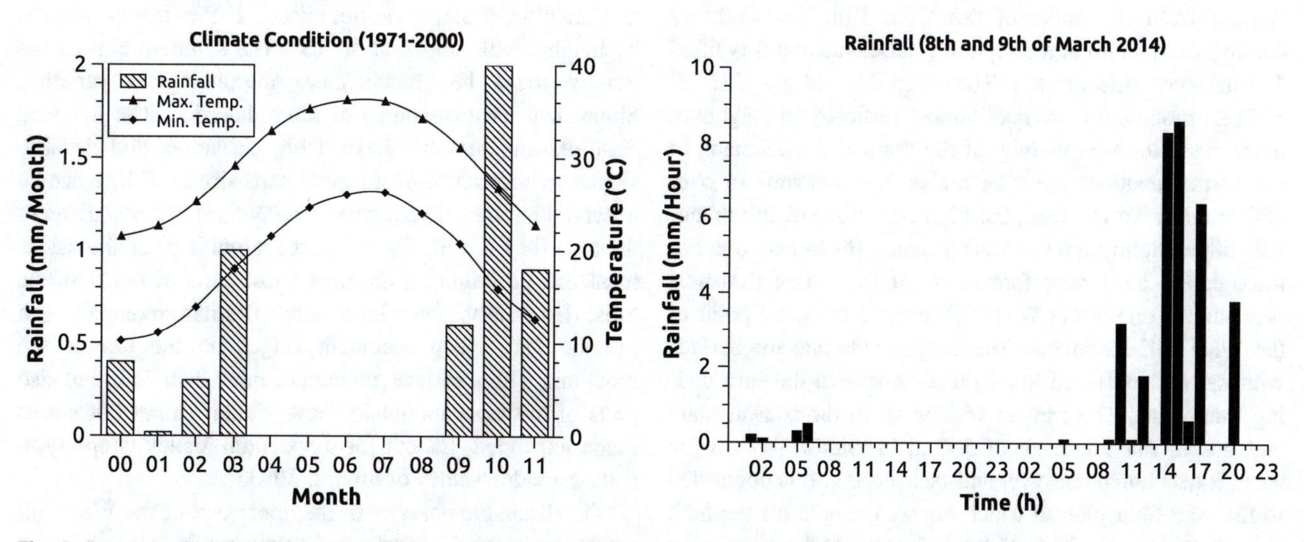

Fig. 3 Long-term (1971-2000) average climate conditions in Hurghada after WMO (2013) (left) and hourly rainfall during the flash flood event March 2014 (right) modified after Hadidi (2016)

dominated by strong winds coming from the north–west. Average wind speed in the Hurghada region is about 5.5 m/s (NOAA 2015). Especially in March and April gusty winds and sandstorms are present. Many sunshine hours, high temperatures, high wind velocities, and low humidity are causing high potential evaporation rates which are measured by about 3700 mm/A (Saleem 1990; Masoud 2007).

2.3 Morphology

The following observations were obtained using a digital elevation model. The basis is the global Shuttle Radar Topography Mission (SRTM) (USGS 2006) dataset available in a resolution of 30 m in the square.

The basin of Wadi Bili is part of the Red Sea Cuesta, which covers almost the whole Egyptian Red Sea coast on the western side. The graben structure of the Red Sea Rift Valley elongates in northwest–southeast direction from Suez up to the border of Sudan. Between Ras Ghareb and Hurghada, the Cuesta is present in several morphological stages with the same general direction like the whole Rift Valley formation. The catchment of Wadi Bili is morphologically structured in 4 units: The coastal plain between El Gouna at the coast and the mountain ridge, Esh El Mellaha mountain ridge, the Abu Shar plateau, and the high mountain formations in the western part of the catchment (Fig. 4). The coastal plain is a fan with smoothly increasing elevations from sea level up to 50 m amsl at the steep cliff of the mountain ridge. The average gradient of the plains′ surface is 0.006 in a relatively homogenous distribution. Shallow flow channels of former flash flood events are cutting into the plain in direction southwest–northeast orthogonal to the elongation of the Rift Valley. The most distinct channel with the most significant depression is pointing toward El Gouna coming from the outlet of the Wadi Bili. The highway coming from Cairo is a morphological element and is lifted 1–2 m above the ground. The steep cliff of the Esh El Mellaha mountain ridge rises almost vertically to heights of more than 200 m in vicinity of the Wadi Bili catchment, to the north elevations are even higher with amounts of over 400 m amsl. To the west, the high elevations of the mountain ridge slightly are decreasing again. Steep canyons run through the hard rock formations of the ridge; the most pronounced one is the Wadi Bili canyon the pour point of the Wadi Bili catchment. The canyon cuts into the surface with more than 100 m in height difference to the surrounding mountains. The canyon widens up to the coastal plain and has in maximum about 500 m in width. Toward the west, it gets much narrower and meandering. It is connected to the Abu Shar plateau which covers the area till the high western mountains about 25 km in length. At the connection point to the valley, the plateau has its lowest point with about 100 m amsl; at the mountain site elevations increase up to 300 m amsl. The average gradient of the surface is slightly higher than the ones of the coastal plain and adopt values of 0.008 at the central channel. Most of the catchments area is part of the high mountain ridge of the Red Sea hills which follow the regional trend of direction of the Red Sea Rift Valley. It covers an area of about 590 km^2 which is two third of the total area of the catchment. Four major mountains are surrounding the western catchment. The northern Gebel Kurm El-Asmar mountain has a height of about 1600 m. It is separated from the southern mountain complex by an about 2.5 km wide valley. The more southern Gebel Abu Marwa is about 1,600 m in height as well. It surrounds a wide valley to the south which is about 8 km in width. The distal southernmost valley elongates in north–south direction and is surrounded by two mountains, the Gebel Abu Dukhan (1800 m amsl) and the Gebel Shavid El Banat at the most southern border of the catchment which has the highest elevation of the catchment with almost 2050 m amsl (Figs. 1 and 4).

The distribution of the steepness of slopes follows, in general, the same structure of the distribution of mountain ridges and the coastal plain and the central plateau (Fig. 4). There is a strong relation between slope and lithological distribution of rocks in the catchment. A detailed description of the geological features is given in Sect. 3.

The steepest slopes can be found in areas where the basement is exposed on the surface. This includes the Red Sea hills and a small area around the Wadi Bili canyon (Fig. 4). The tertiary rocks of the Esh El Mellaha mountain ridge have a mediate strength of slope in comparison to the high mountains and the flat Abu Shar plateau. The Wadi Bili canyon is characterized by steep slopes along the valley with measures up to 45°. The mountain ridge is generally more smooth and undulated, the northern part of it is relatively flat and broad parts slopes do not exceed 1–5%. It is permeated by heights with slopes up to 15°. The southern part of the tertiary rocks has much more morphological varieties. Slopes can have steepness of more than 30°. The flat Abu Shar plateau does not have a big curvature, and usually, slopes do not exceed 3°. In small parts slopes of 10% can be observed but they do not have a big portion in comparison to the area they cover. The steepest morphology of the catchment can be found on the uppermost parts of the Red Sea hills. Especially, the Gebel Abu Dukhan mountain can provide very steep mountain ridges on the top of the mountain with a surface gradient of more than 70%, but also parts of the other mountains have similar slopes but not as much expanded. Except for the Central Valley, slope typically provides values of around 30–35°.

The discharge network of the upper part of the Wadi Bili catchment was calculated using geoprocessing algorithms on the basis of the SRTM digital elevation model. The main

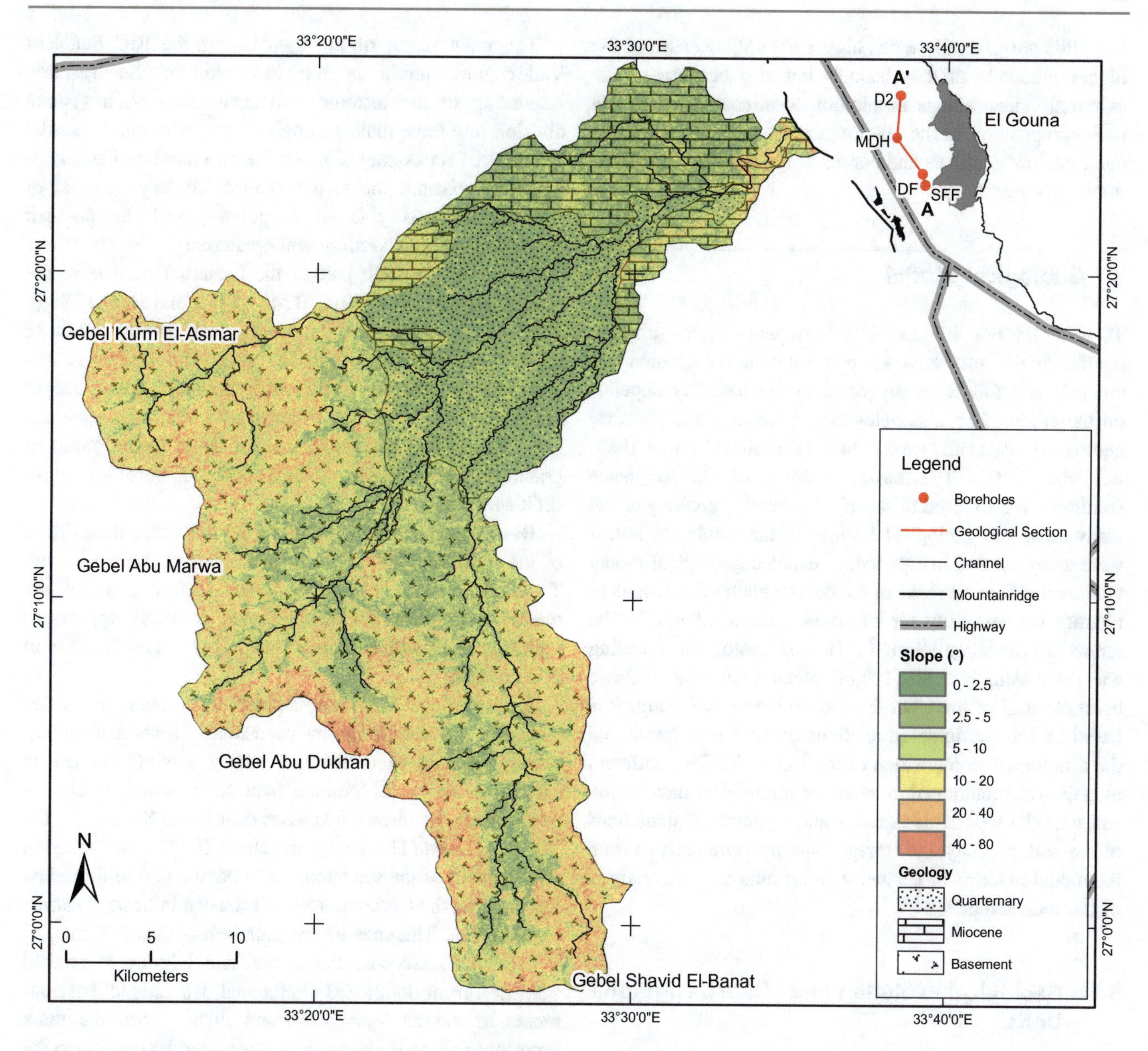

Fig. 4 Distribution of slopes in the Wadi Bili catchment and relation to the geological situation

discharge structures were calculated. Flow distributions on the coastal plain of El Gouna are not included and are part of the Sect. 4, as the flow directions are very sensitive to the accuracy of the digital elevation model and chance over time according to sedimentation processes. Observations prove that besides the flash flood events almost no water is running through the channels.

The pour point of the catchment is the mouth of Wadi Bili. It is also the location where measurements of the discharge during the flash flood of March 2014 were taken. The sub-catchment of the northern Esh El Mellaha mountain ridge discharges separately from the rest of the catchment with a main channel of about 13 km length. The average gradient can be estimated with 0.006. To the west of this junction, the flow channel divides into 2 different pathways. The northern channel divides again into 2 channels coming from the main valleys in the Red Sea Hills. Along the Abu Shar plateau, the resulting 2 northern channels have a length of 18 km and 23 km, respectively, until the steep parts of the Red Sea Hills. As a result, the gradients are about 0.009–0.01. To the mountains the gradient increases to 0.02 for the northern stream and 0.014 for the southern stream. A similar condition exists for the southern channel which follows the surface gradient to the south into the Gebel Shavid El Banat mountain ridge. On the Abu Shar plateau average values of about 0.009 can be estimated along the channel. The steep mountain area provides an average gradient of around 0.018. According to these values, the gradient doubles in the mountainous area in comparison to the plateau. It can be estimated that flow velocities in the distal areas of the Red

Sea Hills are generally much higher not only because of the higher gradients of the channels but also because of the extremely steep slopes in addition to almost impermeable rock composition of the outcropping basement (Sect. 3). In contrast, low gradients and porous sediments slow down the flooding water.

3 Geological Model

To study the possibilities of the flood water recharge to the aquifers in the study area, a geological model of the area was created. The following suggested model basically depends on the study of rock samples from boreholes and surveys, geological maps and tectonic maps (Klitsch and Linke 1983; EGSMA 1981). The basic concepts of the sequence stratigraphy were used to specify the detailed geology of the study area. The geological features of the whole catchment were analyzed and interpreted. A detailed geological model was created for soft rocks of the coastal plain of El Gouna to identify the capability for infiltration and as a basis for the numerical modeling (Sect. 4). The 3D geological modeling was done using indicator kriging method with the available borehole data of the wells located on the coastal plain. It is based on the stratigraphic relations of the layers found and the lithological composition of the soft rocks. Two different models were elaborated, a model of lithological distribution and a model separating aquifers and aquitards. Calculations of the bulk porosity and storage capacity were derived from the model in terms of geometry and results of sieve analysis of the rock samples.

3.1 Geological Evolution and Lithostratigraphic Units

Dominant morphological and tectonic feature of the Red Sea underground composition is the East African Rift Valley. The southwest–northeast extension of the Red Sea had a huge impact on the study area and caused uplifting of the basement to the surface, and hence makes the tectonic lineaments using satellites images visibly clear. On the other hand, these old rocks reflect not only recent tectonic activities but also old ones up to the Precambrian due to old dykes and extinct shear zones. The recent erosion of the uplifted crystalline basement and carbonate platform resulted from the Red Sea Rift System activities and caused an accumulation of huge amounts of clastic sediments as alluvial fans. Derived from the recent stress field the extension direction of the rift valley during the Miocene was oriented N55° E (Bosworth and Taviani 1996) and rotated during the late Pleistocene to recent direction of N10° E.

The main phase of the uplifting of the Red Sea Rift Valley took phase in the Miocene, middle Tertiary. According to the tectonic movement activity, a general division into three main geological categories can be made: The pre-rift rocks consisting of the basement and sedimentary and volcanic rocks, the syn-rift Tertiary deposits of silici-clastics and younger evaporates, and the post-rift clastics of the late Tertiary and Quaternary. The crystalline basement in the north part of the Eastern Desert is of the Neoproterozoic Era (600–570 Ma) (Stern and Hedge 1985). The rocks are remains of the orogeny and mainly consist of acidic magmatic rocks (Greiling et al. 1988). Two formations of the basement are present in the study area: younger granitoids and volcanics dissected by fractures. Granite, granodiorite, and adamellite can be found in the group of granitoids, the volcanics mainly consist of porphyric rocks (EGSMA 1981).

Basement outcrops are mainly the high mountain ridges of the Red Sea Hills in the western part of the catchment. They cover more than 50% of the surface geology. As mentioned in Sect. 2.3 the basement generally appears as high elevations with extremely steep slopes between 500 and 2000 m amsl

Paleozoic and Mesozoic deposits in northeastern Africa generally are dominated by sandstones (Alsharhan 2003), thick sequences of continental poorly fossiliferous quartz sandstones form the ‘Nubian Sandstone’ which is disconnected from the huge thicknesses of it in the Sahara and the Western Desert (Thorweihe and Heinl 2002). The formation environment of the sandstones was continental to a shallow marine; just the Carboniferous carbonates indicate a marine environment. Thicknesses are usually less than 200 m.

During Cretaceous, Paleocene, and Eocene, a gradual transition from dominated continental depositional environments to marine took place and finally form the thick limestones of the Eocene series while later deposits up to the Miocene were dominated by continental conditions and consist of sandstones, siltstones, and conglomerates.

Miocene syn-rift strata are shallow marine sediments and coastal deposits. Conglomeratic fan deltas describe the sedimentation along the ancient coastline and change their composition into marls and evaporitic Sabhka sediments according to the distance of this line. The intense faulting during rifting caused the creation of blocks with discordances to the underlying pre-rift strata. Post-rift Pliocene and Quaternary sediments are widely spread and can have thicknesses over 1500 m (Bosworth 1995).

The dominating lithologies are gravels and sands, interposed by thin layers of evaporites. Quaternary deposits include Aeolian sands and Quaternary alluvium, which cover most of the depressions in the catchment, in addition to the coastal plain deposits, which consist of loose to moderately consolidated coarse clastics derived from topographic heights.

3.2 Hydrostratigraphic Units

According to the evolutional separation of the lithological units, aquifers can be divided similarly. Four major aquifer systems are present in the area of the Wadi Bili catchment: The basement aquifer, the pre-rift aquifer, a syn-rift aquifer, and the shallow post-rift aquifer. The basement is a fractured aquifer with good hydraulic conductivities according to the presence of petroleum reservoirs observed in the Gulf of Suez (Alsharhan 2003; Salah and Alsharhan 1996). In the Wadi Bili catchment, this secondary porosity could not be determined, but observations of Bedouins and one old well located on major fracture show water availability close to the Abu Shar plateau. Before the 1990s during flood natural springs occur.

Nubian sandstone, the Cretaceous, and the Paleogene formations mold an aquifer system. In the Eastern Desert data on these aquifers are not available especially for the study area. (Thorweihe and Heinl 2002) researched freshwater reservoirs in these formations for the Western Desert but after Said (1990) in the Eastern Desert, the aquifers are reduced according to the rise of the Red Sea Mountains. There are no outcrops of the formations in the catchment, and the distribution is not clear until now.

On Abu Shar plateau, evaporitic formations of the middle-late Miocene have been washed out and eroded, as limited outcrops could be noticed. This could be an important reason for elevated salt content in the post-Miocene aquifers, i.e., Pleistocene and Quaternary alluvial fans. The lithology of the carbonatic aquifer is limestone dolomite. Karstification is observed by Hadidi (2016). The groundwater of the aquifer has a relatively high temperature.

The post-rift aquifer covers all plains and central parts of the Abu Shar Plateau. It consists of Pliocene and Quaternary alluvial deposits and lagoon sediments (Hadidi 2016). On the coastal plain of El Gouna, three overlaying aquifers are presently divided by less permeable mud layers and lenses. Measurements of the electric conductivity during drilling showed different salt contents of the groundwater increasing with depth. The groundwater of this aquifer is the source of the drinking water production of El Gouna by desalination of the brackish water. The shallow aquifer with less salt content is used which is not exceeding 30 m in depth close to the city. In the second aquifer at this location salinities higher than the Red Sea were observed in 2017.

3.3 Lithological 3D Model

Four borehole and four liner samples were collected to study the detailed lithology of the study area and correlate the stratigraphy of the underground sediments (Fig. 5). The boreholes D2 and DF (location in Fig. 4) are about 75 m in depth, therefore the source for data between 25 and 75 m below sea level. The boreholes MDH and SFF are shallow with depth up to 25 m. The liner samples gave an idea of the layers thickness and in combination with sieve analysis an idea of the porosity of the layers. Liner data proved the existence of different layers thickness from few centimeters and up.

A 3D lithological model has been created using the software RockWorks14®. The model represents the clastic lithology of the alluvial fans with a depth of less than 100 m, with a rectangular surface of an area of 56 km^2. The horizontal model dimensions are 8 km $\times$ 7 km, with a horizontal spacing of 100 m. The vertical dimension ranges from −75 to 30 m amsl with a vertical spacing of 1 m.

Sequence stratigraphic correlation was used to understand the distribution of lithology of the upper formations and is described in detail in Hadidi (2016). The sedimentation reflects different scenarios of sea level rise and fall. The upper part of the aquifer represents a typical high variability in thicknesses of alluvial fans. Coarse deposits of gravel layers and pebbles are present. Generally, grain sizes decrease to the bottom of the investigated layers. Lithologically the layers of sand and sandy gravels become more dominant to the lower parts. All layers are highly separated by muddy thin deposits indicating a low-speed sedimentation environment. On the basis of this correlation and additional evaluation of boreholes, the 3D geological model was set up (Fig. 6). Important continuous strata are two mudstone and clay layers which are proposed to be present over the whole investigation area (Hadidi 2016). Those layers separate the aquifers (Fig. 6). The chemical composition of groundwater observations correlates with this hydrostratigraphical separation showing lower salinities in the upper aquifer ($\sim$25 mS/cm) and higher in the lower ($\sim$35 mS/cm) measured directly below the mud layer in borehole MDH. Similar observations were done at boreholes SFF and D2. The final hydrostratigraphic model contains three aquifers separated by two aquitards. The first aquitard located at a depth of $\sim$20 m below sea level, and the second located at a depth of 45 m below sea level (Fig. 6).

Quantifying hydrological important parameters was done by analyzing the porosities of different lithology found in boreholes (Table 2).

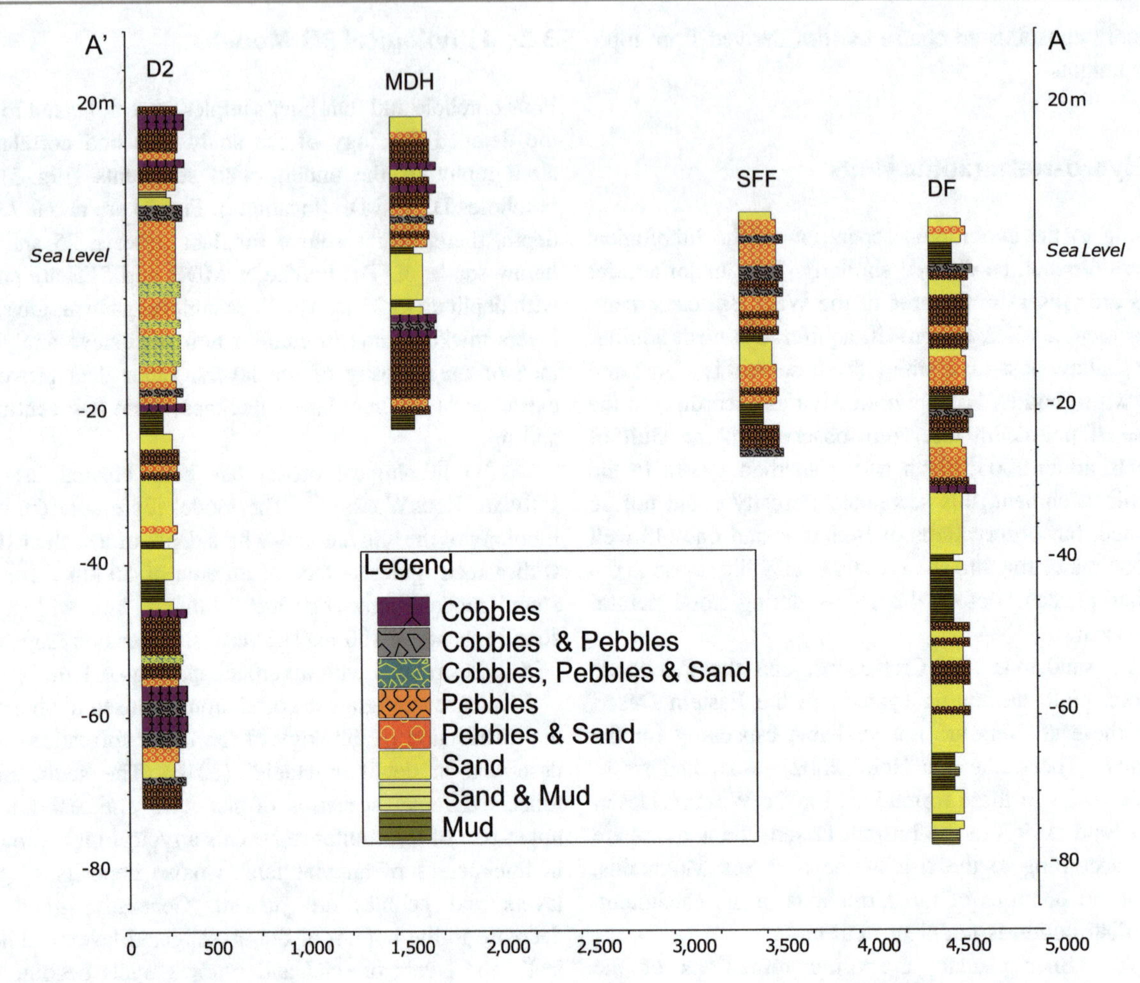

Fig. 5 Stratigraphic correlation of boreholes located at the coastal plain El Gouna modified from Hadidi (2016), location in Fig. 4

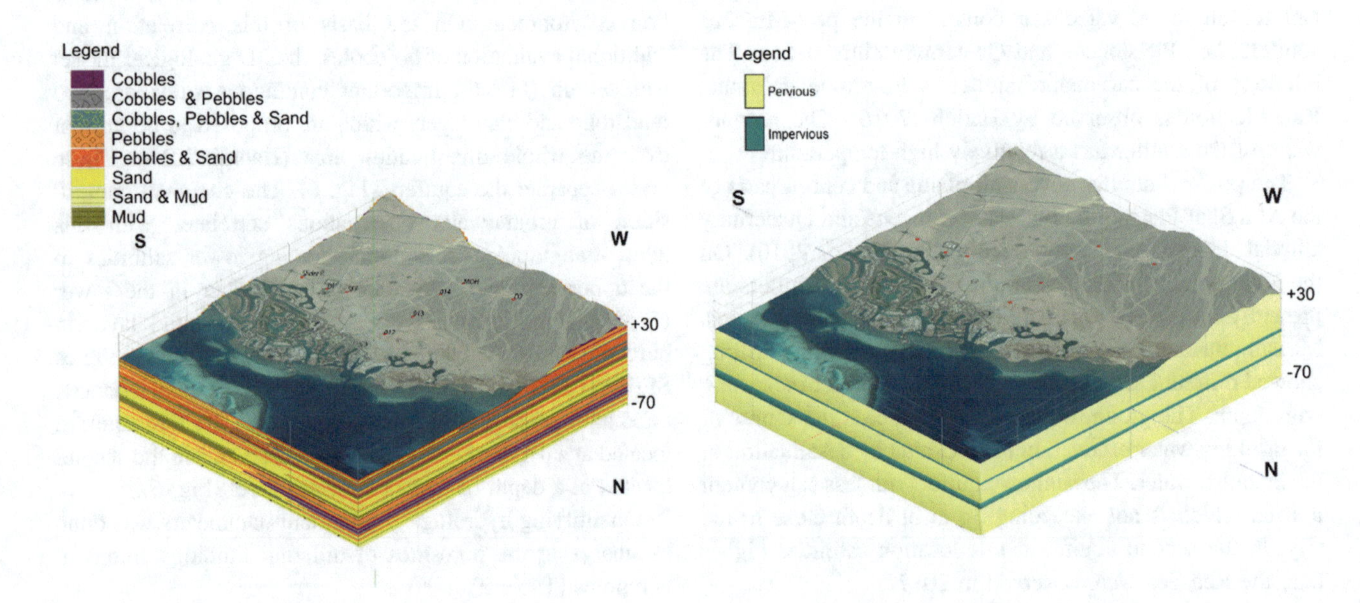

Fig. 6 3D geological modeling of the coastal plain El Gouna and hydrostratigraphic distribution modified after Hadidi (2016)

Table 2 Porosities of upper strata obtained by sieve analysis (modified after Hadidi (2016))

Lithology	Calculated total porosity (%)	Measure total porosity (%)	Drainable porosity (%)
Pebbles & sand	29	27	3
Cobbles, pebbles & sand	26	19	9
Pebbles, sand & mud	30	24	4
Sand with minor pebbles	32	28	2
Sand with pebbles	31	32	2
Pebbles & coarse sand	37	29	5
Pebbles, cobbles, sand & mud	28	30	7
Pebbles with sand	26	19	9

The grain size distribution represents an alluvial fan with many different lithologies. Porosities are taken from sieve analysis and further calculations and measurements to determine the capability for water storage. Therefore, total porosities represent the maximum amount of water which is present in the aquifer, and the drainable porosity is the minimal volume which can be extracted. Average drainable porosity is about 5%. It is much smaller than the total porosity with slightly more than 20% of the total porosity. Regarding the permeable units in the model, the volume of the upper permeable unit is 1858 million m^3 and the volume of the second permeable unit is 925 million m^3. The total porosity volume of the upper unit is 530 million m^3 including the layers up to the ground surface, and it is 289 million m^3 for the second aquifer. The total drainable porosity volume of the upper unit is 99.8 million m^3, and it is 31.7 million m^3 for the second aquifer.

4 Flash Flood Model

After investigations of hydrological, morphological (Sect. 2), and geological (Sect. 3) conditions of the study area, a two-dimensional flash flood model can be used to further study the different processes during flash floods and their interactions as well as possible mitigation measures to protect the study area from severe damages. Flash flood models represent an important tool to better understand the flow processes during and after heavy rainfall events. They can be applied to investigate different scenarios regarding increased rainfall intensities on the one hand and structural measures such as dams, canals and retention basins on the other hand to find suitable solutions to mitigate flood damages. Flash flood models should also be part of early warning systems and be used to create hazard and risk maps raising the awareness regarding zones which are particularly prone to flash floods so that the communities can be better prepared. Also, methods to effectively store and use the fresh water amounts supplied from flash floods can be investigated by combining a flash flood model and a geological model as introduced in Sect. 3 to study the surface flow field and the conditions of the subsurface the same time.

Classically, hydrological catchment models are used to simulate rainfall-runoff processes. They consider the runoff generation including processes such as precipitation, infiltration, evaporation, and retention as well as the runoff concentration regarding the temporal arrival of water from different origin generating the runoff hydrograph at the outlet of the catchment. In those types of models, the consideration of hydrodynamic processes is usually strongly simplified. In contrast, hydrodynamic models are used to simulate more detailed local flow processes in river sections, lakes, or coastal areas and can also be used to study the two-dimensional propagation of flash floods generated by heavy rainfalls. Such models provide water depths, flooding areas, and flow velocity fields for the whole computational domain. Different structural measures to protect settlements and cities against flash floods can be implemented in the model to investigate their effectiveness and to find the best solution. They can also be coupled with transport models to simulate the concentrations and distribution of substances and sediments. Robust numerical methods enable the simulation of flow with very small water depths over complex topography, the propagation of wet–dry fronts, and flow transitions. Together with the continuously increasing computer performance, effective scaling methods, and high-resolution data, these schemes are suitable and recommended to be used for rainfall-runoff and flash flood modeling. Due to the relatively small water depths in such events, two-dimensional shallow water models instead of three-dimensional models are considered to be appropriate as flash flood models (Cea et al. 2008; Tügel et al. 2017a). Processes of evapotranspiration can often be neglected in flash flood simulations because the considered time scale is relatively small, and the evapotranspiration is usually reduced due to cloudiness and decreased temperatures

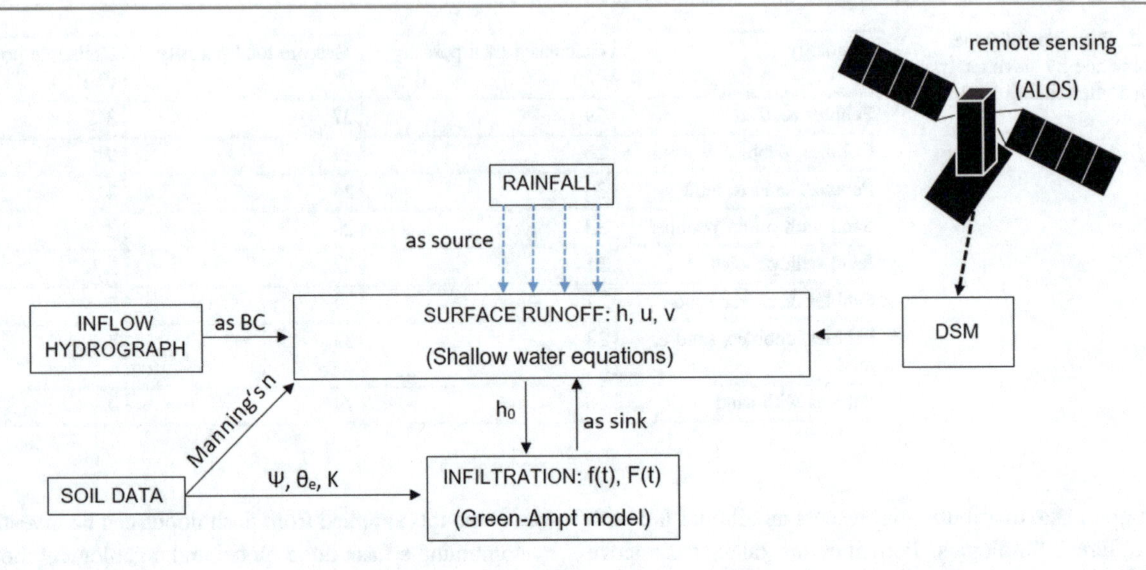

Fig. 7 Flow chart of the modeling procedure (BC: Boundary condition, DSM: Digital surface model; denotations of parameters and variables are given in Sect. 4.1)

during and after heavy rainfalls. In contrast, infiltration processes can have a strong influence on the runoff behavior during flash floods and should be taken into account.

As demonstrated in Table 1, the Eastern Desert of Egypt was often affected by flash floods and many cities are not well protected by mitigation measures or early warning systems. In the following, a flash flood model for the region of El Gouna in the Red Sea Governorate of Egypt is presented as an exemplary study case. Similar models could be set up also for other regions. An introduction of the used methods is followed by the description of the model setup and the discussion of different simulation scenarios considering infiltration processes and mitigation measures. Figure 7 shows a flow chart of the modeling procedure representing the input data and simulated processes with the corresponding parameters and variables as well as their interactions.

4.1 Hydroinformatics Modeling System (HMS)

HMS is a Java-based, object-oriented software framework that has been developed at the Chair of Water Resources Management and Modeling of Hydrosystems, Technische Universität Berlin (Simons et al. 2014; Busse et al. 2012; Simons et al. 2012). The flow is calculated with the two-dimensional shallow water equations, where rainfall and infiltration are considered in source/sink terms in the mass balance equation. Further implemented processes are tracer and sediment transport.

HMS has been designed with a general setup to enable the implementation of single and multiple processes in different spatial and temporal resolutions as well as interactions with geospatial information databases (Özgen et al. 2014).

4.1.1 Governing Equations

The general form of the 2D conversation law can be expressed as

$$\frac{\partial \mathbf{q}}{\partial t} + \frac{\partial \mathbf{f}}{\partial x} + \frac{\partial \mathbf{g}}{\partial y} = \mathbf{s} \tag{1}$$

where $\mathbf{q}$ is the vector of conserved flow variables and t is the time; $\mathbf{f}$ and $\mathbf{g}$ denote the flux vectors in x- and y-direction, respectively, and the vector $\mathbf{s}$ represents the source/sink terms.

The two-dimensional shallow water equations can be written by deploying the vectors given in (2) into the general conservation law (1):

$$\mathbf{q} = \begin{bmatrix} h \\ uh \\ vh \end{bmatrix}, \ldots \mathbf{f} = \begin{bmatrix} uh \\ uuh + \frac{1}{2}gh^2 - \upsilon_t \nabla(uh) \\ uvh \end{bmatrix}, \ldots \mathbf{g} = \begin{bmatrix} vh \\ vuh \\ vvh + \frac{1}{2}\upsilon_t \nabla(vh) \end{bmatrix}, \ldots \mathbf{s} = \begin{bmatrix} r \\ -gh\frac{z_B}{x} - f_x \\ -gh\frac{z_B}{y} - f_y \end{bmatrix} \tag{2}$$

with water depth h, velocity vector components in the x- and y-directions u and v, respectively, and the bottom elevation above datum z_B. r is a mass source/sink term, g is the gravitational acceleration, the turbulent viscosity is denoted with υ_t, the density of water with ρ and the external forces with f which includes the bottom friction. In this study, the turbulent viscosity υ_t is set to zero, and the bottom friction is calculated with Manning's law.

4.1.2 Infiltration

Infiltration processes are considered with the Green–Ampt model which is implemented in HMS. Here, the cumulative

Table 3 Green–Ampt parameters after Rawls et al. (1983)

Soil texture class	Effective porosity θ_e	Wetting front capillary suction head Ψ (cm)	Hydraulic conductivity K (cm/h)
Sand	0.417	4.95	11.78
Loamy sand	0.401	6.13	2.99
Sandy loam	0.412	11.01	1.09
Sandy clay loam	0.330	21.85	0.15

infiltration is calculated with Eq. (3), and the infiltration rate is derived with Eq. (4).

$$F(t) = Kt + (h_0 - \psi)\Delta\theta \ln\left(1 + \frac{F(t)}{(h_0 - \psi)\Delta\theta}\right) \quad (3)$$

$$f(t) = K\left(1 + \frac{(h_0 - \psi)\Delta\theta}{F(t)}\right) = \frac{dF}{dt} \quad (4)$$

where $F(t)$ denotes the cumulative depth of infiltration, $f(t)$ the infiltration rate, and K the hydraulic conductivity at residual air saturation, which is assumed to be 50% of the saturated hydraulic conductivity K_s (Whisler and Bouwer 1970). Ψ is the wetting front capillary suction head, h_0 the ponding water depth, and $\Delta\theta$ the increase of moisture content, which is the difference between the effective porosity θ_e and the initial moisture content θ_i. The effective porosity, wetting front capillary suction head, and hydraulic conductivity are the Green–Ampt parameters and depend on the given soil texture class.

Table 3 shows exemplary the empirically derived Green–Ampt parameters for four soil texture classes after Rawls et al. (1983).

4.1.3 Numerics

By using a cell-centered Finite-Volume method for space discretization and an explicit forward Euler method as time discretization, the conserved variables **q** can be calculated for the new time level $n + 1$ with Eq. (5).

$$\mathbf{q}^{n+1} = \mathbf{q}^n - \frac{\Delta t}{A}\sum_k \mathbf{F}_k^n \cdot \mathbf{n}_k l_k + \Delta t \mathbf{s}^n \quad (5)$$

where $n + 1$ and n denote the new and the old time level, respectively, and **F** the flux vector over edge k and k is the index of a face of the considered cell. The time step is denoted with Δt, and A is the area of the considered cell; **n** is the normal vector pointing outward of a face, and l is the length of a face.

To ensure numerical stability in this explicit scheme, the time step is constrained by the Courant–Friedrichs–Lewy (CFL) criterion (Simons et al. 2014). In this work, the infiltration rate after Green–Ampt was calculated for each time step of the flow calculation. Equation (3) is solved with a second order MUSCL scheme (Hou et al. 2013). Discontinuities in the flow field like they appear in hydraulic jumps or at wet–dry fronts can be handled with the implemented HLLC Riemann solver to compute the fluxes over the cell edges. The application of a total variation diminishing scheme avoids spurious oscillations (Hou et al. 2013; Özgen et al. 2014). To ensure robustness at wet–dry fronts, a cell is defined as dry and taken out of the calculation if a certain water depth threshold (here: 10^{-6} m) is undershot. Furthermore, the scheme switches from second order to first order accuracy at wet–dry fronts and high gradients (Hou et al. 2013; Murillo et al. 2009).

4.2 Model SetUp—Computational Domain, Initial and Boundary Conditions

The model should be used to simulate the flow field between the cross section where the runoff hydrograph from Wadi Bili catchment during the event of March 9, 2014 was measured (see Sect. 2) at the city of El Gouna. Therefore, a model area of approximately 8 km × 11 km was chosen (Fig. 7). The grid consisted of 397,296 rectangular cells of 15 m × 15 m. The digital surface model ALOS WORLD 3D (AVE) was used to represent the topography of the model area. The cell size of this freely available digital surface model (DSM) is 30 m (JAXA 2017). To reach a smoother transition between the cells and a higher resolution of the DSM, an inverse distance weighted interpolation with an output cell size of 10 m was conducted by using SAGA GIS. A Manning coefficient n of 0.033 $m^{-1/3}$•s was used to represent the friction of a natural channel with no vegetation (Chaudhry 2008). The initial water depths in the whole system were considered to be zero, neglecting the initial water depths inside the lagoons of El Gouna. All model boundaries were defined as open boundaries except for 30 m at the western model boundary where the inflow hydrograph from Wadi Bili was implemented according to the runoff measurements from March 9, 2014, which are published in Hadidi (2016). The total period of measured discharge was approximately 18 h, and the peak discharge of about 42 m^3/s was already reached after 1 h and 45 min. For simplicity, the accumulated measured rainfall of 34 mm was considered as equally distributed over a period of 8 h where the measurements showed the highest intensities. Thus, a constant rain intensity of 4.25 mm/h was implemented as a source term in each cell of the domain over the first 8 h of simulation time. A total period of 72 h was simulated to investigate the flow propagation and flooding areas inside the domain (Tügel et al. 2017b).

4.3 Results

4.3.1 Reference Case

The reference case represents the simulated flow processes caused by rainfall inside the domain as well as the inflow from the upstream wadi catchment into the computational domain. The results include fields of water depths and flow velocities for each hour of the total simulation time to analyze the propagation of the flood and the development of flooding areas in space and time. Figure 7 shows exemplary the simulated flow velocity magnitudes after 11 h of simulation plotted on a map of El Gouna. After passing the narrow valley of Wadi Bili at the western model boundary, the flood spread into different streams on the plane between the plateau and the Red Sea. Some of the streams propagated to the north and northeastern directions flowing out of the model domain and others to the southeast, crossing parts of El Gouna before draining into the lagoons and the Red Sea. In addition to the flood wave from the wadi catchment, the rainfall inside the domain accumulated in different flow paths and depressions. Some areas in the northern part of El Gouna are not connected to the flow paths coming from the inflow of Wadi Bili and the streams and ponds in this area were only generated by the local rainfall. Maximum flow velocities up to 0.9 m/s were reached at some locations, mainly in a curve before the area of the highest inundations close to the old desalination plant (see point C in Fig. 8).

Figure 8a shows the simulated water depths after 11 h representing the time step where the maximum water depth occurred inside the computational domain and over the total simulation time. Figure 8b depicts the remaining water depths after 72 h of simulation time. The main southern stream coming from Wadi Bili crossed the road of the connection Hurghada–Cairo and led to inundations up to 0.6 m at one of the crossroads (point A). East of the road, the flood reached some hotel and club areas causing inundations up to 1 m in some local depressions (Point B). The highest water depth of 1.49 m occurred after 11 h in an area around the old desalination plant (Point C). East of point B, the stream crossed El Gouna Park and drained into one of the lagoons connected to the Red Sea. The area around the TU Berlin Campus El Gouna was only affected by rainfall inside the domain and water depths up to 0.49 m occurred after 72 h (D). It could not be derived in detail, to which extent the land areas around the lagoons were affected, because the resolution of this model was too coarse. For such purposes, the mesh inside the city area must be refined to better resolve the flow and water depths around the buildings (Tügel et al. 2017b).

4.3.2 Case 1: Infiltration

Infiltration was considered with the Green–Ampt model which is implemented in HMS (see Sect. 4.1). Considering

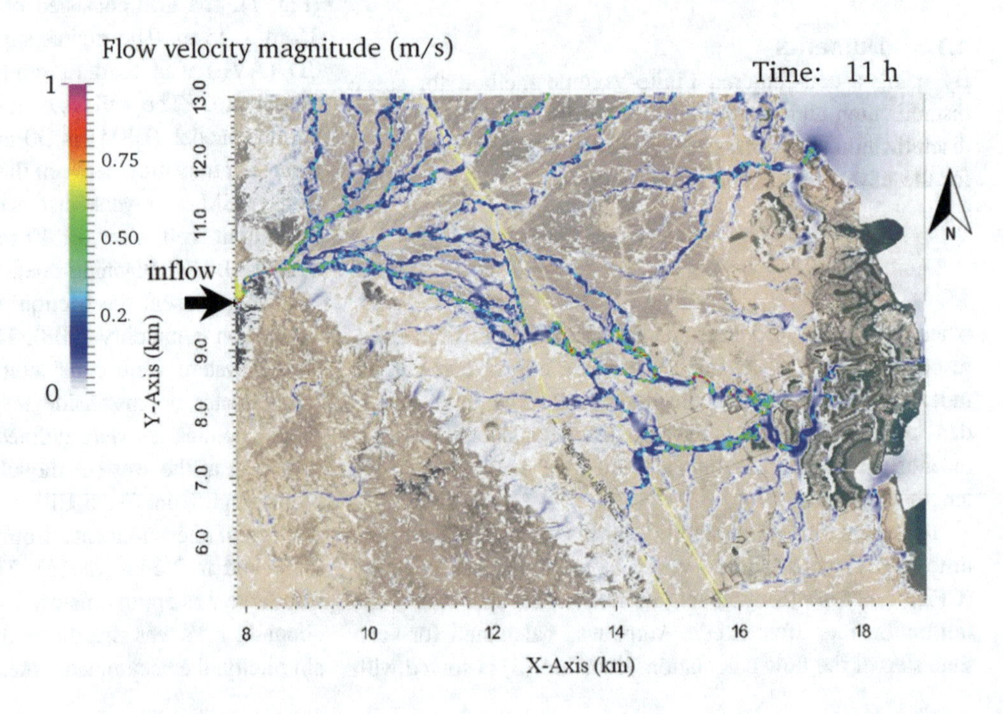

Fig. 8 Field of simulated flow velocities in m/s after 11 h of simulation, plotted on a map of El Gouna. ©Google (2017)

the Green–Ampt parameters for loamy sand after Rawls et al. (1983) (see Sect. 4.2) and initial moisture content of zero assuming a completely dry soil before the event, all the water was infiltrated before reaching the city of El Gouna. Considering the parameters for a sandy clay loam and an initial moisture content of 4%, the simulated water depths were also strongly decreased compared to the simulation without infiltration, but surface runoff still occurred also in the city area. Figure 9 shows the simulated water depths inside the domain after 11 h for sandy clay loam with (a) the average Green–Ampt parameters after Rawls et al. (1983) and (b) a reduced soil suction head of 4.42 cm instead of 21.85 cm. The value of 4.42 cm corresponds to the minimum soil suction head measured for a sandy clay loam by Rawls et al. (1983), while 21.85 cm represents the average soil suction head of all conducted measurements for this soil texture class. As expected, the reduced soil suction head led to higher water depths and further propagation of the flood. As long as the soil is not saturated, the infiltration rate under consideration of a lower soil suction head will be decreased compared to the infiltration with a higher soil suction head. Figure 10 shows the water depths over time at the location of the maximum water depth in the reference case for both variants of infiltration compared to the reference case and scenarios with retention basins, which will be discussed in the next section. The peak water depth was significantly damped and shifted if infiltration was considered. For sandy clay loam with $\Psi = 21.85$ cm the peak water depth was 0.42 m after 32 h, and with $\Psi = 4.42$ cm a peak water depth of 1.0 m occurred after 24 h compared to 1.49 m after 11 h in the reference case. Thus, a higher soil suction head caused a much stronger effect of damping and shifting of the peak water depth.

Field measurements of infiltration rates showed a significant spatial variability inside the considered domain, and at some locations, the sedimentation of fine material from previous flood events reduce the infiltration rate at the surface. As mentioned, for example, in Mein and Larson (1971), the effect of surface clogging due to raindrop compaction during heavy rainfalls has also an important influence on the infiltration behavior. Furthermore, reduced infiltration capacities in the city area due to a higher percentage of less permeable surfaces lead to increased surface runoff. In future, spatially variable infiltration parameters can be implemented in the model to study the effects of the heterogeneous infiltration behavior on the flood development in the study area. To calibrate the infiltration parameters accurately, field measurements during an event itself are necessary.

The temporal development of the modeled infiltration rates at the location of the maximum simulated water depth in the reference case is represented in Fig. 10. Regarding the two cases of infiltration without basins, it can be seen, that during the first hours the infiltration rates were equal to the rainfall intensity of 4.25 mm/h, so that no surface runoff occurred. For the case with a lower soil suction head of $\Psi = 4.42$ cm (blue triangles), the infiltration rate started to decrease after 2 h leading to surface runoff as the rain intensity was higher than the infiltration rate. With the higher soil suction head of $\Psi = 21.85$ cm (black rhombuses), the infiltration rate had the same magnitude as the rainfall intensity during the total rainfall event of 8 h, so that all water from the local rain infiltrated immediately. The rainfall event stopped after eight hours, and the infiltration rate for the soil with $\Psi = 21.85$ cm dropped abruptly to zero because no more water was available. However, for the case of $\Psi = 4.42$ cm, the decreased infiltration led to surface runoff from the surrounding area, as not the whole rainfall could infiltrate immediately. Therefore, there was still water available to infiltrate but with a continuously decreasing rate. In the case of $\Psi = 4.42$ cm, the flood wave reached the considered location after 12 h of simulation time resulting in a rise of the infiltration rate up to a maximum value of 7.4 mm/h after 19 h. With $\Psi = 21.85$ cm, the flood wave reached the location after 20 h, which can be seen in the abrupt rise of infiltration rate from zero to 6.1 mm/h at this time step. After reaching a maximum, the infiltration rate was continuously decreasing in both cases and would gradually approach a similar value, which would finally correspond to the saturated hydraulic conductivity.

As the Green–Ampt model calculates the infiltration rate iteratively, the computational effort increases significantly depending on the number of cells, the time step size resulting from the flow calculations, and the chosen convergence criteria. The computational time for the simulation with infiltration considering sandy clay loam was about three times higher compared to the simulation without infiltration.

4.3.3 Case 2: Retention Basins

Different structural protection and mitigation measures such as dams, canals, and basins can be implemented in the model, and the results can be analyzed regarding the effectiveness of the measures. Several options of protection measures were investigated within the framework of a master thesis (Marafini 2017) and are included in Marafini et al. (2018). Figure 12 shows exemplary the water depths after 11 h when one large and two small retention basins were implemented in the model for the cases (a) without infiltration and (b) with infiltration considering sandy clay loam with $\Psi = 4.42$ cm and $\theta_i = 4\%$. The large basin (green box in Fig. 12a) collected almost the total volume of the inflow from the large wadi catchment, while the two smaller ones (pink and red boxes in Fig. 12a) should collect some of the water in the streams which are generated only by rainfall inside the area. The water can be stored in the basins to prevent high water depths around buildings and

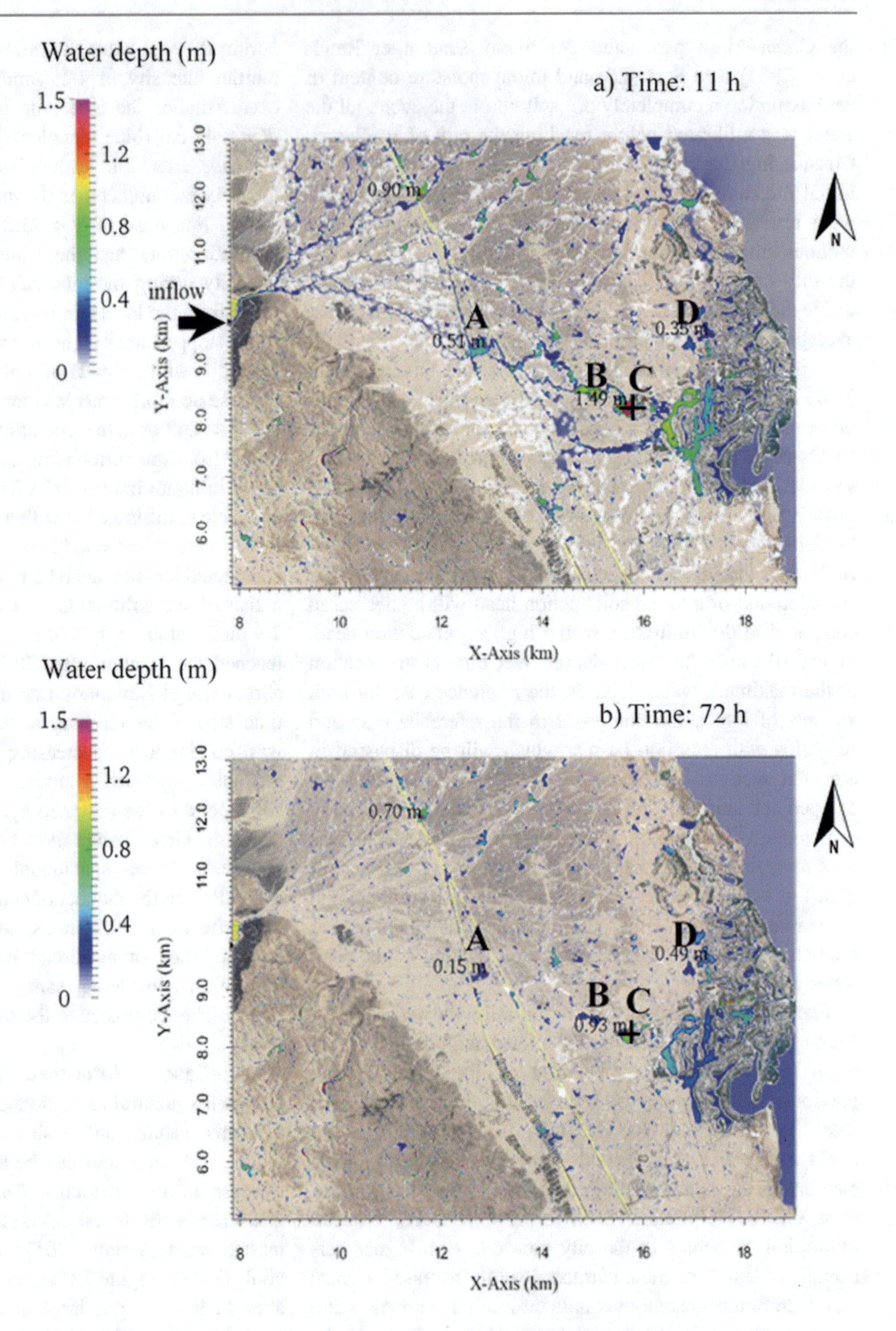

Fig. 9 Simulated water depths plotted on a map of El Gouna. ©Google (2017) **a** after 11 h of simulation, **b** after 72 h of simulation. A: crossroad, B: local depression, C: old desalination plant, D: area around TU Berlin Campus El Gouna. The black cross at point B indicates the location of the plot of water depth over time in Fig. 10

infrastructure in the city area further downstream. In Fig. 11, it can be seen that for the case with retention basins and without infiltration (purple circles) the peak water depth at the considered location (see the black cross in Fig. 8) was damped from 1.49 to 1.07 m and occurred 8 h later compared to the reference case without retention basins. But the value of 1.07 m is still very high and other cases should be considered to reach a more effective protection in terms of a stronger damping of the peak water depth. In city areas, the maximum water depth should not exceed 0.20 m to prevent stronger damages to buildings and infrastructure. For the case with consideration of basins and infiltration a much stronger effect of the retention basins was obtained. Figure 10 shows that the maximum water depth at that location dropped from 1.0 m considering infiltration but no basins (blue triangles) to 0.07 m if infiltration and basins were

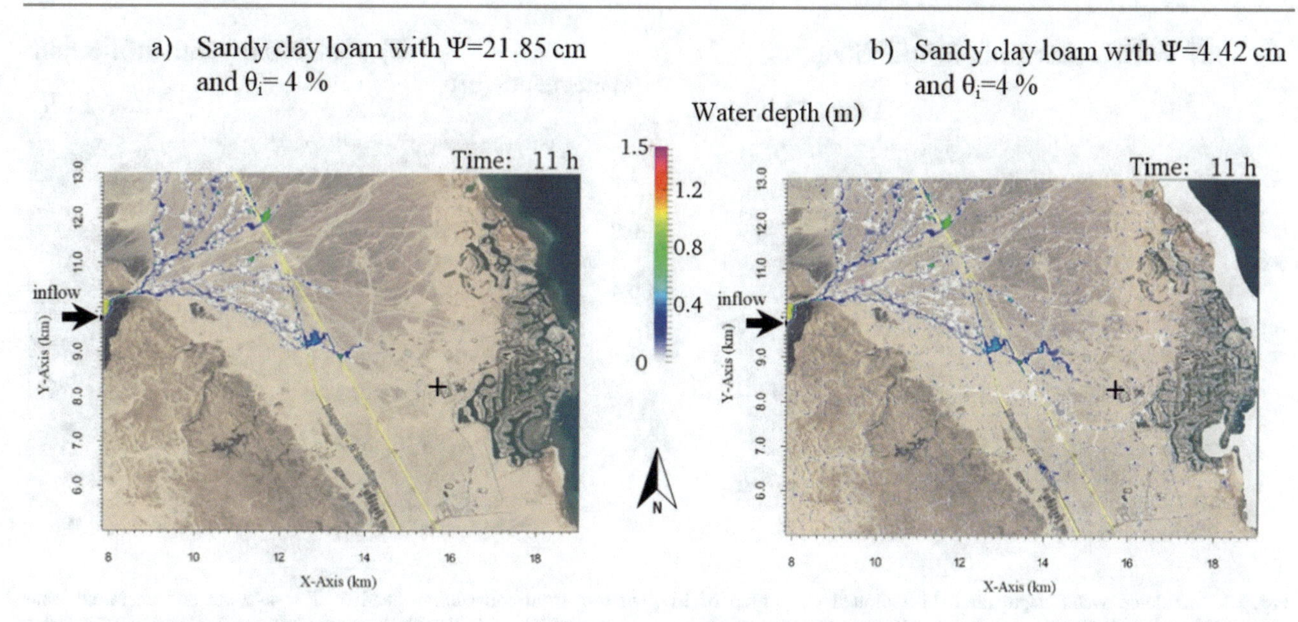

Fig. 10 Simulated water depth after 11 h plotted on a map of El Gouna ©Google (2017) with consideration of **a** sandy clay loam with an initial water content of θ_i = 4% and wetting front soil suction head of Ψ = 21.85 cm and **b** sandy clay loam with θ_i = 4% and a reduced wetting front soil suction head of Ψ = 4.42 cm. The black cross indicates the location of the plots over time of water depth and infiltration rate in the Figs. 10 and 11

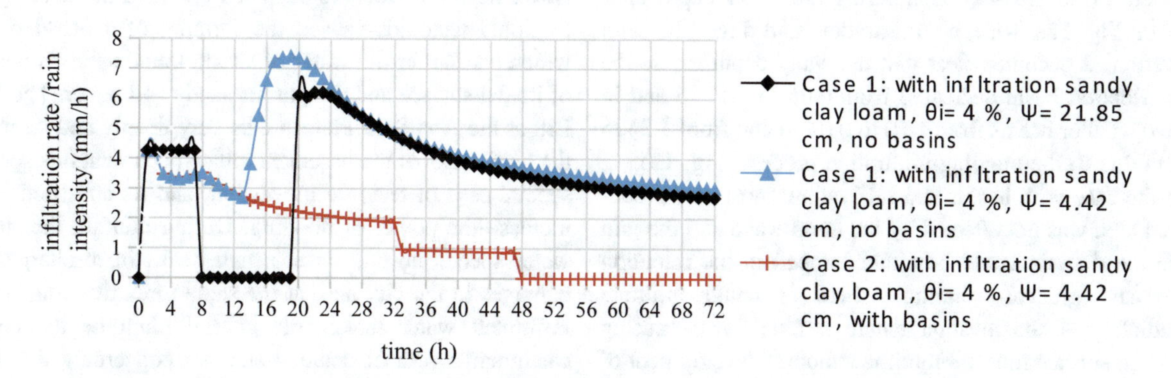

Fig. 11 Temporal development of rain intensity and infiltration rate at the location of maximum water depth (see the black cross in Fig. 8) for sandy clay loam with θ_i = 4% and Ψ = 21.85 cm or Ψ = 4.42 cm, respectively, and with and without retention basins

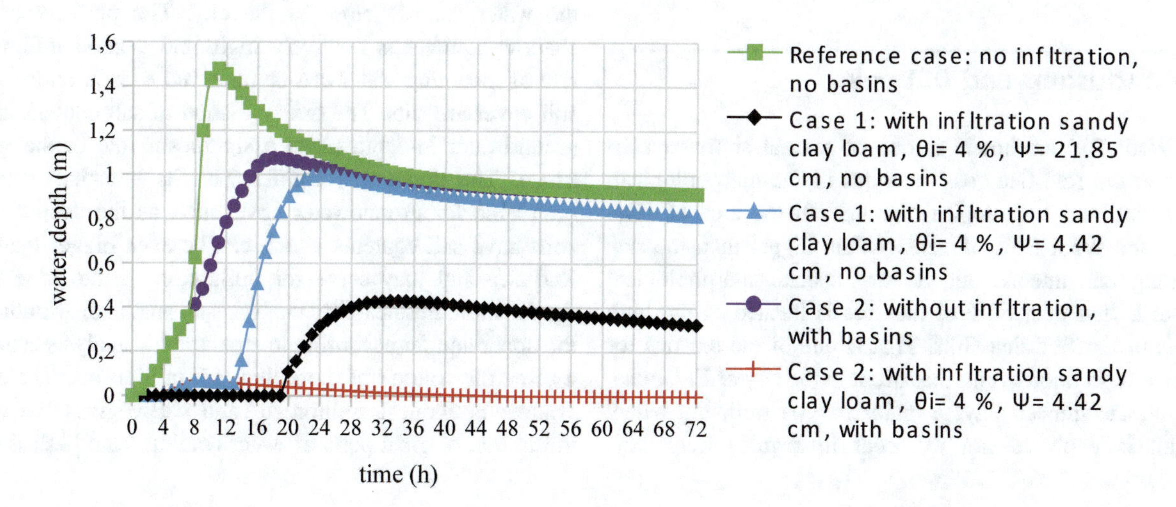

Fig. 12 Water depth over time at the location of maximum water depth in the reference case (see the black cross in Fig. 8) comparing the results of the reference case with different scenarios considering infiltration and retention basins

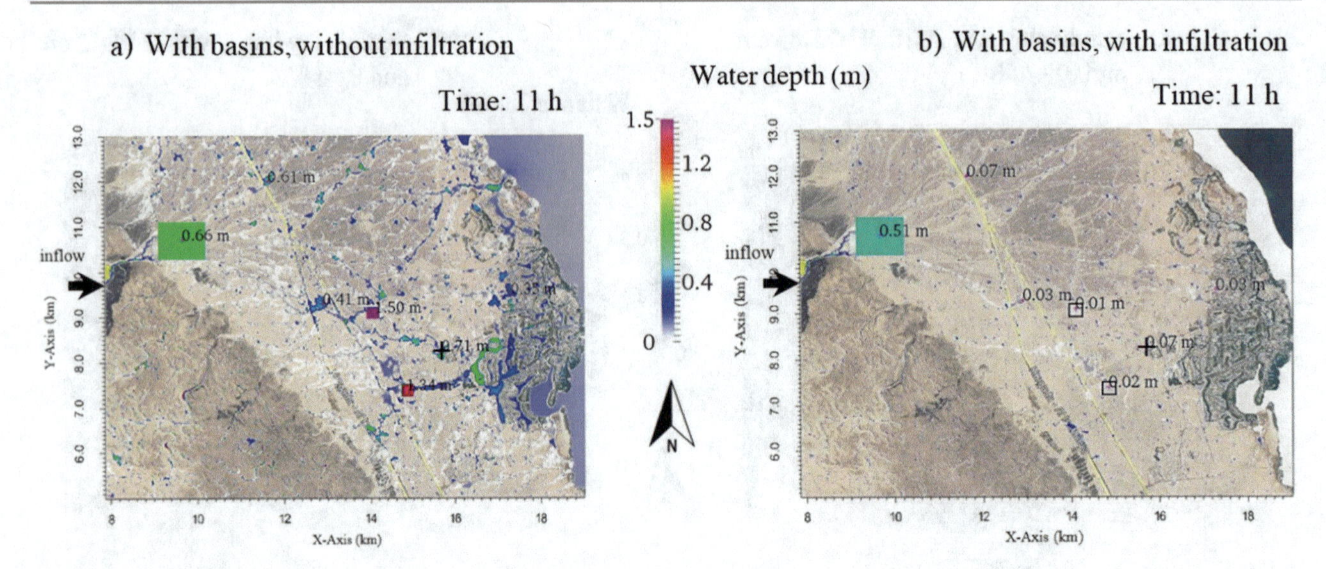

Fig. 13 Simulated water depth after 11 h plotted on a map of El Gouna ©Google (2017) with **a** one large and two small retention basins (green, pink, and red rectangle) and **b** one large and two small retention basins and infiltration considering sandy clay loam with a reduced wetting front soil suction head of $\Psi = 4.42$ cm and an initial water content of $\theta_i = 4\%$. The black crosses indicate the location of the plots over time of water depth and infiltration rate in Figs. 10 and 11

considered (red crosses). Comparing the water depth after 11 h in Fig. 12a with no infiltration and Fig. 12b with infiltration, it becomes clear that the water depth inside the large retention basin decreased from 0.66 to 0.51 m and in the two smaller basins from 1.50 to 0.01 m and from 1.34 to 0.02 m due to the ongoing infiltration process (Fig. 13).

In the case with basins and infiltration, most of the considered area was not affected by the flood wave and the rain inside the domain could be either trapped in the retention basins and gradually infiltrate there or could infiltrate immediately at the location where it falls down causing almost no surface runoff. From the temporal development of the infiltration rate in Fig. 10 it becomes clear, that after 48 h no more water was available in that area, which is why the infiltration rate dropped to zero (red crosses).

5 Conclusions and Outlook

The Wadi Bili catchment is one of several similar catchments at the Red Sea coast in terms of the morphological, geological, and hydrological features. In the area of Hurghada, the Wadi Bilis' dimensions are bigger than the surrounding catchments, and a very special morphological feature is that over 50% of the area is located in the high ranges of the Red Sea Hills. That is one of the reasons for extreme flash flood events that occur in the city of El Gouna. The climate impact plays a major role as well, but heavy rainfall is expected not to occur in regular frequency. Nevertheless, it seems to happen more often in the last years. Regional tectonics caused the forming of a broad plateau which can buffer the water of strong thunderstorms because of its flat surface and the porous sandy soil cover. The Wadi Bili at the Abu Shar Plateau cuts very deeply and steep into the hard rocks of the basement which occurrs at this spot is a special case of tectonic movement, and its constriction can increase the power of flooding. During rainfall, the surface water speeds up and causes high peaks in discharge. The damages in the city and, at the same time, the usage of the infiltrated water makes the coastal plain of the eastern catchment worth of detailed analysis concerning the distribution of the flooding water as well as the storage capability of the underlying aquifers. The effect of saltwater intrusion is measurable at the wells of the desalination plant and limits the water sources close to the city. The porosity of the shallow aquifers is relatively high, and a good infiltration rate of incoming water can be observed even if parts of the soil cover are silts. The differentiation of salt content in the groundwater in-depth plays a significant role in the water usage. The uppermost aquifer with its brackish water is preferable for groundwater abstraction as the deeper horizons have salt contents which can be even bigger than the Red Sea and may serve for infiltration of the brine after desalination. Permeability analysis and spatial distribution of the upper aquifer according to stratigraphic analysis estimate a reservoir volume of more than 1.8 million m^3. The drain volume is about 1 million m^3 and partly gives space for infiltration of great parts of water coming from flash floods.

In spring 2014, the amount of water flowing through the Wadi Bili was about 1 million m^3 which partially can be stored underground for long-term usage.

Recent construction activities in the area of El Gouna covering the prevention of damages caused by flash floods were undertaken. After the experiences of the 2014 flooding, a protection dam was constructed in the southern part of El Gouna where most of the destruction took place. Flash flood models can be applied to study the effectiveness of structural mitigation measures. They also enhance the understanding of flow processes during and after heavy rainfalls and should be used to support the risk analysis.

The represented results of the flash flood model of El Gouna include fields of water depths and flow velocities in the whole simulated domain at different times. The discussion of the modeling results was focused on the analysis of the flood propagation and maximum water depths in the domain as well as on the impact of infiltration processes and retention basins. Due to missing measurement data, the model could not be calibrated, but the results are plausible in terms of flooding areas, ranges of water levels, and flow velocities as well as the temporal development of infiltration rates for different scenarios. The representation of the model results facilitates the understanding of effects from infiltration and retention and their interactions in the overall considered domain. It was shown that the retention basins combined with infiltration processes could completely prevent critical water depths in and around the city area, while the effects of retention and infiltration were significantly less distinct if they were considered separately. In future work, the impact of spatially variable infiltration rates based on measurements as well as other scenarios of structural measures in terms of canals, dams, or combinations of different structures can be investigated with the model. Also, scenarios with higher rain intensities and the resulting flooding areas can be studied. In the future, setting up an early warning system would be desirable by coupling weather observations, a hydrological catchment model, and a flash flood model with a sufficiently fast computational effort to get real-time flood predictions providing information to the community in time to take necessary steps before the flood reaches the city. To use flash flood models for more detailed planning of structural protection measures, more accurate data of topography as well as a higher mesh resolution of the model would be necessary.

Additionally, the construction of protection and infiltration dams in the Wadi Bili itself seems to be a reasonable action to face both challenges, prevention of destruction as well as enhancement of the water recourses. The porous deposits of the Wadis' soils can catch relevant amounts of water. Those underground storage may help to face the increasing amount of water needed by the city of El Gouna. The study can also be seen as a role model of bigger cities like Hurghada which has the same problems like El Gouna. The Eastern Desert has even more impacts with flash floods which are analyzed more and more recently as the density of inhabitants rises. The flash flood event of Ras Gharib 2016 shows dramatically the results of unprotected areas and comes more and more into consideration of urban planners and governmental institutions along the Red Sea coast. Prediction of flash floods and their impact, therefore, are a substantial concern of urban planning and can avoid casualties and in the second place material losses.

Future investigations can foster the knowledge about the principles of flash floods in Wadi Bili and surrounding catchments. Numerical modeling of surface flow in the whole catchment can help to predict the impact of flash floods but also can calculate the amount of water and its location during the flood. In combination with an underground flow model, a holistic water balance can be designed, and the usage of water at surface and underground be quantified. Construction of dams for protection and artificial recharge can be built on the basis of the model results. Nevertheless, the quality of a numerical model strongly depends on the amount and quality of data. In the catchment of Wadi Bili many investigations are ongoing concentrating on modeling surface flow and the quality and origin of groundwater. Most of the research concentrates on the coastal plain of El Gouna. However, the distal parts of the catchment are not very well recognized by measurements yet. Future plans are focusing on basic investigations of the catchment starting with the installation of weather stations and measurements of infiltration rates on the Abu Shar Plateau.

References

Alsharhan, A. S. (2003). Petroleum geology and potential hydrocarbon plays in the Gulf of Suez rift basin Egypt. *AAPG Bulletin, 87*(1), 143–180.

Ashbel, D. (1938). Great floods in Sinai Peninsula, Palestine, Syria and the Syrian Desert, and the influence of the red sea on their formation. *Quaterly Journal of The Royal Meteorological society, 65,* 635–639.

Bosworth, W. (1995). A high-strain rift model for the southern Gulf of Suez (Egypt). *Special Publication Geological Society of London, 80,* 75–112.

Bosworth, W., & Taviani, M. (1996). Late Quaternary reorientation of stress field and extension direction in the southern Gulf of Suez, Egypt: Evidence from uplifted coral terraces, mesoscopic fault arrays, and borehole breakouts. *Tectonics, 15*(4), 791–802.

Busse, T., Simons, F., Mieth, S., Hinkelmann, R., & Molkenthin, F. (2012). HMS: A generalised software design to enhance the modelling of geospatial referenced flow and transport phenomena. In *Proceedings of the 10th International Conference on Hydroinformatics—HIC 2012*. Hamburg.

Cea, L., Puertas, J., Pena, L., & Garrido, M. (2008). Hydrologic forecasting of fast flood events in small catchments with a 2D-swe model. Numerical model and experimental validation.

Chaudhry, M. (2008). *Open-channel flow* (2nd ed.). New York, USA: Springer Science & Business Media LLC.

Dayan, U., Ziv, B., Margalit, A., Morin, E., & Sharon, D. (2001). A severe autumn storm over the middle-east: Synoptic and mesoscale convection analysis. *Theoretical and Applied Climatology, 69,* 103–122.

De Vries, A. J., Tyrlis, E., Edry, D., Krichak, S., Steil, B., & Lelieveld, J. (2013). Extreme precipitation events in the Middle East: Dynamics of the active Red Sea Trough. *Journal of Geophysical Research: Atmospheres, 118*(13), 7087–7108.

EGSMA. (1981). *Geologic map of Egypt. Scale 1:2000000.* Cairo: Egyptian Geological Survey and Mining Authority (EGSMA).

Google. (2017). Retrieved 07, October, 2017, from https://www.google.de/maps/place/el-Guna.

Greiling, R. O., El Ramly, M. F., El Akhal, H., & Stern, R. J. (1988). Tectonic evolution of the northwestern Red Sea margin as related to basement structure. *Tectonophysics, 153*(1–4), 179–191.

Hadidi, A. (2016). *Wadi Bili Catchment in the Eastern Desert—Flash Floods, geological model and hydrogeology. Dissertation.* Berlin: Fakultät VI—Planen Bauen Umwelt der Technischen Universität Berlin.

Hou, J., Liang, Q., Simons, F., & Hinkelmann, R. (2013). A stable 2D unstructured shallow flow model for simulations of wetting and drying over rough terrains. *Computers & Fluids*, 132–147.

Huffman, G., Adler, R. F., Bolvin, D. T., Gu, G., Nelkin, E. J., Bowman, K. P., et al. (2007). The TRMM multi-satellite precipitation analysis: Quasi-global, multi-year, combined-sensor precipitation estimates at fine scale. *Journal of Hydrometeorology, 8*(1), 38–55.

IFRC. (2017). *Emergency plan of action. Egypt: Floods emergency plan of action.* IFRC.

JAXA. (2017). *Japan Aerospace Exploration Agency Earth Observation Research Center, ALOS Global Digital Surface Model "ALOS World 3D—30 m" (AW3D30).* Retrieved 20, February, 2017, from http://www.eorc.jaxa.jp/ALOS/en/aw3d30/index.htm.

Klitsch, E. & Linke, H. W. (1983). *Photogeological interpretation map.* CONOCO coral Inc. Cairo, A.R.E.

Krichak, S. O., & Alpert, P. (1998). Role of large scale moist dynamics in November 1–5, 1994, hazardous Mediterranean weather. *Journal Geophysical Research, 103,* 453–468.

Krichak, S. O., Alpert, P., & Krishnamurti, T. N. (1997a). Interaction of topography and tropospheric flow—A possible generator for the Red Sea Trough? *Meteorology and Atmospheric Physics, 63,* 149–158.

Krichak, S. O., Alpert, P., & Krishnamurti, T. N. (1997b). Red Sea Trough/cyclone development—Numerical investigation. *Meteorology and Atmospheric Physics, 63,* 159–169.

Krichak, S. O., Breitgand, J. S., & Feldstein, S. (2012). A coceptual Model for the Identification of Active Red Sea Trough synoptic events over southeastern Mediterranean. *Journal of Applied Meteorology and Climatology.*

Krichak, S. O., Tsidulko, M., & Alpert, P. (2000). November 2, 1994 severe storms in the southeastern Mediterranean. *Atmospheric Research, 53,* 45–62.

Marafini, E. (2017). Investigation of mitigation measures against wadi flash floods. Master thesis. Rome: Ingegneria Civile per la Protezione dai Rischi Naturali Idraulica—Roma Tre, Università degli studi.

Marafini, E., Tügel, F., Özgen, I., & Hinkelmann, R. L. (2018). Flash flood simulations based on shallow water equations to investigate protection measures for El Gouna, Egypt. In *Palermo: 13th International Conference on Hydroinformatics*.

Masoud, M. H. (2007). *Hydrogeological evaluation of quaternary and tertiary aquifers in some areas between Ras Gharib and Hurghada, Red Sea coat—Egypt*. Alexandria: Alexandria University.

Mein, R. G., & Larson, C. L. (1971). *Modeling the infiltration component of the rainfall-runoff process*. Minneapolis, Minnesota: Water Resources Research Center.

Milewski, A., Sultan, M., Yan, E., Becker, R., & Abdeldayem, A. (2009). A remote sensing solution for estimating runoff and recharge in arid environments. *Journal of Hydrology, 373,* 1–14.

Murillo, J., García-Navarro, P., & Burguete, J. (2009). Conservative numerical simulation of multi-component transport in two-dimensional unsteady shallow water flow. *Journal of Computational Physics*, (15), 5539–5573.

NOAA. (2015). *Hurguada (Hurghada) Climate Normals 1961–1990.*

Özgen, I., Simons, F., Zhao, J., & Hinkelmann, R. (2014). *Modeling shallow water flow and transport processes with small water depths using the Hydroinformatics modelling system.* New York: CUNY Academic Works.

Rawls, W., Brakensiek, D., & Miller, N. (1983). Green-Ampt infiltration parameters from soils data. *Journal of Hydraulic Engineering, 1,* 62–70.

Said, R. (1990). *The geology of Egypt*. Rotterdam: Belkema.

Salah, M. G., & Alsharhan, A. S. (1996). Structural Influence on Hydrocarbon Entrapment in the Northwestern Red Sea Egypt. *AAPG Bulletin, 80*(1), 101–118.

Saleem, M. S. (1990). *Geography of the Egyptian deserts: Eastern desert Part II* (Arabic ed.). Cairo, Egypt: Dar Al-Nahda Al-Arabeeah.

Shahin, M. (2007). *Water resources and hydrometeorology of the Arab region. Water science and technology library* (p. 601).

Simons, F., Busse, T., Hou, J., Özgen, I., & Hinkelmann, R. (2012). HMS: model concepts and numerical methods around shallow water flow within an extendable modeling framework. In *Proceedings of the 10th International Conference on Hydroinformatics*, Hamburg.

Simons, F., Busse, T., Hou, J., Özgen, I., & Hinkelmann, R. (2014). A model for overland flow and associated processes within the Hydroinformatics modelling system. *Journal of Hydroinformatics, 16*(2), 375–391.

Stern, R. J., & Hedge, C. E. (1985). Geochronologic and isotopic constraints on late Precambrian crustal evolution in the eastern desert of Egypt. *American Journal of Science, 285,* 97–127.

Thorweihe, U., & Heinl, M. (2002). *Groundwater resources of the Nubian Aquifer system, NE-Africa. Modified synthesis submitted to: Observatoire du Sahara et du Sahel* (p. 23). Paris: OSS.

Tügel, F., Marafini, E., Özgen, I., Hadidi, A., Tröger, U., & La Rocca, M. &. (2017b). Flash flood simulations considering infiltration processes and protection measures for El Gouna, Egypt. In *3rd International Symposium on Flash Floods in Wadi Systems*, Muscat, Oman.

Tügel, F., Özgen, I., Hadidi, A., Tröger, U., & Hinkelmann, R. (2017a). Modelling of flash floods in wadi systems using a robust shallow water model—case study El Gouna, Egypt. In *SimHydro2017 Conference*, Nice, France.

USGS. (2006). *Shuttle Radar Topography Mission, 1 Arc second scene.* Maryland: Global Landcover Facility, University of Maryland.

Vanderkimpen, P., Rocabado, I., Cools, J., & El-Sammany, M. (2010). FlaFloM—An early warning system for flash floods in Egypt. In D. De Wrachien (Ed.), *WIT transactions on ecology and the environment* (p. 314), WIT Press.

Wagdy, A., El-Zawahry, A., Hussein, K., Imam, Y., & El-Gamal, M. (2003). *Rainfall report II. Developing Renewable Ground Water Rercources in Arid Lands, Task 12.h—JY/03/4*. Egypt: Ministry of Water Resources and Irrigation.

Whisler, F., & Bouwer, H. (1970). Comparison of methods for calculating vertical drainage and infiltration for soils. *Journal of Hydrology, 1,* 1–19.

WMO. (2012). Management of flash floods. In W. M. Oranization, *Integrated flood management tools series* (p. 44), World Meteorological Oranization.

WMO. (2013). *Weather information for Hurghada.*

Environmental Approaches in Flash Flood

Environmental Flash Flood Management in Egypt

Rasha El Gohary

Abstract

Flash Floods are a common natural disaster occurring in many parts of the arid regions. The flashfloods severely threaten the infrastructures, human lives destroying their livelihood, moreover affecting the country's business, economy, and threaten the archeological areas. Egypt economy depends on archeological tourism because it represents one of the very important resources of national income in Egypt. Natural disasters (e.g., floods, landslides, sand/dust storms, etc.) have become during the past four decades widespread dangerous problems in many parts of the world, including the Arab countries, thus contributing to increase in poverty and threatening food security. Today's society becomes rapidly vulnerable to natural disasters due to climate changes, growing urban populations, severe human degradation of marginal lands, and disturbance of formerly stable ecosystems, as well as a lack of natural resource management, land use planning, and preparedness. Several technical reasons connected to population pressures are aggravating the intensity of flooding in the Arab countries becoming as such tragedies and bringing disasters in their wake. Flooding disasters have created, during the last years, widespread environmental, social, and economic destruction in several Arab countries. Floods cause severe degradation reshaping natural landscapes, sedimentation in reservoirs and waterways, disruption of lake ecosystems, and contamination of drinking water.

Keywords

Flash flood • Environmental action plan • Disaster risk reduction-flood • Risk management strategies • Climate change and disaster risk assessment

R. El Gohary (✉)
Central Laboratory for Environmental Quality Monitoring, National Water Research Center, NWRC, Cairo, Egypt
e-mail: rm.elgohary@yahoo.com

A. M. Negm (ed.), *Flash Floods in Egypt*, Advances in Science, Technology & Innovation,
https://doi.org/10.1007/978-3-030-29635-3_6

1 Introduction

Natural hazards and floods are part of nature. They have always existed and will continue to exist. Except for some floods caused by dam failures or landslides, floods are climatic phenomena affected by geomorphology, relief, soil, and vegetation. Meteorological and hydrological operations can be rapid or slow and can lead to sudden flooding or slow-growing predictable floods, also called river floods. Several studies have been conducted to identify possible measures to avoid the risks posed by floods. The engineers have developed mechanisms for harvesting floodwater. These water can be used directly to meet part of the aquatic needs or recharged to the shallow aquifers. Disasters of all kinds strike every nation of the world. Do not differentiate between rich and poor nations. However, developing countries suffer the greatest hazard, often suffering from subsequent internal civil conflicts leading to complex humanitarian emergencies.

Floods have negative effects on human health and the ecology, causing damage to infrastructure, livestock, and plants. Floodwater can cause a range of significant health impacts, including deaths and or injuries, drinking water pollution, loss of electricity, community disruption, and displacement.

This chapter will represent the Egyptian national environmental action plan, and the natural environmental hazards (earthquakes–flash floods–dust and sand storm). Moreover, it will highlight the environmental and disaster risk reduction in Egypt and Arab Countries, and present the tools for implementing disaster risk reduction in environmental hazards zones. The chapter will explain the Strategic Environmental Assessment (SEA): Flood risk management strategies, SEA objectives and assessment method, environmental assessment of the flood risk management, and its mitigation and monitoring.

Finally, it will demonstrate the Climate change, disaster risk assessment and management in Egypt and Arab

Countries, Egypt's national strategy for adaptation to climate change and disaster risk reduction (National strategy Goals —Climate change projections of sea level and impacts on coastal zones and northern lakes), and Climate change adaptation for human health. The discussions in this chapter aim to address the complexity of environmental flash flood risks in Egypt and the Arab region and provide some tools that can be used by local governments, civil society, and other institutions working in the field of environment and disaster risk reduction.

2 Egypt National Environmental Action Plan

The National Environmental Action Plan (NEAP) represents Egypt's agenda for environmental actions for the coming fifteen years. It complements and integrates with sectoral plans for economic growth and social development. NEAP is the basis for the development of local environmental initiatives, actions, and activities. It is designed to be the framework that coordinates for future environmental activities in support of the sustainable development of Egypt.

2.1 Natural Environmental Hazards (Earthquakes–Flash Floods–Dust and Sand Storm)

2.1.1 Earthquakes

Sudden movements along geological faults in rocks comparatively near the surface of the earth result in earthquakes Selishi (2008). Most movements are preceded by the slow build-up of tectonic strain that progressively deforms the crustal rocks and produces stored elastic energy. When the imposed stress exceeds the strength of the rock, it fractures, usually along a pre-existing fault. The point of sudden rupture, known as the focus, can occur anywhere between the surface of the earth and depth of 600–700 km. Shallow-focus earthquakes (less than 40 km below the surface) are the most damaging events, accounting for about 75% of the global seismic energy release. The source point for earthquake measurement is the epicenter, which lies on the surface of the earth directly above the focus.

The geology of Egypt and its tectonic setting and seismic records indicate that there are at least three main lines of seismic activity: (i) Northern Red Sea–Gulf of Suez–Cairo-Alexandria trend; (ii) Eastern Mediterranean–Cairo-Fayoum depression trend; and (iii) Gulf of Aqaba trend. In addition to these trends, there are several areas known to be active, such as Southwest of Aswan. Time series data record 83 noticeable earthquakes in and around Egypt, causing damage of variable degrees.

Earthquakes have been traditionally considered as natural disasters. However, it recognized that human activities could enhance the occurrence and impacts of natural hazards. Human-made earthquakes have known since Carder in 1945 documented the occurrence of about 860 local tremors during the 10 years following the formation of Lake Mead, in Arizona and Nevada, by Hoover Dam in 1935. Other human activities that have enhanced the occurrence of earthquakes are the injection of fluids into deep wells and underground nuclear tests.

2.1.2 Flash Floods

Flash floods occurring due to short-period heavy storms are among the sources of environmental damage, especially in the Red Sea area and southern Sinai. Floodwater velocity depends mainly on the topography of the basin (height, slope, and drainage network capacity), and its soil type and characteristics. In 1979, a flash flood over ElQuseir and Marsa Alam region led to destructing both the Red Sea Coastal Road and Qena-El-Quseir Road. Because of the flood, authorities reported 19 deaths. In 1991, another flash flood hit Mersa Alam where about 37,000 m^3 of water was received in a very short period. In 1993, Alexandria City received a heavy storm with losses including 21 deaths. In November 1994, the Governorate of Assiut suffered one of the most severe and dangerous floods during this decade in which the heavy storm caused fires and loss infrastructure as well as the loss of life.

Many studies have been undertaken to determine possible measures to avoid hazards that flash floods cause. Engineers have developed mechanisms to harvest flash floods water. This water could be directly used to meet part of the water requirements or recharge the shallow groundwater aquifers. Based on statistics, an estimated one BCM of water on average can be utilized annually by harvesting flash floods.

2.1.3 Dust and Sand Storms

Among significant natural sources of air pollution in Egypt are dust and sand storms. Dust and land storms can result in high concentrations of particulate matters affecting visibility, contributing to increased road accidents, and negatively affecting air travel. Dust and sand storms are a common phenomenon in Egypt during the spring and late winter seasons.

3 Environmental and Disaster Risk Reduction in Egypt and the Arab Countries

3.1 Tools for Implementing Disaster Risk Reduction in Environmental Hazards Zones

The total area of the Arab region is 14.2 million km, 90% of which is located in arid, semi-arid, and dry sub-humid areas, in addition to social and economic vulnerabilities that make the region especially vulnerable to climate change.

The region can be portrayed by the cruel condition, and fragile biological systems Ngigi (2009). It is noted that ecological concernsdecrease in the Arab contrasted with districts, for example, the Americas, Africa, and Asia. However, ecological preservation and debacle hazard decrease in this locale are no less deserving of attention, but instead regularly overlooked amidst the monetary stagnation, political clashes, and complex crises. This is clear since water lack is a challange among the most noteworthy challenges for the future. Normal dangers, and natural concerns often lie behind the political conflicts, UNISDR-ROAS (2011) (Fig. 1).

Rapid population growth of about 3% per year (among the highest in the world) alongside the changing utilization examples and ways of life have prompted expanded food request and accelerated land degradation in this arid environment. Land degradation in the Arab region due to widespread abuse is at an accelerated rate. The failure of resource management policies is exacerbated by overgrazing UNISDR-ROAS (2011).

Overexploitation of water and land resources, over-cultivation of marginal lands, deforestation, and the use of inappropriate technologies. The years of intense drought and desertification impact on rural areas, agricultural sectors, urban centers, and economic development. Before the end of the last century, despite national, regional, and international efforts to combat desertification and eliminate the effects of drought and desertification is as yet one of the major natural issues in the Arab Region Abahussain et al. (2002). Over the years, Egypt, Tunisia, Algeria, Syria, and Jordan have experienced frequent waves of drought that have led to a decline in agriculture and food shortages. Recent droughts in Jordan and Syria have been the worst in 30 years

Sudden flooding has killed many people and has extremely influenced the livelihoods of thousands of humans throughout the region. In March 2003, floods in the Jordan River caused extensive damage to agricultural land, resulting in loss of crops for the entire season. In October 2008, Algeria, Morocco, and Yemen encountered exceptional flooding. In Yemen, floods have displaced nearly 25,000 people and destroyed more than 3,000 irreparable homes. Algeria and Morocco were hit by the most exceedingly bad flooding for a considerable length of time influencing in excess of 12,000 families. It is likewise worth referencing that the area experienced its first typhoon in 2007 at a speed of 170 km/h, substantial rain, and high tides. The biggest impact was recorded in the Sultanate of Oman, where floods and landslides caused millions of dollars due to economic damage Zeitoun et al. (2010).

Fig. 1 Destruction after floods in upper Egypt, climate change is causing an increase in the frequency and intensity of extreme weather events source, UNISDR-ROAS (2011)

After droughts and floods, earthquakes pose the greatest threat to human life in the Arab region. The earthquakes affected 1.3 million people and caused economic damage of more than $11 billion in the Arab region between 1980 and 2008. Various earthquakes have been recorded in the region throughout the most recent years, Jordan (February 2004) and Lebanon (February 2008), both with insignificant harms to life or infrastructure. However, the 1960 earthquake in Morocco killed nearly 12,000 people, and the last mission in February 2004 killed 600 people and left more than 30,000 homeless. The 1980 earthquake in Algeria killed about 2,500 people, and in 2003; another earthquake caused more than 500 deaths and 4,500 injuries. The 1992 earthquake in Egypt left more than 370 dead and at least 3,300. In addition to earthquakes, satellites can cause secondary effects, indirect surface processes that are directly related to the earthquake, including tsunamis, landslides, land liquefaction, and ground cracking. In mountain areas, landslides often follow a reaction to earthquakes. This incorporates rock falls, landslides, slow earth flows, and side branches of the soil. If rocks and landslides are not dealt with, they can end up covering the fertile land and thus decrease agricultural production, UNISDR-ROAS (2011).

Also, the region contains the active seabed spread in the Mediterranean, Red Sea, and the Gulf of Aden, posing a threat from the tsunami as well as active tectonics in the Atlas Mountains of the Maghreb Seawater Foundation (2010). A tsunami can be simplified to be long-lasting ocean waves, usually caused by seafloor movements during the earthquake. This causes waves to rise when they reach shallow water. In the worst scenarios, the waves will strike the earth with such a force that nearby livelihoods will be affected. The Indian Ocean tsunami in December 2004 was an example, affecting Oman, Yemen, and Somalia in the Arab region. The Mediterranean is also vulnerable to earthquakes, and tsunamis have occurred throughout the region's history and must be considered a natural hazard. A tsunami that destroys coral reefs, mangroves, seagrasses, coastal, and coastal wetlands can be severely affected, depleting the natural resources on which communities depend on survival. Food security is directly linked to the marine resources of many coastal communities in the region. To ensure that the fisherman can rebuild his economy and people can eat, reefs, wetlands, and marine protected areas must be restored after the tsunami. Environmental experts must work with local officials to rebuild the missing scientific and management capacities needed to restore health-affected ecosystems. In any case, early warning and risk reduction systems can reduce the potential negative impacts of a tsunami on the Arab region. The schematic distribution of the most common hazards is shown in Fig. 2. UNISDR-ROAS (2011).

As disasters are close to threats and vulnerable conditions, and vulnerabilities have increased due to environmental degradation, settlement patterns, livelihood options, and behavior, environmental dimensions must address disaster risk reduction measures. Environmental degradation compounded the actual impacts of hazards and limited the capacity of the region to absorb those impacts, which in turn reduced overall resilience to the effects of hazards and disaster recovery. Understanding the linkages between environmental degradation, vulnerability, and disaster risk and recognition will enable the development of comprehensive approaches to disaster risk reduction and community protection. At the same time, it should be recognized that, to the extent that environmental degradation is a per se, the environment is at risk in times of disaster and should be considered in the assessment of damage and the reconstruction phase of a post waste recovery process. Figure 3 shows a diagram showing the links between vulnerability, environmental degradation, and disaster risk. (AFED 2009).

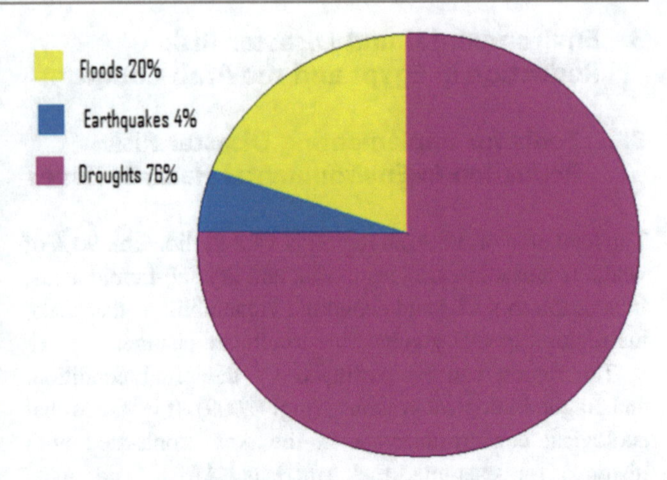

Fig. 2 Number of the total affected population in Arab Countries due to disasters caused by natural hazards between 1980 and 2008, (AFED 2009)

Climate change projections demonstrate that by 2025, the water supply in the Arab area will be just 15% of levels in 1960. Via 2030 the overwhelming impacts of climate change will incorporate a decrease in precipitation, an extraordinary ascent in temperatures, and an expansion in seawater interruption into seaside aquifers as sea levels rise and groundwater overexploitation proceeds UNDP (2004, 2006, 2009), UNDP-UNEP (2011). With expanded urbanization, the urban heat island impact is anticipated to increase evening time temperatures by 3 °C. Climate change will likewise have unbalanced ramifications for women, poor, and marginalized communities who are particularly in danger because of their reliance on natural resources Metwalli (2010). Seeing how climate risks meet and associate with more extensive difficulties of improvement and emergency avoidance is turning into a top need for all nations in the region. All communities in the Arab area are somehow or another vulnerable to climate change UNDP (2018).

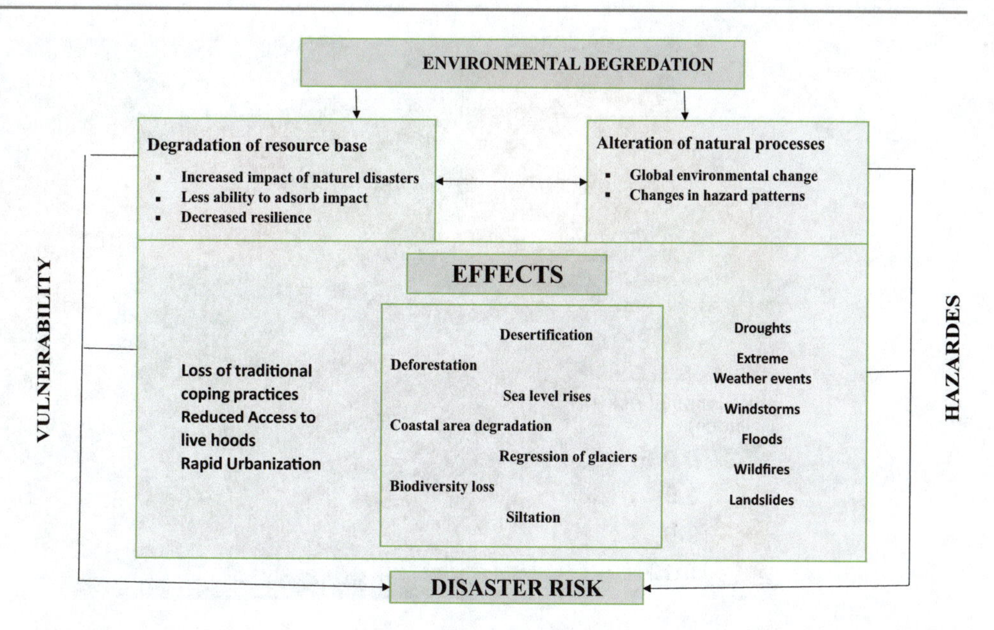

Fig. 3 Linking environmental degradation, vulnerabilities, and disaster risk. *Source* Living with Risk: A global review of disaster reduction initiatives. United Nations, International Strategy for Disaster Reduction (ISDR), Geneva, 2004

The Arab region encompasses populations heavily dependent on climate-sensitive, subsistence farming, fishing, and pastoralism. A number of the monetary hobbies are in the flood-prone river or coastal zones or in Arid and Semi-Arid Land (ASAL) drought-prone areas. In such areas at risk of climate trade, there's a powerful hyperlink between poverty and negative administration of natural resources ACSAD (2007). For example, Yemen has almost a countrywide water scarcity due to over-pumping of groundwater resources to support the water-intensive cultivation of the narcotic. Over 2.7 million people throughout Somalia are expected to face quandary and emergency meals shortages via June 2018 as a result of an incapacity to arrange for and adapt to consecutive seasons of negative rainfall. It's as a result indisputable that pressing, extensive-scale coherent action is required to increase the resilience of the millions of livelihoods in Arab states at severe danger to the impacts of climate change UNDP (2018).

In 2003, the total population of the Arab region was 305 million (4.7% of the world population). Over the past two decades, the population grew at an annual rate of 2.6%, with an increase in the total urban population from 44% to nearly 54%. At the same time, the development and poverty situation in the region is very uneven, and poverty is a serious problem in many Arab countries. Nearly 85 million people live below the $2 a day poverty line, which is about 30% of the total population of the region in 2000. Figure 4 provides an overview of the population at risk of climate change impacts in the region World Bank (WB) (2005, 2007, 2010).

Coastal areas in the Arab region are of great importance. The total length of the coastal area in the Arab region is 34,000 km, of which 18,000 km are inhabited. Most of the major cities and economic activities in the region are located in coastal areas. The fertile agricultural land is largely located in low-lying coastal areas such as the Nile Delta. Popular tourism activities depend on marines and coastal assets, such as coral reefs and associated animals. AFED (2009) (Fig. 5).

The Nile Delta shoreline extends from Alexandria to the west to Port Said to the east with a total length of about 240 km and is typically a smooth wide coast. This zone consists of sandy and silty coasts of greatly varying lateral configurations, depending on where the various old branches of the Nile have had their outlets. The coastline has two promontories, Rosetta and Damietta. There are three brackish lakes connected to the sea: Idku, Burullus, and Manzala. Also, there are several harbors located on the coast, including Alexandria, Edku fishing harbor, Burullus fishing harbor, Damietta commercial harbor, El Gamil fishing harbor, and Port Said commercial harbor. Two main drainage canals, Kitchener and Gamasa, discharge their water directly to the sea within this zone.

The Nile Delta region is presently subject to changes, including shoreline changes, due to erosion and accretion, subsidence, and sea level rise as a result of climate changes Paul and David (2001). Agrawala et al. (2004) surveyed specific large economic centers of Alexandria, Rosetta, and Port Said and obtained quantitative estimates of vulnerable areas and an expected loss of employment in case of no action. They concluded that the Nile Delta coastal zone is highly vulnerable to the impacts of sea level rise through direct inundation and salt-water intrusion. Low elevation coastal zones constitute high-risk areas due to potential damage of sea protection from earthquakes or human activities, AFED (2009).

Egyptian delta coasts are also vulnerable to subsidence NBI (2006). Tide gauges data from the Coastal Research Institute of Alexandria revealed land subsidence of

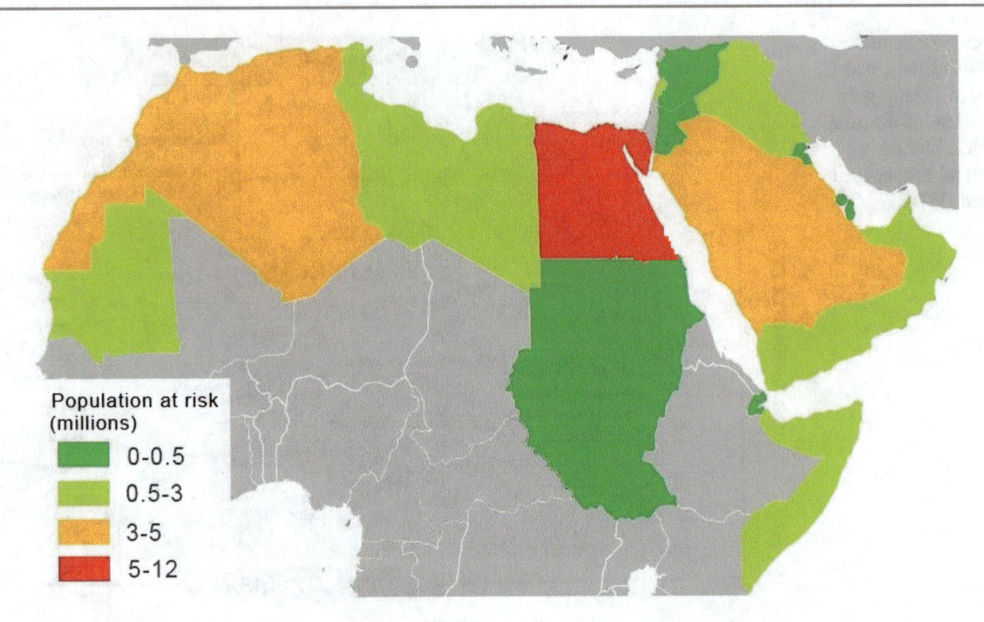

Fig. 4 Arab World indicating population at risk of the impact of climate change in colors, (AFED 2009)

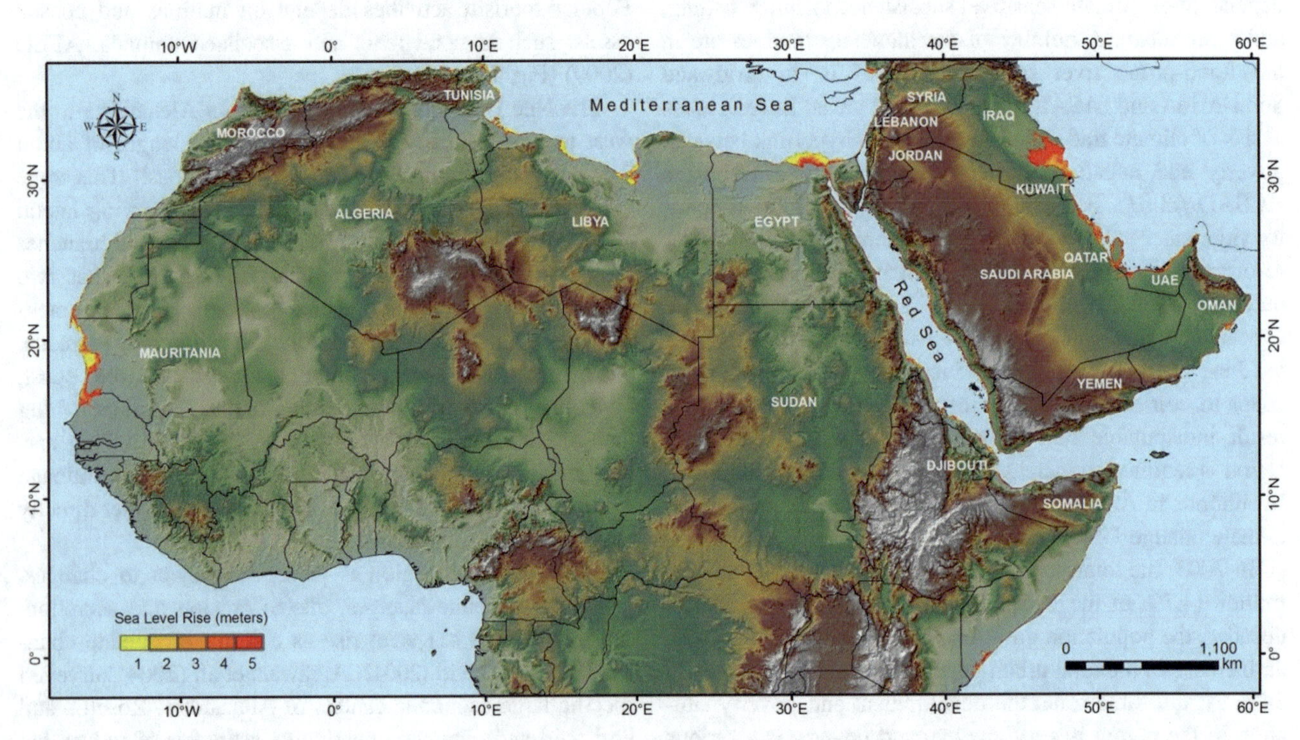

Fig. 5 An overview of most vulnerable coastal areas of the Arab region due to sea level rise. *Source* (Special analysis carried out for AFED Report by E. Ghoneim at the Center of Remote Sensing, Boston University (CRS-BU, E. Ghoneim—AFED 2009 Report)

about 1.6 mm/year at Alexandria (Fig. 6), 1.0 mm/year at Al-Burullus, and 2.3 mm/year at Port Said. However, survey measurements carried out have revealed rates greater than 4 mm/year at Port Said and about 2 mm/year at Alexandria for the Holocene period. (AFED 2009) (Fig. 7).

Recently, work carried out on the Damietta region using Radar satellite interferometry has revealed rates of subsidence that may reach up to 8 mm/year. This prompted carrying out radar image interferometry for Alexandria, which has revealed rates that vary between 5 and 9 mm/year in some specific areas of the city.

From the analysis of the three satellite data, it was found that about 6.5–6.9% of the city is subsiding at rates that vary between 5 and 9 mm/year. These areas are associated with

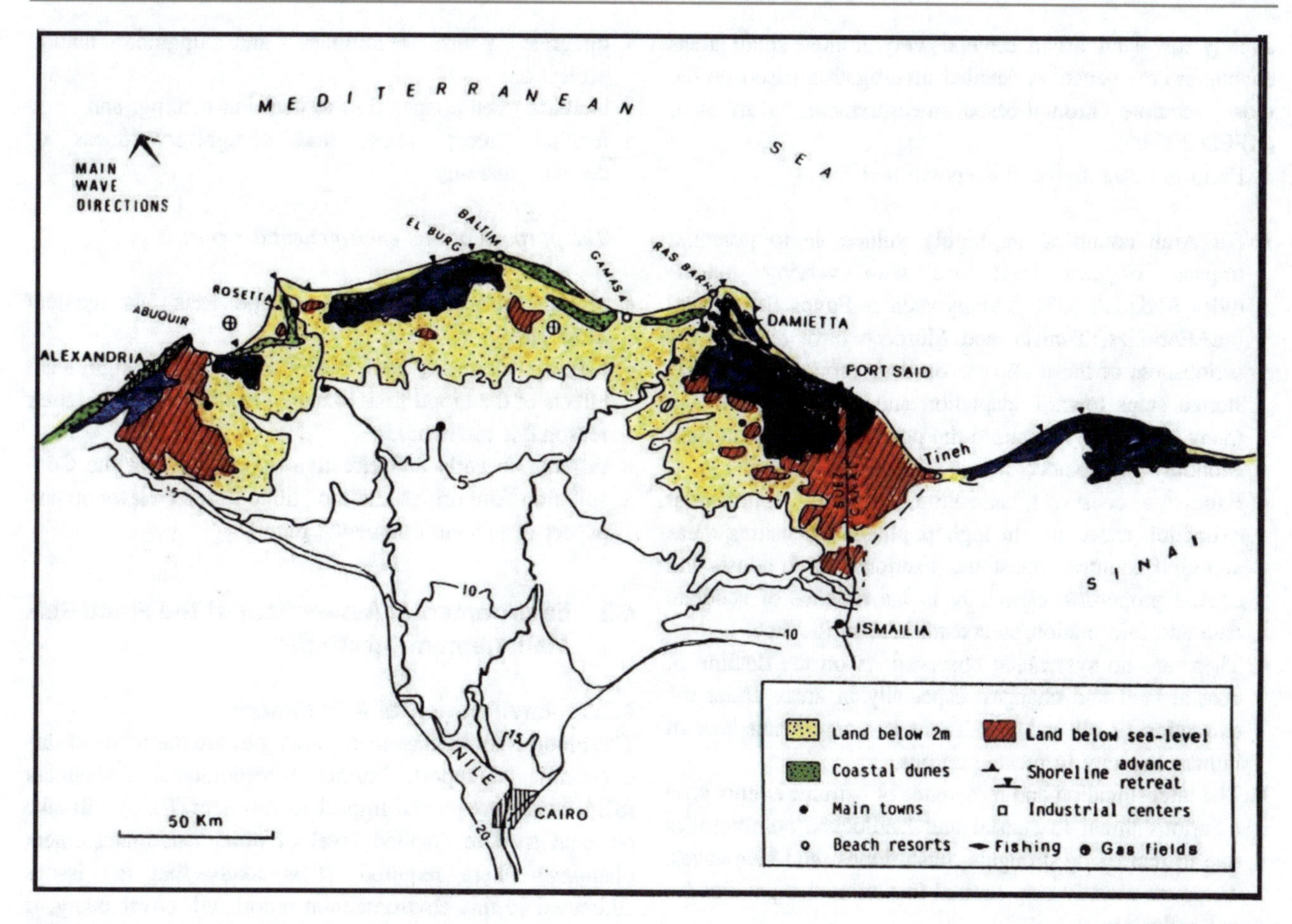

Fig. 6 General topography of the Nile Delta indicating areas below mean sea level in red and areas below 2 m contour level in yellow, (AFED 2009)

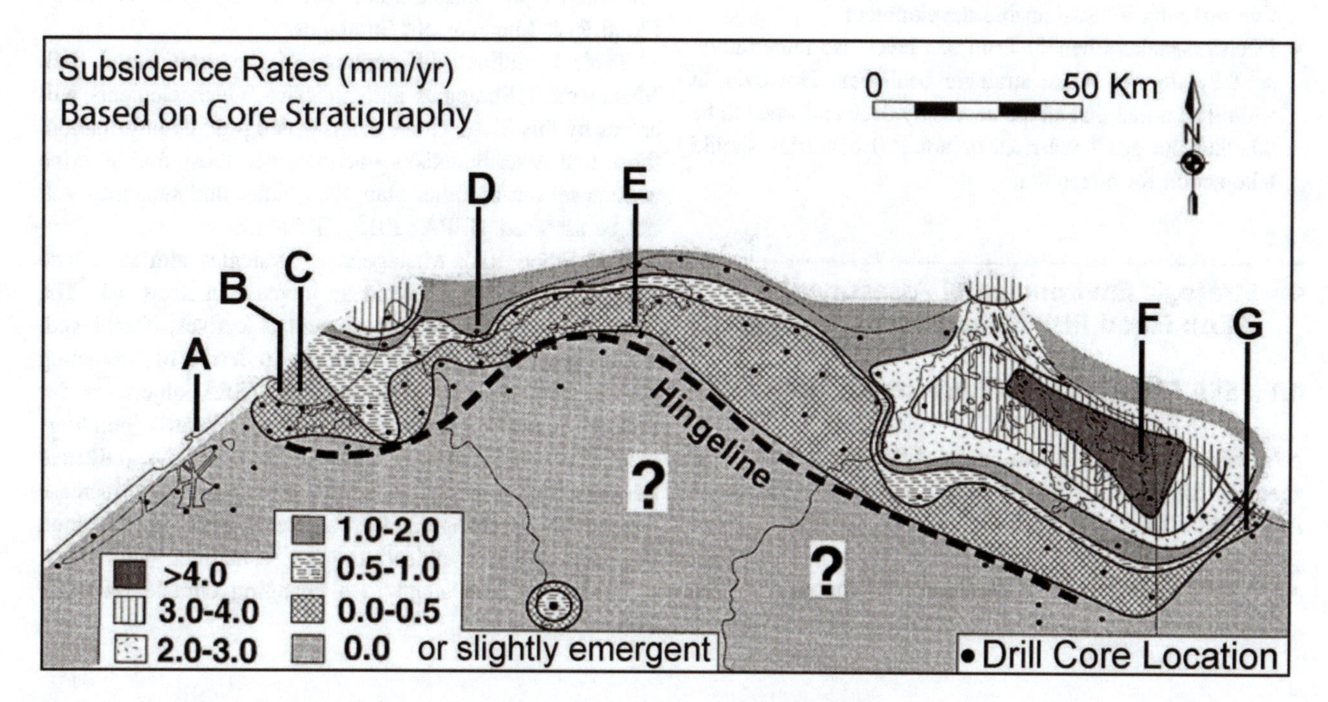

Fig. 7 Land subsidence rates map in the Holocene period as measured at Alexandria, Burullus, and Port Said, (AFED 2009)

already low land urban centers. Very limited small areas encounter emergence. A detailed investigation based on the most accurate ground-based measurements warranted, (AFED 2009).

From the data above, it is shown that

- All Arab countries are highly vulnerable to potential impacts of sea level rise with varying magnitudes Abdallah (2007). Many such as Egypt, Saudi Arabia, Emirates, Tunisia, and Morocco have realized the seriousness of these impacts on their economy and have started steps toward adaptation and risk reduction, and many have not yet realized the potential impacts on their economy and socioeconomic conditions.
- Excessive coastal urbanization, oil and groundwater extraction especially in high population densities areas and Gulf countries constitute a serious risk to coasts and coastal properties, especially in the absence of accurate data and information on coastal land subsidence.
- There are no systematic observations on the decline of coastal land and changes, especially in areas under the extraction of oil and gas. There is a significant lack of human capacity in most countries.
- The intensification and recurrence of extreme events pose a serious threat to coastal and landlocked communities due to increasing droughts, flash floods, and heatwaves. There are already very limited institutional capacities for risk reduction.
- The lack of systematic observation systems, lack of awareness, and weak environmental systems are fundamental obstacles to the proper implementation of proactive planning for sustainable development.
- Decision-makers benefit from sea level rise uncertainty, so they do not make strategic decisions. However, it should be noted that all the necessary decisions need to be whether our sea level rises or not. All countries should take action for adaptation.

4 Strategic Environmental Assessment (SEA): Flood Risk Management Strategies

4.1 SEA Objectives and Assessment Method

As a major aspect of the readiness of the Flood Risk Management Strategies, SEPA is doing a Key Environmental Assessment (SEA). SEA aims to

- improve ecological elements into the arrangement of and decision-making for plans, programs, and systems;
- integrate plans, techniques and upgrade natural protection;
- increase open cooperation in decision-making; and
- facilitate receptiveness and straightforwardness of decision-making;

The purpose of this environmental report is to

- provide information on the Flood Risk Management Strategies;
- identify, describe, and evaluate the likely significant effects of the Flood Risk Management Strategies and their reasonable alternatives;
- provide an early and effective opportunity for the Consultation Authorities and the public to offer views on any aspect of this environmental report.

4.2 Environmental Assessment of the Flood Risk Management Strategies

4.2.1 Environmental Assessment

The Flood Risk Management Strategies are the focus of this environmental report. Further environmental assessment (SEA or Environmental Impact Assessment (EIA)) will take place at a more detailed level of flood risk management planning, where required. It is likely that the issues addressed in this environmental report will cover many, if not all, of the significant environmental effects of the Local Flood Risk Management Plans.

The environmental report focuses on two spatial scales. The main report makes a national assessment across all 14 Flood Risk Management Strategies.

Table 1 outlines the contents of the draft Flood Risk Management Strategies and indicates which elements will assess by this SEA. Those parts, which provide information, those that describe existing actions, and those that describe actions set out by other plans, programs, and strategies will not be assessed. SEPA (2015) (Table 2).

The Flood Risk Management Strategies aim to reduce overall flood risk and avoid an increase in flood risk. The aims of the flood risk management objectives, if achieved, should deliver significant benefits in terms of protecting people, property, and infrastructure (SEA objectives for **human health** and **material assets**). Historic buildings may also benefit from a reduction in flood risk (**cultural heritage**). Without action, flood risk is expected to increase due to climate change and urban creep: therefore, the flood risk management objectives are consistent with SEA objectives to help adapt to a changing climate (**climatic factors**).

Table 1 Elements of the draft flood risk management strategies assessed in this SEA SEPA (2015)

The content of draft flood risk management strategies	Has this been assessed in this SEA?
Local Plan District overview Defines the Local Plan District, the public bodies involved, and a summary of flood risk for the area. The Potentially Vulnerable Areas for the Local Plan District also identified	No, this provides information about the Local Plan District
Local Plan District objectives and actions Further information on flood risk management objectives and actions for this Local Plan District that span more than a single Potentially Vulnerable Area. These are in relation to flood warning, surface water management planning, land use planning, and generic initial objectives and actions	Yes, new potential actions have assessed
Local Plan District: delivery plan A delivery plan developed by the lead local authorities that contains information on the timing and funding arrangements for the Local Plan District level actions. (This delivery plan is optional and may not produce for all Local Plan Districts)	No, this information is prepared by the lead local authority as part of the Local Flood Risk Management Plans and so does not fall within the scope of this SEA. The lead local authorities will determine whether the Local Flood Risk Management Plans require an SEA
Surface water flooding The impacts of surface water flooding summarized for each Local Plan District, alongside existing actions to manage these risks. Future impacts due to climate change and links to river basin management described	No, this provides information about surface water flooding and existing actions
River flooding River flooding in the Local Plan District is described here. The impacts described alongside existing actions to manage river flooding in individual chapters. Future impacts due to climate change, the potential for natural flood management, and links to river basin management also described	No, this provides information about river flooding and existing actions
Coastal flooding Coastal flooding impacts summarized alongside an explanation of the coastal processes that influence flooding. Existing actions to manage risk and future impacts due to climate change are summarized, and the potential for natural flood management and links to river basin management also described	No, this provides information about coastal flooding and existing actions
Potentially Vulnerable Areas: Characterization Describes and summarizes the likelihood and impact of all types of flooding for Potentially Vulnerable Areas, including the local history of flooding and existing actions to manage floods	No, this provides information about flooding in each Potentially Vulnerable Areas and existing actions
Potentially Vulnerable Areas: Objectives and actions Further information on flood risk management initial objectives and actions for each Potentially Vulnerable Area	Yes, we will assess the likely significant effects of the initial objectives and potential actions
Potentially Vulnerable Areas: Delivery plan A draft six-year delivery plan that has been developed by the lead local authorities that contains information on proposed funding and timing of potential actions	No, this information prepared by the lead local authority as part of the Local Flood Risk Management Plans, and so does not fall within the scope of this SEA. The lead local authorities will determine whether the Local Flood Risk Management Plans require an SEA

4.3 Mitigation and Monitoring

The Environmental Assessment requires a clarification of "the measures imagined to anticipate, reduce, and as completely as conceivable counterbalance any critical unfriendly impacts on the earth of implementing the arrangement or program." Table 3 sets out any ecological issues that are likely to result from the usage of the Flood Risk Management Strategies and proposes measures for the counteractive action, decrease and balance of huge unfavorable impacts. These measures are recommendations that should be taken forward at progressive point by point dimensions of surge hazard management arranging, for example, at possibility and configuration stages. Different associations, especially local specialists, will lead to stages that are progressively detailed. We will set out how our recommendations on alleviation will take forward in the post-adoption statement SEPA (2015).

4.3.1 Monitoring

Environmental assessment requires the responsible authority (SEPA) to monitor the significant environmental impacts of

Table 2 SEA objectives and assessment questions, SEPA (2015)

SEA topic	SEA objective	Do the flood risk management strategies
Population and human health	*Protect the human health, reduce health inequalities, and promote healthy lifestyles*	• Improve the health and living environment of people and communities • Reduce flood risk • Improve opportunities for healthy lifestyles
Biodiversity, fauna, and flora	*Conserve and where appropriate enhance species, habitats and biodiversity, and habitat connectivity*	• Avoid adverse effects on, and improve protected species and habitats • Avoid adverse effects on and improve wider biodiversity • Support healthier ecosystems • Help promote habitat connectivity
Soil	*Protect and where appropriate enhance the function and quality of the soil resource*	Safeguard soil quality, quantity, and function, including valuable soil resources such as agricultural land and carbon-rich soils
Water	*To prevent deterioration, protect, and where appropriate enhance the water environment*	• Protect and enhance the overall water environment • Avoid adverse effects on the status of water bodies • Avoid adverse effects on sensitive coastal areas and the marine environment
Climatic factors	*Contribute to mitigation of and adaptation to climate change*	• Improve adaptability to the effects of climate change • Contribute to reducing greenhouse gas emissions
Material assets	*Contribute to protecting property and infrastructure. Minimize waste and energy consumption and promote resource efficiency*	• Protect material assets, e.g., infrastructure, properties • Promote resource efficiency, including energy, waste, water, and minerals
Cultural heritage	*Protect and where appropriate enhance the character, diversity, and special qualities of cultural heritage and the historic environment*	• Protect the historic environment and its setting • Enhance or restore historic features and their settings • Improve the quality of the wider built environment
Landscape	*Protect and where appropriate enhance the character, diversity, and special qualities of landscapes*	• Protect, enhance, or restore landscape quality • Avoid adverse effects on protected landscapes

implementing flood risk management strategies. This must be done in a manner that also identifies the unforeseen adverse effects and takes appropriate remedial action.

Monitoring should inform the impacts of flood risk management strategies themselves rather than wider trends. SEPA's water environment is widely used, and we suggest that you take advantage of existing activities instead of making any new monitoring. The monitoring activities for the proposed strategic environmental assessment are presented in Table 4. The impacts of individual projects will be monitored in accordance with the plans developed as part of the EIA at the project level SEPA (2015).

5 Climate Change and Disaster Risk Assessment and Management in Egypt and Arab Countries

5.1 Development Strategies and Crisis/Disaster Risk Reduction

Development can increase the level of crises and disasters. There are so many examples of the economic development and social improvement approach toward creating new crises and disasters, such as the quick urban transformation PCC (2007a). It is known that the growth of slums and poor areas in the middle of cities leads to creating unstable environments. These slums are mainly built in narrow alleys or over steep slopes, or plains and valleys subject to floods and torrential rains, or beside transportation lines and hazardous industries.

Life in rural areas endangered by the huge diversity in the economic and social structure and its interaction with the environment. There are special recurrent features for sustainable development in forming disasters and crises in rural areas. Poverty in the countryside is one of the main factors causing many crises and disasters (https://docplayer.net/53116069-National-strategy-for-crisis-disaster-management), SEPA (2015).

5.1.1 Risks of Crises, Disasters, and Planning for Sustainable Development

Reduction of crises and disaster risks can be incorporated into sustainable development planning UNESCO-IHE (2009). It is quite certain that the repetition of natural disasters and crises should put their elimination on top of the priorities set forth by planners of sustainable development. These plans differentiate between two kinds of crisis and disaster management:

Table 3 Measures envisaged for the prevention, reduction, and offsetting of any significant negative effects, SEPA (2015)

SEA objective	Potential negative effects of the flood risk management strategies	Proposed mitigation measures and recommendations
Population and human health: Protect the human health, reduce health inequalities, and promote healthy lifestyles	No significant adverse effects	
Biodiversity, fauna, and flora: Conserve and where appropriate enhance species, habitats and biodiversity, and habitat connectivity	Storage, conveyance and control actions, river defenses, and coastal defenses could damage ecosystems such as wetlands, native floodplain woodlands, and coastal habitats that are already fragmented/degraded	Potential negative effects can be mitigated through the identification of impact, sympathetic design, and timing of works to avoid or minimize the effects on habitats and wildlife, along with consultation with relevant organizations
	All structural actions could have significant negative effects on designated nature conservation sites, for example, by altering patterns of river flow or coastal processes or through disturbance	Potential negative effects on protected sites will be assessed by SEPA as part of the Habitats Regulations Appraisal for the Flood Risk Management Strategies and mitigation applied where required. At more detailed levels of planning, Habitats Regulations will also apply during which the responsible authority will need to take steps to mitigate negative effects on protected sites
Soil: Protect and where appropriate enhance the function and quality of the soil resource	Storage, conveyance, and control actions can alter natural processes and lead to increased erosion of carbon-rich soils or agricultural land	Modeling of natural processes can help to predict better and mitigate potential negative effects: this should be addressed during feasibility and detailed design stages
Water: To prevent deterioration, protect and where appropriate enhance the water environment	Storage, conveyance and control actions, river defenses, and coastal defenses could lead to potential degradation of beds and banks of rivers and the coastline	The potential negative effects can be mitigated by minimizing potential habitat loss and including habitat creation in flood risk management schemes. Negative effects should be addressed during feasibility and detailed design stages Actions that can affect the freshwater environment (such as river defenses or storage actions) regulated under The Controlled Activities Regulations, which aim to protect the water environment. Mitigation considered as part of the authorization process Some actions, particularly those deemed as development, regulated under the land use planning system: environmental effects will be addressed through project level Environmental Impact Assessments
Climatic factors: Contribute to mitigation of and adaptation to climate change	Storage, conveyance and control actions, river defenses, and coastal defenses could lead to potential loss or degradation of habitats (e.g., wetlands, woodlands, coastal) that help to mitigate and adapt to a changing climate	The potential negative effects can be mitigated by minimizing potential habitat loss and including habitat creation in flood risk management schemes. Negative effects should be addressed during feasibility and detailed design stages
Material assets: Contribute to protecting property and infrastructure. Minimize waste and energy consumption and promote resource efficiency	No identified negative effects. Effects on waste, energy, and resource efficiency uncertain at this stage	Opportunities to minimize waste and resource use should be examined during feasibility and detailed design stages
Cultural heritage: Protect and where appropriate enhance the character, diversity, and special qualities of cultural heritage and the historic environment	No significant negative effects identified (although assessment is uncertain as effects depend strongly on the type of action and its location)	Potential negative effects can be mitigated through the identification of any heritage assets (including archeology) and the early engagement of heritage interests during feasibility and detailed design stages

(continued)

Table 3 (continued)

SEA objective	Potential negative effects of the flood risk management strategies	Proposed mitigation measures and recommendations
Landscape: Protect and where appropriate enhance the character, diversity, and special qualities of landscapes	Coastal defenses (and also storage, conveyance and control actions, and river defenses) could lead to landscape degradation	Potential negative effects should be addressed early during feasibility and detailed design stages. Consultation with SNH, National Park Authorities, and affected communities recommended

Table 4 Proposed SEA monitoring program, SEPA (2015)

What is being monitored	Data source, the frequency of monitoring	Timescale and responsibility
Flood risk to people and properties, cultural heritage, and designated environmental sites	SEPA National Flood Risk Assessment and baseline flood risk data updated every 6 years	SEPA, as part of the National Flood Risk Assessment update in 2017 and in the second cycle of flood risk management strategies in 2021
Status of the water environment	WFD classification data; monitored via the river basin management plans (6 yearly publication cycle)	SEPA, as part of the third river basin management plans in 2021

- Prospective management of crisis and disaster risks should be incorporated in sustainable development planning. Development programs and projects should be reviewed to identify their potentials in eliminating vulnerability and risks or their aggravation (medium and long term).
- Compensatory management of crisis and disasters such as preparation and confrontation should be compatible with developmental planning. It mainly focuses on reducing vulnerability and the natural risks that incremented during the past development plans. Compensatory policies are necessary for eliminating current risks (short term).

This combination of disasters and crises risk reduction require three main steps:

1. Collecting basic data on natural disasters risks and developing planning tools to track the relationship between the development policy and the risks of crises and disasters.
2. Collecting and publishing the best practices in development planning and policies in the field of natural crises and disaster risk reduction.
3. Mobilizing the political will to redirect the development sectors and to eliminate the risks of disasters and crises (https://docplayer.net/53116069-National-strategy-for-crisis-disaster-management).

5.2 The Egyptian Legislation and Laws Concerned with Crisis and Disaster Management

Any national system for crisis and disaster management needs enforceable laws and legislation, accurately identifying the measures and plans, to not only manage crises and disasters but also for the three phases of the pre, during, and post-crisis. These laws and legislation shall accurately state the tasks, levels, and measures required to deal with the crisis or disaster MWRI (1997, 2010), NWRP (2005), MOALR (2009).

5.2.1 First: The Current National Legislation Concerned with Crisis and Disaster Management

By reviewing, the current legislation concerned with crisis and disaster management, a number of laws and presidential and ministerial decrees organizing work in the areas related to crisis and disaster management demonstrated, mainly:

1. **Law no. 179/1956 on civil defense:**

The said law aims to prevent civilians and secure facilities and transportation against air raids and military operations. Also to secure the organized flow of work through organizing the alarm and firefighting systems and the exchange of relief aid among different cities and other kinds and means of rescue at the time of air and military operations.

The second item in article three of the said law deals with the civil defense authority, ministry of interior competence in facing public disasters, that is under a cabinet decree. It may directly request from any ministry to provide the required personnel, resources, and tools. It may also use civil defense forces. The law hereto is composed of 25 articles, organizing all the works of the civil defense. Provisions of this law amended several times in law 10/1965, law 20/1974, law 175/1981, and law 107/1982. It is also noteworthy that the said law had included establishing a higher council for civil defense presided by the Minister of Interior.

2. **Presidential Decree no. 300 on establishing the Search And Rescue Centre at the Ministry of Defense**:

In light of this Decree, the Centre of "Search and Rescue" established at the Ministry of Defense in Cairo. The decree specified its tasks and objectives.

The law stated on forming the committee of search and rescue, comprising a number of representatives of the concerned ministries.

3. **Law no. 4/1994, by adopting a law for the protection of the environment and its executive regulations**:

The law aims to protect the environment from pollution, maintain its natural resources, and achieve the balance required between development and environment in light of the concept of sustainable development. In this respect, the law includes a number of articles concerned with crisis and disaster management, such as article 25 that identifies the competencies of the Egyptian Environment Affairs Agency (EEAA), in terms of the preparation of an emergency plan to manage environment disasters. Article 26 also deals with securing the availability of all required resources in all public and private bodies to manage environment disasters. Articles 53, 54, and 55 address accidents of oil ships, which can or may lead to polluting the regional sea or the economic zone of Egypt.

5.2.2 Second: The Current National Legislation that Contributes to Risk Reduction of Crises and Disasters

Since global approaches give major significance to the measures of disaster risk reduction in the pre-crisis/disaster stage, there are currently a number of national laws and decrees that contribute to the reduction of crisis/disaster risks such as

1. **Law no. 4/1994, for enacting a law on the protection of the environment and its executive regulations:**

As mentioned previously, the said law meant to protect the environment against pollution, to maintain natural resources, and achieve a balance between development and the environment under the concept of sustainable development.

The said law also included some articles that curbed down the risks of crises and disasters, namely; no. 19,20,21,22, and 23 in chapter one dealing with development and the environment, within the scope of part one on protecting the environment from pollution.

These articles organize the process of requesting all the new development facilities to prepare their environmental impact assessment for the facility applying for a license. The law hereto covers all the new expansions or renewals of the existing facilities under the same provisions stated in the said articles.

2. **Presidential Decree no. 153/2001, for establishing the National Centre for the Land Use Planning at the country level**:

Recognizing the importance of establishing a national policy to organize and plan land use at the country level. The country initiated in May 2001 establishing the national center for land use planning at the country level. The center coordinates several competencies with the concerned bodies, such as

- Counting and assessing the out of zone state lands and preparing the general planning for developing and using them in light of the general country policy.
- Preparing a map of country land off border usages for all purposes, in coordination with the Ministry of Defense.
- Handing each ministry, a map of the lands allocated for its activities.
- Counting the annual programs for the development and use of land for each ministry and the revenue and expenses budget of development.
- Participating in choosing and identifying the sites required for the main new development projects in the country.
- Holding the technical and environmental studies and research required for the use of off border country lands in coordination with the concerned ministries and bodies.

3. **Prime Minister's Decree no. 1537/2009 of constituting the Cabinet—National Committee for Crisis/Disaster Management and Disaster Risk Reduction**:

In the framework of the country's concern with developing the national system for crisis/disaster management and DRR, the Prime Minister's decree n. 1537/2009 issued constituting the National Committee for Crisis/Disaster Management and Disaster Risk Reduction. The committee mandated with activating the national institutional framework aligned with UNISDR or what are called National Platforms for disaster risk reduction. Concurrently, the decree sets out the committee's objectives, jurisdictions, and tasks, representation level of ministries, governorates, concerned entities, and other national agencies. The decree stipulates as well constituting the Advisory Committee comprising selected prominent scholars and specialists to provide the NCCDMDRR with technical and academic assistance.

5.2.3 Third: Evaluating the Current Legislative Situation

It has observed that the current laws had issued years ago lacking the following basic elements:

1. The absence of a legislative integration to deal with crises and disasters.
2. Most of the current laws and decrees deal with crisis or disaster confrontation, and they do not focus on the measures that should be taken pre or post the crisis/disaster.
3. The absence of measures for DRR according to international developments.
4. Most of the legislation should be updated according to the new national, regional, and international developments.

5.2.4 Action plan

The government should work on achieving the following:

(1) Conducting national baseline evaluations of the status of disaster risk reduction and disseminating their results periodically through structured scientific studies in accordance with the capacity and needs of the country. The studies will also be conducted based on specific indicators aimed at monitoring the implementation of the relevant action plan of crisis/disaster management and DRR. When appropriate, such information should be shared with the relevant local, regional, and international authorities as well as citizens and experts concerned;

(2) The National Committee for Crisis/Disaster and Disaster Risk Reduction implements tasks and duties of the National Platform for Disaster Risk reduction as stipulated in the UN International Strategy for Disaster Reduction (UNISDR) and priorities of Hyogo Framework (Article Three of the Prime Minister's decree no. 1537/2009);

(3) Following-up on (or developing) institutional systems responsible for early warning data and indicators as well as on the availability of their periodic data in the speed and quality required. Furthermore, following-up on DRR systems, ensuring their validity, efficiency, and calibration, and issuing periodicals about them;

(4) Following-up on integrating DRR aspects related to current and potential climate changes in the Strategies for Disaster Risk Reduction. Also, adapting to climate change and ensuring that the relevant disaster administration is giving due consideration to the disaster at hand—natural and climatic—for example, earthquakes, landslides, and flash floods in education programs according to geographic location. At the same time, evaluating and following-up on the extent by which influential segments, like decision-makers, and most affected people, like women, children, and the elderly are responding;

(5) Publishing a digest on DRR national programs and success stories related to the Hyogo Framework for Action, provided that it contains analyses of ongoing costs, benefits, observations and evaluations of vulnerable areas, and risks related especially to those areas subject to climate changes, earthquakes, and floods, as deemed relevant;

(6) Conducting a comprehensive review to evaluate the risks involved in national and major industrial projects to ensuring their compatibility with the industrial and environmental safety standards as well as issuing certificates proving such compatibility;

(7) Following-up on and studying data pertinent to road accidents and identifying vulnerability areas as well as causes. Furthermore, attempting to increase road safety by conducting scientific studies to improve roads, vehicles, and highway administrations.

6 Egypt's National Strategy for Adaptation to Climate Change and Disaster Risk Reduction

6.1 National Strategy Goals

Egypt faces many challenges in various areas of sustainable development (social; financial; and environmental), and on unique levels (countrywide; regional; and international) that have got to be tackled in an effort to be competent to speed up and reinforce the developmental impact Strzepek and Yates (2000). For example, at the national stage, water shortage is a major venture within the light of populace growth, creation, and consumption needs NEEDS (2010). Issues related to mitigation and adaptation to climate change additionally present a valuable mission that has to be addressed. Also, the proportion of informality in Egypt's economic climate is excessive and growing. Considering we will manage what we measure, reaching the Sustainable Development Goals (SDGs) would require considerable efforts to formalize the informal sector through setting up appropriate incentive structures. These efforts are currently underway, and a countrywide technique has been adopted to increase and formalize the casual sector. Additionally, while Egypt has done commendable development in regards to female's empowerment, mainly ladies' entry to education, drastic discount within the practice of feminine genital mutilation among younger ladies, and a historical milestone in feminine representation within the parliament, there are nonetheless limitations impeding women from realizing their talents as strong dealers of social and fiscal development. MIC-ARE (2016).

The National Strategy for adaptation to climate change and disaster risk reduction aims to achieve the following objectives NSACC (2011), Sayed and Nour El-Din (2001, 2002) , Sayed (2003, 2004, 2008):

1. Increasing the resilience of Egyptian society in dealing with the risks and disasters resulting from climate change and its impact on different sectors. These sectors include coastal areas, water resources, irrigation, agriculture, health, urban areas, housing and roads, and tourism. This objective can be achieved through an in-depth analysis of the current situation in different sectors of society. These facilities are available and required to raise the level of readiness for confrontation and flexible interaction with developments.
2. Enhanced capacity to absorb and contain climate-related risks and disasters: This can be achieved through the development of specialized sectoral programs and action plans to meet the needs of society as a whole and adapt to new conditions. Through various means, ranging from the basics to the use of the latest technology. In this way, systems are created to adapt to potential climatic changes, namely, increasing temperature and water scarcity, and negative forecasts of increase, precipitation, and sea level rise.
3. Disaster reduction associated with climate change: This is possible through accurate scientific calculations. Field and theoretical observation of various sectors of society; appropriate support for existing projects; choice of sites and designs that are most appropriate and appropriate for new projects; and enhancement of infrastructure in a way that helps to reduce climate change-related disasters, MWRI (2013).

To achieve these objectives, the following measures should be taken:

- Identify the risks and crises associated with climate change, taking into account the precise scientific treatment of uncertainties about the current potential impacts and their spread in terms of geographical scope and time scale.
- Integration of adaptation plans for different sectors into five-year plans and national development programs.
- Building a culture of "safety first" and raising community awareness, taking into account that climate change is a long-term phenomenon.
- Promote community participation at all levels (governmental, non-governmental, and civil society).
- To strengthen regional and international cooperation and to consolidate current initiatives to adapt to climate change.
- Monitoring, evaluation, and development.

6.2 Climate Change Projections of Sea Level and Impacts on Coastal Zones and Northern Lakes

In general, the impact of risks, disasters, and crises resulting from climate change on the sectors and areas covered by the strategy is negative. It is important to highlight that these impacts will occur if no special adaptation measures taken to reduce climate change risks, disasters, and crises. Naturally, there will be a positive outcome if the necessary preventive measures are carried out MWRI (2013).

<u>The Coastal Zone</u>:

Risks, disasters, and crises caused by climate change, sea level rise, and extreme events:

Climate change and sea level rise are associated with direct impacts, which include the erosion and inundation of sandy beaches, and thus, the gradual regression of shorelines Zanella (2010). Another impact is the possible recurrence of storms and hurricanes, which would have a negative impact on multiple sectors of the coastal zone, MWRI (2013).

This depicted in detail as follows:

1. **The Coastal Zone of the Mediterranean Sea**

The gradual rise in sea level may lead to increased erosion rates for the low-lying coastal area (0–1 m above sea level), especially in endangered areas. This may cause erosion of the shore at the narrow, low-lying sand barrier separating the sea from the northern lakes, which will result in gradual integration of these lakes with the sea, as expected for Lake Manzala. In such cases, the ecological aspects of marine life and the natural and chemical properties of the lake will change. If the shore goes inland toward agricultural land, this may increase the salinity of groundwater, leading to the salinization of agricultural land in the area. It is also possible that some of the sites will be flooded by flooding the low-lands unless the engineering protection structures are strengthened. These installations operate efficiently with current sea level. Special measures should also be taken to anticipate the expected climate change, given the multiple negative impacts of increasing frequency and intensity of storms, hurricanes, and tsunamis.

2. **The Coastal Zone of the Red Sea**

The coastal area of the Red Sea is likely to affect climate change and sea level rise, particularly in lowlands, wetlands, and beaches. Coral reefs may be affected by sea level rise, especially if the sea level is higher than coral reef growth. Coral bleaching may also increase due to environmental stresses. In any case, the rocky composition of the coast which distinguishes the Red Sea, slightly above sea level from 2 to 5 meters is a natural barrier against rising sea levels in these areas. It is worth mentioning that frequent tsunami may cause many negative effects.

3. **Beaches**

The beaches are defined as sand-covered coastal areas and serve as the front lines of coastal areas because of their direct contact with the sea. They are among the most important natural environments, not only as of the first line of defense against sea attacks but also because of their environmental, esthetic, and tourist importance. However, beaches are increasingly vulnerable to flooding and progressive loss of their sand (due to erosion), especially if sea level rises above the standard rate and when the strength of waves and sea currents is affected by strong winds and storms. It is therefore expected that, according to scenarios adopted in this strategy, all low-lying beaches with levels ranging from 0 to 1 m will be flooded. The shores of the Mediterranean Sea and the Red Sea with higher levels of sea level are expected to be safe, depending on the topographical characteristics of each region El-Raey (2009).

4. **The Condition of the Sea and Extreme Natural Events**

In general, it is noticed that the rise and strength of the sea and ocean waves have increased over the past decades. Climate change can double the power of storms and hurricanes and change their paths, increasing their destructive power in coastal areas OECD (2012). At the local level, none of the studies confirmed any significant abnormalities in prevailing natural events, including temperatures, wind, waves, hurricanes, and winter storms. However, many projections suggest that the frequency and intensity of storms, hurricanes, and extreme events may increase in the future.

5. **Social and Economic Structure of Coastal Zones**

There will be a clear negative impact on the socioeconomic aspects of coastal areas. The intensity of this impact can be reduced by measures to reduce the negative impact of climate change and the expected rate of sea level rise. These measures can be implemented as precautionary measures, or as a rehabilitation process for some areas. Under unusual circumstances, these measures could be implemented to transfer some activities to areas less vulnerable to the negative impact of climate change and sea level rise.

Climate projections

Met Hardley Centre now produces confusing parameter sets to form an individual model known as HadCM3C, to explore uncertainties in physical and biochemical feedback processes. The results of this analysis will become available next year and will complement the projections of multi-model environmentally sound management (CMIP5), providing a more comprehensive set of data to help advance understanding of climate change in the future. However, many studies have been used to address the effects of climate model output CMIP3. For this reason, and because it is still the most widely used range of projections, the CMIP3 result for temperatures and precipitation, for the A1B emission scenario, shows below Egypt and the surrounding region.

Summary of temperature change in Egypt

Figure 8 shows the percentage change in the average annual temperature by 2100 from the baseline climate of 1960–1990, with an average of more than 21 CMIP3 models. All CMIP3 models in the CMIP3 project have increased temperatures in the future, but the size of each pixel indicates the compatibility of the models with the size of the increase.

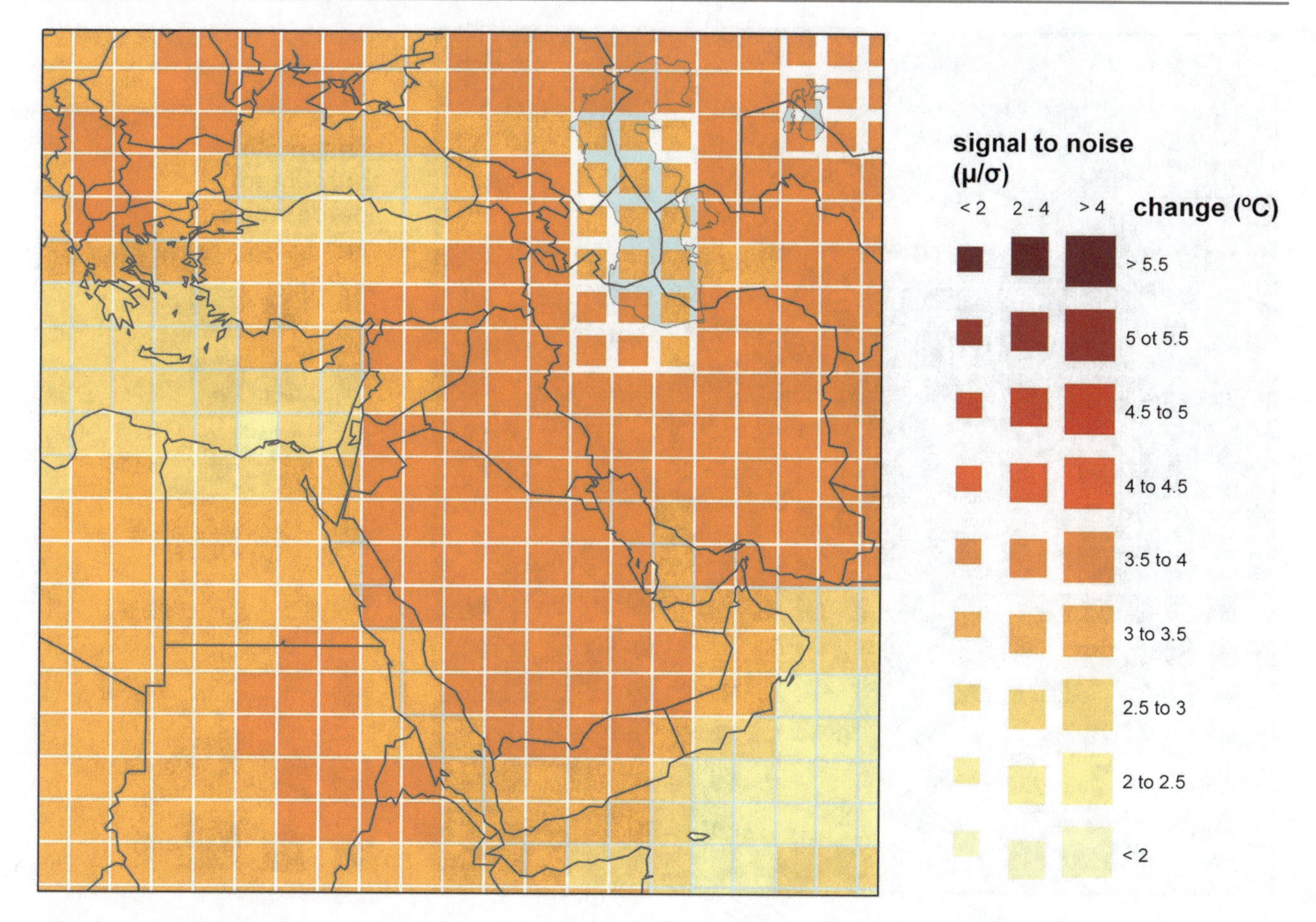

Fig. 8 Percentage change in average annual temperature by 2100 from the 1960 to 1990 baseline climate, averaged over 21 CMIP3 models. The size of each pixel represents the level of agreement between models on the magnitude of the change, UNFCCC (2007)

Expected increases in temperatures in Egypt range from about 3–3.5 °C with consistently good agreement between models in the Middle East region in general, UNFCCC (2007).

Summary of precipitation change in Egypt

Figure 9 shows the percentage change in the average annual precipitation by 2100 from the baseline climate of 1960–1990, with an average of more than 21 CMIP3 models. Unlike temperature, models sometimes differ on whether precipitation is increasing or decreasing in an area, so in this case, the size of each pixel indicates the percentage of the models in the group that agrees to the change in deposition mark. Egypt is expected to experience a major reduction in precipitation, in common with the wider Mediterranean region and the majority of the Middle East region. More than 20% projected declines in the west of the country, with a strong group agreement. Smaller changes are expected toward the southeast, UNFCCC (2007).

Figure 10 presents the results of this analysis for Egypt. The Nile and its tributaries are the only surface water bodies included in this analysis. To this end, the remainder of Egypt is characterized as "no appreciable flow" Teshome (2009). Most of the population and economic activity take place close to the river. In other parts of the country, water resources are provided from groundwater-fed springs and oases. Egypt also benefits from favorable water sharing arrangements, which result in it receiving the biggest share of rainfall resources of all ten countries in the Nile Basin. Hence, because of this analysis, the level of threat to human water security is moderate to low. Figure 10 generally reflects the dominance of surface water usage via the Nile in Egypt; indeed, 97.6% of Egypt's total renewable water resource is from surface water, whereas 2.4% is from groundwater recharge, UNFCCC (2007).

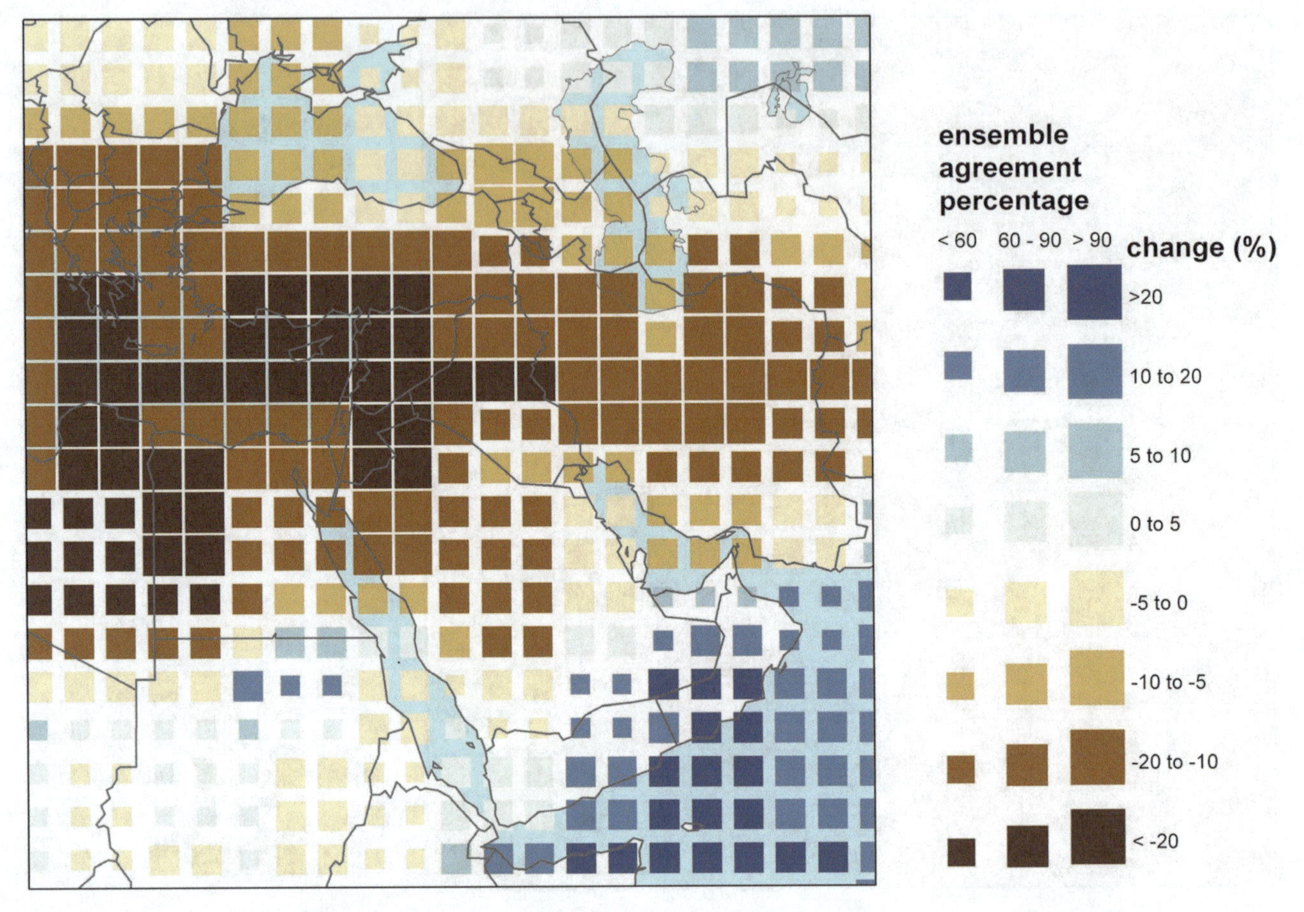

Fig. 9 Percentage change in average annual precipitation by 2100 from the 1960 to 1990 baseline climate, averaged over 21 CMIP3 models. The size of each pixel represents the level of agreement between models on the sign of the change, UNFCCC (2007)

7 Floods: Climate Change and Adaptation for Human Health

7.1 Retrospective Perspectives on Health Risks Associated with Flooding

It is clear that there is little data of relatively good quality available on the health risks associated with flooding. For many types of effects, for example, leptospirosis, meningitis, and the dangers of bad floods, the effects are relatively rare, so the risks need to maintain the context. However, even small numbers of effects, such as deaths and injuries, can contribute useful information to the formulation of prevention programs ECIDSSC (2010). Full flood effects may not be reported because everyone will not seek medical assistance. In this context, more epidemiological data can be obtained through interviews or from local newspaper reports.

Drinking water: Sources of drinking water contaminated. Thus, you may need to cut for a certain period after a flood occurs. Watersheds are known and can be easily drawn

Sewage system: In Cologne, the Rhine infiltration into the sewerage system Todd and McNulty (1976). As a result, valves are installed.

Animal diseases: They can be transmitted to humans by touching the water (not necessarily drinking water), Janev Holcer et al. (2015).

The problems of predicting an event are extremely difficult for floods; this is mainly due to difficulties in predicting where heavy rainfall is likely. Moreover, there are more difficulties in educating the public to respond appropriately in anticipation of or during flooding. Public service announcements have proven to be very ineffective in helping people who are under stress. Reducing risk behavior is a clear need. A total of 40% of the health effects of flooding were estimated to be directly related to the wrong behavior. Evidence from the 1999 storms in France indicates that from 17 deaths, there were 8 cases due to risky behavior, including the death of two people to repair the roofs of buildings and the occurrence of two people for jogging at the time of the storm. Many people in London asked how they would respond in the event of a flood when they said

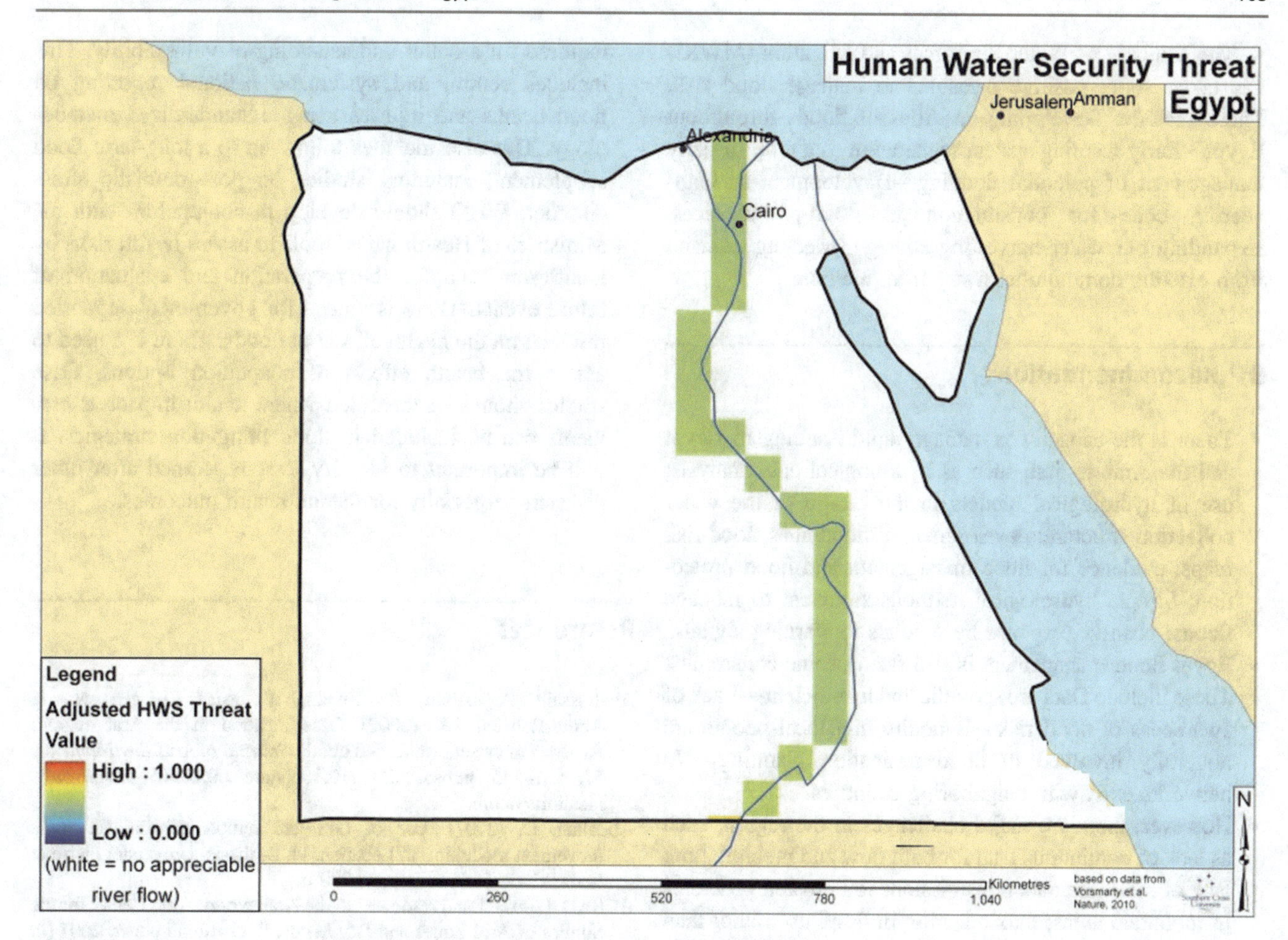

Fig. 10 Present adjusted Human Water Security threat (HWS) for Egypt UNFCCC (2007)

they would go underground to the public transport system. In California, seismic preparation programs were largely successful and could serve as a model for flood events schemes.

There is a need to distinguish between recurrent and extreme events. Structural defenses should be directed to minimize the effects of recurrent events. The basic requirement is the need to develop long-term infrastructure. This is a major issue independent of the potential impacts of climate change TearFund (2010). In Europe, there are strict regulations on nuclear and chemical plants, but consideration of potential health effects is not part of these laws. In France, it is sometimes difficult to implement guidance at the local level. For urban stormwater events, management may not be ready. Required weakness maps, including consideration of successive effects. Safety is one of the logical reasons for warnings; however, warnings do not include mental health and other health outcomes. Risks should focus on health effects and not just on economic damage, as is the case now.

8 Conclusions

Egypt suffers from a water shortage that gradually increasing far from the Nile, especially in the Eastern Desert and Sinai. Renewable water resources are limited due to limited rainfall that sometimes causes a rapid flooding. If these floods are properly managed, they will serve the needs of the fragile communities in the desert. Moreover, social stability and economic development are strongly linked to the sustainable existence of water resources.

Socioeconomic development in the Sinai in terms of infrastructure (roads, natural gas, oil pipelines, airports, ports, resorts, settlements, etc.), sudden flooding causes severe damage to human life and infrastructure.

The Egyptian experience with the floods has shown that much remains to be done to improve the use of water resources and to protect our infrastructure that crosses dangerous areas.

The Ministry of Water Resources and Irrigation (MWRI) has taken some positive measures to manage flood risk. These include: Developing an Atlas of floods throughout Egypt—Early warning system registration as a tool for daily management of potential flooding—Development of engineering code for construction in flood-prone areas, Expanding our water harvesting strategy, Meeting demand from existing communities away from the Nile

9 Recommendations

- There is the capacity to manage rapid flooding in Egypt and the Arab region, such as hydrological data analysis; use of hydrological models in the design of the water collection structure; development of flood atlas flood risk maps; evidence for flood management and flood protection; Simple hydrological methods sufficient to manage floods; positive response by officials to warning signals.
- Egypt flood management is still facing some constraints. These include Data are sporadic and insufficient—Lack of awareness of flood risk—Bedouins and local people are not fully involved in flood mitigation planning—No networks exist with neighboring countries.
- However, there are major challenges in the region, such as lack of continuous data for both flow and precipitation; lack of real-time data transmission; soil erosion problems in mountain areas; maximization of flood use rather than flood risk management; the absence of a flash flood early warning system; the absence of a disaster risk management plan; there is little interest in the users of the flood; the availability of a model to describe the hydrological conditions of the Arab region; social gaps in terms of minimal awareness of frequent catastrophic events; communication technology is very unreachable and expensive to upgrade; the absence of an integrated approach to rapid flood management, not only the warning system but the full cycle and continuity; the lack of methodology for risk assessment; the lack of design standards for precipitation network; inter-state technical and administrative conflict, there is no flood insurance policy, lack of access to data collected by previous studies through international partners; conflict and disputes among institutions responsible for flood management, the absence of a legal framework for preventive construction; the view of rapid flood as a source and blessing rather than risk; there is a knowledge gap that begins with data, information, and data analysis, human skills; attitude (do not give up as a professional) and the value of data., there is a need to train trainers for all aspects of rapid flood aspects.
- More and better quantitative data are needed on health impacts associated with all flood categories. This data is required for a better understanding of vulnerability. This includes central and systematic national reporting on flood deaths and injuries using a standardized methodology. This also includes follow-up to a long-term flood supplement, including studies on post-traumatic stress disorder. WHO should develop in conjunction with the Ministries of Health MOH tools to assess health risks by identifying data for the preparation and evaluation of future events. There is a need for government-supported research on the health effects of floods. There is a need to assess the health effects of adaptation options. Case studies should be identified where health impact assessments can be included in flood mitigation strategies. It will be important to identify lessons learned after other disasters, especially for mental health outcomes.

References

Abahussain, A., Abdu, A., Zubari, W., Alaa El-Deen, N., & Abdul-Raheem, M., (2002). Desertification in the Arab region: Analysis of current status and trends. *Journal of Arid Environments, 51*, 521–545. https://doi.org/10.1006/jare.2002.0975, http://www.idealibrary.com.

Abdallah, C. (2007). Use of GIS and remote sensing for mass movement modelling in Lebanon. Ph.D. thesis, Université Pierre et Marie Curie Paris 6, France, 300 p.

ACSAD. (2007). Land resources in the Arab world. Arab Center for the Studies of Arid Zones and Dry Lands. Background paper, 2007 (in Arabic).

AFED. (2009). Arab forum for environment and development arab environment. Impact of climate change on the Arab countries. (AFED) Report 2009, Published with Technical Publications and Environment & Development magazine, http://www.afedonline.org, ISBN: 9953-437-28-9.

Agrawala S, Moehner A, El Raey M, Conway D, Van Aalst M, Hagenstad M, Smith J (2004). Development and climate change in Egypt: focus on coastal resources and the Nile. *Organization for Economic Co-operation and Development* COM/ENV/EPOC/DCD/DAC (2004)1/FINAL.

El-Raey, M. (2009). Mapping areas affected by sea-level rise due to climate change in the Nile Delta Until 2100, in Hexagon Series on human and environmental security and peace (HESP), H. G. Brauch (ed.), Free University of Berlin, UNU-EHS and AFES-PRESS.

Janev Holcer, N., Jeličić, P., Bujević, M. G., & Važanić, D. (2015). Health protection and risks for rescuers in cases of floods. *Arh Hig Rada Toksikol, 66,* 9–13. https://doi.org/10.1515/aiht-2015-66-2559.

Metwalli, N. (2010). *Climate change risk management program (CCRMP)-Egypt*. UNDP/MDG: Mid Term Review.

Ministry of International Cooperation MIC-ARE. (2016). National voluntary review sustainable development goals. http://www.miic.gov.eg/English/Resources/Publications/NonPeriodical/DOCUMENTS/ENGLISH.PDF_915201625622PM.PDF.

MOALR. (2009). Sustainable agricultural development strategy towards 2030. Ministry of Agriculture and Land Reclamation, Cairo, Egypt.

MWRI. (1997). The Egyptian Water Strategy Till 2017. Planning Sector—Ministry of Water Resources and Irrigation, Cairo, Egypt.

MWRI. (2010). Water Resources Development and Management Strategy in Egypt—2050. In *Arabic, Ministry of Water Resources and Irrigation*, Cairo, Egypt.

MWRI. (2013). Proposed climate change adaptation strategy for the Ministry of Water Resources & Irrigation. Joint Programme for Climate Change Risk Management in Egypt.

NEEDS. (2010). Egypt national environmental, economic and development study for climate change; Egyptian Environmental Affairs Agency/Climate Change Central Department, Under the United Nations Framework Convention on Climate Change on Climate Change (UNFCCC).

Ngigi, S. N. (2009). Climate change adaptation strategies Water resources management options for smallholder farming systems in sub-saharan Africa. The Earth Institute At Columbia University and the MDG Centre. http://hdl.handle.net/10919/68902.

Nile Basin Initiative. (2006). Baseline and needs assessment of National Water Policies of the Nile Basin Countries: A Regional Synthesis. Water Resources Planning & Management Project, Water Policy Component, Addis Ababa, Ethiopia.

NSACC. (2011). Egypt's national strategy for adaptation to climate change and disaster risk reduction. Egyptian Cabinet/Information and Decision Support Center/UNDP.

NWRP. (2005). National water resources plan, facing the challenge. Ministry of Water Resources and Irrigation Integrated Water Resources Management Plan for 2017.

OECD. (2012). Recent OECD work on adaptation to climate chang. Organization for Economic Co-Operation Development. www.oecd.org/env/cc/adaptation.

Parry, M., Arnell, N., & Dodman, D. (2009). *Assessing the costs of adaptation to climate change*. London: Imperial College.

Paul, D., David, K. (2001). Africa and global climate change. *Inter-Research, Climate Research*, CR-Special-8, *17*(2).

PCC. (2007a). Climate change impacts, adaptation and vulnerability. contribution of working group ii to the fourth assessment report of the intergovernmental panel on climate change. Cambridge, UK: Cambridge University Press.

Sayed, M. A. A., (2003). Lake Nasser flood and drought control/integration with climate change uncertainty and flooding risks (LNFDC/ICC). In *AMCOW Conference*, Adiss Ababa.

Sayed, M. A. A. (2004). Impacts of climate change on the nile flows. Ph.D. thesis, Ain Shams University, Cairo-Egypt.

Sayed, M. A. A. (2008). Eastern Nile planning model, integration with IDEN projects to deal with climate change uncertainty and flooding risk. *Nile Basin Capacity Building Network Journal*.

Sayed, M. A. A., & Nour El-Din, M. M. (2001). Adaptation of water policy of Egypt to mitigate climate change impacts. In *Proceedings of the Stockholm Water Symposium*.

Sayed, M. A. A., & Nour El-Din, M. M. (2002). Adaptation of water policy in Egypt to Mitigate climate change effects. In *Stockholm Water Symposium*, Stockholm.

Seawater Foundation. (2010). Seawater foundation, Retrieved November 23, 2010, from http://www.seawaterfoundation.org.

Selishi, B. et al. (2008). A review of hydrology, sediment and water resources use in the Blue Nile Basin. In *Working Paper 131—International Water Management Institute*, IWMI.

SEPA. (2015). Scottish environment protection agency environmental assessment of the flood risk management strategies. Heriot Watt Research Park, Edinburgh. https://www.sepa.org.uk/media/219209/sea_post_adoption.pdf.

Strzepek, K. M., & Yates, D. N. (2000). Responses and thresholds of the Egyptian economy to climate change impacts on the water resources of the Nile River. *Climatic Change, 46,* 339–356.

TearFund. (2010). How to integrate climate change into national-level planning in the water sector. UK: TearFund. http://www.adaptationlearning.net/guidance-tools/how-integrate-climate-change-adaptation-national-level-policyand-planning-water-sect.

Teshome, W. (2009). Transboundary Water Cooperation in Africa: The case of the Nile Basin Initiative (NBI). Review of General management, issue: 2. www.ceeol.com.

The Egyptian cabinet information & decision support center. (2010). The national strategy for crisis/disaster management and disaster risk reduction.

Todd, D. K., & McNulty, D. E. O. (1976). Polluted groundwater—A review of the significant literature. publisher, Water Information Center.

UNDP. (2004). Adaptation policy frameworks for climate change—Developing strategies, policies and measures, Cambridge University Press.

UNDP. (2006). Beyond scarcity: Power, poverty, and the global water crises. UNDP.

UNDP. (2009). Project document. adaptation to climate change in the Nile delta through integrated coastal zone management.

UNDP. (2018). Climate change adaptation in the Arab States: Best practices and lessons learned from country experiences. www.undp.org, https://www.uncclearn.org/sites/default/files/inventory/arab-states-cca.pdf.

UNDP-UNEP. (2011). Mainstreaming climate change adaptation into development planning: A guideline for practitioners. UNDP-UNEP

UNESCO-IHE. (2009). IWRM as a tool for adaptation to climate change. Training Manual and Facilitator Guide.

UNFCCC. (2007). United nations framework convention on climate change, Climate change: impacts, vulnerabilities and adaptation in developing countries. https://vdocuments.mx/technology/climate-change-effects-on-the-uk.html.

UNISDR-ROAS. (2011). United Nations Office for disaster risk reduction—Arab network for environment and development (RAED). An overview of environment and disaster risk reduction in the Arab Region: A community perspective. https://www.unisdr.org/we/inform/publications/23612.

World Bank (WB). (2005). Arab Republic of Egypt Country environmental analysis (1992–2002). MENA Region, Water, Environment, Social & Rural Development Department.

World Bank (WB). (2007). Assessing the impact of climate on crop water needs in Egypt. Global Environment Facility. Policy Research Working Paper, World Bank. WPS4293. 35 pp.

World Bank (WB). (2010). World development report 2010. Development and climate change. Washington, DC: The World Bank.

Zanella, D. (2010). Seawater forestry farming: An adaptive management strategy for productive opportunities in "barren" coastal lands. M.A. dissertation, California State University, Fullerton, United States California. Retrieved November 22, 2010, from Dissertations & Theses: Full Text (Publication No. AAT 1470817).

Zeitoun, M., England, M., Hodbod, J. (2010). Water demand management and the water-food-climate nexus in the Middle-East & North Africa Region. Regional Water Demand Initiative, IDRC-CRDI-IFAD, www.idrc.ca/wadimena.

Flash Flood Management and Harvesting Via Groundwater Recharging in Wadi Systems: An Integrative Approach of Remote Sensing and Direct Current Resistivity Techniques

Mohamed Attwa, Dina Ragab, Mohammed El Bastawesy, and Ahmed M. Abd El-fattah

Abstract

An integrated approach is carried out to study the hydrogeological conditions in wadi systems. In this chapter, the use of remote sensing and direct current (DC) resistivity techniques was considered to manage the flash flood and explore groundwater in desert lands. As case studies, two wadis at the Eastern and Western Deserts, Egypt, are presented to show the efficiency of using the suggested integrated study. Based on satellite images, remote sensing (RS) and geographic information system (GIS) techniques are used to identify the regional geology, geomorphological features, lineaments, and active channels regarding flash flood events in desert lands. Consequently, a hydrograph and runoff modeling can be estimated for wadi systems. Furthermore, the calculated flash flood total discharges and the storage capacity of existing mitigation measures can be recognized. From a geophysical point of view, the DC resistivity method can be applied to identify the subsurface layer distributions, image the near-surface lateral heterogeneities, subsurface structures, and potential groundwater zones. Based on geophysical data inversion results, this chapter shows that it is possible to recognize sites for successful dam construction and groundwater bearing zones exploration. Accordingly, the results of these case studies represent the importance of the integrated approach for the flash flood hazards management and its harvesting.

Keywords

Remote sensing • GIS • DC resistivity • ERT • Flood hazards mitigation • Groundwater exploration

M. Attwa (✉)
Structural Geophysics Group (SGG), Faculty of Science, Geology Department, Zagazig University, Zagazig, 44519, Egypt
e-mail: attwa_m2@yahoo.com

M. Attwa · M. El Bastawesy
National Authority of Remote Sensing and Space Sciences (NARSS), Cairo, Egypt

D. Ragab
Geological Sciences Department, National Research Centre, Dokki, Cairo, Egypt

D. Ragab
Geosciences Department, Boise state University, Idaho, USA

A. M. Abd El-fattah
Geology Department, Faculty of Science, Aswan University, Aswan, Egypt

1 Introduction

The flood is a water overflow that usually submerges drylands in wadi systems. After heavy rainfall, large water volumes are discharged in dry riverbeds. Further, the flood hazards refer to a phenomenon causing societal damage from the natural environment. Accordingly, flash flood hazards assessment is a key component of natural risk management. The natural hazards can be considered within a hydro-meteorological and geological frame, where floods, earthquakes, landslides, storms, volcanoes, droughts, and tsunamis are the major types (Alcantara-Ayala 2002; Youssef et al. 2009). Such natural hazards occur worldwide, but their environmental impact is greater in developing countries regarding geographic location and existence of different types of social, cultural, and economic vulnerabilities. Additionally, limited water resources and increasing population in such countries increase the freshwater demand for sustainable development.

It is worth mentioning that the sustainable development in the arid and semi-arid region depends mainly on groundwater exploration and utilizing the available water resources for favoring human demands. Specifically, the flash flood water can be considered as an important source for renewal water resources in desert lands to meet a part of water demand for sustainable development. There are many authors who investigated the groundwater recharge and

A. M. Negm (ed.), *Flash Floods in Egypt*, Advances in Science, Technology & Innovation,
https://doi.org/10.1007/978-3-030-29635-3_7

rainfall-runoff processes for wadi systems applying hydrogeological models (e.g. Gheith and Sultan 2002) and flash flood modeling approaches (e.g. El Bastawesy et al. 2009) using geographic information system (GIS), remote sensing (RS), and digital elevation models (DEM). Against the ongoing climate change, runoff water harvesting is considered as a supplementary action worldwide for (i) flash flood hazards management and (ii) water resources development.

RS has become an important tool for collecting multi-temporal, spatially extensive, and cost-effective data. Satellite RS offers a chance for more systematic analysis and better inspection of various geomorphologic units. Further, GIS is a powerful tool for analysis and integrating multi-thematic layers. GIS provides an excellent tool for integrating the information on the controlling parameters. Accordingly, RS and GIS become very important techniques for hydrogeological investigations (e.g. Tweed et al. 2007; webhelp.esri.com/arcgiSDEsktop/9.1/index.cfm?…An%20overview%).

In general, the electrical geophysical methods are not routinely integrated with RS datasets. Direct current (DC) resistivity method is one of the most popular electrical methods for groundwater exploration and geoenvironmental investigations (e.g. Al-Manmi and Rauf 2016; Attwa et al. 2009. In high conductive soils, penetration depth and resolution of DC resistivity method are less than that of resistive medium and resistivity imaging under conductive soils. They can be quite challenging if not impractically used (Attwa and Günther 2013; Reynolds 2011). Consequently, DC resistivity method, especially when combined with RS and GIS, can be capable of categorizing and characterizing areas prone to geohazards.

Recently, Egypt faces several flash flood events in the coastal areas and the Nile wadi system. The total economic damage was estimated at about 1.2 billion USD/year from 1975–2014. From the inspection of downstream features in Egypt, it can be noticed that the wadis discharge the floodwater into coast areas (Red sea Wadis, e.g., W. Ambagi and W. Safaga) and/or Nile plateaus (Eastern desert (ED) wadis, e.g., W. Atfih and W. Abadi). On the other hand, Nile water is currently insufficient especially after the establishment of large-scale land reclamation projects in desert fringes around the Nile valley. Consequently, the development and management of groundwater and floodwater have recently become extremely important. As a consequence, the mitigation of flash flood hazards and effective use of floodwater for groundwater recharge become vital and urgent.

With the aim to mitigate the flash flood hazards, this chapter attempts to present an integrative approach of GIS, RS, and DC resistivity data to manage the flash flood in wadi system. Furthermore, this approach aims to assess the water resources in arid and semi-arid areas making recommendations concerning sustainable development. Here, two case studies will be presented in two wadis in Egypt in which we used an amazing integrative approach for the investigationand management of flash floods.

2 Methods

This section presents an overview of the methods used to achieve the objective of this chapter based on the methodology shown in Fig. 1. The present integration strategy aims to combine geological, hydrogeological, and geophysical data in order to evaluate the hydrogeological conditions and to manage the flash flood hazards wadi systems. The suggested strategy depends on using the surface geological data as a key while interpreting geophysical data which assess many geological controls that enhance subsurface modeling and reduce the geological uncertainty. The integration strategy workflow is summarized in Fig. 1.

2.1 GIS and Remote Sensing

Recently, the water resources and flash flood hazards management is carried out using GIS and RS techniques. In the present integrative methodology, GIS, RS, and DC resistivity collected data were integrated to assess the hydrogeological conditions in arid and semi-arid areas (e.g., wadis). The GIS and RS are applied to construct the hydrological model according to four steps.

2.1.1 A Digital Elevation Model (DEM)

DEM is a digital file of terrain elevations or a terrain's surface 3D representation created from ground elevation data usually for a planet including Earth (Fig. 2). A DEM can be shown in the form of a raster (a squares grid) of a continuous surface where each cell value in the raster represents elevation. DEMs are constructed using RS techniques and land surveying. DEMs are used in GIS and are the basis for digitally relief maps production. Shuttle Radar Topography Mission (SRTM images) which is produced in 2003 with a spatial resolution of 90 m is used. To get accurate results from the surface analysis, the data must have been projected.

In general, the constructed DEMs are processed using ArcGIS software to fix a few errors. The grid involves little errors because of anomalies values which usually represent either sinks or peaks (maybe dams and wells or pits). A cell with indeterminate drainage direction, where no surrounding cells are lower, is known as a sink. On the other hand, a cell where no neighboring cells are higher is known as a peak. These errors are often as a result of data resolution, missing or insufficient data, or elevations rounding to the nearest integer value. Furthermore, sinks lead to stop the drainage

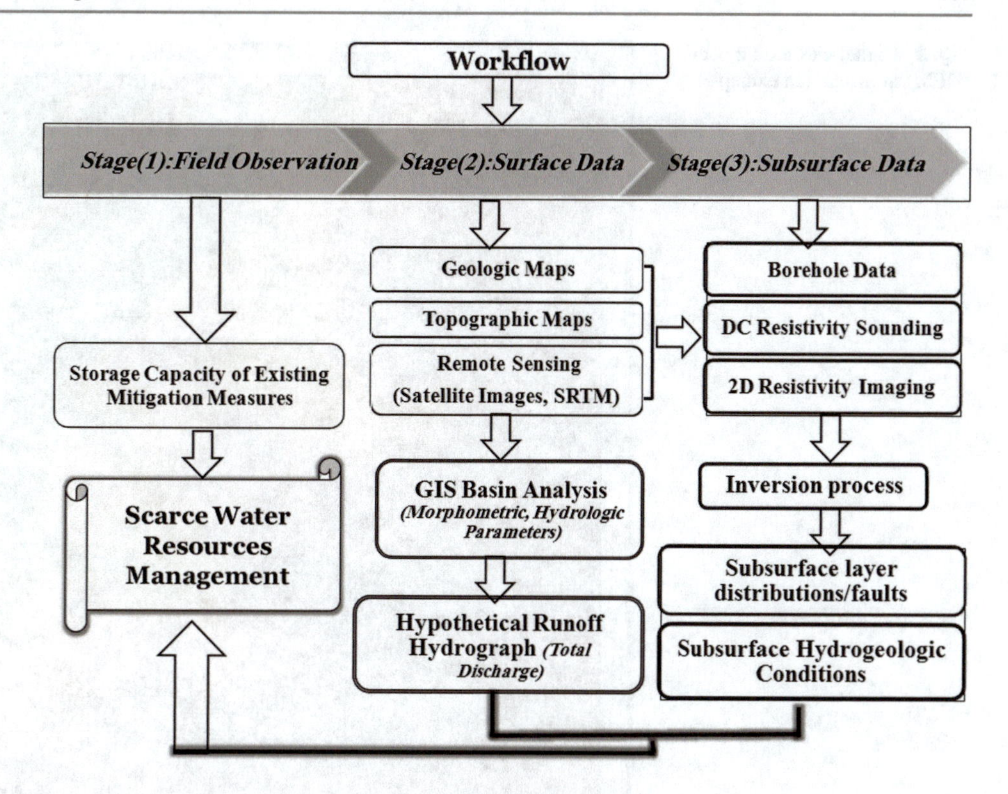

Fig. 1 Workflow of an integrated approach for hydrogeological assessment and flash flood management in wadi systems

network, separate a part of the basin (Attwa et al. 2014), and finally give undependable results while flow directions calculation. Therefore, filling sinks should be applied to guarantee accurate delineation of watersheds and streams. The tool can be applied to remove peaks, i.e., cells with an elevation greater than would be expected. Because sinks and peaks are filled, others can be formed at the boundaries of the filled areas, which are removed in the next iteration.

2.1.2 Flow Direction Determination

The flow direction is circumscribed by the direction of flow from each cell in the raster to its neighbor of the steepest downslope or its maximum drop. This can be calculated as follows:

$$\text{Maximum Drop} = \text{Change\_In\_Z} - \text{Value} / \text{Distance} * 100$$

The distance is delineated between cells'centers; therefore, if the two cells are orthogonal, the distance between the two of them is 90 m. If the two cells are diagonal, the distance between the two of them is about 127 m. This suggests that each extracted stream will be ≥ 90 m in case of using one cell as a threshold for delineation of the stream.

The approach of the D-8 flow model (Tarboton 2000) has been used to get the direction of flow out of each cell where eight output directions are valid; flow could pass through the eight bordering cells. When the steepest descent direction is found, the output pixel is coded with the value characterizing that direction (Fig. 3).

2.1.3 Calculating Flow Accumulation and Delineating Watersheds

The flow accumulation is calculated as the accumulated weight of the whole cells flowing into each downslope cell. The cell value of the output raster represents the cells' number flowing into each cell. Cells with 0 value of flow accumulation are considered restricted topographic highs and can be used for identifying ridges; the higher the flow accumulation of the cells, the more the concentration of the flow in these areas. Therefore, the main channels can be identified. In the following graphic, the image placed in the bottom left direction shows the travel direction from each cell and the image placed in the bottom right direction shows the cells'number of this flow into each cell.

Watershed delineation means creating a boundary that represents the contributing area for a particular control point or outlet. For example, the value of the outlet point of a watershed located in the central Eastern desert, west Red Sea coast, and Al-Quseir city has been determined, and the boundary of it is delineated (Fig. 4). The pixel value (flow accumulation) of this point is 16979 representing the number of cells contributing to the watershed.

2.1.4 Extracting Stream Networks, Orders, and Automatic Vectorization

In this step, stream networks have been detected by applying a value of threshold to the output of the flow accumulation process using these null tool. Then, the stream link tool has been applied to assign distinctive values to all the links in a

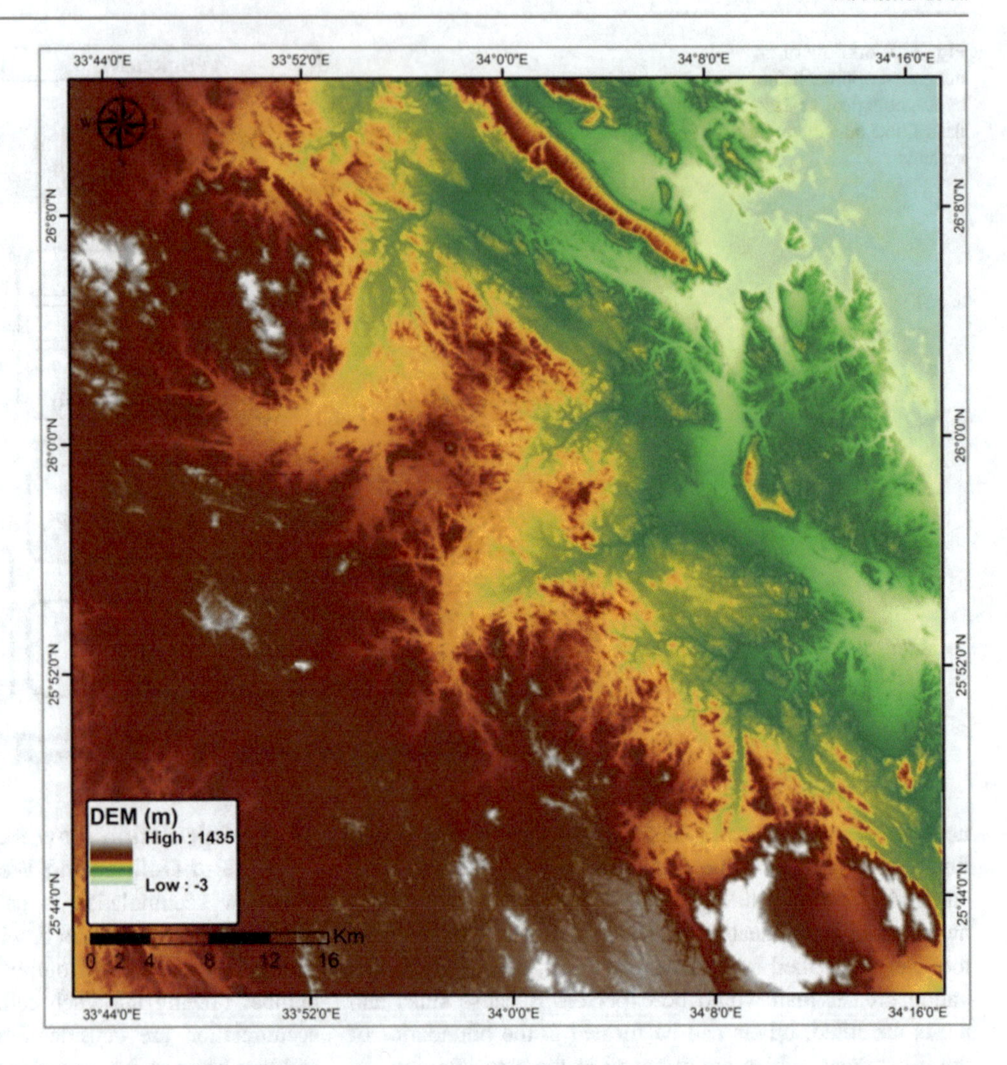

Fig. 2 Digital elevation model (DEM) in meters (an example)

linear network of a raster between intersections where links are the stream channel segments linking two consecutive junctions, the outlet and a junction, or the drainage divide and a junction. Stream ordering is a technique to derive main stream channel and its tributaries. The stream order tool has two methods you can use to assign orders. There are two methods to assign the stream order proposed by (Jenson and Domingue 1988; StrahlerAN 1957).

Flow length is known as the length of the longest overland flow path within a given basin to the upstream or the downstream direction. It can also be o defined as the weighted distance along the flow path for each cell. In GIS, the flow length can be determined by the summation of the progressive distances along the flow path from each pixel center-to-center from the designated pixel to the outlet one (https://www.slideshare.net/BUGINGOAnnanie/exercise-advanced-gisandhydrology). The tool can be used to create distance–area diagrams of hypothetical rainfall and runoff events using the weight raster as impedance to movement downslope. Once the flow direction for each cell is obtained, the grid of flow direction is reclassified to get the flow length through each cell. The moment that both the flow velocities and lengths are determined for each cell, the travel time of the flow in each cell is acquired by dividing the length by the flow velocity estimated for each cell using the Manning equation (El Bastawesy et al. 2009). Once the velocity and the length of the flow within each cell are estimated, the travel time for each cell in the catchment is possible to be estimated. The flow length function in ArcInfo can determine the accumulated downstream flowlength. In this time grid, each one of the cells possesses a value that represents the time needed (in seconds) for the surface runoff generating within or transferred to each cell to arrive at the basin outlet. Therefore, the watershed can be divided into several time–area zones (in hours) detached by isochrones. Transmission loss, eliminations because of storage or loss along the flow path, are neglected, and the surface runoff created from any time–area zone will arrive at the catchment outlet in a certain time (ShreveRL 1966). The runoff will pursue the same paths with the same velocity despite surplus rainfall depth.

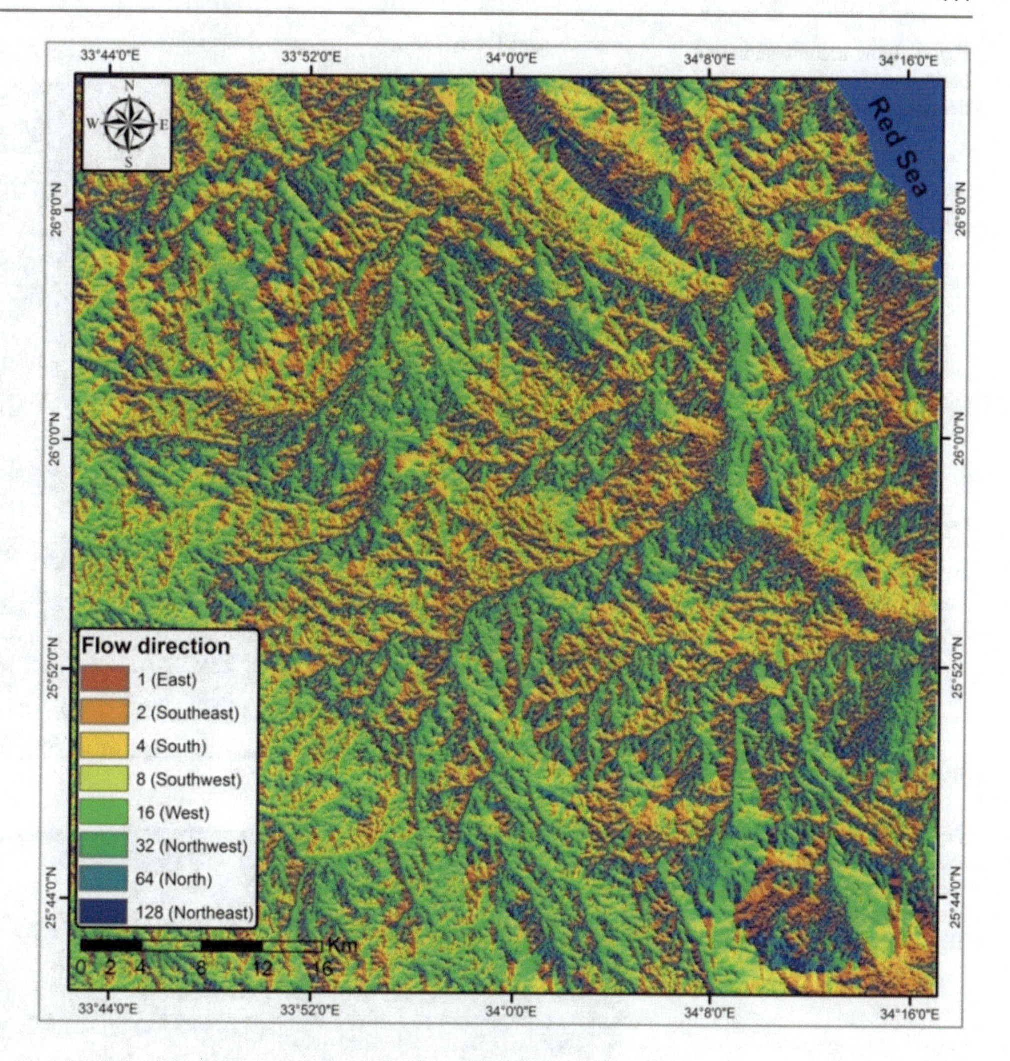

Fig. 3 Flow direction distribution (an example)

2.2 DC Resistivity Method

DC resistivity method is one of the oldest geophysical survey techniques used until today. It aims to determine the distribution of subsurface's resistivity based on the principle of applying electric current to the earth through two electrodes and measuring the potential difference between two or more other electrodes (Fig. 5). The true resistivity of the subsurface can be estimated from these measurements.

The electrical resistivity of a geological formation is a physical characteristic depending on the flow of electric current in the formation. Subsurface resistivity is related to several geological parameters and varies with these parameters such as the texture of the rock, nature of mineralization, and conductivity of electrolyte contained within the rock. Further, resistivity not only changes from formation to formation but even within a particular formation (Maidment 1993). Accordingly, there are a few parameters that influence the ground resistivity such as mineral and fluid content, porosity, and the degree of water saturation in the rock. These parameters make the resistivity method an ideal technique for geotechnical, archeological, environmental, and hydrogeological investigations (e.g., Sharma 1997; Attwa and Günther 2012; Attwa and El-Shinawi 2017).

DC resistivity measurements involve four active electrodes (stainless steel bar). These electrodes are planted into the ground. Two current electrodes (A and B) are used to inject an electrical current into the ground (earth), Fig. 5. We measure the ability of that material to resist the current (I) by recording the resulting voltage difference (ΔV) at two points on the ground surface measured by two potential electrodes (M and N), shown in Fig. 5. The potential difference is then given by

$$\rho_a = K\left(\frac{\Delta V}{I}\right)$$

where K is the geometrical factor having the dimension of length (m), which depends on the arrangement of the four electrodes.

If the ground is heterogeneous, the resistivity is called the apparent resistivity. The apparent resistivity (ρ_a) depends on

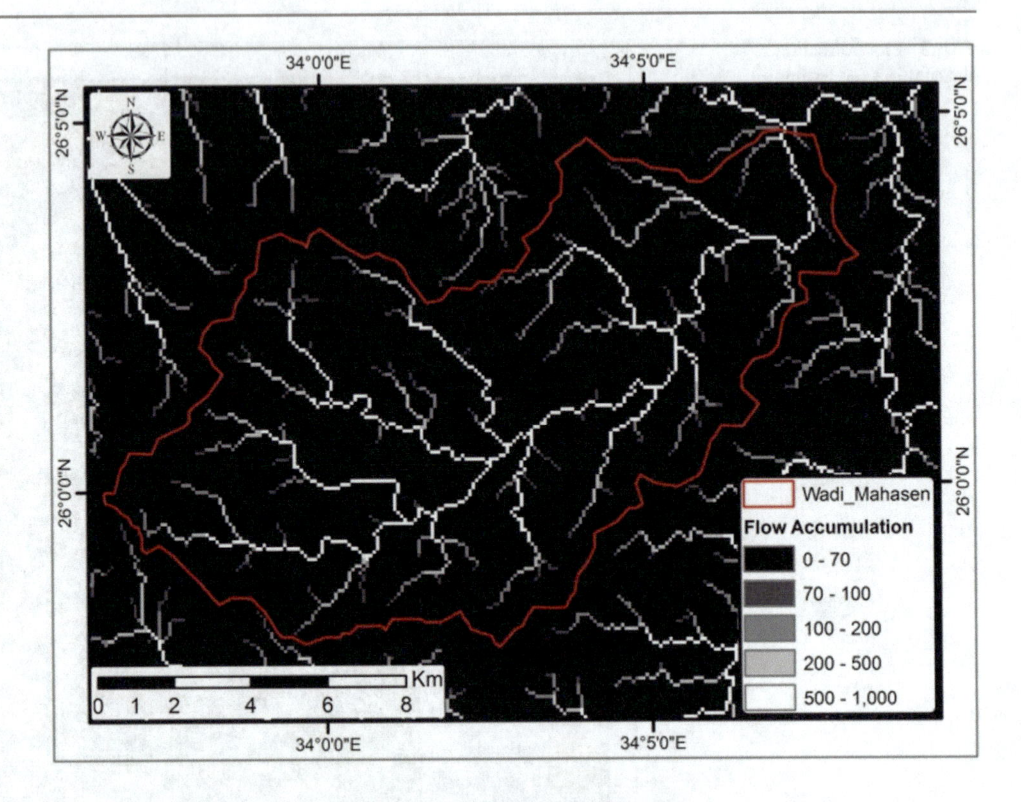

Fig. 4 Flow accumulation and basin boundary after watershed delineation (an example)

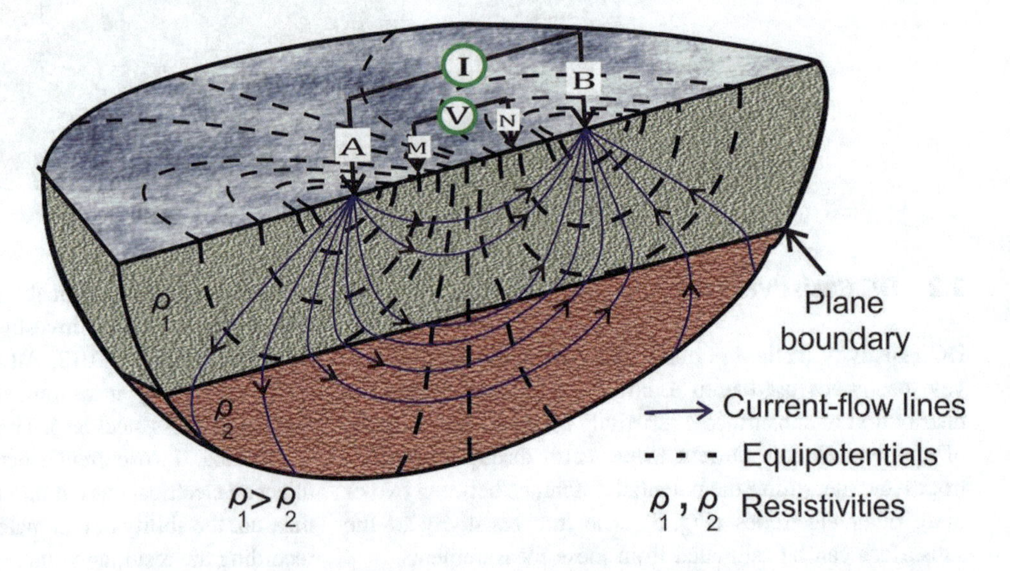

Fig. 5 Principle of resistivity measurements with a four-electrode configuration (modifies after, Attwa and Henaish 2018)

the element geometry and resistivity constituting the given geologic medium.

The ground resistivity variation scan causes divergences in the current flow and the measured voltage differences. As its origin, DC resistivity method is known as resistivity sounding. This is a one-dimensional (1D) survey where only vertical changes are taken into account. For lateral heterogeneity imaging, 2D-electrical resistivity tomography (ERT) survey can be carried out using different configurations of electrodes on the ground surfaces. There are frequent configurations/ arrangements for placing the potential and current electrodes. Some of the common electrode arrays for geologic studies are Schlumberger (SC), Wenner (MN), dipole–dipole (DD), and pole–dipole (PD) arrays (Fig. 6).

In desert lands, the condition of measurements is extremely difficult in most locations; this difficulty can be attributed to the ground resistance contact and inhomogeneities of the ground surface layer, which composed of dry loose sands mixed with gravel, large boulders, and rock fragments Fig. 7. To overcome the high resistance, digging a

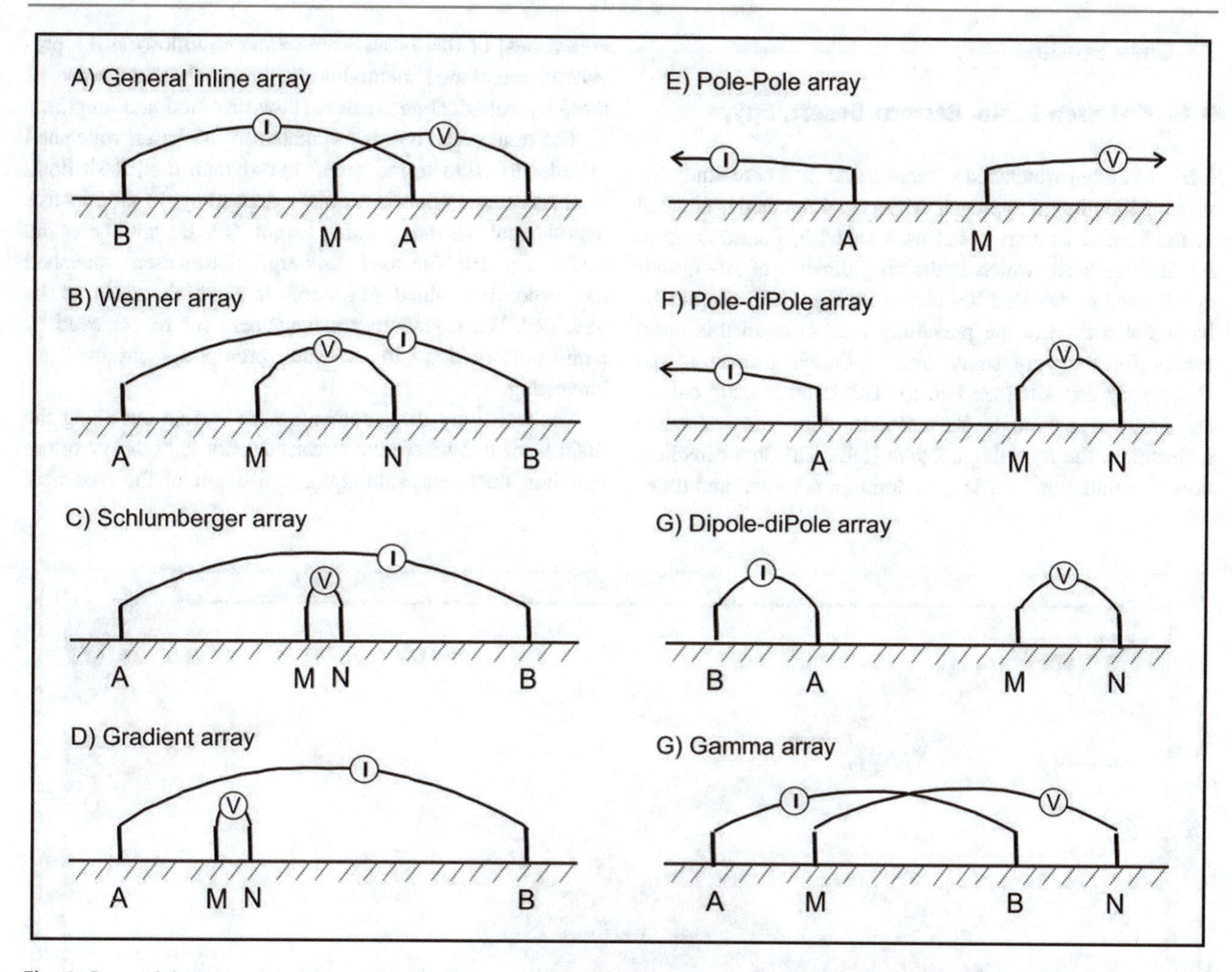

Fig. 6 Some of the most commonly used electrode configurations. The letters A and B denote the current electrodes and the letters M and N denote the potential electrodes

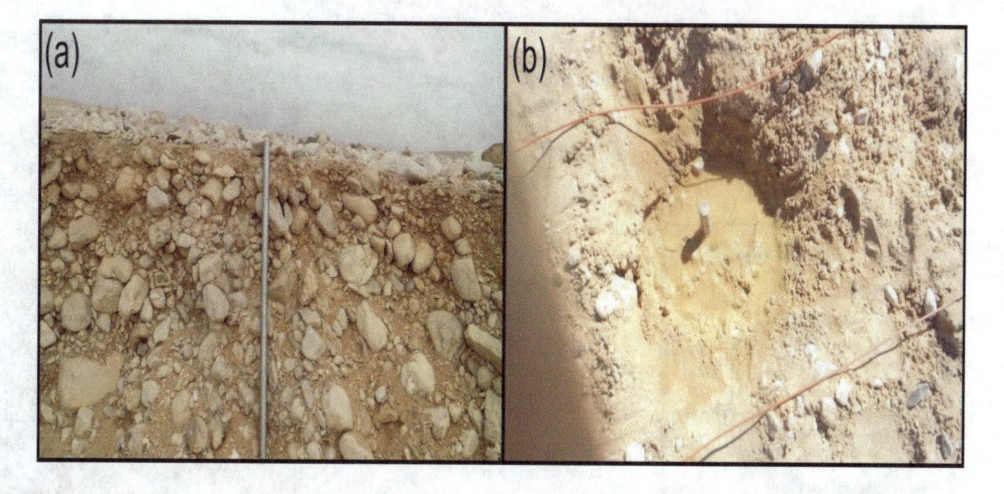

Fig. 7 The difficult conditions due to the resistance contact and inhomogeneity of the surface cover. **a** Surface cover composed of gravel, large boulders, and rock fragments. **b** Cone-shaped hole is filled by bentonite

hole for each electrode and adding water in the hole were performed or two electrodes can be set in parallel at each point to decrease this value. However, in most cases with difficult conditions, we dug a cone-shaped hole with a depth of 50 cm and fill it by drilling mud (bentonite) and then adding water after plugging electrode.

3 Case Studies

3.1 Mahasen Basin, Eastern Desert, Egypt

Here, Mahasen watershed is represented as a case study for proving the present approach. Mahasen watershed is situated in the central Eastern desert as a subbasin contributing to Al-Ambagi basin which is draining directly in Al- Quseir city located on the Red Sea coast (see Fig. 8). Geologically, basement rocks are the prevailing rock units in this basin except for the very small area of Quseir formation and Quaternary deposits (see Fig. 8). This characteristic enhances the composition of flash floods. After the watershed delineation, the hydrologic layers (DEM-fill-flow direction, flow accumulation flow length, drainage network, and time–area zones) of that basin were extracted following the previously mentioned methodology. Figure 9 shows some of these hydrological parameters (flow direction and length).

The time–area zone is calculated for Mahasen watershed in order to estimate the runoff hydrograph (i.e., flash flood discharge curve) that is acquired depending on an effective hypothetical 10 mm rainfall event (El Bastawesy et al. 2009), Fig. 10. The total discharge of Mahasen watershed has been determined (1427868 m^3) which needs to be managed. Water resources management can be achieved by dam construction in a site promoting groundwater harvesting.

The resistivity measurements were carried out along the main wadi bed where the stream density is high to explore alluvium thickness, saturation, and depth of the basement

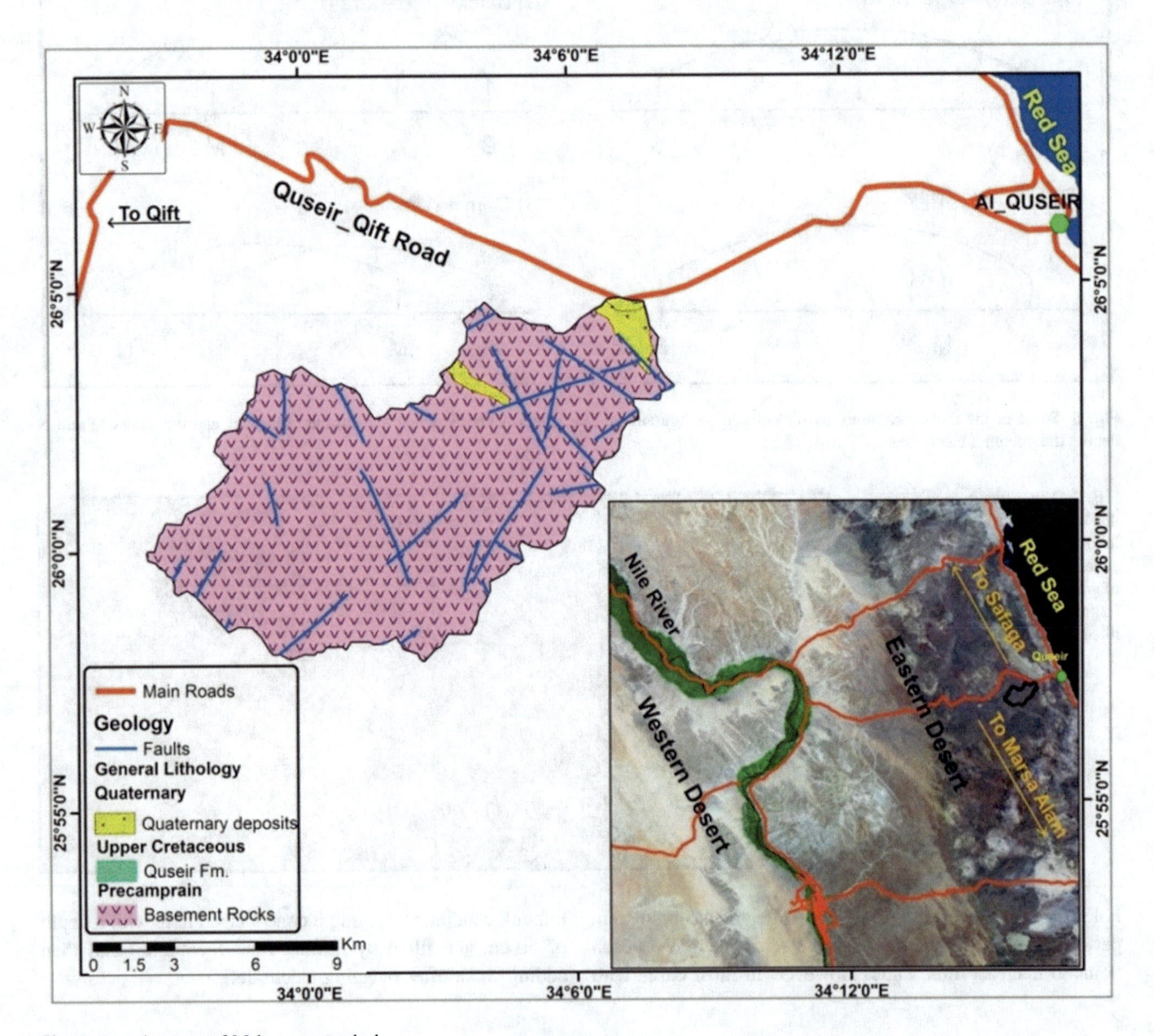

Fig. 8 Location map of Mahasen watershed

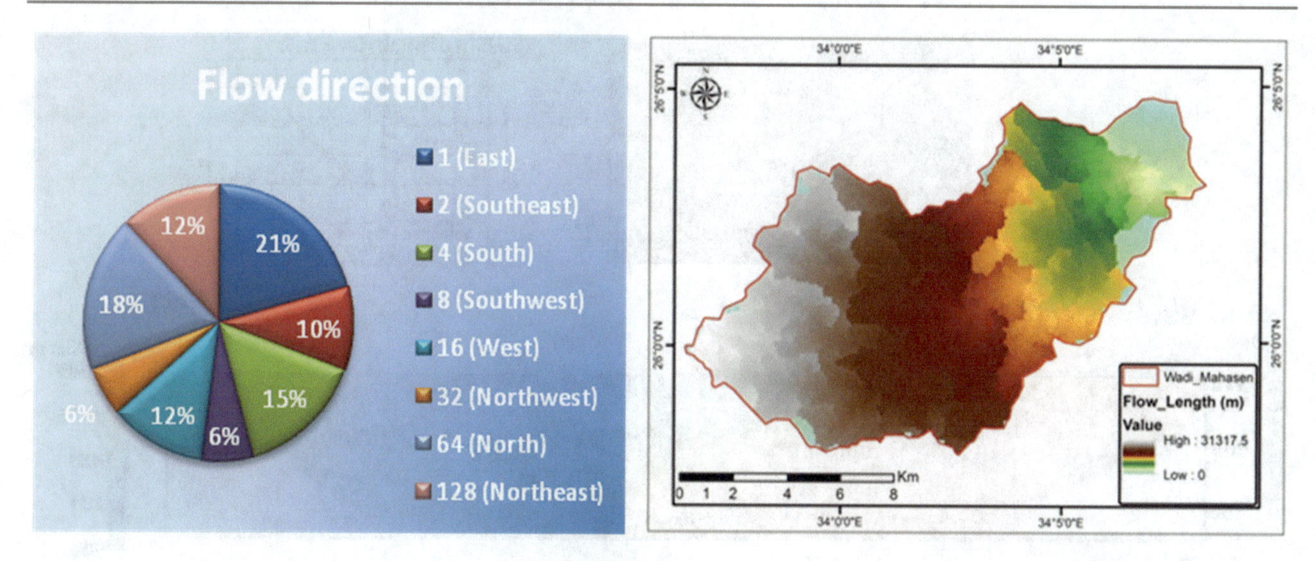

Fig. 9 (left) Flow direction and its distribution and (right) flow length in meters of Mahasen basin

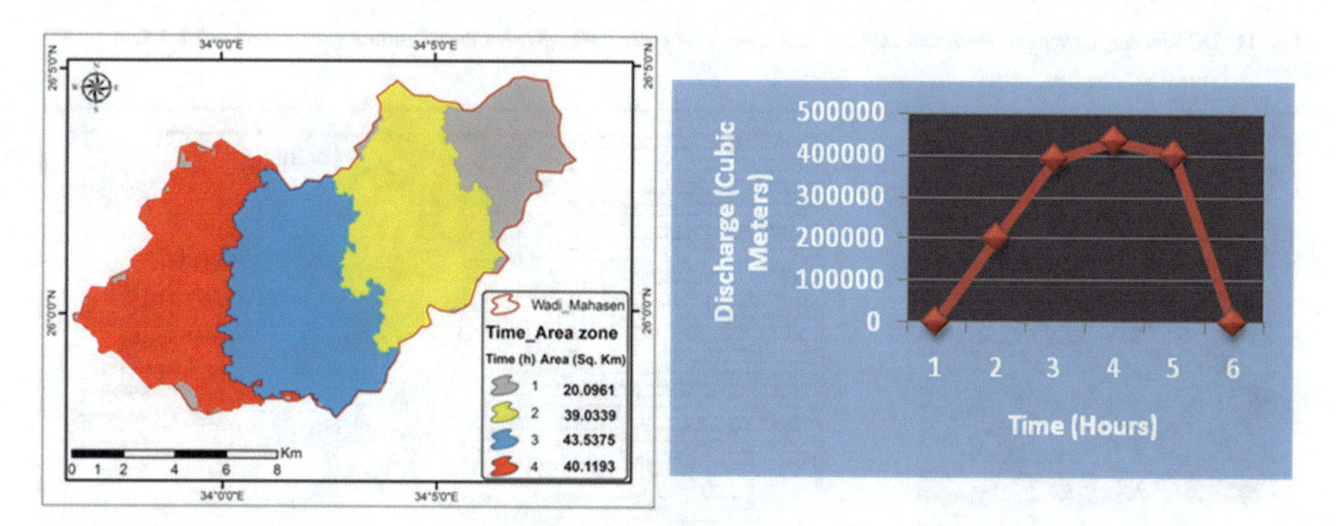

Fig. 10 (left) Time–area zones and (right) hypothetical runoff hydrograph of Mahasin basin

rocks. The inversion result of the 2D profile (i.e., 2D-ERT) showing the subsurface geology and inferred faults is shown in Fig. 11. After the 2D profile had been interpreted, it was found the same location of the profile is an optimum site for constructing a dam as a tool for scarce water management because of the chance given to the runoff water to be recharged and also the underground structure location managing the movement of the groundwater. Furthermore, the great thickness of sediments located between the two basement blocks is very useful for runoff recharge. The recommended dam should be constructed by special dimensions to control more than the entire runoff from the whole catchment. The proposed dam will be expected to exceed the needed total storage capacity as it will control about 2.6 million cubic meters.

3.2 Wadi Esna, Western Desert, Egypt

Wadi Esna watershed is represented as a second case study for proving the present approach. The investigated area is located on the west bank of the River Nile, west Esna city. It situated within the transitional zone between the western desert and the Nile valley; it is bounded from the west by Limestone Eocene plateau and from the east by the river Nile. The surface land of the area is characterized by a change of terrains and elevations. The average ground elevation varies from about 69 m above mean sea level (AMSL) in the eastern part to about 606 m AMSL in the western part.

In the present case study, morphometric analysis of the Esna basin (Fig. 12), based on several drainage parameters

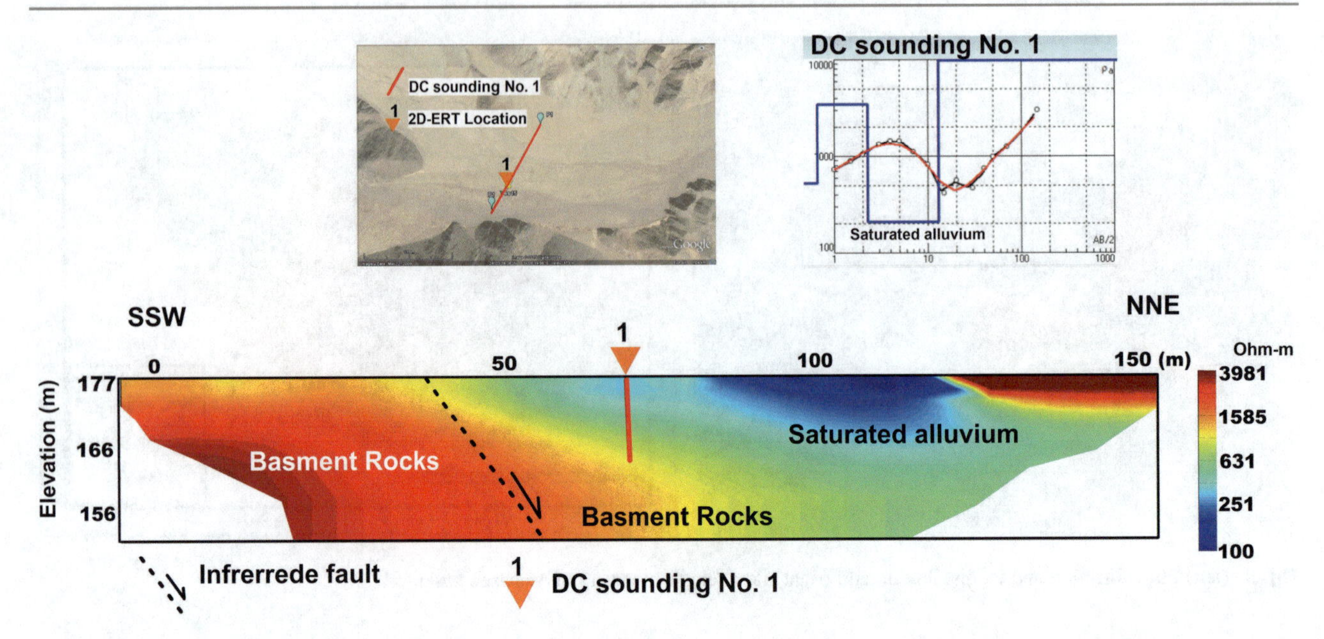

Fig. 11 DC soundings inversion results and 2D-ERT carried out at Mahasin basin showing the subsurface geology and inferred fault

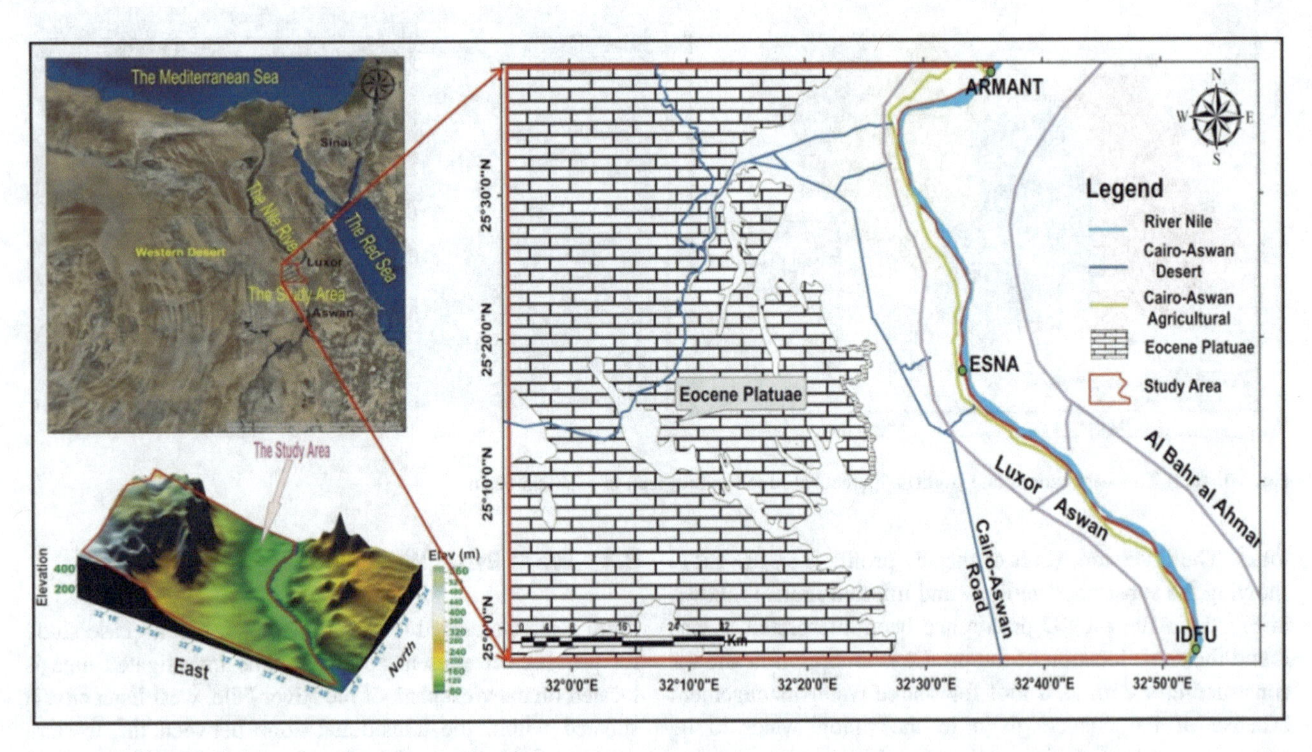

Fig. 12 Location map of Esna, Western Desert, Egypt

applying RS satellite data and latest GIS techniques for drainage analysis, has been represented. It is observed that the present wadi can be classified into six basins (Fig. 13). Detailed morphometric analysis of all basins indicates dendritic to sub-dendritic drainage forms, which indicate homogenous lithology and variations of R_b values among the basins related to the difference in elevation and geometric development. The maximum frequency of stream order is noticed in the case of first-order streams and then for the second order. Accordingly, it is observed that there are stream order increases as the stream frequency decreases and vice versa.

Runoff, especially flash flood is a real danger for the development in these areas, and is considered being the main reason to lose large quantities of freshwater besides the destruction of life stocks and infrastructure in the desert.

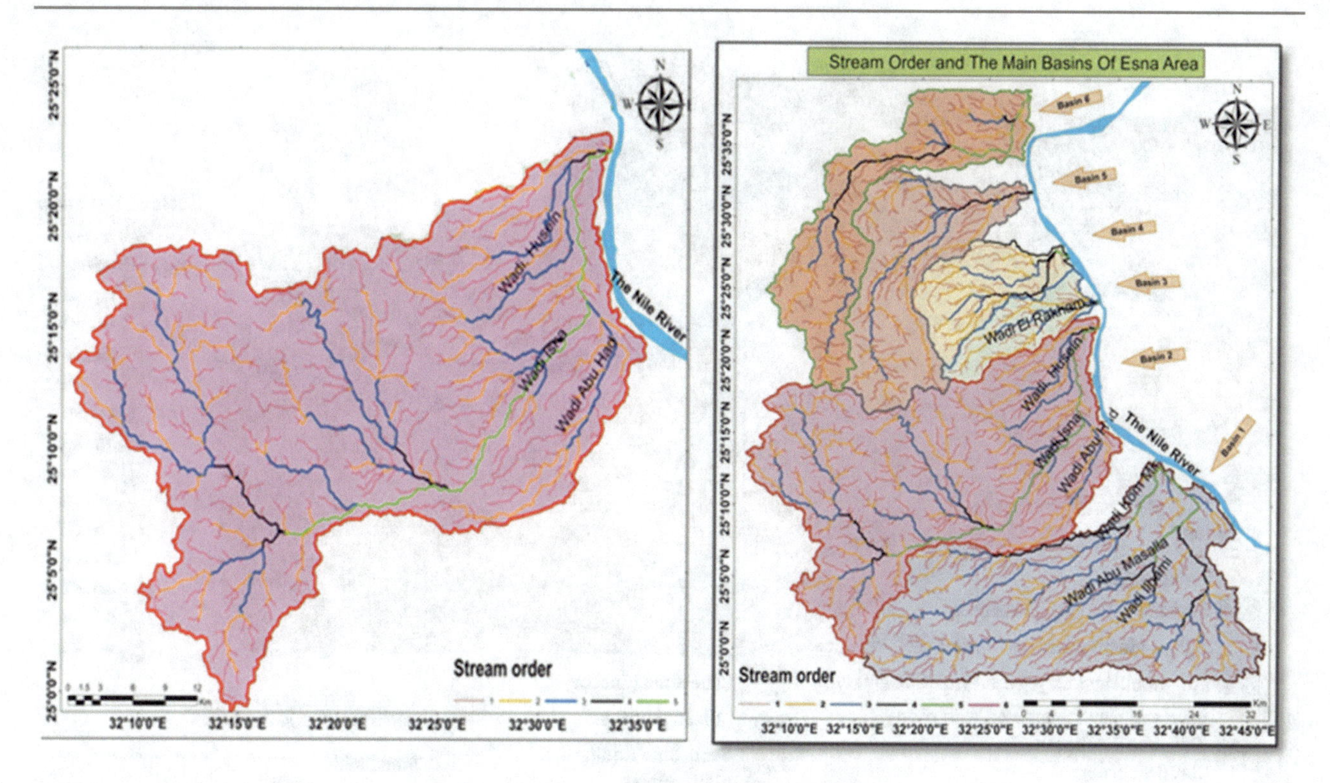

Fig. 13 Drainage network map of main basins at Esna basin, Western Desert, Egypt

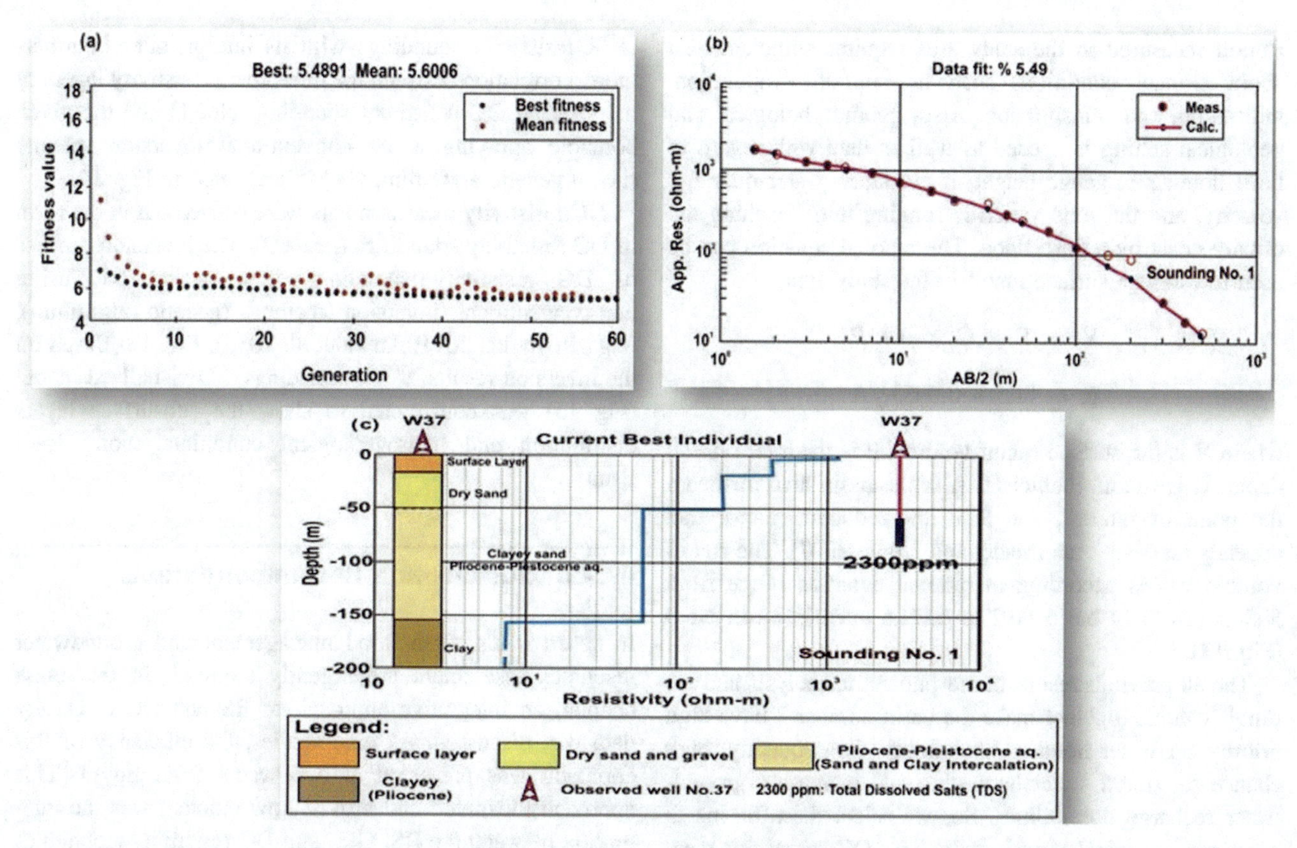

Fig. 14 The inversion results of DC resistivity soundings using a genetic algorithm (GA); **a** Variations of mean and best misfit versus generation, **b** a comparison between the measured and calculated data at the best misfit value, **c** a comparison between the model response and borehole information

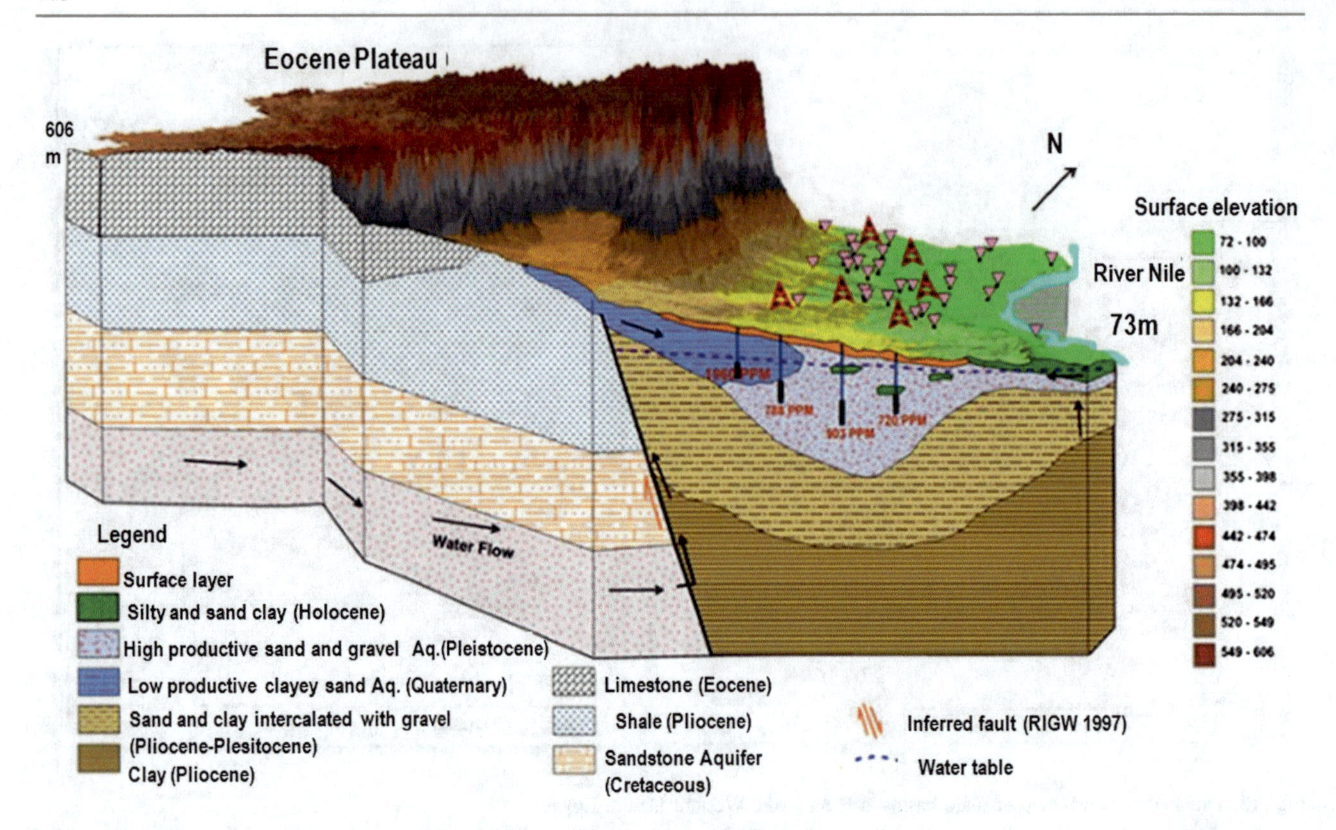

Fig. 15 3D visualized model showing the subsurface layers distribution and hydrogeological conditions at Esna basin, Western Desert, Egypt

Runoff measured in the study area requires sufficient data about climate conditions such as rainfall, evaporation, infiltration, and transpiration. Also, geomorphological and geological setting is needed as well as data with regard to flash flood, i.e., gauge height, if available, water quantity, velocity, and the time which is running until reaching the estuary coast by a flash flood. The rational equation can be used to estimate surface runoff in the study area:

$$V = R * C * A * P$$

$$R = 1.05 - 0.0053\sqrt{A}$$

where V is the surface runoff volume, P is the total rainfall depth, C is runoff coefficient, A is the basin area upstream the point of interest, and R is the reduction factor that depends on basin area (Seidel and Lange 2007). The runoff volume values according to rational equation range from 30549.4 m^3/h in basin no.2 to 6571.5 m^3/h in basin No.3 (Fig. 13).

The all previous results of morphometric analysis and the runoff volume together make the basin number 2 have high priority for water flooding potentiality. Consequently, high chance for runoff water harvesting and increasing groundwater recharge possibilities. So, we select these basins to carry out the geoelectrical study (i.e., DC resistivity measurements). Figure 14 shows an example of the measured DCR resistivity soundings with its interpretation. Furthermore, correlation between the interpreted resistivity layers of the selected DC resistivity sounding (No. 1) and the given borehole applying a non-conventional inversion scheme (i.e., a genetic algorithm, GA) is presented in Fig. 14.

DC resistivity measurements were carried out in the form of DC resistivity soundings (i.e., 1D). The inversion process of DC resistivity soundings was carried out using non-conventional inversion method (genetic algorithm), (e.g., Reynolds 2011; Attwa et al. 2016), Fig. 13. Based on the inversion results of DC soundings, 3D visualized model (Fig. 15) was constructed to show the subsurface layers distribution and hydrogeological conditions along wadi Esna.

4 Conclusions and Recommendations

In desert lands, flash flood management and groundwater resources assessment are urgently required. In this book chapter, an integrative approach of RS and DC resistivity data was discussed. As case studies, the efficiency of this approach was presented at two wadis from Egypt. This approach illustrated and proved how efficient was the integration between the RS, GIS, and DC resistivity techniques for scarce water resources management in wadi systems. The proposed approach was useful not only for the mitigation of flash flood hazards but also for replenishing aquifers in an attempt to urban development even in small scales. RS and GIS techniques can be applied to delineate the basin of a high chance for runoff water harvesting and groundwater recharge possibilities. On the other hand, the DC resistivity method can be used to delineate/image the subsurface layer distributions and aquifers extension. Accordingly, the successful applications of the present integrative approach open the way for sustainable development in wadi systems by decision-makers to (i) overcome water scarcity problem and (ii) reduce the flash flood hazards.

References

Alcantara-Ayala, I. (2002). Geomorphology, natural hazards, vulnerability and prevention of natural disasters in developing countries. *Geomorphology*, *47*(2–4), 107–124.

Al-Manmi, D. A. M., & Rauf, L. F. (2016). Groundwater potential mapping using remote sensing and GIS-based, in Halabja City, Kurdistan, Iraq. *Arabian Journal of Geosciences, 9*(5), 357.

Attwa, M., & El-Shinawi, A. (2017). An integrative approach for preliminary environmental engineering investigations amidst reclaiming desert-land: a case study at East Nile Delta, Egypt. *Environmental Earth Sciences, 76*, 304. https://doi.org/10.1007/s12665-017-6627-4.

Attwa, M., & Günther, T. (2012). Application of spectral induced polarization (SIP) imaging for characterizing the near-surface geology: an environmental case study at Schillerslage, Germany. *Australian Journal of Applied Sciences, 6,* 693–701.

Attwa, M., & Günther, T. (2013). Spectral induced polarization measurements for environmental purposes and predicting the hydraulic conductivity in sandy aquifers. *Hydrology and Earth System Sciences (HESS), 17,* 4079–4094.

Attwa, M., & Henaish, A. (2018). Regional structural mapping using a combined geological and geophysical approach—A preliminary study at Cairo-Suez district, Egypt. *Journal of African Earth Sciences, 144,* 104–121.

Attwa, M., Günther, T., Grinat, M., & Binot, F. (2009). Transmissivity estimation from sounding data of holocene tidal deposits in the North Eastern Part of Cuxhaven, Germany. In *Near Surface 2009-15th European Meeting of Environmental and Engineering Geophysics*, Extended Abstract. https://doi.org/10.3997/2214-4609.20147099.

Attwa, M., Akca, I., Basokur, A. T., & Günther, T. (2014). Structure-based geoelectrical models derived from genetic algorithms: a case study for hydrogeological investigations along Elbe River coastal area, Germany. *Journal of Applied Geophysics, 103,* 57–70.

Attwa, M., Gemail, K. S., Eleraki, M. (2016). Use of salinity and resistivity measurements to study the coastal aquifer salinization in a semi-arid region: A case study in northeast Nile Delta, Egypt. *Environmental Earth Sciences, 75*(9) https://doi.org/10.1007/s12665-016-5585-6.

El Bastawesy, M., White, K., & Nasr, A. (2009). Integration of remote sensing and GIS for modelling flash floods in Wadi Hudain catchment, Egypt. *Hydrological Processes: An International Journal, 23*(9), 1359–1368.

Gheith, H., & Sultan, M. (2002). Construction of a hydrologic model for estimating Wadi runoff and groundwater recharge in the Eastern Desert Egypt. *Journal of Hydrology, 263*(1–4), 36–55.

Jenson, S. K., & Domingue, J. O. (1988). Extracting topographic structure from digital elevation data for geographic information system analysis. *Photogrammetric Engineering and Remote Sensing, 54*(11), 1593–1600.

Maidment, D. R. (1993) *Developing a spatially distributed unit hydrograph by using GIS* (pp. 181–192), IAHS publication.

Reynolds, J. M. (2011). *An introduction to applied and environmental geophysics*. Wiley.

Seidel, K., Lange, G. (2007). Direct current resistivity methods. In Knödel, K., Lange, G., Voigt, H. J. (eds.), *Environmental geology* (pp. 205–238), Chap. 4. Springer BH.

Sharma, P. V. (1997). *Environmental and engineering geophysics*. Cambridge: Cambridge University Press.

Shreve, R. L. (1966). Statistical law of stream numbers. *The Journal of Geology*, *74*(1), 17–37.

Strahler, A. N. (1957). Quantitative analysis of watershed geomorphology. Eos, *Transactions American Geophysical Union*, *38*(6), 913–920.

Tarboton, D. G. (2000) TARDEM, a suite of programs for the analysis of digital elevation data. http://www.engineering.usu.edu/cee/faculty/dtarb/tardem.html#Acknowledgements.

Tweed, S. O., Leblanc, M., Webb, J. A., & Lubczynski, M. W. (2007). Remote sensing and GIS for mapping groundwater recharge and discharge areas in salinity prone catchments, southeastern Australia. *Hydrogeology Journal, 15*(1), 75–96.

webhelp.esri.com/arcgiSDEsktop/9.1/index.cfm?…An%20overview%.

https://www.slideshare.net/BUGINGOAnnanie/exercise-advanced-gisandhydrology.

Youssef, A. M., Pradhan, B., Gaber, A. F. D. Z., & Buchroithner, M. F. (2009). Geomorphological hazard analysis along the Egyptian Red Sea coast between Safaga and Quseir. *Natural Hazards & Earth System Sciences*, *9*, 751–766.

Environmental Monitoring and Evaluation of Flash Floods Using Remote-Sensing and GIS Techniques

Noha Donia

Abstract

Egypt receives very little rainfall apart from on its coastal areas where average annual rainfall is 50–250 mm on its west coast and 50–150 mm on its east coast—rainfall increases from west to east. As a result of such low levels of rainfall Egypt might be considered safe from flooding. However, such an assumption would be dangerous because Egypt is exposed to high rainfall rates every 100 years. Therefore, studies must be carried out on all Egypt's basins to study their characteristics and calculate probable flood volumes for such events, thereby allowing the provision of flood protection. As traditional methods of flood response are put under pressure by dynamic changes experienced in Egypt, this chapter presents a model for the application of advanced technologies such as remoyte-sensimg and GIS techniques capable of identifying and monitoring flash floods and watershed delineation and flood estimation in different areas in Egypt.

Keywords

Flash floods • GIS • Remote sensing • Egypt • Model • Rainfall

1 Introduction

A flood is generally defined as the location of a flow of water; geologically it is the flow of water over the ground. Floods are governed by a number of factors, the most important being rainfall, evaporation, and leakage. Meteorologically, a flood is defined as the result of heavy rainfall —associated with the occurrence of sudden atmospheric instability over mountainous areas during the cold, winter season. The three major flood types are flash floods, rapid onset floods, and slow onset floods.

Flash floods result from heavy rain, dam breaks, or snowmelt. They occur over very short timeframes (2–6 h, sometimes in a matter of minutes). They can also be caused by intense rainfall from slow-moving thunderstorms. Flash floods are destructive and can be fatal—commonly taking people by surprise. Often they arrive with no warning, allowing no preparation, with impacts both swift and devastating.

Rapid Onset Floods take longer to develop than a flash flood and tends to last only for a day or two. Such floods can be managed by people but are very destructive.

Water bodies overflowing their banks cause slow onset floods. They are slow to develop but can last for days or weeks. They usually spread over many kilometers and tend to occur in flood plains (fields prone to flooding in low-lying areas). They bring with them disease and malnutrition as well as an increase in the number of snakebites—rising water levels drive animals into closer proximity to people.

Floods arguably represent the most devastating, widespread, and frequently occurring natural hazards globally. Flash floods are formed from excess rain falling on upstream watersheds, rapidly flowing downstream with great speed and force. Often, they are sudden and appear without warning. Therefore, such floods have considerable consequences, with damage becoming especially pronounced when water flows through human settlements and concentrations of infrastructure.

Remote sensing and geographic information systems (GIS) provide a wide range of tools for determining areas affected by flash floods and delineating watersheds. A GIS can be used to assemble information from different maps, aerial photographs, satellite images, and digital elevation

N. Donia (✉)
Department of Environmental Engineering, Institute of Environmental Studies and Research, Ain Shams University, Cairo, Egypt
e-mail: Noha.samir@iesr.asu.edu.eg

A. M. Negm (ed.), *Flash Floods in Egypt*, Advances in Science, Technology & Innovation,
https://doi.org/10.1007/978-3-030-29635-3_8

models (DEMs). Remote-sensing technology along with GIS has become the key tool for load hazard and risk maps for vulnerable areas. El Bastawesy et al. (2008, 2009, 2012) and Yousif et al. (2013) applied GIS and remote-sensing techniques for watershed delineation.

This chapter attempts to synthesize a database in a spatial framework to create a flood risk map using remote-sensing and GIS techniques. This study also focuses on the identification of factors affecting flash flood risk.

2 Methodology

A watershed is an area of land that contributes runoff water to a common point, it is a natural physiographic or ecological unit composed of interrelated parts and functions (Al-Jabari et al. 2009), determined by the topography of the land around it. Identifying a watershed is necessary to calculate runoff depth and volume; drainage networks are necessary for water-harvesting activities. In this study, drainage networks and watersheds are identified using a DEM and ArcGIS 10.1 software from Esri—an American company specializing in GIS. This represents the latest computer technology and can be used for processing and spatial analysis in 3D. It permits the processing and interpretation of enormous spatial and descriptive data and provides outputs in the form of maps and reports. The program can also be used to extract basin watersheds, drainage networks, stream order, flow, and direction, and other basin characteristics using ArcHydro, part of Arc Toolbox. The software is comprised of a collection of tools for basin analysis and hydrological applications (O'Callaghan and Mark 1984).

2.1 Watershed and Drainage Network Delineation Using a Geographic Information System

2.1.1 Digital Elevation Model

A DEM represents a 3D view of the Earth's terrain and has significant applications in science and other fields. DEMs depend on aerial images, global signature systems, and digital signage. They use either a geographic grid with spacing along parallels and meridians—in which case data spacing changes because of the curvature of the Earth—or they use a projected grid like Universal Transverse Mercator (UTM)—in which case the merging of datasets eventually becomes a problem because of the Earth's curvature. Small-scale DEMs generally use geographic grids, whereas large-scale DEMs can use either projected or geographic grids. The DEM used in this study was based on data from the Shuttle Radar Topography Mission (SRTM)—a space survey of Earth by NASA, the National Geospatial-Intelligence Agency (NGA), DLR (Germany's national aeronautics and space research center), and ISA providing radar data (wavelength of 56 mm) with 90-m spatial resolution (obtained from the United States Geological Survey website, http://www.usgs.gov.)

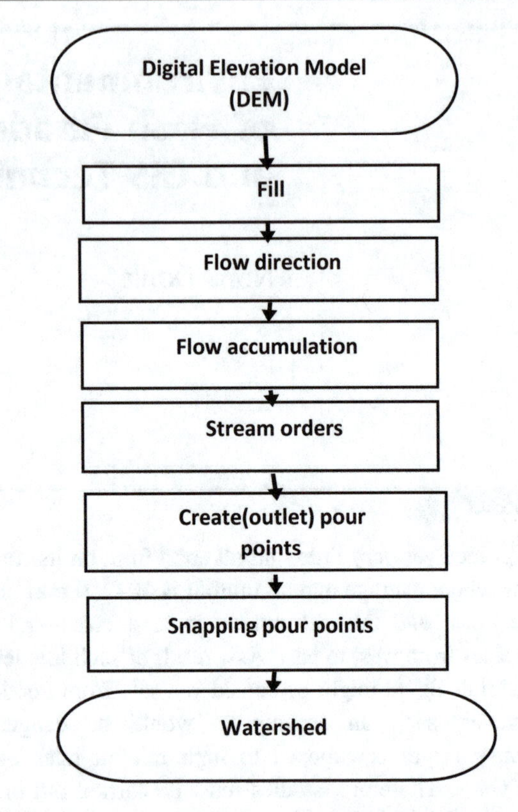

Fig. 1 Hydrological processing using ArcGis 10.1 software

The drainage network and watershed were delineated from DEM inputs (Jones 2002). The hydrological processing steps are shown in Fig. 1 and were completed using ArcGis 10.1 software in Arc Hydro in the main interface of Arc Toolbox (ESRI 2010).

STEP 1 Fill operation

This operation was used to remove all possible and abnormal values—either sink or peak errors—due to data with no resolution or a rounding of elevations to the nearest integer, to ensure proper watershed and stream delineation.

STEP 2 Flow direction

The flow direction for each well was determined using the ARCGIS grid processor. Additionally, the D-8 drainage model was used to derive flow directions between the central cell and its eight adjacent neighbors.

STEP 3 Flow accumulation

Cells with the greatest accumulated flow were determined and a network of high-flow cells created. Flow accumulation can be used to determine how much rain has fallen within a given watershed—flow accumulation output, representing the amount of rain flowing through each cell, was used to determine the drainage pattern (defined in terms of cumulative contribution counts for drained water into outlets).

STEP 4 Stream orders

The stream order of a watershed reflects the ability of a stream to erode and deposit sediment and is therefore linked to soil erosion and flooding. In this study, Strahler's method was used to identify and classify stream type based on tributary numbers.

STEP 5 Create outlet (pour) points

Pour point placement is an important step in the process of watershed delineation. Pour points should be positioned within an area of high flow accumulation to calculate total contributing water flow.

STEP 6 Snapping pour points

The Snap Pour Point tool snaps pour points created in Step 5 to the closest cell of high flow accumulation to account for any error during placement.

STEP 7 Watershed delineation

The watershed was delineated using a flow-direction raster.

2.1.2 The Main Morphometric Parameters of the Watershed and Drainage Networks

The area and perimeter of the watershed, the order of its streams, and the sum of stream lengths within drainage networks were calculated using ArcGIS 10.1 software.

Drainage density is an important factor related to geomorphology and hydrology. It reflects the method of surface water flow according to geology, gradient, plant cover, and the quantity and intensity of precipitation. Drainage density was calculated using the following equation:

$$LDD = SL/A$$

where *SL* is the sum of stream lengths; *A* is the area of the watershed (Rashash et al. 2015); and *LDD* the longitudinal drainage density.

2.1.3 Sub-watersheds

The main watershed was divided into sub-watersheds depending on land use/cover and the positions of cultivated areas between dykes through wadis to determine the effect of geomorphology on surface runoff, study the effect of spillway height on water distribution through wadis, and calculate the lost runoff volume to the sea.

2.2 Estimation of Surface Runoff

Water is one of the most important resources in arid and semi-arid regions and is essential to their development. Runoff estimation is especially helpful in areas lacking measuring stations (Al-Gamdi 1991). Surface runoff is the flow of precipitated water in a catchment area through a channel after all surface and sub-surface losses (Bansode and Patil 2014). In this study, an estimate of surface runoff was based on watershed characteristics, the natural components associated with land use/land cover, hydrological soil group (HSG), monthly rainfall, and antecedent moisture conditions using the Soil Conservation Service (SCS) curve number method and GIS. The runoff estimation methodology is shown in Fig. 2.

The SCS of the US Department of Agriculture developed the curve number (SCS-CN) method, also known as the Hydrologic Soil Cover Complex Method. This procedure has been used widely for runoff estimation. Rainfall amount and curve number are needed for this method and depend on the HSG of an area, land use, and hydrological conditions. Normally the SCS model computes direct runoff with the help of the following relationship:

$$S = (25400/CN) - 254$$

$$I_a = 0.2S$$

$$CN = \Sigma(CN_i timesA_i)/A$$

$$Q = (P - I_a)^2/(P - I_a) + S$$

where:

- CN is the runoff curve number of a hydrological soil cover complex;
- S is the potential maximum retention of water by the soil (mm);
- I_a is the initial abstraction which represents all losses before runoff begins (mm);
- P is the total storm rainfall (mm);
- Q is the actual direct runoff (mm);
- CN_i is the curve number from i = 1 to the number of areas;

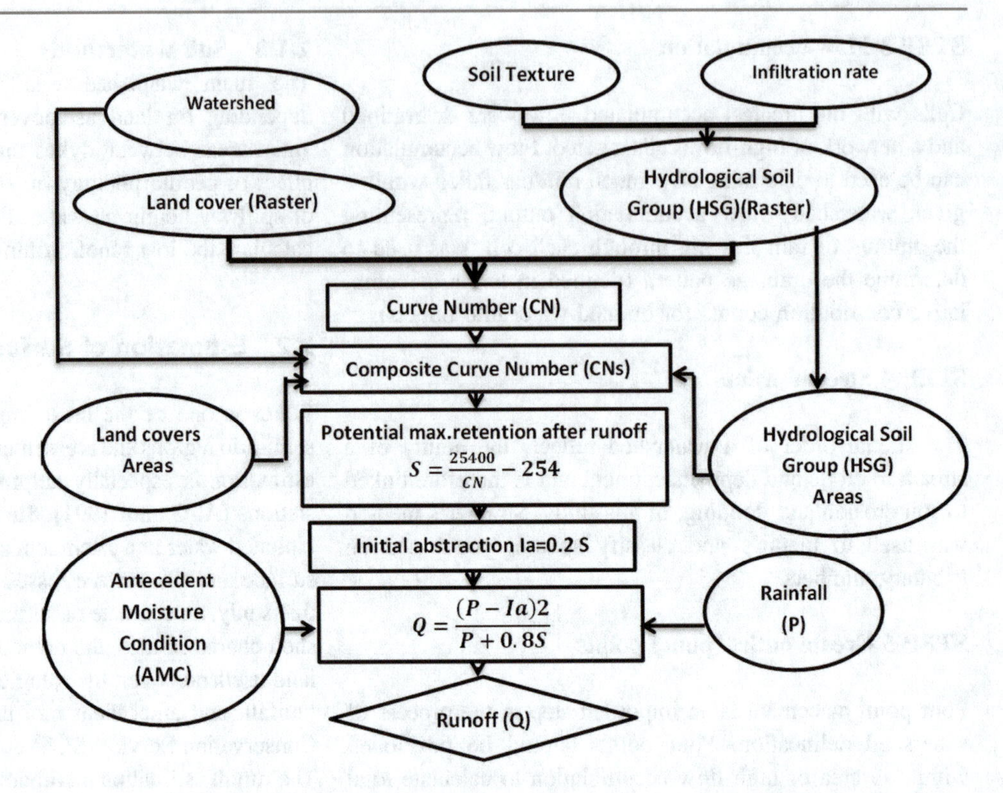

Fig. 2 Runoff estimation methodology

A_i is the area with curve number CNi (m^2); and
A is the total area of the watershed (m^2)

3 Study Areas

3.1 Wadi Firan

Wadi Firan is about 90 km east of Abou Redis, in the Sinai governorate. It includes 47 principal settlements and 22 secondary settlements and has a population of 7000. It contains many historical sites (Dir el Banat) and includes two churches (Nabi Mouus and Anba Dinamous). Monagat Mountain is 1 km west of Wadi Firan village. In addition, the region of Briga and the historical site of Nabtia are about 3 km from the village. Tahouna Mountain (the location of Alia Nabi Church and Actafious Church) is also nearby, as is Serbal Mountain—in the Agla Seil region—and the historical site of Nawawis. The area of Firan is illustrated in Fig. 3 and its geological map is provided in Fig. 4. The watershed model for the area, created using GIS, is shown in Fig. 5. Figures 6, 7, 8, and 9 show the results of a GIS analysis of the watershed.

3.2 Wadi Hashim

The area of Wadi Hashim is located about 50 km east of the city of Marsa Matrouh, bounded by latitudes 31°8′0″N and 31°9′0″N and longitudes 27°37′30″E and 27°38′30″E (Fig. 10). The drainage networks and watershed were extracted using ArcGIS 10.1 software (Fig. 10). The watershed was linked with an attribute table and its area calculated to be 3.305 km^2.

3.2.1 Land Use

The locations of different land use classes were determined during a field survey using GPS devices. Their boundaries were digitized using ArcGIS 10.1, attribute tables were assigned, and areas calculated (Table 1 and in Fig. 11).

The watershed was found to comprise three classes of land use with bare soil covering the greatest area (1.834 km^2, representing 55.5%) and cultivated land covering the least area (0.255 km^2, representing 7.7%).

3.2.2 Soil Texture

The particle size distribution of eight samples of soil from four sites, incorporating all land use types, was determined according to the United States Department of Agriculture (USDA) classification and soil textures were determined with the USDA soil texture triangle. The results are given in Table 2 and show that the soil texture of all samples is sandy loam.

3.2.3 Soil Infiltration Rate

Infiltration rates for soils, at the same sampling sites used to ascertain soil texture, were determined using a double-ring

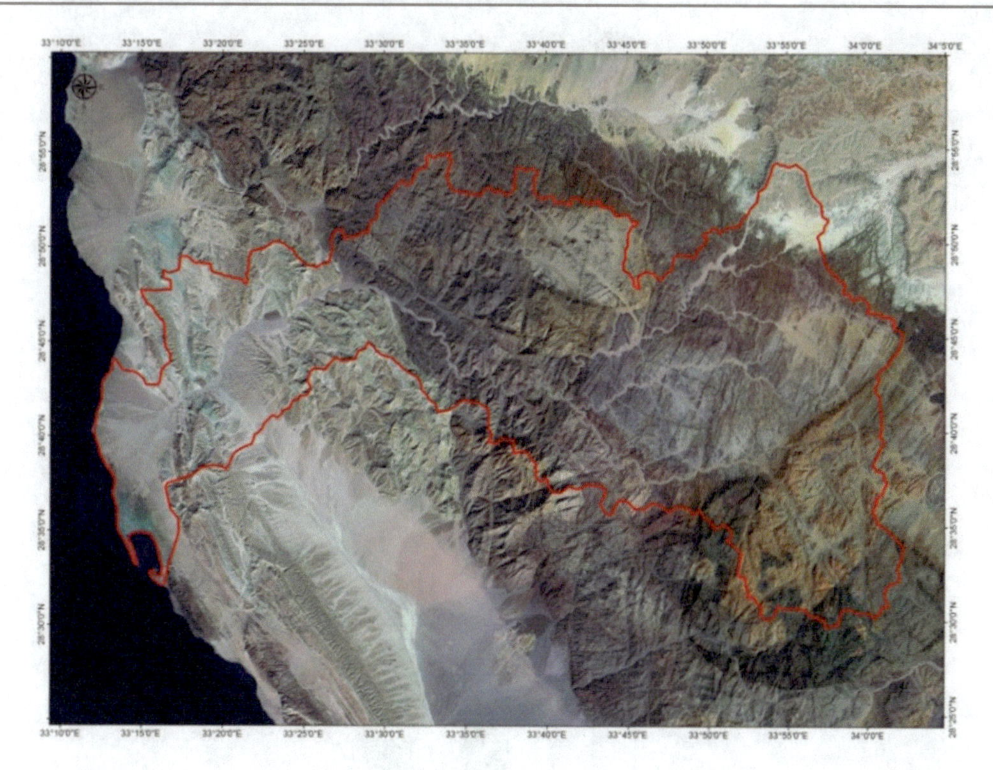

Fig. 3 Study area

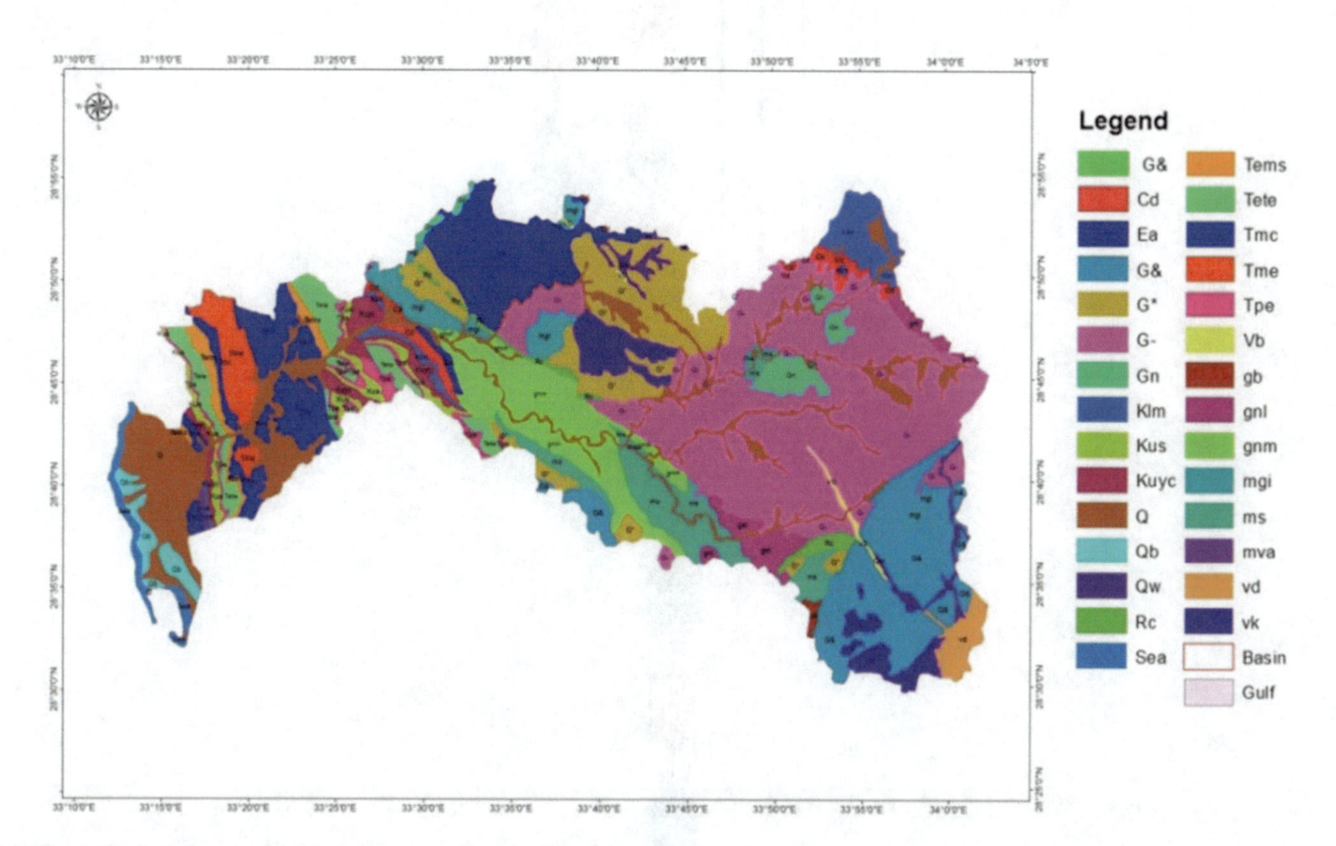

Fig. 4 Geological map of the study area

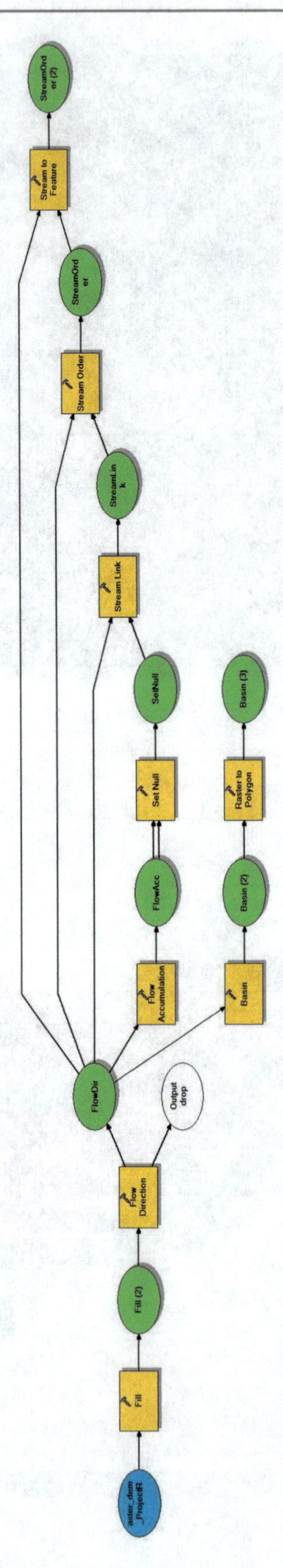

Fig. 5 Watershed model

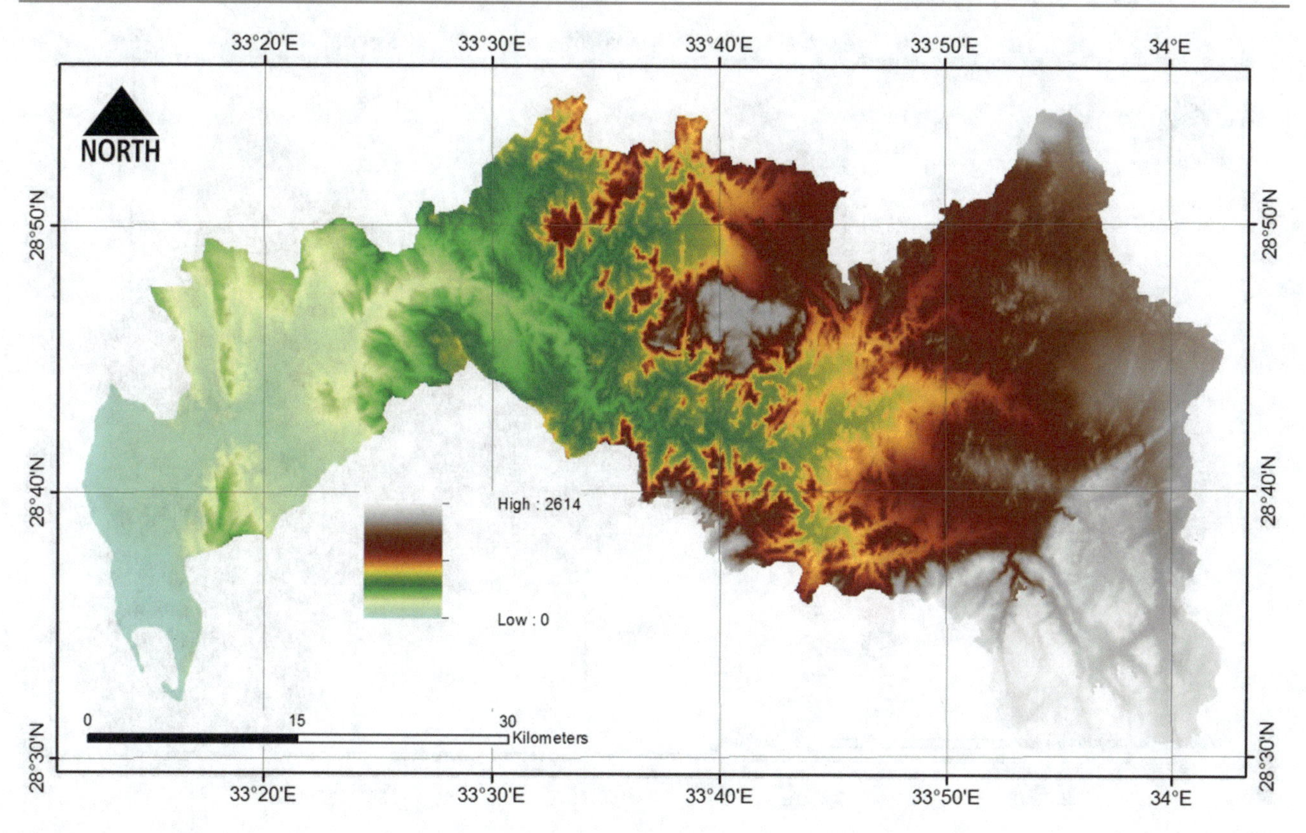

Fig. 6 DEM of the study area

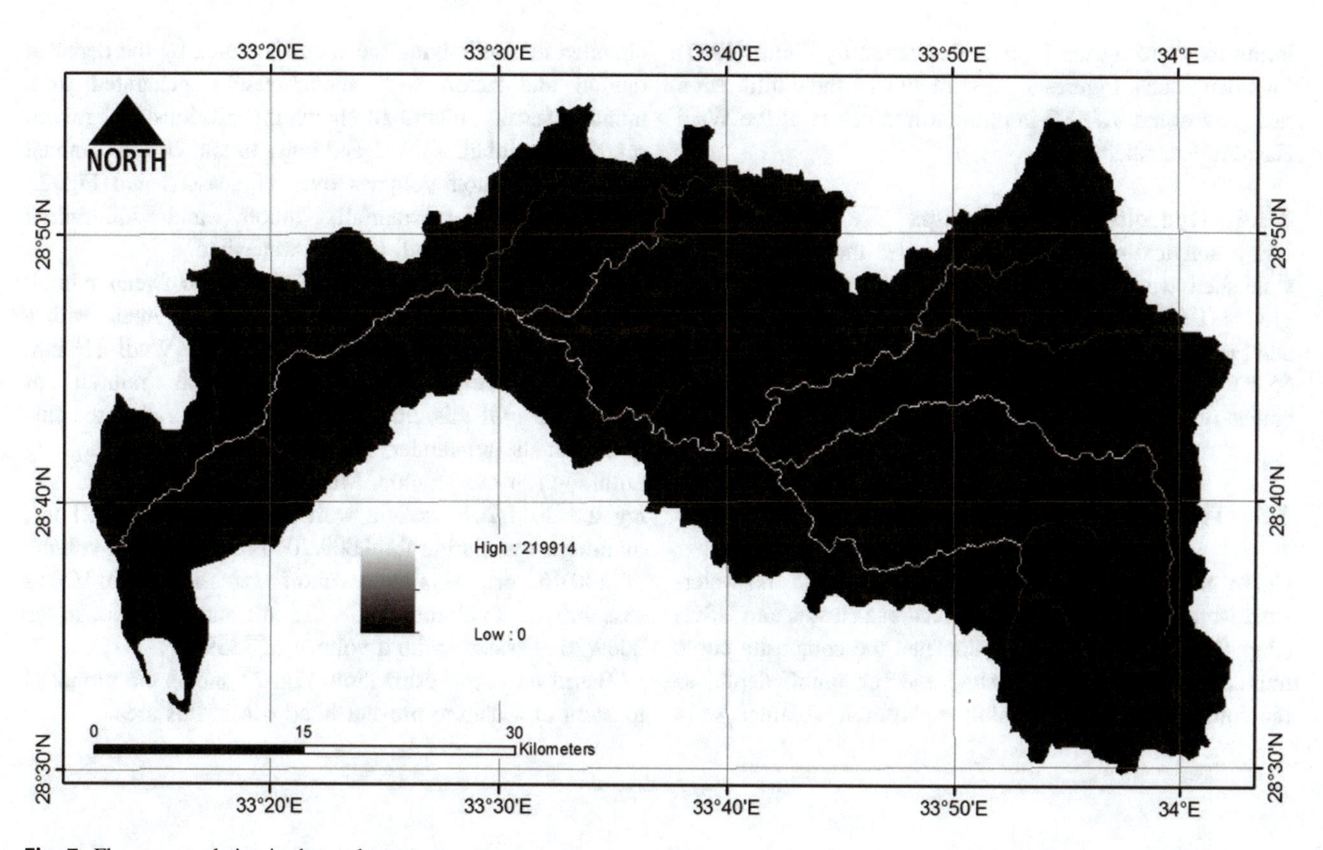

Fig. 7 Flow accumulation in the study area

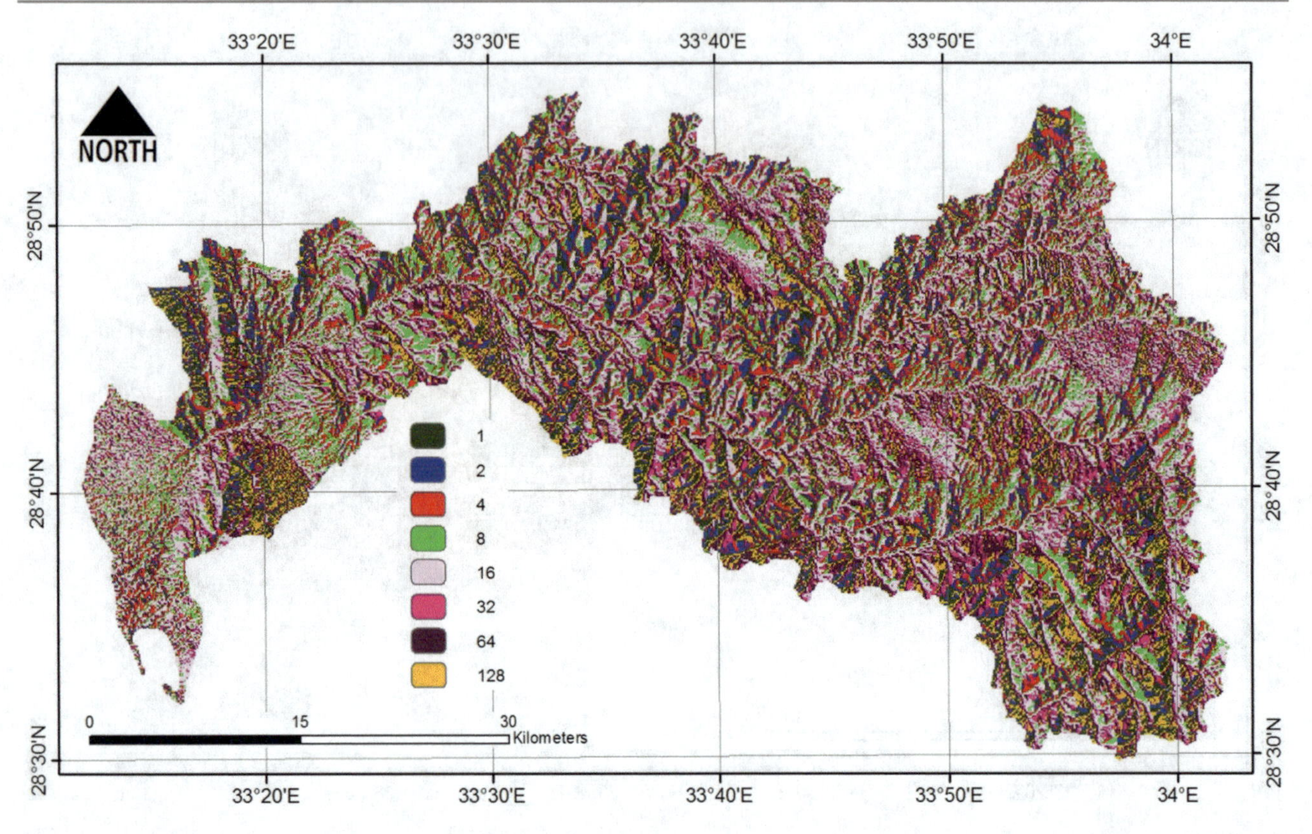

Fig. 8 Flow directions in the study area

infiltrometer using the formula developed by Philip (1957) for calculations. Figures 12, 13, 14 and 15 show infiltration rate curves and Table 3 infiltration rate values at the Wadi Hashim watershed.

3.2.4 Hydrological Soil Groups

Using soil texture and infiltration rates the Wadi Hashim watershed was broken down into three hydrological soil groups (HSGs) (according to USDA): A, C, and D (Table 4 and Fig. 16). The results show that HSG A is dominant, with 55.5% coverage, followed by type C and type D with coverages of 36.8% and 7.7% respectively.

3.3 The Depth and Volume of Runoff

Curve numbers (CNs) for different land uses in the watershed were obtained according to land use classes and HSGs (Fig. 17), from which were calculated the composite curve number (CNc) of the watershed and the runoff depth, as previous detailed. The rainfall and runoff volumes were obtained by multiplying the watershed area by the depth of rainfall and runoff with annual results generated from monthly results. Figure 18 shows the relationship between runoff and rainfall. Table 5 and Figs. 19 and 20 show annual rainfall and runoff volumes over 17 seasons and Fig. 21 compares average rainfall, runoff, and infiltration/evaporation at the Wadi Hashim watershed.

The results show that the relationship between rainfall and runoff at the Wadi Hashim watershed is linear with a correlation coefficient (R) of 0.9227. The Wadi Hashim watershed receives an average annual rainfall of 336,616 m^3. Of this, 56314 m^3 (16.7%) runs over the surface with the remainder, 280,302 m^3 (83.3%), undergoing infiltration or evaporation. Maximum rainfall occurred during the 2011/2012 season with a volume of 545,821 m^3; minimum was during the 2009/2010 season with a volume of 140,463 m^3. Maximum runoff was in the 2011/2012 season, with a volume of 138,452 m^3; minimum was in the 2009/2010 season, with a volume of 3359 m^3.

Based on runoff estimation, Fig. 22 shows the proposed location of a dam to prevent flood risk in this area.

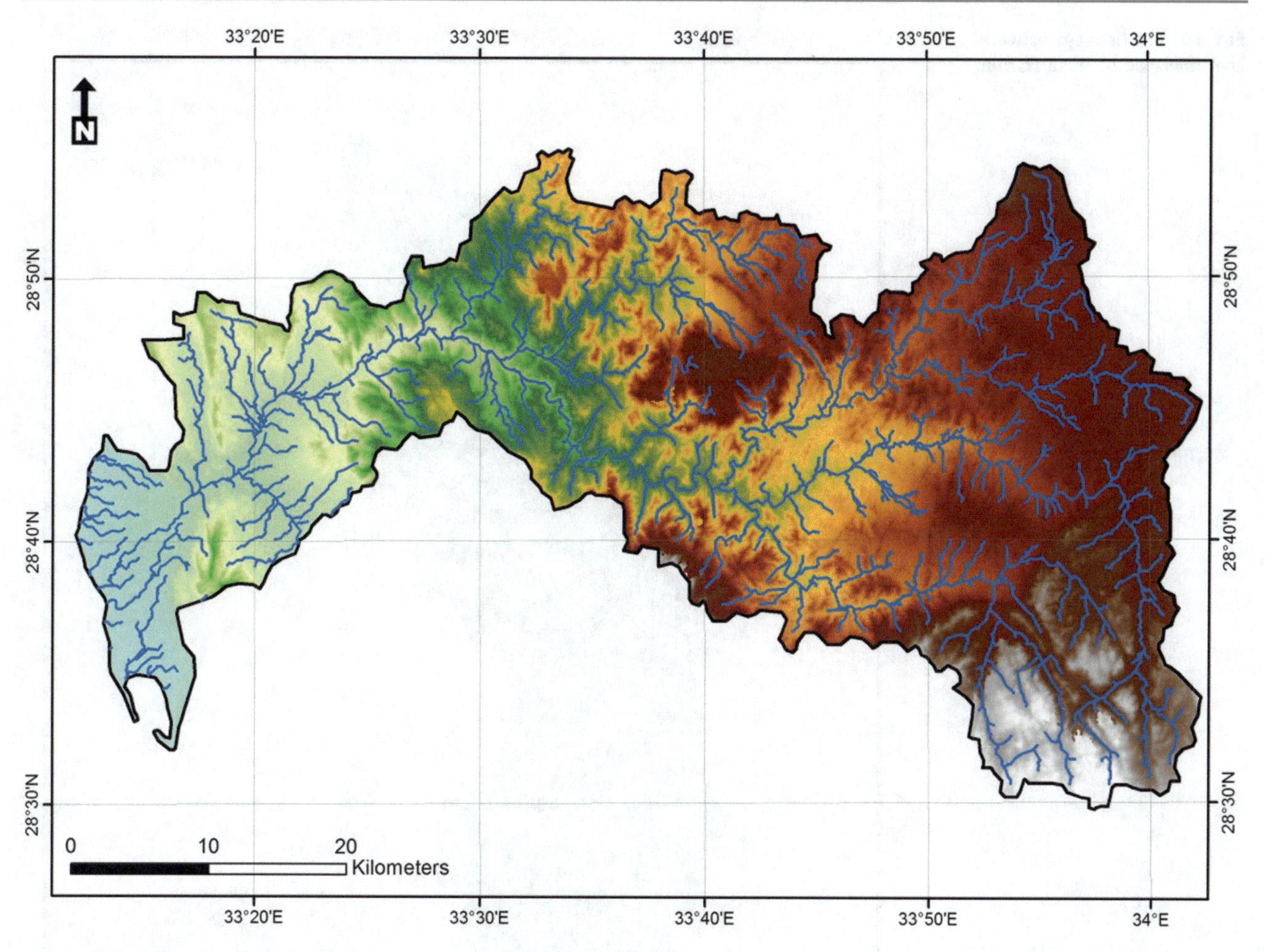

Fig. 9 Flood mapping in the Firan catchment and drainage area

4 Conclusions

This chapter demonstrates how remote-sensing techniques and GIS technology can be used effectively to extract a watershed and monitor flash flooding. Drainage flow directions, (upslope) areas, and catchments were extracted from DEMs in areas that have suffered flash flooding. Areas at risk of flooding were delineated. Such methodologies were applied to case studies in the Firan catchment, Sinai, and Wadi Hashim on the northern coast of Egypt.

In this study the GIS-based SCS-CN method was used to estimate runoff from the watershed. It is reasonable to conclude that land use planning and watershed management can effectively and efficiently be completed using the SCS-CN method and GIS. The study demonstrates that the combination of SCS-CN method and GIS is a powerful tool for estimating the runoff of ungauged watersheds, facilitating better watershed management and conservation. The methodology followed in this study can be applied to all wadi watersheds, but runoff and rainfall percentages will depend on watershed area and physical characteristics.

Our results suggest that the following would be beneficial:

1. Installing meteorological stations in all watersheds to provide spatial and temporal rainfall data that can be used for hydrological modeling.
2. Carrying out field measurements of runoff volumes at as many positions as possible and comparing results with the estimated values so that the SCS-CN method can be calibrated.

The results indicate that accurate watershed analysis can be developed successfully using remotely sensed data and currently available open source software. This approach saves time, effort, and cost. Finally, validation of the current technologies for watershed modeling and flash flood mapping is required through further research.

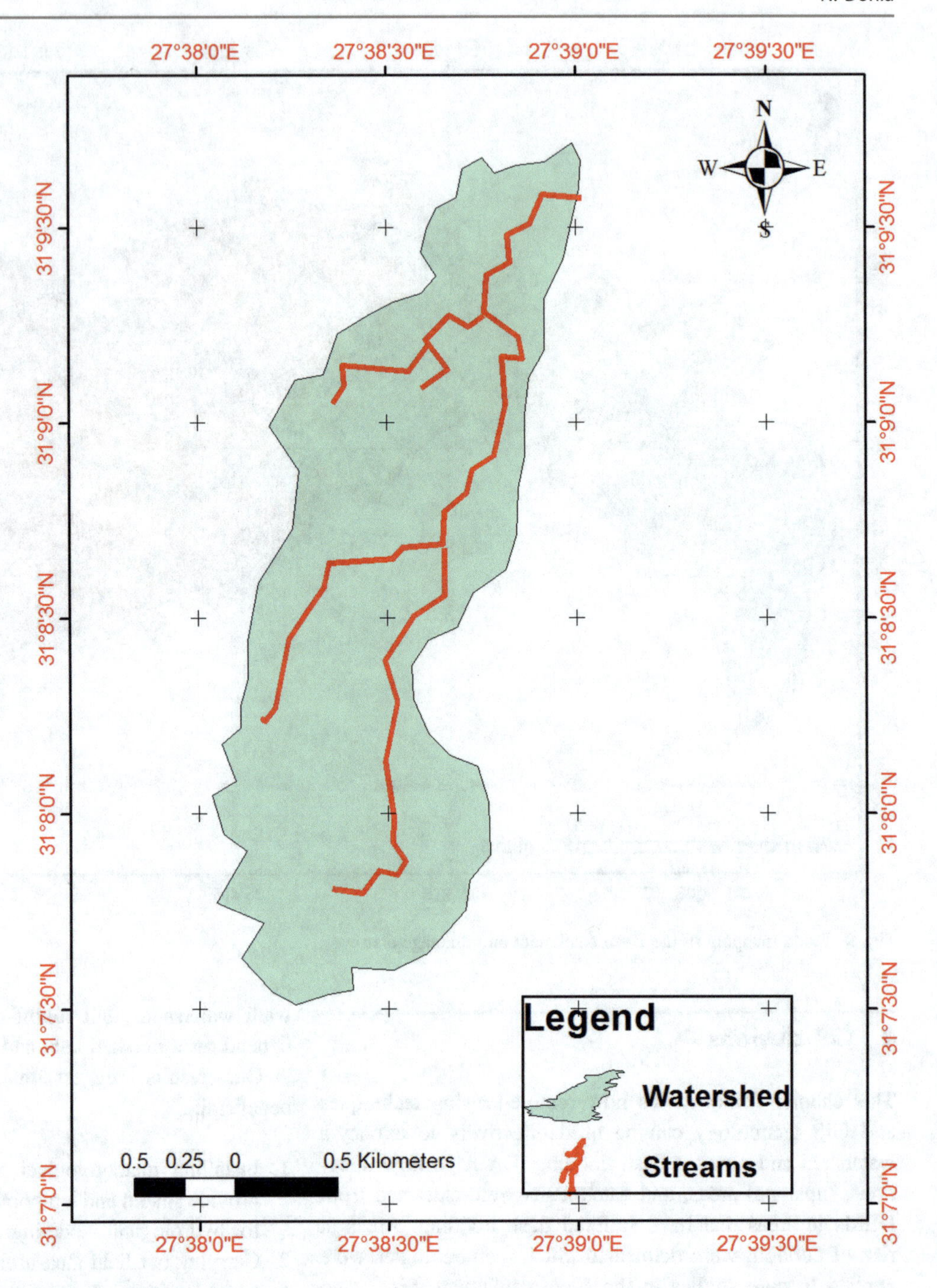

Fig. 10 The drainage networks and watershed of Wadi Hashim

Table 1 Classes of land use in the Wadi Hashim watershed

Land use	Area (m^2)	Percentage of area
1. Rock covered with thin hard crust	1.216	36.8
3. Cultivated area	0.255	7.7
3. Bare soil	1.834	55.5
Total	3.305	100

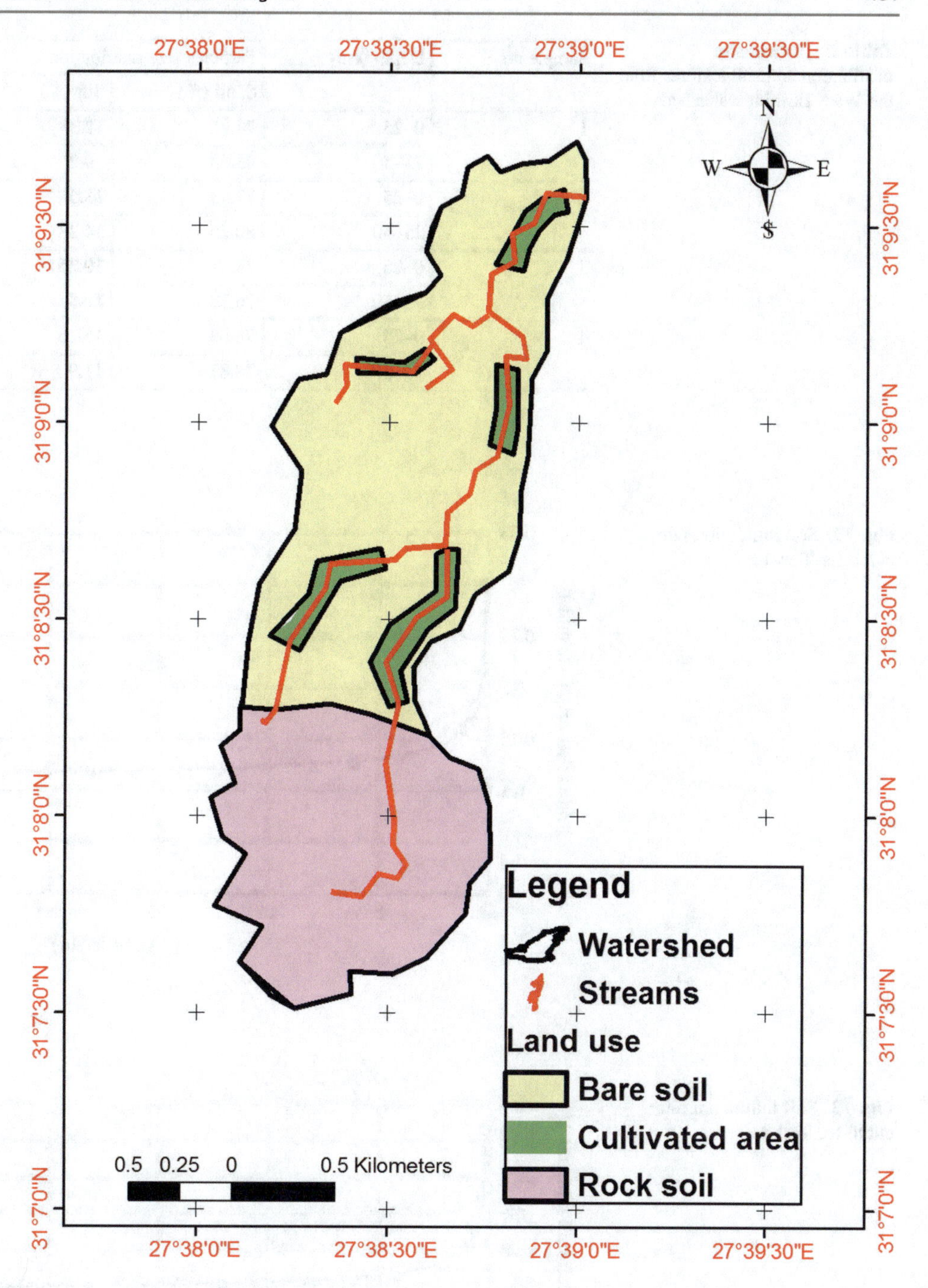

Fig. 11 Land use in the Wadi Hashim watershed

Table 2 Particle size distribution and soil textures from the Wadi Hashim watershed

Sample no.	Depth (cm)	Particles size distribution			Soil texture
		Sand (%)	Silt (%)	Clay (%)	
1	0–25	74.93	12.33	12.74	Sandy loam
	25–50	83.68	9.2	7.12	Sandy loam
2	0–25	73.41	18.5	8.09	Sandy loam
	25–50	80.26	14.28	5.46	Sandy loam
3	0–25	73.09	10.26	16.65	Sandy loam
	25–50	76.32	12.54	11.14	Sandy loam
4	0–25	78.03	15.26	6.71	Sandy loam
	25–50	78.87	11.9	9.23	Sandy loam

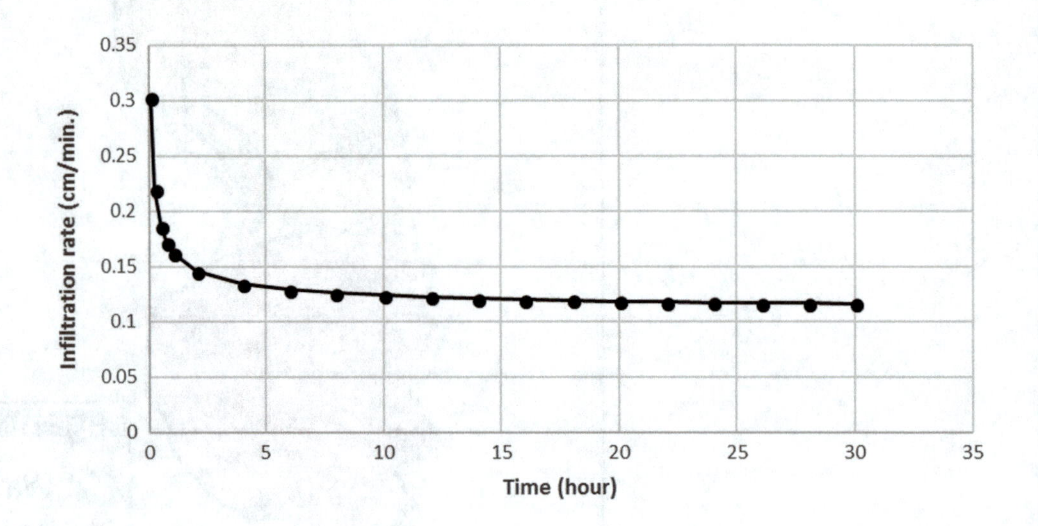

Fig. 12 Soil infiltration rate curve for Test 1

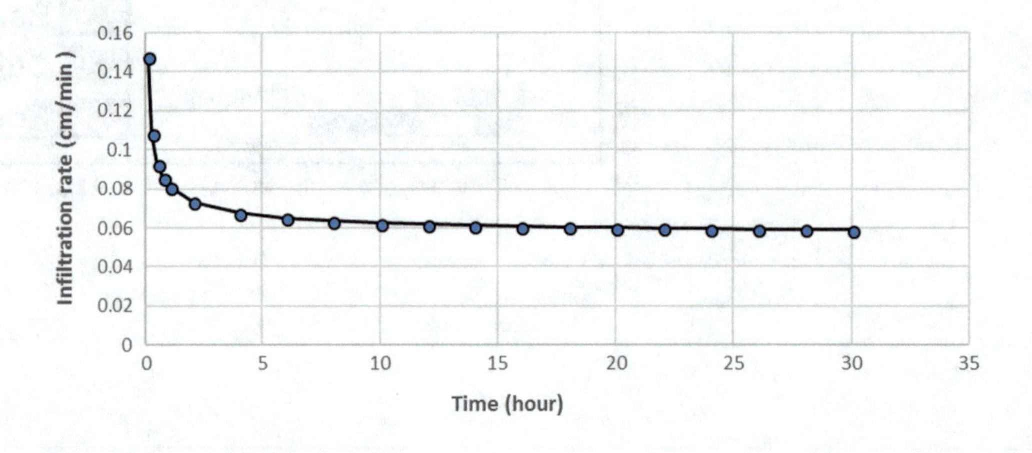

Fig. 13 Soil infiltration rate curve for Test 2

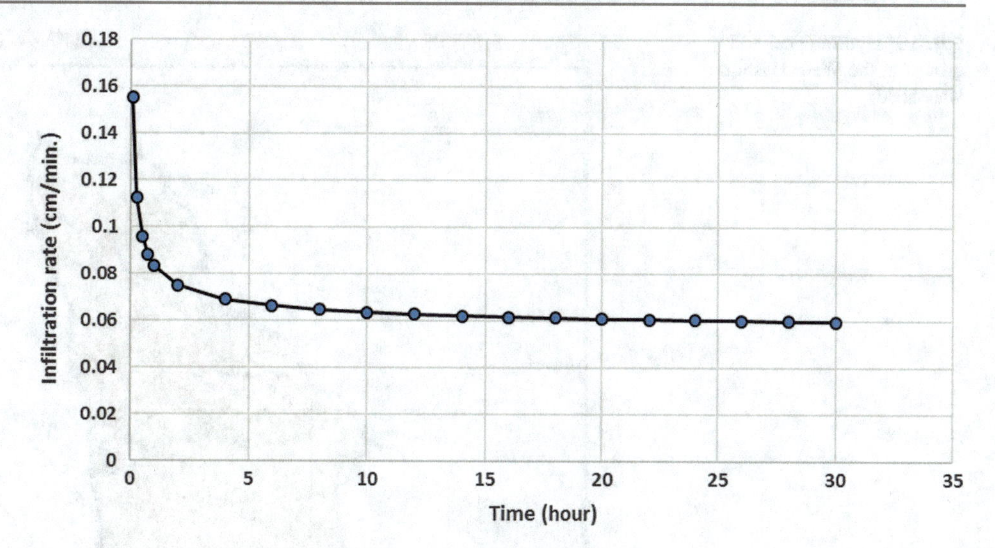

Fig. 14 Soil infiltration rate curve for Test 3

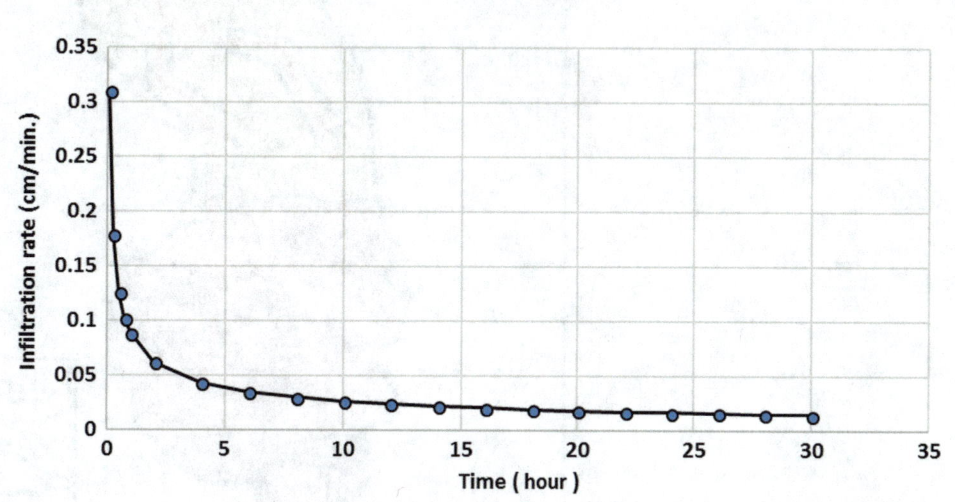

Fig. 15 Soil infiltration rate curve for Test 4

Table 3 Soil infiltration rates at different sites in the Wadi Hashim watershed

Test no.	Infiltration rate (cm/min)
1	0.12
2	0.06
3	0.06
4	0.02

Table 4 Hydrological soil groups at the Wadi Hashim watershed

Land cover	HSG	Area (km^2)	Percentage of area
1. Rock covered with thin hard crust	C	1.216	36.8
2. Cultivated area	D	0.255	7.7
3. Bare soil	A	1.834	55.5
Total		3.305	100

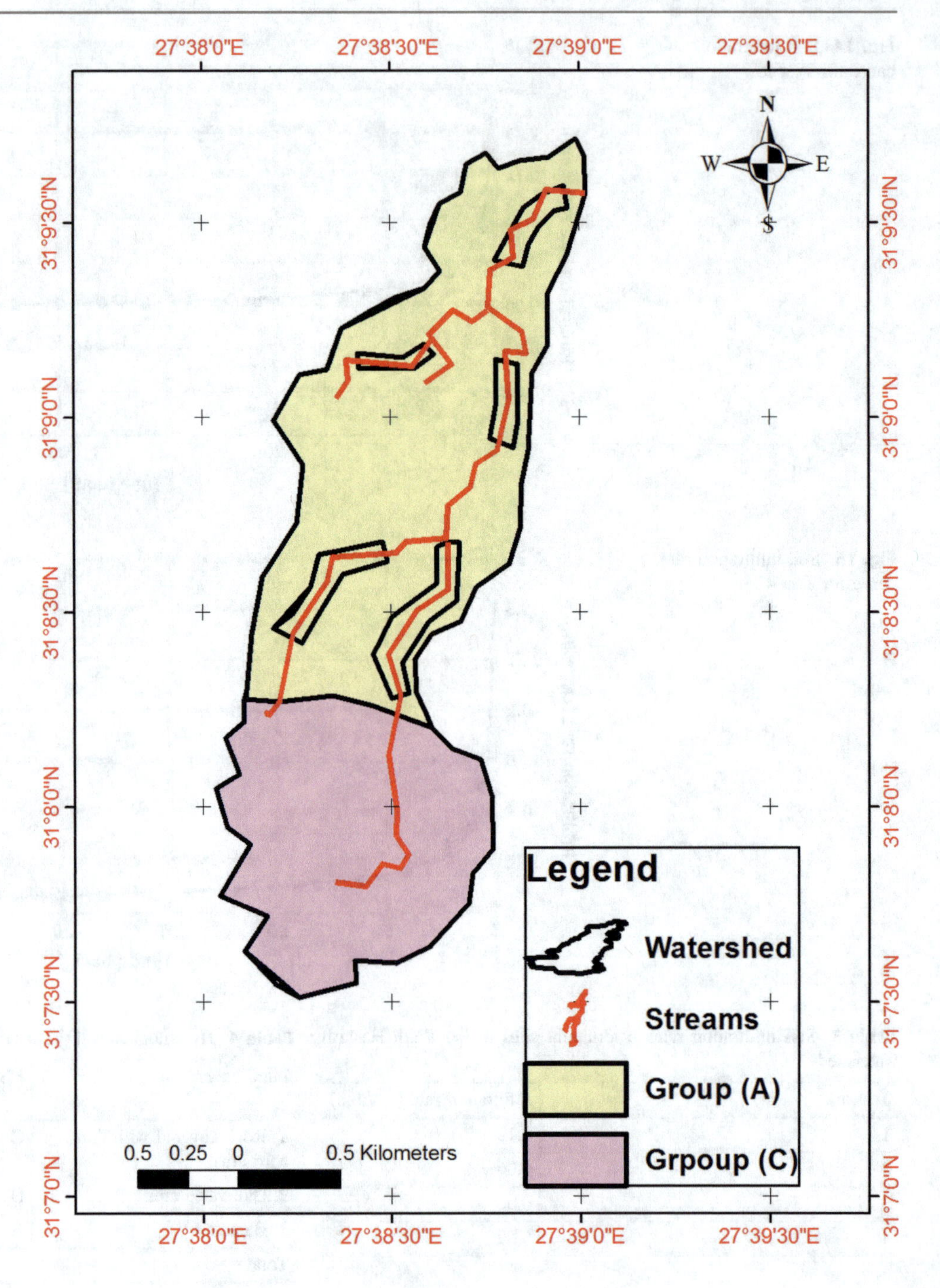

Fig. 16 Hydrological soil groups at the Wadi Hashim watershed

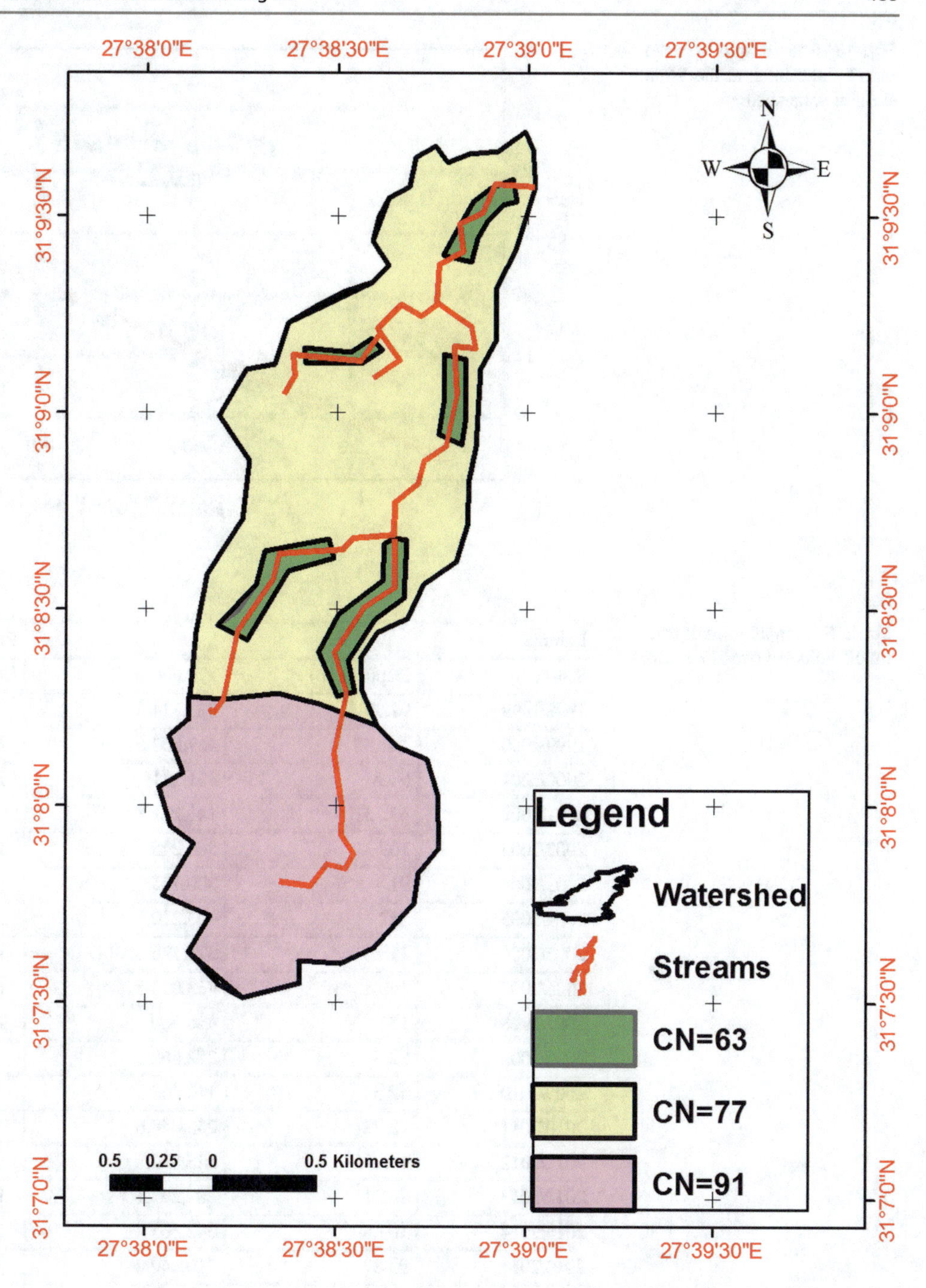

Fig. 17 Land use curve numbers (CN) at the Wadi Hashim watershed

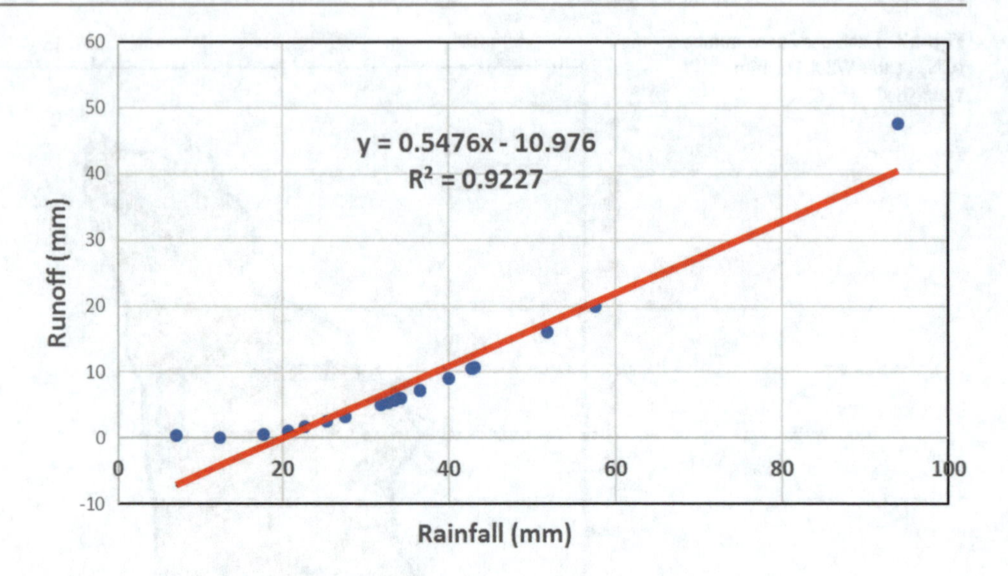

Fig. 18 Relationship between runoff and rainfall at the Wadi Hashim watershed

Table 5 Annual rainfall and runoff volumes over 17 seasons

Rainfall			Runoff	
Season	Depth (mm)	Volume (m^3)	Depth (mm)	Volume (m^3)
1998/1999	97.95	323,724.8	18.18	60,083.2
1999/2000	141.32	467,062.6	48.61	160,667.8
2000/2001	94.2	311,331	14.39	47,559.77
2001/2002	43.35	143,271.8	5.42	17,926.15
2002/2003	109	360,245	19.68	65,026.24
2003/2004	91	300,755	13.14	43,433.72
2004/2005	122	403,210	20.79	68,713.86
2005/2006	113.2	374,126	17.42	57,583.84
2006/2007	140.1	463,030.5	19.79	65,417.94
2007/2008	132	436,260	28.92	95,569.93
2008/2009	82.5	272,662.5	8.12	26,838.4
2009/2010	42.5	140,462.5	1.02	3,358.732
2010/2011	73.72	243,644.6	2.39	7,907.296
2011/2012	165.15	545,820.8	41.89	138,451.8
2012/2013	86.01	284,263.1	11.01	36,380.26
2013/2014	103.59	342,365	16.53	54,622.61
2014/2015	93.87	310,240.4	2.36	7,800.101
Total	1731.46	5,722,475	289.66	957,341.7
Average	101.8506	336,616.2	17.04	56,314.22

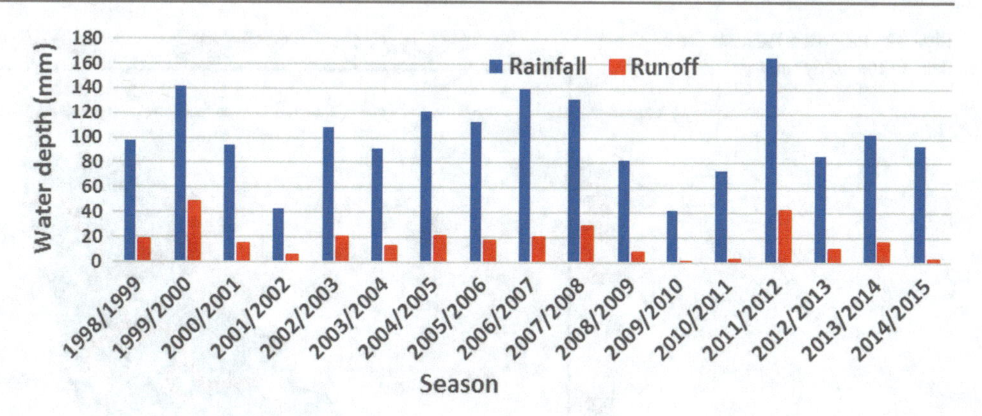

Fig. 19 Annual rainfall and runoff depths at the Wadi Hashim watershed for 17 seasons

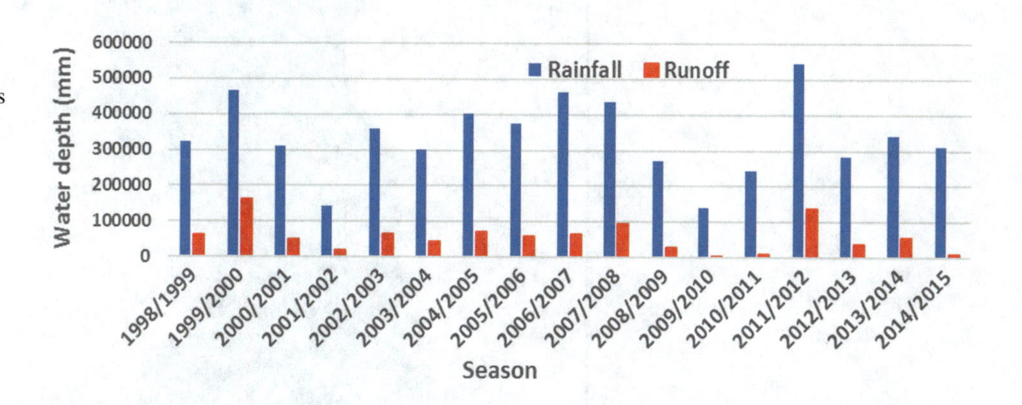

Fig. 20 Annual rainfall and runoff volumes at the Wadi Hashim watershed for 17 seasons

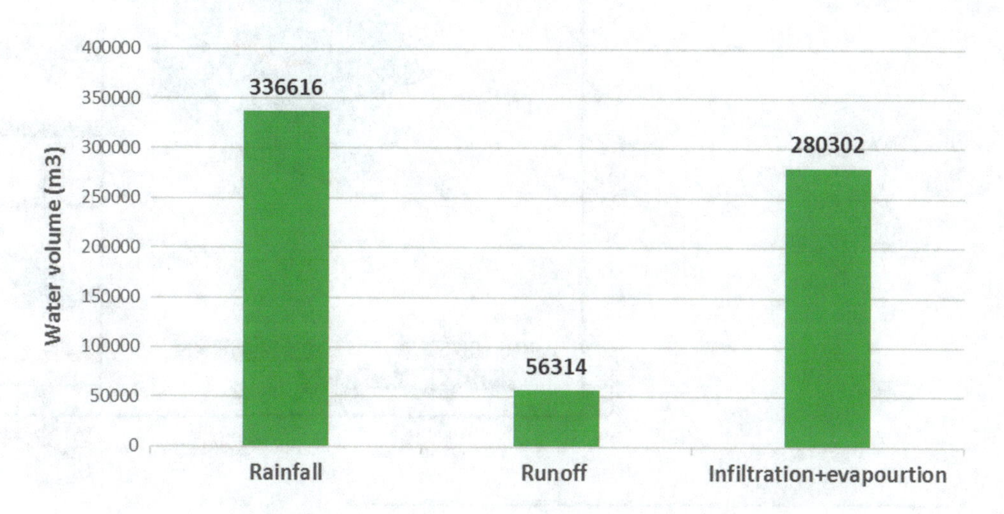

Fig. 21 Average rainfall, runoff, and infiltration/evaporation at the Wadi Hashim watershed

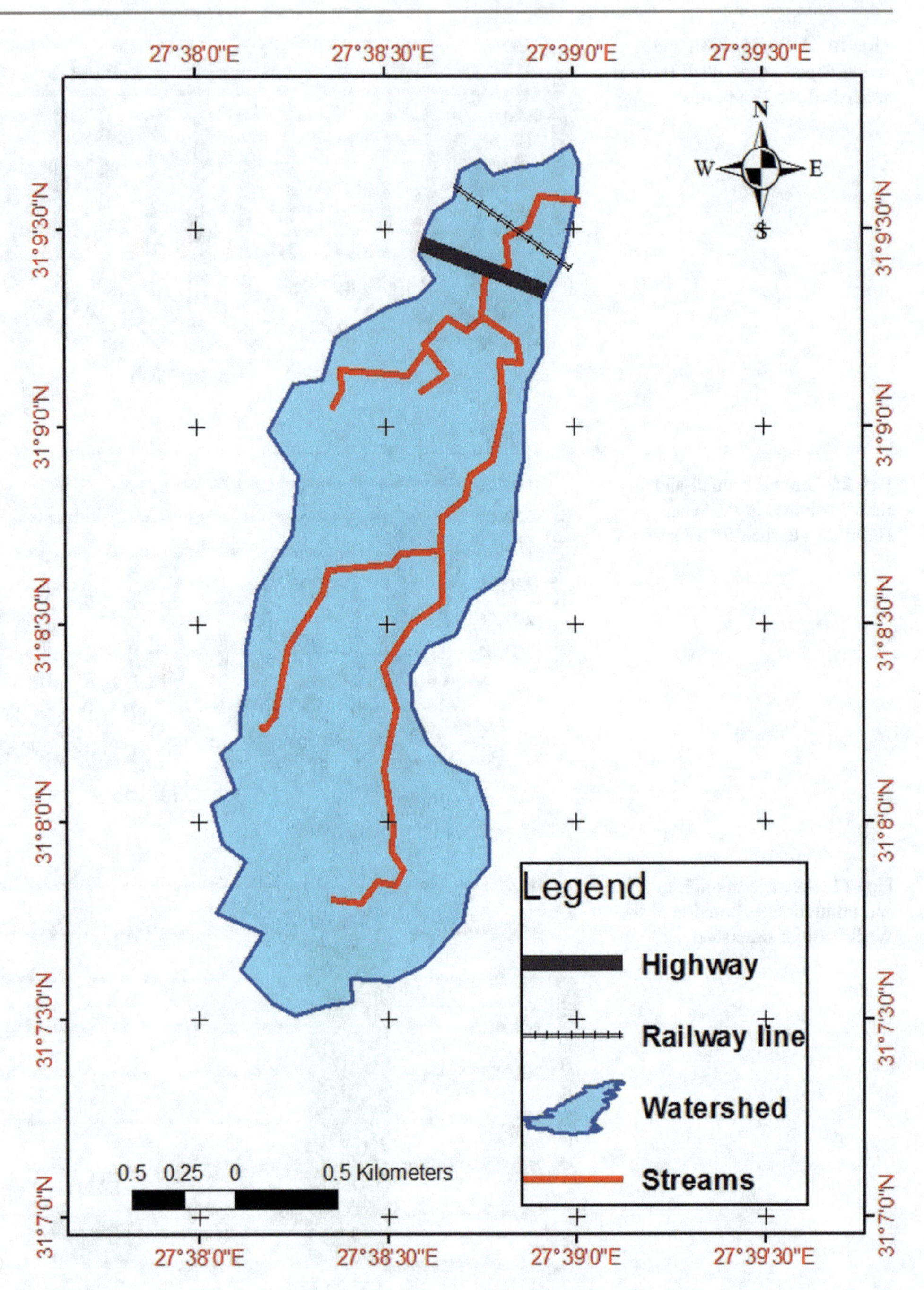

Fig. 22 Proposed location for a dam in the study area

References

AL-Gamdi, S. (1991). Estimating runoff curve numbers of the soil conservation service in arid and semi-arid environments using remotely sensed data. A dissertation Submitted to the Faculty of the University of Utah, USA.

Al-Jabari, S., Sharkh, A. M., & Al-Mimi, Z. (2009). Estimation of runoff for agricultural watershed using SCS curve number and GIS. In *Thirteenth International Water Technology Conference, IWTC 13 Hurghada*, Egypt.

Bansode, A., Patil, A. K. (2014). Estimation of runoff by using SCS curve number method and Arc GIS. *International Journal of Scientific & Engineering Research*, *5*. ISSN 2229-5518.

El Bastawesy, M., Ali, R. R., Nasr, A. H. (2008). The use of remote sensing and GIS for catchments delineation in Northwester Coast of Egypt: an assessment of water resources and soil potential. The *Egyptian Journal of Remote Sensing* and *Space* Sciences, *11*, 3–16.

El Bastawesy, M., White, K., & Nasr, A. (2009). Integration of remote sensing and GIS for modelling flash floods in Wadi Hudain catchment. *The Egyptian Hydrological Processes*, *23*, 1359–1368.

El Bastawesy, M., El Harby, K., & Habeebullah, T. (2012). The hydrology of Wadi Ibrahim Catchment in Makkah City, the Kingdom of Saudi Arabia: The interplay of urban development and flash flood Hazards. *Life Science Journal, in an Arid Environment. Journal of Hydrology*, *292*, 48–58.

ESRI. (2010). *Environmental systems research institute (ESRI) Press*. Redlands, California.

Jones, R. (2002). Algorithms for using a DEM for mapping catchment areas of stream sediment samples. *Computers & Geosciences 28*, 1051–1060.

O'Callaghan, J. F., & Mark, D. M. (1984). The extraction of drainage networks from digital elevation data. In *Computer vision, graphics and image processing* (Vol. 28, 323–344).

Philip, J. R. (1957). The theory of infiltration. 2–4 Soil Science, p. 34, 83, 85,163 and 257, 28.

Rashash Ali, A., Mohamed, E. S., Belal, A., El Shirbeny, M. (2015). GIS spatial model based for DAM reservoir on dry Wadis ACRS 2015. In *36th Asian Conference on Remote Sensing: Fostering Resilient Growth in Asia, Proceedings*.

Yousif, M., Abd, E. S. E., & Baraka, A. (2013). Assessment of water resources in some drainage basins, northwestern coast, Egypt. *35TApplied Water Science35T*, *3*, 35TI. 235T, pp. 439–452.

Sustainable Development of Mega Drainage Basins of the Eastern Desert of Egypt; Halaib–Shalatin as a Case Study Area

Hossam H. Elewa, Ahmad M. Nosair, and Elsayed M. Ramadan

Abstract

This book chapter focuses on using effective tools of monitoring and management of natural resources, based on the integration of remote sensing (RS) and geographic information systems (GIS) techniques with a field survey in surface and groundwater resources evaluation. It is anticipated to provide operational and effective systems of investigation, management and protection of the available natural resources, and improve the livelihood of the surrounding population. This work depends on the previous expertise and overwhelmed researches of the National Authority for Remote Sensing and Space Sciences (NARSS) and addresses the key challenges for the sustainable development in this remote area. Sustainable water supply is vital for the development of communities in arid regions, such as that of the South Eastern Desert of Egypt. The economic importance of the area is enormous, besides the fact that it has long been a target zone for mineral resources excavation and mining. One of the challenges facing this arid area is the limited water resources needed for agricultural, industrial, mining, or domestic uses. Bedouin depend mainly on rainwater, which constitutes the main source feeding their hand-dug wells and fracture springs. Rainwater harvesting (RWH), as a historical and worldwide trend, could fulfill the gap of water scarcity in arid or semi-arid regions. This proposed work is to use the modern techniques of RS, geographic information systems (GIS), and watershed modeling systems (WMS) to provide a plan for the RWH. RWH is the accumulation and storage of rainwater for reuse before it reaches the aquifer system (Groundwater). Multi-spectral remote sensing (MSRS) and geographic information systems (GIS) are vital tools to optimize the surface water usage of episodic rainfalls, where the concept of runoff water harvesting (RWH) in promising watersheds should be applied. (Elewa et al. in Am J Environ Sci 8:42–55, 2016). GIS and digital elevation models (DEM) enable the development of hydrological models to investigate every ancient terraced field in a non-invasive manner, without disturbing the archaeological remains (Bruins et al. in J Environ 166:91–107, 2019). The RWH could be used also for maximizing the recharge possibilities of groundwater. As a non-conventional water resource, RWH could provide water for gardens, livestock, irrigation, mining, cleaning of bathrooms as in the first flush, etc. In many places with similar climate conditions, the collected water is redirected to a deep pit with percolation to recharge the groundwater for later use and protection, especially in structurally controlled groundwater accumulations. The harvested water could be used as drinking water, if the storage is a tank that can be accessed and cleaned when needed. The work recommendations will be a good source for the up-to-date databases, which could be used effectively by the decision-makers, researchers, executive authorities, planners, and related governorates. The **objective** of this book chapter is to assess the South Eastern Desert of Egypt for the RWH capabilities, with the determination of their optimum methods and techniques. The overall **goal** is to assist in poverty alleviation, Bedouin and urban allocation, supporting animal husbandry, accelerating agricultural development, improved agricultural and food production for local inhabitants, combating desertification, resolving unemployment problems, and raising individual incomes. Bedouin and natives as the main end users will be a major target of

H. H. Elewa (✉)
Engineering Applications & Water Division (EAWD), National Authority for Remote Sensing & Space Sciences (NARSS), Cairo, Egypt
e-mail: elewa.hossam@gmail.com; ; hossam.eliewa@narss.sci.eg

A. M. Nosair
Geology Department, Faculty of Science, Zagazig University, Zagazig, Egypt

E. M. Ramadan
Water Structures Department, Faculty of Engineering, Zagazig University, Zagazig, Egypt

A. M. Negm (ed.), *Flash Floods in Egypt*, Advances in Science, Technology & Innovation,
https://doi.org/10.1007/978-3-030-29635-3_9

the work. Innovative ways to improve the capture, storage, and use of rainwater will have their own bearing on the sustainable and profitable production of dry season vegetable crops in South Eastern Desert. According to the worldwide trends and techniques in RWH, which is applied aggressively in many neighboring countries, Egypt should enter the era of catching every water droplet for domestic and agricultural development. The results of the present research work could establish a good example to be applied in other parts of the country as well as worldwide.

Keywords

Egypt • Eastern desert • Remote sensing • GIS spatial modeling • Drainage systems • Hydromorphometric analysis • Watershed modeling • Water harvesting

List of Acronyms and Abbreviations Listed in the Text

ANOVA	Analysis Of Variance
ASTER	Advanced Spaceborne Thermal Emission & Reflection Radiometer
BA	Basin Area
BL	Basin Length
BS	Basin Slope
Dd	Drainage Density
DEM	Digital Elevation Model
E	Degree of Effectiveness
EIA	Environmental Impact Assessment
ETM	Enhanced Thematic Mapper
Fm	Formation
GIS	Geographic Information System
GSA	Global Sensitivity Analysis
HS	Halaib–Shalatin region
IF	Basin Infiltration Number
masl	Meter above sea level
MCDSS	Multi-Criteria Decision Support System
MFD	Maximum Flow Distance
OFD	Overland Flow Distance
RS	Remote Sensing
RWH	Runoff Water Harvesting
SAM	Spatial Analysis Model
LMP	Land use Master Plan
ST	Total effect, Sensitivity index
VAF	Volume of Annual Flood
V_i	A partial variance
W	Wadi (dry valley)
W_c	Criterion Weight
WMS	Watershed Modeling System (WMS Software); Water Management System
WSPM	Weighted Spatial Probability Model

1 Introduction

In Halaib–Shalatin (HS) area (about 23,615 km^2), the natural springs and hand-dug wells are the main dependable water resources from the native dwellers. The area is suffered for several decades from intensive aridity, overgrazing, and desertification. The rainless periods may continue for 4, 5, or even 8 years, which leads to great tragic starvation for the native people, and for the animal husbandry.

Serious strategies for water exploration and development must be established to fully utilize the surface and groundwater resources. The success of these strategies is dependent on the availability and reliability of basic hydrogeologic and hydrologic information and related subjects.

Recently, the area has got more attention as a promising region for different developmental activities as tourism, fishery, animal husbandry, and mining and for its importance as a trading route between Egypt and Sudan. The growth of such activities requires a simultaneous strategy for using and development of the available water resources of the study area to meet the water demands.

Remote sensing (RS) techniques were used as an important tool, among others, for investigation of surface geological, structural, and hydrological (drainage systems) features. These surface features were used, with ground investigations to study the surface and groundwater resources.

1.1 Aim and Scope of the Present Work

The objectives of the present work are to evaluate HS area for the runoff water harvesting (RWH) and groundwater recharge possibilities. It also include determining the suitable sites for control works implementation either for RWH and/or groundwater recharge using optimum methods and techniques. The overall goal is to construct a developmental and management system for water resources in this important area to increase and improve agricultural activity, food production, and improve the standard of living for natives. Additionally, the present work includes construction of RWH system helps in the mitigation of flood hazards that are frequently threatening the different developmental activities. The specific objectives of the present work include drainage mapping of main wadies, construction of a digital database for runoff volumes, delineation of watersheds, determining the suitable sites for implementing water harvesting control works.

1.2 Location

HS area is a strategic region located at the far southeastern (SE) corner of Egypt and lies between longitudes 34°30′–37° 00′ E and latitudes 22°00′–23°30′ N (Fig. 1). The study area

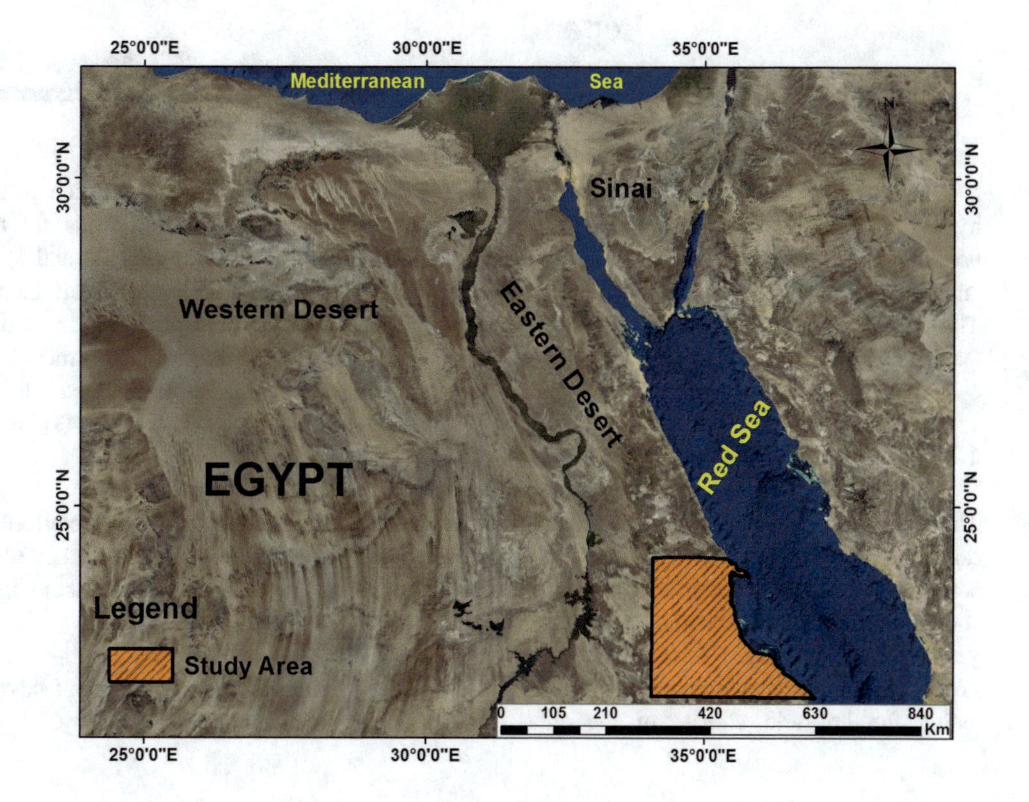

Fig. 1 Location map of the study area

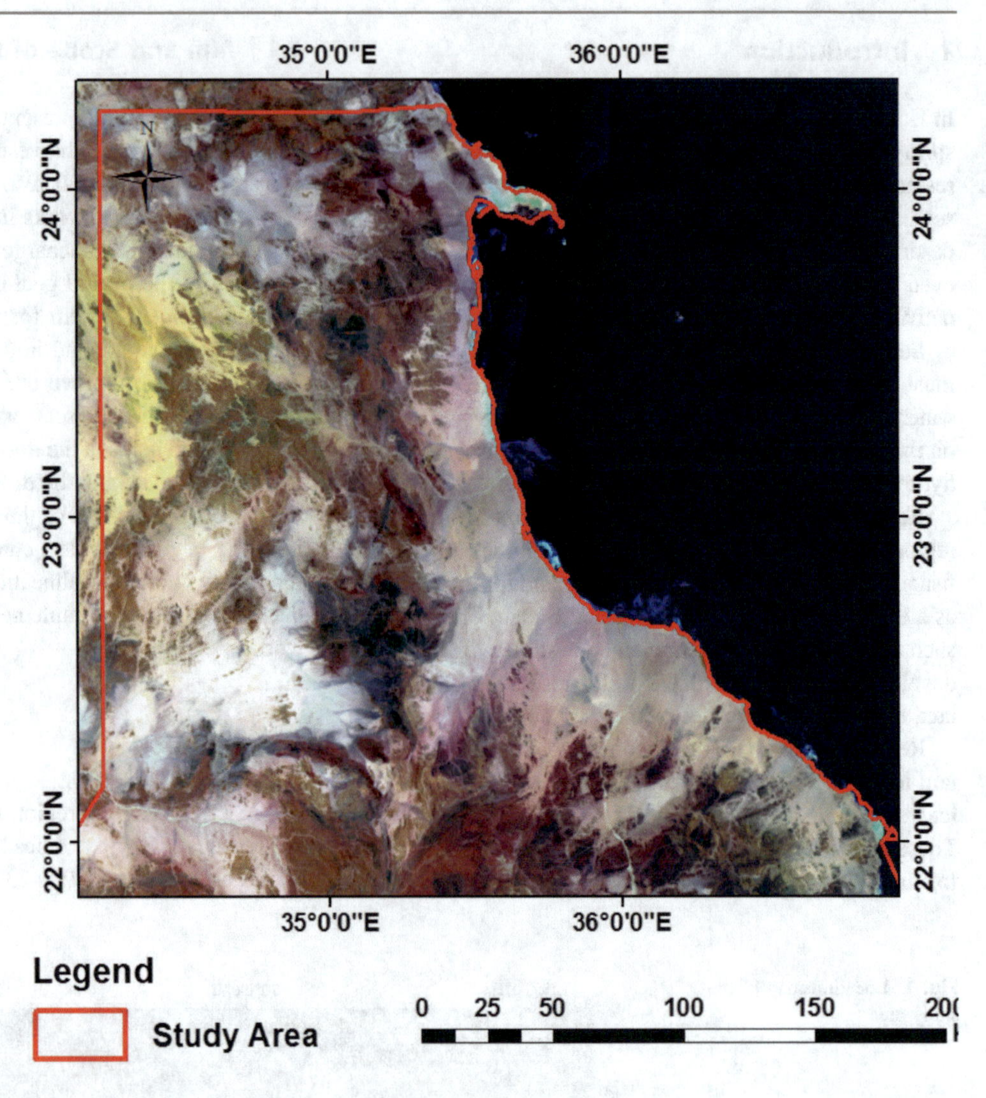

Fig. 2 A satellite ETM + 8 image showing the study area

is bounded by the Red Sea coast to the east, the Nile Valley hydrographic basin to the west, and the Egyptian–Sudanese border to the south (lat. 22°00′ N). This area is accessible and traversed by a number of paved roads and desert tracks. The area is covered by Landsat (ETM + 8) image scenes (see Fig. 2).

1.3 Climatic Conditions

The study area is situated within the arid belt where sporadic rainfall may occur from time to time and accompanied by some flash floods. Sometimes an extremely rainless or shortage in rainfall may continue for years, reaching 6 or 8 years, this is impressed through the desert varnish phenomena formed in many places of the study area and indicating the hot arid climate of the area, (Fig. 3). The inhabitants depend on their own ways of storing rainwater and drill hand-dug wells for their domestic use and for animal drinking (Figs. 4 and 5).

The climatic parameters include temperature, solar radiation, rainfall precipitation, and wind velocity. The Desert Research Institute constructed a complete meteorological station in the study area (in Wadi (W.) Rahaba)(Desert Research Institute 1998) (Table 1).

1.3.1 Temperature

The study area is characterized by a hot summer with an average temperature of about 32 °C, and cool winter with an average temperature of about 19 °C.

1.3.2 Humidity

The net humidity is about 43–49%, while the evaporation rate is about 16.8 mm/day.

Fig. 3 Field photo showing the desert varnish phenomena in W. Hodein indicating the arid climate of the study area

Fig. 4 A field photo showing a hand-dug well with hoisting via robes (Bir Iqeet)

1.3.3 Rainfall

The rainfall is scarce over most of the year. The average rainfall precipitation is about 2.35 mm/year. The maximum rainfall precipitation is about 24.8 mm (recorded in October 1997), while the minimum rainfall precipitation is about 0.76 mm (recorded in April 1998).

1.3.4 Wind Velocities

The minimum average wind velocity is about 30.39 km/h (recorded in November 1997) and the maximum average wind velocity is about 38.93 km/h (recorded in June 1997). The maximum wind velocity is about 41.30 km/h (recorded in February 1997), which represents the beginning of

Fig. 5 A field photo showing protection of the hand-dug wells from sanding up by flash floods, by building up a high masonry wall around it (in W. Rahaba)

Table 1 Meteorological parameters, recorded at Shalatin Station (the period from October 1997 to October 1998), after Desert Research Institute (1998)

Month	Third of month	Average Temperature (°C)	Solar radiation K.W./M^2	AV. Wind velocity	Rainfall (mm)
October 1997	2nd	28.8	0.253	31.80	24.849
	3rd	26.4	0.22	26.56	–
November 1997	1st	25.3	0.219	32.88	–
	2nd	24.8	0.219	32.88	–
	3rd	23.6	0.195	29.19	–
December 1997	1st	22.4	0.192	29.65	–
	2nd	20.5	0.177	39.18	–
	3rd	20.4	0.173	29.81	–
January 1998	1st	20.3	0.169	39.8	–
	2nd	18.0	0.198	33.6	–
	3rd	18.9	0.208	28.3	–
February 1998	1st	20.7	0.220	27.4	–
	2nd	19.0	0.226	31.6	–
	3 rd	20.4	0.236	41.3	–
March 1998	1st	20.3	0.246	34.2	–
	2nd	21.5	0.231	34.0	–
	3rd	21.0	0.226	28.2	1.27
April 1998	1st	22.7	0.305	40.3	–
	2nd	27.5	0.295	36.1	0.76
	3rd	25.2	0.279	37.2	–
May 1998	1st	28.5	0.308	33.7	–
	2nd	27.8	0.324	38.9	1.27
	3rd	29.7	0.317	40.0	–
June 1998	1st	28.5	0.315	36.6	–
	2nd	30.4	0.323	41.1	–
	3rd	29.8	0.320	39.1	–

(continued)

Table 1 (continued)

Month	Third of month	Average Temperature (°C)	Solar radiation K.W./M^2	AV. Wind velocity	Rainfall (mm)
July 1998	1st 2nd 3rd	31.7 32.4 32.9	0.319 0.297 0.302	36.3 35.8 35.5	– – –
August 1998	1st 2nd 3rd	33.2 35.4 35.8	0.314 0.285 0.310	36.4 36.1 35.0	–
September 1998	1st 3rd	28.2 29.0	0.317 0.320	31.50 31.40	– –
October 1998	1st 3rd	28.4 28.0	0.248 0.211	31.2 25	

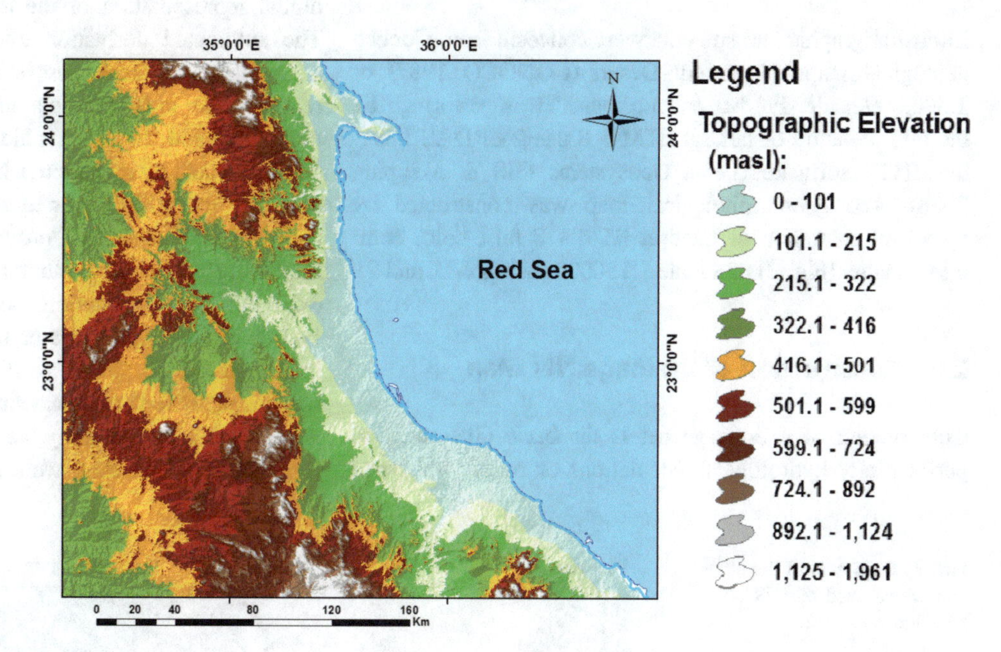

Fig. 6 Topographic elevations of the study area (based on ASTER DEM of 30-m resolution)

El-Khamsin storm. Prior to the present arid nature of the climate, semi-arid or even wet climatic conditions were dominant in Egypt. Butzer (1959) and Said (1990), give some details of that period.

1.4 Topographic Features

The topography of the area ranges from gently sloping coastal plains to rugged mountainous and hilly lands, at the western and southwestern (SW) parts of the study area and heights up to 1961 masl (Fig. 6).

2 Materials and Methods

To accomplish the work objectives, the following tasks were performed which address the undertaken methods and techniques:

2.1 Satellite Image Collection, Preparation, and Processing

The ETM + 8 satellite image (acquired in 2014) was used, which has been designed to provide long-term continuity of data collection but with successive improvements in the technical capability and performance of the sensing systems involved.

The image is calibrated into geographic latitudes/ Longitudes, GRS 1980, and transformed from .dat format to .img format through the import module of Erdas Imagine 9.0© software and then converting them into Universal Transverse Mercator (UTM), WGS 1984, to be compatible with the different Geographic Information System (GIS) thematic layers (i.e., compatible with digital elevation model (DEM), enhanced thematic mapper images (ETM + 8), geological map for soil data, etc. (ERDAS Field Guide 2005).

2.2 Construction of Base Map

The map was constructed by using the published and validated maps of the Egyptian General Authority for Civil Survey (EGACS 1989) with multi-scales series and Google Earth maps and Satellite ETM + 8 images (Fig. 2). The constructed base map of geographic and drainage basins' data of the study area comprises the watershed boundaries of drainage basins shown in Fig. 7.

2.3 Construction of Geologic and Geomorphologic Maps

Lithostratigraphic and soil data were collected from Conoco geological maps of eastern Desert (CONOCO 1987) of 1:500,000 scale. Further enhancements' was performed by the interpretation of landsat ETM + 8 using ERDAS Imagine 10.1© software (Leica Geosystems GIS & Mapping 2008). The geomorphological map was constructed by visual interpretation of Landsat ETM + 8 false-color compiled image, (Fig. 2) of a scale 1:250000 (bands 4, 3, and 2).

2.4 Construction of Drainage Net Map

Construction of a drainage net is the basic GIS entity to perform any hydrological calculations or runoff watershed modeling practices. In modern research methods, the reliance on DEMs and satellite imagery with high precision for the extraction of drainage networks and the boundaries of their basins are common (Jenson and Domingue 1988; Mark 1984; Moore et al. 1991; Martz and Garbrecht 1992; Tribe 1992).

Hydrologic process and water resource issues are commonly investigated by using distributed watershed models. These watershed models require physiographic information such as the configuration of the channel network, location of drainage divides, channel length and slope, and sub-catchment geometric properties. Traditionally, these parameters are obtained from maps or field surveys. The digital representation of the topography is called a DEM. The automated derivation of topographic watershed data from DEMs is faster, less subjective, and provides more reproducible measurements than traditional manual techniques applied to topographic maps (Tribe 1992). Digital data generated by this approach also have the advantage that they can be readily imported and analyzed by the GIS. The technological advances provided by the GIS and the increasing availability and quality of DEMs have greatly expanded the application potential of DEMs to many hydrologic, hydraulic, water resources, and environmental investigations (Moore et al. 1991). In the present work, the runoff calculations and assessing the potential for the RWH, was performed by dividing the study area into 52 watersheds (Fig. 8), which in turn, were subdivided into smaller sub-

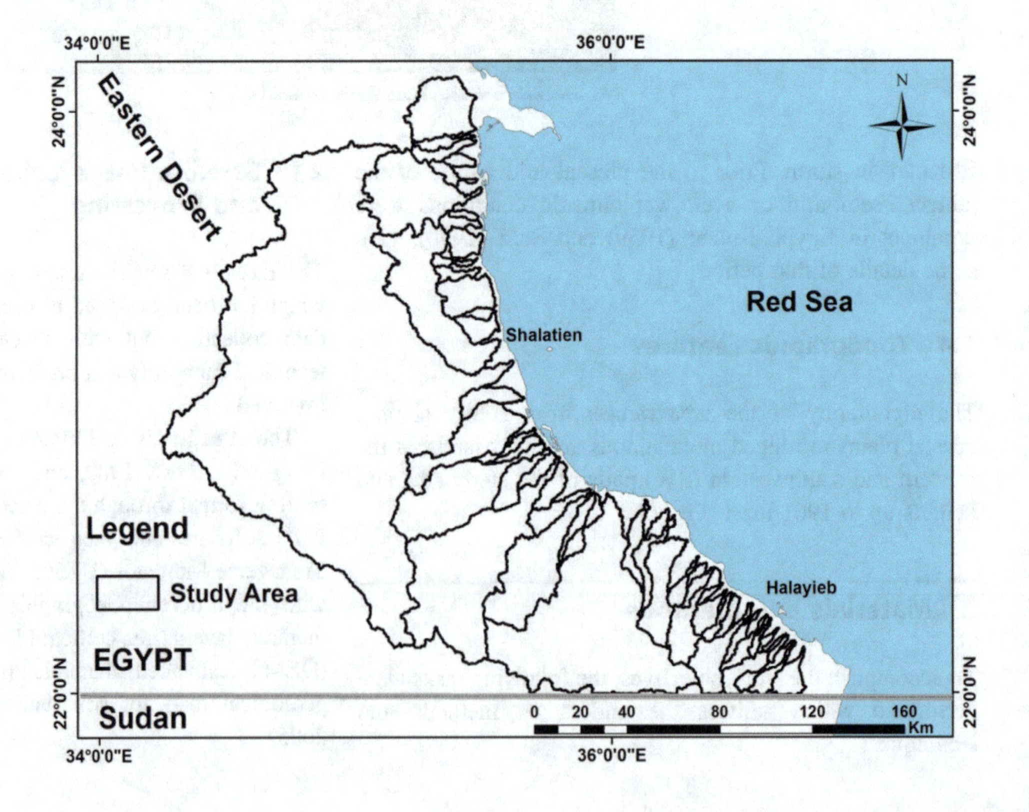

Fig. 7 Base map of the study area constructed by GIS techniques

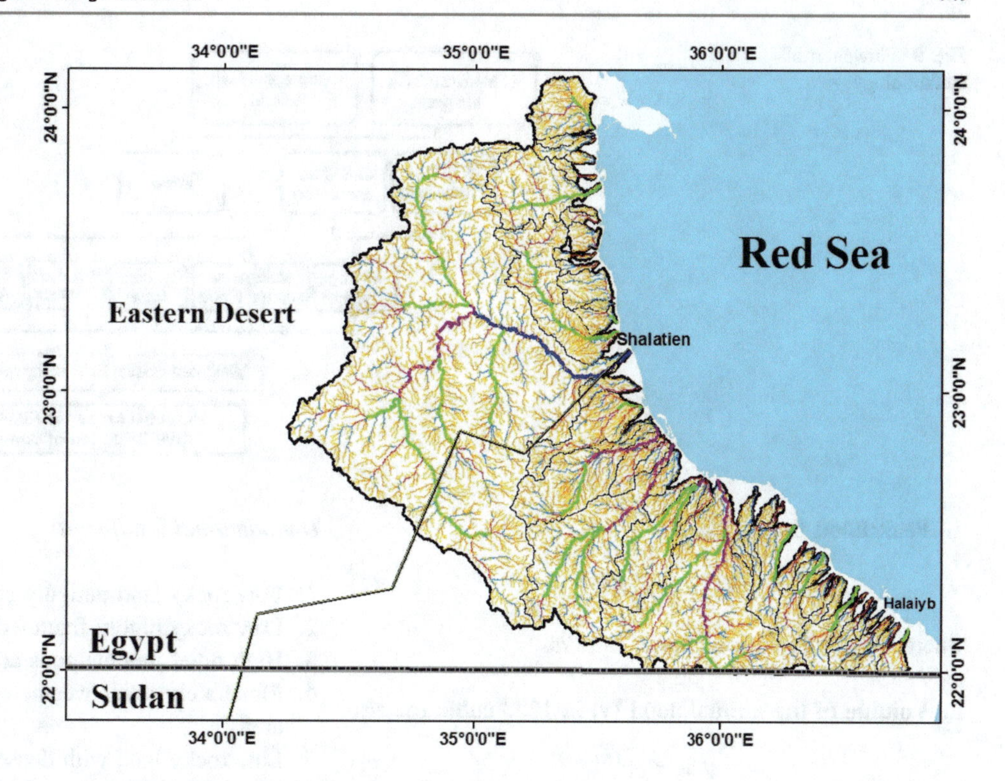

Fig. 8 Drainage net of study area comprising the: watersheds boundaries and drainage network

and sub-watersheds. The present work describes a process for determining site characteristics and developing an integrated approach including RS, GIS, and watershed modeling system software (WMS 8.0©) for performing such objectives.

DEM with a 30-m resolution of the study area has been obtained from the ASTER (Rabus et al. 2003), which was subsequently enhanced by the topographic contours and Landsat heights. One program that has been developed for the automatic delineation of watersheds and stream arcs on a DEM is the Topographic Parameterization (TOPAZ©) software (Garbrecht and Martz 1997). A special version of TOPAZ has been created for using within WMS software platform, which only requires an elevation grid as input and produces a flow direction grid and a flow accumulation grid as outputs. After defining basins with a DEM, the results are converted to drainage coverage for easier data storage and manipulation.

2.5 Runoff Calculations and Watershed Modeling

The hydromorphometric parameters of Halaib and Shalatin watersheds were determined using watershed modeling systems (WMS 8.0©) software (Aquaveo 2008), which differentiated the basins and provided multiple watershed characteristics. Accordingly, several thematic maps, viz, volume of annual flood (VAF), average overland flow (OFD), maximum flow distance (MFD), infiltration number (IF), drainage density (Dd), basin area (BA), basin slope (BS), and basin length (BL) were integrated as input layers for the weighted spatial probability model (WSPM) to perform a determination for the efficient sites suitable for the RWH. The WMS 8.0© software calculated the hydromorphometric characteristics for each watershed value used in the WSPM. These values are provided for each of the delineated watersheds. These layers are generated in steps, i.e., digitization, editing, building a topological structure, and finally polygonization in ArcGIS 10.1© Spatial Analyst module (ESRI 2007).

Finkel (1979) method was used through the present work to calculate the volume of the annual flood which was run inside the WMS 8.0© software (AQUAVEO 2008). Finkel (1979) used his method for the Araba Valley, which has similar climate conditions to Sinai Peninsula and arid regions in Egypt. It is a simple graphical method to determine the probability or frequency of occurrence of yearly or seasonal rainfall.

The Finkel empirical method (Finkel 1979) uses the following parameters (Eqs. 1 and 2):

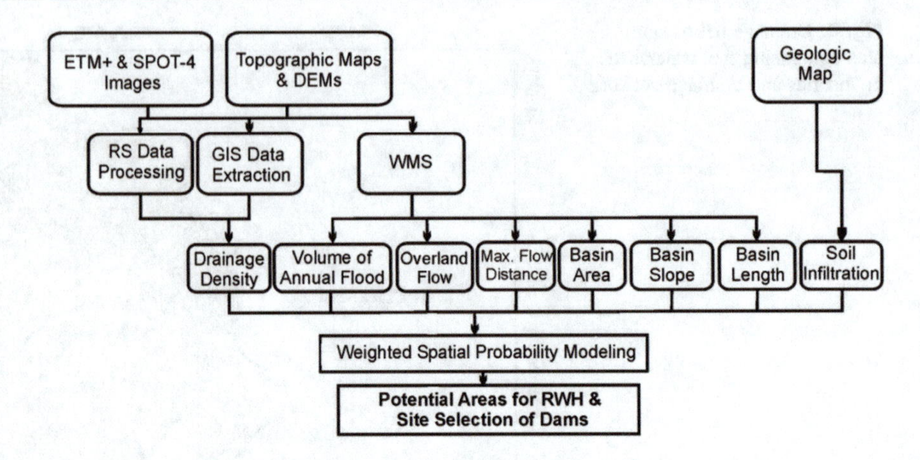

Fig. 9 Flowchart of methodology

1. **Peak flood flow (Q_{max})**

$$Q_{max} = K_1 A^{0.67} \tag{1}$$

where Q_{max} = Peak flood flows, in m^3/s.

2. **Volume of the annual flood (v) in 1000 cubic meters**

$$V = K_2 A^{0.67} \tag{2}$$

where A is the area of the basin in km^2, and K_1 and K_2 are the constants depending on the probability of occurrence:

Probability of occurrence in a given year	K_1	K_2
10%	1.58	26.5

Here, we used 10% because it is very suitable for the developmental conditions.

The overall flowchart of the methodology is given in Fig. 9.

3 Geomorphological Setting, Geological Setting, and Soil Characteristics

3.1 Geomorphologic Features

Geomorphology is the face of the terrain, which reflects the geological setting of the area and has an important effect on accumulation, transportation, and percolation of rainfall. The investigated area comprises different types of landscapes including exogenic and endogenic landforms. The main geomorphologic units and features are:

3.1.1 Geomorphological Units

The study area can be classified into the following geomorphological units (Fig. 10):

Denudational landforms:

1. Low rocky land partially covered with sand and gravel
2. Low rocky highly fractured land
3. High relief mountainous area with sharp peaks
4. Moderately relief weathered and fractured mountainous area
5. Low rocky land with dense dyke swarms
6. Nearly flat sedimentary high land
7. Piedmont
8. Isolated hills
9. Hilly country.

Accumulated origin landforms:

1. Alluvial fan-like delta
2. Sand dunes
3. Wadi deposits.

Coastal plain:

1. Beach
2. Sabkhas
3. Sand sheets.

In the following paragraph, a brief description of some of the major geomorphologic units recognized in the study area will be given.

Denudational Landforms

They are represented by nine geomorphic units (Fig. 10) formed chains of mountains that cover most parts of the study area. These mountains possess different slopes, but they are mostly ranging from steep to moderate slopes. Several pointed peaks are found in the study area such as those of Gebel Elba (1417 m), Gebel Shindeb (1334 m), and

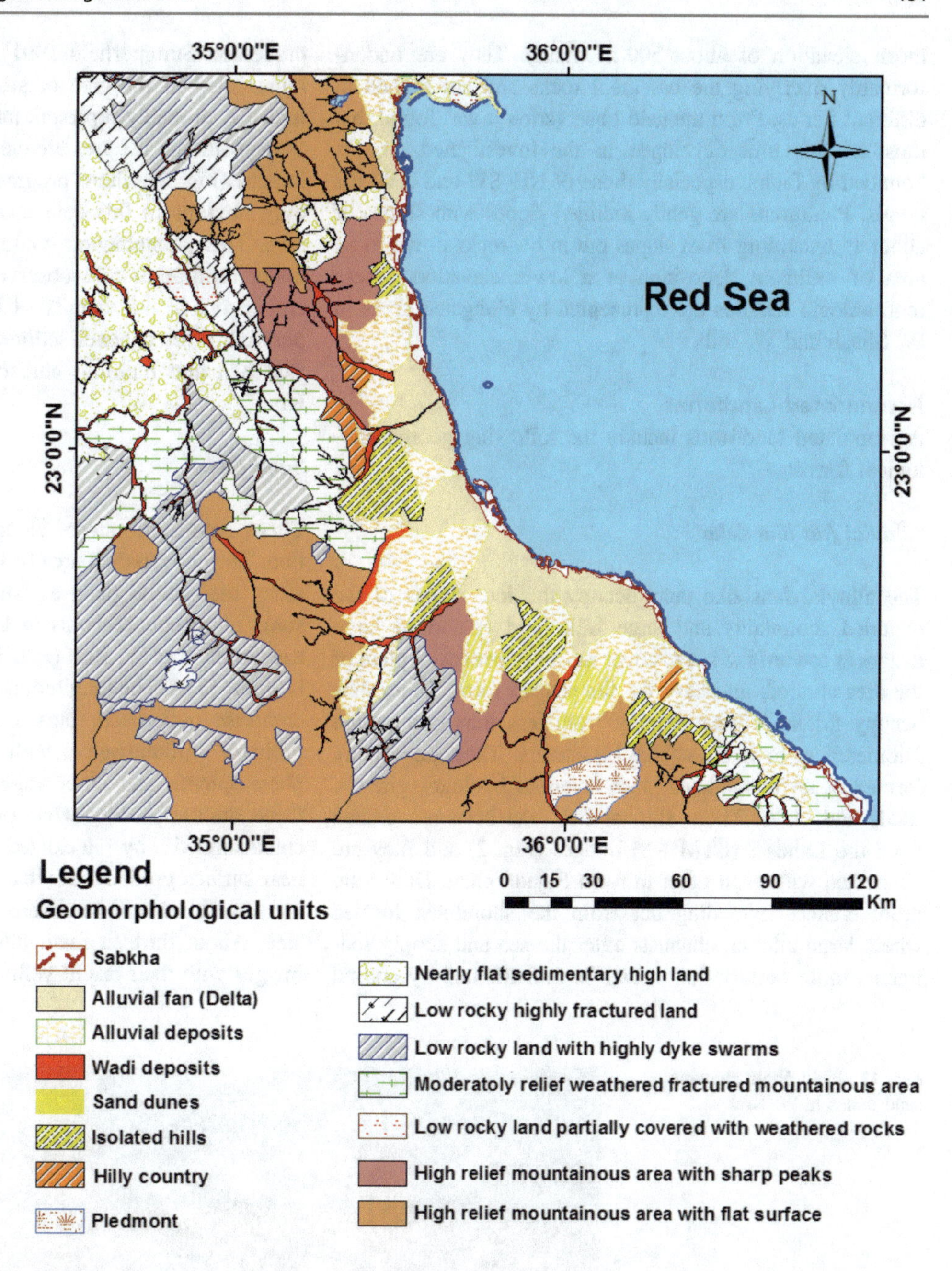

Fig. 10 Geomorphologic map of the study area

Gebel Iss (1050 m). The eastern boundary of the mountainous chains is partially delineated by NNW to NNE faults, while the NE, NW, and ENE faults mainly shape the mountains. The hills range in topographic elevations from about 100 m to 400 m. They are formed of igneous, metamorphic, and sedimentary rocks, and range in size from tens to hundreds of meters. Major NE and NW faults as well as dykes of many trends and major wadis trace them. Most of these hills have gentle slopes and rounded peaks. Wadis have generally separated these hills and drain toward the Red Sea.

Some of these hills comprise some of the main economic minerals in the study area, such as magnesites, chromite, and iron oxides. These hills are mainly exposed at the area around Gebel Elba, W. Kraf, and the area between W. Hodein and W. Shaab deltas. The nearly flat sedimentary high land is well stratified and characterized by low to moderate topographic relief. They are represented by the Nubia Sandstone beds located at the northwestern (NW) part of the study area overly the basement rocks. They are composed of sandstones with clay intercalation and are highly weathered and highly fractured with a consistent mean elevation of about 500 m (masl). They are unconformably overlying the basement rocks and are faulted in different trends. From the field observation it was found that most of this unit developed in the investigated area is bounded by faults, especially those of NE–SW and NW–SE trends. Piedmonts are gently inclined slopes with slopes of about 1° extending from slopes cut in bedrocks down to an area of sediment deposition at a lower elevation. These morphologic features are represented by elongated areas in W. Shaab and W. Ibib.

Accumulated Landforms

Accumulated landforms include the following geomorphological forms:

Alluvial fan-like delta

The alluvial fans like delta occupy the foot slopes of the denuded mountains and large hills, and extend to large distances toward the sea to cover most of the coastal plain of the area studied, and have the fan-shaped form. These fans occupy different areas ranging from less than one square kilometer up to many square kilometers. They are usually formed of unconsolidated loose layers of boulders, gravels, sands and clays. These alluvial fans had been recognized from the Landsat (ETM + 8) images (Fig. 2) and they are connected with each other to form Bajada plain. Deltas are protuberances extending out from the shorelines formed where large alluvial channels enter the sea and supply sediments more rapidly than it can be redistributed by coastal processes (Summerfield 1991). This features are somewhat represented in the area of study by the delta of W. Kraf which is probably represent an early stage developed delta, as depositional features are clear on the Landsat (ETM + 8) images (Fig. 2) where progradation of the coastal plain is only obvious in this sole characteristic wadi of the study area. This phenomenon could be attributed to the relative lower infiltration rate observed from the infiltration tests carried out in the vicinity of the wadi which gives a high possibility for channel sedimentation by the wadi into the Red Sea and building characteristic fan-like delta of W. Kraf.

Sand dunes

These sand dunes are of elongated type, of NE–SW direction. They covered an area of low land of granodiorite rocks along the coastal plain as well as located at the edges of some high mountains, as in Gebel Elba area, or along the eastern flank of W. Kraf (Fig. 11). The height of these dunes is about 13 m. The longitudinal dunes found in W. Kraf may comprise underneath buried paleo-drainage systems with probable groundwater bodies. On the coastal dunes, phreatophytes and other vegetation types are growing on, these dunes lie downwind of large sand supply and are characterized by localized vegetation sustained by near-surface groundwater that may be present. These dunes are sometimes called nebkhas and formed of dry sand grain size. About thirteen main dunes are interpreted from TM images with their bright yellow color.

Fig. 11 Field photo showing sand dunes in W. Kraf

Wadi and alluvial deposits

The wadis of the area of study are formed of bedrock and alluvial channels, the later flow into alluvial channels formed in unconsolidated sediments. These sediments may range in caliber from boulders (at the upstream channels) to clay-sized material at the downstream channels). Alluvial channels are “self-formed” and their morphology arises from the mobilization, transportation, and deposition of sediment and represents an adjustment to prevailing hydrological and sedimentological conditions. Alluvial channels differ significantly from bed rock channels which can usually change only slowly and whose morphology is dominated by structural and lithological controls.

Coastal Plain

The coastal plain of the study area is composed of different geomorphic features; beach, sabkhas, sand sheets, and alluvial terraces.

Sabkhas

Sabkhas occupy wide areas near the shoreline of the Red Sea from Abu Ramad in the north to Ras Halaib in the south. They are composed mainly of salts, clay, and silts with few grains of sand. They are wetlands and partially covered by seawater and salt. Some people extract NaCl salt from these sabkhas. They exhibited low topographic relief and were formed in the Quaternary age. The sabkhas are replaced by northward of the latitude of Abu Ramad by Beach sands which constitute quantitatively the most significant accumulations of subarially exposed sediment along the Red Sea coast.

Sand Sheets

Sand sheets are sand accumulations covered some small areas along the Red Sea coast. These sheets rarely give rise to significant topographic features but domes. The geomorphologic map (Fig. 10) may differentiate the infiltration capabilities of different areas as units characterized by high infiltration possibilities in terms of nearly flat surfaces, weathered surfaces, low rocky lands, low to moderately relief weathered fractured areas, nearly flat sedimentary high land, sand dunes and sand sheets, and areas characterized by low infiltration capabilities such as alluvial fans (deltas), wadi and alluvial deposits, sabkhas, high relief mountainous area with sharp peaks, piedmonts, isolated hills, and hilly country. The precipitation falling on a slope either run off on the surface, is held in surface depressions (surface detention), or infiltrates through the slope surface. Except on the slope of the impermeable rock where some proportion of the precipitation reaching the slope almost invariably infiltrates the surface and either percolate down to the water table.

3.2 Geological Setting

The southern part of the Eastern Desert of Egypt represents a part of the Arabo-Nubian Shield. The area is almost occupied completely (exceptionally the area of Gebel Abraq and the upstream portion of W. Hodein are made of Nubia Formation) by basement rocks which form the mountainous ranges of high to medium topographic relief. The sedimentary rocks cover is a relatively small in area and forming moderately weathered bedded hills.

Quaternary deposits are present to cover the Red Sea Coastal plain and partly cover the sedimentary and Precambrian rock units. A geological map (Fig. 12) was prepared using Landsat ETM + 8 composite image (Fig. 2) followed by intensive field work where the geological contacts were sharpened and checked in the field and by matching it with the published Conoco geological map (scale:1:500,000), Sheet of Bernis).

3.2.1 Description of Rock Units

The following are different rock units from older to younger which could be identified in the study area:

- ***Precambrian basement rocks***:

– Gneisses and Migmatites
– Ultrabasic rocks
– Metasediments rocks
– Metavolcanic rocks
– Gabbro–diorite Group
– Older Granitoide rocks
– Younger Granites
– Alkaline Granite
– Dykes
– Ring Complexes.

- ***Cretaceous sedimentary rocks***:

– Abu Aggag Formation
– Timsah Formation
– Umm Barmil Formation.

- ***Miocene carbonates rocks***:

– Gebel El-Rusas Formation.

- ***Quaternary deposits***:

– Wadi deposits
– Sabkhas

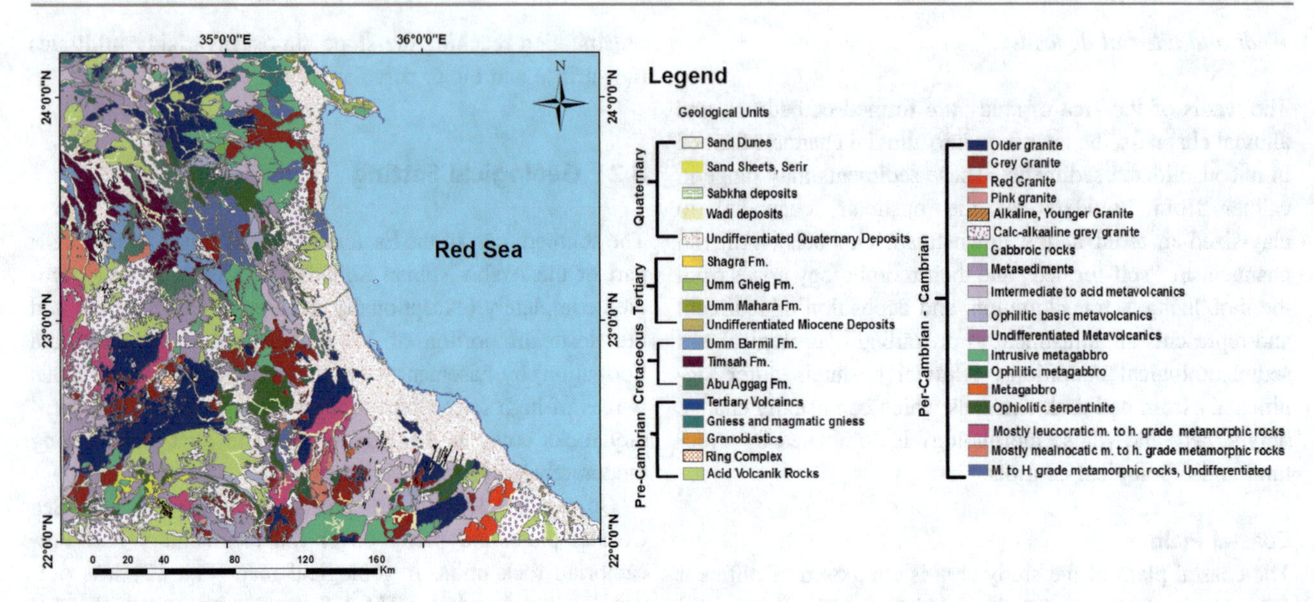

Fig. 12 Geological map of the study area (after Conoco 1987)

- Aeolian deposits
- Gravel plain.

These rock units formed the host environment for the groundwater occurrences as they provide favorable conditions for the groundwater aquifers and affect its hydrochemical quality. The following is a brief description of these units.

Precambrian Basement Rocks

It is accepted now that Egyptian basement rocks in the Eastern Desert were developed during the Pan African progeny and were formed by the accretion of intraoceanic island arcs. These basement rocks essentially consist of ultrabasics, metamorphic complex, and granitic masses. The metamorphic sequence of interbedded metasediments and metavolcanics is presented in most localities. The granitic masses belong to late to post-orogenic. The following is a short description for the Precambrian basement rocks.

Gneisses and Migmatites

Gneisses and migmatites are the oldest outcrops of the basement rocks in the Eastern Desert of Egypt. Low topographic relief, highly weathered characterizes them, gneissose texture and highly fractured. They are present as isolated small hills partly covered by sands and exposed due to the NW part of W. Hodein. Gneiss and migmatite rocks are highly deformed with a high grade of metamorphism, protruding melanosoum bands of hornblende and leucosoum. The latter is mainly quartz. Garnet crystals are abundant in these rocks.

Ultrabasic and Related Rocks

The ultramafic rocks occur at Sol Hamed area, as a belt of serpentinites striking in NE–SW trend. They show high topographic relief and exhibit green color. They are fractured in different directions and are also thrusted. The marginal parts formed talc carbonate rocks with brown color, highly weathered, and weak in hardness. These rocks are composed mainly of olivine and pyroxene relics, antigorite, and comb-shape crysotile veinlets. They contain iron oxides and chromite bodies. There are some quartz veins of E–W trend cutting the rocks. Also the white and yellow color carbonates minerals filling the joints. Pale green talcoses surfaces are abundant in the eastern side of this serpentinites belt, with some aggregates of olivine and pyroxene.

Metasediments

These rocks are characterized by pale gray in color and schistose texture, flaky, and of bedding fabric, highly affected by tectonic structure processes and highly weathered. They show pencil-shape weathering products. The metasediments consist of green schist facies where they are composed mainly of actinolite and tremolite with rare chlorite, epidote, and iron oxides. These rocks are exposed at different localities, such as the elongated areas around W. Hodein, W. Kraf, and great area at the SW part of the study area.

Metavolcanic Rocks

Nuweir and El-Sharkawi (1978) suggested that anywhere in the Eastern Desert of Egypt, the metavolcanic rocks are always associated and interbedded with the metasediment rocks. These rocks are of schistose greenschist facies composed mainly of actinolite, tremolite with few chlorites, and epidote with iron oxides and quartz veinlits. Many dykes in different directions cross them. These rocks are exposed to the south of Gebel Sul Hamid.

Gabbro–diorite Group

Akkad and Ramly (1960) gave the term of the gabbro–diorite group. These rocks are outcropped in Sol Hamed area and other greater parts of the study area. They are of medium to coarse grained in size, highly weathered, and highly fractured. These rocks are composed mainly of hornblende, calcic plagioclase, and iron oxides with few pyroxene and biotite crystals. There are chlorite and epidote in the joints.

Older Granitoid Rocks

These rocks are widely spread as isolated small hills with low topographic relief in the area of study. They show pale to deep gray colors and highly weathered with exfoliation fabric. These facies are named as gray granites by El-Ramly (1972) and previously named as syntectonic granites by Sabet (1972). The older granitoide rocks are medium to coarse grained and composed mainly of plagioclase, hornblende, biotite, and quartz and are cut by younger granites. They range in composition from granodiorite to adamllite.

Younger Granites

These granites were previously named as pink granites by Akkad and Ramly (1960). Sabet (1972) named them as late-tectonic granites. They are exposed as high topographic relief bodies with sharp contacts with the all other older basement rocks. They are composed mainly of orthoclase, microcline, quartz, and plagioclase, and they are of medium to coarse grained. Some quartz aggregates are present particularly at their contact with other rocks.

Alkaline Granites

These rocks are most probably belonging to Cretaceous age. They are of alkaline composition and composed mainly of alkaline feldspars (microcline and orthoclase) and quartz, of medium to coarse grain. These rocks are mainly represented in Gebal Elba with very high topographic relief. They are white in color and slightly weathered. They are jointed in vertical system, which filled with iron oxides; also there are some pockets of aggregated quartz.

Dykes

The Precambrian basement rocks were injected by a group of acidic and basic dykes. These dykes are penetrating the older basement rocks, which led to the creation of longitudinal ridges. The study area is affected intensively by these dykes which are responsible for the presence of several water points as, sometimes block the water in the wadis and form natural dams. Therefore, it was found that several water points were dug in the wadi beds. These wells are usually shallow, with depths rage from 8 to 10 m. The intensive role of these dykes was observed in several localities and was responsible for the presence of their water points as in: W. Rahaba, W. Hodein (W. Fiqo and W. Gimal, W. Ibib, W. kraf, and W. Meisah). Most of these dykes are sriking in NE–SW direction with vertical walls.

Ring Complex

Gebels Meshbeh and Naqroub El-Fokani is a typical ring complex of syenite rocks in the southern Eastern Desert of Egypt. They have a ring shape of very high topographic relief. They are composed mainly of syenite, trachyte, and younger granite rocks. They are not cut by any dykes but are cutting by NE–SW faults and belong to Cretaceous age. They show pale and deep red color of fine to medium grain size.

Cretaceous Sedimentary Rocks (Nubia Sandstone)

The Cretaceous rocks in the area of study are composed of alternating sequences of fine- to coarse-grained sandstone intercalated with silt or clayey beds. These rocks are exposed in the NW part of the study area, the region of W. Hodein, in Gebel Abraq and Abu Saafa area (Fig. 13).

The Nubia Sandstone succession is composed of the following formations, from older to younger.

Fig. 13 Field Photo showing the Nubia sandstone outcrops at W. Abu Saafa area

Abu Aggag Formation

El-Naggar (1970), noted that Abu Aggag Formation is composed of conglomeratic to coarse-grained kaolinitic sandstone. The predominant structures are large to small-scale trough cross-bedding and ripple laminations. Superimposed channel sandstones can be assigned to a low-sinuosity or braided river environment. The upper part of the formation shows fining upward sequences of lens-shaped channel sands and thick paleosoles. In the type area, it has a thickness of 30–40 m. The bottom of this formation is not exposed or reached by drilling in the area of study.

Timsah Formation

The Timsah Formation is composed of nearshore marine to deltaic sequences of silt and fine-grained sandstone with thick shale intercalations. Two or three oolitic iron ore beds occur in this formation (El-Naggar 1970). The thickness of this formation ranges from 5 to 130 m. Timsah Formation conformably overlies the Abu Aggag Formation and disconformably overlain the fluvial section of Umm Barmil Formation. The age of Timsah Formation ranges from Coniacian to Santonian (El-Naggar 1970).

Umm Barmil Formation

The Umm Barmil Formation is composed of coarse to medium-grained sandstones of large-scale tabular and trough cross-bedded originated from a low sinuosity fluvial environment (El-Naggar 1970). Paleo-current direction of forests points to N–NW. In the upper part, the sequence contains fining-upward cycles with intercalation of gray siltstone and ripple-laminated sandstone.

The thickness of this sequence at the type area is about 40 m, but may exceed 260 m in Wadi Dif area (Abdel Razik 1972; Endriszewitz 1988). Accordingly all the drilled groundwater wells in the area of study especially, in Abu Saafa and Dif areas did not encounter the base of this formation. In the subsurface, this formation consists of fine to coarse-grained sandstones, white to gray in color, mostly silty and clayey and slightly ferruginated. Umm Barmil Formation overlies Timsah Formation with an erosional contact and its age ranges from Santonian to Lower Campanian (Klitzsch et al. 1986; Hendricks and Luger 1987).

Miocene Carbonates Rocks

The Miocene sediments are found as isolated hills located to the west of Abu Ramad and Halaib areas in W. Serimtai, W. Aideib, W. Daaet, W. Kraf, W. Darera, and the region of W. Hubal. They mainly consist of alternating limestone and marl beds, and they are equivalent to the beds of Gebel El-Rusas Formation. The rocks of Gebel El-Rusas Formation are composed of alternating beds of marls and limestones interbedded with clay layers, which are covered by gravel, and boulders detached from the surrounding rocks. The Gebel El-Rusas Formation is characterized by the presence of a great number of manganese and calcite ore-bearing veins. These veins are affected by a group of faults generally striking NW–SE direction, i.e., parallel to the Red Sea. Also another group of faults having the general direction of NE–SW with a lateral displacement affects them. This displacement led to the appearance and disappearance of the Miocene rocks with its ore veins in some places in the study area.

The Quaternary Sediments

The Quaternary sediments comprise gravel plain, wadi deposits, sabkhas, aeolian sediments, etc. The following is some details about these units:

Gravel Plain

The coastal plain of the investigated area is mostly covered by gravel sheet (s) depending on the topography of the plain. The western side of the plain faces the high range of the basement rocks from which the separated blocks and boulders are rested nearby this side. As far as, at the shoreline of the Red Sea, the size of these blocks and boulders decreases gradually to flint and pebble size.

Wadi Deposits

These deposits formed terraces like islands, especially near the downstream of the wadis, and they are composed of gravels and pebbles graded in different sizes and interbedded with thin bands of coarse-grained sands.

Sabkhas

These deposits are extending along the Red Sea coast and occupy different pans according to the topography of the area. These deposits are formed of mud horizons intermingled with sands and salt. Some pans are partially covered by seawater depending on the extent of the tide, while others are permanently covered by water. The salt formed in these pans may reach in thickness to 5 cm.

Aeolian Deposits

These sediments are composed of windblown sands and generally form the serir feature. Also, these blown sediments form the longitudinal or crescent-shaped dunes in the area. The heights of these sand dunes range from 5 to 10 m above the ground surface. The length, height, and width of these dunes are increasing toward the Red Sea coast, where some of these dunes may reach 10 km in length. These movable sand dunes formed parallel groups of ridges beginning from the basement rocks to the near of the Red Sea coast. Some of the sand dunes are not movable and form stabilized sand dunes. This type of dunes is found along the eastern flank of W. Kraf. These dunes may be trapped in a local structural control and wetness of the ground. The importance of these dunes area is that they may comprise a buried paleo-drainage system along with W. Kraf. On the other hand, this aeolian sand may contain black sand derived from the Precambrian basement rocks.

3.2.2 Tectonics

The study area, as the other parts of the Eastern Desert, was subjected to different deformation processes, which affected the Precambrian basement and the overlying sediments. The faulting systems are one of the main factors, which control the occurrences of groundwater and its flow. The main structural elements affecting the study area are:

Faults

Many wadis and tributaries in the study area were channeled their preferred pass routes on the faults or fractures zones. The promising faulting system which affecting the study area can be summarized in the following sets:

Red Sea faulting system: These are the main faults, and having the trend N 25°–35° W, which is parallel to the Red Sea rifting system. In this system, the faults may extend more than 50 km. They possess strike–slip movements but some of them are of the normal type with their downthrown side toward the east.

Gulf of Aqaba (Dead Sea) system: This is the main secondary trend, which affected the study area. It has a direction of N 20° E.

The third set of faults have the N 50°–60° E trend which is considered as the oldest trend (Cretaceous) which affected the basement rocks. This trend is encountered in Gebel Elba at the southern part of the study area.

Fractures

Most of these structural features were formed as an echo of the major faulting systems, and others are primary structures; hence, they are numerous and intensified along the whole areas, with relatively shorter lengths than the main fault trends. These fractures are very important in groundwater studies as they are responsible for the presence of many hand-dug wells and springs in the study area, as will be discussed later.

Dykes

The different basement rock units outcropped in the investigated area is injected by swarms of dykes of acidic, intermediate, and basic composition. Generally, these dykes are arranged in parallel groups forming swarms of different trends and of longitudinal ridges. Also, these dykes are intersecting each other. As mentioned before, the rock types encountered in the area of study are highly fractured and jointed. They are together with the dyke swarms play an important role in preserving water. Where the former (fractures and joints) acts as passage ways and when they are intersecting with each other's tubular water bodies are formed among them (Davis 1965), while the later (sealed dykes) preserve and block water, especially when they are intersected with each other's or intersected with fractures and faults. The dyke swarms are easily detected on the Landsat images. The trends and orientations of these dykes are very important in groundwater movement, occurrences, and exploration, as when they are crossing the wadis they capture the groundwater at the upstream side of the wadis. Such case was observed in the field, where these dyke swarms (masked and not seen on the Landsat ETM + 8 images) are responsible for the occurrence of the famous water point of Bir Gahelia, in W. Rahaba.

3.3 Soil Characteristics

The present pedological setting of the study area is based on the interpretation of satellite images, field work survey, and analytical data. HS area is characterized by many types of soil as described in the following terms:

3.3.1 Soil Types in Halaib–Shalatin Area

Soil of the Wadies

These soils are sedimentary movable, sandy textures in some areas with a crumbs of rock and gravel in the elevated areas changed to loamy sand in moderate elevation areas and there are Celtic crust layers with thickness of 1 cm approximately cover the surface of the soil shows in Fig. 14, while at the end of the wadi, there is compacted soil which shows signs of water erosion (Figs. 15 and 16).

The morphological features of soil profiles representing this unit show that the dry soil color is very pale brown (10YR7/4) to brownish yellow (10YR 6/6) and it is pale brown (10YR 6/3) to yellowish brown (10YR 5/6) in moist condition was determined with the aid of Munssel Color Charts, (Anon 1975), the texture class is sandy loam for the surface layers. Soil profiles are characterized by deep soils, soil structure is sub-angular blocky and massive in the surface layer. The consistency is soft to slightly hard, non-sticky to slightly sticky, non-plastic to slightly plastic. Based on the morphological feature of the representative's profile of this unit and chemical analysis, the soils are classified as typic Torrifluvents.

Soil of the Delta and Alluvial Fans

These soils are sedimentary movable with water or wind, soil profiles are characterized by deep soils and moderately deep, soil texture is sand, loamy sand, and sandy loam.

The morphological features of soil profiles representing this unit show that the dry soil color is very pale brown (10YR7/4) to yellowish brown (10YR5/4) and it is pale brown (10YR 6/3) to dark yellowish brown (10YR 4/4) in moist condition, and the texture class is sandy loam to loamy sand for the surface layers. Soil structure is sub-angular blocky and massive in the surface layer. The consistency is

Fig. 14 Loamy crust on surface soil of the wadi

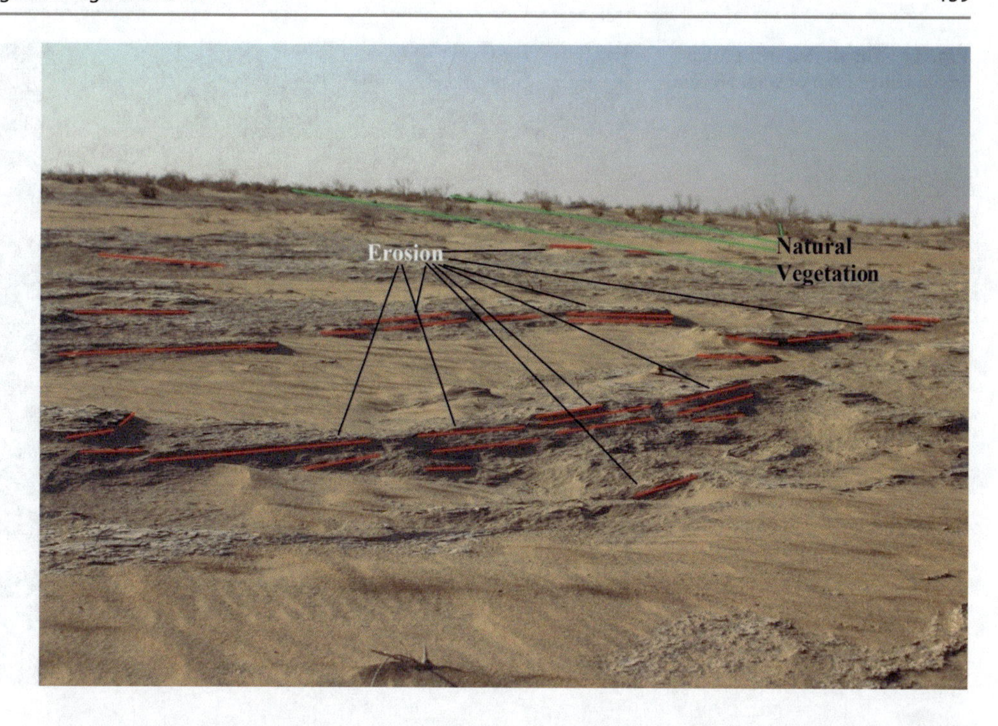

Fig. 15 Compacted soil in the upstream parts of the wadi

Fig. 16 Water erosion in the upstream parts of the wadi

soft to slightly hard, non-sticky to slightly sticky, non-plastic to slightly plastic. The horizon boundary is clear smooth. The morphological features of soil profiles representing this unit show that, most of the studied soil profiles contain a small percentage of salts, gypsum, and calcium carbonate. The depth of drilled soil profile is moderate to deep (Fig. 17). Based on the morphological feature of the representative's profile of this unit and chemical analysis the soils are classified as Typic Torrifluvents.

– **Soil of Sabkhas**

These soils are sandy and too salty which is flat to undulating and found in the low lying areas and extended with the

Fig. 17 The drilled soil profile in the alluvial fan of wadi Hodein

coastline, exist mainly in depressions and originated as a result of sea intrusion which rises water table. The soil profile is shallow and water table depth of 30–80 cm. Soil profile contains mostly on salic horizon or gypsic horizon.

The morphological features of soil profiles representing this unit show that the dry soil color is very pale brown (10YR 7/4) to brownish yellow (10YR 6/6). It is yellowish brown (10YR 5/6) in moist condition, the texture class is coarse sand. Soil profiles are characterized by shallow to moderately deep soils (30–80 cm), soil structure is single grained structure and very friable in the surface layer. The consistency is slightly hard, no sticky and no plastic. The horizon boundary is a clear smooth and diffuse smooth boundary. This unit is shallow deep water table. Salt is increased with depth. It is mainly present in fine salt crystals forms for the surface layer and the accumulation of salts for soil. Gypsum content present in the studied layers is in the form of small shiny crystals. The soils of this unit are classified as Typic Aquisalids.

Soil of Sand Sheet

Soil profiles are characterized by deep soils, soil texture is sandy. The morphological features of soil profiles representing this unit show that the dry soil color is very pale brown (10YR7/4) to brownish yellow (10YR 6/6) and it is pale brown (10YR 6/3) to yellowish brown (10YR 5/6) in moist condition, the texture class is sandy for the surface layers. Soil structure is single grained and massive in the surface layer. The consistency is soft to slightly hard non-sticky, non-plastic. The horizon boundary is clear smooth. Most of the studied soil profiles are a small percentage of salts, gypsum, and calcium carbonate. The soils of this unit are classified as Typic Torripsamments.

3.3.2 Soil Classification of Halaib–Shalatin Area

The American Soil Survey Staff system (USDA 2010) was applied down to the sub-great group level for the studied soil mapping unit. Table 2 shows the calculated area for each

Table 2 Soil classification according to USDA (2010) for HS area

Soil order	Soil sub-order	Soil great group	Soil sub-great group	Units
Aridisols	Salids	Aquisalids	Typic Aquisalids	Wet and dry Sabkhas
Entisols	Psamments	Torripsamments	Typic Torripsamments	Sand sheet and sand dunes
	Fluvents	Torrifluvents	Typic Torrifluvents	Wadi, Alluvial fans and Delta
	Orthents	Torriorthents	Typic Torriorthents	Alluvial deposits

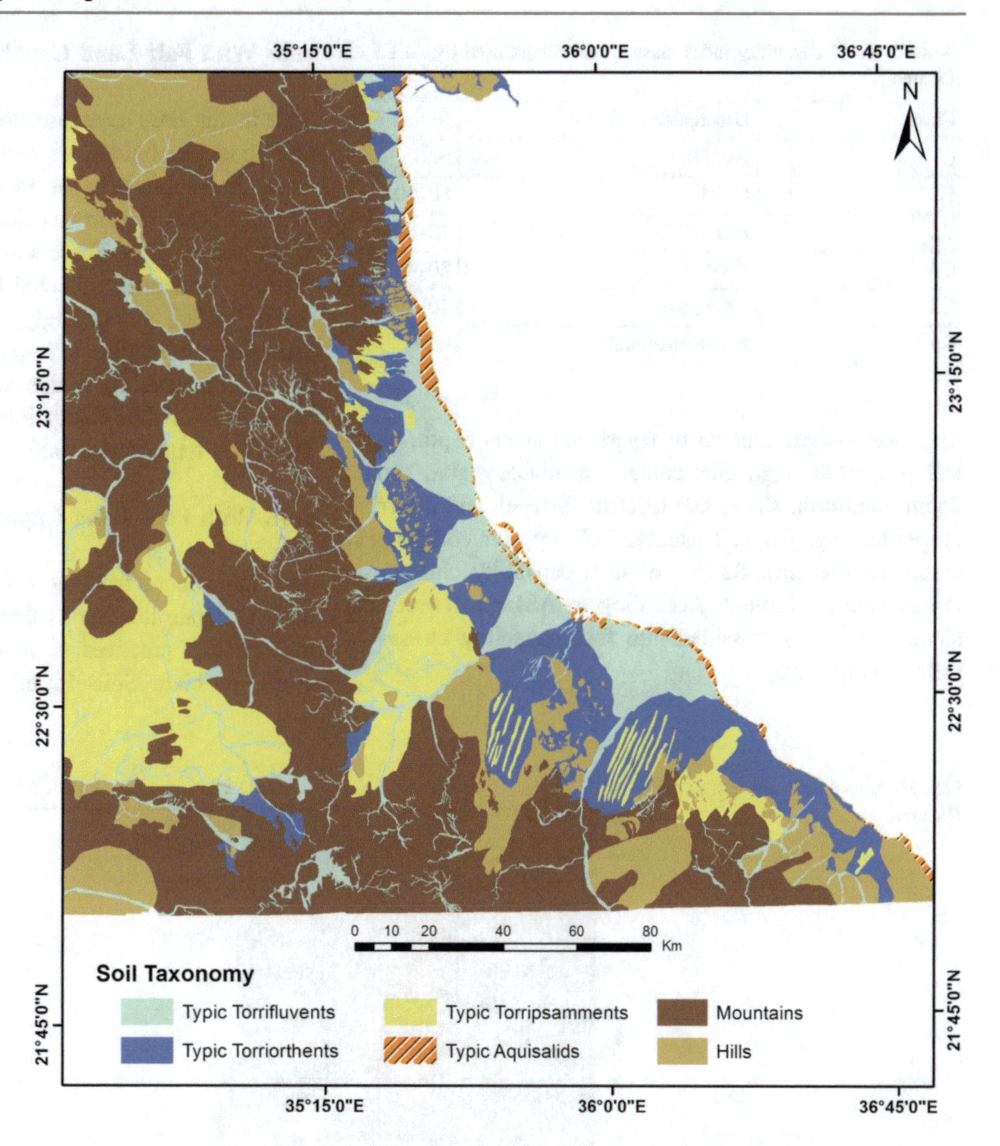

Fig. 18 The soil map of HS area

soil classification unit. The spatial distribution of these soil great groups is illustrated in Fig. 18.

The defined diagnostic horizons of Aridisols are salic and gypsic horizons. Salic horizon usually exists in soils close to salt marches and poorly drained area such as salt marshes are also common in these areas. The process of soil salinization is more often connected with rising of water level and high evaporation rate. These areas mostly located in the northern parts near the coast.

Diagnostic horizons (salic and gypsic) are different in presence with each other in accordance to the conditions associated with the configuration in the arid and semi-arid regions due to poor rainfall and relatively high evaporation which, in turn results in the accumulation process. Precipitations of salts arise from processes of evaporation and thus, at least soluble salt (carbonate) precipitates first followed by the most soluble salt (gypsum), where calcic horizon exists are down the profile and gypsum above calcic horizon. Either in the presence of groundwater level (such as salt marshes), the movement of the water to down, and, thus, it consists of salic horizon followed by gypsic horizon.

The soil of the study area contains two orders of soil classification; the Entisols, which includes Psamments, Orthents, and Fluvents suborders and the Aridisols, which include Salids suborders.

3.3.3 Land Evaluation

Land Capability Classification

Land capability classification was carried out using the Applied System of Land Evaluation (ASLE software) which has been developed by (Ismail et al. 2005). ASLE software is inserting for soil database and calculates possible indices combinations between the major factors. The soil properties factor was used to calculate land capability. This property is

Table 3 Land capability index classes and ratings used by ALES of HS area

Class	Description	Rating (%)
C1	Excellent	>80
C2	Good	60–80
C3	Fair	40–60
C4	Poor	20–40
C5	Very poor	10–20
C6	Non-agricultural	<10

irrigation system, number of layers and layers depth, physical properties (e.g., clay content, available water, profile depth, landform, slope, and level of surface), and chemical properties (e.g., pH, soil salinity, CEC, gypsum content, and carbonate content). Ranges of land capability classes are represented in Table 3. According to ASLE software, the studied area was classified into four capability classes as follows (Fig. 19).

Soils With Fair Land Capability (C3)

Soils in this class have more than slight limitations and more than moderate limitations that require moderately intensive management practices or moderately restrict the range of crops, or both. Soils in this class have fair soil index between 42.64 and 57.47%. These soils are deep to moderately well drained, moderately affected by alkaline and electric conductivity (EC). Accordingly, these soils need slightly good management practices to improve their current situation. So, the current capability of this soil map unit can be changed to be good with moderately intensive management practices. This class represented wadi and alluvial fans.

Soils With Poor Land Capability (C4)

Soils in this class have more than slight limitations and more than moderate limitations that require moderately intensive management practices or moderately restrict the range of crops, or both. Soils in this class have poor soil index

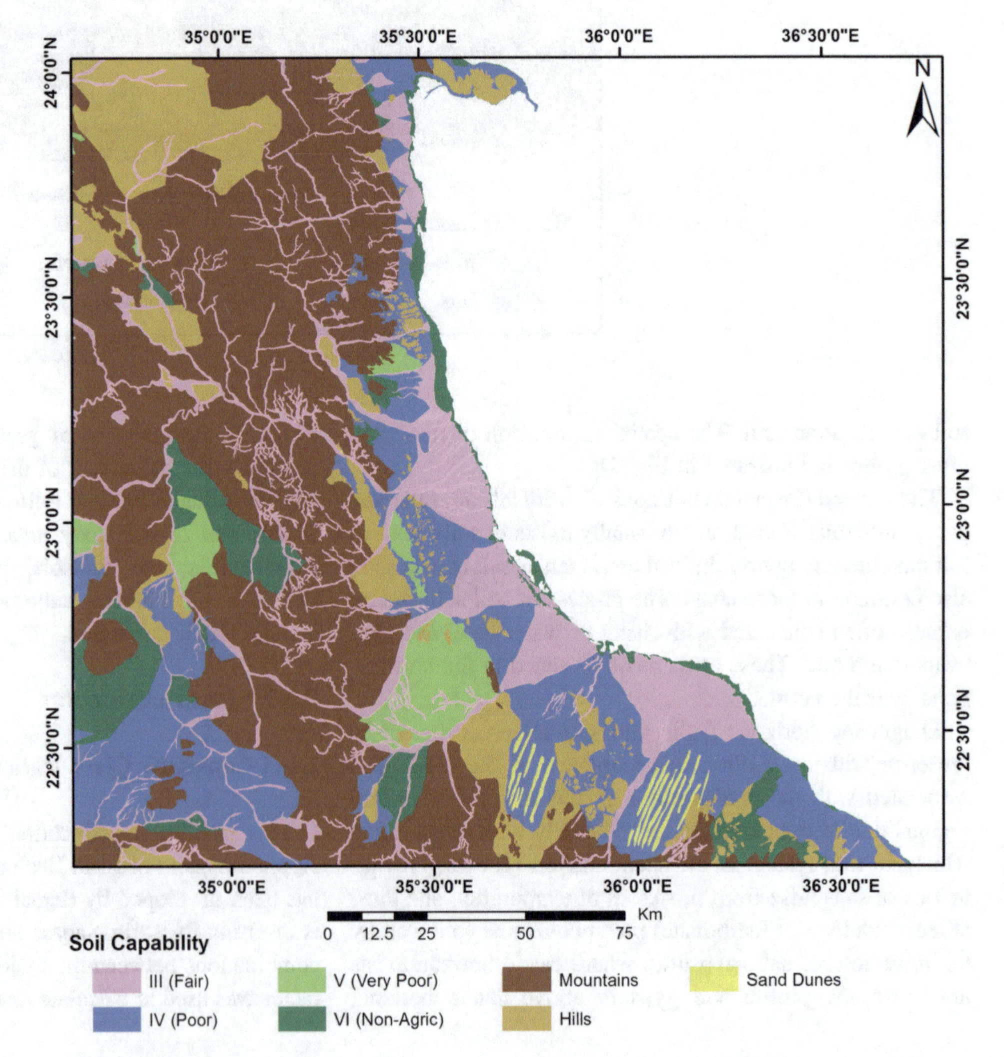

Fig. 19 Capability soil index of HS area

between 20.3 and 38.3%. Accordingly, these soils need good management practices to improve their current situation. So the current capability of this soil map unit can be changed to be fair with moderately intensive management practices. This class represented some soils of alluvial deposites.

Soils With Very Poor Land Capability (C5)

Soils in this class have one or more severe limitations that exclude the use of land, or with one or more severe limitations that require special management practices or severely restrict the range of crops, or both. These soil units have some limitations such as salinity, soil depth, and content of $CaCO_3$ because it has low soil index (10.37–18.32%). These lands require good and proper management. These soil units will be improved to be fair or good. They are moderately high in productivity for a fair range of crops. This class is represented on the plain with rock out crop.

Soils With Non-Agriculture Land Capability (C6)

Soils in this class have severe or very severe limitations that cannot be corrected such as texture, high salinity, high contents of gypsum, and very shallow soil profile. It has low soil index ranging 1.39%. This class is represented by sabkhas.

4 Results of Watershed Modeling and Discussions

4.1 Generalities

HS area is classified into 52 watersheds, each has its geographical location and own hydromorphological parameters (Fig. 20 and Table 4).

4.2 Criteria of Runoff Water Harvesting (RWH) Potentiality Determination in Halaib–Shalatin Area

RWH could play a vital role in reducing flood proneness and providing household consumption (Sepehri et al. 2018). The RWH potentiality of the study area was determined by spatially integrating eight thematic layers, which represent the most important hydrographic and hydromorphometric parameters or criteria for determining the RWH potentiality. After defining the watershed attributes with the DEMs inside the platform of WMS 8.5© software, the multi-criteria decision support (MCDSS) layers are to be converted into data coverage for easier data storage and manipulation. The input criteria (layers) and their ranges used in the construction of the weighted spatial propability model (WSPM) are

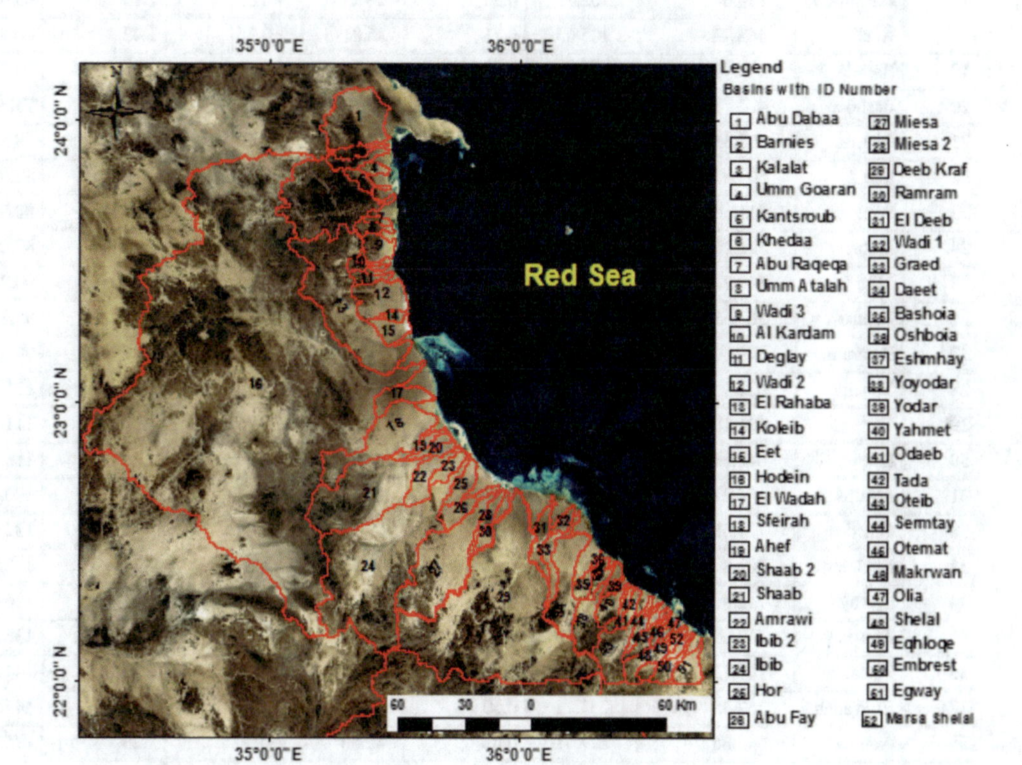

Fig. 20 Landsat satellite ETM + 8 image showing HS Watersheds

Table 4 WMS 8.0© software hydrographical output criteria used for demarcating the watersheds' characteristics of HS area

Basin ID	Watershed	Volume of annual flood (VAF) (10^3 m^3) Finkel method	Peak flood flow (Q_{max}) (m^3/s) Finkel method	Overland flow distance (OFD) (km)	Max. flow distance (MFD) (km)	Basin infiltration No. (IF)	Drainage density (Dd) (km^{-1})	Basin Area (BA) (km^2)	Basin slope (BS) (m/m)	Basin length (BL) (km)
Halaib–Shalatin watersheds										
1	Olia	218.10	36.72	0.63	14.11	0.97	1.25	23.24	0.04	12.61
2	Mersa Shalal	225.11	38.50	0.72	15.91	0.88	1.43	24.37	0.02	12.65
3	Yodar	225.77	38.67	0.72	14.58	0.77	1.45	24.47	0.01	13.16
4	Abu Raqeqa	249.07	44.77	0.59	14.21	0.67	1.18	28.34	0.15	12.21
5	oshboiya	254.37	46.20	0.70	17.54	1.00	1.39	29.24	0.02	13.95
6	Makrawan	282.15	53.93	0.57	20.14	0.84	1.14	34.13	0.05	16.74
7	Shaab 2	418.66	97.19	0.72	30.39	0.77	1.43	61.51	0.06	23.84
8	Kolieb	302.00	59.69	0.66	30.48	0.91	1.32	37.78	0.01	14.53
9	Al kardam	335.48	69.83	0.53	17.33	0.72	1.05	44.20	0.01	18.25
10	El Laqd	364.60	79.07	0.58	17.08	0.67	1.15	50.05	0.08	14.77
11	Odaeb	395.16	89.16	0.59	20.60	0.92	1.18	56.43	0.01	21.37
12	Eshmhay	396.39	89.58	0.69	24.95	0.92	1.38	56.69	0.20	20.10
13	Tada	401.15	91.19	0.68	25.12	0.80	1.36	57.71	0.02	18.68
14	Deglay	411.04	94.56	0.52	24.74	0.72	1.05	59.85	0.08	17.99
15	Sermatay	413.38	95.37	0.57	21.12	0.79	1.14	60.36	0.21	25.64
16	Umm Atalah	424.41	99.19	0.50	22.00	0.72	1.00	62.78	0.16	16.45
17	Kanstroub	434.67	102.79	0.56	24.22	0.83	1.13	65.06	0.18	16.45
18	Ahef	442.72	105.65	0.71	36.81	0.75	1.43	66.87	0.01	28.06
19	Barnies	449.91	108.22	0.57	29.74	0.63	1.13	68.49	0.12	24.19
20	Bashoia	472.76	116.52	0.60	27.18	0.80	1.20	73.75	0.03	22.03
21	Otemat	497.33	125.68	0.59	35.63	0.80	1.18	79.54	0.08	28.55
22	Ibib 2	507.81	129.65	0.58	27.54	0.75	1.16	82.06	0.02	20.26
23	Eqhloqe	511.50	131.06	0.63	38.89	0.75	1.27	82.95	0.11	29.64
24	Embrest	528.26	137.52	0.55	33.95	0.62	1.10	87.04	0.10	27.05
25	Misa 2	543.27	143.39	0.64	42.76	0.66	1.28	90.75	0.01	33.35
26	Kalalat	563.65	151.49	0.55	37.78	0.66	1.09	95.88	0.17	25.93
27	Ramram	579.00	157.69	0.63	41.45	0.75	1.27	99.81	0.01	41.50
28	Yahemt	622.99	175.90	0.60	28.95	0.72	1.20	111.33	0.01	21.36
29	Hor	625.16	176.82	0.65	27.90	0.70	1.29	111.91	0.10	21.45
30	Eet	642.03	183.99	0.61	27.45	0.74	1.22	116.45	0.01	18.99
31	Abu Mad	656.27	190.11	0.57	21.12	0.74	1.15	120.32	0.14	15.05
32	Graed	701.58	210.03	0.60	63.24	0.67	1.20	132.93	0.02	41.50
33	El Deeb	703.34	210.82	0.62	51.54	0.64	1.24	133.43	0.02	35.58
34	Egway	712.90	215.11	0.61	31.52	0.79	1.22	136.15	0.05	23.98
35	Abu Fay	713.09	215.20	0.46	38.38	0.24	0.92	136.20	0.02	28.35
36	Shelal	740.71	227.75	0.56	40.15	0.72	1.13	144.15	0.15	33.61
37	El Wadah	752.35	233.12	0.60	35.81	0.73	1.20	147.54	0.01	25.97
38	Amrawi	845.35	277.41	0.62	45.80	0.90	1.23	175.57	0.02	32.22
39	Kelibtab	1109.47	416.25	0.60	39.35	0.81	1.19	263.45	0.09	25.13

(continued)

Table 4 (continued)

Basin ID	Watershed	Volume of annual flood (VAF) (10^3 m^3) Finkel method	Peak flood flow (Q_{max}) (m^3/s) Finkel method	Overland flow distance (OFD) (km)	Max. flow distance (MFD) (km)	Basin infiltration No. (IF)	Drainage density (Dd) (km^{-1})	Basin Area (BA) (km^2)	Basin slope (BS) (m/m)	Basin length (BL) (km)
40	Yoyodar	1118.38	421.25	0.56	52.23	0.73	1.11	266.61	0.11	40.72
41	Otieb	1147.92	437.97	0.58	60.46	0.79	1.16	277.19	0.18	40.14
42	Daeet	1427.07	606.09	0.61	53.74	0.74	1.22	383.60	0.04	42.13
43	Abu Dabaa	1678.51	772.20	0.79	40.10	1.68	1.58	488.73	0.06	28.52
44	Sfeirah	1900.70	929.63	0.64	72.93	0.81	1.27	588.37	0.05	53.70
45	Meisa	2357.61	1282.18	0.54	90.19	0.69	1.09	811.51	0.07	64.02
46	Khedaa	2439.80	1349.47	0.47	68.63	0.59	0.95	854.09	0.20	49.57
47	Shaab	2547.82	1439.61	0.60	100.17	0.77	1.19	911.15	0.09	68.16
48	Rahaba	2834.56	1688.00	0.57	101.65	0.76	1.13	1068.3	0.09	64.56
49	Ibib	4115.46	2944.85	0.52	129.72	0.70	1.04	1863.8	0.13	92.44
50	Kraf (from Egypt)	5563.66	4618.49	0.56	146.27	0.75	1.12	2923	0.09	82.23
51	Hodien	13989.60	18288.53	0.34	303.83	0.19	0.69	11575	0.07	143.64
52	Um Goran	342.22	71.94	0.68	24.88	0.9	1.36	45.5	0.20	12.6

Table 5 Ranges of input criteria used in the WSPM for HS area

Watershed RWH Criteria	Very high	High	Moderate	Low	Very Low
Volume of annual flood (1000 m^3)	>11590	11580–8603	8602–5512	5511–2476	<2475
Overland Flow Distance (km)	>0.6218	0.6217–0.5733	0.5732–0.5125	0.5124–0.4392	<0.4391
Maximum Flow Distance (km)	>247.2	247.1–175.1	175–112.3	112.2–61.03	<61.02
Basin infiltration number	>1.169	1.168–0.7715	0.7714–0.6022	0.6021–0.3804	<0.3803
Drainage density(km^{-1})	>1.244	1.243–1.147	1.146–1.026	1.025–0.8782	<0.8781
Basin Area (km^2)	>9279	9278–6455	6454–3586	3585–1309	<1308
Basin Slope (m/m)	>0.1368	0.1367–0.1038	0.1037–0.07783	0.07782–0.04807	<0.04806
Basin Length (km)	>117.6	117.5–91.8	91.79–64.56	64.55–38.82	<38.81

given in (Tables 4 and 5). Integration of these criteria in the GIS-based WSPM will result in the production of comprehensive maps determining the efficient sites suitable for RWH, with a number of classes. These thematic layers, which represent the WSPM model inputs, include: average overland flow distance (OFD), VAF, BS, Dd, BL, BA, basin infiltration number (IF), and MFD (Table 4). The following is a short description of these themes:

– **Average Overland Flow Distance (OFD)**

The average OFD within the basin is computed by averaging the overland distance traveled from the centroid of each triangle to the nearest stream. The overland flow is the water that flows over the slopes of the drainage basin and is then concentrated into stream channels. Also, it is known as surface flow. It is affected by the type of soil and surface topography that govern the erosion rates caused by the overland flow (Montgomery and Dietrich 1989). High–very high OFDs (>247.2 km) are represented by the eastern side of the study area constitute a strip parallel to the Red Sea coast. It covers most of the small watersheds that are close to the coast such as (Embrest, Egway, Mrasa Shelal, El-Deep, El-Laqd, El-Wadah, Sfeirah, Abu Dabaa, Baranies, and Kalalat Watersheds) (Fig. 21). Additionally, it covers the downstream parts of lager watersheds such as (Hodein, Ibib,

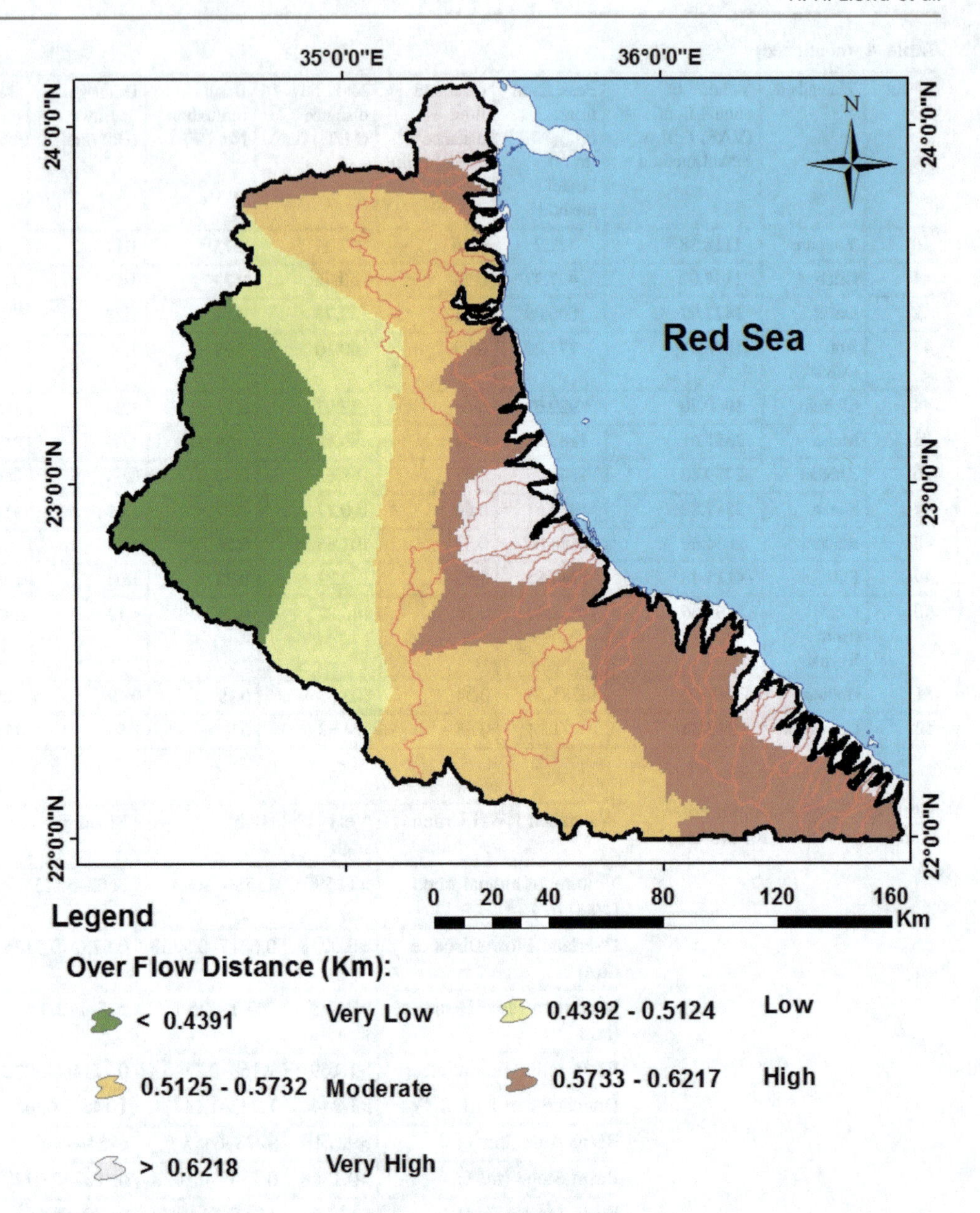

Fig. 21 The OFD thematic layer used in the WSPM

and Deep Kraf. The thematic layer of the OFD indicates a pronounced decrease in value toward the western side of the study area reaching to very low (<0.4391 km) at the upstream parts of Hodein Watershed (Fig. 21 and Table 5). The moderate class of OFD (0.5732–0.5125 km) was occupied by the most parts of Miesa, Khedaa, Wadi 3, and Ibib Watersheds in addition to some parts of Hodein, Deep Kraf, and Sfeirah Watersheds. This layer was assigned a weight of 12.5 in the WSPM (Table 6).

– **Volume of Annual Flood (VAF)**

The availability of a pronounced VAF in a drainage basin is one of the most important parameters for the success of the RWH (Elewa and Qaddah 2011). The VAF reflects the quantity of water available for harvesting. The values of VAF were calculated and listed in Tables 4 and 5 in units of 1000 m^3.

The VAF was calculated by Finkel's method (Finkel 1979). Figure 22 shows the classes of VAF, where the high–very high classes (>11590) occur mostly in the western corner of study area mainly in W. Hodein Watershed at its upstream parts (Fig. 22). The moderate class (8602–5512) of the VAF occurs as a small strip in the central part of W. Hodein Watershed in addition to a very small part in Shaab Watershed at its upstream. The low and very low VAF classes (<2475) comprises large parts of the study area represented mainly by the eastern side parallel and close to

Table 6 The first WSPM scenario (equal weights to criteria), ranks, and degree of effectiveness of themes used in the RWH potentiality mapping of HS area

Thematic layer (Criterion)	RWH potentiality class	Average rate (Rank) (R_c)	Weight (W_c)	Degree of effectiveness (E)
Volume of annual flood (VAF)	I (Very high) II (High) III (Moderate) IV (Low) V (Very low)	0.9 0.7 0.5 0.3 0.1	12.5	11.25 8.75 6.25 3.75 1.25
Average overland flow distance (OFD)	I (Very high) II (High) III (Moderate) IV (Low) V (Very low)	0.9 0.7 0.5 0.3 0.1	12.5	11.25 8.75 6.25 3.75 1.25
Maximum flow distance (MFD)	I (Very high) II (High) III (Moderate) IV (Low) V (Very low)	0.9 0.7 0.5 0.3 0.1	12.5	11.25 8.75 6.25 3.75 1.25
Basin infiltration number (IF)	I (Very high) II (High) III (Moderate) IV (Low) V (Very low)	0.9 0.7 0.5 0.3 0.1	12.5	11.25 8.75 6.25 3.75 1.25
Drainage density (Dd)	I (Very high) II (High) III (Moderate) IV (Low) V (Very low) V (Very low)	0.9 0.7 0.5 0.3 0.1	12.5	11.25 8.75 6.25 3.75 1.25
Basin area (BA)	I (Very high) II (High) III (Moderate) IV (Low) V (Very low)	0.9 0.7 0.5 0.3 0.1	12.5	11.25 8.75 6.25 3.75 1.25
Basin slope (BS)	I (Very high) II (High) III (Moderate) IV (Low) V (Very low)	0.9 0.7 0.5 0.3 0.1	12.5	11.25 8.75 6.25 3.75 1.25
Basin length (BL)	I (Very high) II (High) III (Moderate) IV (Low) V (Very low)	0.9 0.7 0.5 0.3 0.1	12.5	11.25 8.75 6.25 3.75 1.25

the Red Sea coast extending from northeast to southeast (Fig. 22).These classes are represented by all watersheds with different areas ranging from small parts as in Hodein Watershed to hole watershed area as in the other watersheds (Fig. 22). In general, the spatial distribution of the VAF classes was degraded from the very high to very low classes passing by high, moderate, and low classes from the western to the eastern parts of the study area (Fig. 22). This layer was assigned a weight of 12.5 in the WSPM for the RWH (Table 6).

– **Basin Slope (BS)**

The BS of the drainage basin is a key factor for the selection of water-harvesting locations in order to get the maximum storage capacity in the channel. It is the average slope of the triangles comprising this basin (Horton 1945; Leopold and Maddock 1953). Reasonable care should be taken into account when determining this parameter, as the peak discharge and hydrograph shape is sensitive to the slope value used for the basin (Jones 1999). In the present work, the

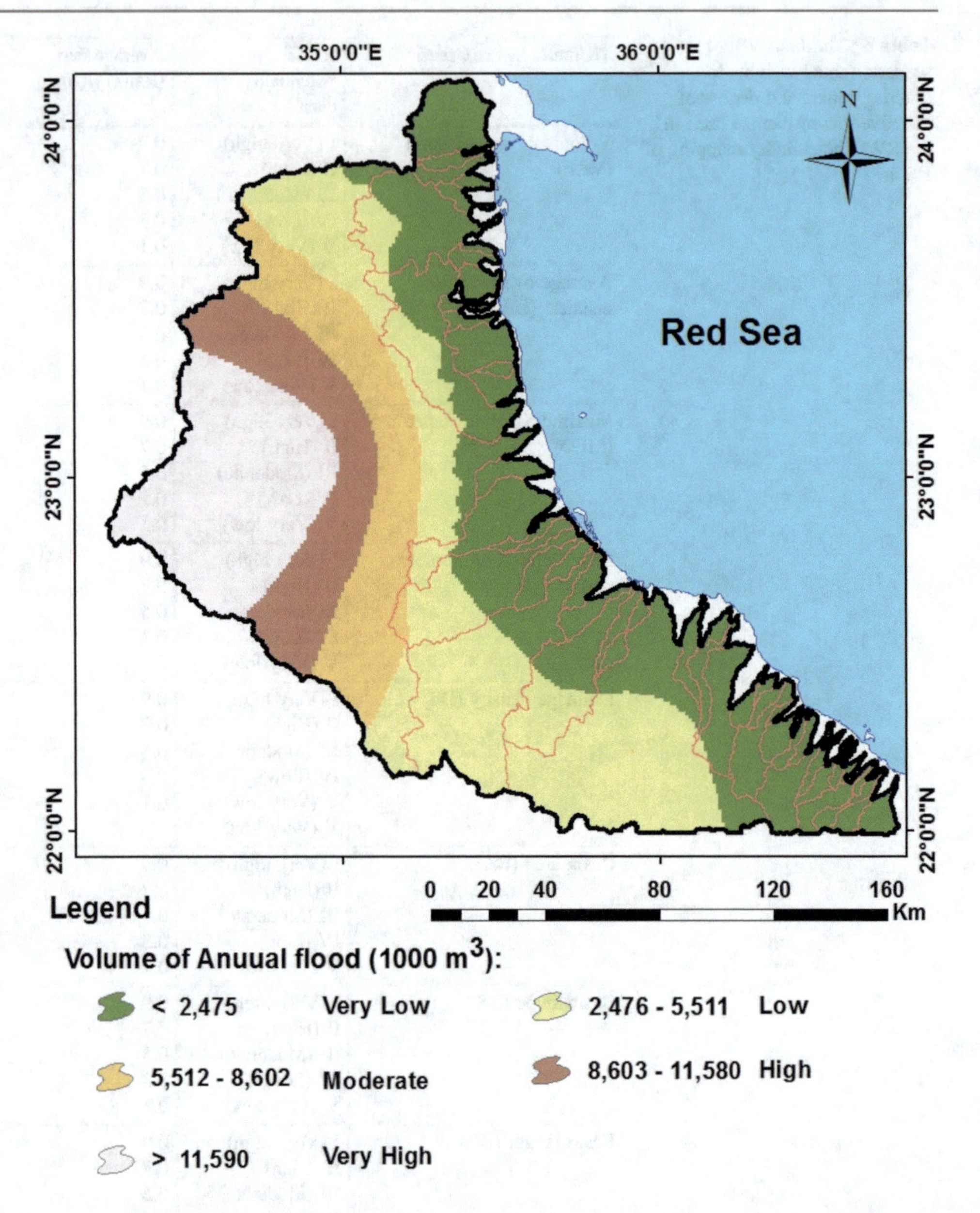

Fig. 22 The VAF (calculated by Finkel's method) thematic layer used in the WSPM

slope map is generated from the ASTER DEM. Five slope classes were generated (Fig. 23). The slope map was merged with the basin map to create the slope attributes of each drainage basin. The thematic layer of the BS indicates that high–very high slope (>0.1368) are represented by the SE corner of the study area occupying the upstream parts of many watersheds such as (Marsa Shelal, Egway, Embrest, Eqhloqe, Shelal, Deep Kraf, and Otieb Watersheds). These classes also represented by Umm Goaran, Kantsroub, Khedaa, Abu Ragea, and Umm Atalah Watersheds in the northeastern (NE) side of the study area in addition to some parts of Hodein, Rahaba, Kelibtab, and Deglay Watersheds (Fig. 23). Moderate class of BS (0.1037–0.07783 m/m) are represented by a considerable strip in southern parts of the study area covering the upstream parts of Shaab, Ibib, Miesa, and Deep Kraf Watersheds in addition to some parts of Hodein, Rahaba, Kelibtab Watersheds in the northern parts of the study area (Fig. 23). Moderate class is decreased to low–very low BS classes(<0.04806 m/m) to the eastern side mainly in the central parts of the study area extending from central parts to the outlet of many Watersheds (Fig. 23) which doubles the possibilities of the RWH. The possibility of RWH is higher in gentle or medium-sloped basins parts. This layer was assigned a weight of 12.5 in the WSPM (Table 6).

– **Drainage Density (Dd)**

The Dd was classified into five classes and were ordered as: >1.244, 1.243–1.147, 1.146–1.026, 1.025–0.8782,

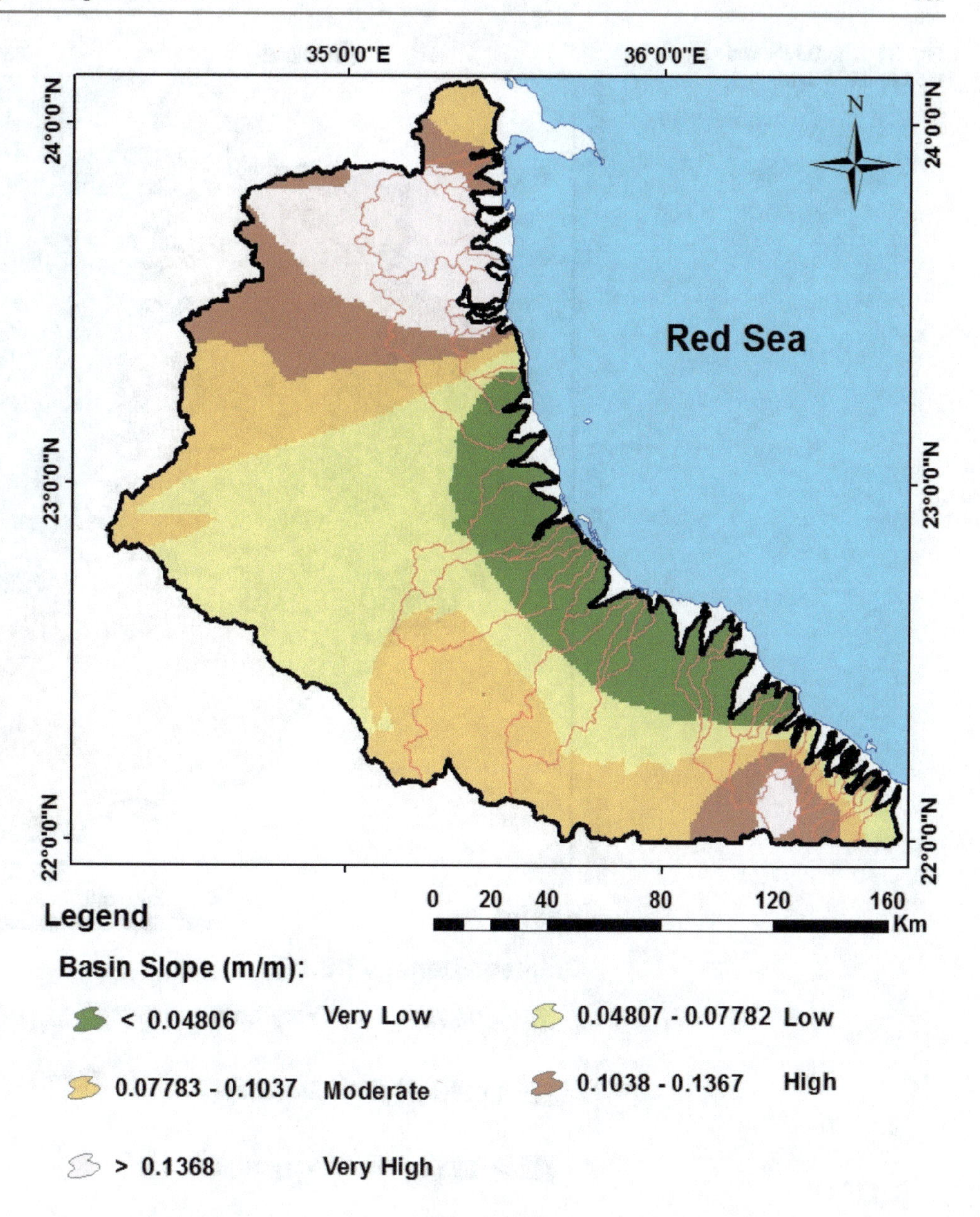

Fig. 23 The BS thematic layer used in the WSPM

and <0.8781 km^{-1} for the very high, high, moderate, low, and very low potential for the RWH, respectively (Table 5; Fig. 24). The general decrease in the Dd values due to the western parts (watershed upstream areas). The thematic map of Dd arranged the Dd classes from very low, low, moderate, high, and very high, respectively, from west to east parts of the study area (Fig. 24). It is also observed that there are many watersheds in the eastern side of the study area are totally covered by high–very high Dd classes as in Abu Dabaa, Baranis, Kalalat, Amrawi, Ibib 2, Shelal, Eqhloqe, Embrest, Egway, Marsa Shelal, Oteib, and Otemat Watersheds (Fig. 24). This layer had been assigned a weight of 12.5 in the WSPM (Table 6).

– **Basin Length (BL)**

The BL is defined as the distance which cut the basin into two similar parts (Horton 1945). The longer the BL the lower the chances that such a basin will be flooded; this is because, the longer the basin, the lower its slope, and hence the higher the possibilities for the RWH. The longer basins in the study area (High–very high BL classes; >117.6 km) are represented only by Hodein Watershed mainly in its central and upstream parts (Fig. 25). High–very high flowed by moderate BL class and represented by considerable parts of Shaab, Ibib, Miesa, and Deep Kraf watersheds. The general decrease in BL is due to the east parts of the study

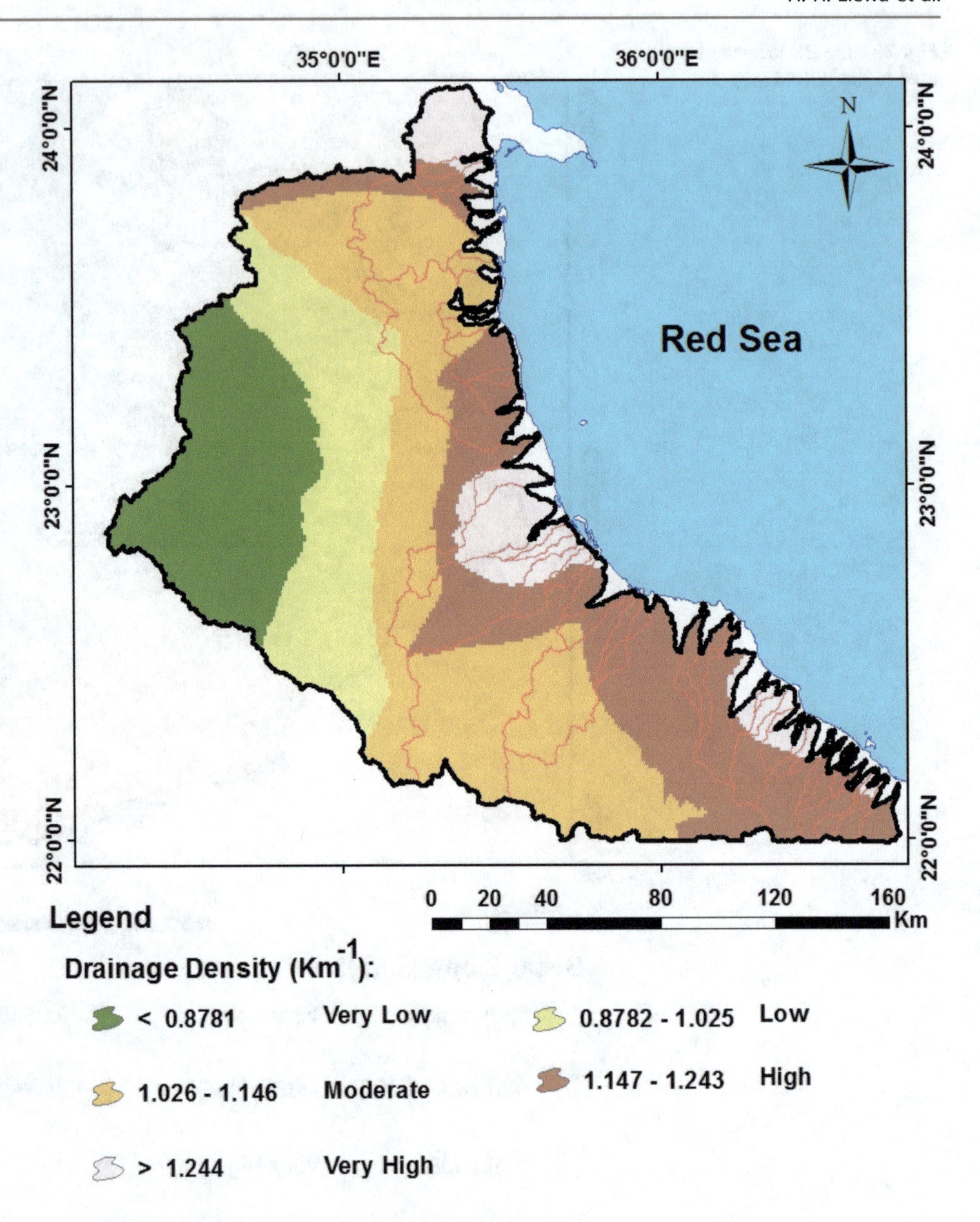

Fig. 24 The Dd thematic layer used in the WSPM

area, whereas the low–very low classes totally covered the micro-watersheds extending along the eastern side in addition to some parts of larger watersheds (Fig. 25). Micro-catchment RWH practices are more successful in shorter BLs (i.e., Abu Dabaa, Baranis, Kalalat, Amrawi, Ibib 2, Shelal, Eqhloqe, Embrest, Egway, Marsa Shelal, Oteib, and Otemat Watersheds), whereas the macro-catchment procedures are more applicable in the longer BLs (i.e., Hodein, Ebib, and Miesa Watershds). This is because, the longer the basin the lower its slope and, hence, the higher the possibilities for the RWH. This layer was assigned a weight of 12.5 in the WSPM (Table 6).

– **Basin Area (BA)**

The BA is the total area in square kilometers enclosed by the basin boundary (Horton 1945). BA is the most important of all the morphometric parameters controlling the catchment runoff volume and pattern. This is because, the larger the size of the basin the greater the amount of rain it intercepts and the higher the peak discharge that result (Morisawa 1959; Verstappen 1983). Another reason for the high positive correlation between BA and the discharge is the fact that the BA is also highly correlated with some of the other catchment hydromorphometric characteristics, which

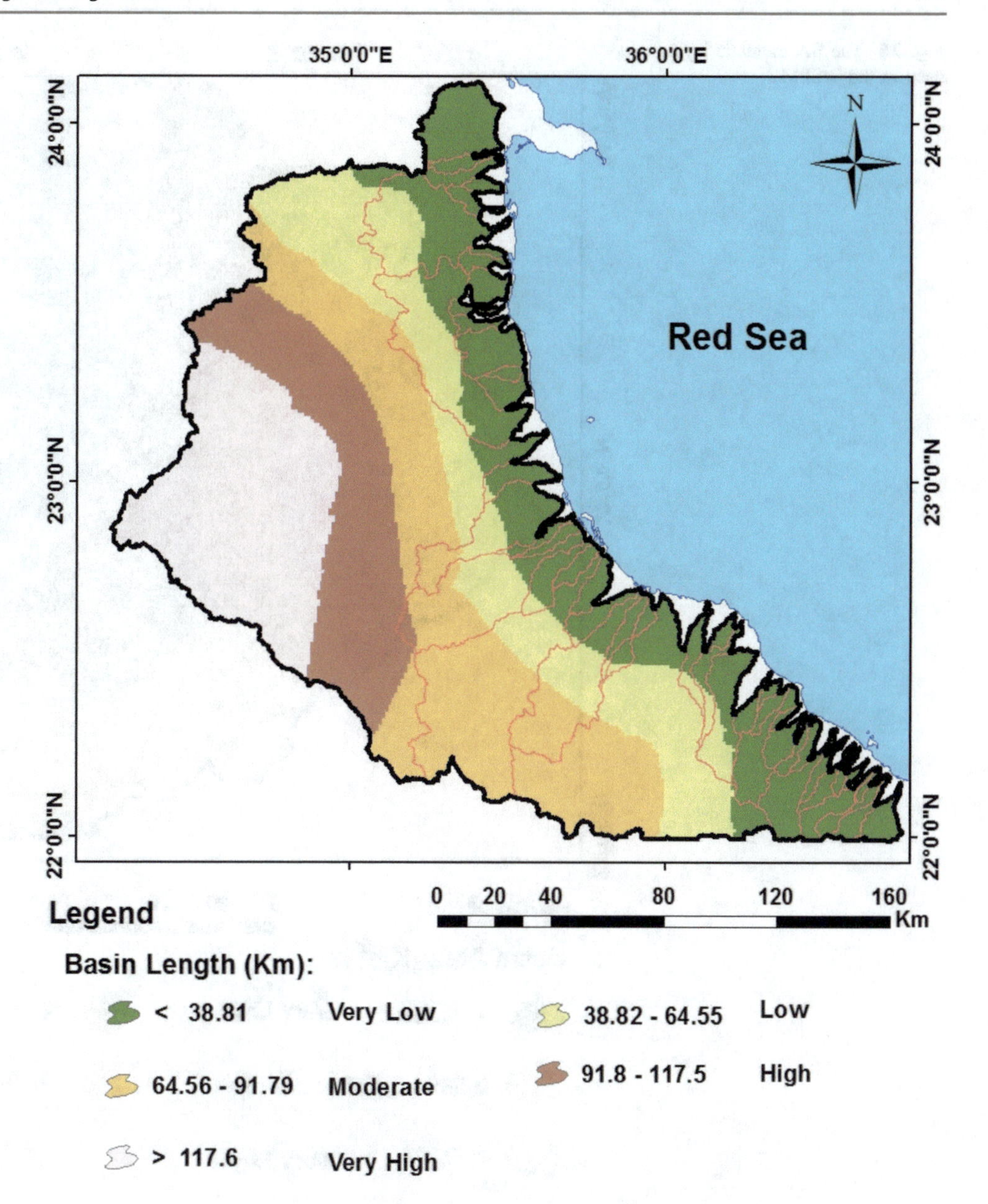

Fig. 25 The BL thematic layer used in the WSPM

influence runoff, such as BL (i.e., the larger the BA the longer its length), average OFD, and the MFD (Gregory and Walling 1973; Jain and Sinha 2003). A thematic layer of the BA with five classes was generated (Fig. 26). The very high–high BA classes (>9279 km^2) occur in W. Hodein upstream Watershed. The moderate BA class (6454–3586 km^2) is represented by a curved zone in the central part of Hodein Watershed in addition to a very small part of W. Shaab Watershed (Fig. 26). The low–very low classes (<1308 km^2) are represented by the other parts of the study area, which occur in the eastern and the central parts. This layer was assigned the weight of 12.5 in the WSPM (Table 6).

– **Basin Infiltration Number (IF)**

The IF is defined as the product of Dd and drainage frequency (Df) (Faniran 1968). The higher the infiltration number the lower will be the infiltration and consequently, higher will be the runoff. This leads to the development of higher Dd. IF gives an idea about the infiltration of a drainage basin. The IF thematic map reflects that most of the study area is represented by moderate IF class (0.7714–0.6022) which are characterized by moderate RWH potentialities and infiltration capabilities. The low–very low IF classes (<0.3803) are represented by the upstream and central parts of W. Hodein Watershed in addition to very small

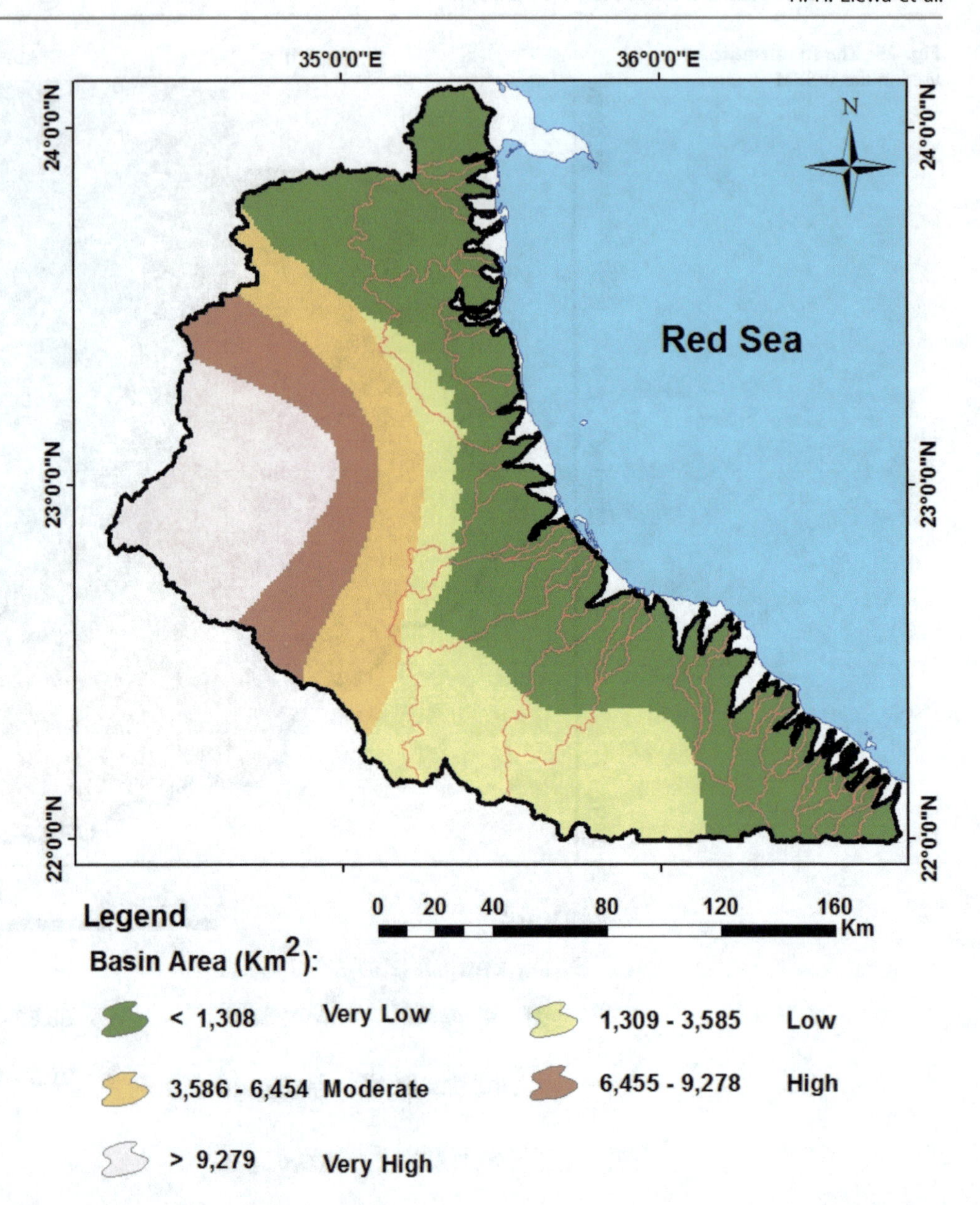

Fig. 26 The BA thematic layer used in the WSPM

parts of Ibib, Hor, Miesa, and Khedaa Watersheds (Fig. 27). The very high IF class only occurs in Abu Dbaa watershed at the NE part. High IF class is distributed as scattered areas in northern, central, and southern parts (Fig. 27). This layer was assigned a weight of 12.5 in the WSPM (Table 6).

– **Maximum Flow Distance (MFD)**

The MFD of a basin includes both overland and channel flow (Horton 1945). It is the maximum length of the water's path in the drainage basin in kilometers. This factor is important in determining the RWH capability of a drainage basin, as the higher the MFD the higher the RWH possibilities, and vice versa. The high–very high MFD classes (>247.2 km) occur mainly in central and upstream parts of W. Hodein Watershed. The moderate MFD class (175–112.3 km) occurs as longitudinal zone extending from north to south through the central and SW parts of the study area occupying central and upstream parts of Hodein, Rahaba, Sfeirah, Shaab, Ibib, Miesa, and Deep Kraf Watersheds. The moderate MFD class if decreased further to the eastern strip of the study area reaching to very low class (<61.02 km) covering mostly micro-watersheds close to Red Sea coast and downstream parts of larger watersheds (Fig. 28).

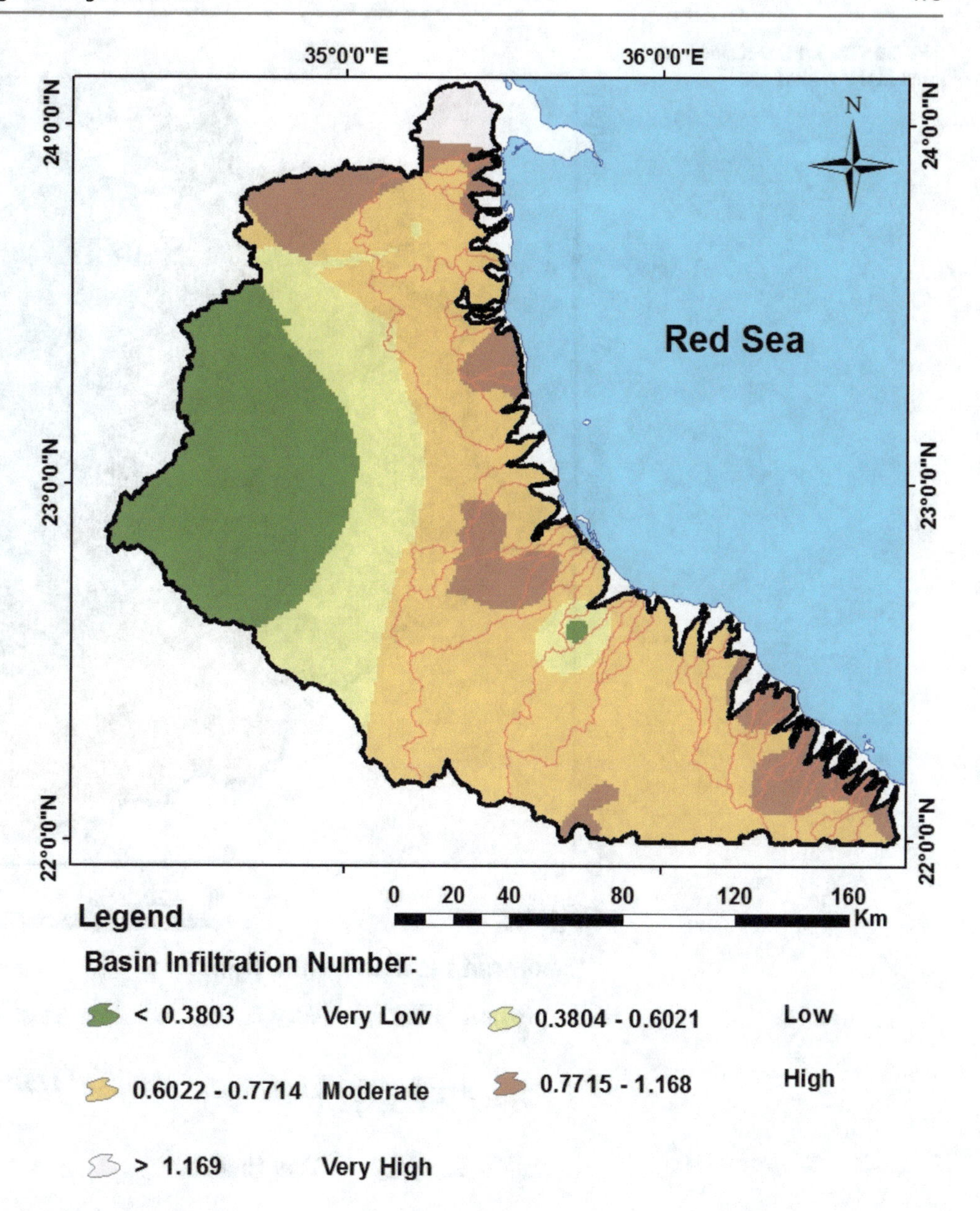

Fig. 27 The IF thematic layer used in the WSPM

4.3 Weighted Spatial Probability Models (WSPMs) for RWH in Halaib–Shalatin Area

The MCDSS (Malczewski 1996, 2006) represented by the previously discussed eight thematic layers, were ranked according to their magnitude of contribution to the RWH, thus they were categorized from the very high to very low contribution, and the same classes were used in the RWH potentiality mapping.

Three weighted spatial probability models' scenarios (WSPM) were generated, where the model was running three times by using criteria of: (1) Equal weights, (2) Weights proposed by the authors' judgments, and (3) Weights justified by the sensitivity analysis. The model's running implied the integration of all criteria as thematic layers in the WSPM. Accordingly, an output map for each scenario with several classes indicating the categories of RWH potentiality (i.e., very high, high, moderate, low, and very low) was obtained.

4.3.1 WSPM's Scenario I (Equal Weighting of Criteria) for Halaib–Shalatin Area

In this model's scenario, the previously discussed criteria had been proposed to have the same magnitude of contribution in the RWH potentiality mapping (Table 6; Fig. 29). However, some criteria work positively, while others work negatively in the RWH potentiality mapping. For example, the BS criterion works negatively in the RWH potentiality mapping, whereas the VAF, OFD, BA, BL, Dd, and MFD, work positively. The weights and rates of the previously

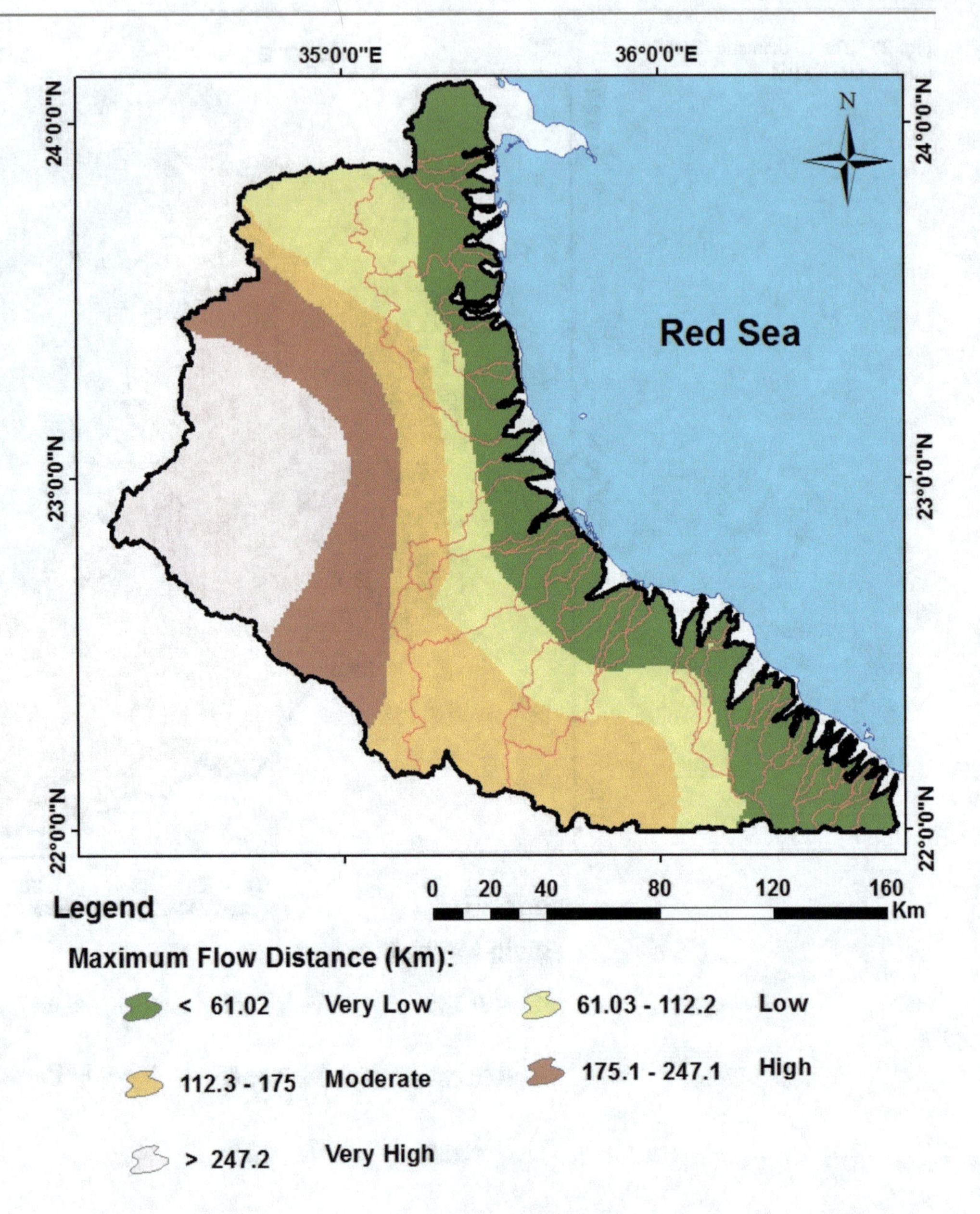

Fig. 28 The MFD thematic layer used in the WSPM

discussed MCDSS were assumed and optimized here to have equal weights of contribution in the RWH potentiality mapping (Table 6), where the integrated criteria were given an equal weight of 12.5% with a summation of 100% for all data themes. After proposing criteria weights, categorization was applied to each of the five classes among each criterion. For example, the classes were graded from I (very high potentiality) up to V (very low potentiality) in the RWH potentiality mapping (Table 6; Fig. 29). Taking 100% as the maximum value for the rank, thus for the five classes, ranks will be classified as: 100–80, 80–60, 60–40, 40–20, and 20–0%, respectively. Consequently, the average ranking for each class will be 0.90, 0.70, 0.50, 0.30, and 0.10% for classes from I to V, respectively (Table 6).

The degree of effectiveness (E) for each thematic layer was calculated by multiplying the criterion weight (W_c) with the criterion rank (R_c). For example, if the weight of VAF equals to 12.5% and this is multiplied by the average rank of 90 (for class I), the E will be 11.25 (Eq. 3).

$$E = W_c x R_f = 0.125x90 = 11.25 \quad (3)$$

According to this method of data manipulation, the assessment of the effectiveness of each decision criterion provides a comparative analysis among the different thematic layers. Therefore, it is clear from Table 6 that the class I in all data themes (i.e., E = 11.25) represent the most effective criteria in performing the RWH potentiality mapping, compared to the least effective class V (i.e., E = 1.25)

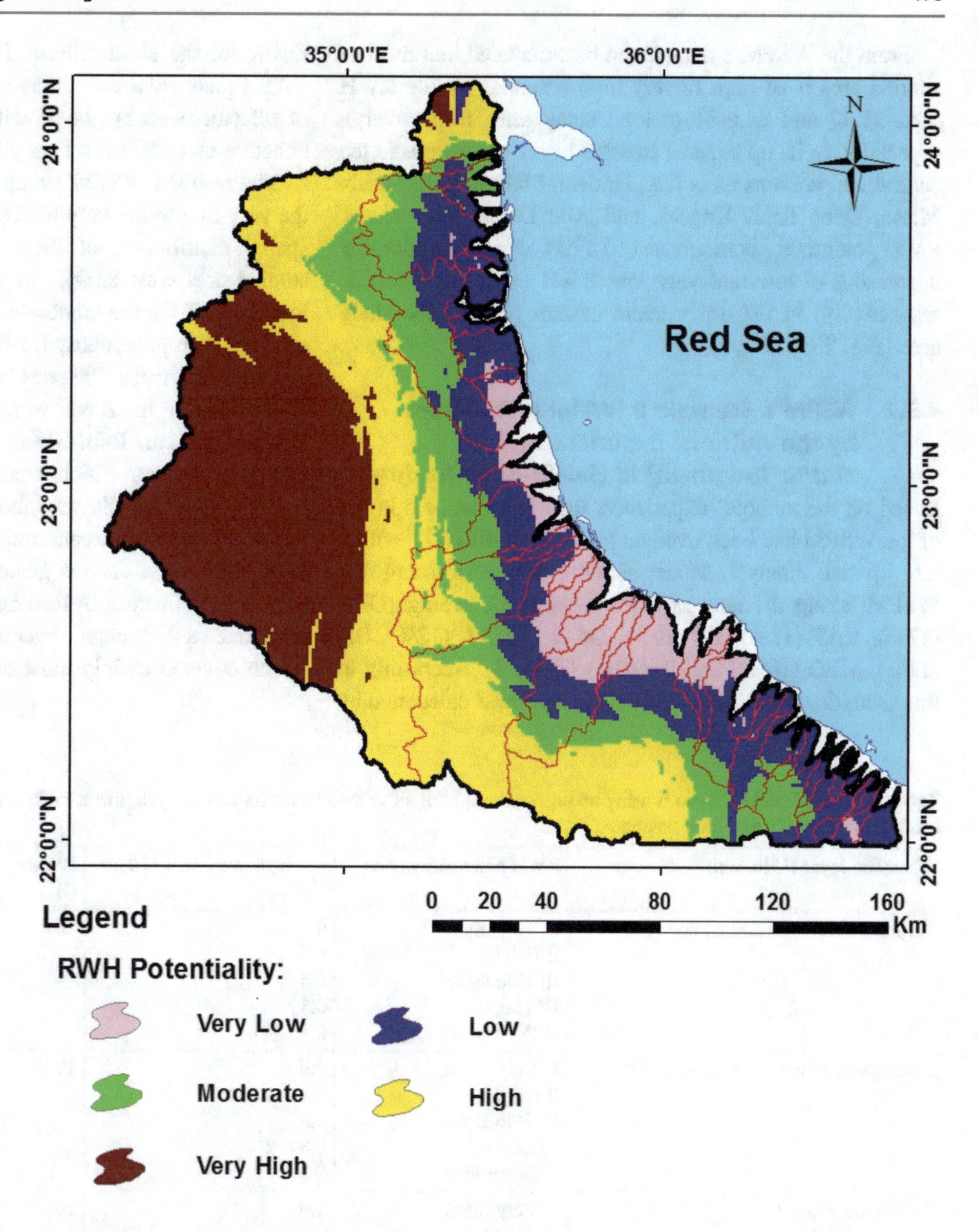

Fig. 29 WSPM's Scenario I map showing the potential areas for the RWH in HS area

Table 7 Areas of RWH potentiality classes (WSPM Scenario I) in HS area

RWH potentiality map					
RWH potentiality class	Very low	Low	Moderate	High	Very high
Area (km^2)	2091.55	4841.74	4207.38	7989.81	6261.96
Area (in % relative to the total studied area) (Total study area: 25392.4 km^2)	8.24	19.07	16.57	31.47	24.66

in all criteria. Therefore, an arithmetic overlay approach was built within the ArcGIS 10.1© Spatial Analyst Model Builder to conduct the WSPM. The overlay processing manipulates both continuous and discrete grid layers and the derived data are continuous grid data layer. Thus, the WSPM output map for the RWH potentiality with five classes ranging from the very low to very high potentiality was produced (Fig. 29).

The percentage of the spatial distribution of the resulting RWH potentiality classes to the total study area of the study area: 8.24 (very low), 19.07 (low), 16.57 (moderate), 31.47 (high), and 24.66% (very high) (Fig. 29; Table 7).

From the WSPM's map, it can be concluded that most of studied area is of high to very high potential for the RWH (i.e., 31.47 and 24.66% of total study area, respectively), especially, in its up streams areas of larger watersheds in the central and western parts (i.e., Hodein, Sfeirah, Ibib, Shaab, Miesa, Deep Kraf, Khedaa, and Abu Dabaa Watersheds. RWH potentiality is moderate (16.57%), which is noticeably decreasing to low and very low RWH (19.07 and 8.24%, respectively) in the downstream eastern parts of the study area (Fig. 29; Table 7).

4.3.2 WSPM's Scenario II (Weights Assigned by the Authors' Experience and/or Judgment) in Halaib–Shalatin Area

Based on the authors' experience, the eight thematic layers of the WSPM had been overlain by the ArcGIS 10.1© within the Spatial Analyst Model Builder for performing the WSPM taking the new assigned weights as: average OFD (17%), VAF (16%), BS (12%), Dd (12%), BL (12%), BA (11%), MFD (10%), and IF (10%) (Table 8). According to this scenario, the E for each thematic layer was calculated by Eq. 1, for the eight criteria. For example, if the weight of VAF equals 16% and this is multiplied by the average rank of 90 (for class I), the E will be 14.4, whereas the least effective class V has a E of 1.6 (Table 8).

The resulted WSPM's map has five classes ranging from the very low to the very high potentiality for the RWH. The spatial distribution of these classes relative to the total studied area was: 8.09% for the very low, 16.09% for the low, 18.27% for the moderate, 33.10% high, and 24.44% for the very high potentiality for the RWH (Fig. 30; Table 9). It is observed from the resulted map that the promising watersheds for the RWH were represented by watersheds of Hodein, Sfeirah, Ibib, Shaab, Miesa, Deep Kraf, Khedaa, and Abu Dabaa, which characterized by high–very high RWH potentiality classes (about 59% of total studied area), especially in its upstream and central parts (Fig. 30). High–very high RWH classes generally decreased to the eastern side in the direction of Red Sea coast, where it followed by moderate RWH class reaching to low–very low classes which covered mainly most of the micro-watersheds.

Table 8 The WSPM scenario II using an unequal weighting of criteria based on authors' judgment; ranks and degree of effectiveness of themes used in the RWH potentiality mapping

Thematic layer (Criterion)	RWH potentiality class	Average rate (Rank) (R_c)	Weight (W_c)	Degree of effectiveness (E)
Average volume of annual flood (VAF)	I (Very high)	0.9	16	14.4
	II (High)	0.7		11.2
	III (Moderate)	0.5		8
	IV (Low)	0.3		4.8
	V (Very low)	0.1		1.6
Average overland flow distance (OFD)	I (Very high)	0.9	17	15.3
	II (High)	0.7		11.9
	III (Moderate)	0.5		8.5
	IV (Low)	0.3		5.1
	V (Very low)	0.1		1.7
Maximum Flow Distance (MFD)	I (Very high)	0.9	10	9
	II (High)	0.7		7
	III (Moderate)	0.5		5
	IV (Low)	0.3		3
	V (Very low)	0.1		1
Basin infiltration number (IF)	I (Very high)	0.9	10	9
	II (High)	0.7		7
	III (Moderate)	0.5		5
	IV (Low)	0.3		3
	V (Very low)	0.1		1
Drainage density (Dd)	I (Very high)	0.9	12	10.8
	II (High)	0.7		8.4
	III (Moderate)	0.5		6
	IV (Low)	0.3		3.6
	V (Very low)	0.1		1.2
Basin area (BA)	I (Very high)	0.9	11	9.9
	II (High)	0.7		7.7
	III (Moderate)	0.5		5.5

(continued)

Table 8 (continued)

Thematic layer (Criterion)	RWH potentiality class	Average rate (Rank) (R_c)	Weight (W_c)	Degree of effectiveness (E)
	IV (Low)	0.3		3.3
	V (Very low)	0.1		1.1
Basin slope (BS)	I (Very high)	0.9	12	10.8
	II (High)	0.7		8.4
	III (Moderate)	0.5		6
	IV (Low)	0.3		3.6
	V (Very low)	0.1		1.2
Basin length (BL)	I (Very high)	0.9	12	10.8
	II (High)	0.7		8.4
	III (Moderate)	0.5		6
	IV (Low)	0.3		3.6
	V (Very low)	0.1		1.2

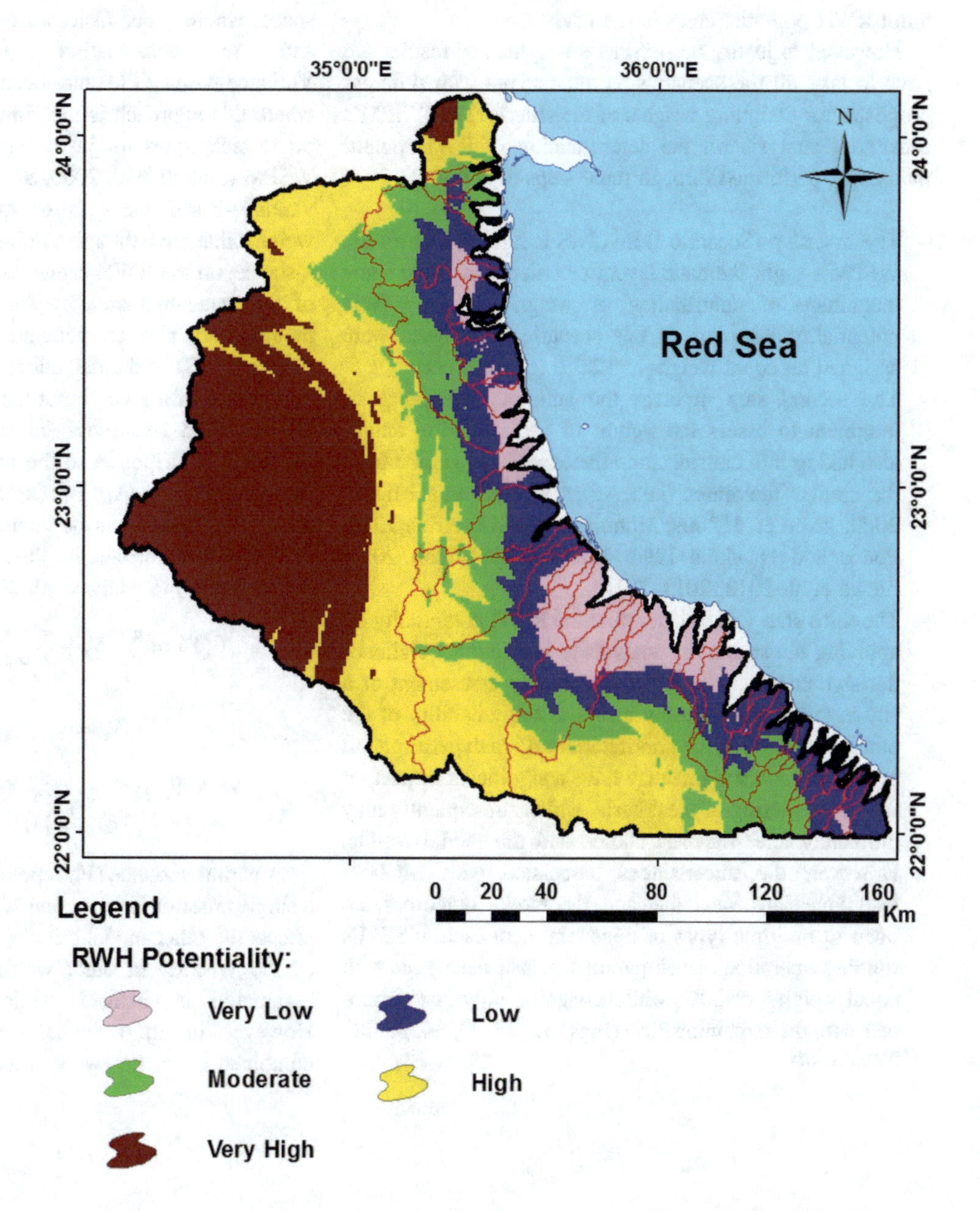

Fig. 30 WSPM's Scenario II map showing the potential areas for the RWH in HS area

Table 9 Areas of RWH potentiality classes (WSPM Scenario II) in HS area

RWH potentiality map					
RWH potentiality class	Very low	Low	Moderate	High	Very high
Area (km^2)	2055.03	4085.24	4639.91	8405.47	6206.81
Area (in % relative to the total studied area) (Total study area: 25392.4 km^2)	8.09	16.09	18.27	33.10	24.44

4.3.3 WSPM's Scenario III (Justified Weights by the Sensitivity Analysis) in Halaib–Shalatin area

In the third scenario, a sensitivity analysis (Van Griensven et al. 2006) was performed to justify the weights of the WSPM's criteria in order to attain a more justified or optimum RWH potential areas in the study area.

However, to justify the WSPM's weights and results, we have to take all the scenarios as alternatives with different proposals for assigning weights of the criteria. The WSPM's sensitivity analysis for the determination of RWH potentiality was performed through three steps as follows:

- The first step (Scenario I) involves assuming that all the WSPM's eight thematic layers or criteria have the same magnitude of contribution or weights in the RWH potentiality mapping. In this scenario, all criteria were assigned an equal weight of 12.5% (equal effect).
- The second step involves the authors' experience or judgment to assess the weight of each criterion and/or also taking into consideration the experience gained from the similar literatures (i.e., Adiga and Krishna Murthy 2000; Javed et al. 2009; Montgomery and Dietrich 1989; Ponce and Hawkins 1996; Elewa and Qaddah 2011; Elewa et al. 2012, 2013, 2014).
- The third step was to determine the RWH potentiality by applying the sensitivity analysis to justify the weights of decisive criteria. The RWH potentiality assessment of a site involves the necessity to assess the reliability of the parameters used in the prioritization. A small perturbation in the decision weights may have a significant impact on the rank ordering of the criteria, which subsequently may ultimately alter the best choice and the model results. However, the uncertainties associated with MCDSS techniques are inevitable and the model outcomes are open to multiple types of uncertainty. In each WSPM's running operation, seven parameters had been kept with equal weights of 10%, while assigning only one parameter with the remaining 30% (Figs. 31, 32, 33, 34, 35, 36, 37 and 38).

The previously discussed WSPM's running practice is necessary to apply the variance-based global sensitivity analysis (GSA) or what is called the ANOVA (ANalysis Of VAriance), which subdivides the variability and apportions it to the uncertain inputs (Ha et al. 2012; Feizizadeh et al. 2004). GSA is based on perturbations of the entire parameter space, where input factors are examined both individually and in combination (Ligmann-Zielinska 2013). Variance-based GSA has been used in the present work, where this approach is identified as one of the most appropriate techniques for justifying the factors' weights of the WSPM (Saltelli et al. 2000; Saisana et al. 2005). The goal of variance-based GSA is to quantitatively determine the weights that have the most influence on model output, in this instance, on the RWH categorization computed for each cell of the watershed area by the decisive factors. With this method we aim to generate two sensitivity measures: first-order (S) and total effect (ST) sensitivity index. The importance of a given input factor (X_i) can be measured via the so-called sensitivity index, which is defined as the fractional contribution to the model output variance due to the uncertainty in (X_i). For (k) independent input factors, the sensitivity indices can be computed by using the following decomposition formula for the total output variance ($V(Y)$) of the output (Y) (Saisana et al. 2005) (Eqs. 4–6):

$$V(Y) = \sum_i V_i + \sum_i \sum_{j>i} V_{ij} + \ldots + V_{12,\ldots k} \tag{4}$$

$$V_i = V_{xi}\{E_{x-i}(Y|X_i)\} \tag{5}$$

$$V_{ij} = V_{xixj}\{E_{x-ij}(Y|X_i, X_j)\} - V_{xi}\{E_{x-i}(Y|X_i)\} - V_{xj}\{E_{x-j}(Y|X_j)\} \tag{6}$$

A partial variance (V_i) represents the repeated variation of a single criterion (i) (i.e., one of the eight model criteria) that affects the other model criteria, which constitutes the inputs of the WSPM. In other words, one of the WSPM eight parameters is changed, while the rest remain constant. However, in Eq. 4, higher order effects (V 1, 2, p) are combined effect for two or more inputs. The partial effects

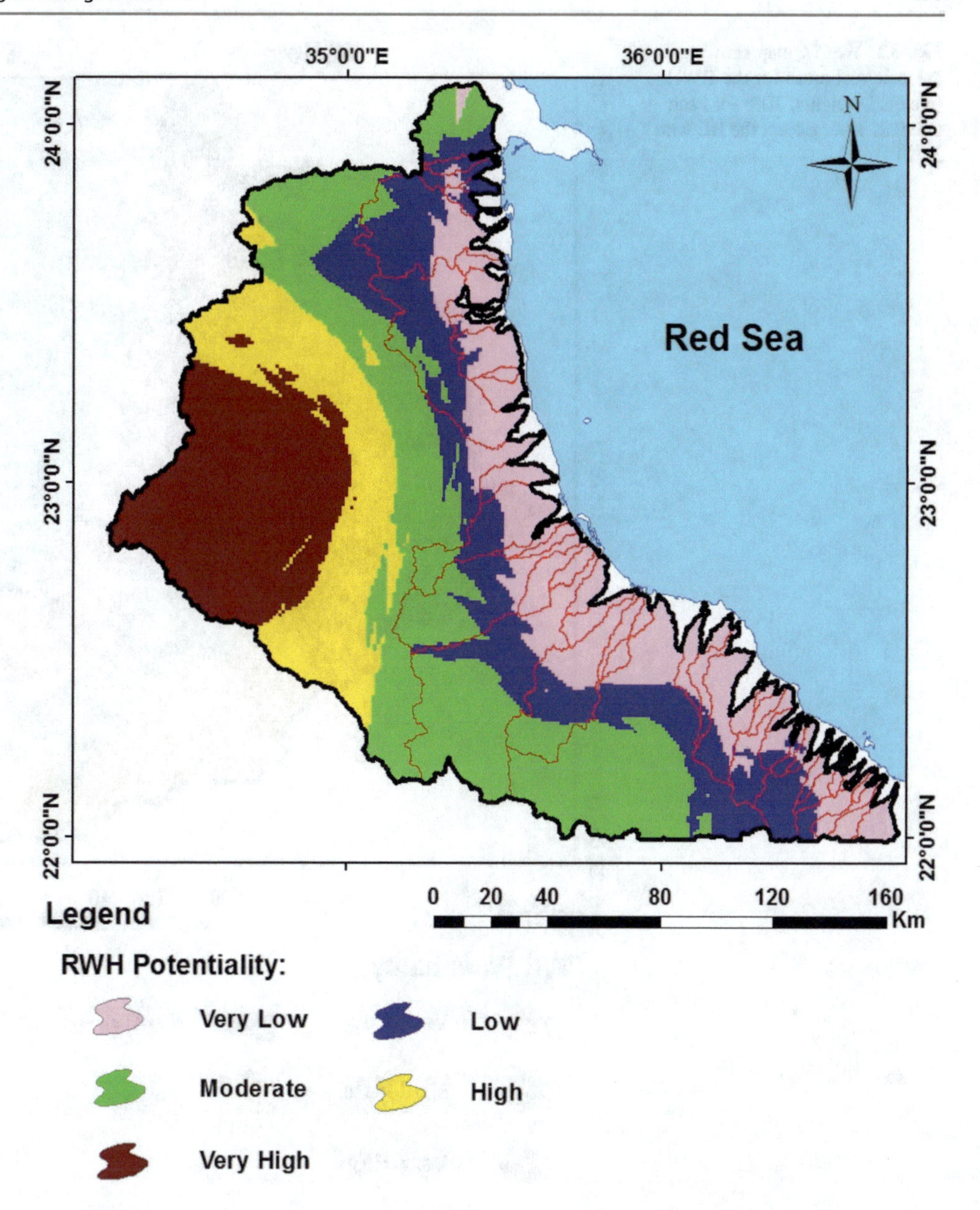

Fig. 31 WSPM map showing the potential areas for the RWH (unequal weights: 10% for each thematic layer except the BA with 30%)

can be calculated with special sampling schemes that are often computationally demanding (Saltelli et al. 2000). Accordingly, the effect of each model criterion was calculated by comparing its effect on the summation of classes that have high and very high RWH potentiality, in case that the criterion was assigned a unique weight value of 30% compared to the first scenario of equal model weights.

Figure 39 shows the classes of the high–very high RWH potentiality and their summation area (green columns), which resulted from scenario I (equal weighting) and the Scenario III, in which all criteria have equal weights of 10% unless only one criterion with a weight of 30%.

Figure 40 represents the percentage of variance in the total area of high and very high potentiality classes for the RWH, which had been resulted from scenario III comparable to the scenario I. So, it could be noticed that the length of overland flow has a higher effect value, followed by volume of annual runoff and the smallest effect resulted from MFD and IF.

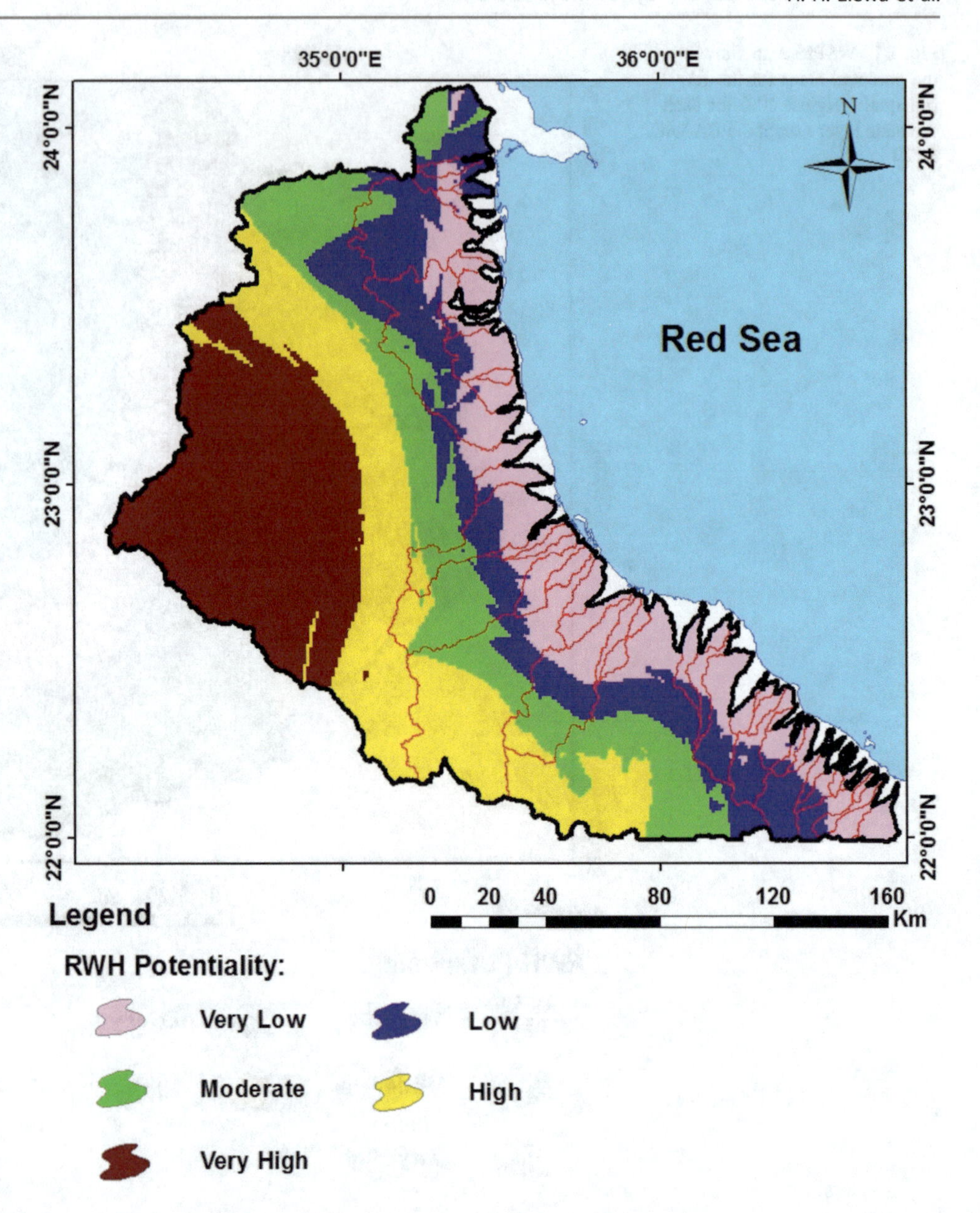

Fig. 32 WSPM map showing the potential areas for the RWH (unequal weights: 10% for each thematic layer except the BL with 30%)

From Fig. 40 and Table 10, the summation of all variance ratios in the high–very high RWH potentiality classes in scenario III for each criterion with respect to their areas in scenario I is 272.96%.

Accordingly, the justified weight of each criterion was calculated by dividing the variance ratio shown in Fig. 40 and Table 10 by the sum of all variance ratios. For example, the justified or optimum weight of the length of over land flow distance (OFD) could be obtained by dividing the variance ratio (60%) by the summation of all variance ratios (272.96), which is equal to 21.99% (Table 10).

Depending on the justified or optimum weights of thematic layers, another run for the WSPM was carried out taking into consideration these new weights as: OFD (21.99%), VAF (12.54%), BS (7.65%), Dd (21.99%), BL (7%), BA (17%), IF (1.95%), and MFD (9.88%).

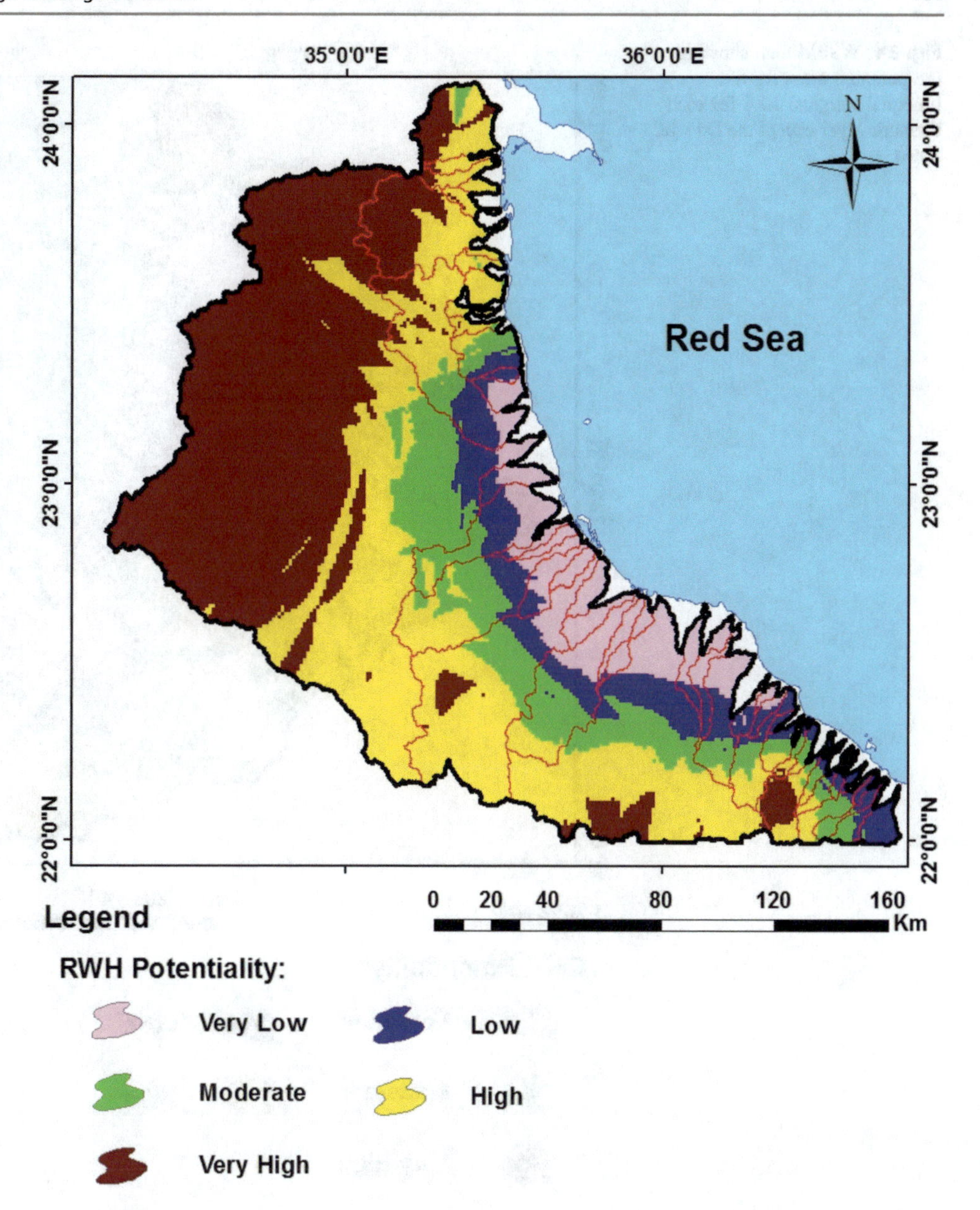

Fig. 33 WSPM map showing the potential areas for the RWH (unequal weights: 10% for each thematic layer except the BS with 30%)

The WSPM output map with five classes ranging from very low to very high potentiality was obtained (Fig. 41). The spatial distribution of these classes relative to the total studied area was: 5.62% for the very low, 18.26% for the low, 28.9% for the moderate, 29.51% for the high, and 17.7% for the very high potentiality for the RWH (Fig. 41; Table 11).

From the justified WSPM's map, it could be concluded that about 47.2% of the study area was represented by high–very high RWH potentiality distributed mainly in larger watersheds at its upstream and central areas (i.e., Hodein, Sfeirah, Ibib, Shaab, Deep Kraf, and Abu Dabaa watersheds). Moderate RWH potentiality class was represented by 28.91% decreasing to low–very low RWH potentiality classes due to downstream eastern parts.

In conclusion, the WSPM's map of scenario III, which had been constructed by the justified weights, was compared with scenarios I and II, and with the field verification carried out by the research team. Hence, a good correlation among the three scenarios was noticeable, where the moderate, high, and very high classes are dominating the study area. Accordingly, the justified scenario III could be considered as

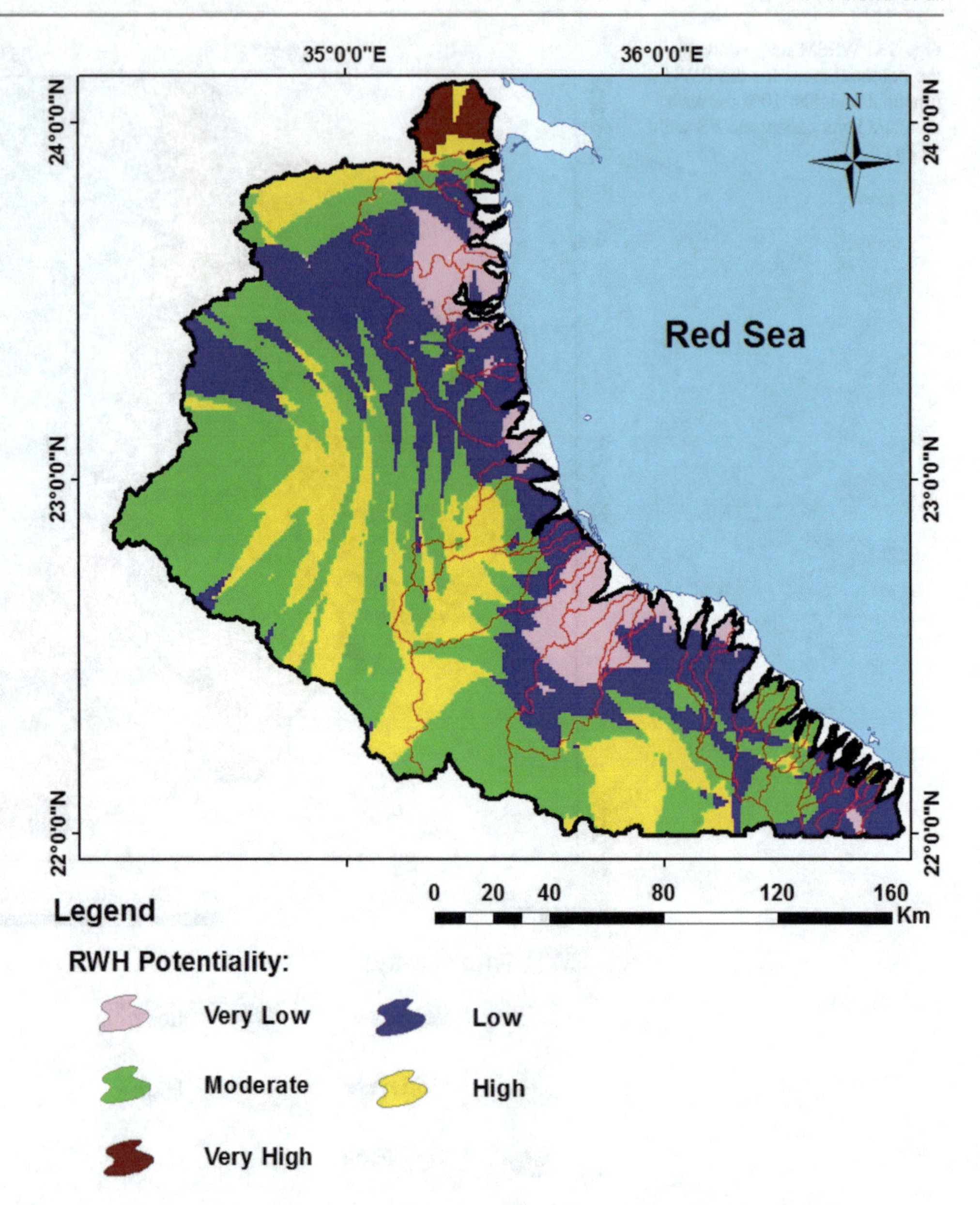

Fig. 34 WSPM map showing the potential areas for the (unequal weights: 10% for each thematic layer except the Dd with 30%)

a product of high reliability for determining the RWH potentiality, where it also coincides with the local inhabitants' experience and needs.

4.4 Proposals of Surface Storage Projects

4.4.1 Proposed R.W.H Construction in Halaib–Shalatin Watersheds

RWH technique in the basins of the studied area of HS on the coast of Red Sea in the SE of Arab rubablic of Egypt was made by the construction of 11 small dams in the main basins of that area. The water-harvesting system in the study area also comprises a six ground cisterns cross the basins main stream in additional to the infiltrated water into the soil surface. This volume of water would be sufficient for some human activities and drinking water for local habitants of the study area. The locations of these dams and cisterns were shown in Fig. 42 and Tables 12 and 13.

In present work, the criteria used for the site selection of proposed dams include:

Collection of runoff water at the outlets of HS basins, which are characterized by adequate VAF.

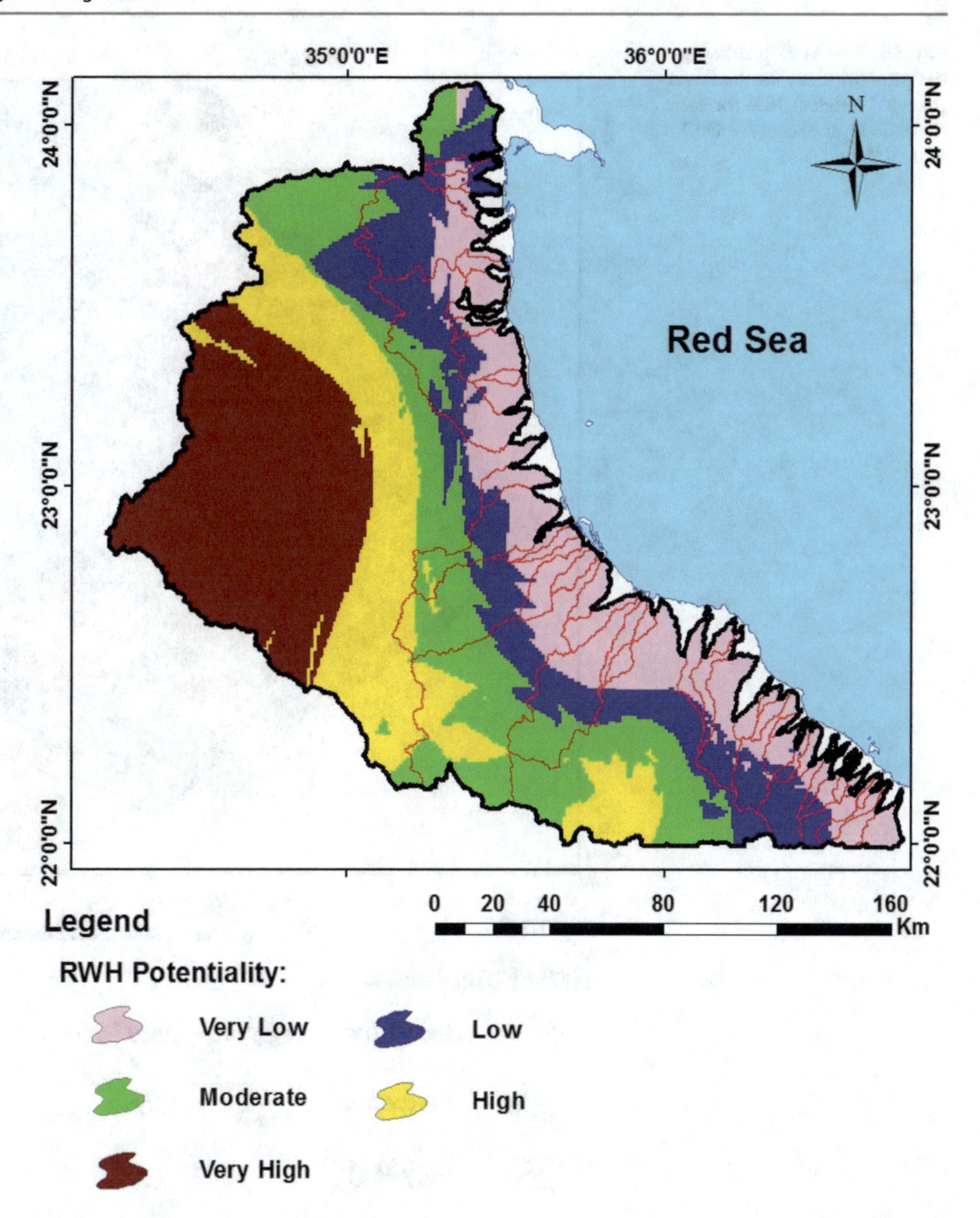

Fig. 35 WSPM map showing the potential areas for the RWH (unequal weights: 10% for each thematic layer except the MFD with 30%)

The results of the WSPM for determining RWH potentialities (moderate to very High).
The soil characteristics, which will provide a good environment for agriculture (alluvial or wadi deposits).
Existing land use pattern, which should be outside the present inhabited areas. The harvested runoff water will provide new areas suitable for the settlement of new communities.
Surface topography in terms of side slopes, which provide shoulders for the proposed dams to maintain reasonable stability for the installed proposed dams.

The successful design, construction, and operation of a reservoir of a dam project over a full range of loading require a comprehensive site characterization, a detailed design of each feature, and continuous evaluation of the project features during operation (Mcmahon 2004).

The proposed dams were aligned with respect to their heights to be straight or of the most economical alignment fitting to the topography and founding conditions. Additionally, the dams were designed to satisfy the basic design criteria of crest levels, minimum top widths, in addition to

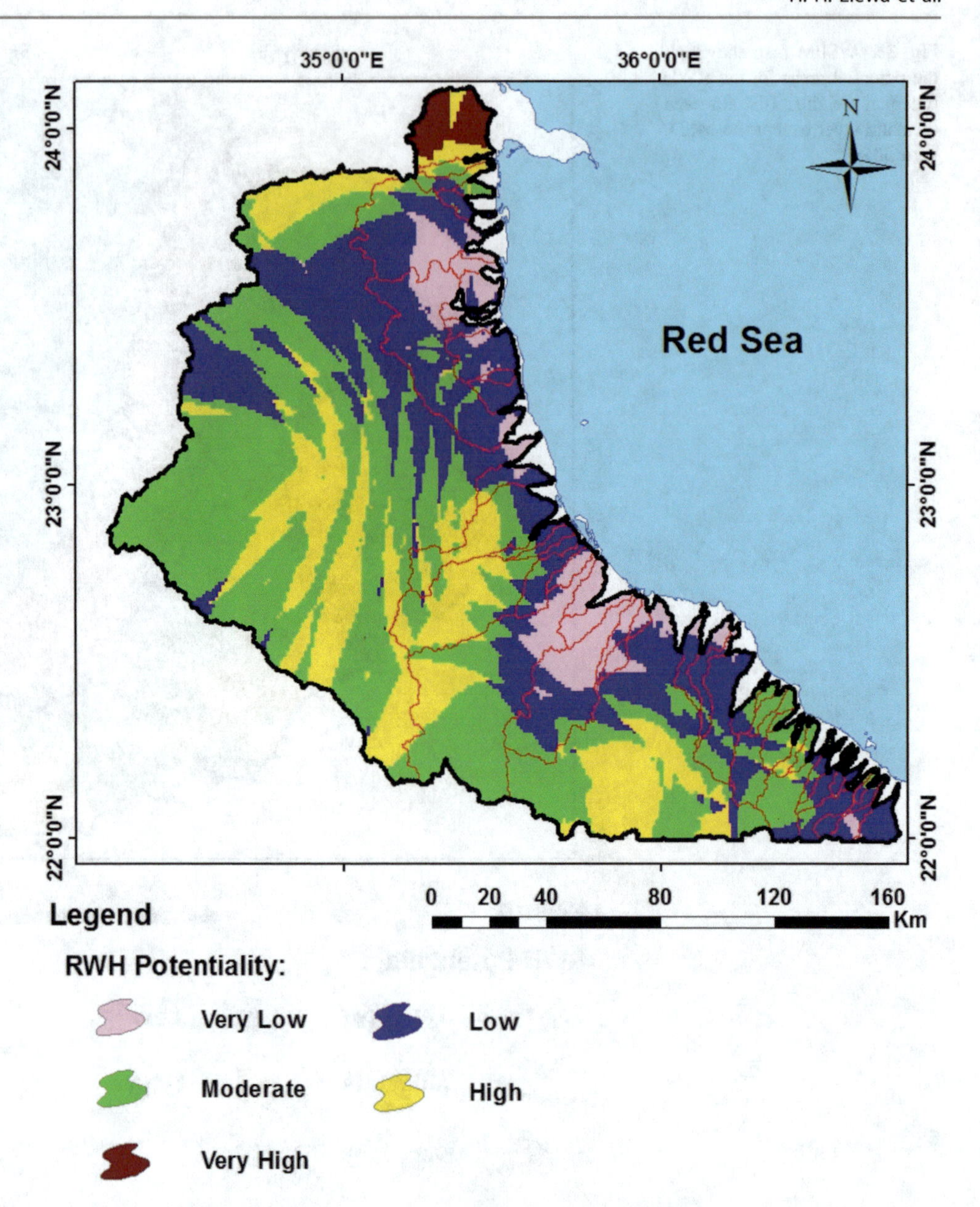

Fig. 36 WSPM map showing the potential areas for the RWH (unequal weights: 10% for each thematic layer except the OFD with 30%)

the basic technical and administrative requirements of an embankment dam (Greimann 1987), to meet the dam safety requirements (i.e., the dam foundation and abutments stability under all static or dynamic loading conditions, seepage control, freeboard, spillway, and outlet capacity, etc.).

The proposed dams in HS basins are embankment dams of the rock-fill type. The rock-fill dam can be classified into few groups by the configuration of dam sections (Kunitomo 2000). The selected rock-fill type consists of various layers of rock materials with an inclined core of impervious materials. The main body of the rock-fill dams, which should have a structural resistance against failure, consists of rock-fill shell and transitional zones, core and facing zones, which have a role in minimizing the leakage through the embankment. Filter zone should be provided in any type of rock-fill dams to prevent loss of soil particles by the expected erosion resulting from the seepage flow through the embankment.

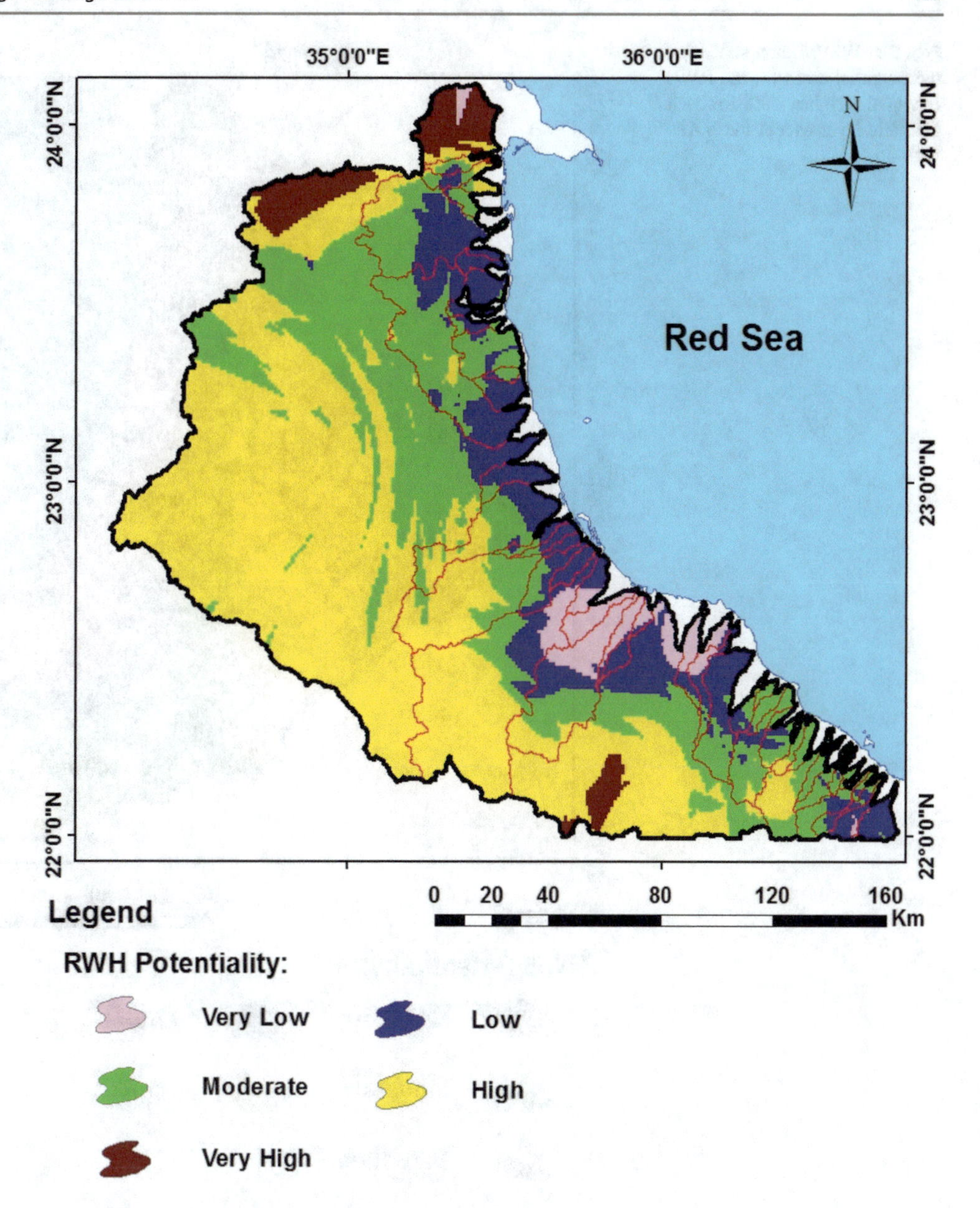

Fig. 37 WSPM map showing the potential areas for the RWH (unequal weights: 10% for each thematic layer except the IF with 30%)

Figure 43 shows a vertical cross section through the proposed dams in HS basin. The proposed dam allows its reserved water flow through a pipe which has an adequate inclination.

The economic feasibility is the major constraints that control the dam construction. Therefore, the types of proposed dams were selected to satisfy the economic feasibility by using, to a large extent, the available local building materials.

With reference to the proposed dams' cross sections illustrated in Fig. 43, it is shown that they consist of different building materials in layers (i.e., well-faced rock surface using cement mortar, previous layer of rolled fill material, and core layer from impervious rolled fill material). The cross sections of dams were designed according to the dam layout and topographical conditions with respect to the dam heights throughout each cross section.

Construction Stages of Proposed Dams

Upon beginning the construction stages of dams, additional detailed work is necessary, which includes:

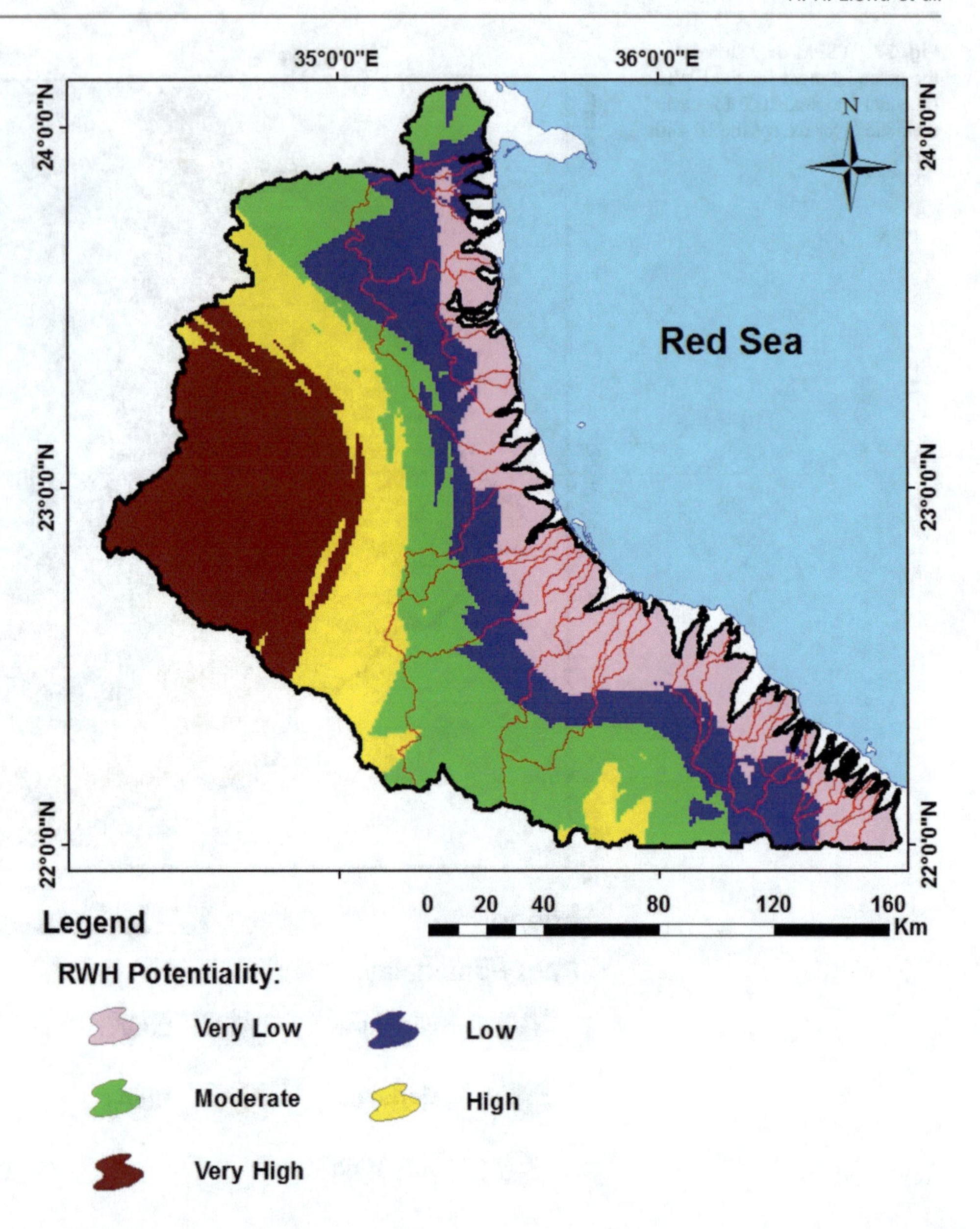

Fig. 38 WSPM map showing the potential areas for the RWH (unequal weights: 10% for each thematic layer except the VAF with 30%)

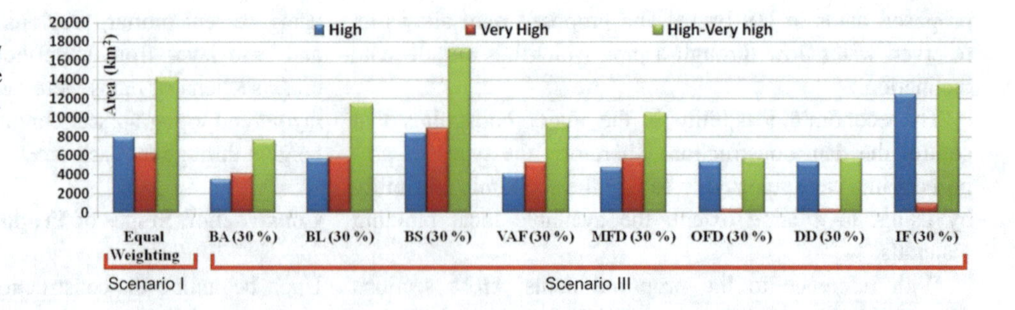

Fig. 39 Areas (in km^2) of the high–very high RWH potentiality classes and their summation in the two WSPM scenarios in HS area

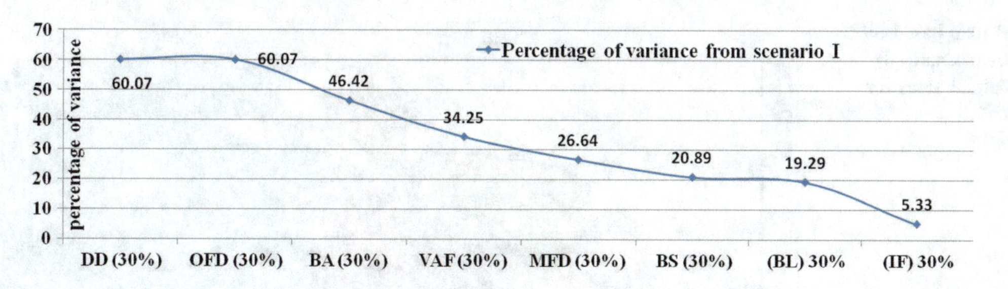

Fig. 40 A graph indicating the variance ratios of the high–very high classes in scenario III with respect to their areas in scenario I in HS area

Table 10 The variance ratios and justified weights of the WSPM's criteria used in the RWH potentiality mapping in HS area

WSPM Criterion	Overland flow distance	Volume of annual flood	Basin slope	Drainage density	Basin length	Basin area	Basin infiltration number	Max. flow distance
Variance ratio (%)	60.07	34.25	20.89	60.07	19.29	46.42	5.33	26.64
Justified weight (%)	21.99	12.54	7.65	21.99	7	17	1.95	9.88

Complete Design and Site Investigations:
Data collection (i.e., geotechnical works).
Selection of construction method.
Complete structural design of dams' cross sections and outlets.

Preparation of dam layouts taking into consideration the required facilities (i.e., roads, bridges, etc.).

Site Preparation:

Validation of dams' layouts and design details with the actual in situ conditions and locations.

Preparation of dams' foundations.

Fulfilling the compaction requirements of the base soil.

Follow-up Stages:

Construction of different structural elements of water outlets, pipe instalment, inlets and outlets of pipes.

Construction of different dam layers according to the details of cross sections.

Smoothen and finishing the surface of the dam with a layer of cement mortar and construction of different dam layers according to the details of cross sections.

Smoothen and finishing the surface of the dam with a layer of cement mortar and allocating the road pathways in the vicinity of the reservoirs.

The proposed six ground cisterns cross the basins will store yearly storage of 3,000 m^3. The location of these cisterns was shown in Table 13 and Fig. 42. Also the proposed cisterns will increase the protection from the dangerous effects of flooding on the Red Sea coastal road.

Figures 44 and 45 show a typical cross section in the elevation view and complete plan view for the proposed ground cisterns, respectively. These cisterns have a storage capacity of 500 m^3 for each, where it consists of 8-m-long rectangular side and storage height of 4 m.

5 Land Use Master Plan

5.1 Generalities

All the previously discussed issues about the developmental parameters of HS triangle area constitute a fruitful foundation for constructing an overall Land Use Master Plan (LMP). The objective of this plan is ensuring a sustainable vision for the development of the HS area according to the development capabilities of this area.

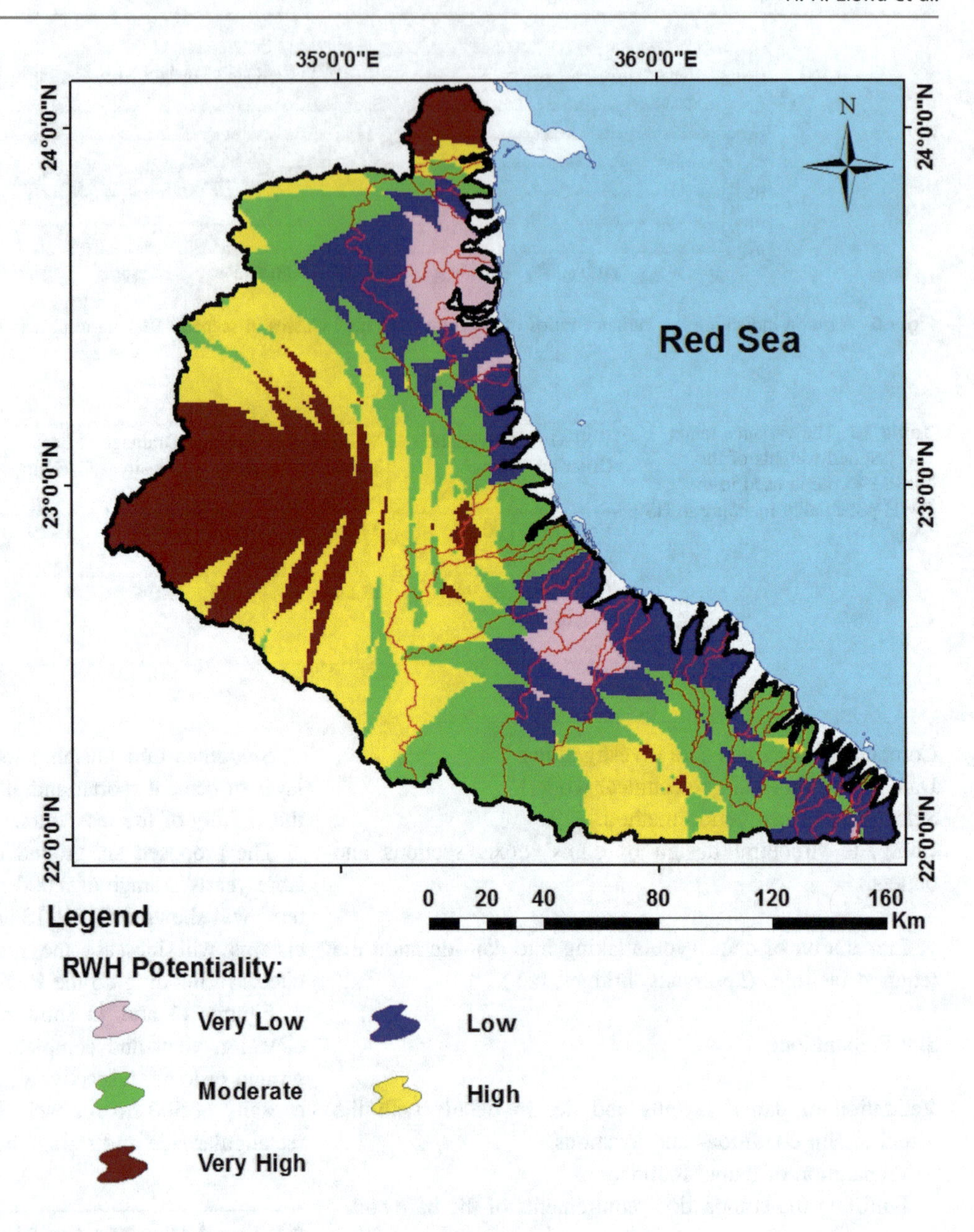

Fig. 41 WSPM map based on sensitivity results (scenario 3) showing the potential areas for RWH in HS area

Table 11 Areas of RWH potentiality classes (WSPM Scenario III; based on the results of sensitivity analysis) in HS area

WSPM's map for the RWH potentiality classification					
RWH potentiality class	Very low	Low	Moderate	High	Very high
Area (km^2)	1425.99	4637.7	7340.02	7493.48	4495.276
Area (in % relative to the total studied area) (Total study area: 25392.4 km^2)	5.62	18.26	28.91	29.51	17.70

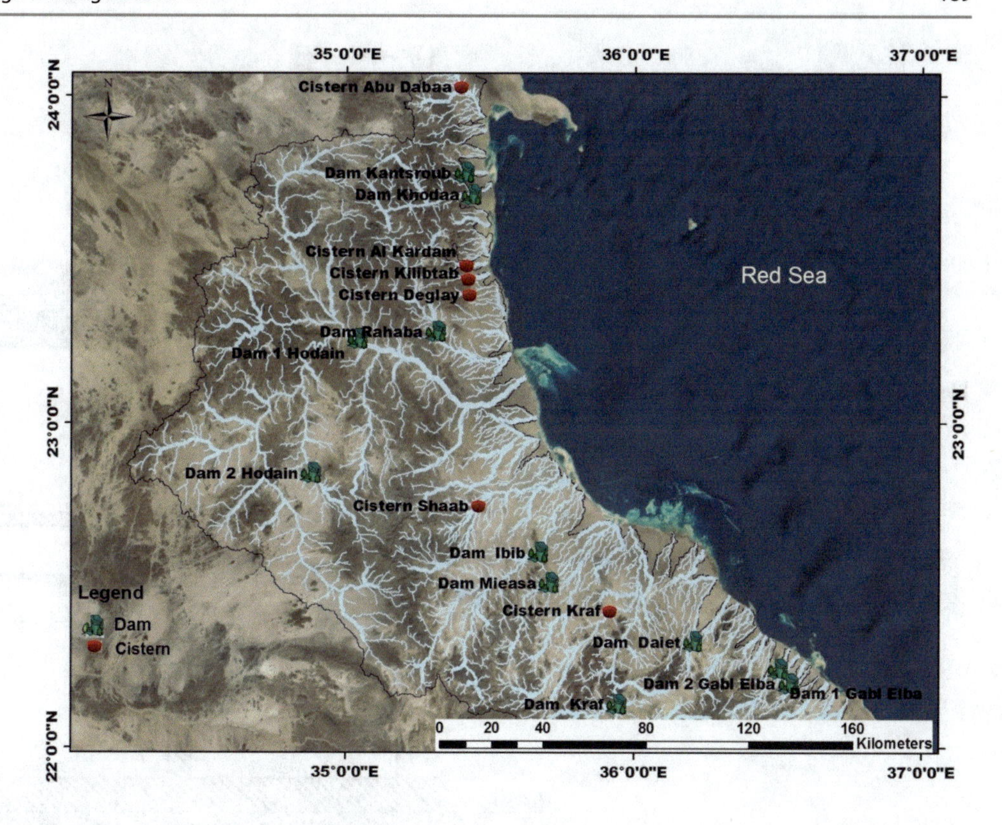

Fig. 42 Location of proposed dams and cisterns in HS watersheds

Table 12 Location of proposed dams in HS basins

Dam no.	Location	Lat.	Long.
1	Gabl Elba (1)	22°14′38.56″ N	36°30′07.10″ E
2	Gabl Elba (2)	22°11′ 56.70″ N	36°32′ 22.87″ E
3	Hodain	23°15′ 23.87″ N	35°02′ 24.96″ E
4	Hodain (2)	22°50′ 43.44″ N	34°52′ 53.10″ E
5	Daiet	22°19′ 36.08″ N	36°12′ 20.79″ E
6	Ibib	22°36′ 10.71″ N	35°39′ 55.95″ E
7	Kantsroub	23°45′ 54.95″ N	35°24′ 12.90″ E
8	Khodaa	23°41′ 55.56″ N	35°25′ 45.26″ E
9	Kraf	22°08′ 03.34″ N	35°56′ 25.46″ E
10	Mieasa	22°30′ 25.69″ N	35°42′ 09.09″ E
11	Rahaba	23°16′ 48.72″ N	35°18′ 23.31″ E

Table 13 Location of proposed cisterns in HS basins

Cistern No.	Location	Lat.	Long.
1	Abu Dabaa	24°01′44.55″ N	35°23′30.79″ E
2	Al kardam	23°29′04.33″ N	35°04′43.55″ E
3	Deglay	23°26′33.75″ N	35°25′09.02″ E
4	Kilibtab	23°23′36.11″ N	35°25′20.19″ E
5	Kraf	22°25′23.33″ N	35°54′58.64″ E
6	Shaab	22°44′49.40″ N	35°27′20.51″ E

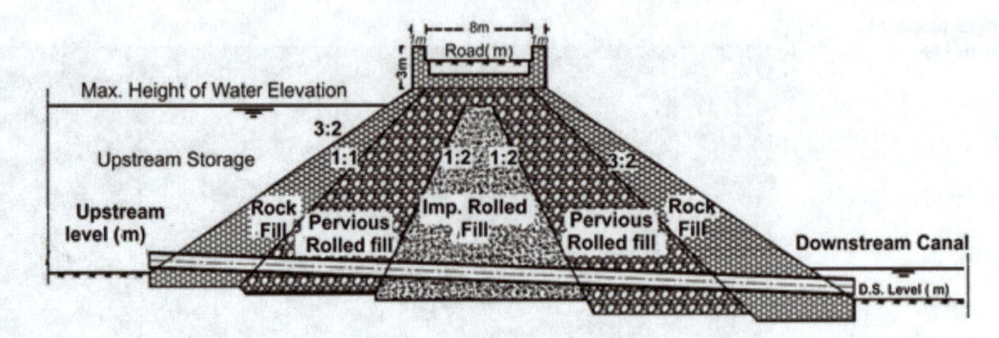

Fig. 43 Typical cross section of the proposed dams in HS basins (Elewa et al. 2013)

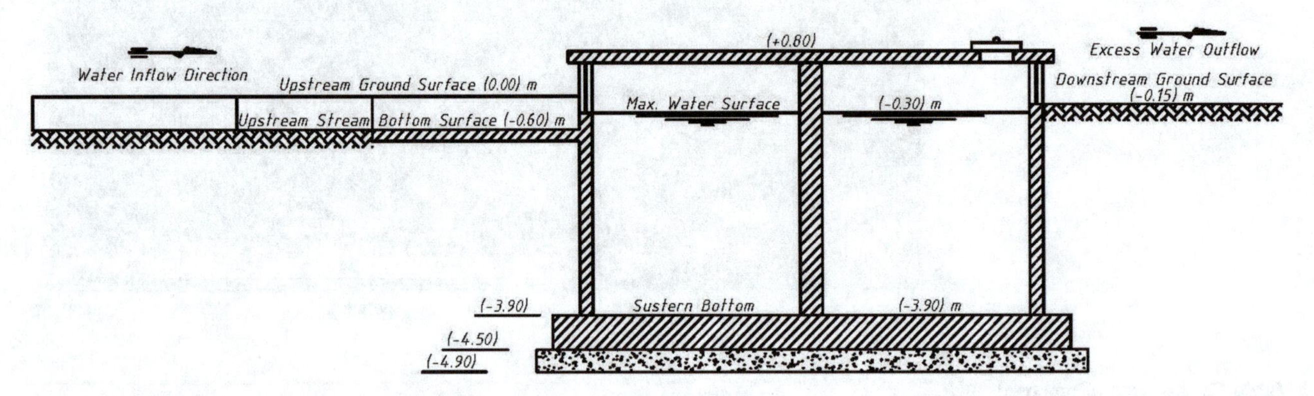

Fig. 44 Typical cross-sectional elevation in proposed cisterns in HS basins

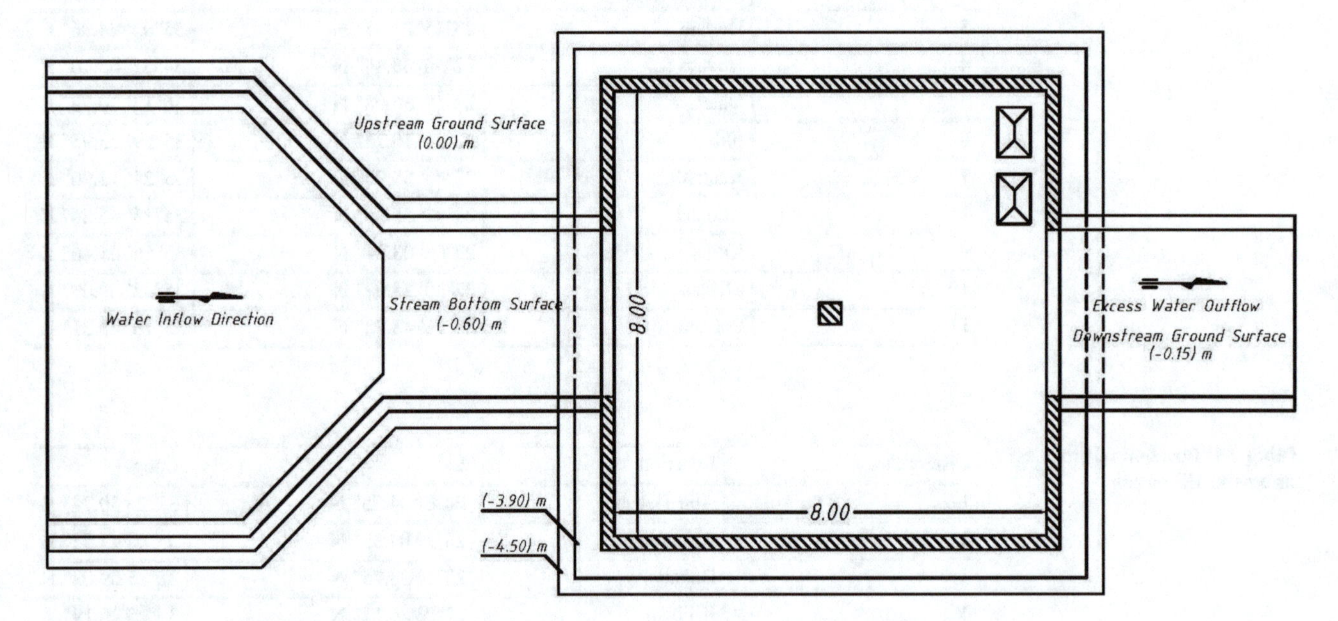

Fig. 45 Typical plan view in proposed cisterns in HS basins

The data used for the constructing the LMP were collected from:

- Different thematic maps and results of the present work.
- Ministry of Agriculture and Land Reclamation.
- Groundwater resources field inventory from 1992 uphill 2016.
- Previous works (Dissertations, Reports, articles, Web, etc.
- Geological data from EGS and Conoco Coral, 1987.
- General Authority for Rehabilitation Projects and Agricultural Development (GARPAD).
- National Water Research Center (NWRC), Egypt.
- National Authority for Remote Sensing & Space Sciences (NARSS).

5.2 Master Plan Objectives

The overall LMP has been elaborated to achieve, among the others, the following main objectives:

- To design the LMP for the whole HS region;
- To support a growth pattern for the area based on sustainable Green Economy;
- To promote the integration among different sectors and economic activities, as they are interfingering each other;
- To promote the integration among the economic activities to be developed, the infrastructure system, and the urban settlements;
- To promote the creation of clusters and urban settlements composed by a mix of economic and urban functions, guarantying the integration with the physical environment;
- To preserve the natural beauties and the heritage assets as a unique and inalienable patrimony of HS;
- To promote variety in use, therefore flexibility in planning according to the possible changes in socioeconomic conditions;
- To integrate within the LMP design the outcomes of all thematic maps previously discussed and the recommendations received from the research team involved in the present work;
- To build and design the LMP on the outcomes of the land suitability carried out for the HS region;
- To establish development criteria for sustainable use of the region:

HS area has great potential due to its geographic position on the Red Sea coast, which allows internal connectivity across the in-desert roads to the Nile Valley territory and sophisticated airports, toward Cairo and the delta area, to the North, and international connectivity toward the rich countries of the Arab Peninsula, to the west. This connectivity is facilitated by the presence of:

- Two international airports: Hurghada and Mersa Allam to the south and Luxor to the West;
- The HS region has a surface of almost 23,000 km^2; it includes a wide range of potential resources, well known since Pharaohs and Romans' times;
- Mineral resources, located in the mountains in an area between 10 and 80 km from the Red Sea coast; mineral resources include gold, granite, manganese, zinc, tantalum, and construction materials;
- Agriculture, due to the water scarcity in this area, most of its crops are exported from other areas located along the Nile River valley. However, small areas in the desert in Abu Saafa area (NW reaches of W. Hodein) were used as a pilot agricultural project in the form of greenhouses on limited groundwater resources. These significant water systems need to be matched by more efficient irrigation systems targeting water at the roots of each plant;
- Tourism, located mainly along the Red Sea coastline from north Shalatin to Halaib. This portion of the Red Sea coast presents a great potential for the tourism development to the uncontaminated coastline and reefs;
- Ecotourism–heritage tourism; the area presents a bulk of potentially exploitable sights in the old part of the Abraq and Abu Saafa, Abu Ramad, and along the historic connection routes, where roman ancient graffiti engravings in G. Abraq survives.

5.3 The Infrastructure System is Simple and Not Yet Efficient

One main road is connecting Shalatin to Halaib with minor paved and unpaved roads connecting its villages. Red Sea main highway is the principal road for accessibility. The local roads include about 45-km crosscut road between Shalatin and Bir El-Gahelia and then about 30-km road toward Bir Abraq. These roads are local rather than primary or secondary.

The main urban centers are the main cities of Shalatin, Abu Ramad, and Halaib.

These cities are located on the Red Sea corridor. Their location is favorable for the international market for export/import trading and for tourism; thanks to the international airports of Hurghada and Mersa Allam.

Within this system there are isolated nomadic nodes, separated from each other by a gap of 40 km of empty coastline and separated from the Nile Valley by a nearly about 220 km of mountains and desert zone.

5.4 Land Suitability Analysis

5.4.1 The Physical Environment

The present work carried out an assessment of the HS physical environment in the frame of the objectives and available baseline data. During the LMP construction, the work utilized the information previously collected in order to identify the main features of the area in view of the identified development opportunities. Following such an approach, the study areas/features have been identified and mapped:

1. **Map 1** (see Fig. 46): "Proposed Protected Areas" within HS area. The map includes:

– Proposed protected Areas by the present work, which include G. Elba, W. Gimal, G. Abraq, Abu Saafa, rock engravings in G. Abraq and Abu Saafa, coral reefs, and mangroves sites along the Red Sea shoreline.

2. **Map 2** (see Fig. 47) "Slope," which indicates the areas with slope in range under 5% and between 5% and 30% and those over 30% (using "DEMs"). The map shows the level of constraining in the mountain area to the west of the coastal zone.
3. **Map 3** (see Fig. 48): "Wadis and flood areas," which shows the wadi network, and the areas subjected to flash floods. The map shows the wadi systems:

– Nine main wadis bed from the western mountain area toward the Red Sea, the average length of 40–80 km; the lengthiest ones exist in the vicinity of Shalatin area, while the shortest ones are characterizing those of G. Elba and Halaib areas.
– The wadi width varies and there is no fixed width; the length of the wadies does not give an indication on the hazard, which depends on many factors. In the LMP, the indication for any development will be included (i.e., RWH capabilities, flood hazards, groundwater, etc.).

5.4.2 Analysis of Exploitable Natural–Heritage Resources for Economic Activities

The natural and heritage resources identified in the area are listed and analyzed in the following maps:

1. **Map 4** (see Fig. 49): shows proposed "Agriculture areas," Agriculture areas are located mainly in the lower parts of the major wadi beds directly connected to Red Sea. Salt tolerant crops with low water requirements (i.e., medical herbs) are highly encouraged for small areas close to the Red Sea coastline. RO water desalinization plant in Shalatin area (3600 m^3 daily capacity) could be used as a minor source for pilot cultivated areas of low water consumption. In-desert pilot agricultural areas are proposed depending on the seasonal harvested flood water, which was determined feasibly in the present work.

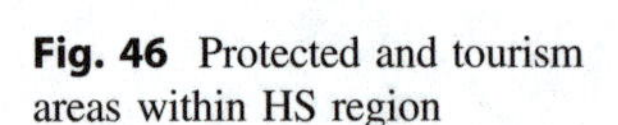

Fig. 46 Protected and tourism areas within HS region

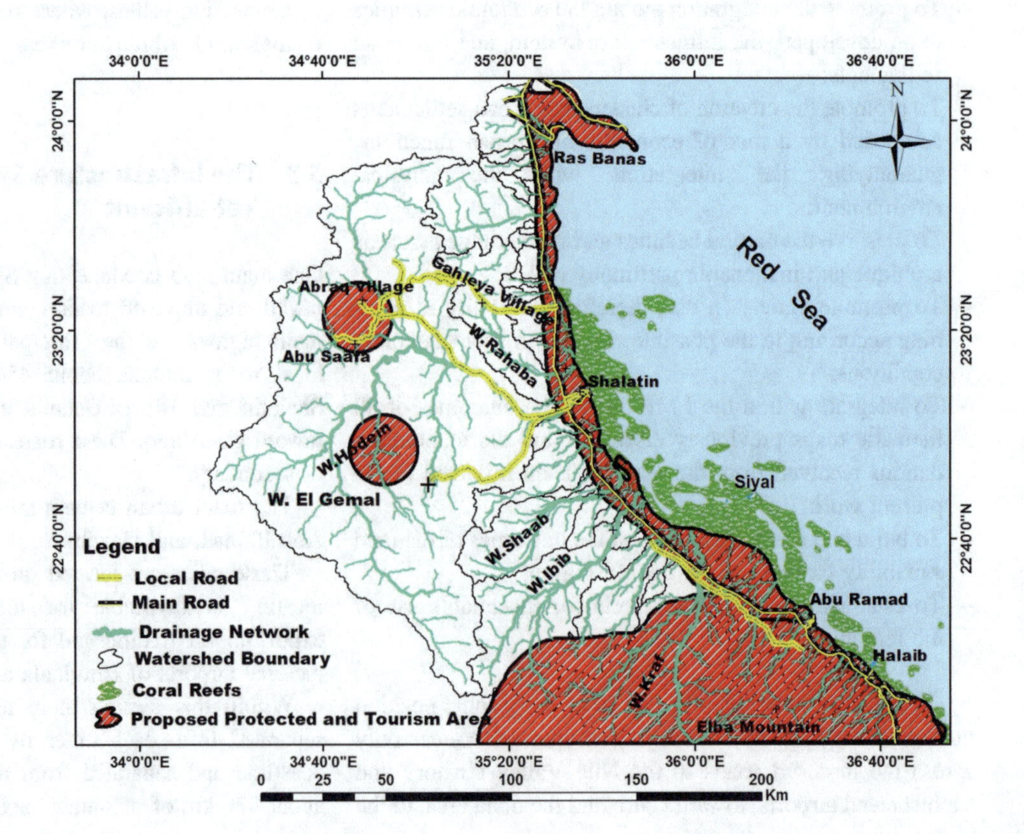

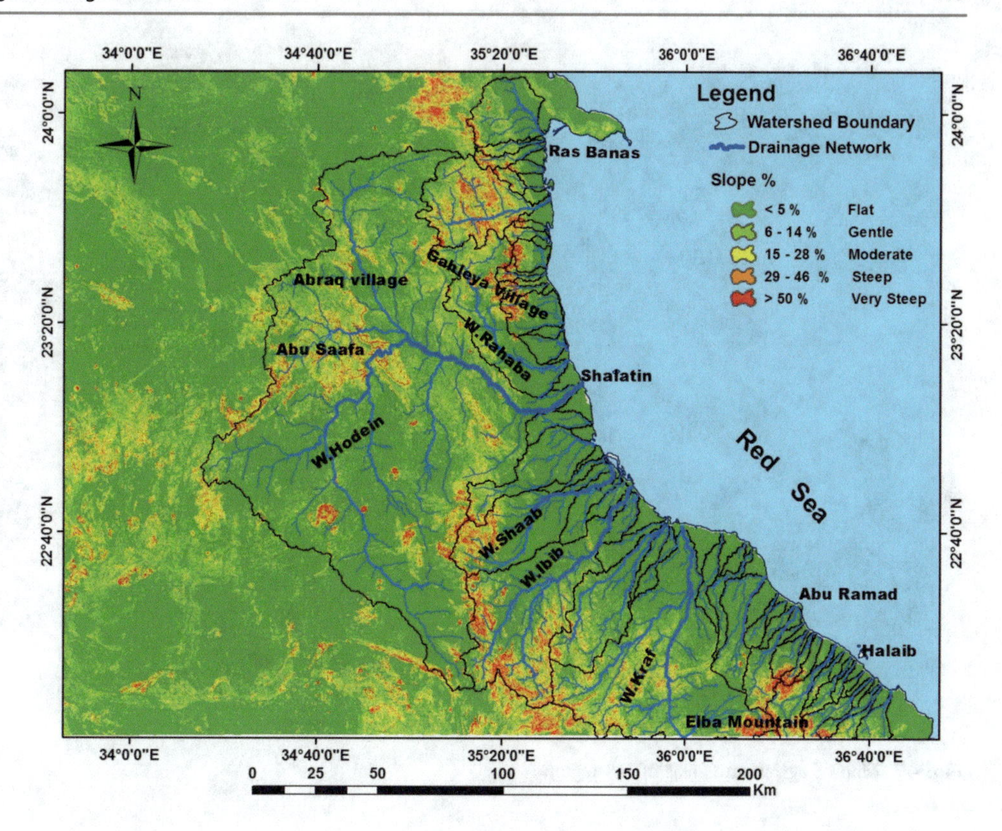

Fig. 47 Slope map of HS area (based on a 30-m resolution ASTER DEM)

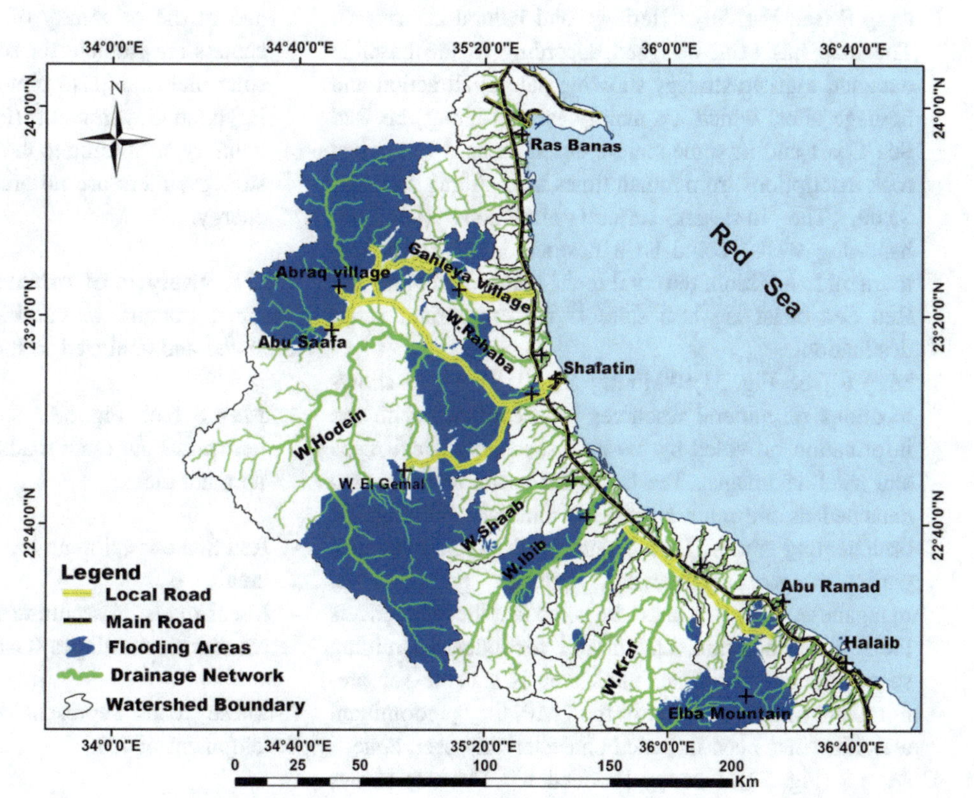

Fig. 48 Drainage network and flooding areas of HS region

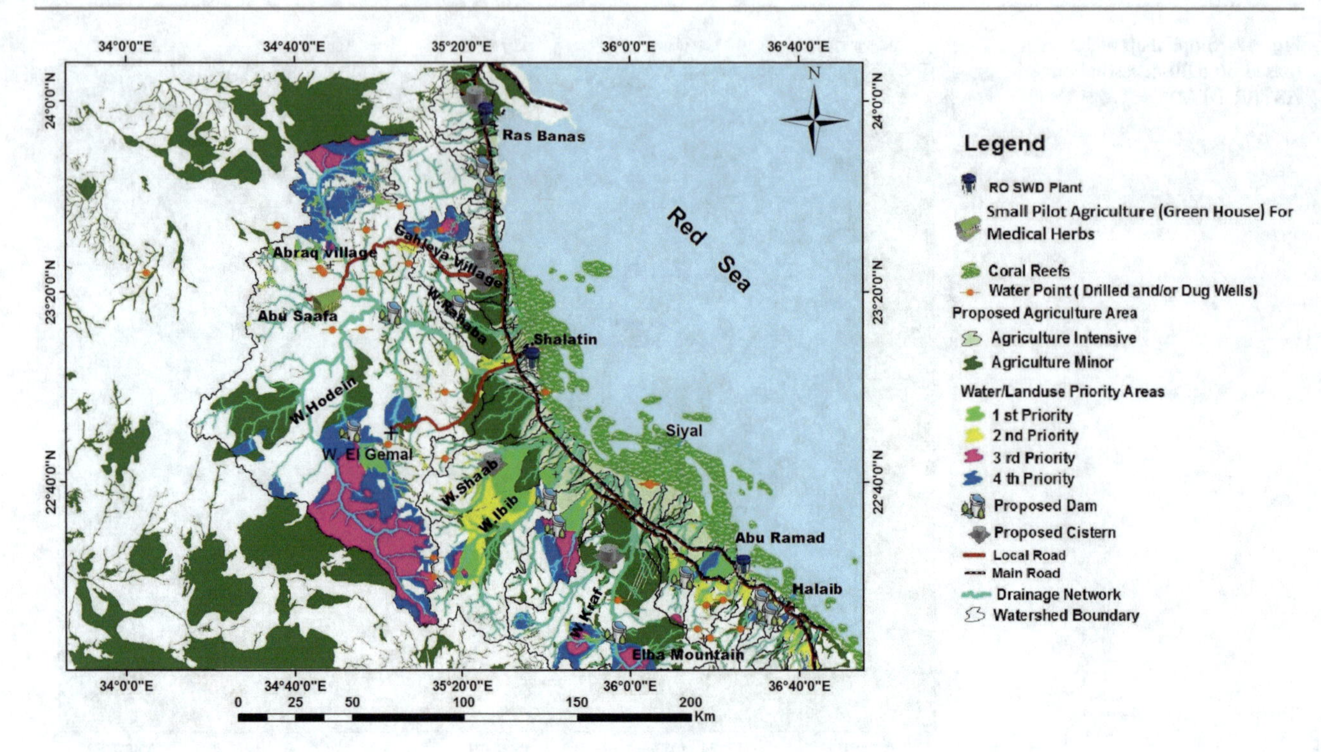

Fig. 49 Map of agriculture areas of HS region

2. **Map 5** (see Fig. 50): "Heritage and natural attraction". The map has been designed according to the baseline data and tourism strategy showing natural attraction and heritage sites, which are mainly located along the Red Sea Coast and in some remote desert areas with ancient rock inscriptions from roman times in G. Abraq and Abu Saafa. The in-desert cultural attractions (i.e., old hand-dug wells) could be a destination of safari ecotourism. In addition, the coral reefs, mangroves along the Red Sea coast are also suitable for ecotourism safari destination.
3. **Map 6** (see Fig. 51): "Mining Sites". The map shows locations of mineral resources in accordance with the information provided by the published geological maps and satellite images. The basement rocks territory was detached as the main source for minerals, building, or ornamenting stones. These minerals or rocks include, in particular, zinc, muscovites, granites, granodiorites, manganese, or gold. It has to be noted that the map reflects the regional baseline data suitable for industrial mining sector strategy for which such area is suitable for predominant mining activity. In the LMP, the "predominant activities" has been indicated characterizing each zone.
4. **Map 7** (Fig. 52): "Energy Production". The map shows the location of areas for energy production suitable for solar energy plants: all the other areas outside mountains and in the proximity of main power trunks and urban centers are suitable for solar energy plants. The planned solar plants are also plotted on the map (ongoing by the Egyptian Government) (PI self-communications). On the contrary, according to the baseline data and energy sector strategy, there are no areas extremely suitable for wind energy.

5.4.3 Analysis of Infrastructures and Network

The infrastructures and utilities networks existing in the area are listed and analyzed in the following maps:

1. **Map 8** (see Fig. 53): shows the current infrastructure network. Four main roads bound HS territory connecting its main cities:

- Red Sea coastal main road connecting Shalatin to Halaib area.
- Local roads crosscutting the main coastal Red Sea road toward local villages (Gahelia, Abu Ramad, Hodein).

These roads represent a potential for the integrated development of HS.

2. **Map 9** (see Fig. 54): shows the current utilities networks.

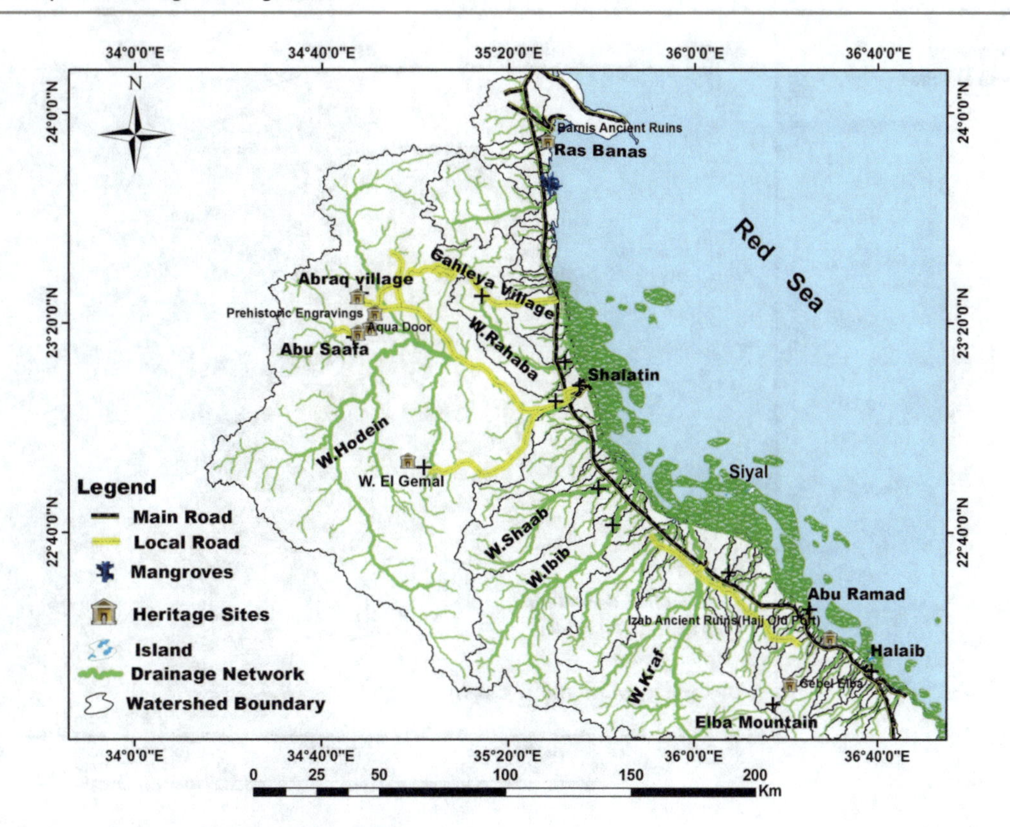

Fig. 50 Map of Heritage and natural attraction sites in HS area

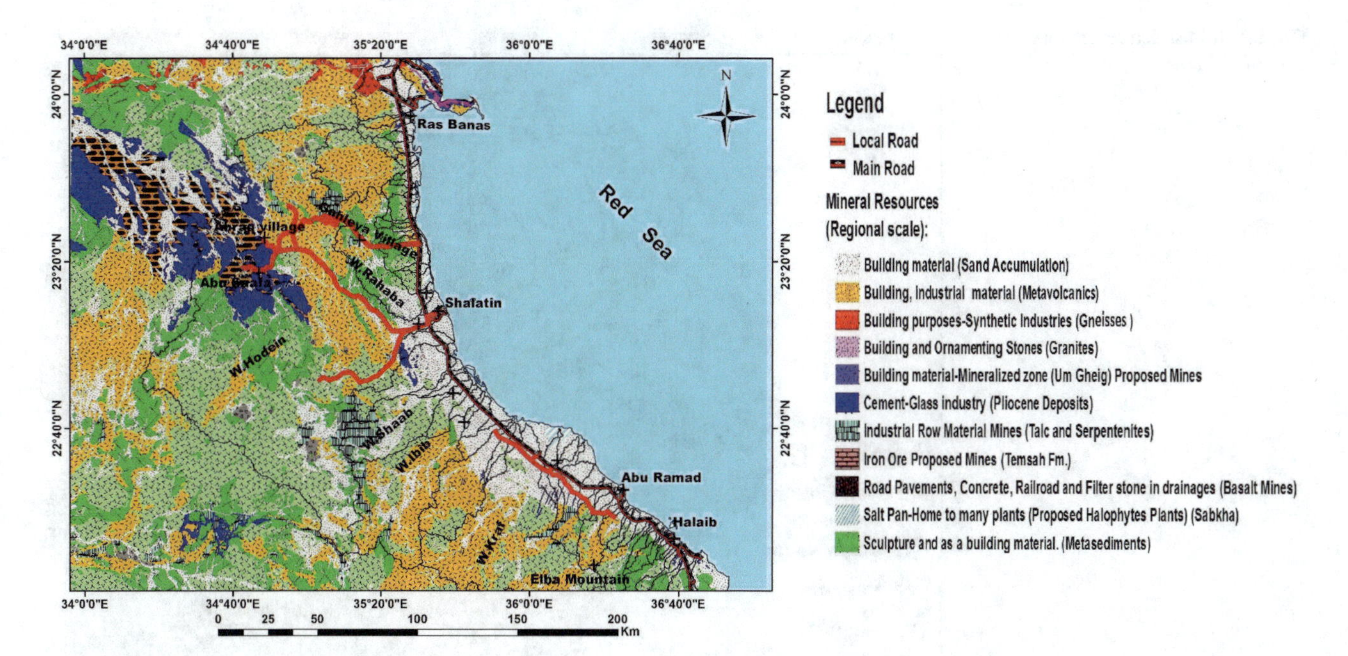

Fig. 51 Map of mining sites in HS area

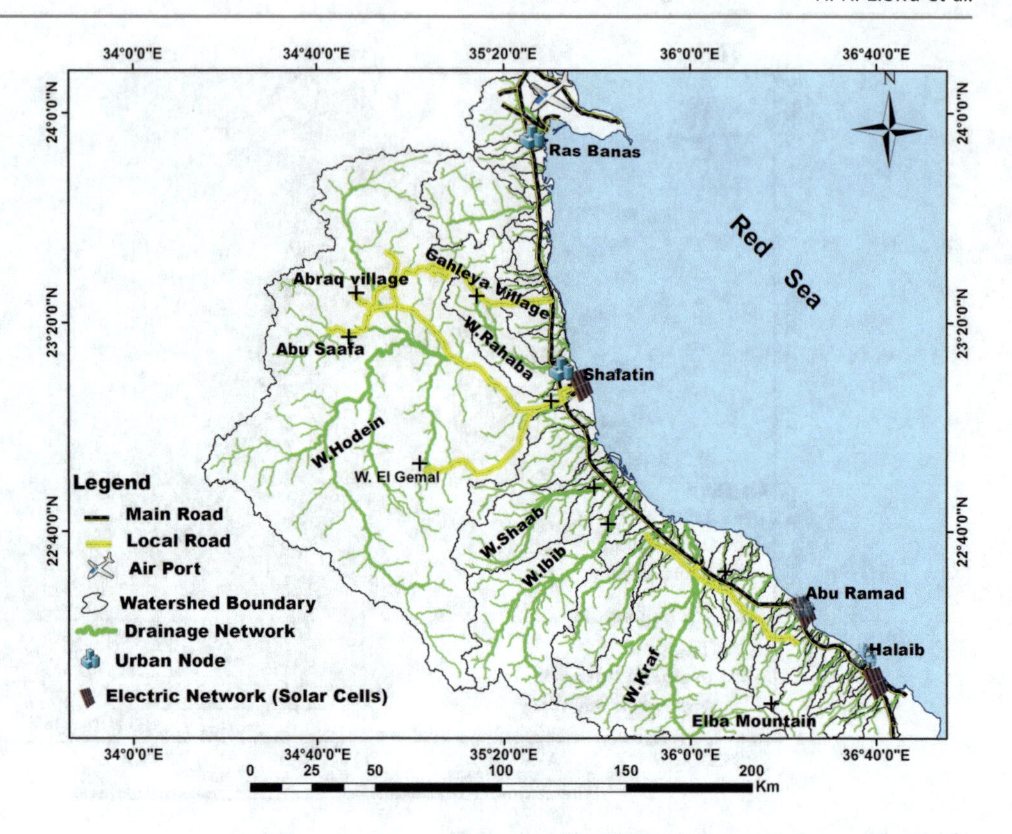

Fig. 52 Map of energy production sites in HS area

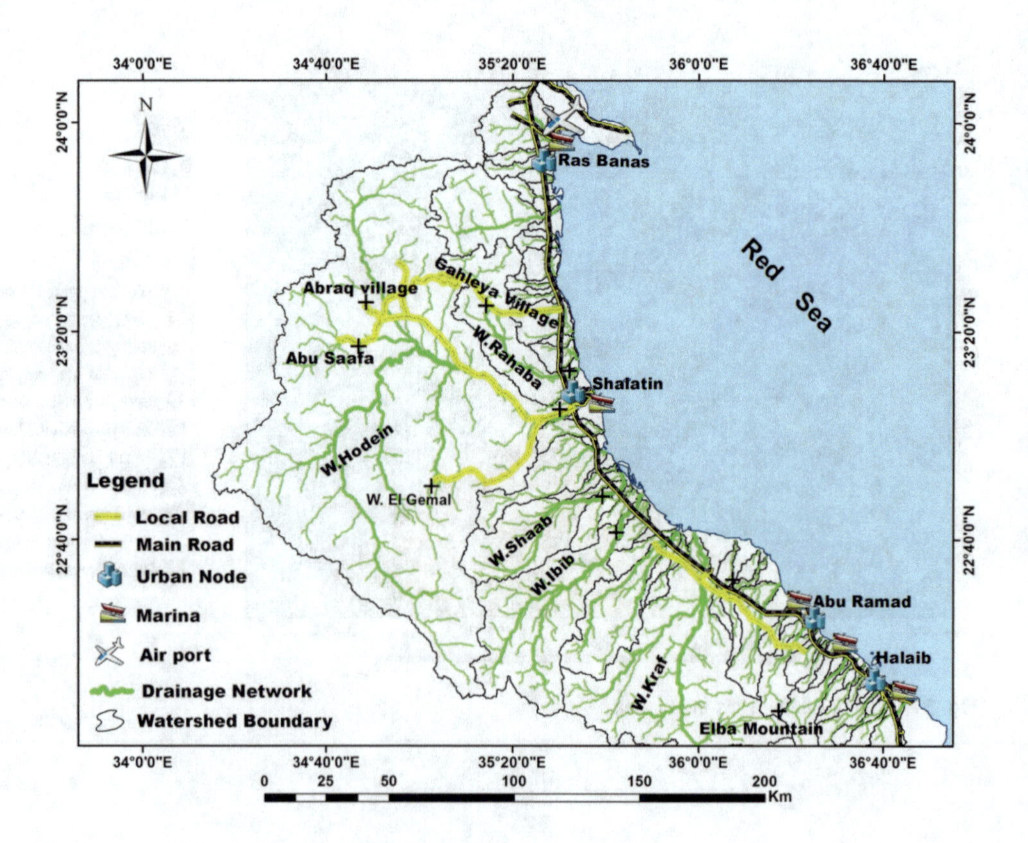

Fig. 53 Infrastructure network map of HS area

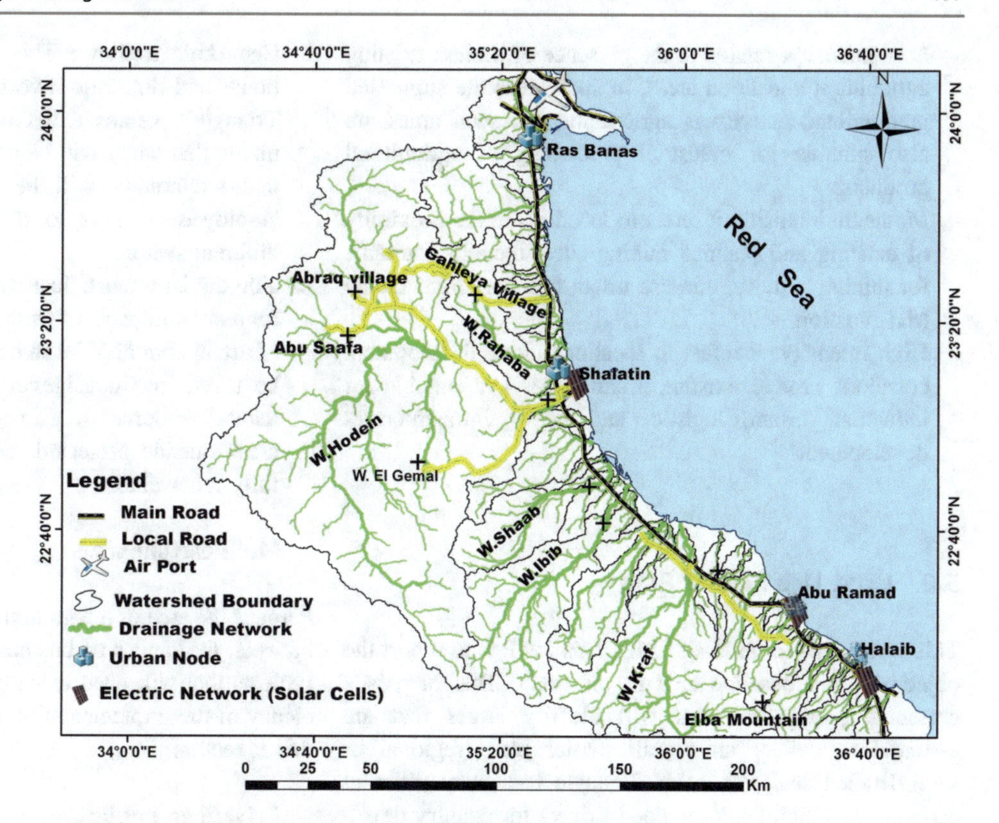

Fig. 54 Utilities networks map of HS area

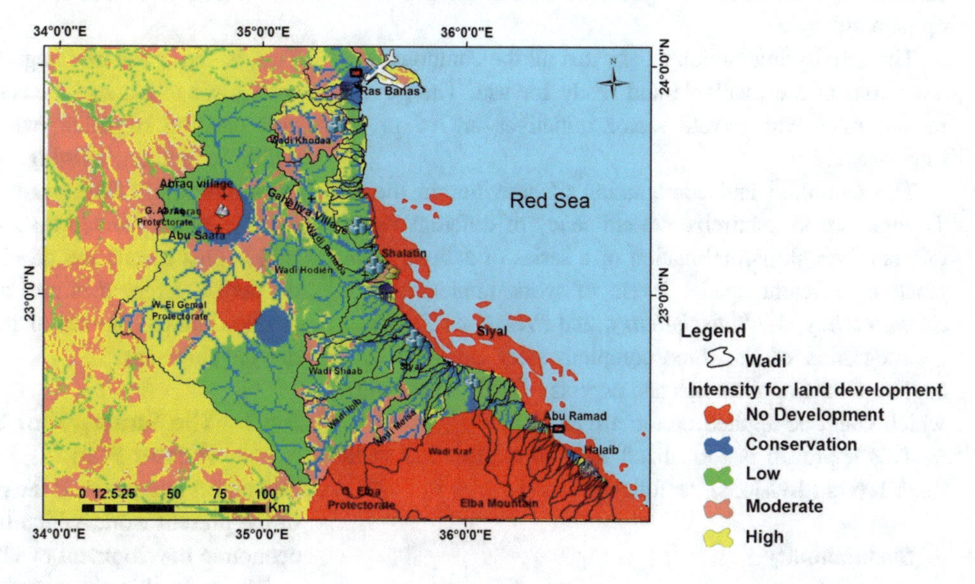

Fig. 55 Land suitability map of HS area

5.5 Land Suitability Map

This map (Fig. 55) combines the outcomes of the physical environment analysis and the results of the analysis of natural–heritage resources for economic activities and provides information on the suitable use of the territory in relation to its hazard and its resources. The map shows the areas of:

- No development: due to hazards, protected sites, natural sites, in this category are recommended no development;
- Conservation: due to the presence of heritage sites, it allows the development of important activities for tourism like restoration and renewal of high sights and infrastructures development;

– Low intensity: related to the presence of limited existing agricultural and flood areas, in such areas the suggested predominant activity is agriculture, but such areas are also suitable for industrial processing of agricultural products.
– Moderate intensity: it refers to locations in the proximity of existing and planned mining sites, therefore suitable for mining activity, mix-use urban function, light industrial function.
– High intensity: it refers to locations along development corridors, around existing urban nodes and suitable for industrial, urban, logistic, and heavy infrastructures development.

5.6 Land Use Master Plan

This work aims to build an LMP. This LMP is to report the objectives and baseline data in one map. This map is a crosscutting approach that will identify issues that are essential to deliver an overall master plan (regional or semi-detailed scale). The HS Triangle has many different economies, which could be good drivers for country development strategies.

The activity integration means that all the components of development are available and ready for use. The governmental rules and private sector initiatives are of pivotal importance.

The definition and combination of activities in the HS Triangle are to maximize the efficiency of different sectors outputs through a combination of a series of actions taking place at different spatial levels of work implementation: ***Sustainability***, ***Multidisciplinary,*** and ***Private or Public***.

Integration of the three complementary divisions is to capture all the applications and aspects of LMP formulation, which could be applied on the ground.

This approach is formalized with an explanation of the three layers (divisions), as follows:

– **Sustainability**

The present work is underpinned by the principle that all activities should take place under the broad umbrella of sustainability, where NARSS has chosen to interpret it as the development of a **Green Economy**; this will rotate around three parameters:

a. **People**: by being conscious that every aspect of the development plan must recognize the resource presented by the people resident in the HS Triangle, as well as those living in adjacent territories.
b. **Economic activity**: This will mean targeting international and domestic investors prepared to share the HS Triangle's values and develop their businesses recognizing that water will be priced as a limited resource and that technology will be deployed to enable directly employed workers to be economically productive in different sectors.
c. **The environment**: Just by being conscious of the environmental impacts of economic activity and the presence of urban communities in otherwise virgin lands. It should be taken into consideration the limited availability of natural resources (e.g., water, agricultural lands, mines areas outside protected areas), whereas solar energy is fully renewable.

– **Multidisciplinary**

Perhaps the area of integration that we all think about most is between the ranges of business units and the infrastructural projects that will need to be put in place to ensure the efficiency of the implementation of the master plan's economic driver sectors.

– **Private or Public**

This is the level of integration that allows private sector enterprises to manage the achievement of efficiency and the best creation of value, within a collection of enterprises within a sector. **Mining, ecotourism, trading, animal husbandry, trading, industry, agro-industry, fisheries, etc**. are being planned on a short-, medium-, and long-term bases. This means the development of integrated activities on both a large, medium, and small scales; which would enable different classes of investors to participate in activities development.

5.6.1 The Strategy for Upgrading Land Use Master Plan

The report integrates all the previously discussed parameters of the present work, which has addressed more in detail the economic development of HS Triangle region.

The main objectives of the Master Plan are the following:

– The urban nodes for workers employed in mining, fisheries, industrial, and tourism sectors have been combined, to form a stronger cluster;
– The urban nodes for mining, tourism, fisheries, and industrial sectors have been located considering the linear urban development principle (development along connecting roads and infrastructures).
– The cities of Shalatin and Halaib participate in the urban growth associated to the economic development of

trading, mining, fisheries, tourism, and industrial sectors. This, because some industrial activities will be located in proximity of the two cities, due to technical and logistic requirements as expressed by specific sector strategies, as well as, the proximity of some mining sites to the two cities, than the creation of new urban nodes.
- The proposed new urban nodes at the deltas of main wadis to capture all planned or expected urban growths in future, and to overcome the demographic vacuum and to build man power for the planned developmental sectors.

In other words, the LMP of HS triangle region has been studied for each main area, that includes the following:

1. Proposed or current protected areas.
2. Conservation areas.
3. Agriculture (Green House or Medical Herbs).
4. Fisheries.
5. Energy production.
6. Tourism use.
7. Mining use.
8. Industrial use.
9. Logistics.
10. Ports.
11. Airports.
12. Development corridors.
13. Urban nodes—Existing.
14. Urban nodes—Proposed.

5.6.2 Developing the Master Plan

Reasons Behind the Construction of LMP

The reason behind the construction of LMP has been based on the following principles and objectives:

1. To promote integrated and sustainable development of HS Triangle.
2. To better exploit the physical characteristics and natural resources of HS Triangle.
3. To add value to the socioeconomic development of the country, in terms of attractiveness for population, employment created, attractiveness for public or private enterprises, and economic activities.
4. To better use the territory in a balanced and sustainable way, with long-term view.
5. To better sustaining an efficient and internal interconnectivity between land uses and urban nodes are existing and proposed.
6. To better planning for future land use parallel to urban and economic growth.
7. To create attractive urban nodes as clusters for the sustainable and integrated development for the region and the country.

Land Use Units

Based on the methodology outlined above, land use zones have been identified.

Each zone is defined by the following characteristics:

- Predominant use.
- Allowed use.

The identified land use units are the followings:

1. **Proposed Protected Areas**

These areas have been recognized as environmentally sensitive and they have been envisaged as future planned protected areas by the Egyptian Environmental Affairs Agency (EEAA). Intervention is subject to environmental impact assessment (EIA) and to the approval of a specific development project.

2. **Conservation Areas**

These are natural parks and wadies subject to conservation; aiming at forming a green natural vegetation belt between cities and productive areas, and flood water harvesting theater, ecotourism and safari tourism platform, nomadic use, and animal husbandry, etc.

- Predominant use: including the sustainability principles, are providing public natural zones to balance the industrial–urban development.
- Allowed use: tourism, ecotourism and infrastructures, subject to the approval of specific development project.

3. **Agriculture-Intensive Use**

This area is the main wadi deltas nearby urban nodes with adequate distances from the Red Sea; it is characterized by intensive agriculture on desalinized water mixed with urban villages (existing or proposed urban nodes), in detail:

- Predominant use is agriculture for sustaining principal crops to the residents and nomadic communities;
- Allowed uses are: tourism, mix-urban use (artisanal, commercial, offices, residential, services), urban (residential–public services).

4. **Agriculture Minor Use**

These areas include the wadi beds of the drainage basins and the mountainous desert areas, and between the Red Sea coast and the mountain areas; the area is suitable for agriculture extensive use, according to the strategy of the specific agriculture sector (i.e., medical herbs, green houses on groundwater in Abu Saafa), in details:

- Predominant use is agriculture;
- allowed uses are urban (residential–public services) and tourism.

5. **Energy Production Use**

Areas that are suitable for solar energy production.

- Predominant use: energy production;
- Allowed uses: Tourism, Urban Park, Infrastructure.

6. **Tourism Use**

These are the areas with peculiar environmental attractiveness for tourism use; they include the entire Red Sea coast, mountains, desert wadis, heritage areas within remote deserts and the existing urban centers (particularly Shalatin, Halaib, Abu Ramad, Abu Saafa, Gahelia, and Abraq) and other heritage areas identified in the suitability analysis, in details:

- Predominant use: tourism;
- Allowed uses are urban (residential–public services) and agriculture;

7. **Mining Use**

These are the areas with a concentration of suitable mining sites in the mountains outside the environmentally protected areas, as emerged in the suitability analysis, in details:

- Predominant use is mining;
- Allowed uses are: tourism, infrastructures, facilities associated with the mining activity.

8. **Industrial Use**

These are the areas where products will be processed; products related to the following sectors:

- Fisheries (fishing, canning, exporting).
- Mining sector.
- Agriculture sector (Medical herbs preparation, drying, and canning).
- Construction sector.
- Energy.
- Other industrial sectors:
- The areas are confined along development corridors, in proximity of:
- Available human resources–work force;
- Available water resource;
- Available infrastructure–road–transport network.

These land use units are also in proximity of urban centers, in order to reduce the commuting time and costs, planned in a way not to affect the urban center with pollution. Accordingly, a set of mitigation measures should be taken into consideration, among which we mention: maintain a minimum distance of 5–10 km between urban and industrial sites; study of the local microclimate in particular wind directions; introducing a "green belt-urban park" as a buffer between the urban area and industrial one. In details:

- Predominant: industrial use;
- allowed use: Logistics, infrastructures, support services, public services;

9. **Logistics**

These are the areas with logistic value, as per specific sector plan. They include dry ports. In details:

- Predominant use is logistic;
- Allowed uses are industrial, port, infrastructures, public services;

10. **Ports**

These are the areas with port use, as per specific Sector Strategy.

- Predominant use is port;
- Allowed use is logistics, infrastructures, public services.

11. **Development Corridors—Mix-Urban Use**

These are the areas with a vocation for mix use, which includes handicraft, commercial, offices, residential, services. Consequently, they are located along development corridors-connecting roads, to link/complement urban areas with economic development zones. In details:

- Predominant use is a mix use (handicraft, commercial, offices, residential, services);
- Allowed use: tourism, infrastructures, public services.

12. **Urban Use Existing**

These are the existing urban centers in Shalatin, Abu Ramad, and Halaib, plus minor centers in Abraq, Abu Saafa, and Gahelia villages along connecting roads. In details:

- Predominant uses are residential, commercial, retail, offices, public services, tourism;
- Allowed uses are tourism, handicrafts, and agriculture;

13. **Urban Use Proposed**

Areas that are planned urban centers or nodes at the peripheries of main wadis deltas in W. Shaab, W. Meisah, W. Kraf, etc. These proposed urban nodes could be used as attractive areas for new residents, especially with the expansion of economic projects in all sectors.

As a final product for the previous discussed peculiar maps, the final land/use master plan for HS region was constructed successfully (Fig. 56). Units of the proposed land use unit were calculated and listed in Table 14.

6 Summary and Conclusions

HS area is considered to be a promising region for different developmental activities as tourism and mining. The present work presented an overview of the optimum use for the natural resources in HS area.

The topography of the area ranges from gently sloping coastal plains to rugged mountainous and hilly lands. Geologically, the basement rocks occupy the area completely (except the area of Gebel Abraq and the upstream portion of W. Hodein), that covers vast areas forming high to medium topographic reliefs. The sedimentary rocks cover a relatively smaller area forming moderately weathered-bedded hills.

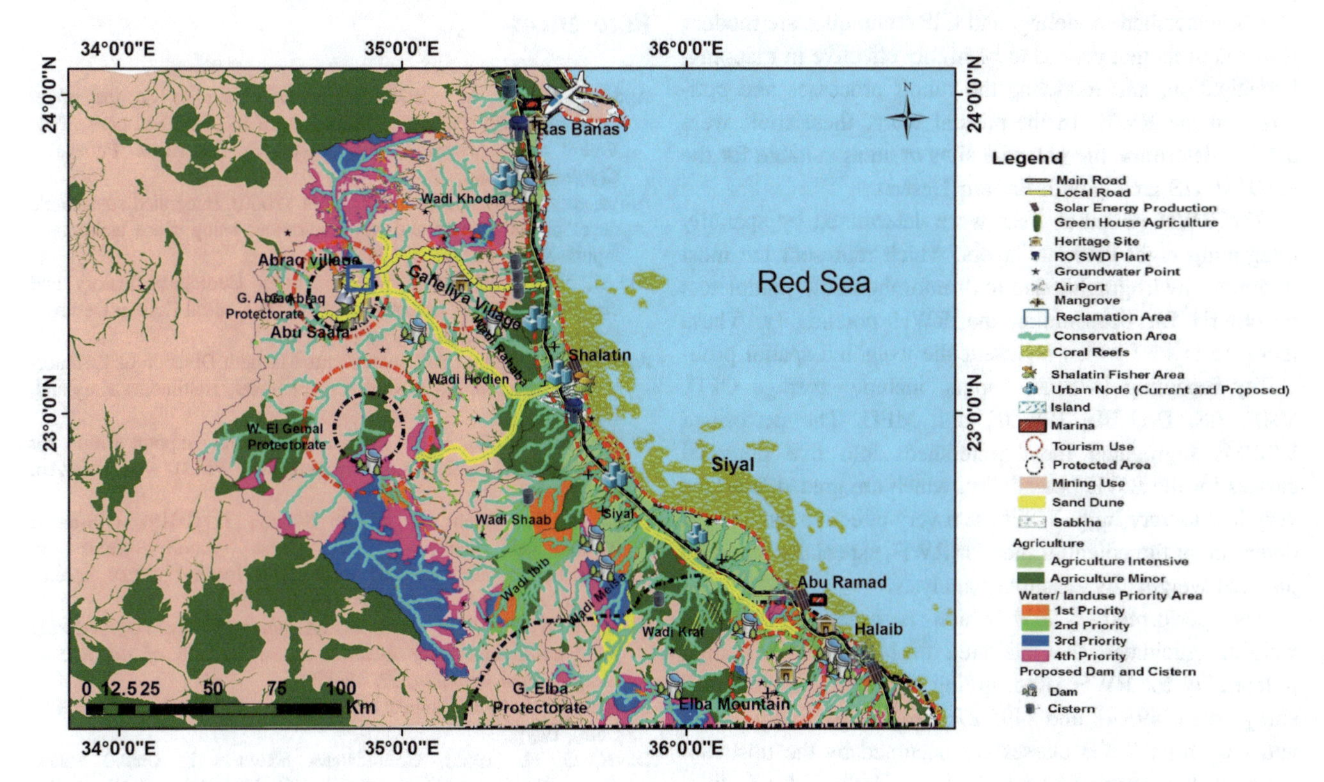

Fig. 56 Map of Land use master plan of HS area

Table 14 Land use units of HS area

Land use unit	Area (km^2)
1st priority area for water/land use	1164.6
2nd priority area for water/land use	1293.5
3rd priority area for water/land use	1792.7
4th priority area for water/land use	2231.7
Agriculture intensive	1582.6
Agriculture minor	10489.7
Mining use	35681.6
Conservation area	5031.0
Tourism area	7638.2
Protected area	7030.2
Sand dunes	407.1

Quaternary deposits are present mainly along the Red Sea coast and partially cover the sedimentary and Precambrian rock units.

Drainage net map and hydromorphometric parameters are introduced in details for all HS watersheds through the present work.

RS, watershed modeling, and GIS techniques are modern research tools that proved to be highly effective in mapping, investigation, and modeling the runoff processes and optimization the RWH. In the present work, these tools were used to determine the potential sites or areas suitable for the RWH in HS area (South Eastern Desert).

The RWH potential areas were determined by spatially integrating eight thematic layers, which represent the most important hydrographic and hydromorphometric parameters or criteria for determining the RWH potentiality. These thematic layers, which represent the weighted spatial probability modeling (WSPM) inputs, include: average OFD, VAF, BS, Dd, BL, BA, IF, and MFD. The performed WSPMs segregated these watersheds into five potential classes for the RWH potentiality, which are graded from the very low to very high. WSPM is a very effective method for determining the potential sites for RWH, especially based on justified weights by sensitivity analysis.

The performed WSPM which is based on justified weights elucidated that the area of high and very high potentiality for RWH is occupying about 47% of total HS study area (7493.48 and 4495.276 km^2, respectively). High and very high RWH classes are occupied by the upstream parts of the larger watersheds (i.e., Hodein, Kraf, Ibib, El-Rahaba, and Shaab) decreased to low and very classes in the downstream parts. A LMP was developed by the present work to foster the sustainable development of HS area.

Acknowlegment The authors wish to express their deep gratitude for the National Authority for Remote Sensing & Space Sciences (NARSS) for funding the project entitled “Planning for the Sustainable Development of Halaib-Shalatin Area, South Eastern Desert, Using RS, GIS, and other Techniques,” where the present work was performed.

References

Abdel Razik, T. M. (1972). Comparative studies on the upper cretaceous-early Paleocene sediments of the Red Sea coast, Nile Valley and Western Desert, Egypt. In *8th Arab. Petroleum Congress, Algiers*, 23p.

Adiga, S., & Krishna Murthy, Y. V. N. (2000). Integrated sustainable development of land and water resources using space technology inputs. *Space Forum, 5,* 179–202.

Akkad, M. K., El-Ramly, M. F. (1960). Geological history and classification of the basement rocks of the Central Eastern Desert of Egypt. *Geological Survey of Egypt*, 24p.

Anon. (1975). Munsel soil color charts Macbeth Division of Kollmorgen Corporation. 2441 North Cavert Street, Baltimore, Maryland, USA.

AQUAVEO. (2008). Water modeling solutions. Support forum for sub-watershed modeling system software (WMS). www.aquaveo.com.

Bruins, J. H., Guedj, H. B., & Svoray, T. (2019). GIS-based hydrological modelling to assess runoff yields in ancient-agricultural terraced wadi fields (central Negev desert). *Journal of Arid Environments, 166,* 91–107.

Butzer, K. W. (1959). Environment and human ecology in egypt during pre-dynastic and early dynastic times. *Bulletin of the Society Geographical Egyptian, 32,* 42–88.

CONOCO. (1987). Geological map of Egypt, scale 1:5000,000. EGPC, Cairo, Egypt.

Davis, G. H. (1965). Groundwater networks in United States. *International Association of Scientific Hydrology*, Publ. no. 68, 433p.

Desert Research Institute. (1998). *Meterological Data Recorded at Shalatin station*. Cairo, Egypt: Installed by Desert Research Institute.

EGACS. (1989). Egyptian general authority for civil survey. Topographic sheets, scales 1:500000–1:250000.

Elewa, H. H., Qaddah, A. A. (2011). Groundwater potentiality mapping in Sinai Peninsula, Egypt, using remote sensing and GIS-watershed-based modeling. *Hydrogeology Journal*, (Springer, Heidelberg, Berlin), *19*, 613–628. https://doi.org/10.1007/s10040-011-0703-8.

Elewa, H. H., Qaddah, A. A., & El-Feel, A. A. (2012). Determining potential sites for runoff water harvesting using remote sensing and geographic information systems-based modeling in Sinai. *American Journal of Environmental Sciences, 8,* 42–55. (Science Publications, USA).

Elewa, H. H., Ramadan, E. M., El-Feel, A. A., Abu El Ella, E. A., & Nosair, A. M. (2013). runoff water harvesting optimization by using RS, GIS and watershed modelling in Wadi El-Arish, Sinai. *International Journal of Engineering Research & Technology (IJERT), 2,* 1635–1648.

Elewa, H. H., Ramadan, E. M., El Feel, A. A., Abu El Ella, E. A., & Nosair, A. M. (2014). Runoff water harvesting optimization by using RS, GIS and watershed modelling in Wadi El-Arish, Sinai. *International Journal of Engineering Research & Technology (IJERT), 2,* 1635–1648.

Elewa, H. H., Ramadan, E. M., Nosair, A. M. (2016). Spatial-based hydro-morphometric watershed modeling for the assessment of flooding potentialities. *Environmental Earth Sciences*, *75*, 927.

El-Naggar, Z. R. M. (1970). On a proposed lithostratigraphic subdivisions for the late cretaceous-early paleogene succession in the Nile Valley, Egypt. In *7th Arab Petroleum Congress*, Kuwait.

El-Ramly, M. F. (1972). A new geological map for the basement rocks in the eastern and south western deserts of Egypt, scale 1: 1000,000. *Annals of the Geological Survey of Egypt, 2,* 1–18.

Endriszewitz, M. (1988). Gliederung der Nubischen Serie in Südostägypten. Auswertung von Gelände-Und Fernerkund dungs daten-Berliner geowiss. Berlin, Ah. A, 79, 141p.

ERDAS Field Guide. (2005). Geospatial imaging, LLC Norcross, Georgia, United States of America.

ESRI. (2007). ArcGIS 9.2® Software and user manual. *Environmental Systems Research Institute*, Redlands, California 92373-8100, USA. http://www.esri.com.

Faniran, A. (1968). The index of drainage intensity—A provisional new drainage factor. *Australian Journal of Science, 31,* 328–330.

Feizizadeh, B., Jankowski, P., & Blaschke, T. (2004). A GIS based spatially-explicate sensitivity analysis approach for multi-criteria decision analysis. *Computers Geosciences, 64,* 81–95.

Finkel, H. H. (1979). Water resources in arid zone settlement, a case study in arid zone settlement. In G. Colany (ed.), *The Israeli Experience*, Pergamon.

Garbrecht, J., Martz, L. W. (1997). TOPAZ: An automated digital landscape analysis tool for topographic evaluation, drainage identification, sub-watershed segmentation and sub catchment parameterization; TOPAZ Overview, U.S. Department of Agriculture, Agricultural Research Service, Grazing lands Research Laboratory, El Reno, OK, USA, ARS Publication No. GRL 97-3.

Gregory, K. J., & Walling, D. E. (1973). *Drainage Basin form and process: A geomorphological approach* (p. 456p). London: Edward Arnold.

Greimann, B. (1987). Design of small dams, (3rd edn), United States Department of the Interior, Bureau of Reclamation, a water resources technical publication, (pp. 287–312).

Ha, W., Lu, Z., Wei, P., Feng, J., & Wang, B. (2012). A new method on ANN for variance based importance measure analysis of correlated input variables. *Structural Safety, 38,* 56–63.

Hendricks, F., & Luger, P. (1987). The Rakhyiat formation of the Gebel Qreiya area: Evidence of middle campanian to Early Maastrichtian syn-sedimentary tectonism-Berliner geowiss. *Abh. A., 75,* 83–96.

Horton, R. E. (1945). Erosional development of stream and their drainage basins; hydrophysical approach to quantitative morphology. *Geological Society of America Bulletin, 56,* 275–370.

Ismail, H. A., Bahnassy, M. H., Abd El-Kawy, O. R. (2005). Integrating GIS and modelling for agricultural land suitability evaluation at East Wadi El-Natrun, Egypt. *Egyptian Journal of Soil Science, 45,* 297–322.

Jain, V., & Sinha, R. (2003). Evaluation of geomorphic control on flood hazard through geomorphic instantaneous unit hydrograph. *Current Science, 85,* 26–32.

Javed, A., Khanday, M. Y., & Ahmed, R. (2009). Prioritization of sub-watersheds based on morphometric and land use analysis using remote sensing and GIS techniques. *Journal of Indian Society of Remote Sensing, 37,* 261–274.

Jenson, S. K., & Domingue, J. O. (1988). Extracting topographic structure from digital elevation data for geographical information system analysis. *Photogrammetric Engineering and Remote Sensing, 54,* 1593–1600.

Jones, J. A. A. (1999). *Global hydrology: Processes* (p. 399p). Resources and Environmental Management: Longman.

Klitzsch, E., List, F. K., Pohlmanm, G. Handley, R., Hermina, M., Meissner, B. (1986). Geological map of Egypt, scale 1:500,000, 20 sheets, Cairo (Conoco/EGPC).

Kunitomo, N. (2000). *Design and construction of embankment dams* (pp. 2–5). Aichi Institute of Technology: Department of Civil Engineering.

Leopold, L. B., Maddock, T. (1953). The hydraulic geometry of stream channels and some physiographic implications: U.S. Geological Survey Professional Paper 252, 52p.

Ligmann-Zielinska, A. (2013). Spatially-explicit sensitivity analysis of an agent-based model of land use change. *International Journal of Geographical Information Science, 27,* 1764–1781. https://doi.org/10.1080/13658816.2013.782613.

Malczewski, J. (1996). A GIS-based approach to multiple criteria group decision-making. *International Journal of Geographic Information Science, 10,* 955–971.

Malczewski, J. (2006). GIS-based multi criteria decision analysis: A survey of the literature. *International Journal of Geogrphic Information Science, 20,* 703–726.

Leica Geosystems GIS & Mapping, LLC. (2008). *ERDAS field guide* (Vol. 1), Norcross, GA: Leica Geosystems Geospatial Imaging, LLC.

Mark, D. M. (1984). Automatic detection of drainage networks from digital elevation models. *Cartographica, 21,* 168–178.

Martz, L. W., & Garbrecht, J. (1992). Numerical definition of drainage network and Sub-catchment areas from digital elevation models. *Computers & Geosciences, 18,* 747–761.

Mcmahon, J. R. (2004). *General design and construction considerations for earth and rock-fill dams, Department of the Army* (pp. 2–8). Washington: U.S. Army Corps of Engineers.

Montgomery, D. R., & Dietrich, W. E. (1989). Source areas, drainage density, and channel initiation. *Water Resoures Research, 25,* 1907–1918.

Moore, I. D., Grayson, R. B., & Ladson, A. R. (1991). Digital Terrain modeling: A review of hydrological, geomorphological and biological applications. *Hydrological Processes, 5,* 3–30.

Morisawa, M. E. (1959). Relation of morphometric properties to runoff in the Little Mill Creek, Ohio Drainage Basin, (Columbia University, Department of Geology), Technical Report, 17, office of Naval Research, Project NR 389-042.

Nuweir, A. M., & El-Sharkawi, M. A. (1978). The Um Had-Um Effein metamorphic aurede, Central Eastern Desert, Egypt. *Bulletin, Faculty of Science, 51,* 285–301. Cairo University.

Ponce, V. M., & Hawkins, R. H. (1996). Runoff curve number: Has it reached maturity? *Hydrologic Engineering, 1,* 11–19.

Rabus, B., Eineder, M., Roth, A., & Bamler, R. (2003). The shuttle radar topography mission- a new class of digital elevation models acquired by space-borne radar. *Journal of Photogrammetry and Remote Sensing, 57,* 241–262.

Sabet, A. H. (1972). On the stratigraphy of the basement rocks of Egypt. *Annals of Geological Survey of Egypt, 2,* 79–101.

Said, R. (1990). *The geology of Egypt* (p. 734p). Rotterdam: A.A. Balkema.

Saisana, A., Saltelli, A., & Tarantola, S. (2005). Uncertainty and sensitivity analysis techniques as tools for the quality assessment of composite indicators. *Journal of Royal Statistical Society, 168,* 307–323.

Saltelli, A., Chan, K., & Scott, E. M. (2000). *Sensitivity analysis*. New York: Wiley.

Sepehri, M., Malekinezhad, H., Ilderomi, A. R., Talebi, A., & Hosseini, S. Z. (2018). Studying the effect of rain water harvesting from roof surfaces on runoff and household consumption reduction. *Sustainable Cities and Society, 43,* 317–324.

Summerfield, M. A. (1991). Global geomorphology. An introduction to the study of landfornis. xxii + 537 pp. Harlow: Longman; New York: John Wiley Inc. Price £17.99 (paperback). ISBN 0 582 30156 4 (Longman), 0 470 21666 2 (Wiley).

Tribe, A. (1992). Automated recognition of valley heads from digital elevation models. *Earth Surface Processes and Landforms, 16,* 33–49.

Van Griensven, A., Meixner, T., Grunwald, S., Bishop, T., Diluzio, M., & Srinivasan, R. (2006). A global sensitivity analysis tool for the parameters of multi-variable catchment models. *Journal of Hydrology, 324,* 10–23. https://doi.org/10.1016/j.jhydrol.2005.09.008.

Verstappen, H. (1983). The applied geomorphology, International Institute for Aerial Survey and Earth Science (I.T.C.), Enschede, Netherlands, Amsterdam, Oxford, New York.

Torrents Risk in Aswan Governorate, Egypt

El-Sayed E. Omran

Abstract

Identification of areas vulnerable to torrent hazard is the basis of decision-making to take necessary management plan and mitigate the risk the effects of flood risk to reduce losses as possible. Historical records indicate that Aswan Governorate was and continues to be exposed to the geomorphological torrents hazards. The present study aims to identify stream networks and the drainage patterns in Aswan Governorate by the integration between remote sensing and GIS data for determining the directions of running water and the probable sites for storing it. Slope was extracted from Digital Elevation Model (DEM-30 m) data and stream network is established by hydrological model. The results investigated that the lowest elevation is more risky than the highest, where Eastern Nile basins in the area between Edfu and Aswan cities are very risky, particularly in the area of Kom Ombo and East of Aswan city. Results also show that the streams that flow into Lake Nasser represent the minimum risk of torrents as there are no urban communities. This study provides basic information for preliminary results of risk torrent assessments and hazard mapping. The results recommended that avoiding the risks of torrents and benefit from them.

Keywords

Torrents • Risk • Hazard • Aswan • Remote sensing • GIS • Vulnerable areas • Flooding

E.-S. E. Omran (✉)
Soil and Water Department, Faculty of Agriculture,
Suez Canal University, Ismailia, 41522, Egypt
e-mail: ee.omran@gmail.com

E.-S. E. Omran
Institute of African Research and Studies, Nile Basin Countries,
Aswan University, Aswan, Egypt

1 Introduction

On the one hand, torrents are one of the most serious natural disasters that threaten many countries in the world, affecting urban structures, road networks, and infrastructure in arid and semi-arid desert areas. Due to heavy rainfall and carrying all the strength of mud, rocks, sand, and cover all the trees, houses, causing loss of human lives and properties damage. The risk degree is depending on torrents' strength and the layout readiness of the city or village to face and manage that risk (Abd-Allah 2007). On the other hand, flood is an overflowing of water over land that usually is dry or not usually submerged (Olajuyigbe et al. 2012). Flood risks could accelerate as a result of increase in global climate change and rapid urbanization (Karakuyu 2004; Teng et al. 2006; Nirupama and Simonovic 2007). In the time between 1972 and 2014, heavy rainfall and torrents commonly hit many regions in Egypt causing damage to properties and infrastructures, collapse of the ways, and loss of lives and environmental systems. In May 2014, flash flood which inundated Taba–Nueiba area, killed people, and left infrastructures damage (Eliwa et al. 2015).

Torrents of Aswan arise as a result of rainfall on the mountains east of the government where water is heading in a group of valleys west to the River Nile. Most vulnerable areas are:

Wadi Al-Sarraj area (about 13 km south of Edfu), Wadi Ajam area (about 1 km north of Aswan), Umm Habbal area (about 55 km southeast of Aswan), and Wadi Haymour Allaqi area (about 200 km southeast of Aswan). The streams that flow into Lake Nasser represent the minimum risk of torrents as there are no urban communities. Eastern Nile basins in the area between Edfu and Aswan cities are very risky, particularly in the area of Kom Ombo and east of Aswan city. In May 1979, the flood flow disrupted the railway lines and affected Edfu, Kom Ombo, and Aswan centers. About 300 families have been displaced and the falling of torrential rocks on some parts of the agricultural

A. M. Negm (ed.), *Flash Floods in Egypt*, Advances in Science, Technology & Innovation,
https://doi.org/10.1007/978-3-030-29635-3_10

road and railway lines. Floods were repeated in 1980 and 1987, 2005, 2010, and 2014 (ASRT 1993; Saber et al. 2017). In 2010, some villages in Aswan city were severely hit by that torrent. About 500 families were evacuated from their houses that at risk, and lost their livestock and harvest but they were indemnification by the government through the donation account they have created for the Egyptian flash floods (Al-Momani and Shawaqfah 2013).

Several studies developed models for flood risk management and planning by integrated remote sensing and GIS techniques (Todini 1999; El Bastawesy et al. 2009; Dawod et al. 2011; Al-Momani and Shawaqfah 2013; Hamdy et al. 2014; Saber et al. 2017; El-Shaikh 2019). Management of torrents and delineation of area under risk are critically important due to urban development on hill slopes, which has continuously expanded rapidly, and lacking data addressing flood risk assessment (Saber et al. 2017). In order to advance a flood mitigation approach for Aswan, the effects of all the complex mechanisms must be taken into account. To mitigate flood damages and efficiently harvest the highly needed fresh water, the chapter's objective is to manage and mitigate the flood hazard and minimize their effect. In order to study the main objective, the following specific objectives are identified to draw the vulnerability map for several wadis in Aswan, and to determine the flood hazard occurrence on the vulnerable areas of Aswan.

2 Study Area

Aswan is the gateway to Egypt from the south, and represents a point of contact between Egypt and Africa as it the last governments of Upper Egypt in the south, and bordered to the north by Luxor, the Red Sea in the east, and El Wadi El Gedid Governorate to the west. Aswan Governorate is located 879 km from the capital Cairo, and it rises about 85 meters above sea levels. Aswan Governorate (Fig. 1) is located between Lat. 22° 00′ to 25°40′ N and Long. 31°20′ to 33°30′ E. Aswan Governorate occupies an area 62.766 km^2, representing about 30.4% of the total area of the Upper Egypt. The government consists of five administrative centers (Table 1): Aswan, Edfu, Kom Ombo, Nasr El-Nuba, and Daraw. The governorate includes ten cities in addition to the new city of Aswan, and 97 villages. The city of Aswan is the capital of the province.

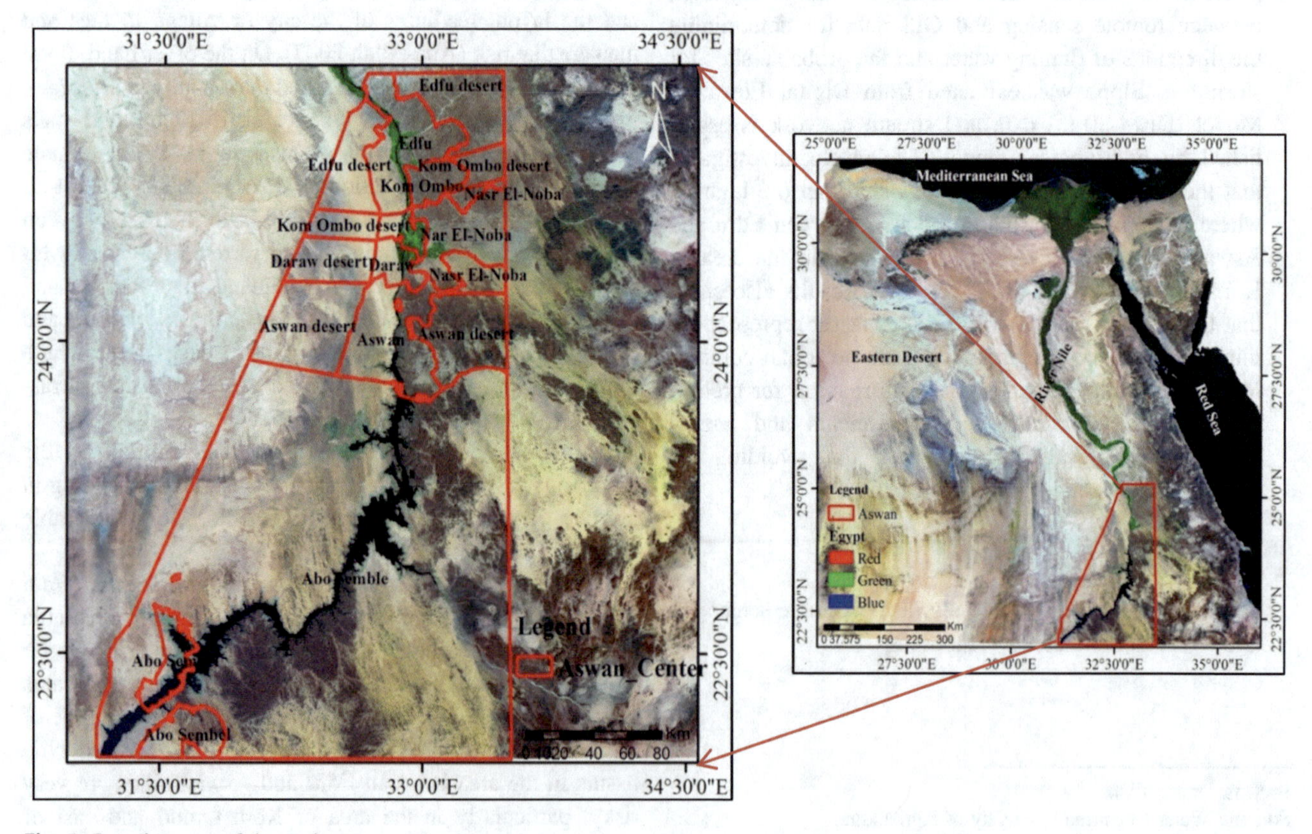

Fig. 1 Location map of the study area

Table 1 Area of Aswan administrative centers

Area (Km2)	Center
11947.6	Aswan
17921.7	Edfu
23895.7	Kom Ombo
2987.0	Nasr El-Nuba
5974.0	Daraw

3 When Does Change Occur? the Crisis Is in Minds, not in Torrents

The vast majority of officials and mass media deals with torrents as a matter of fate, and ignored all those saying that torrents are a possible natural phenomenon, and they have specific paths since the age of the time. Even if a torrent has not passed a century ago, ordinary people can see the course of the torrent that has missed an opportunity that has been lost, and opportunities are lost because of lack of awareness and mismanagement. There is no difference between the disaster caused by torrents, although it is a natural phenomenon and the incident of the train. Both are wrong.

Both reveal insistence on the absence of scientific management. Torrents must be seen as a blessing from God in a country that is now suffering from water poverty and faces threats to sustain its share of the Nile River. For example, rain falls in separate areas on its high mountains, with narrow valleys running each year. The higher the mountain and the greater the valley, the faster and heavier the hill. People in Aswan know the dates of the torrential rain, and know that the rains are raining so often about every 5–10 years as they did this year. Depending on where rain falls, the path of the torrents is determined. Although there are 12 drains exist on the Nile River (Table 2).

Torrents in the region have been running out of loss of life. However, after several years in which the rain did not fall, the usual inaction took place in the Egyptian administration, and the devices became engaged when rain fell, increasing losses.

In Aswan, the disaster was even greater. Because local administrations have been lulled for many years, citizens have been building their mud houses in the torrents, and they have not been warned. Torrents are rarely given. These local administrations have not bothered to find an alternative place for these citizens. The problems there are many. No one wants to worry about another problem. That is why when the torrents came, it changed from grace to curse. The issue is not in torrents. But it is in the style of management and performance. According to the plans that the torrents should not be surprised, cause great losses, and make use of the water. These plans require first lifting the state of emergency immediately after receiving the first notification from meteorology of the possibility of the occurrence of torrents, and then reviewing the plans in the main crisis center in the governorate. When the warning time approaches, the likely passage of the torrents over some of its parts is supposed to be closed. Direct sends the accommodation for the potential areas before rain falls causing torrential rains. But the meteorology warned. But the governorate's agencies have not taken measures, and the plans at the crisis Management Center may not have been opened, and the center itself may no longer exist. The torrents came as a surprise to officials as citizens were surprised. No difference.

Table 2 The drain name and the distance from the high dam in Aswan Governorate arranged according to geographic location (Ahmed 2015)

No.	Drain Name	Location on the Nile river	Distance from the high dam (Km)
1	Aswan Drain	On the right side	10500
2	Abu Sbiera Drain	On the right side	23000
3	Fatera Drain	On the right side	57000
4	Wady elnoqra Drain	On the right side	63000
5	North Seelwa Drain	On the right side	73000
6	Hager El-metamia Drain	On the left side	122250
7	Wdady abady Drain	On the right side	123750
8	Domaria El-Balad	On the right side	124000
9	North Domaria Drain	On the right side	124500
10	North Nagh Hilal Drain	On the right side	138500
11	El-Mahamid Drain	On the right side	139000
12	El-Sharawna Drain	On the right side	143000

The case is not in torrents. But it is in the minds that are responsible for dealing with it. The administration's dominant pattern has nothing to do with science. So government always finds itself in a state of reaction to all events, even political ones. If the scientists had seen us, we would have dams to reduce their overdrive and increase the overstock to be a stock that we would ever need.

If we do not want to hurt ourselves, it will be necessary to create an independent authority with genuine scientific management to deal with crises and disasters, linked to a group of sub-centers in all governorates and cities, whose task will develop perceptions of the risks expected in each region, both natural and non-natural. Scientific plans are put to face every danger in case it occurs, and all the capabilities of the state's authorities are under their control in case of emergency. Without this measure, we will continue to cry in newspapers and satellite channels whenever we are in a disaster, then we forget and wait for the next disaster, and we will have chosen ignorance on the way and failure as a target.

4 Identification of the Risk and Vulnerable Areas of Torrents

Figure 2 shows the DEM map and stream networks. As illustrated in Fig. 2a, slope map, which extracted from DEM, shows the areas vulnerable to flooding (Maidment 2002; Ozdemir and Bird 2009). Figure 2b shows the drainage pattern that is helpful, for decision makers, in determining the directions of running water and the probable sites for storing it, also, suitable sites for dams construction (El-Shaikh 2019).

Identification of drainage networks can be achieved using DEM data. Risk areas of Abouelreesh village, Aswan were identified by Hamdy et al. (2014) using DEM (2012, 30 m) map analyses (Fig. 3a). It was investigated that the lowest elevation which is more risk than the others, contains the most urban area under risk about 48.8% of the total risk area that categorized to three risk zones. About 21% of urbanization and 62% of agriculture land (Fig. 3b) are the vulnerable areas in Aswan City (Saber et al. 2017).

5 Causes and Consequences of Torrents in Aswan

Egypt has been hit by a flood in October, 2019, which the country has not seen for 50 years in terms of increasing rates. This year's flood began on average in mid-August, and then increased in October, and the rains on the Ethiopian plateau significantly increased above all previous rates. The high dam is designed to have a maximum reserved water level of 183 m in front of it, with a storage lake capacity of 169 billion cubic meters. Storage capacity is divided into 31.6 billion cubic meters of dead-load storage capacity. Storage capacity of 89.7 billion cubic meters, which included an average annual disposal rate of 84 billion cubic meters was distributed between Egypt and Sudan. The share is divided into 55.5 billion cubic meters for Egypt and 28.5

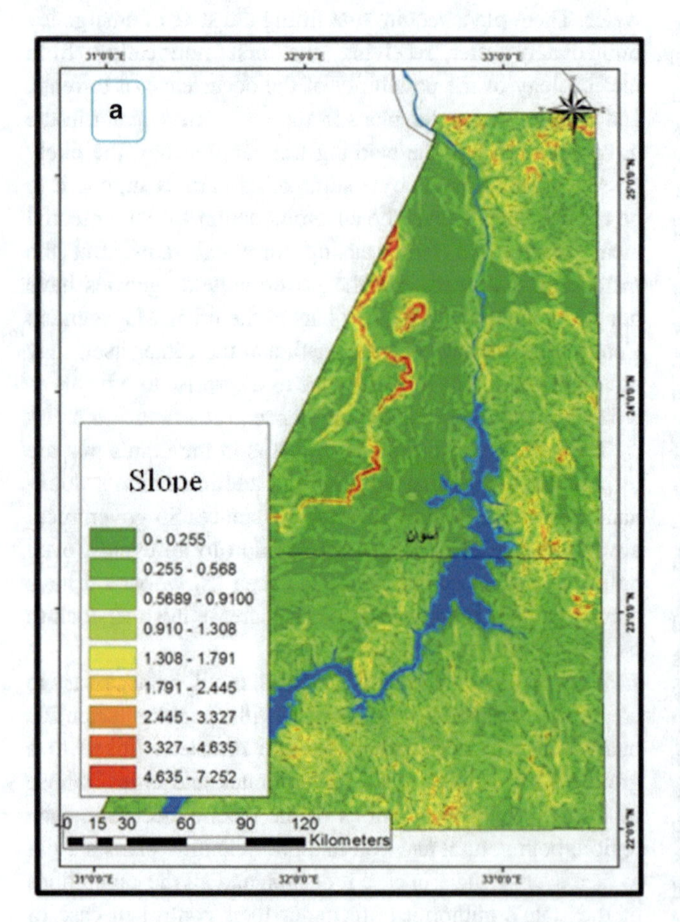

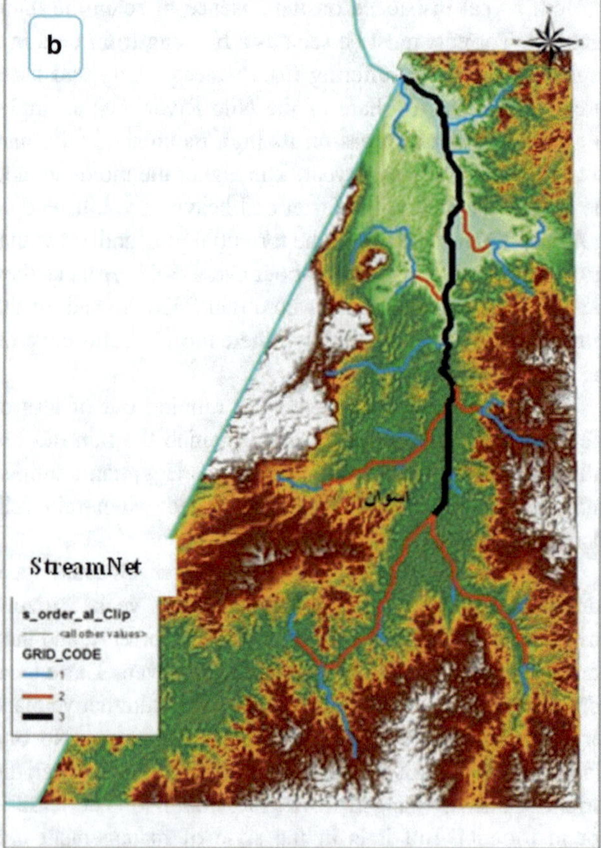

Fig. 2 Slope **a** and Stream networks **b** of Aswan Governorate (El-Shaikh 2019)

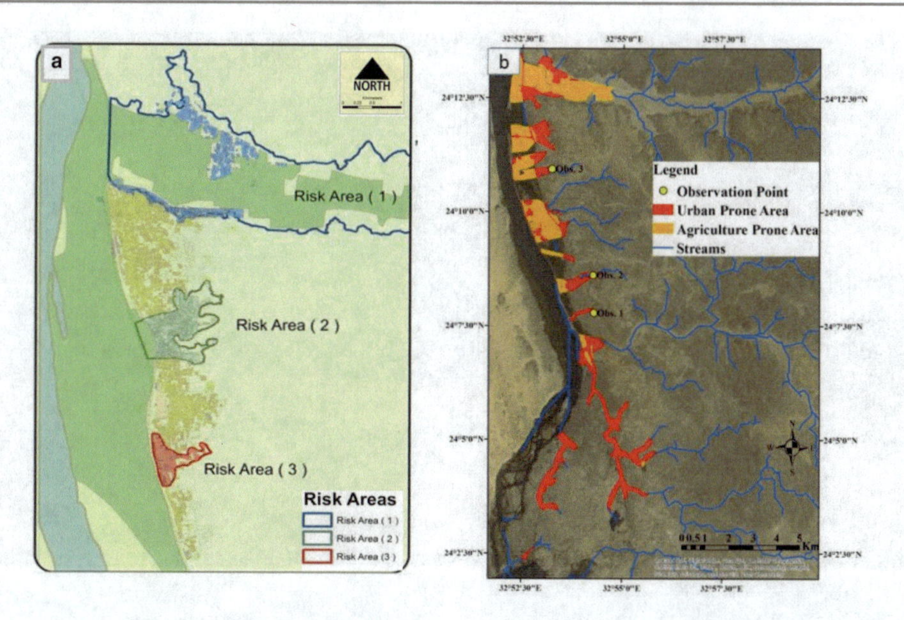

Fig. 3 Risk and vulnerable areas (urban and agriculture) for torrents, **a** Abouelreesh village (Hamdy et al. 2014). **b** Aswan city (Saber et al. 2017)

billion cubic meters for Sudan. A storage capacity of 77.7 billion cubic meters was allocated for flood protection.

The Toshka Lake was established at the end of 1981 to prevent the country from being in danger of high floods and the possible large-scale landslides and destruction of its water installations. A flushing has been carried out to drain the lake if it rises above the maximum stored level of 183 m. The water can be drained if its level increases in front of the high dam above 178 m through the Toshka to low level in the Western Desert, south of the high dam, by about 250 km. It is allowed to pass 200 million cubic meters per day. The water entered the Toshka Lake for the first time on November 15, 1996, when the water level in front of the high dam reached 178.55 m. The Toshka benefit has not been opened since 2001.

This year 2019, torrents and rain contributed to the increase in the level of Lake Nasser. This led to the first time that workers were seen in the port of the high dam (a river port on Lake Nasser, a gateway for passengers and commercial goods between Egypt and Sudan. The construction of the port began in 1964 after the transformation of the Nile River into the construction of the high dam) the presence of high water in a pavement side area that has long been totally dry (Fig. 4—upper).

The most important factor affecting the danger of flash flooding is the inappropriate use of land for urbanization where the wadis' paths have been blocked. Flash floodwaters in Aswan come from the city's south and southeast drainage basins and hills. Runoff water runs northwest along the original roads, passing through the town. The rate of flash floods has increased due to extreme precipitation events, causing the region to be at risk of flooding. Extreme flooding resulted in loss of life, damage to urban areas and infrastructure (main roads), damage to hundreds of vehicles, and erosion on roads and wadi sediment. It took people away from their homes.

Figure 4 shows the impacts of floods that occurred in 2019, which inundating the southern district of Aswan at Abu Simbel at a depth of more than 50 cm. Field and data analysis indicated that a number of factors contributed significantly to these flash flood events, including:

1. unprecedented severe rainfall;
2. inadequate wadi paths due to anthropogenic interference (urban expansion and encroachment across the paths);
3. insufficient drainage systems used to drain the floodwater (a design problem);
4. factors related to manmade infrastructure such as roads, highways, agricultural areas, and earth dikes that obstructed the path of the water; and
5. other factors related to the absence of detailed hydrologic studies, early warning systems, and engineering solutions for the upper reaches of the wadis.

6 Benefit from the Flood Water

Rainwater falling when running in the valleys evaporates part, and the other part runs below the surface to feed the reservoir and the rest of the water goes from it, the greater part to the Red Sea. So, this water can be utilized either to feed the aquifer or store some of it can resort to the following methods:

Fig. 4 Field photos showing large portions of the port of the high dam, which completely inundated with water during flood events of 2019 (upper) and (below) flood water inundating the southern district of Aswan at Abu Simbel, which attacking the district at a depth of more than 50 cm

- Construction of aggregate dams from rocks in the stream of the valley and its tributaries.
- Upstream and downstream to reduce water flow and increase the chance of feeding the aquifer.
- The establishment of reservoirs to receive surface water, which is running in valleys, known as Cisterns, lined with reinforced concrete or plastic laminates with special specifications covered with a cement layer to prevent water leakage. Such tanks used in drinking and some other purpose and depend on the size of the tank.
- Study the possibility of establishing relatively large dams on some of the main valleys where the nature of the land morphology allows for the storage of large amounts of rainwater on a seasonal basis.
- Groundwater accumulates in the sedimentary valleys characterized by high permeability of water flowing to the sea. It is only stopped by the presence of a natural barrier, whether an impermeable layer, a breaker or a splitter.

7 Recommendations to Address the Risks of Flooding in Aswan

Flood risk recommendations and how to address them can be categorized into two parts; the first dealing with floods in their streams, the second deals with the precautions to be taken in establishing urban facilities, agricultural land and others, taking into account that most villages in Aswan are actually built adjacent to hell slopes and without pre-planning (General Organization for Urban Planning 2009).

1. The recommendations made based on this study dealing with floods in their streams are:

- Conducting analytical studies of the drainage basins, especially the recurrent flow.
- Drilling flood streams and lined them with limestone blocks to reach the Nile River.
- Constructing dams of cumulative obstruction of the rock masses in the drainage basins of upstream or incomplete obstruction dams to the both sides of the valleys allowing water to pass through a winding road to reduce the speed and give a chance to leak some to feed the aquifer and receive water away from urban areas, such methods are low cost.
- Construction of an industrial drain on the margins of the flood plain to store water and absorb capacity amounts of excess water to places where runoff can be exploited.
- Construction of a set of surface tanks (Crabs) on the roofs flood fans surrounded by stone arches so that they reserve deposits and purifies the water from them. The reservoirs reach the ground clean and therefore maintain the storage capacity of the reservoirs, which they can be used to irrigate agricultural land.
- It is necessary to create bumpers for torrents and set up concrete dams in the wadies.
- Construction of artificial lakes to receive water.
- Building an advanced early warning system.
- It also recommended specific road specifications and routes in these areas, and identified areas for the construction of dams with special specifications for water reserving and conversion to the aquifer in the region that are suffering from water scarcity.

2. Precautions to be taken in the selection of urban areas:

- Conducting geotechnical studies to select the most suitable places for building houses and away from slum areas.
- Construction of bridges at the intersection of floods with roads.
- Spreading awareness among citizens by avoiding building in areas threatened by serious torrents only after recourse to the decision maker authorities. Even these areas were not previously exposed to these risks of torrents and stay away from dangerous places taking.
- Optimal land planning for high risk areas to mitigate the effects of flood risk to reduce losses as possible.
- Prioritize the implementation by building a conversion channel group of flood hazard prevention and protection plans. These developments will take place in phases, taking into consideration the urban areas vulnerable to flood hazards.
- The results of this study point out that no authority shall be permitted to divide, plan, develop, or use any land within or outside the boundaries of urban development unless it receives approval from the Ministry of Municipal and Rural Affairs for the land scheme and subsequent approval of hydrographic studies carried out to prevent the risk of flooding due to any such project.
- This study suggests that it is important to apply the decision of the Council of Ministers setting out the controls and procedures to be followed in response to flood damage, and to adopt hydrological and engineering designs necessary for flood drainage before adopting housing or farming plans.
- It is also suggested that work permits in the valleys should not be granted unless the work is done in coordination with the relevant authorities.
- Environmental sustainability is one of the conditions for successful development, and this will only be accomplished through the preservation and restoration of the valleys by using water and energy to contribute to the protection of water, food, and housing safety with a view to achieving sustainable urban development and minimizing negative environmental impacts.

References

Abd-Allah, M. M. A. (2007). Modeling Urban dynamics using geographic information systems. In *Remote sensing and Urban growth models*.

Ahmed, A. Z. (2015). *Assessment report for Aswan*. 18 pp.

Al-Momani, A. H., & Shawaqfah, M. (2013). Assessment and management of flood risks at the city of Tabuk, Saudi Arabia. *The Holistic Approach to Environment, 3*(1), 15–31.

ASRT, Academy of Scientific Research and Technology. (1993). Flood Risk in Egypt.

Dawod, G. M., Mirza, M. N., & AlGhamdi, K. A. (2011). GIS-based spatial mapping of flash flood Hazard in Makkah City, Saudi Arabia. *Journal of Geographic Information System, 3*(3), 225–231.

El Bastawesy, M., Kevin, W., & Ayman, N. (2009). Integration of remote sensing and GIS for modelling flash floods in Wadi Hudain catchment, Egypt. *Hydrological Processes, 23*(9), 1359–1368.

Eliwa, H., Mamoru Murata, M., Ozawa, H., Koza, T., Adachi, N., & Nishimura, H. (2015). Post Aswan High Dam flash floods in Egypt: Causes, consequences and mitigation strategies. In *School education research bulletin, Naruto University of Education* (Vol. 29, pp. 173–186).

El-Shaikh, A. A. (2019). The use of GIS and remote sensing in the study of geomorphology of the Aswan region. *JSSA, 5*(20), 544–587.

General Organization for Urban Planning. (2009). Environmental perspective of the urban development strategy of southern Upper Egypt (p. 189).

Hamdy, O., Zhao, S. H., Salheen, M. A., Eid, Y. Y. (2014). *IOP Conference Series: Earth* and *Environmental Science*, *20*, 012009.

Karakuyu, M. (2004). The effects of Urbanization on the climate change and flood. *Marmara Geographical Journal*, (6), 1–12 (Istanbul).

Maidment, D. R. (2002). *Arc hydro: GIS for water resources* (Vol. 1, p. 140). Redlands, CA, USA: ESRI Press.

Nirupama, N., & Simonovic, S. P. (2007). Increase of flood risk due to Urbanization; A Canadian Example. *Natural Hazards, 40,* 25–41.

Olajuyigbe, A. E., Rotowa, O. O., & Durojaye, E. (2012). An assessment of flood Hazard in Nigeria: The case of Mile 12, Lagos. Mediterr. *The Social Science Journal*, *3*(2), 367–377.

Ozdemir, H., & Bird, D. (2009). Evaluation of morphometric parameters of drainage networks derived from topographic maps and DEM in point of floods. *Environmental Geology, 56,* 1405–1415.

Saber, M., Kantoush, S., Abdel-Fattah, M., & Sumi, T. (2017). Assessing Flash Floods Prone Regions at Wadi basins in Aswan, Egypt. In *DPRI annuals* (No. 60, pp. 853–863).

Teng, W. H., Hsu, M. H., Wu, C. H., & Chen, A. S. (2006). Impact of flood disasters on Taiwan in the last quarter century. *Natural Hazards, 37*(1–2), 191–207.

Todini, E. (1999). An operational decision support system for flood-risk mapping. *Forecasting and Management, Urban Water, 1*(13), 1–143.

Hazards, Risk, Harvesting and Utilization of Flash Floods

Egypt's Sinai Desert Cries: Flash Flood Hazard, Vulnerability, and Mitigation

El-Sayed E. Omran

Abstract

Water demand will increase and augmenting freshwater resources from the rainwater and flash flood will decrease the gap between supply and demand. It revolves around the preparation of flood hazard maps for the vulnerable areas and proposed early warning system to mitigate it. Sinai is identified as a flood-prone area where flash floods were recorded and resulted in significant infrastructure damages, population displacement, and sometimes loss of lives. The establishment of different dams leads to the presence of communities around these dams to work in agriculture and grazing. In addition to reducing water losses, reducing the speed of floods is the main factor to protect the soil from water erosion. Floods such as the floods in El-Arish (2010) may not be due to flooding alone, but due to the nature of the randomness of the construction of buildings in the corridors of floods and without the work of a previous geological study of the area on which the various facilities will be built. One effective way to reduce the risk of flash floods lies in the implementation of an early warning system. The early warning information system (EWIS) consists of a number of components, linked and activated through an automatic platform. The EWIS sends warnings based on user-defined hazard thresholds. The alert may vary from a straightforward message to a map displaying the vulnerable areas. The warning is given to decision-makers as an external warning. This provides lead time for decision-makers to react and take action to prevent (or minimize) damage.

Keywords

Flash flood • Remote sensing • Hydrologic model • Land cover • Hazard • Vulnerability • Mitigation • Sinai

1 Introduction

Flash floods are considered one of the worst kinds of hazard in Sinai Peninsula. Flash floods, which define as "a sudden local flood of great volume and short duration" are the consequence of high precipitation rates holding on a relatively long time (a couple of a few hours) (Doswell et al. 1996). Meanwhile, in response to the loss of 26 lives across three governorates in 2016 and 7 lives in Alexadria in 2015, the problem of our ability to address climate change consequences remains unanswered. Is Egypt actually ready?

Flash floods in hot deserts (e.g., Sinai) are portrayed by high speed and low length with a sharp discharge peak (Ashour 2002). Sinai is identified as a flood-prone area where flash floods were recorded and resulted in significant infrastructure damages, population displacement, and sometimes loss of lives (Abuzied and Mansour 2018; Mohamed and El-Raey 2019). Catastrophic flash flooding has become an unfailing event in Sinai resulting in great disaster and extensive loss of life and property. This situation handicaps human use and development in Sinai, which calls for the use of modeling for environmental prediction. Several flash floods were recorded in Sinai, which brought about noteworthy infrastructural damages, population movement, and sometimes loss of lives. Large sediment loads might be conveyed by floods threatening settlements in the wadis and people who are living there. Wadis are especially vulnerable to this kind of event. A barrier is that observation data are generally scarce, and model results are excessively coarse to allow accurate predictions. The significance of the challenge is only likely to increase since the frequency and influence of flash floods are relied upon to

E.-S. E. Omran (✉)
Soil and Water Department, Faculty of Agriculture, Suez Canal University, Ismailia, 41522, Egypt
e-mail: ee.omran@gmail.com

A. M. Negm (ed.), *Flash Floods in Egypt*, Advances in Science, Technology & Innovation,
https://doi.org/10.1007/978-3-030-29635-3_11

develop because of climate change. It is critically essential to precisely predict the occurrence of flash floods in terms of both timing and magnitude. Remote sensing (RS) and geographic information system (GIS) are utilized to provide improved spatial considerate of basin response to storm rain events (Moawad 2013) and flood monitoring. Therefore, the first chapter's objective is to determine the flood hazard occurrence on the vulnerable areas of Sinai.

Flash floods in Sinai are inadequately understood feature due to a lack of accurate environmental and hydrological data, which are challenging and expensive to develop and manage in such region. Furthermore, hydrological extremes are more typical than in humid climates among others because of the accompanying: (1) high potential evaporation, (2) low runoff volume on a yearly basis and runoff occurs as short discontinuous flash floods, (3) precipitation is low on a yearly basis; however, it falls as high-intensity storms of regularly restricted spatial extent, and (4) short duration and high-intensity rainfall, characterized by spatial heterogeneity (Lázaro et al. 2001; Wheater 2008). These features are considerably more noticeable in regions with topographic complexity (Wilson and Guan 2004) such as mountain ranges in Sinai. A hydrological model is driven mainly by information on the land cover distribution (derived from RS), and soil properties (derived from field measurement), which was used to predict risk areas from large peak flows associated with flash flooding in wadis located in Sinai. In the wadis area, runoff calculation is very difficult because of a lack of direct runoff data. Consequently, the issue is remaining unsolved until long-term runoff data will be accessible. It might be useful to utilize different strategies to estimate runoff in the area such as morphometric analyses and climatic data evaluation. The morphometric analysis of the drainage networks, basins, and alluvial fans, which are extracted from topographic maps and digital elevation model (DEM) is considered as main considerations affecting the flash floods. One of the keys for diminishing losses is to offer reliable information through flood vulnerability maps, which can predict future flood-prone areas (Cook and Merwade 2009). Wadis offer the most appropriate routes for roads and infrastructure; leading to increasing pressure for construction in flood-prone areas (Elsayad et al. 2013). Therefore, analyzing of wadi characteristics were used to assess the area at risk to create a final risk zones map. Several wadis in southern Sinai are famous for their activities during flood such as Wadi Watir, Wadi Dahab, and Wadi Kid in the eastern part of southern Sinai and Wadi Ras Suder, Wadi Werdan, Wadi Feiran, Wadi Sedr, Wadi Gharandal, and Wadi Meiar in the western part of southern Sinai. Therefore, the second objective is to draw the vulnerability map for several wadis in southern Sinai.

It is important to understand that factors (Ţîncu et al. 2018) and risks are determined not only by the climate and weather events, i.e., the hazards, but also by the exposure and vulnerability to hazards (Azmeri and Isa 2018), which have been induced by human activity (IPCC 2012). Flash flood hazard mapping is a supporting component of non-structural measures for flash flood prevention (Abdelkarim et al. 2019). "Therefore, effective adaptation and disaster risk management strategies and practices also depend on a rigorous understanding of the dimensions of exposure and vulnerability, as well as a proper assessment of changes in those dimensions" (https://www.ipcc.ch/site/assets/uploads/2018/03/SREX-Chap2_FINAL-1.pdf). Urban planners and highway engineers become insensitive to the potential even though infrequent dangers of dry wadi beds and closed basins, which must carry and store excess surface water. Alluvial fans of wadis area boarding the mountains are of main interest in this respect. On the one hand, they strongly attract human occupancies because they represent preferred sites in terms of local topography, accessibility, and, mainly, water supply. On the other hand, the alluvial fan is located at the outlet of a mountain watershed into a lower foreland such as a rift valley or a major valley, threatening Bedouin villages. In order to advance a flash flood mitigation approach for Sinai, the effects of all the complex mechanisms must be taken into account. Precipitation is a key variable in the hydrological circle, regardless of the climate region. To mitigate flash flood damages and efficiently harvest the highly needed fresh water, the third chapter's objective is to manage and mitigate the flood hazard and minimize their effect.

2 Description of Sinai Peninsula

2.1 Topography of Sinai

The Sinai Peninsula covers an area of approximately 61,000 Km^2 (Al-Gamal and Sadek 2015). The length of the Sinai coast is around 700 km; nonetheless, the length of Egypt's coast is around 2400 km. It is triangular with its apex found by the junction of the Gulf of Aqaba, and the Gulf of Suez in the south and its base by the Mediterranean Sea coastline between Port Said in the west and Rafah in the east (Fig. 1). The Sinai's topography has an elevation difference of 2640 m between its highest point and its lowest point. The elevation of the northern part of Sinai ranges from 1626 m above sea level to the lowest level of the Mediterranean Sea.

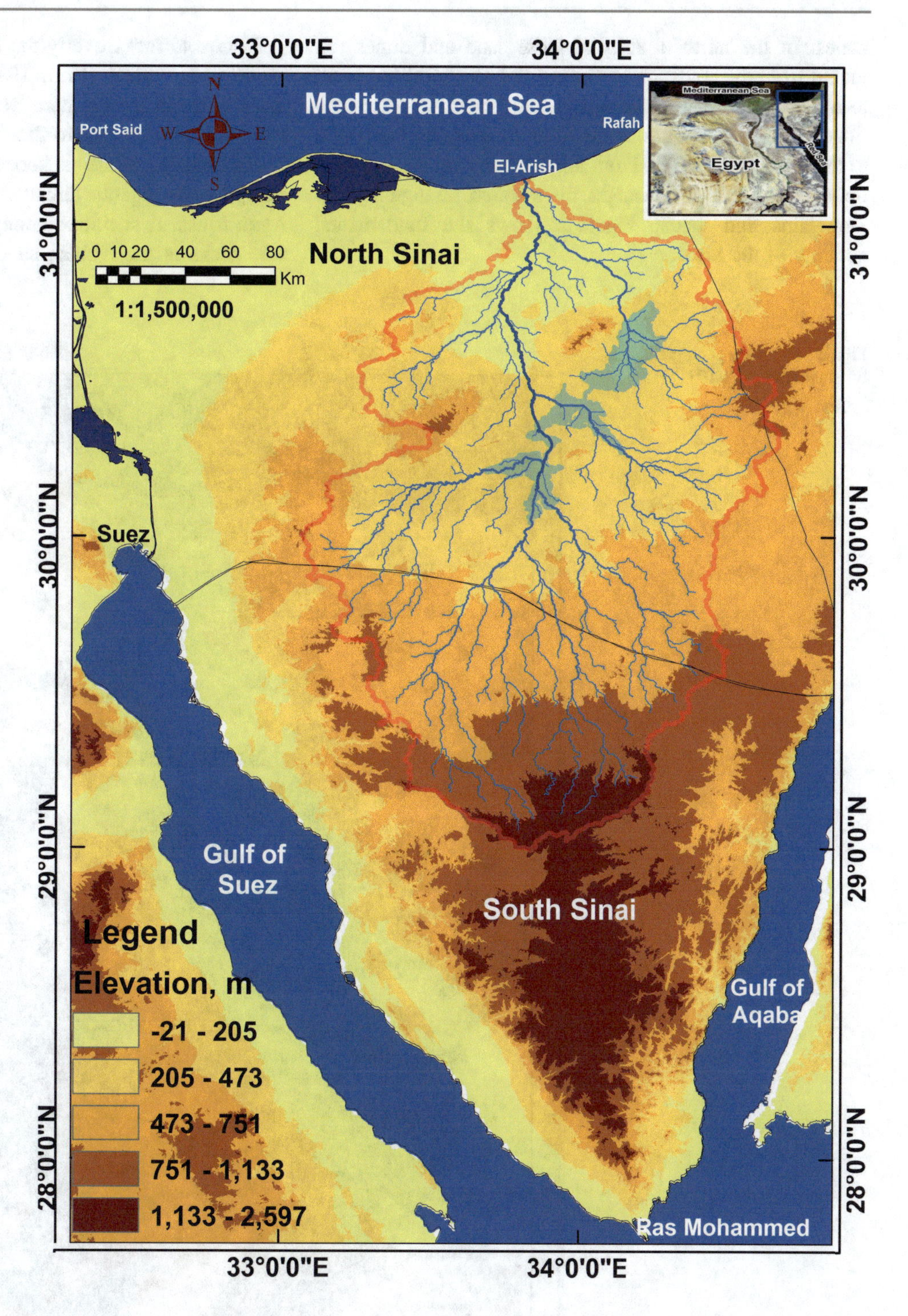

Fig. 1 A digital elevation model representing the topography of Sinai

The elevation of the southern part of the Sinai ranges from 2640 m above sea level to the lowest level at the Gulf of Suez and Gulf of Aqaba.

In arid regions like Sinai, the geology and soils are the most dominant factors in the estimation of the resulting runoff. It is also known that in such regions, the evaporation rate is very high and there is neither land use nor vegetation cover.

2.2 Geological Setting

Groundwater resources and watershed areas of the Sinai are enormously controlled by the related geological formations. Therefore, it is essential to give general establishment information about these formations. The general geology of Sinai was delineated by Said (1962) and Shata (1956). Geologically, the peninsula is divided into various distinct zones. In the north, a strip of loose sand and dunes run inland from the coast for 16–32 km and afterward levels off to a flat, barren plain. This gravel and limestone plain continues for about 241 km, rising at its southern farthest point to the Plateau of El-Tih. From this level toward the southern tip of the Sinai, it is caught by a rough arrangement of mountains and wadis. Figure 2 shows the fundamental geology of the Sinai.

Paleozoic rocks overly the Precambrian basement in the southwestern Sinai (Kora 1995; Omara 1972). Mesozoic strata crops in the northern Sinai, where a nearly complete sequence from Triassic to Cretaceous is known (Kerdany and Cherif 1990), create a subsurface section achieving an exceptionally gigantic thickness (955 m for Jurassic rocks) at Ayun Musa. A standout among the most imperative Mesozoic rocks is the Nubian sandstone of Lower Cretaceous,

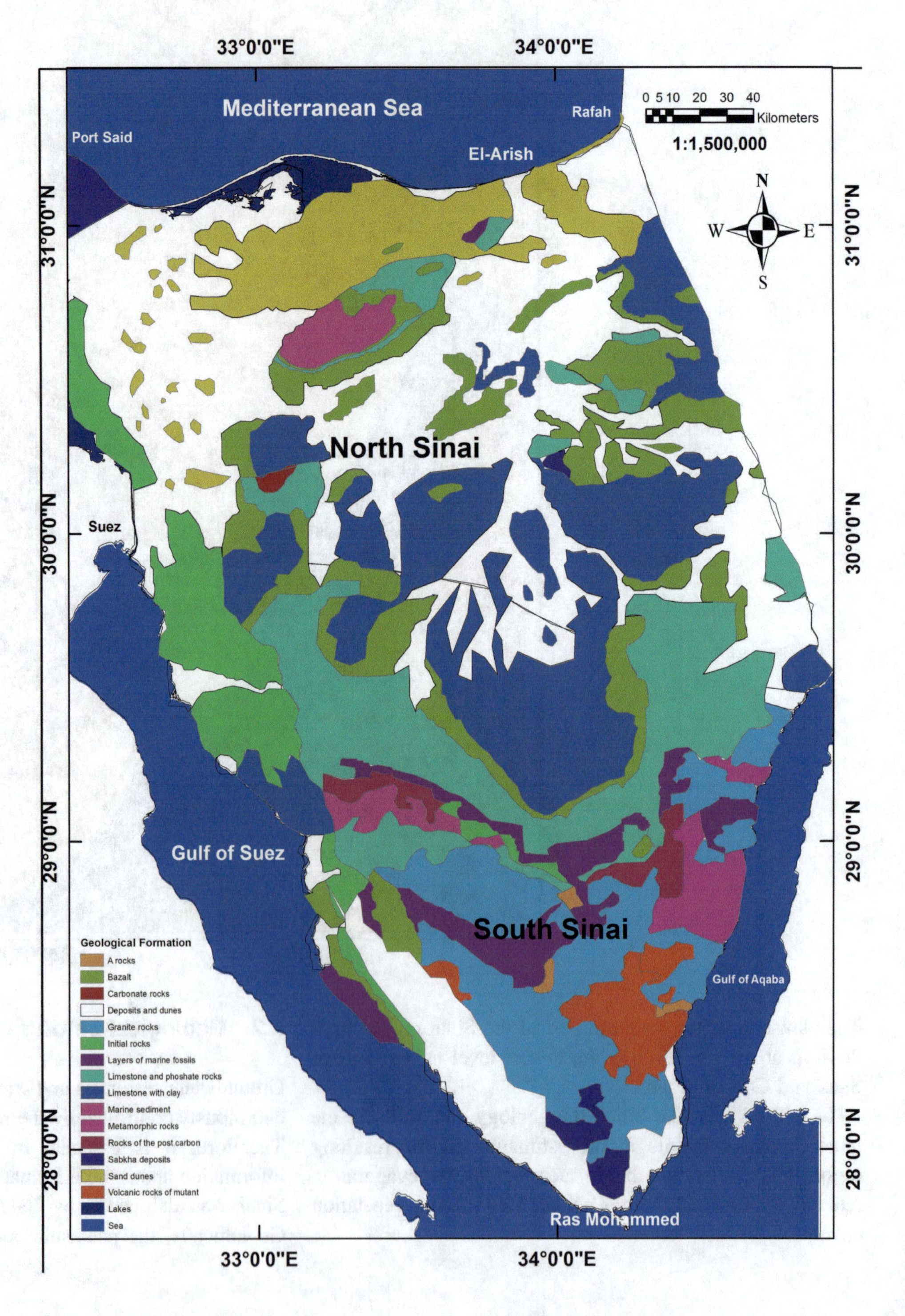

Fig. 2 The main geological units of Sinai (EGSMA 1981)

which represent the main water-bearing unit in the region. It achieves the highest thickness at around 500 m, while at central Sinai, it is comprised of 70–130 m (Kora 1995). Sinai was classified geologically into 15 formations: e.g., basalt, carbonate rocks, granite rocks, limestone and phosphate rocks, metamorphic rocks, and sand dune (Fig. 2).

The high mountains are complex, such as the Gabal Catherina (2641 m above mean sea level—amsl), Um Shomar (2586 m amsl), and Serbal (2070 m amsl), occupies the southern part of the Sinai. Toward the north of this mountain mass, happens the great El-Egma limestone plateau, which inclines from more than 1000 m downward the Mediterranean Sea. The southern mountains are exceedingly dismembered by watersheds draining either to the Gulf of Suez or to the Gulf of Aqaba. Most of the drainage basins of the northern plateau are debouching northward to the Mediterranean Sea (Fig. 2). Two gulfs represent two grabens, which continue along the Suez Canal and Wadi Araba.

In western Sinai, the Nubian sandstones (Issawi 1973) lay on Carboniferous limestone. These rocks comprise of continental sandstones with thin beds of marine limestones, and marls. This sandy unit relates to continental deposits (Schütz 1994). The prerift marine formations that cover the Nubian sandstones contain a thick Cretaceous to Eocene succession of marine deposits that are subsequently overlain by a conglomerate and evaporite synrift series in the center of the trough. The uncovered sandstones in southwestern Sinai marginal to the Gulf of Suez are altogether or incompletely of Carboniferous age. This sandstone unit is represented by medium to coarse-grained sandstones. These sandstones are uncovered along the scarp of Gebel El-Tih. The Cretaceous rocks are broadly distributed in Sinai. They are all around uncovered along the eastern and the western sides of the Gulf of Suez as well as at the northern part of Sinai.

2.3 Soils of Sinai

The soil of Sinai differs in its characteristics according to the method of composition, origin, and geographical location, most of which are sedimentary transmitted either by wind, water, or both (Omran 2017). Sandy soil transported by wind predominates in the northern and eastern parts of Sinai. Sedimentary soils are found in the bottom of the valleys and flood plains, covering the greater part of Tina plain and scattered parts of the eastern lakes. Most of the soils of the northeastern and middle parts of Sinai have high calcium carbonate content, as most of them were formed from the limestone rocks that predominate with El-Egma and Tina Plain. While most of the soil of South Sinai is very poor in calcium carbonate except for the soil of Sedr valley, and the valley of Gharandal and El-Qaa plain, due to the dominant of igneous and metamorphic rocks in South Sinai.

3 Flash Flood Hazard Mapping

On November 2, 1994, Egypt suffered flash floods, which cause serious problems including excessive life and property losses. Several areas are vulnerable to flash flooding, which hampered development and resource exploitation activities. One important process in risk assessment is to characterize the area subject to a flash flood. This process mainly involves two main topics: information to be collected and tools to be used. On the one hand, the information to be collected has to provide scientific data for hazard, vulnerability, and risk analysis and to assist decision-makers during the subsequent process of planning of measures. The information collected should include details about the Geography of the catchments, hydrological and hydraulics information, and vegetation, land use, and historical analysis of local flood events. On the other hand, there are three main tools are to be considered to characterize the area subjected to flash floods: (1) a database for storage of general information; (2) a GIS for graphical representation of maps and spatial analysis; and (3) a set of computer programs for data processing (e.g., hydrological and hydraulic models).

3.1 History of the Flash Flood in the Area Concerned

History of strong flash floods in Sinai during the period extending from 1975 to 2010 are presented in Table 1.

Runoff causes several problems to land and infrastructure, including flooding, erosion, and pollution. Figure 3 highways, bridge, and village in the study area were destroyed by a flash flood in January 2010 at El-Arish.

3.2 Climatic Conditions

The climatic conditions of the Sinai Peninsula are like those, which describe desert areas in different parts of the world. They incorporate extreme aridity, long hot and rainless summer months, and a mild winter. During the winter months, some areas of Sinai encounter brief but the intensity of rainfall that makes wadi beds overflow and sometimes causes severe flash floods, which damage the roadways and, sometimes, human lives (JICA 1999). The climate of any particular place is influenced by several interacting factors. These include latitude, elevation, nearby water, topography, vegetation, and prevailing winds.

The average annual precipitation based on TRMM 3B42. v7A 3-h data acquired (1998–2013) over Sinai (Mohamed 2015) is generally under 200 mm (Fig. 4). In the far northern zone of Rafah and El-Arish, the yearly rainfall in the lower south of the area of Ras Mohamed is less than 20 mm.

Table 1 History of strong flash floods in Sinai during the period extending from 1975 to 2010

Flood event date[a]	Location	Flash floods description and damage
20–25/2/1975	Wadi El-Arish	A unique rain event occurred in February 1975 at El-Arish drainage basin and brought an extreme flood flow to the coastal city. The rain strengthened during the early morning hours of 20 February and continued throughout the day (http://ajer.org/papers/v6(05)/V0605172181.pdf). The water submerged the coastal area and extended to about 200 km till central Sinai and formed a deep lake with a length of 8 km and width of 3 km (may be El-Halal strait). The rain amounts were measured by rain gauges in the area: St. Catherine Monastery (south of the basin), 73 mm; at the east of the basin 68 mm; Nekhel 48 mm; and only 8 mm in El-Arish in the northern part of the basin. It is estimated that the average rain over the basin was 40–50 mm. The total water volume on the basin was estimated to be around 800×10^6 m^3, the peak flow reached 1650 m^3/s, and the total flood volume was 120×10^6 m^3. The results were as follows: 17 persons died and 105 inhabitants became homeless because about 200 homes were destroyed (http://ajer.org/papers/v6(05)/V0605172181.pdf)
18/10/1987	Wadi Watir	Destroy and damage of Nuweiba and Ras El-Naqab road as led to the injury of 27 persons with snake Boswell as destroying parts of coastal road between Nuweiba and Taba especially these parts, which cross the mouths of wadis. This flash flood led to the injury of 27 persons with snake poison. The flooding took away the cars in the roads and buried some of them with everything inside, i.e., people and their belongings
6/1/1988	Wadi Sudr	The flood in a wadi close to Ras Sudr caused the death of 5 passengers inside a car. The flood happened 60 km far from Sud town
13/10/1991	Wadi Feiran	The main road following Wadi Feiran and some houses in El-Tarfa village had been partly destroyed. The flash flood originated in St. Catherine area
20–22/3/1991	Wadi El-Aawag	The big flash flood in Wadi El-Aawag destroyed some houses in El-Wadi village, and some animals died. The flood developed in Wadi Meiar and Wadi El-Mahash
1991, 1993, 1994, 1997, and 25–27/10/2004	Wadi Watir	The runoff in Wadi Watir destroyed the highway to Nuweiba to 40%. The flood developed in the whole basin resulted in a big runoff
17–18/01/2010	Wadi Watir	The high flash flood developed in the whole basin resulted in a big runoff. The rainfall was 11–30 mm, which destroyed the highway to Nuweiba
17–18/01/2010	Wadi El-Arish	Flash flood was hated El-Arish city after heavy rainfall. It was the heaviest rainfall in the last 30 years. According to a report published by the Crisis Management Centre in north Sinai (2010), the flood left 780 homes totally destroyed, 1076 submerged and the area suffered material losses of over US $25.3 million. The destroyed homes would cost the government $3.5 million in compensation. The report also said the floods ruined 59 km of roads, killed 1838 animals and felled 27,820 (mostly olive) trees. The more important was the recorded human death

[a]There are historical records of flash floods in Wadi El-Arish (Klein 2000): October 1925 (very strong), December 1928 (strong), December 1930 (strong), October 1931 (medium), December 1933 (strong), October 1935 (strong), October 1937 (very strong), October 1938 (medium), October 1940 (medium), December 1942 (strong), March 1943 (weak), March 1945 (very strong), March 1947, February 1948, December 1949, December 1950, March 1951, December 1951, February 1952, March 1953, March 1954, 17 November 1964, 11 December 1964, 14 December 1964, 12 January 1965, 27 March 1965, 19 February 1975, February 1979, 2 November 1994, 18 January 2010. Also, there are historical records of flash floods in Wadi Watir: 16 October 1987, 20 December 1987, 1 April 1988, 17 October 1988, 12 March 1990, 23 October 1990, 22 March 1991, 1–2 January 1994, 2 November 1994, 17–18 November 1996, 14 January 1997, 17–18 Octoberuary 1997, 15 January 2000, 9 December 2000, 27–31 October, 3 November 2002, 15 December 2003, 5 February 2004, 29 October 2004, 24 October 2008, 17–18 January 2010 (https://mafiadoc.com/an-early-warning-system-for-flash-floods-in-hyper-arid-egyp)

However, the mountain region (high mid-southern zones) has the yearly precipitation from 50 to 150 mm (Mohamed 2015). Meanwhile, the amount of yearly precipitation falling on the southern Sinai heights between 50 and 75 mm. Central Sinai is viewed as the driest zone in the Peninsula, where the highest yearly precipitation adds up to about 30 mm. The average yearly precipitation ranges from 80 to 100 mm. Despite the lack of overall rain in Sinai, it is the most plentiful rain compared with the eastern and western deserts.

Figure 4 shows the sequential Meteosat images displaying the rainstorm activities every half hour throughout the flash flood in the Southern Sinai on 17 November 1996. It can be seen from Fig. 4 that the rainy storms can be followed to forecast the flash flood from satellite images to avoid the flood hazards. It can be utilized to calibrate rainfall estimates from other satellites. South Sinai is characterized by arid climatological conditions. Values of the average monthly climatic elements recorded at the Ras Sudr meteorological station (Egyptian Meteorological Authority) for the period 1978–1993 show that the highest average monthly maximum temperature is recorded in July (35.3 °C) and the lowest average monthly minimum temperature is recorded in

Fig. 3 On 17 and 18 January 2010, severe flash floods hit Sinai owing to heavy rains from which flood of Wadi El-Arish was the worst. The water level reached 2 m above the ground in some locations. The water washed away cars, auto-trucks, trees, roads, electric towers, and water lines. Soils, plants, and infrastructure affected by runoff of floodwater event

December (6.5 °C) (http://curresweb.com/mejas/mejas/2015/209-222.pdf). Highest rates of evaporation are recorded in June, where the average value was 15.70 mm. Rainfall is frequently occurring during the autumn (October, November, and December) and winter (January, February, and March) seasons. The highest average monthly value of rainfall is recorded in January (4.11 mm). Rainfall may cause severe flash floods, which have a noticeable destructive impact on the asphaltic roads and residential areas (http://curresweb.com/mejas/mejas/2015/209-222.pdf).

3.3 Hydrogeological Setting

Estimating transmission losses in arid environments is difficult because of the variability of surficial geomorphic characteristics and infiltration capacities of soils and near-surface low-permeability geologic layers. Losses occur as flow infiltrates channel bed, banks, and floodplains. Transmission losses in ephemeral channels are nonlinear functions of discharge and time and change spatially along the channel reach and with soil antecedent moisture conditions (Sharma et al. 1994). Without the correct administration of this water resource, the extra precipitation can be rapidly lost to the high evaporative environment or lost from the watershed by means of runoff.

In general, as the soil of the wadis is dry preceding winter storms with insignificant evaporation, the alluvium underlying the wadi beds are expected to viably transmit infiltration losses down to the subsurface alluvial aquifers. Wadi bed infiltration also has a significant influence on flood spread as the flood waves travel downstream to the catchment outlet. Hence, wadi bed infiltration may be the prevalent procedure hidden groundwater recharge. Quantification of the infiltration loss is accordingly vital.

The runoff analysis worries with the amount of runoff produced by watershed for a given rainfall pattern. This influences the runoff volume resulted from a rainfall during the storm. Correlation between meteorologic parameters and the morphologic features of the basin enables to characterize the risk areas of the basin, which may threats the economic regime of the national industrial projects area.

According to Badawy (2013), the total runoff volume of Wadi El-Arish was estimated as 124×10^6 m^3 (Table 2), which constituted about 18.8% of the total rain volume (Vr). Runoff volume (Fig. 5) ranged between 1.77×10^6 m^3 for Wadi El-Azariq and 50×10^6 m^3 for Wadi El-Ruaq. However, the total surface losses were estimated at 541.4×10^6 m^3, which reveal the effect of CN coefficients in reducing the potential runoff. The rain falls majority at very low intensity (<1 mm/h) owing to light rain (Fig. 5), which may have fallen during a shorter period than an hour, but the radar technology assembled the images only every 6 h. "High rain intensities were localized mostly in the western and central watercourses of the catchment during 17 January, rather than the southern watercourses" (https://www.tandfonline.com/

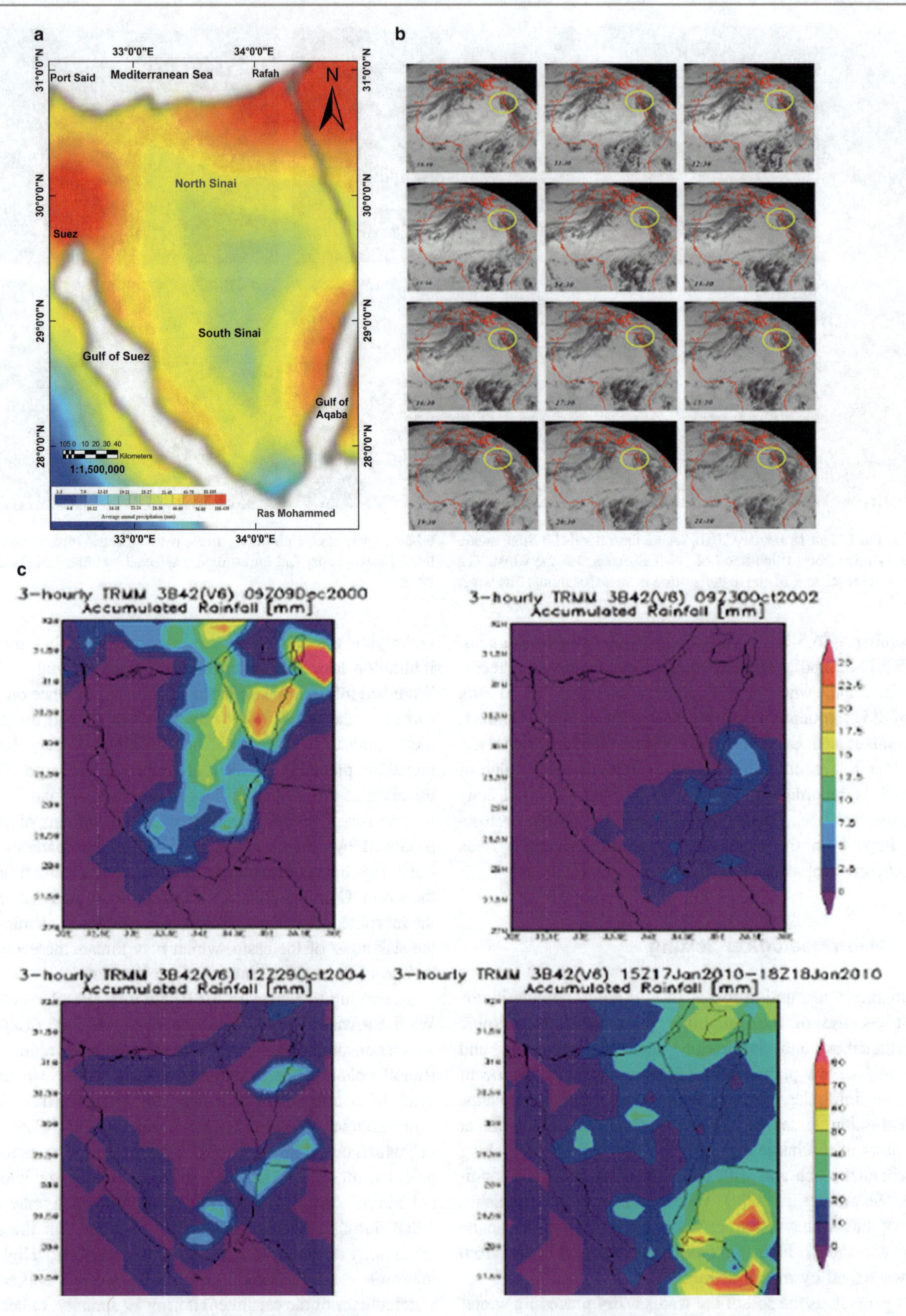

Fig. 4 **a** Average annual precipitation extracted from TRMM 3B42 (v7) A 3-h data acquired (1998–2013) over Sinai (Mohamed 2015). **b** Sequential Meteosat images are showing the rainstorm movements every half hour during the flash flood in the Southern Sinai. **c** Selection of 3-h rainfall estimates from TRMM for the events in December 2000, October 2002, October 2004, and January 2010. Images have been extracted with NASA's Giovanni (http://giovanni.gsfc.nasa.gov) and (https://mafiadoc.com/an-early-warning-system-for-flash-floods-in-hyper-arid-egyp)

Table 2 Flow estimation (Badawy 2013)

Wadi	Vr ($m^3 \times 10^6$)	Q ($mm \times 10^3$)	Vq ($m^3 \times 10^6$)
El-Ruaq	94.5	45.9	50
El-Aqabah	39.7	30.9	28.37
El-Bruck	191	4.8	3
El-Qariya	57.8	20.0	21.26
Abu El-Matamir	5	0.2	0.15
El-Gaifi	37.6	3.5	1.97
Awlad Ali	6.7	0.6	2.13
El-Azariq	5.5	1.5	1.77
El-Hassana	158	10.4	10.91
Lower El-Arish	69.6	5.5	4.42
El-Arish	665.4	123.3	124

Vr, Rain volume (m^3); Q, Surface runoff (mm); Vq, Runoff volume (m^3)

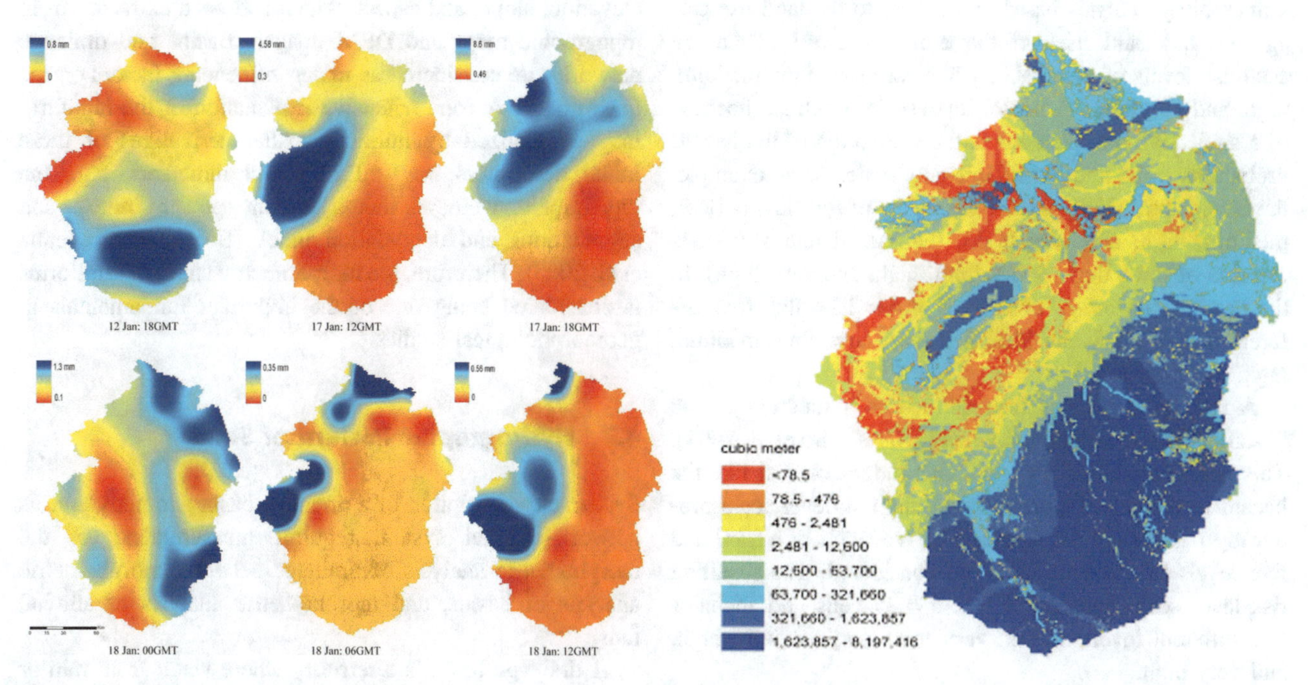

Fig. 5 Six-hour rain intensity and runoff volumes of Wadi El-Arish (Badawy 2013)

doi/full/10.1080/19475705.2012.731657). This resulted in the accumulation of a large amount of water and contributed to more runoff from the central watercourses.

4 Identifying Susceptible Areas Prone to Flash Floods

4.1 Vulnerability Analysis and Risk Assessment

During the process of identifying susceptibility areas prone to flash floods, some decision concerning the choice of the detail level of the analysis will be taken that includes, e.g., map scale and hazard intensity scale. In addition, the definition of hazard scenarios and the construction of a basic map of hazard will be made. Based on the "Hydrology and hydraulics" information collected, in particular, flooding frequency and water levels of a certain time of occurrence, a scale of probability levels of the hazard scenarios should be defined. A probability level is then assigned to each hazard scenario.

There are various methods for determining vulnerability. One method is to use a vulnerability scale, which is defined via the consideration of the defined hazard scenarios and the information available regarding the damage to the population and environment. For instance, we can build our

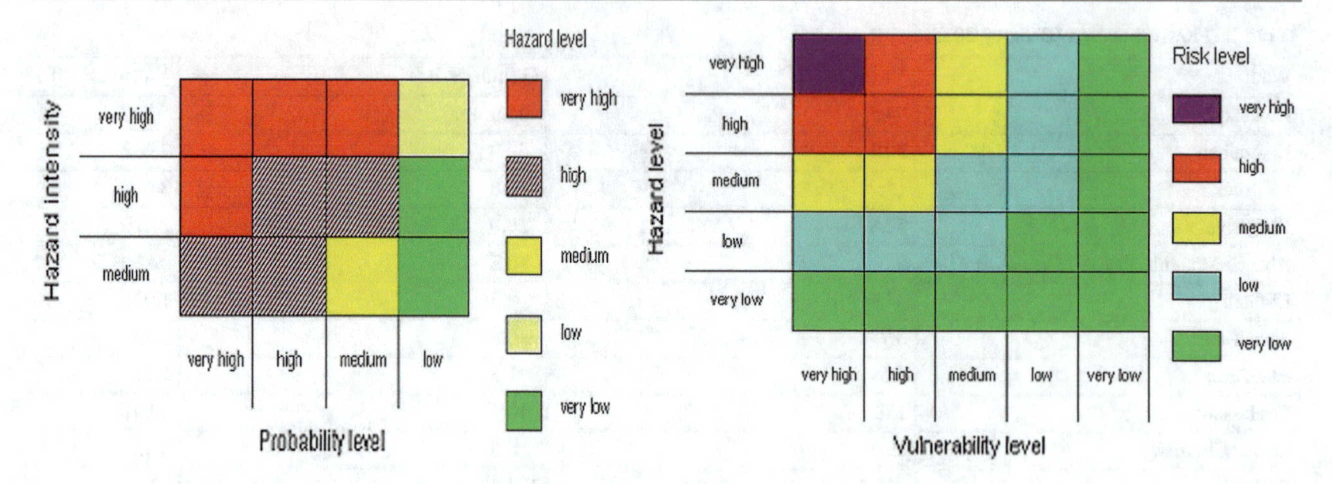

Fig. 6 An example of a hazard- and risk-level scale (Alessandro et al. 2002)

vulnerability analysis based on damage to the land use categories and land uses of the area concerned. Then we establish levels of vulnerability like very low, low, medium, high, and very high. A hazard-level scale is defined because of a subjective interpretation and combination of the hazard probability scale and hazard intensity scale. As an example, the hazard probability scale can consist of four levels (low, medium, high, and very high); the hazard intensity scale consists of three degrees (medium, high, and very high). In the resulting hazard-level scale ($3 \times 4 = 12$ cells), five different hazard levels are identified (very low, low, medium, high, and very high).

A risk-level scale is then defined as a function of the hazard level and vulnerability level (Alessandro et al. 2002). This scale is obtained by subjective judgment, just like the hazard-level scale. Figure 6 shows the risk-level scale produced to assess the risk in a case of five levels of hazard and five levels of vulnerability were considered. The resulting risk-level scale consists of $5 \times 5 = 25$ cells and includes five different levels of risk: very low, low, medium, high, and very high.

South Sinai is exposed to floods through some valleys, especially the active ones like Wadi Ferran, Wadi Dahab, Gharandal and Wadi Sedr. Figure 7 shows the hazardous (risky) areas and the most vulnerable to floods, which caused agricultural and infrastructure projects destroyed, and sometimes may extend the damage to human life.

Because of the hydrological analysis, runoff volumes assessed for design storm of 50-year return period are presented in Fig. 8. This runoff volume ranges from 2 to 132 million m^3 (Elsayad et al. 2013). The watersheds discharge range from 26 to 1717 m^3/s (Fig. 8).

In addition, the defined flash flood hazard map area is shown in Fig. 8.

The morphometric study of the drainage networks begins with the extraction of basic components of relief, such as an elevation, slope, and aspect, which had been extracted from topographic maps and DEM images. Basins and drainage networks are considered as major influencing factors on the flash floods. A comprehensive explanation of the landform may be realized by utilizing spatial derivatives of these initial descriptors, as well as useful indicators, e.g., the topographic wetness index, stream power index, and aggradations and degradation index (Bolongaro-Crevenna et al. 2005). Therefore, the morphometric study of landforms is considered being one of the important fundamentals in geomorphological studies.

4.2 Hydrographic Pattern of Sinai

Systematic description of a drainage basin geometry and its network–channel system requires measurements of the morphometric analysis of drainage network, morphometric analysis of basins, and morphometric analysis of alluvial fans.

A drainage basin is a territory where water from rain or snow melt drains downhill into a waterway, e.g., a valley, river, lake, dam, estuary, wetland, or sea. The drainage basin incorporates both the streams that convey the water and in addition, the land surfaces from which water drains into those channels. The drainage basin acts as a funnel—gathering all the water inside the basin area and diverting it into a waterway. Each drainage basin is isolated topographically from nearby basins by a ridge, hill, or mountain, which is known as a water divide.

The morphometric analysis of the main basins in the area depends mostly on measurements of these basins from topographic maps (large-scale), areal photographs, photo maps, satellite images (TM, ETM, SPOT, and SRTM), and field work measurements to make a data base, which can be analyzed mathematically, geometrically, and statistically by

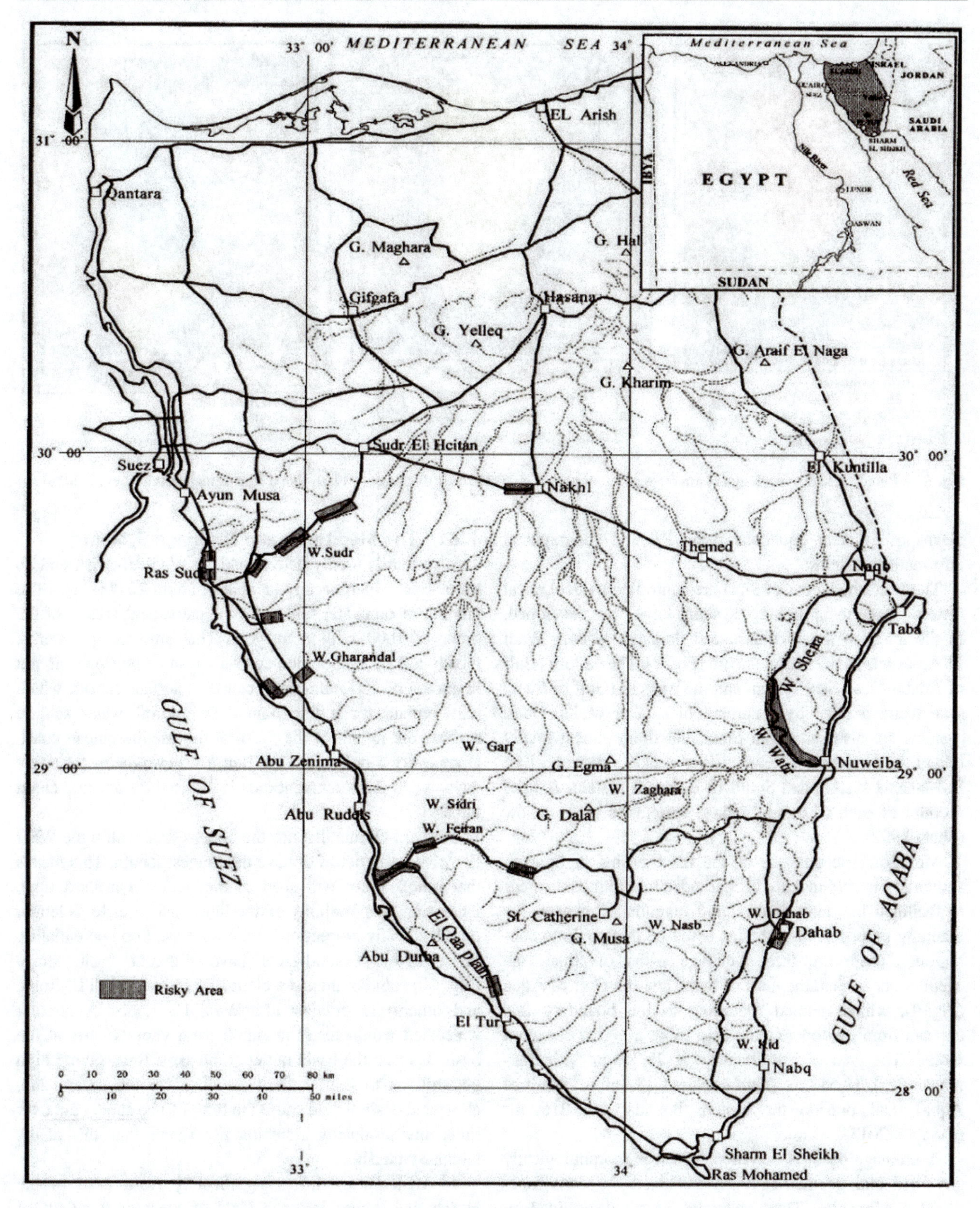

Fig. 7 The hazardous and risky areas most vulnerable to floods in Sinai (JICA 1999)

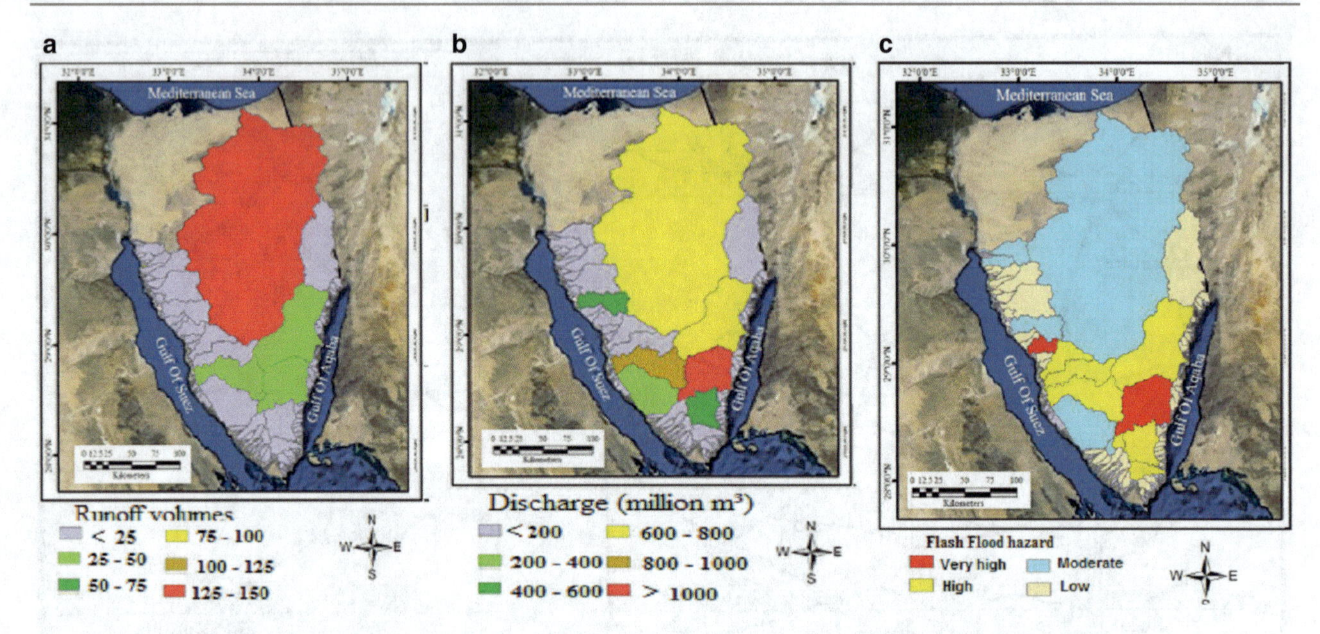

Fig. 8 **a** Runoff volumes for design storm 50-year return period. **b** Watersheds discharge. **c** Flash Flood hazard map (Elsayad et al. 2013)

means of different equations using RS and geographical information systems.

The watershed areas of Sinai are controlled by geological factors. Three main drainage systems have been developed. In the north of the Mediterranean drainage system, Wadi El-Arish acts as the master stream (Fig. 9). The eastern (Gulf of Aqaba–Dead Sea) system and the western (Gulf of Suez) system are drained by a number of smaller wadis. Other systems are also known in Sinai, but drain almost totally inland and comprise East Bitter Lakes system, East El-Manzala system and South El-Bardawil system. A brief account of each of these drainage systems is given below (Shata 1992).

Morphometric analysis of the main basins in Sinai is shown in Fig. 9. Sinai has been divided into four basin areas to facilitate the analysis of runoff potential. However, for planning purposes, these basins could be grouped into sub-regions. There are three drainage paths of Sinai, the Mediterranean Sea, the Gulf of Suez, and the Gulf of Aqba (Fig. 9), which defined according to the boundary that appears from the topographic map using a digital elevation model. The area of this drainage is 23,749 m^2 (Mediterranean Sea), 18,542 m^2 (Gulf of Suez), 11,360 m^2 (Gulf of Aqba), and outside the eastern boundary is 2160 m^2 (ElSayed 2013).

A drainage basin is a system, which is morphologically governed and geometrically characterized by some functional relationships. These relationships are determined by measuring the different elements, which describe the basin as following: basins area, basins dimensions (length, width, and the perimeters), basins shape, and basins surface.

4.2.1 The Mediterranean Drainage System

This system is totally determined by Wadi El-Arish and its tributaries, which occupy an area of about 20,000 km^2. The amount of rainwater falling on the catchment area is of the order of 1000 million m^3/year. This amount of water is mostly lost by evaporation and only a small portion of it, not in excess of 100 million m^3 causes a surface runoff, which may remain for a few hours. The runoff water seldom reaches the mouth of the Wadi at the Mediterranean coast. Once every 5 or 7 years, the runoff water remains for a few days, when the catchment area is affected by unusual cloud bursts.

Wadi El-Quraia that has the biggest area within the Wadi El-Arish basin does not have the longest length. This means that runoff water will shed to the outlet in a short time, increasing the possibility of flooding. Basin area to its length ratio is directly proportion with basin flash flood potentiality. Wadi El-Arish sub-basins have different basin shape (Fig. 10) some of them are elongated such as Wadi El-Ruaq, and others are circular like Wadi El-Quraia. A circular watershed would result in runoff from various parts of the basin to reach the basin outlet at the same time, giving high possibilities to initiate flash flooding, on the other hand, elongated basin would cause the runoff to be spread out over time, thus producing a smaller flood peak than that of the circular watershed.

El-Arish Basin differs significantly from most of the basins that empty into the Gulf of Suez or the Gulf of Aqabah. Firstly, because of its generally low gradient, low drainage density, and the natural controls, the El-Arish Basin can be expected to have a lower runoff yield per km^2 of the

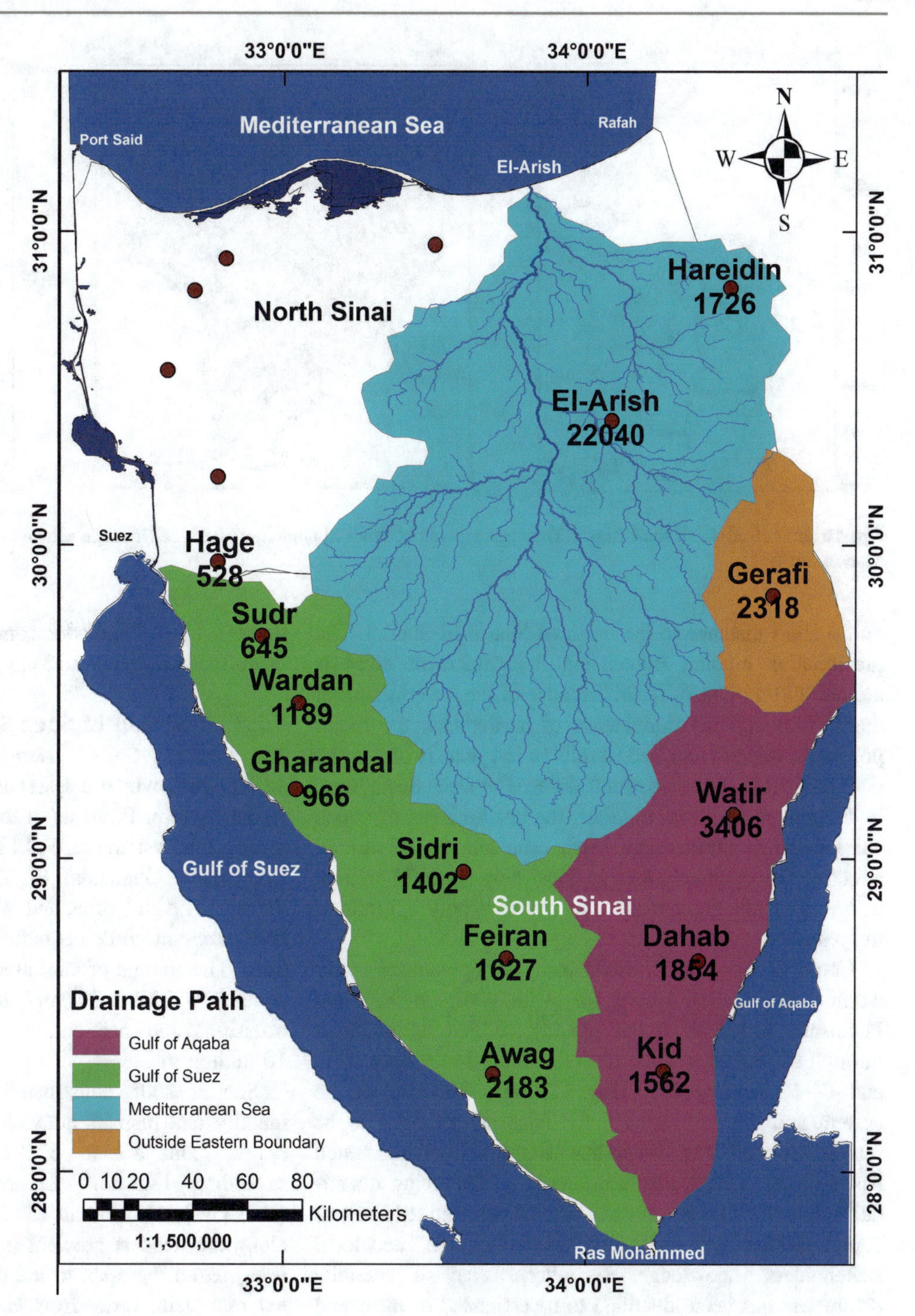

Fig. 9 Hydrographic Basins and drainage pattern of Sinai

catchment area, at least as measured at the lower end of the basin. Secondly, because of the presence of the natural controls on Wadi El-Arish, the average sediment load of floods on the wadi, as measured on the lower part of the basin, should be significantly less than would be expected from floodwaters in the wadis draining to the Gulf of Suez or the Gulf of Aqabah (Dames and Moore 1985).

4.2.2 The Gulf of Aqaba–Dead Sea System

The Gulf of Aqaba–Dead Sea system occupies an area of about 8500 km^2, characterized by severe ruggedness and steep surface slopes. The main wadis dissecting the surface in this area are; from north to south: Wadi Wasit which empties at Nuweiba, Wadi Dahab, which empties at Dahab, Wadi Kid and Wadi Umm Adawi which debouches in the

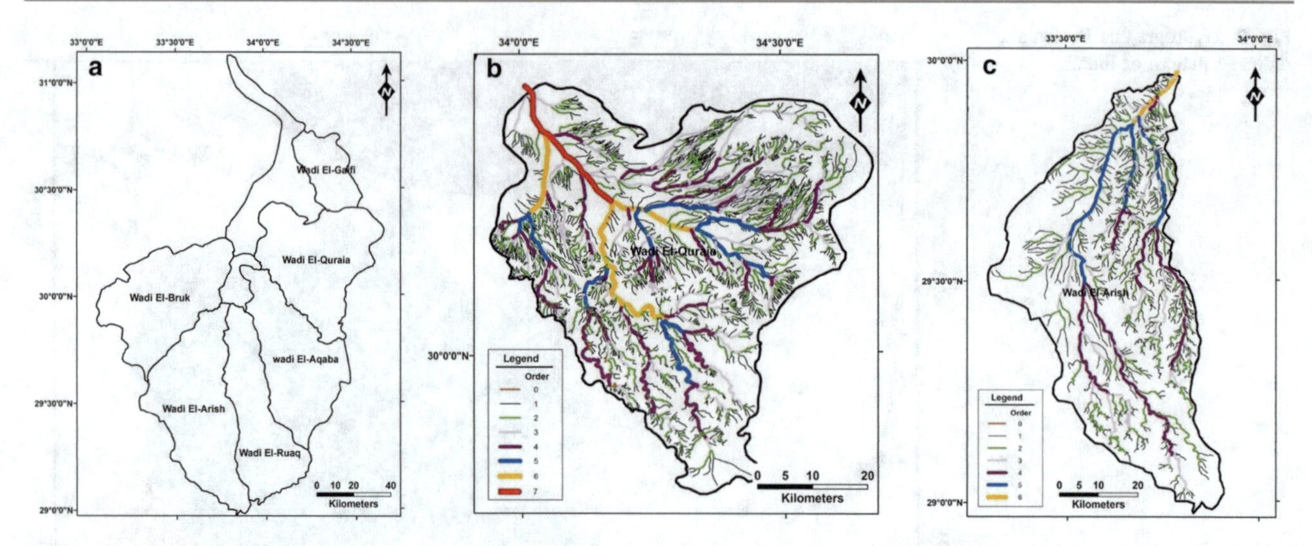

Fig. 10 **a** Wadi El-Arish sub-basins. **b** Drainage network of Wadi El-Quraia sub-basin. **c** Drainage network of Wadi El-Arish sub-basin (Abd-El Monsef 2010)

gulf a short distance to the north of Sharm El-Sheikh. The precipitation amount influencing the catchment areas is around 200 million m^3/year. Because of the steep nature of the surface and the domination of hard rocks, the major portion of this amount (±350 million m^3/year) is lost in the Gulf of Aqaba as surface runoff water. The Dead Sea system is determined by Wadi El-Giraf, the uptake areas of which are located in northeastern Sinai. The amount of rainfall affecting the catchment area in Sinai may reach 75 million m^3/year, but the amount of runoff water is only 2.5 million m^3/year.

Cools et al. (2012) established an operational early warning information system for Wadi Watir on the Sinai Peninsula in Egypt. It has already verified its potential through the forecast of the flash floods of 24 October 2008 and 17–18 January 2010 (Fig. 11). Yet, the skills of the system and the (in)tolerance for false alarms need to be further explored. The system has been developed and tested based on the best available information, this being quantitative data (field measurements, simulations, and RS images) complemented with qualitative "expert opinion" and local stakeholders' knowledge. Secondly, a set of essential parameters has been identified to be estimated or measured under data-poor conditions (https://mafiadoc.com/an-early-warning-system-for-flash-floods-in-hyper-arid-egyp). These are: (1) a list of past important rainfall and flash flood events, (2) the spatial and temporal distribution of precipitation events (3) transmission and infiltration losses, and (4) thresholds for issuing warnings. Over a period of 30 year (1979–2010), only 20 significant rain events have been measured. Nine of these resulted in a flash flood. Regional storms and four by local convective storms caused five flash floods (https://mafiadoc.com/an-early-warning-system-for-flash-floods-in-hyper-arid-egyp).

4.2.3 The Gulf of Suez System

The Gulf of Suez system occupies an area of about 12,000 km^2, which a lesser steep surface compared to the Aqaba system. From north to south, the main wadis determining this system are W. El-Raha, W. Lahata, W. Sudr, W. Wardan, W. Gharandal, W. Tayiba, W. Baba, W. Sedr, W. Feiran, W. Abu Durba, and W. El-Qaa. Most of these wadis have pronounced deltas before terminating into the Gulf of Suez. The amount of rainfall affecting the catchment areas is about 240 million m^3/year, but the amount, which causes occasional and short-duration surface runoff hardly, exceeds 10 million m^3.

Sherief (2008) constructed digital morphometric maps to identify and map 22 alluvial fans (see Fig. 12). Fans area ranges from 2.1 to 69.3 km^2 with an average value exceeding 18.5 km^2. Largest mapped fans are Feiran (69.3 km^2) and Habran (58.5 km^2). Fan length, measured along the steepest gradient (i.e., at a right angle to contour lines), from the apex to the distal boundary contacting the external plain, varies from Hadahid with 1.8 km to Habran with 11.4 km. Within the mapped features, altitude ranges from sea level at Araba west fan to 520 m at Timan fan. Total relief within the fans ranges from 35 m for the Hadahid fan, to more than 280 m for the Habran fan. Low values of total relief are frequently linked to the smallest fans (Hadahid, Araba west, and Taghda). The spatial distribution of relative relief and slopes of the alluvial fans in the study area ranges between 0.5° in Araba east fan and 2.1° in Timan fan, respectively, and reaches to 0.6° in Feiran fan.

Fig. 11 Wadi Watir: topography, sub-basins, rain gauges, and damage resulting from the flash flood of 24 October 2008

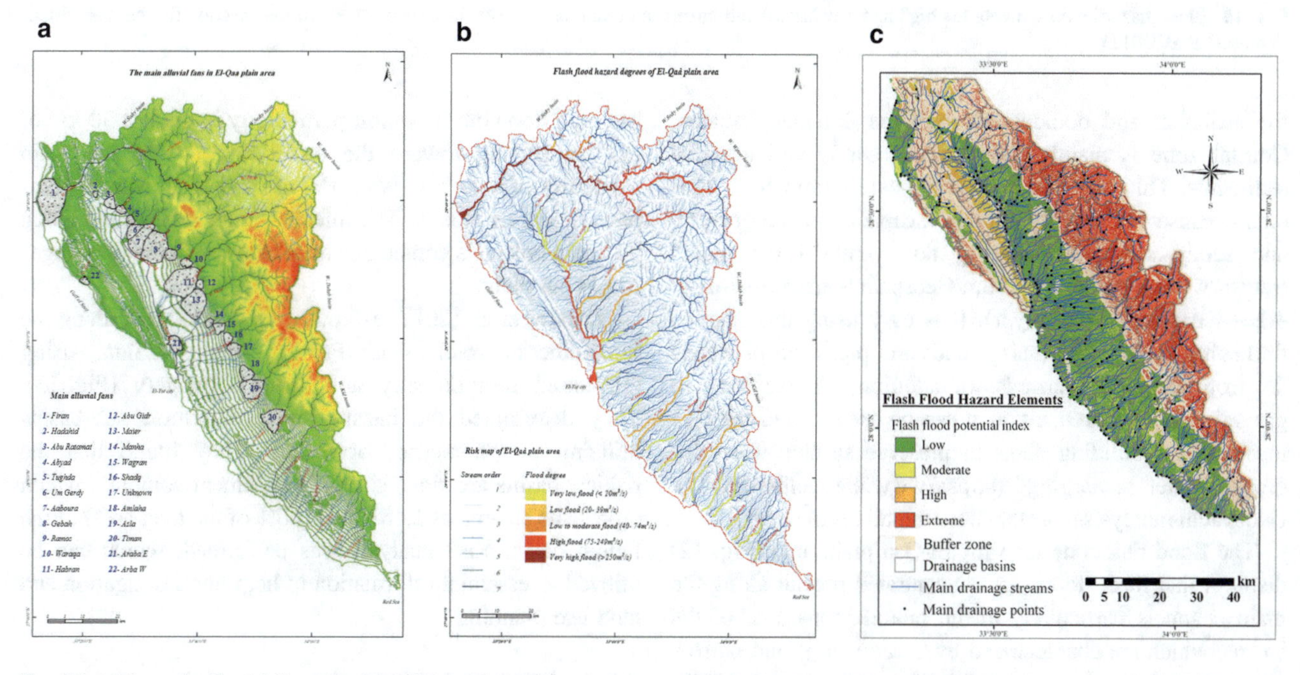

Fig. 12 **a** The main alluvial fans (Sherief 2008) and **b** the risk zones map and flash flood hazard degrees of El-Qaà plain area. **c** Flash flood hazard map of El-Qaa plain

Fan size, shape, location, and dominant morphological processes are controlled by drainage basin features such as size, slope, erodibility of the exposed rocks, rainfall, and flood combined with sediment yield.

Wahid et al. (2016) model spatially the runoff amount and density related to flash flood development (Fig. 12). They create a flash flood hazard map of El-Qaa plain as an example of the coastal plain in a desert environment with the large and complex hydrologic setting. ASTER images are used to develop a DEM and land use/land cover (LULC) data sets of El-Qaa plain area. GIS was used to perform runoff and flash potential flood analyses of the created databases and to show distributed runoff and flooding potential in spatial maps. The main two factors controlling runoff amounts and flash flood potential in such kinds of areas are the slope and soil types. The eastern side is the most area subjected to flash floods where "the slope is very steep, the soils are impervious and flooded water can flush

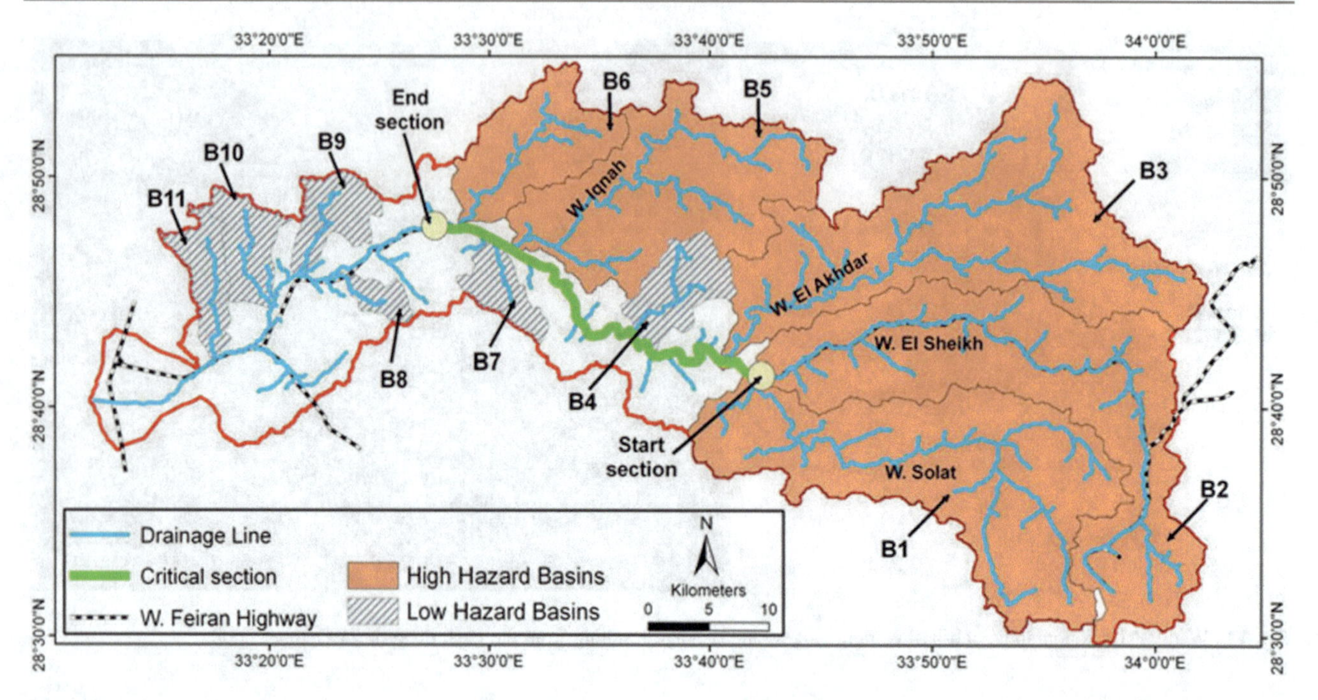

Fig. 13 Flood hazard map showing the high and low hazard sub-basins in Feiran basin. Note the critical sensitive road section for the flash floods (Youssef et al. 2011)

the sediment and boulders through the drainage streams. Coastal zone is mainly composed of sandy and gravelly sediments. This makes this zone the safest from flash hazards because of the high infiltration capacity of the gravelly and sandy sediments forming this gentle slope plain" (https://www.thefreelibrary.com/Geospatial+analysis+for+the+determination+of+hydr). It is easy using this map to find suitable sites for running roads and pipes (oil or water) far from the flood hazards. In addition, "it can help in groundwater management, as it can be used to find the best locations for building dams to preserve surface water for ground water recharging" (https://www.thefreelibrary.com/Geospatial+analysis+for+the+determination+of+hydr).

The flood risk zone map for El-Qaà plain area (Fig. 12) displays that flash floods are concentrated prevail along the main channels particularly in the mountainous area of the basins, which are characterized by a steep slope and narrow channel. At the main streams, the risk is strongly limited. By contrast, some basins such as Wadi Feiran and Wadi El-Aawag basin have wideness channels and comprise some big valleys with broad valley bottoms with Bedouin villages and cultivated areas. Consequently, these cultivated areas and villages often have been destroyed during flash floods. Wadi Feiran basin with the villages (El-Tarfa and Feiran villages) is considered the most dangerous. The flash flood of it is classified as high because of the same classification in the three main tributaries. Along a 50-km distance of Wadi Feiran toward Suez Gulf, the flash flood with of the main channel is classified as very high. Wadi El-Aawag basin also has high flood during runoff particularly in the last 30 km of the mainstream toward the outlet area, while the main tributaries of it such as Wadi Habran and Wadi Meiar basins have moderate floods. The affected areas are limited along the main channels sometimes attacking the Bedouin villages in El-Qaà plain.

Youssef et al. (2011) estimated flash flood risk along the St. Katherine road (Wadi Feiran), southern Sinai, using GIS-based morphometry and satellite imagery (Fig. 13). They determined the hazard level of diverse sub-basins utilizing morphometric parameters. They found that the riskiest basins are situated on the basement complex, which represents an area of 1,467 km^2 (80% of the total area) of the Feiran basin. Risk analysis was performed, which can be utilized as essential information to help flood mitigation and land use planning.

4.2.4 Internal System

These are represented by:

1. The south Bardawil system occupies an area of about 6,000 km^2 and is crossed by a number of wadis, which are lost in the dune sand area. The rainfall amount affecting this area is of the order of 450 million m^3/year, which infiltrates almost totally into the dune sands and forms local occasional surface runoff (2 million m^3/year).
2. The east El-Manzala system occupies an area of about 2000 km^2. It forms a portion of the east side of the Nile Delta basin. The rainfall amount affecting this system is

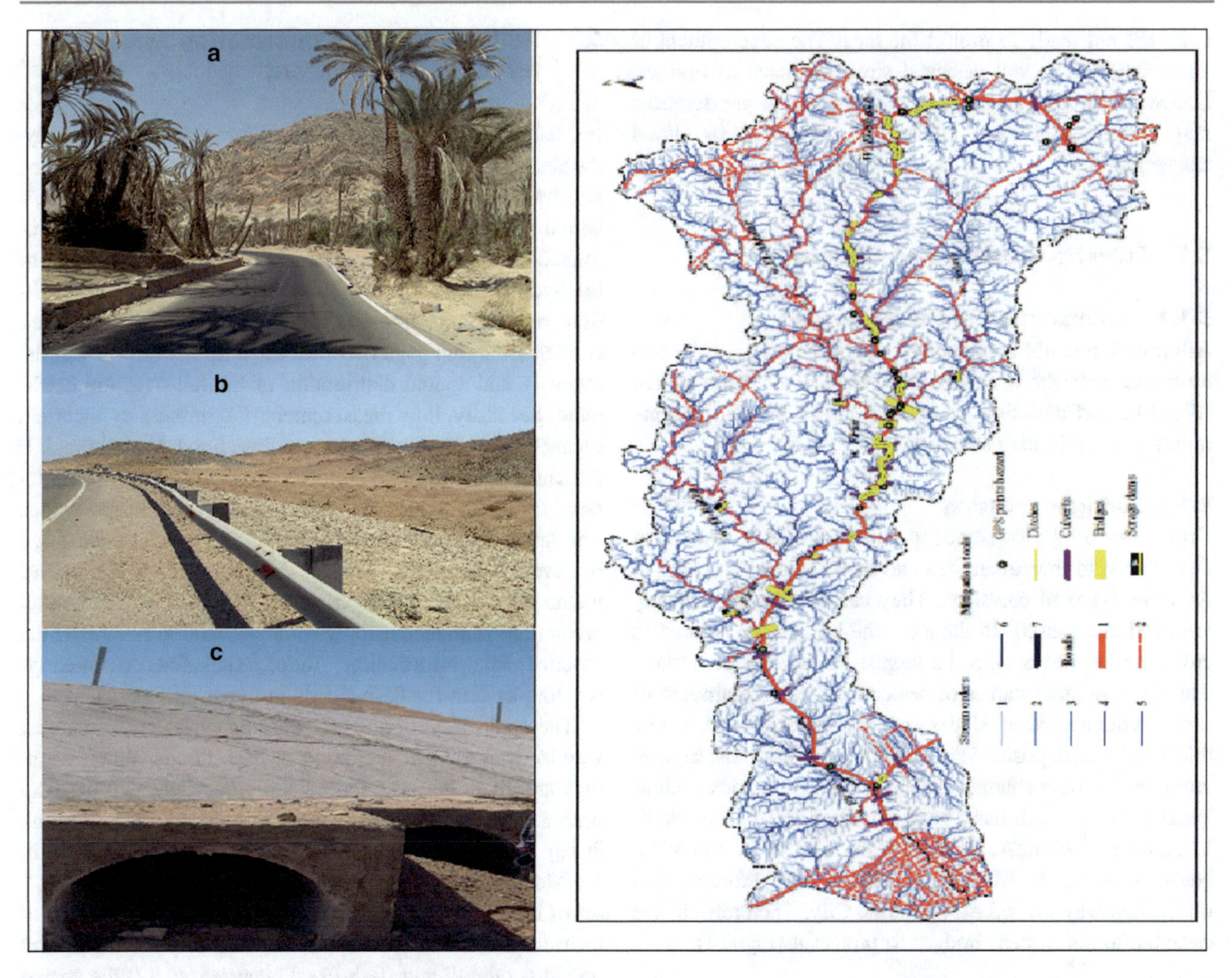

Fig. 14 Photo **a** Afforestation of the main channel of Wadi Feiran by palm trees. Photo **b** Ditches and road dam of Suez–Sharm El-Sheikh Highway. Photo **c** Construction of diversion the main channel of Wadi Feiran. **d** Suggestion methods to mitigate flash flood hazard in Wadi Feiran basin

of the order of 40 million m^3/year. Because the surface is so flat, no real drainage lines are distinguished, and the rainwater accumulates in situ, mostly in the salt flats.

3. The east Bitter Lakes system occupies an area of about 3000 km^2 and is characterized by two main drainage lines, which terminate in the dune sand area. The southern line is known as Wadi El-Giddi, and the northern is known as Wadi Um Khisheib. The rainfall amount affecting this area is of the order of 60 million m^3/year, apportion of which may account for a short-duration surface runoff (50,000 m^3/year).

The sum up, the amount of rainfall affecting the different drainage system in Sinai is of the order of 2000 million m^3/year. The amount of possible runoff water is only about 150 million m^3/year (about 7.5%).

5 Flash Flood Mitigation in Sinai Peninsula

Although, the government had built some culverts crossing (Fig. 14) and bridges (Fig. 14b–c) to mitigate and protect the economic activities in the area, these tools up to this point are not ready to protect the area. The development of unsuitable and unwell designed structures such as bridges, floodwalls, groins, and dike as well as human interferences may have significant effects on flooding or flood exaggeration.

5.1 Culverts Crossing and Bridges

5.1.1 Afforestation of the Watershed

Afforestation could be attempted at chosen areas in the two upstream watersheds of the main channel of Wadi Feiran (Fig. 14). Afforestation conditions affect the infiltration–runoff process in the catchment.

5.1.2 Bridges Crossing

Ordinarily, bridge construction causes the least amount of disturbance to the stream bed and banks when compared to the other types of crossings. They can also be immediately removed and reused. In the area, the bridges are limited to cross the highways with the largest basin channels, which usually have large values of peak runoff, a large amount of runoff volume, steep slopes, narrow streams, and a few thicknesses of deposits. Wadi Feiran basin has most bridges crossing the main channel and their main tributaries such as Wadi Solave, which has a peak runoff of 157.1 m^3/s, Wadi El-Akhdar 159.7 m^3/s, Wadi El-Raha 73.3 m^3/s, and Wadi Nesreen 30.2 m^3/s. All these tributaries have a direct effect on the highway toward St. Catherine City. Therefore, it was essential to build these bridges to protect it (Fig. 14).

5.1.3 Ditches and Roads Dam

The ditches are considered one of the main tools used to mitigate flash foods and help to drain the water far from the road. They are constructed along the highways and distributed along St. Catherine Highway in Wadi Feiran and Sharm El-Sheikh highway in Wadi El-Aawag and other wadis (Fig. 14).

5.1.4 Construction of Diversion Channel

Diversion channel is considered an important tool to protect strategic places, which lie in the flood channel, and deltas. The Egyptian Government used this tool to protect the oil center above the delta of Wadi Feiran (Fig. 14). The diversion channel of the main channel of Wadi Feiran may be not effective, especially during a high flash flood. It is obvious that the wall, which was constructed to change the channel, may be destroyed by a high flood. Therefore, the best method may be the deepening of the main channel of Wadi Feiran upstream of the delta.

5.2 An Early Warning Information System for Flash Flood Hazard

For data-poor areas, the challenges are exacerbated. Firstly, the absence of accessible information is a prime reason for the constrained comprehension of the flash flood dynamics, which thus inhibit the calibration and validation of hydrological and hydraulic models. In addition, huge numbers of the hydrological models are worked for more humid conditions and do not represent arid conditions well. Moreover, conventional rain gage densities often do not well reflect the intensity and spatial distribution of rainfall over the catchment. Secondly, flow measurements are missing or uncertain owing to the damaging power of a flash flood. In addition, it is difficult to quantify and collect field data due to the remoteness, hostile climate, and damaged highways inside wadis. The latter makes it hard to observe and predict flash flood occurrences and encourages the creation of alternative methods for collecting information. The use of RS and rainfall predictions is an increasingly common trend to counteract the absence of information (https://mafiadoc.com/an-early-warning-system-for-flash-floods-in-hyper-arid-egyp).

The application of an early warning system is an efficient way to decrease the likelihood of flash floods. When warnings are given before a flash flood case, there will be extra time to take action to save life and property (Marchi et al. 2010) report a comprehensive evaluation of flash floods in the Mediterranean region. Under data-poor circumstances, a set of basic parameters was acknowledged for evaluation or measurement. These are: (1) an inventory of significant previous rainfall and flash flood occurrences, (2) the spatial and temporal distribution of rainfall events, (3) transmission and infiltration losses, and (4) thresholds for issuing warnings (https://mafiadoc.com/an-early-warning-system-for-flash-floods-in-hyper-arid-egyp).

5.2.1 Operational EWIS

Omran (2016) proposed an early warning information system (EWIS). The EWIS consists of a number of components, linked and activated through an automatic platform (Fig. 15). Predicting rainfall is the first and most important element. Consequently, the rainfall information are converted and aggregated for each subcatchment in the region into spatially averaged catchment rainfall. The rainfall prediction of subcatchment acts as an input for the model of rainfall–runoff. Either using the rainfall–runoff model or a more comprehensive hydraulic model can perform the routing. Finally, the system sends warnings based on user-defined hazard thresholds. The alert can range from a straightforward message to a map displaying at-risk areas and even a full

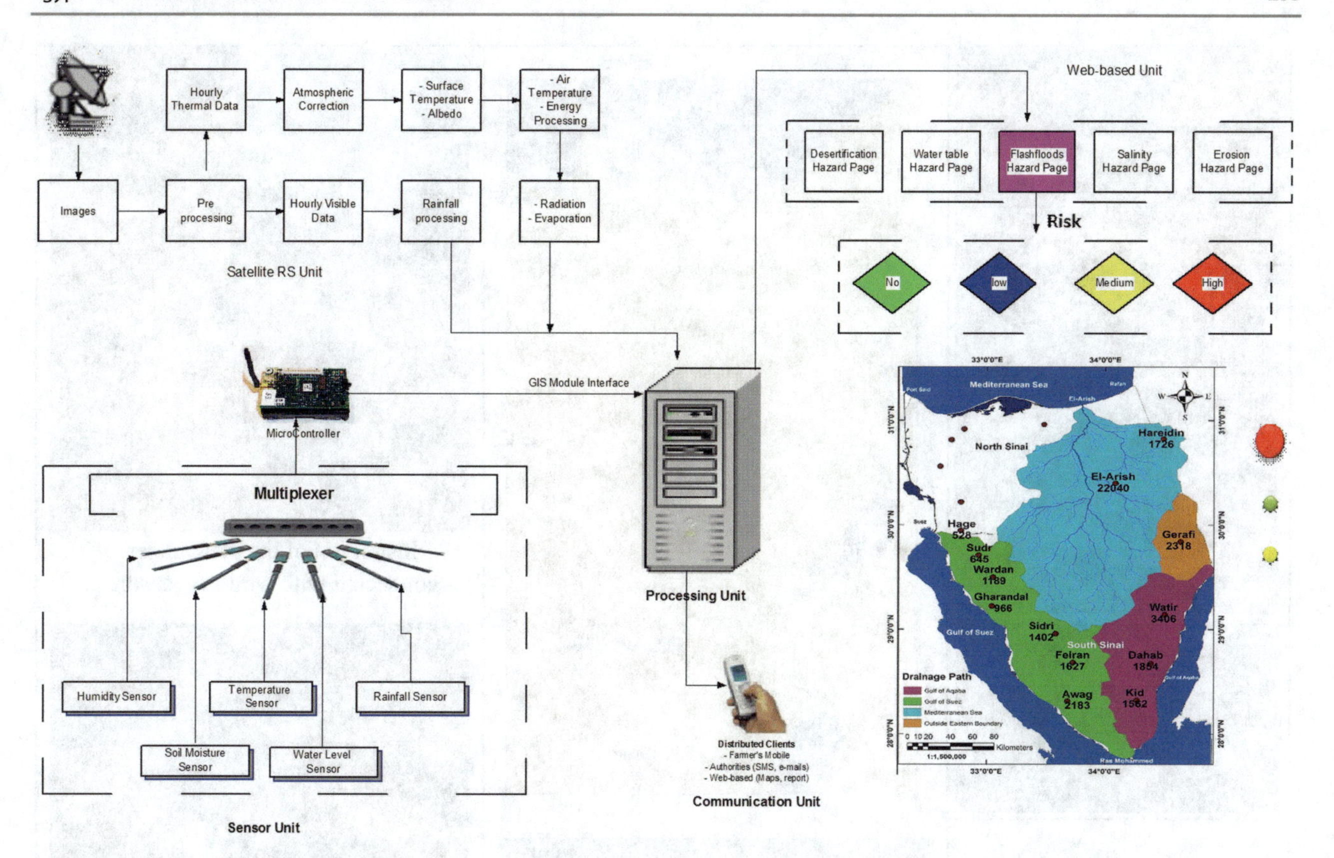

Fig. 15 Concept of real-time outdoor transmission unit, wireless sensor network, WiFi, and control database unit linked with RS and GIS environment for flash flood prediction

(automatically prepared) report. An operator will first handle a warning to exclude false alarms by quickly screening simulation anomalies and communicating with on-field professionals (e.g., based on cloud patterns and Bedouin traditional weather knowledge) (https://mafiadoc.com/an-early-warning-system-for-flash-floods-in-hyper-arid-egyp.). If positive, the warning is submitted as an external warning to decision-makers. This gives decision-makers lead time to respond and take actions to avoid (or minimize) damages.

Figure 15 demonstrates the proposed wireless sensor network, WiFi, and sensor node utilized for flash flood prediction. Different wireless sensors can be used for sensing flash flood hazards. The sensor will utilize the signals, which need to be transmitted to the users. It is not practical to transmit the electrical signal from the field gadget to the customers via wire as it is so long, and we have many wells and sensors in the field at various places. Due to the complexity of the communication scheme and the energy needs of such devices, the use of frequently used wireless technology is also not advisable. It is also not economical to demonstrate power supply or to have a solar panel. Such devices may require high battery energy, and periodic battery charging or substitution is extremely expensive. Alternative option is to supply solar panel energy. The solar panels are costly and are therefore theft after confirmation. A wireless sensor network (WSN) can be regarded to monitor flash flood risks and jointly transfer their information to a primary place through the network (https://mafiadoc.com/an-early-warning-system-for-flash-floods-in-hyper-arid-egyp.).

5.2.2 EWIS Implementation

From an implementation point of view (Fig. 16), sensor nodes are controlled by WSN. The base station is stationary and collects sensor node information. The controller performs duties, processes data, and regulates other component features in the node of the sensor. Sensors are hardware instruments that generate a measurable reaction to changes in a physical situation such as water level, soil moisture, temperature, or precipitation (Fig. 16). An analog-to-digital converter digitizes the easy coherent sign produced by the sensors and is sent to controllers for further processing (https://mafiadoc.com/an-early-warning-system-for-flash-floods-in-hyper-arid-egyp). Many wireless standards have been established. The standards for wireless LAN, IEEE 802.11b ("WiFi") (IEEE 1999), wireless PAN, IEEE 802.15.1 (Bluetooth) (IEEE 2002), and IEEE 802.15.4 (ZigBee) (IEEE 2003), are used in measurement and automation applications more commonly. An 802.11 (WiFi) mesh network consist of high-end nodes, like gateway units.

Fig. 16 Actual real-time implementations of EWIS prototype and field photos along the New Suez Canal in the Sinai

The e-mail/transfer file will be read at this stage and data will be processed within the geographic information systems (GIS). The last values processed will then be set against the threshold values, triggering an early warning if any of the values exceed the threshold limits. The early warning panel would subsequently be hosted on the website (http:/cms.iti-sinai.com), whereby such data would be made available to the public and officials concerned. An alert, triggering and allowing a management reaction, would be given to the appropriate authorities. The map of various risk areas of the website will be updated as the sensor data are constantly calculated. A blinking red light with a constant replay of the 'warning' sound and flash light would show that there is a certain region at high danger, whereas yellow is medium and blue for low risk places. Remarks on the proportion of risk zone and acreage would also appear in the display panel for the built-up places, highways and agricultural regions (https://mafiadoc.com/an-early-warning-system-for-flash-floods-in-hyper-arid-egyp).

Figure 16 shows the actual real-time implementation of EWIS prototype. The operational system collects all data and coordinates computations required to create the different hazard maps. The EWIS system will be used for leadership in everyday activities. Daily automated runs are planned, taking new data and predictions after arrival. Model output can be checked and compared to observations to help forecasters decide on suitable levels of warning. Distributed warnings of the water level are based on model projections and forecasters ' specialist judgement. EWIS has strong graphical tools for viewing information from time series; collecting time series; and editing capacities—copy to—from, e.g., Excel. All models are planned to operate automatically, including data assimilation (https://mafiadoc.com/an-early-warning-system-for-flash-floods-in-hyper-arid-egyp).

6 Conclusions and Study Significance

The central focus of this chapter rotates around determination the flood hazard occurrence, draw the vulnerability map, mitigate the flood hazard, on the vulnerable areas of Sinai. Sinai is progressively suffering from an overwhelming water crisis. Flash flood and runoff water could be an answer to this issue. The produced risk map is useful to know the locations that have a high flood risk in order to avoid loss of life and reduce damages to property. The main watersheds flowing through Sinai are classified into four categories where 4% of watersheds have very high risk, 10% has high risk, 38% has moderate risk, and 48% has moderate to low risk. The aim of any risk study is to minimize the harm caused by flooding. This involves reducing the likelihood of flooding and reducing the impacts when flooding occurs. At the same time, there are underlying pressures that are increasing risks, such as climate change, housing development, or changes in land use. Flood risk assessment and flood mapping will help to show which places are most at risk and in what circumstances. After that, governments can take the correct strategy for flood risk reduction or mitigation. In addition, they can select the suitable locations of weirs and dams for two reasons. The first is to prevent the risk and the second is to collect water in this arid area for Bedouins. The establishment of different dams leads to the presence of communities around these dams to work in agriculture and grazing. In addition to reducing water losses, reducing the speed of floods is the main factor to protect the soil from water erosion.

Floods may not be due to flooding alone, but due to the nature of the randomness of the construction of buildings in the corridors of floods. So, local knowledge of flash floods of Bedouins is needed for the development of the early warning system. Because of the rapidity of flash flood occurrence and its power, flash flood experts recommend the use of early warning systems for reducing vulnerability. Flood risk assessment helps to create flood vulnerable map, and from the historical rainfall data, we can make an early warning system. The early warning system is very important to protect the city by reducing the losses and victims in the region.

However, for future work, relatively simple calculation methods for flash flood hazard mapping is necessary. With the development of computational technologies, the hydrodynamic model can be used to model flash floods to provide more detailed information. With the rapid development of LiDAR and sonar measurement technology, it is becoming easier and more practical to obtain high-resolution DEM data, which can be used to develop more accurate flash flood hazard maps, especially in an area with limit data. Moreover, except for the danger of the life of the population, floods cause also considerable material and financial losses, which need to be explored. Measures for the reducing losses are undertaken depending on the capital value of the threatened objects and the expected consequences from their damage or destruction. Disasters have a negative impact on economic activity, and the associated economic uncertainties hamper investment in long-term commercial relationships. Conversely, particular types of economic activity and a truncated policy focus can increase a country's economic vulnerability to natural disasters.

References

Abd-El Monsef, H. (2010). Selecting the best locations for New Dams in Wadi El-Arish Basin using remote sensing and geographic information system. In *Sinai International Conference for Geology & Development Saint Katherine*, South Sinai, Egypt.

Abdelkarim, A., Gaber, A. F. D., Youssef, A. M., & Pradhan, B. (2019). Flood Hazard assessment of the Urban Area of Tabuk City, Kingdom of Saudi Arabia by integrating spatial-based hydrologic and hydrodynamic modeling. *Sensors, 19,* 1024.

Abuzied, S. M., & Mansour, B. M. H. (2018). Geospatial hazard modeling for the delineation of flash flood-prone zones in Wadi Dahab basin, Egypt. *Journal of Hydroinformatics, 21,* 180–206.

Al-Gamal, S., & Sadek, M. (2015). An assessment of water resources in Sinai Peninsula, using conventional and isotopic techniques, Egypt. *International Journal of Hydrology Science and Technology, 5,* 241–257.

Alessandro, C., Hervas, J., & Arellano, A. (2002). Guidelines on flash flood prevention and mitigation. NEDIES Project—Report EUR 20386 EN, 64 PP.

Ashour, M. (2002). Flash flood in Egypt—A case study of Durunka village–upper Egypt. *Society* of *Geography, 75.*

Azmeri, A., & Isa, A. H. (2018). An analysis of physical vulnerability to flash floods in the small mountainous watershed of Aceh Besar Regency, Aceh province, Indonesia. Jàmbá. *Journal of Disaster Risk Studies, 10,* a550.

Badawy, M. (2013). Analysis of the flash flood occurred on 18 January 2010 in wadi El Arish, Egypt (a case study). Geomatics. *Natural Hazards and Risk, 4,* 254–274.

Bolongaro-Crevenna, A., Torres-Rodrıgueza, V., Sorani, V., Frame, D., & Ortiz, M. (2005). Geomorphometric analysis for characterizing landforms in Morelos State, Mexico. *Geomorphology, 67,* 407–422.

Cook, A., & Merwade, V. (2009). Effect of topographic data, geometric configuration and modeling approach on flood inundation mapping. *Journal of Hydrology, 377.*

Cools, J., Vanderkimpen, P., El Afandi, G., Abdelkhalek, A., Fockedey, S., El Sammany, M., et al. (2012). An early warning system for flash floods in hyper-arid Egypt. *Natural Hazards and Earth Systems Sciences, 12,* 443–457.

Dames & Moore. (1985). Sinai development study—Phase I, Final Report, Water Supplies and Costs. Volume V-Report Submitted to

the Advisory Committee for Reconstruction, Minstry of Development, Cairo.
Doswell, C. A., Brooks, H. E., & Maddox, R. A. (1996). Flash flood forecasting: An ingredients-based methodology. *Weather and Forecasting, 11,* 560–581.
EGSMA. (1981). Geological map of Egypt, Scale 1,2000000. Egyptian Geological Survey and Mining Authority.
Elsayad, M. A., Sanad, A. M., Kotb, G., & Eltahan, A. H. (2013). Flood hazard mapping in Sinai Region. In *Global Climate Change, Biodiversity, and, Sustainability Conference, Arab Academy for Science, Technology and Maritime Transport*.
El Sayed, M. (2013). Risk assesment of flash flood in Sinai. Ms.C thesis, College of Engineering Department, Arab Academy for Science, Technology and Maritime Transport (AASTMT).
IEEE. (1999). Wireless medium access control (MAC) and physical layer (PHY) specifications: Higher-speed physical layer extension in the 2.4 GHz Band. IEEE Standard 802.11b. The Institute of Electrical and Electronics Engineers Inc., 345 East 47th Street, New York, USA.
IEEE. (2002). Wireless medium access control (MAC) and physical layer (PHY) specifications for wireless personal area networks (WPANs). IEEE Standard 802.15.1.. The Institute of Electrical and Electronics Engineers Inc., 345 East 47th Street, New York USA.
IEEE. (2003). Wireless medium access control (MAC) and physical layer (PHY) specifications for low-rate wireless personal area networks (LR-WPANs). IEEE Standard 802.15.4. The Institute of Electrical and Electronics Engineers Inc., 345 East 47th Street, New York, USA.
IPCC. (2012). Managing the risks of extreme events and disasters to advance climate change adaptation. In C. B. Field, V. Barros, T. F. Stocker, D. Qin, D. J. Dokken, K. L. Ebi, M. D. Mastrandrea, K. J. Mach, G.-K. Plattner, S. K. Allen, M. Tignor, & P. M. Midgley (Eds.), *A special report of working groups I and II of the intergovernmental panel on climate change* (p. 582). Cambridge: Cambridge University Press.
Issawi, B. (1973). Nubia Sandstone: Type section. *Bulletin. AAPG, 57,* 741–745.
JICA. (1999). South Sinai groundwater resources study in the Arab Republic of Egypt. Main Report, pp. 220.
Kerdany, M., & Cherif, O. (1990). Mesozoic. In R. Said (Ed.), *The geology of Egypt* (pp. 407–438), Balkema, Rotterdam.
Klein, M. (2000). The formation and disappearance of a delta at the El-Arish river mouth. In *Proceedings of the Jerusalem Conference* (No. 261, pp. 303–310). Jerusalem: IAHS Publishing.
Kora, M. (1995). An introduction to the stratigraphy of Egypt. Lecture notes. In *Geology Department Mansoura University* (116 pp).
Lázaro, R., Rodrigo, F. S., Gutiérrez, L., Domingo, F., & Puigdefabregas, J. (2001). Analysis of a 30-year rainfall record (1967–1997) in semi-arid SE Spain for implication on vegetation. *Journal of Arid Environments, 48,* 373–395.
Marchi, L., Borga, M., Preciso, E., & Gaume, E. (2010). Characterisation of selected extreme flash floods in Europe and implications for flood risk management. *Journal of Hydrology, 394,* 118–133.
Moawad, M. (2013). Analysis of the flash flood occurred on 18 January 2010 in wadi AL-Arish, Egypt. Geomatics. *Natural Hazards and Risk Journal, 4*.
Mohamed, L. (2015). Structural controls on the distribution of groundwater in Southern Sinai, Egypt: Constraints from geophysical and remote sensing observations. Dissertations Paper 593.
Mohamed, S. A., & El-Raey, M. E. (2019). *Vulnerability assessment for flash floods using GIS spatial modeling and remotely sensed data in El-Arish City, North Sinai*. Natural Hazards year: Egypt.
Omara, S. (1972). An early Cambrian outcrop in southwestern Sinai, Egypt. *N.JP Geologica et Palaeontologica, 5,* 306–314.
Omran, E.-S. E. (2016). Early warning information system for land degradation hazards in New Suez Canal region, Egypt. *Modeling Earth Systems and Environment, 2,* 1–13.
Omran, E.-S. E. (2017). Land and groundwater assessment for agricultural development in the Sinai Peninsula, Egypt. A. M. E. Negm (Ed.), (pp. 1–45). Springer, Berlin, Heidelberg.
Said, R. (1962). *The geology of Egypt*. Amesterdam, New York: El-Sevier Publishing Company.
Schütz, K. I. (1994). Structure and stratigraphy of the Gulf of Suez, Egypt. *APPG Mem., 59,* 57–96.
Sharma, K. D., Murthy, J. S. R., & Dhir, R. P. (1994). Stream flow routing in the Indian arid zone. *Hydrological Processes, 8,* 27–43.
Shata, A. (1956). Structural development of the Sinai Peninsula, Egypt. *Bull Inst. Desert* (Cairo) *1*.
Shata, A. (1992). Watershed management, development of potential water resources and desertification control in Sinai. In *Geology and Development of Sinai* (pp. 273–280), Ismailia, Egypt.
Sherief, Y. (2008). Flash Floods and Their Effects on the Development in El-Qaá Plain Area in South Sinai, Egypt, A Study in Applied Geomorphology Using GIS and Remote Sensing. Ph.D. am Fachbereich Chemie, Pharmazie, Geowissenschaften, der Johannes Gutenberg-Universität Mainz.
Ţîncu, R., José Luis, Z., & Gabriel, L. (2018). Identification of elements exposed to flood hazard in a section of Trotus River, Romania, Geomatics. *Natural Hazards and Risk, 9,* 950–969.
Wahid, A., Madden, M., Khalaf, F., & Fathy, I. (2016). Geospatial analysis for the determination of hydro-morphological characteristics and assessment of flash flood potentiality in Arid coastal plains: A case in Southwestern Sinai, Egypt. *Earth Sciences Research Journal, 20,* E1–E9.
Wheater, H. S. (2008). Modelling hydrological processes in arid and semi-arid areas: An introduction to the workshop. In H. Wheater, S. Sorooshian, & K. D. Sharma (Eds.), *Hydrological modelling in Arid and Semi-Arid areas* (pp. 1–20). Cambridge: Cambridge University Press.
Wilson, J. L., & Guan, H. (2004). Mountain-block hydrology and mountain-front recharge. In F. M. Phillips, J. Hogan, & B. Scanlon (Eds.), *Grounwater recharge in a desert environment: The Southwestern United States* (pp. 113–137). Washington, D. C.: American Geophysical Union.
Youssef, A. M., Pradhan, B., & Hassan, A. M. (2011). Flash flood risk estimation along the St. Katherine road, southern Sinai, Egypt using GIS based morphometry and satellite imagery Environmental Earth Sciences. *Environmental Earth Sciences, 62,* 611–623.

Egypt's Sinai Desert Cries: Utilization of Flash Flood for a Sustainable Water Management

El-Sayed E. Omran

Abstract

Water demand will increase and augmenting freshwater resources from the rainwater and flash flood will decrease the gap between supply and demand. Flash floods, which is considered as one of the worst weather-related natural disasters are highly unpredictable following brief spells of heavy rain. The eastern desert and the Sinai Peninsula are subjected to flash floods, where floods from the mountains of the Red Sea and Sinai are causing heavy damage to man-made features. The central focus of this chapter was to achieve a sustainable water resources management in the Sinai Peninsula. The objective of the current chapter is to mitigate and utilize the floods water as a new supply for water harvesting in Sinai. Applying water harvesting of the flash flood will reduce the flood risk at the outlet. In addition, it could be used for recharging the groundwater aquifers, which are the basis for sustainable development in Sinai. Furthermore, Bedouins usually move from place to place searching for fresh grazing for their animals and water for their families. These locations sometimes are hazardous areas as flash floods occur there. This chapter helps to make developed sustainable planning for the Bedouins, as hazard areas were defined. Established places to harvest and store the flooded water would allow the Bedouins to resettle the area. The total amounts of rainfall affecting the different drainage systems in Sinai is of the order of 2000 million m^3/year, and the most of possible runoff water is about 150 million m^3/year rainfall was roughly determined at some sites; for El-Arish region is about 1000 million m^3/year (100 million m^3/year runoff), 75 million m^3/year (4 million m^3/year runoff) for El-Gerafi region, 240 million m^3/year (10 million m^3/year runoff) for the Gulf of Suez region, 150 million m^3/year for Bitter Lakes region, 300 million m^3/year for south El-Bardawil region, and 100–600 million m^3/year for the Gulf of Aqaba.

Keywords

Flash flood • Remote sensing • Water harvesting • Dams • Groundwater harvesting • Water management • Sinai

1 Introduction

Nile water is insufficient to meet all the Egyptian demands, especially in the future due to increasing of the population. Water security and flood hazards are of great environmental, economical, and political importance for arid regions (Tooth 2000), especially in Egypt. The problem of water shortage in Sinai is due to low rainfall and uneven distribution throughout the season, which makes rainfed agriculture a risky enterprise. Rainfall in Sinai is characterized by its instability and scarcity from one year to another. This instability could denote a wave of flash floods with devastating and undesirable environmental and social impacts. Flash floods, as well as floodwater resources (Abdelkarim et al. 2019), are very destructive and have a significant impact on local society, economy, environment, agriculture, and cultural heritage. Flash floods are furthermore not well understood from a hydrological, meteorological as well as from a socioeconomical point of view because they come unexpected and often at a very localized scale (WRRI 2010). It is therefore extremely difficult to forecast flash floods with sufficient lead-time to take emergency actions. In arid regions, flash floods are the main cause of loss in infrastructure, property, and human life. The eastern desert and the Sinai Peninsula are subjected to flash floods, where floods from the mountains of the Red Sea and Sinai are causing heavy damage to man-made features (Abuzied and Mansour 2018; Mohamed and El-Raey 2019). Despite this, floodwater

E.-S. E. Omran (✉)
Soil and Water Department, Faculty of Agriculture, Suez Canal University, Ismailia, 41522, Egypt
e-mail: ee.omran@gmail.com

A. M. Negm (ed.), *Flash Floods in Egypt*, Advances in Science, Technology & Innovation,
https://doi.org/10.1007/978-3-030-29635-3_12

could be used to create a great amount of Sinai water resources, and be utilized to fulfill part of the increasing water demand in areas prone to flooding that are currently experiencing high population growth and economic development. This is especially true in the wadi system of Southern Sinai, where flash floods have become an unlucky annual occurrence.

Flood risk management (Jalilov et al. 2018) can be defined as a combination of comprehensive and continuous societal analysis, assessment, and interventions for a reduction in flood risk (Schanze 2006). However, the magnitude of potential water savings and controlling water in order to sustain life has always been one of the basic difficulties faced by mankind. Water saving is an important objective of Egypt's water strategy to serve a growing population with limited resources, which are a big problem inherent in the opening up of the deserts in Egypt to modern development. The chapter's overall aim was to achieve sustainable water resources management in the Sinai Peninsula through the utilization of flash flood water.

Water resources management is a critical issue in arid regions due to the shortage of precipitation, high evaporation, as well as increasing population, together with restricted water resources and associated increases in the need for water. Flash floods are devastating and represent a big threat to human life and arid environments, but can also be utilized and accomplished appropriately to make such floods valuable as water resources. One of the significant targets is thus to evaluate the flash flood water as an additional water resource.

Flash floods spread violence and fear over the land. They sometimes bring peace and grace. Despite their hazardous impacts, flash floods in Sinai, and different parts of southern Egypt represent a potential for non-conventional freshwater sources. However, the majority of the floodwater goes waste as runoff to the Gulf of Suez and this could meet part of the water demands for a multitude of uses in this area if efficiently utilized. A major challenge in these areas is the wise use of floodwater to permit the sustainable management of water resources. Therefore, the chapter objective is to put the best ways to mitigate and utilize the floods water as a new supply for water harvesting in Sinai. In addition, guidelines suggestion for floods water utilizing, and at the same time reducing and controlling their hazards on the area has been discussed.

2 Description of Sinai Peninsula

The Sinai Peninsula is located in the northeast of Egypt. The Suez Canal and the Gulf of Suez are separated from the rest of the Egyptian territories. It is a triangular shape in the south at Ras Mohammed and its base overlooks the Mediterranean Sea in the north. More than half of it is located between the Gulf of Aqaba and the Gulf of Suez (Fig. 1). Sinai is described as the hinge that leaves around it the continent of Asia in the east and Africa in the west. "The separation of the two continents caused the form and geographical shape of Sinai the way it looks today" (http://www.allsinai.info/sites/geology.htm). Sinai has an area of about 61000 km^2, or about 6 % of the area of Egypt and about three times the area of the Nile Delta and exceeds the area of some countries. The length of Sinai beaches about 700 km, representing nearly one-third of Egypt's coastline.

The Peninsula of Sinai may be divided into three geological districts, namely, the granitic and metamorphic, limestone, and sandstone rocks of which they are composed (Said 1962). Generally, Sinai reflects all geologic columns of Egypt. "The northern part of Sinai mainly consists of sandstone plains and hills. The Tih Plateau forms the boundary between the northern area and the southern mountainous with towering peaks" (http://www.allsinai.info/sites/geology.htm).

The following are the stratigraphic column described in detail by (Said 1962) is divided into four major units:

1. Upper Calcareous Division is composed of sandstone, shales, conglomerates, evaporates, and foraminiferal marl. They range in age from Oligocene to Pleistocene.
2. Middle Calcareous Division mainly consists of marls, chalks, limestones, and shales of Cenomanian to Eocene age.
3. Lower Clastic Division, in the south; these are Carboniferous to Lower Cretaceous sandstones. Toward the north, the rocks are more calcareous. This complex, as especially its upper part, is also known as the Nubian Sandstone.
4. Basement Complex, Precambrian igneous and metamorphic rocks; many dykes occur crossing the basement rocks and are considered of Tertiary age. Oil is produced along the shore of the Gulf of Suez. Production is mainly from Miocene sandstones, and in certain fields, from Carboniferous, Cretaceous, and Eocene formations (http://www.tandfonline.com/doi/pdf/10.1080/02626667109493772).

Climate conditions vary in different regions, where dry climatic conditions are prevalent in the north and severe drought in the center and southwest. Rainfall is concentrated in the northern region, with rainfall decreasing from north to south. The average rainfall in North Sinai ranges from 100 to 200 mm per year. The highest quantity is recorded in the Rafah area, while in the interior it ranges from 20 to 100 mm/year or less. In general, the intensity and intensity of rain vary greatly. Large amounts of rain may fall within a day or even within a few hours, as in the case of thunderstorms.

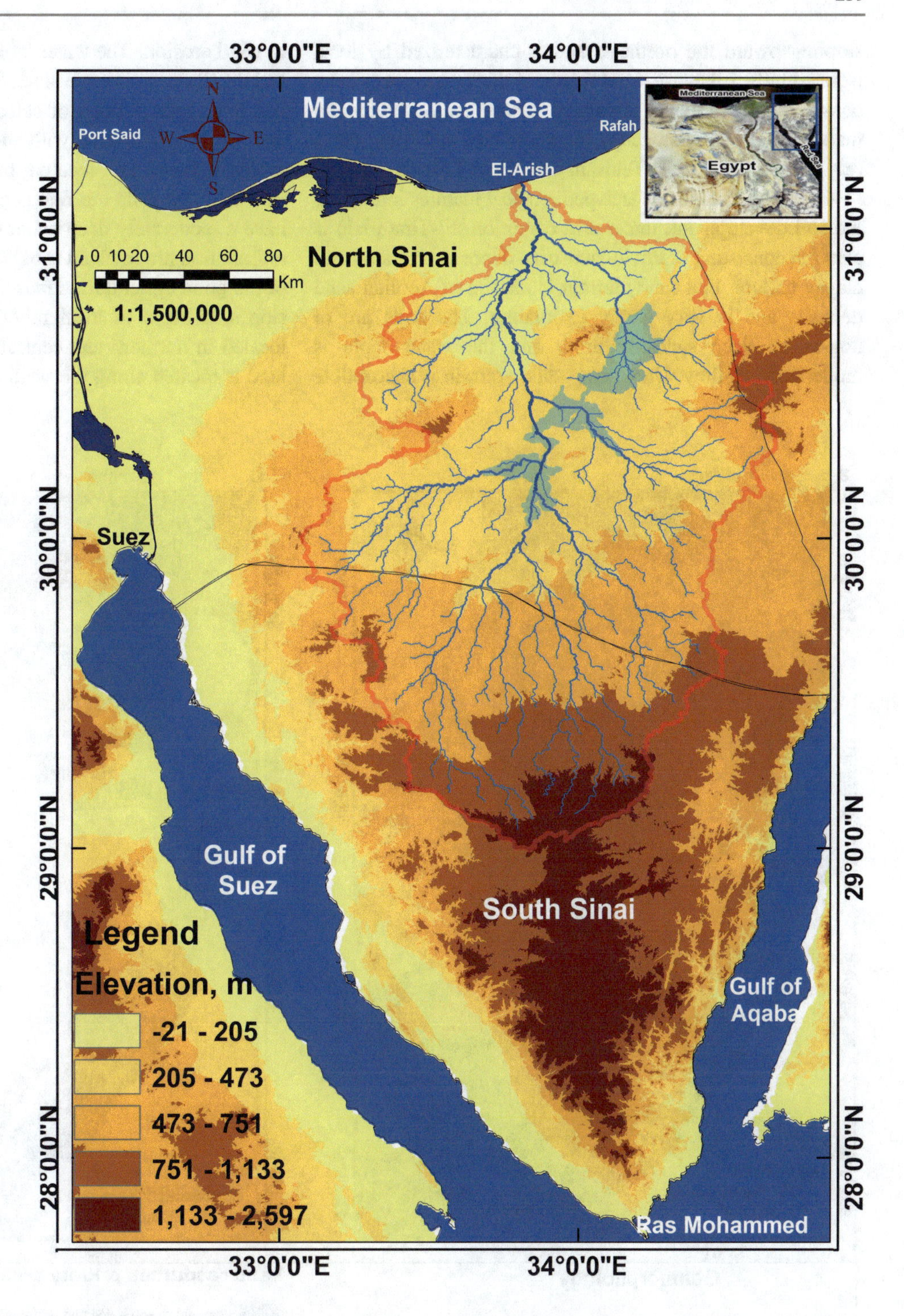

Fig. 1 A digital elevation model representing the topography of Sinai

Soil resources of the Sinai are enormously influenced by the related landform types. The Sinai was classified into six landform units (Dames and Moore 1983), which are Highlands (southern mountains, central tableland, El-Egma tableland, and northern folded blocks); northern piedmont plain; western outwash plain (Feiran, Sidri-Baba, and Sudr-Abu Suweira plains); morphotectonic depressions (El-Qaa plain, Wadi Araba, El-Saghir, Bitter Lakes, Temsah Lake, the area between El-Maghara-Risan Aneiza and the area between Gabal Yelleq and El-Halal); northern coastal plain; and sand dunes. Geomorphologically, the Peninsula includes seven regions (Fig. 2a). The southern elevated mountains area includes the southern part of the Sinai assuming a triangular shape with its apex at Ras Mohamed to the south. The central plateau district includes the central part of Sinai as two principle questas; El-Egma to the southwest, and El-Tieh to the north (Abu Al-Izz 1971). A hilly region lies to the northeast of the Sinai. It is gently sloping toward the northeast and is characterized by local isolated hills. Likewise, sand ridges of thick sandstones set a coastal district of gently undulated surface apart. Muddy and marshy land districts occupy the shorelines and some lakes (e.g., El-Bardaweel, El-Temsah, and Bitter Lakes).

Generally, Sinai soils lack pedagogical features indicating the soil development, under arid condition. El-Tina plain is clay flat consisting of fluvio-lacustrine deposits of loamy and clayey texture. Flat sandy terrain is formed of Aeolian sand deposits and is very gently undulating. The soils are of non-saline deep sandy-textured, and the water table is moderately shallow. Undulating sandy terrain is susceptible to wind erosion. The water table is expected to be very deep and the drainage is excessive. The soils of the Wadi El-Arish bed have been formed of calcareous fluvio-marine deposits and are loamy texture with moderate drainage. Calcareous sandy terraces are located between Rafah and El-Sheik Zuweid. The soils are fine sand or loamy sand texture and have a moderately deep water table. Factors restricting land reclamation are lowland sabkhas and mobile sand dunes. All of the good agricultural lands (Fig. 2b), those having 0–8 % slopes, 0–50 m of local relief, and a deep soil cover, are located in northern and central Sinai. The good agricultural land is located along the wadi channels. The main areas are

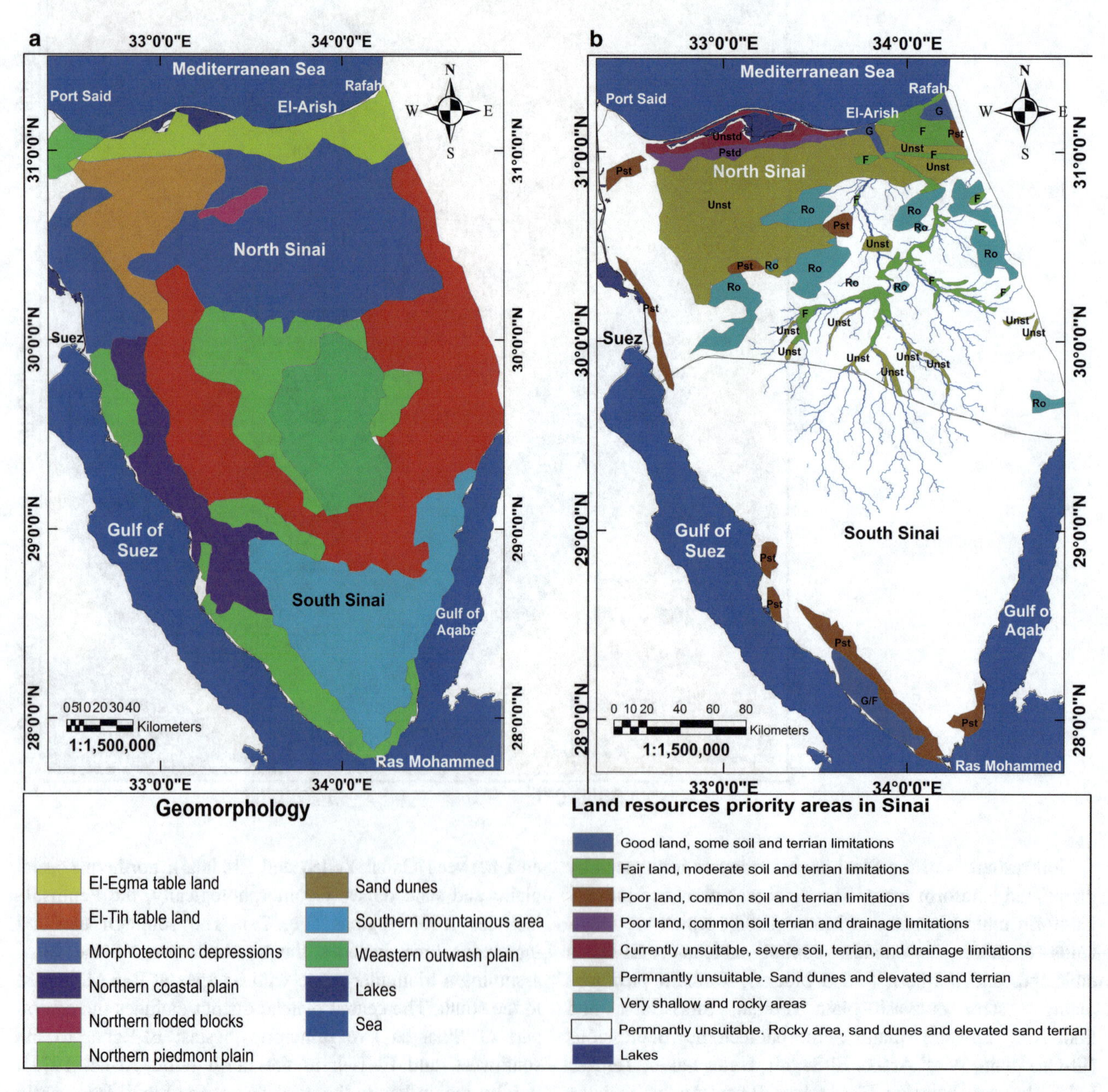

Fig. 2 **a** The main geomorphological units (EGSMA 1981) and **b** the mainland resources of Sinai

along Wadi El-Bruk, extending from Gabal El-Kaib north to Bir El-Thamada, and then east to the confluence of Wadi El-Arish near Gabal Kherim and the Wadi El-Arish main channel and its tributary. Wadi El-Aqabah also forms a good agricultural area, but a smaller and narrower one than the Wadi El-Bruk area. The Wadi El-Arish agricultural area extends from the confluence of Wadi Abu Gidi north to Wadi El-Bruk. There are several significant areas of fair agricultural land in the Sinai (Fig. 2b). On the western, northern, and southern sides of Wadi El-Bruk, between the wadi and the mountains, there is an uneven, narrow band of fair agricultural land. An extensive area of fair agricultural land lies around the base of Gabal Libni. Another large region of fair agricultural land in North Sinai is a 5–20 km strip that trends north–south along the Great Bitter Lakes area. In South Sinai, the largest area of fair agricultural land lies in the El-Qaa Plain. Two extra areas of fair agricultural land are found at Abu Rudeis and Belayim. These areas occupy alluvial fans developed by Wadis Baba and Feiran, respectively. Long narrow strips of fair agricultural land are found in most of the main wide wadis of South Sinai.

3 Maximum Utilization Flash Flood Water as Significant Water Resources

3.1 Existing Versus Potential Sites for Dam Construction with High Flash Flood

Seven dams in Sinai (Table 1) are mentioned in the literature (Shata 1992). Three are masonry dams (Rawafaa Dam, Ein Gudeirat Dam, and Perkins Dam). Rawafaa and Wadi Nefuz (near Feiran Oasis) are currently silted up. The dam built at El-Wadi, north of El-Tor, has been washed away. Ein Gudeirat is masonry and badly damaged. Locations of seven historic dams are shown in Fig. 3, along with sites proposed for future development. Of the seven dams, Rawafaa is the most interesting from several points of view. An arched masonry dam is located on Wadi El-Arish, about 52 km south of El-Arish town. It was built in 1946 and reportedly had an initial capacity of about 3 million m^3. The dam is silted up and it was reduced in capacity from 3.03×10^6 m^3 in 1949 to 2.94×10^6 m^3 in 1958 (equivalent to an average loss of capacity of only 10,000 m^3/year). Assuming that the useful life of the reservoir would end if its capacity dropped to 1×10^6 m^3, and assuming an average siltation rate of 10,000 m^3/year.

It would appear that the reservoir should have a useful life up to the year 2150. In addition to the problem of siltation, the Rawafaa reservoir experienced or is experiencing, both evaporation and seepage losses. Several investigators have noted the presence of well-fractured Eocene limestone underlying the reservoir. This could lead to high seepage losses if the sediment has not been deposited thickly and uniformly enough over the entire reservoir area.

3.2 Water Harvesting

Rainwater harvesting, instead of enabling it to run off, is the accumulation and storage of rainwater for reuse on site. It offers an autonomous water supply during regional water constraints and is often used to complement the primary supply in developed countries. Rainwater harvesting offers water when a drought happens, can assist mitigate flooding in low-lying regions, and decreases usage of wells that can sustain groundwater levels. It also contributes to the supply of drinking water, since rainwater is significantly free of

Table 1 Inventory of existing dams in Sinai (Shata 1992)

Dam location	Description
Rawafaa Dam	Arched masonry dam located on Wadi El-Arish, about 52 km south of El-Arish; capacity about 3,000,000 m^3; reported to be silted up
El-Gudeirat Dam	Masonry dam, located about 9 km west of the Israeli border; built by the Turks in World War I, but presently silted up
Perkins Dam	Masonry dam on Wadi El-Gudeirat located on Wadi Sad, one of the short wadis draining the southeastern flank of Gebel Dalfa (east of Gebel El-Halal)
Wadi Gharandal Dams	Two small diversion dams in Wadi Gharandal, located 10–12 km from the Gulf of Suez, used for irrigation. The uppermost one ponds back 300–400 m of slightly brackish water; presumably an earthen structure in both cases
Wadi Nefuz Dam	Wadi Nefuz flows from Gebel Banat to Wadi Feiran and is located north of the Oasis of Feiran; the dam is located in the upper part of this wadi, where it flows through a red granite canyon, exhibiting springs; the dam is filled with eroded soil
El-Wadi Dam	A large dam built by the Bedouins at El-Wadi, north of El-Tor, for irrigation purposes; it was subsequently washed away
Wadi Shellal Dam	Formerly supplied piped water to Umm Bugma

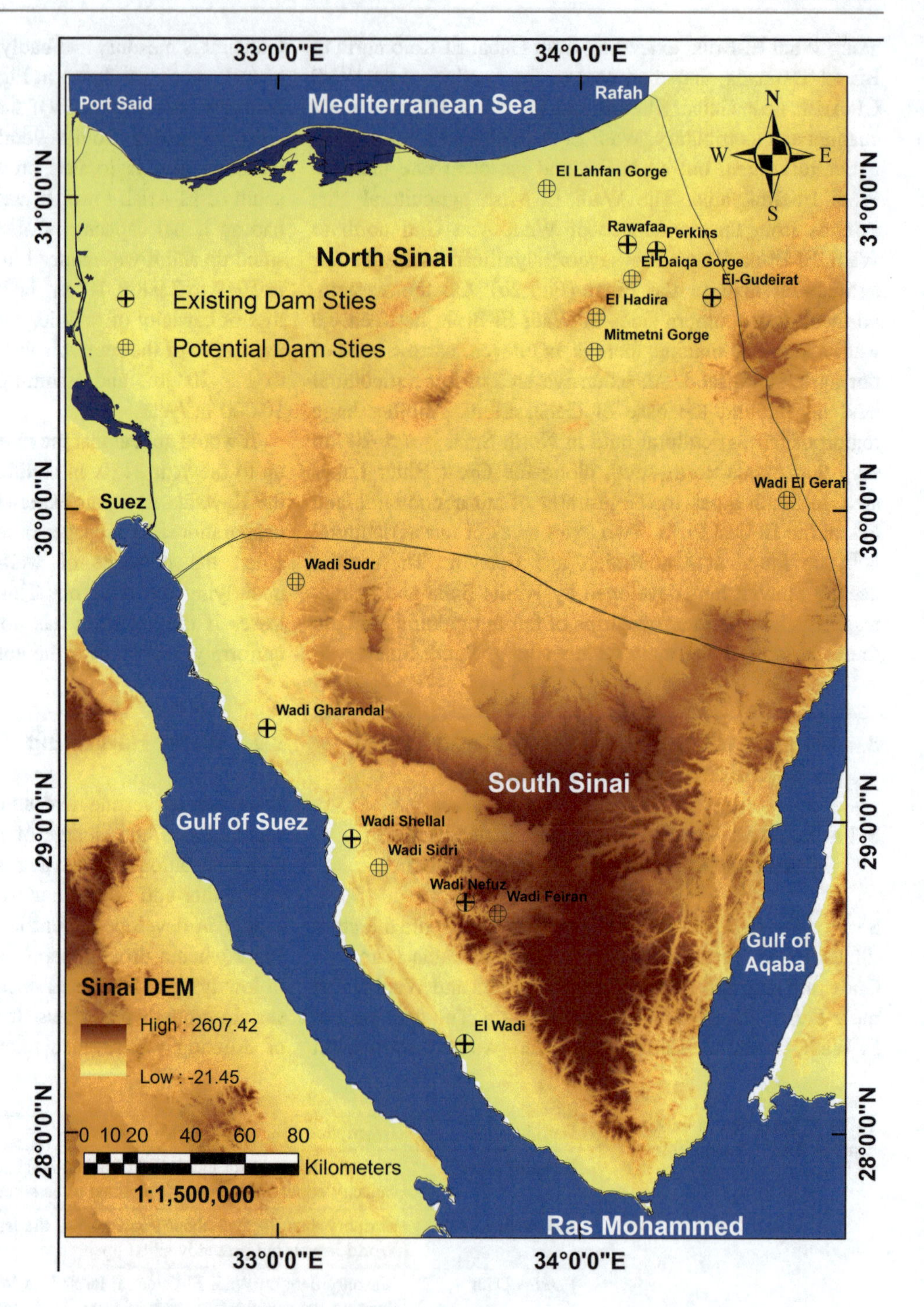

Fig. 3 Existing versus potential dam sites in Sinai

salinity and other salts. Water harvesting, with its different types, is getting a new interest to be evaluated as traditional water management, especially in Sinai. This old technology is gaining new popularity these days. Water harvesting is a traditional water management technology to ease future water scarcity in many arid and semi-arid regions (Shadeed and Lange 2010).

The harvesting of rainwater has numerous positive advantages. It is cheap and extremely decentralized, empowering water management for people and communities. It is secure from the environment and can be used fairly. It offers a reliable renewable resource with limited investment and unique management. It is possible to transport the harvested water with little energy. In agriculture, rainwater harvesting has shown the potential of enhancing food output by 100 % compared to the 10 % rise in food manufacturing from irrigation (Shadeed and Lange 2010).

It is possible to distinguish three groups of water harvesting (Prinz and Singh 2000): (1) Rainwater harvesting is described as a technique for generating, gathering, storing, and preserving local surface runoff for agriculture in arid and semi-arid areas. (2) The collection and storage of creek flow for irrigation purposes can be described as flood water harvesting. Flood water harvesting, also known as "large catchment water harvesting" or "Spate Irrigation." (3) Groundwater harvesting is a rather modern word that is used to cover both traditional and unconventional methods of extracting groundwater. There are few examples of groundwater harvesting methods in Qanat systems, underground dams, and unique kinds of wells.

3.3 Best Locations for Possible Water Harvesting Dams in Sinai

It is critical to construct new dams to protect the Sinai city and its provinces from feature flooding. Abd-El Monsef (2010) uses the precise mapping capabilities of remote sensing and GIS to find the best locations for new dams in Wadi El-Arish to reduce the impact of any future flash flood. Drainage network of Wadi El-Arish Basin was interpreted from Enhanced Landsat TM image. The basin divided into 6 sub-basins. The drainage network of each sub-basin had been analyzed, and their geomorphometric parameters were calculated. Digital elevation model (DEM) was used to determine the slope and aspect of the sub-basins. The geomorphometric parameters of Wadi El-Arish sub-basins reveal their low flash potentiality. The occurrence of a flash flood in the northern part of Wadi El-Arish Basin is mainly due to its high circularity that drives runoff water to come out of the southern sub-basins outlets nearly at the same time. The runoff water is combined at the area between Daikat El-Khorm and Daikat El-Halal. Two sites had been selected, as recommended sites to construct new dams to effectively mitigate the Wadi El-Arish flash flood located at the entrance of Daikat El-Khorm and Daikat El-Halal.

All the Wadi El-Arish sub-basins have low flash potentiality, due to their low drainage density, stream frequency, moderate bifurcation ratio, and their elongation shape (Abd-El Monsef 2010). The possibility of the initiation of the flash flood is still high in the area located between Daikat El-Khorm and Daikat El-Halal. This is due to the high circularity of Wadi El-Arish Basin, that makes the runoff water come through south sub-basins of Wadi El-Arish meet nearly at the same time in this area. The initiation of such flash flood needs a heavy precipitation storm, to allow runoff to go through the long path of the southern sub-basins and overcoming the infiltration capacity of the fractured limestone that covers most of the surface, and the high evaporation rate characterized the area. This might be why the flash flood happened once every two decades in Wadi El-Arish Basin. The idea of developing a retardation dam system through Wadi El-Arish Basin will be faced with a critical objective, in addition to delay the flow time, retardation dams should not allow the runoff water from reaching Daikat El-Khorm at the same time. On Monday, January 18, 2010, the runoff water filled the embankment of El-Rawafaa dam by 5.5 million cubic meters and started to pass the dame toward El-Arish town. The fallen of El-Rawafaa dam to protect El-Arish town raises the high demand of other dams to protect the city.

Two suitable locations for new dams (Fig. 4) were proposed by Abd-El Monsef (2010). The first location is sited at the entrance of Daikat El-Khorm. An embankment for this dam can easily be constructed on the southwest side of the dam where an area of about 80 km^2 and an average depth 8 m compared with the surrounding heights is existing. This embankment can store the runoff water coming from Wadi El-Bruk, Wadi El-Arish, Wadi El-Ruaq, and Wadi El-Aqaba sub-basins. The second location is sited at the entrance of Daikat El-Halal. An embankment for this dam can be constructed in the area located on the south side of the dam where an area of about 77 km^2 and average depth 6 m compared with the surrounding heights is present. This embankment can store the runoff water coming from Wadi El-Quraia sub-basin. This embankment can work as backup reservoir for the first dam.

Sumi et al. (2013) forecast flash floods and propose mitigation strategies in order to reduce the threat of flash floods and water harvesting. A combination of RS data and a distributed Hydrological model so-called hydrological river basin environmental assessment model (Hydro-BEAM) has been proposed for flash flood simulation at Wadi El-Arish, Sinai. Simulation has been effectively done to flash flood occasion that hit Egypt in January 2010. Simulation results present remarkable characteristics, for example, the short time to the maximum peak, short flow duration, and serious damage resulting in difficulty in evacuating people from the vulnerable regions. Six outlet points have been selected based on sub-catchments of the target wadi for this simulation as depicted in Fig. 4. Simulation results of this event show that flash flood characteristics are highly variable from one outlet to another in terms of flow rate and time needed to reach the maximum peak within the whole watershed. Furthermore, at the downstream outlet at Wadi El-Arish catchment, the flow was very severe, at 2864.84 m^3/s. At W. Abu-Tarifieh, one of the sub-catchments of Wadi El-Arish, the discharge was calculated about 240.52 m^3/s, and the discharge of all sub-catchments was also calculated.

Time to peak and flow duration in flash flood simulations have been estimated. Time to peak averaged between 8 h in the upstream regions and 17 h at downstream outlets, which mean that time is very short. In other words, evacuating people are very difficult in such arid regions. In terms of evaluation of sub-catchments water contribution toward Wadi El-Arish, the flow volume of water that can reach the downstream point of each sub-basin has been calculated (Table 2).

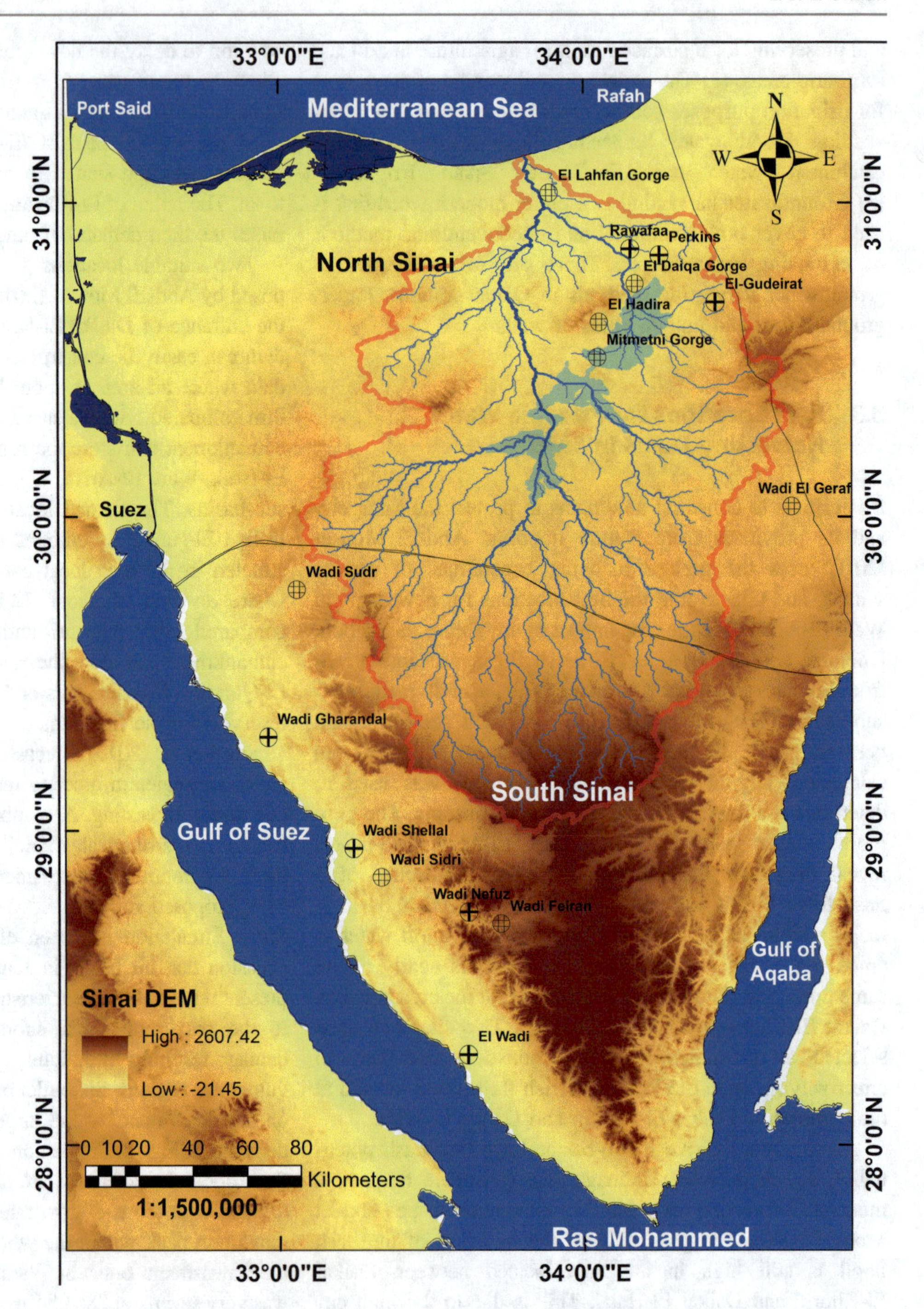

Fig. 4 Suitable locations for new dams of Wadi El-Arish catchment, Sinai

3.4 High Potential for Groundwater Recharging

Masoud (2011) predicted the runoff infiltration volume, and the sites demarcated to conceivably have a high potential for the flash flood in Southern Sinai. The groundwater recharge could enhance the solutions for the flood risk mitigation and the accessibility of water from the wadi systems. Runoff controlling systems could reduce flood vulnerability and save a considerable amount of water in these harsh environments. Located sites for conceivable high flash flood hazards and groundwater recharging appear in Fig. 5. About 50 % of the sites determined to likely have a high potential for groundwater recharging their wells dug for domestic use by the local Beduin in particular in W. Feiran, W. Sidr, St. Catherine area, and Sudr and El-Qaa Plains where shallow Quaternary aquifer underlies the terrain. Among the catchments anticipated having high flash flood potential, the catchment of W. Dahab displayed the highest rank followed by the W. Fieran and W. Watir catchments. W. Kid and W. Al-Awaj had an intermediate potential. The catchments of

Table 2 Simulation results of event (Jan. 18–22, 2010) at Wadi El-Arish

Sub-catchments of Wadi El-Arish Outlets	Time to peak (h)	Peak discharge (m^3/s)	Flow volume (m^3)
Wadi El-Arish1	17	2864.84	3.54×10^8
Wadi El-Arish2	14	1080.38	1.26×10^8
Wadi Griha	13	1050.41	1.26×10^8
Wadi El-Barok	10	885.38	9.9×10^7
Wadi Eqabah	7	386.66	4.5×10^7
Wadi Abu-Tarifieh	8	240.52	2.7×10^7

W. Sidri, W. Gharandal, Ras Nusrani, W. Sudr, W. Samra, W. Tayyibah, W. Lahata, and W. Wardan exhibited low potential. Among the catchments that had high groundwater recharge potential, W. Watir had the highest potential, followed by W. Al-Awaj. The catchments of W. Feiran, W. Dahab, W. Sidri, and W. Samra had intermediate groundwater potential. W. Gharandal, W. Sudr, W. Kid, W. Tayyibah, Ras Nusrani, W. Lahata, and W. Wardan fell inside the low potential class. Ghodeif and Gorski (Ghodeif and Gorski 2001) indicated that the recharge rate (32,000 m^3/day) of the Quaternary aquifer underlying the El-Qaa Plain where W. Al-Awaj flows surpasses the discharge (26,000 m^3/day). Thus, W. Watir could be promising for groundwater exploration as it has significantly higher potential than W. Al-Awaj. The present rainfall-runoff model displayed a rainfall volume sum of 725.5×10^6 m^3. The rainfall distribution over the three components was 307.8×10^6 m^3 (43.4 % runoff volume), 131.5×10^6 m^3 (18.1 % initial loss), and 286×10^6 m^3 (39.4 % transmission loss). The shallow groundwater aquifer recharge was estimated about 57.5 % of the total rainfall. An assessed recharge over the mountainous St. Catherine area resulting from the individual large storm was 135.4×10^6 m^3, which was 50 % of the total precipitation (Masoud 2011).

Numerous studies (Ramli 1982) have shown that a large part of rainwater is lost by evaporation and surface runoff. The total annual precipitation of the Sinai is about 3 billion m^3, most of which is lost by evaporation or surface runoff, and only about 300 million m^3 of groundwater (Dames and Moore 1985). The majority of rainwater falling on the Sinai was lost and not used. In addition to the devastating impact of the floods on the facilities and crops, it also has an impact on desertification of large areas of Sinai. Therefore, it is necessary to develop a scientific plan to harvest the largest amount of rainwater by establishing different dams at the appropriate sites. There are several dams in Sinai, the most famous and the oldest and the largest is the Ruwafa dam.

There are some of the proposed sites for the construction and building of some dams based on the study of satellite images and the field study. The following proposed 37 sites for the construction of dams (Ramli 1982): (1) At the intersection between Wadi Heirdeen with Wadi El-Arish. (2) At Wadi El-Garor. (3) At Wadi Qaria (West South Gebel umm Hadira). (4) At the intersection between Wadi Gadei with Wadi El-Arish (North Gebel Magmel). (5) At Southwest Gebel Al-Khatemia (North Gebel ummKhosheib). (6) Mittla Pass area (the area between Mittla Pass and Wadi Al-Hag). (7) At Wadi Akaba (North and Northeast Qalaet-Nakhel). (8) At Wadi Akaba (West Gebel Al-Reesha). (9) At Awlad Ali area. (10) At Wadi El-Geifi (South El Qussiema). (11) At Talaet Al-Badan (at Al-Rawafa George). (12) At Wadi El-Bruk, at its crossing with Wadi El-Arish, West and Southwest Gebel Khereem. (13) At Wadi El-Hassana (North Gebel El-Monshareh). (14) At the connection between Wadi El-Bruk with Wadi Abou Kanadu, Northeast El-Kuntilla well, and Wadi El-Geifi. (15) At Wadi Al-Tamarni (Southwest El-Kontilla). (16) At Wadi Al-Rawaq. (17) At Wadi Al-Thamad (north Bir Al-Abd). (18) At South of Nakhel (at the crossing of Wadi Abu-Aligana with Wadi Ei-Arish). (19) At Wadi Sudr. (20) At Wadi Werdan (North East Ras Matarma). (21) At North East Wadi Al-Massagid (West Gebel Lobna). (22) At Wadi Graa (West Al-Aqaba Gulf). (23) At Wadi Sidri (East Abu Rudees). (24) At Wadi El-Garaf (East Abou Zinema). (25) At Wadi Al-Akhadar (North East Feiran oasis), and Wadi Feiran (West and Northwest Feiran oasis). (26) At the Wadi which drains in El-Qaa plain from Gebel Qobyliat (North and Northwest El-Tor). (27) At Wadi Berriraq and Abu Tarika (El-Egma plateau). (28) At Wadi Al-Aaeb, Wadi Al-Nasseib, and Wadi Nasaib (draining in the Delta, Dahab). (29) At Wadi Kid, Wadi Umm Adwi Lethi, and Nabaq area. (30) At Wadi Zelefa and Wadi Arada (North and Northwest Gebel Gena). (31) The Wadis which pour in El-Qaa plain from Gebel umm shomer and Gebel Al-Thabat, Wadi Al-Mahash, and umm Qterkha. (32) At Wadi Al-Tamrani and Wadi Al-Qries. (33) At Wadi Zelieq and Wadi El-Biar (Southeast El-Emag plateau). (34) East Sant Khathrine and Gebel El-Banat, Northwest Sant Khathrine, West and Northwest Gebel El-Banat. (35) Al-wadi area, (West Gebel Al-Thabat and Gebel umm Homata), pour in El-Qaa plain. (36) At Al-Wadi area, between Ellatt and North Gebel El-Asafeer (West Gebel El-Barga and Wadi El-Qaseeb area).

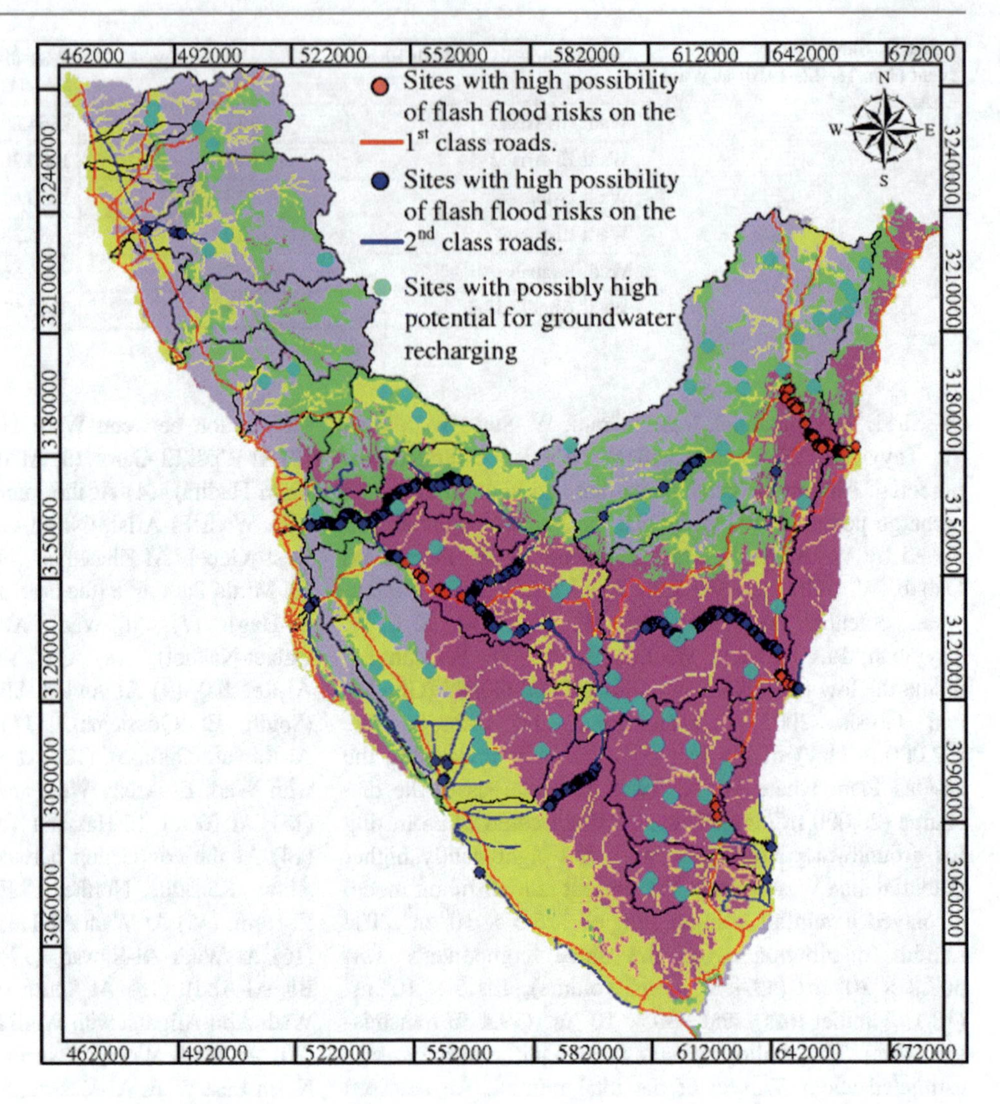

Fig. 5 Located sites for possible high potential for groundwater recharging in the South Sinai catchments (Masoud 2011)

(37) At Wadi Azala area (between Gebel Abu-Khusheib, Gebel El-Safra).

3.5 Floodwater Resources Management

3.5.1 Structural Dams and Artificial Lakes

The aim of floodwater harvesting is to store floodwater in artificial, closed ponds made of different materials. Different low-cost storage mechanisms for floodwater harvesting were identified to suit the different technical and socioeconomic conditions. Firstly, an underground concrete reservoir is one of the most appropriate water harvesting techniques and easily maintained by the Bedouins themselves. A small underground reservoir is capable of offering drinking water for a small Bedouins community and their livestock. As water is sealed in a concrete case, it is protected from evaporation. Secondly, a Haraba is a low-cost alternative to capture floodwater, often used by Bedouins (Dames and Moore 1985). A Haraba is a superficial basin excavated from limestone or rocks. In the case of sedimentary rock (with a high infiltration rate), stones and cement will be used. The main advantage is that in the construction of a Haraba, no building material needs to be brought in. The use of instruments, such as a compressor can, however, facilitate the excavation. The major disadvantage of a Haraba is that it is an open reservoir, and hence water is rapidly lost due to evaporation. Thirdly, one of the potential technically and highly requested by the stakeholders is a low-cost gabions dam with an underground reservoir as the one constructed in Wadi Ghazala (Fig. 6).

The traditional techniques applied comprise the following (Shata 1992):

1. Man-made cisterns or natural caves in the mountain edges adjacent to a Wadi bed (locally known as Harraba).
2. Low earthen or stone dykes in the Wadi beds (locally known as Oqum). They are usually protected by vegetation remains. The objective of these dykes is to reduce the passage of water downstream; not to store it.

Frequently, these dykes are damaged or destroyed by floods before they are stabilized by the substantial deposit of sediments.

3. Masonry dams for the storage of water. They include El-Rawafaa, El-Gudeirat, Perkins, Gharandal, Nefuz, El-Wadi, and Shellal.

El Shamy (1995) mentioned that traditional water controlling systems mostly depend on the construction of huge masonry dams with big lakes behind at the downstream parts of the hydrographic basins, frequently subjected to flash floods. In addition to their costly construction costs, these structures expose many environmental disadvantages such as the deprivation of pervious upstream formations from the only recharge source, the heavy evaporation from the concomitant, the soil and shrub drift by flash floods and the gradual silting of these dams. (El-Shamy 1992) proposed controlling systems, which must be started at the upstream tributaries of the hydrographic system using simple retardation dams. This will promote the recharge to the existing and fractured rock formations and allow a slow controlled runoff into the downstream areas with a consequent high chance for infiltration. The proposed system stabilizes the soil layer, enhances the growth of the plant cover, and gradually improves the local microclimate. The applicability of the system to some hydrographic basins in Sinai; viz, Wadi El-Arish, Wadi Suder-Wardan, and Wadi Watir, is showed in Fig. 7.

Regardless of their dangerous impacts, flash floods in Sinai, and different parts of southern Egypt represent a potential for non-conventional freshwater sources. Dams have been utilized for protection against floods, diversion of streams, storing of water and irrigation of land, each of

Fig. 6 Underground (Haraba) reservoir construction for water harvesting in Wadi Watier near Nuweiba City. **a** Low-cost gabions and obstacle dam constructed in Wadi Ghazala, **b** (WRRI 2010) artificial lake, **c** water trap, **d** well, **e** water storage tank, and **f** for water saving in Feiran Oasis

which is a technique for water control. They offer for agricultural irrigation. They are utilized to store and give tap water, further to more complex purposes, for example, hydroelectric power generation, rendering barren and infertile lands arable, and avoidance of erosion due to floods and accumulation of sand and clay at river-mouths. Water dams and reservoirs have essentially been utilized to serve four capacities [8]: Irrigate crops, provide hydroelectric power, supply water, and control flooding.

3.6 Check Dams Versus Diversion Dams

Check dams are known in Tunisia and west Libya. They consist of the "Tabia," which is the main structure built on the Wadi floor. These dams are suitable for slopy areas in the Gulf of Aqaba and the Gulf of Suez. Diversion dams and water spreading dykes are usually designed to improve pasture land and to protect plant life. These dams are most suitable for north central Sinai.

Fig. 7 **a** Location map of the Sinai' basins, **b** The proposed system for El-Arish mega-basin, **c** Proposed rainwater controlling system for Sudr-Wardan basin, and **d** for Watir basin (El-Shamy 1992)

Some workers (Dames and Moore 1985) referred to eight potential sites for this purpose (Table 3), which include the following: El-Daiqa, El-Mitmetni, Lehfin, and El-Hadhira in Wadi El-Arish Basin, Sudr, and Sidri in west Sinai, Feiran in South Sinai, and El-Gerafi in east Sinai. This type of dams is not suitable for an arid area like Sinai. Several reasons stand for this and include (Shata 1992) erratic rainfall and infrequent runoff water, high evaporation losses, pronounced seepage in the country rocks dominated by fractured carbonates, rapid siltation of the reservoir due to degradation of the surface (10,000 m^3/year), shortage of recharge of the areas downstream the dam site, and relatively high cost of construction of the dam and of water conveyance.

For such reasons, plans aiming at the construction of three new storage dams at El-Karm, El-Maghara, and El-Gudeirat, and increasing the height of El-Rawafaa old dam should be modified. Fourteen dams proposed for Wadi Wasit (Watir) in eastern Sinai can be completed in a similar way to the check dams known in Tunisia and western Libya. For Wadi El-Gerafi, the landscape is most suitable for conversion dams and water spreading.

Appropriate controlling systems must be started at the upstream tributaries of the hydrographic systems using simple retardation dams. This will promote the recharge to the existing previously and fractured rock formations and allow a slow controlled runoff into the downstream areas with a consequent high chance for infiltration. Moreover, the system stabilizes the soil layer, enhances plant growth, and gradually improves the local microclimate (El Shamy 1995).

It is expected that the establishment of adequate number of diversion dams and water spreading dykes in central and north Sinai as well as in the outlets of a number of wadis in west Sinai, (e.g., Wadi Sudr, Wadi Wardan, Wadi Baba, Wadi Sidri, and Wadi El-Qaa) is very important for many reasons. These reasons include increasing the agricultural and grazing activity, minimizing the degree of desertification, contributing to the productivity of the already degraded surfaces, harvesting a lot of water, recharging the underground water in the area, and decreasing the velocity of the runoff water which minimize the soil erosion. These dams also help in improvement of water supply system for irrigated agriculture.

The establishment of the various dams leads to the presence of communities around the areas of these dams to work with agriculture and grazing, as well as to reduce wastage of water, and reduce the speed of the flooding and thereby protect the soil from water erosion. Disasters such as floods disaster in El-Arish in 2010 may not return to the floods alone but as a result of the random nature of the establishment of the buildings in the corridors of the floods and without the work of the previous geological study of the area where the various facilities will be installed on them.

The total amount of rainfall and flash floods that could be used annually is estimated at around 1.3 billion cubic meters. This quantity can be increased to 1.5 billion cubic meters

Table 3 Possible dam sites in Sinai

Location	Possible purpose	Possible problems
El-Daiqa Gorge, El-Arish	Supply of water to El-Arish town; flood protection for El-Arish town	Deprivation of downstream areas; requires long conveyance system; evaporation loss relatively high
Mitmetni Gorge, El-Arish	Supply of water to El-Arish town; flood protection for El-Arish town	Deprivation of downstream areas; requires long conveyance system; evaporation loss relatively high
El Lahfan Gorge, El-Arish	Supply of water to El-Arish town; flood protection for El-Arish town	Possible foundation problems; long dam required; seepage losses could be high
Gebel El-Halal, El Hadira	Supply of irrigation water to El-Arish flood plain south of Gebel El-Halal	Expected yield is relatively low
5 km west of Ain Sudr, Wadi Sudr	Public supply at Ras Sudr; irrigation supply for Wadi Sudr delta; cultivation in bed of reservoir	Rate of siltation expected to be high; evaporation loss relatively high; expected yield is not high
Near Gebel Maghara (South Sinai), Wadi Sidri	Public supply for Abu Rudeis; irrigation water for Wadi Sidri Plain	Rate of siltation expected to be high; evaporation loss relatively high; expected yield is not high
Upstream of the Oasis of Feiran, Wadi Feiran	Flood protection for oasis; irrigation supply to the oasis and northern El-Qaa Plain	Rate of siltation expected to be high; evaporation loss relatively high
Near Kuntilla, Wadi El Geraf	Irrigation supply; groundwater recharge	Rate of siltation and evaporation expected to be high; modest yield; would require negotiations with Israel

upon taking the following measures (https://propertibazar.com/article/egypts-national-strategy-for-adaptation-to-cli):

- Extended development of dams and reservoirs to capture and use this water directly for drinking or farming, or for storage in reservoirs of groundwater.
- Using modern techniques in the field of water harvestings, such as RS and GIS, in order to study the basic properties of the flood-prone areas. This involves studying and analyzing surface runoff as well as identifying basin and soil type features. Avoid the hazards of flash floods by mapping risk assessments for each region and taking suitable precautions to prevent potential risks (https://propertibazar.com/article/egypts-national-strategy-for-adaptation-to-cli).

4 Conclusions and Importance of the Study

Sinai is progressively suffering from an overwhelming water crisis. Flash flood and runoff water could be an answer for this issue. The central focus of this chapter rotates around utilization of the floods water as a new supply for water harvesting in Sinai. Applying water harvesting of the flash flood will reduce the flood risk at the outlet. In addition, it could be used for recharging the groundwater aquifers, which are the basis for sustainable development in Sinai. Furthermore, Bedouins usually move from place to place searching for fresh grazing for their animals and water for their families. These places sometimes are hazardous areas as flash floods occur there. This chapter helps to make developed sustainable planning for the Bedouins, as hazard areas will be defined. Proved locations to get use of the flooded water and store it will encourage the Bedouins to resettle the area. The total amounts of rainfall affecting the different drainage systems in Sinai is of the order of 2,000 million m^3/year, and most of possible runoff water is about 150 million m^3/year.

Structural dams and artificial lakes are highlighted in this chapter to utilize the flash flood.

Because of their features, flash floods are hard to handle through traditional approaches to flood management. The following must be regarded during the development of a management plan for flash floods.

- Flash flood prediction and warning play a key role in flash flood management. However, providing users with precise and timely forecasting and warning data is still a challenge at times. Appropriate technical methods and suitable legal and institutional frameworks are needed to address these issues.
- Appropriate spatial planning can assist decrease exposure and decrease flash flood risks.

References

Abd-El Monsef, H. (2010). Selecting the best locations for New Dams in Wadi El-Arish Basin using remote sensing and geographic information system. In *Sinai International Conference for Geology & Development Saint Katherine*, South Sinai, Egypt.

Abdelkarim, A., Gaber, A. F. D., Youssef, A. M., & Pradhan, B. (2019). Flood Hazard assessment of the Urban area of Tabuk City, Kingdom of Saudi Arabia by integrating spatial-based hydrologic and hydrodynamic modeling. *Sensors, 19,* 1024.

Abu Al-Izz, M. S. (1971). *Landforms of Egypt* (281 p). Dar-A1-Maaref, Cairo, Egypt.

Abuzied, S. M., & Mansour, B. M. H. (2018). Geospatial hazard modeling for the delineation of flash flood-prone zones in Wadi Dahab basin, Egypt. *Journal of Hydroinformatics, 21,* 180–206.

Dames and Moore. (1983). Sinai development study, Final report.

Dames and Moore. (1985). Sinai development study—Phase I, Final Report, Water Supplies and Costs. Volume V-Report Submitted to the Advisory Committee for Reconstruction, Minstry of Development, Cairo.

EGSMA. (1981). Geological Map of Egypt, Scale 1,2000000. Egyptian Geological Survey and Mining Authority.

El-Shamy, I. (1992). Towards the water management in Sinai Peninsula. In *Proceedings of the 3rd Conference on Geology and Sinai Development* (pp. 63–70), Ismailia.

El Shamy, I. (1995). The control of flood. *Egyptian Geographic Society of Egypt* (in Arabic).

Ghodeif, K., & Gorski, J. (2001). Protection of fresh ground water in El-Qaa Quaternary aquifer, Sinai, Egypt. New approaches characterizing flow. Seiler & Wonhnlich, Swets & Zeitliger Lisse. ISBN: 902651-848-X.

Jalilov, S.-M., Kefi, M., Kumar, P., Masago, Y., & Mishra, B. K. (2018). Sustainable Urban water management: Application for integrated assessment in Southeast Asia. *Sustainability, 10.*

Masoud, A. A. (2011). Runoff modeling of the wadi systems for estimating flash flood and groundwater recharge potential in Southern Sinai, Egypt. *Arabian Journal of Geosciences, 4,* 785–801.

Mohamed, S. A., & El-Raey, M. E. (2019). *Vulnerability assessment for flash floods using GIS spatial modeling and remotely sensed data in El-Arish City, North Sinai.* Natural Hazards year: Egypt.

Prinz, D., & Singh, A. (2000). Technological potential for improvements of water harvesting. Study for the World Commission on Dams, Cape Town, South Africa. (Report: Dams and Development).

Ramli, I. (1982). Water Resources in the Sinai Peninsula and its basic and regional development plans in the next 50 years (1982–2032). Desert Research Institute, Ministry of Agriculture, Al-Matareya, Cairo.

Said, R. (1962). *The geology of Egypt*. Amesterdam, New York: El-Sevier Publishing Company.

Schanze, J. (2006). Flood risk management—A basic framework. In J. Schanze, E. Zeman, & J. Marsalek (Eds.), *Flood risk*

management: Hazards, vulnerability and mitigation measures. NATO science series IV (pp. 1–20). Berlin: Springer.
Shadeed, S., & Lange, J. (2010). Rainwater harvesting to alleviate water scarcity in dry conditions: A case study in Faria Catchment, Palestin. *Water Science and Engineering, 3,* 132–143.
Shata, A. (1992). *Watershed management, development of potential water resources and desertification control in Sinai* (pp. 273–280). Ismailia, Egypt: Geology and Sinai Development.
Sumi, M., Saber, M., & Kantoush, S. A. (2013). Japan-Egypt hydro network: Science and technology collaborative research for flash flood management. *Journal of Disaster Research*, *8*, 177–178.
Tooth, S. (2000). Process, form and change in dryland rivers: A review of recent research. *Earth-Science Reviews, 51,* 67–107.
WRRI. (2010). Flash floods in Egypt protection and management, final report. National Water Research Center. Ministry of Water Resources & Irrigation Egypt LIFE06 TCY/ET/000232.

Flash Flood Risk Assessment in Egypt

Ahmed M. Helmi and Omar Zohny

Abstract

The subject of the flash flood risk assessment is an inclusive task that relies on the characteristics of the study area and the nature of previously recorded incidents. The Egyptian Nile Wadies (East Nile, and West Nile) are draining toward the highest population density and associated assets, while the Red Sea and Sinai wadies are draining toward high-density touristic compounds and scattered big cities and connecting roads. The existence of high urban densities and associated assets in the highest discharge locations (at wadies outfalls) without adequate consideration of wadi paths led to a considerable wadies encroachments and catastrophic recorded incidents. All recorded incidents are either due to unplanned urban and agricultural expansion, or insufficient flood mitigation measures, or lack of maintenance. Due to the freshwater stress in Egypt, the rainfall harvesting in the form of dams or artificial lakes should be considered as a top priority flood mitigation measure wherever applicable. The total capacity of all flood protection dams and artificial lakes all over Egypt is about 70 million m^3 (MWRI 2016) that raises the potentiality for more similar measures to increase the rate of investment return from both flood mitigation and reduction in freshwater stress. The available data for this study were sufficient enough to calculate the catchments peak discharge and runoff volume. The 100 year return period was selected for the peak discharge calculations. Many thresholds have been tested for catchment delineation in order to obtain a reasonable number of catchments suitable for such a regional-scale study. The SRTM 90 × 90 DEM file was utilized as an input in the delineation procedure, with selected threshold was set to 50 km^2. Due to the large variance of the catchments peak discharge and runoff volume, the box plot technique was employed to eliminate the ranking outlier values. The catchments were classified into five categories very high risk, high risk, moderate risk, low to moderate risk, and low risk. This categorization was done for the Peak Flow Standardized Risk Factor (PFSRF) and Runoff Volume Standardized Risk Factor (RVSRF) in order to prioritize the flood mitigation measures required for projects. The classification based on the runoff volume can guide the designer accounting for rain harvesting projects that would increase the rate of investment return from both flood mitigation and the reduction of freshwater stress. A two-dimensional HEC-RAS rainfall-runoff modeling is conducted for Ras Gharib city by using updated 30 × 30 DEM files to contain the manmade topographical modifications. The model was verified versus aerial photos for the 2016 incident. In order to assess the effectiveness of the newly constructed culvert (16 vents, 3 m × 3 m box culvert) with attached two dikes, another updated two-dimensional HEC-RAS rainfall-runoff model has been conducted and the results showed significant improvement in flood intensity values in Ras-Gharib city.

Keywords

Peak flow • Runoff volume • Standardized risk factor • Stormwater harvesting • Freshwater stress

A. M. Helmi (✉)
Department of Irrigation and Hydraulics, Faculty of Engineering, Cairo University, Giza, P.O.Box 12613 Egypt
e-mail: ahmed.helmi@eng.cu.edu.eg

O. Zohny
AIEcon. Consultants, Cairo, Egypt

A. M. Negm (ed.), *Flash Floods in Egypt*, Advances in Science, Technology & Innovation,
https://doi.org/10.1007/978-3-030-29635-3_13

Abbreviations and Notations

A	Catchment Area (km^2)
ASRF	Catchment Area Standardized Risk Factor
CN	Curve Number
D	Rainfall duration corresponding to the time step of calculations
D_d	Drainage density
d	Flow depth
DSRF	Drainage Density Standardized Risk Factor
FI	Flood Intensity
F_s	Stream frequency
FSRF	Drainage Frequency Standardized Risk Factor
GIS	Geographic Information System
IF	Intensity Factor
L	Longest flow path in (m)
LSRF	Surface flow Length Standardized Risk Factor
MENA	The Middle East and North Africa
MWRI	Ministry of Water Resources and Irrigation
N_u	Number of streams of order (U)
N_{u+1}	Number of streams of order (U + 1)
O&M	Operation and Maintenance
PFSRF	Peak Flow Standardized Risk Factor
R_b	Bifurcation ratio
RVSRF	Runoff Volume Standardized Risk Factor
S_l	Slope of the longest flow path
SRTM	Shuttle Radar Topography Mission
SSRF	Slope Standardized Risk Factor
TCSRF	Time of Concentration Standardized Risk Factor
T_c	Time of concentration in (min)
T_L	Lag time
T_p	Time to peak discharge
TL_s	Total length of streams (m)
TN_s	Total number of streams
U	Stream orders according to (Horton 1945)
V	Flow Velocity
WSRF	Weighted Standardized Risk Factor
WMO	World Meteorological Organization

1 Introduction

Flash floods are natural phenomenon characterized by its short duration and massive runoff volume with limited time for response. As a natural phenomenon, flash floods depend on climatic conditions, geological nature, and prevailing terrain. The rainfall in Egypt starts at the beginning of the fall season as of mid-October. Falling rains on mountains and large catchment form streams with high velocity in case of the topographical high slopes, carrying sedimentary materials and stones located on the soil surface and leading to severe damage to the assets located at the catchments outlets. In Egypt, the highest population density is centralized around the Nile banks and high-density touristic compounds and scattered big cities with connecting roads are distributed along the whole Red Sea coast, and in the Sinai Peninsula. The high-density urbanization with all associated assets without proper consideration of wadi paths led to a considerable wadies encroachment and catastrophic recorded incidents.

Due to the huge amount and the cost of work required for encroachments assessment and the implementation of proper flood mitigation measures, prioritization of wadies based on its Peak Flow Standardized Risk Factor (PFSRF) has been a critical task since most of the encroachments are located at wadies outlets. Stormwater harvesting can provide a share in reducing freshwater stress associated with a reduction in the required drainage structures accordingly, another catchments prioritization is provided based on Runoff Volume Standardized Risk Factor (RVSRF). The provided RVSRF can guide the designer to give a priority for artificial lakes and dams as a recommended flood mitigation measure for such catchments. The storage capacity of all flood mitigation measures in Egypt is about around 70 million m^3 (MWRI 2016), which raises the potentiality to increase the number of similar flood mitigation measures.

For future planning, 30 × 30 DEM files, and two-dimensional Rainfall-Runoff HEC-RAS can be a useful tool for performing an initial assessment of flood plain. However, this does not replace the topographic survey at project location with reasonable coverage of the surroundings to capture fine terrain details that cannot be achieved by using DEM files, especially in flat terrain areas.

2 The Study Area

Egypt is divided into six hydrological regions based on its hydrological parameters and outfalls.

- Sinai hydrological region.
- Nile hydrological region.
- Red Sea hydrological region.
- Northern coast hydrological region.
- Delta hydrological region.
- Oasis hydrological region.

The study area covers the Nile, Red Sea, and Sinai hydrological regions. The catchments outlets are endangering densely populated areas, as shown in Fig. 1.

3 Flash Flood Characteristics

Flash flood is a natural phenomenon that has a variety of definitions. The WMO states that a flash flood is: "A flood of short duration with a relatively high peak discharge." The American Meteorological Society (AMS) defines flash floods as "A flood that rises and falls quite rapidly with little or no advance warning, usually as the result of intense rainfall over a relatively small area." In terms of warnings, a flash flood is a local hydrometeorological phenomenon that requires: Both Hydrological and Meteorological expertise in real-time forecasting/warning and knowledge of local, up to the hour information for effective warning (24/7 operation). Response time to flash floods is usually less than 6 h (WMO 2012).

The flash flood discharge at the catchment outlet is affected by the rainfall (intensity, duration, distribution, and directions), catchment properties (shape, average slope, stream slope, soil infiltration rate, and the existence of surface depression) as shown in Fig. 2. The danger of flash floods is due to its sudden occurrence associated with a large flow velocity which moves large amounts of debris and sediments. The force of the flow can be high enough to destroy structures and buildings that stand in its path. A flash flood may also result from a failure of dams, embankments, or other hydraulic infrastructures, and when it is associated with a storm event it dramatically increases its negative impacts.

4 Flood Risk Assessment

Definition of flash flood risk is a mandatory step prior to the flood management planning stage. The risk is the potential losses that can occur due to any disaster. In the case of flood, it indicates the impact of a flash flood on human lives, environment, and properties. Three crucial elements are required to characterize the flash flood risks (hazards, exposure, and vulnerability), as shown in Fig. 3. For flash flood consideration the three items can be defined as follows:

Hazard: can be defined as the "flood inundation maps showing the depths of inundation and related water velocities for a storm event corresponding to an agreed return period." The risk assessment of human lives and assets will be properly defined after considering the exposure and vulnerability parameters. The hazard also can be defined as the value of the discharge or volume of runoff as the parameters used to rank the catchments regarding the risk at its outlets and the potential of stormwater harvesting.
Exposure: can be defined as "people, properties, infrastructure, housing, and any other type of assets located in a predefined hazard area." The exposure can be presented in land-use maps, infrastructure maps, population density maps, etc.
Vulnerability: can be defined as "The risk scale determination of an individual, a community, and assets to the impacts of flood hazard" which can be defined by considering the health, physical, social, and economic conditions of the considered items in the hazard area. It can be presented in a map with, the very low, low, medium, high, and very high-risk scale for each considered item.

The water resources in Egypt as in all Middle East and North Africa (MENA) Region are subjected to high stresses due to the population increase, which requires that the flood mitigation measures are not only for protection purpose, but also it has to consider a proper utilization of this freshwater. "That is why it would be of little use to consider Flood Risk Assessment (FRA) by itself without casting it in the framework of flood risk management and water management at large," (Rudari 2017).

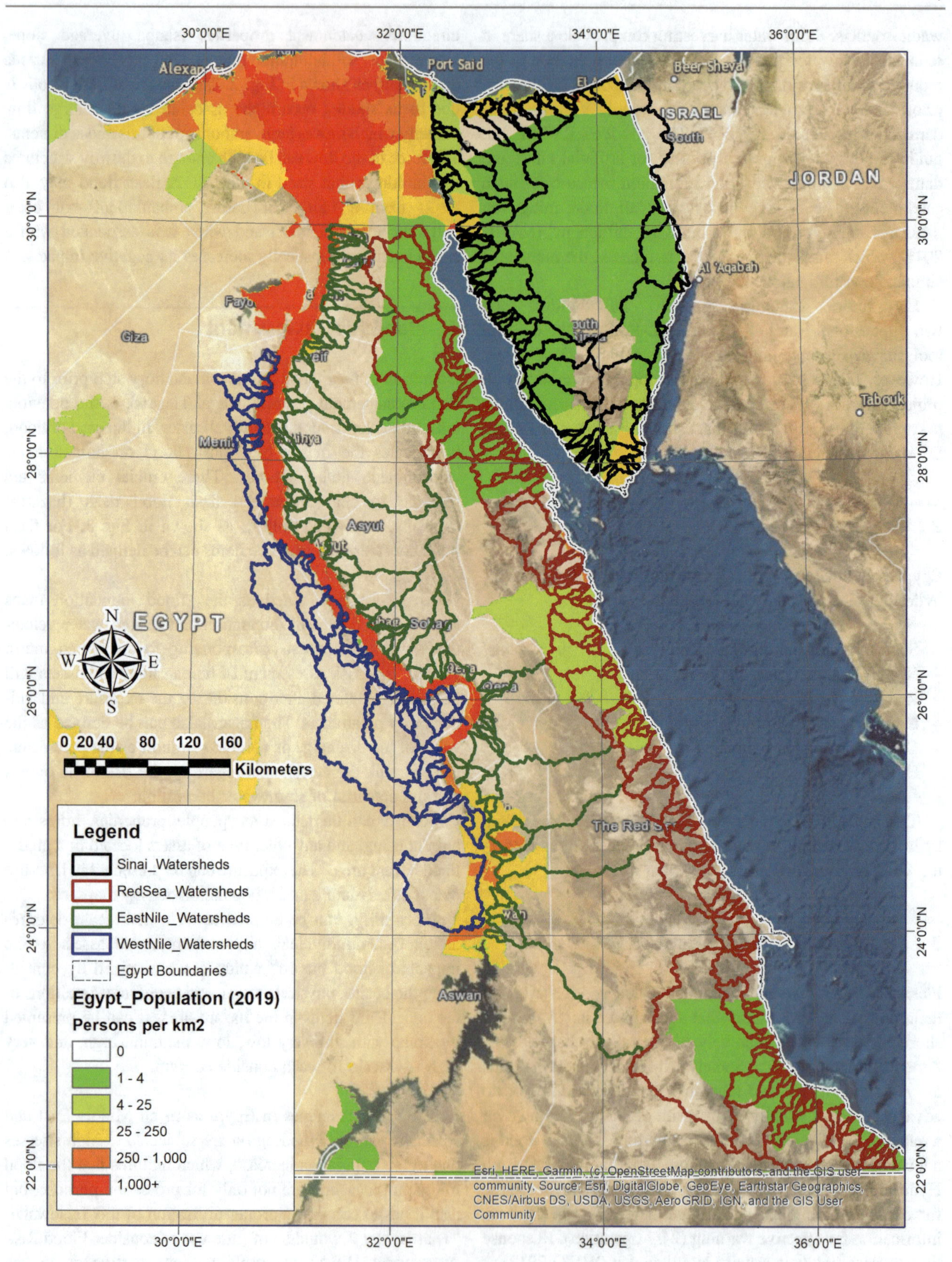

Fig. 1 Distribution of population densities. Adapted from NASA (2019)

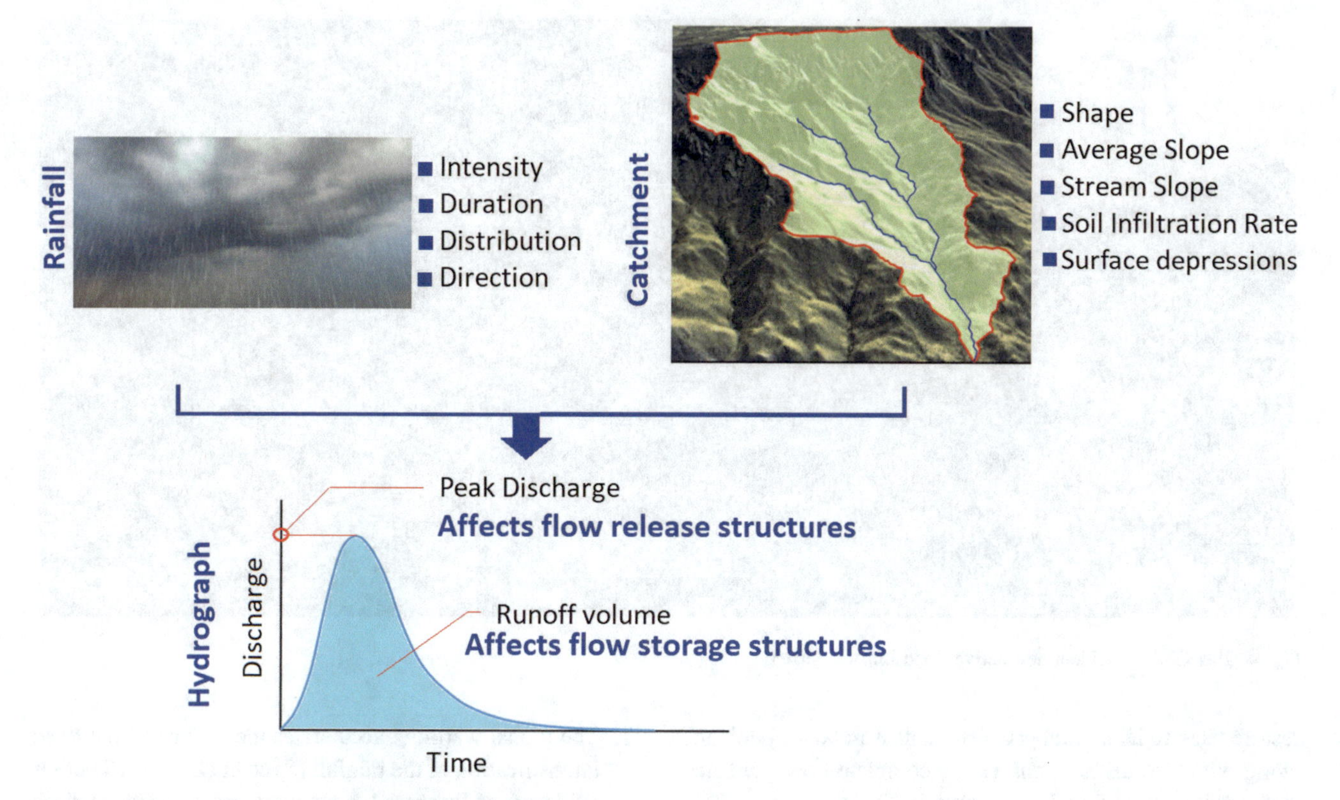

Fig. 2 Factors affecting catchments' outlet hydrographs

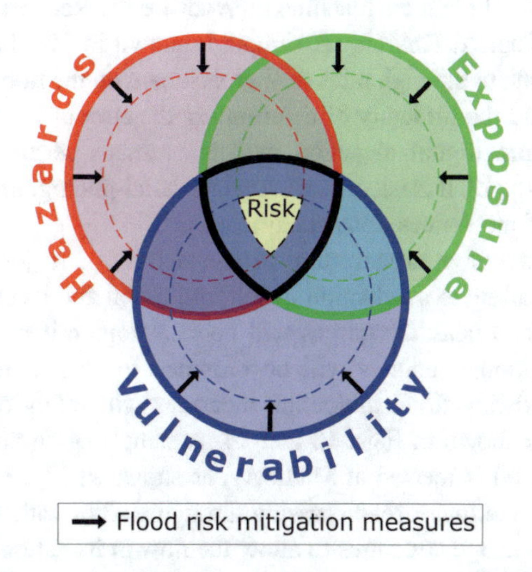

Fig. 3 Flood risk characterization. Adapted from WMO (2006)

High-resolution flood inundation maps are costly and require a very detailed topographical data, and intensive calibration works. It is recommended to divide the procedure of obtaining Flood Risk Plan (FRP) into two stages:

Preliminary Flood Risk Assessment (PFRA): which is utilized to define priority areas using Digital Elevation Models (DEMs), brake lines, and available flood mitigation measures for further detailed assessment.

Final detailed Flood Risk Assessment (FRA): to be applied to the high-risk areas defined in the PFRA in order to maximize the return on investment where many more resources are invested in the topographical survey, modeling, and calibration. The potential areas of development should be taken into consideration, although they have not been shown in preliminary studies (European-Commission 2016).

In addition to the previous procedure, the historical information is an extremely important source of information from the flash flood that has occurred in the past, and its impacts for a specific event is a fact with 100 % confidence.

5 Sample of Previous Flash Flood Incidents

On Friday, October 28, 2016, seven people were killed and 23 others injured after heavy rain hit Ras-Gharib city in the Red Sea governorate as shown in Fig. 4. The main cause of this catastrophic incident was not due to the rainfall depth on Ras-Gharib city; instead, it was due to the diversion of almost 70 % of wadi Had catchment flow toward the city. The diversion of wadi from its natural stream path toward the city was due to the missing drainage structures along Ras-Gharib–El-Sheikh-Fadl road, as shown in Fig. 5. The proposed solution was to provide access to allow the wadi to

Fig. 4 Ras-Gharib incident destructive flood impacts (2016)

restore back to its natural path (16 vents 3 m × 3 m culvert) along with two dikes to ensure the complete flow guidance to the original wadi path, as shown in Fig. 6.

On Tuesday, June 17, 2010, Aswan Governorate witnessed a flood event that swept the home furnishings and people, leaving behind many houses destructions in addition to the damage to thousands of feddans of agricultural land, as well as damaging tons of harvested dates which people had put them under the sun to dry out as shown in Fig. 7. Although the wadi was provided by two dikes to guide the flow toward the drainage channel leading to the River Nile as shown in

Figure 8 due to the lack of maintenance and the use of the channel as a dump site and drain to the houses and agricultural land, the capacity of the drainage channel was reduced and led to the occurred flooding negative impacts (Figs. 9 and 10).

Frequent occurring incidents affect rural roads due to insufficient flood protection measures.

6 General Investigations of Flash Flood Triggers in Egypt

As a global view and according to the collected data from stakeholders, the main causes of flash flood problems can be summarized in the following points:

1. The global warming accelerates the water and led to an intensification of the rainfall (Syed et al. 2010). There is evidence of increased frequency and severity of flash flooding in recent years (USAID 2018).
2. Flood plain encroachment (Wadi Degla, Red sea coast, Wadi Al Khurait). A sample is shown in Fig. 11.
3. Due to the high rates of encroachment on the floodways and the difficulty of determining the current flood path from digital elevation models or maps as shown in Fig. 12. It is necessary to update aerial-photogrammetry of the wadies downstream areas.
4. Excessive urbanization in the vicinity of major roads leading to a reduction in soil infiltration and increase in flood peak discharges will be catastrophic if no attenuation structures will be provided to ensure that the existing flood mitigation measures will safely operate as shown in Figs. 13 and 14. A sample of an artificial pond is located at Madinaty, as shown in Fig. 15.
5. Some major roads passing through wadies without any drainage structures to allow the flow in its natural path, leading to the diversion of wadi paths toward urbanized areas (Ras Gharib–El-Sheikh-Fadl road) as shown in Figs. 5 and 6.
6. There are encroachments by the people on the dam sites and the theft of some dam materials (stones and gabions), which caused the collapse of some dams (Al Ain-Al Sukhna Dams).

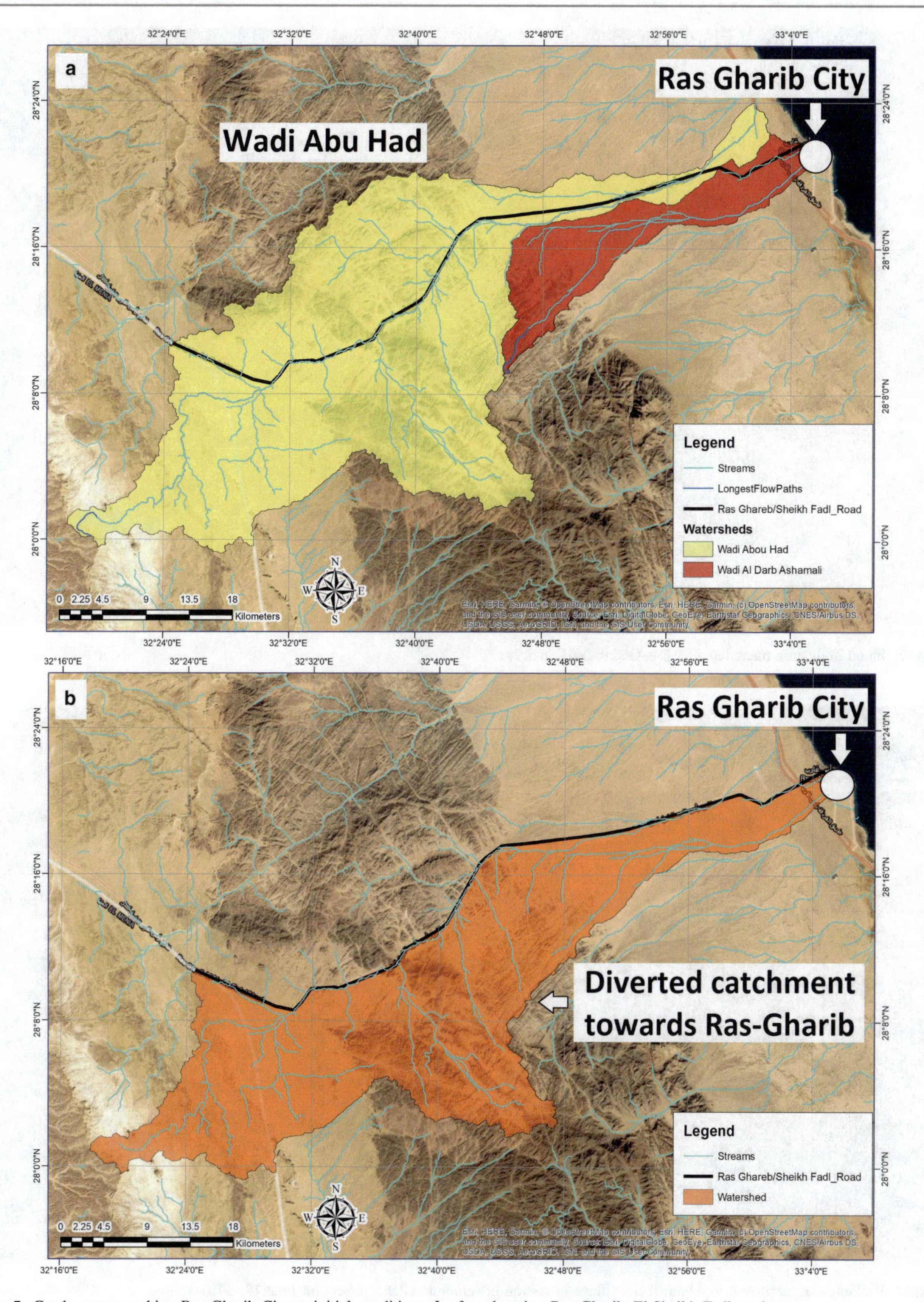

Fig. 5 Catchments attacking Ras-Gharib City, **a** initial conditions; **b** after elevating Ras-Gharib–El-Sheikh-Fadl road

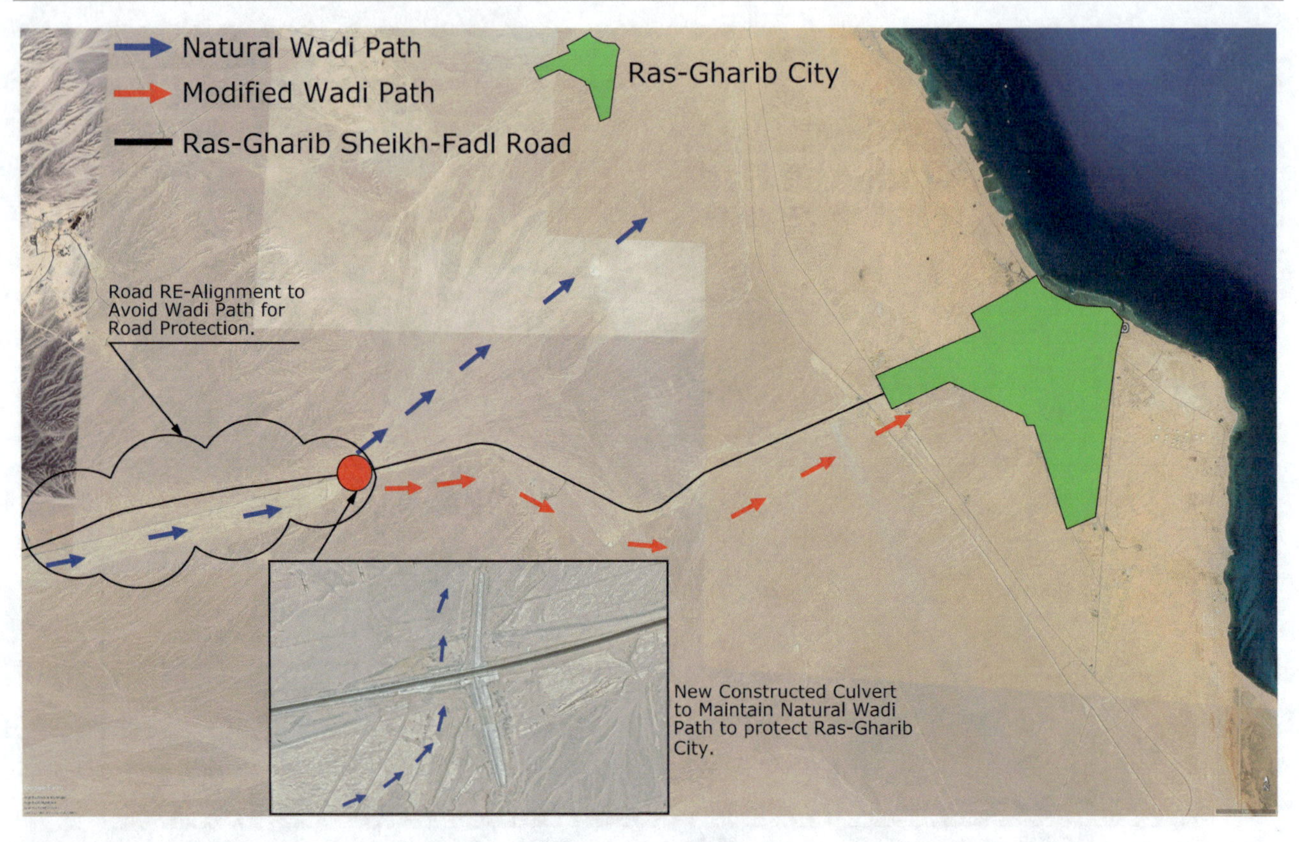

Fig. 6 Flood mitigation measures after Ras-Gharib 2016 incident

Fig. 7 Buildings are destroyed in Abouelreesh village in Aswan governorate after a torrent on June 17, 2010

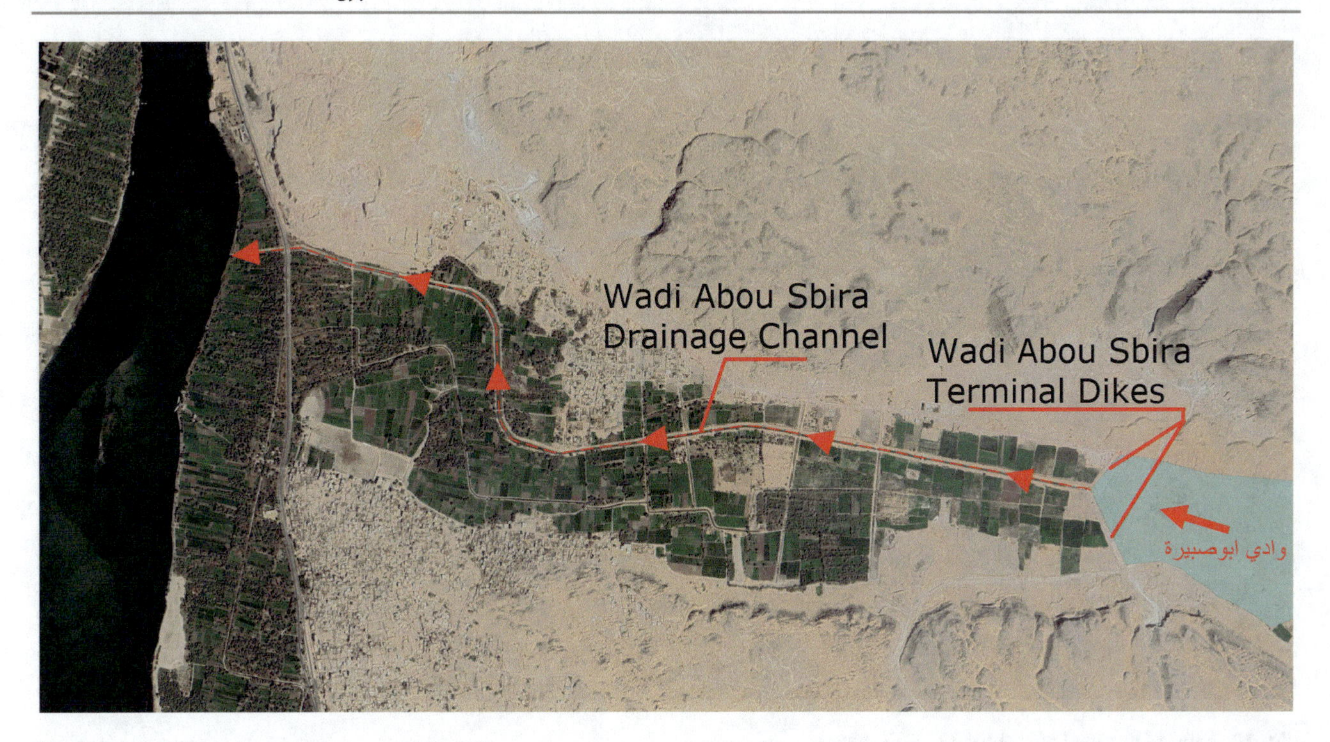

Fig. 8 Wadi Sbira drainage system toward the Nile

Fig. 9 El Sokhna road during April 25, 2018 rains, 2018

Fig. 10 Safaga-Qina road after April 12, 2017 rains

Fig. 11 Wadi Al Khurait encroachments at Aswan governorate (MWRI 2016)

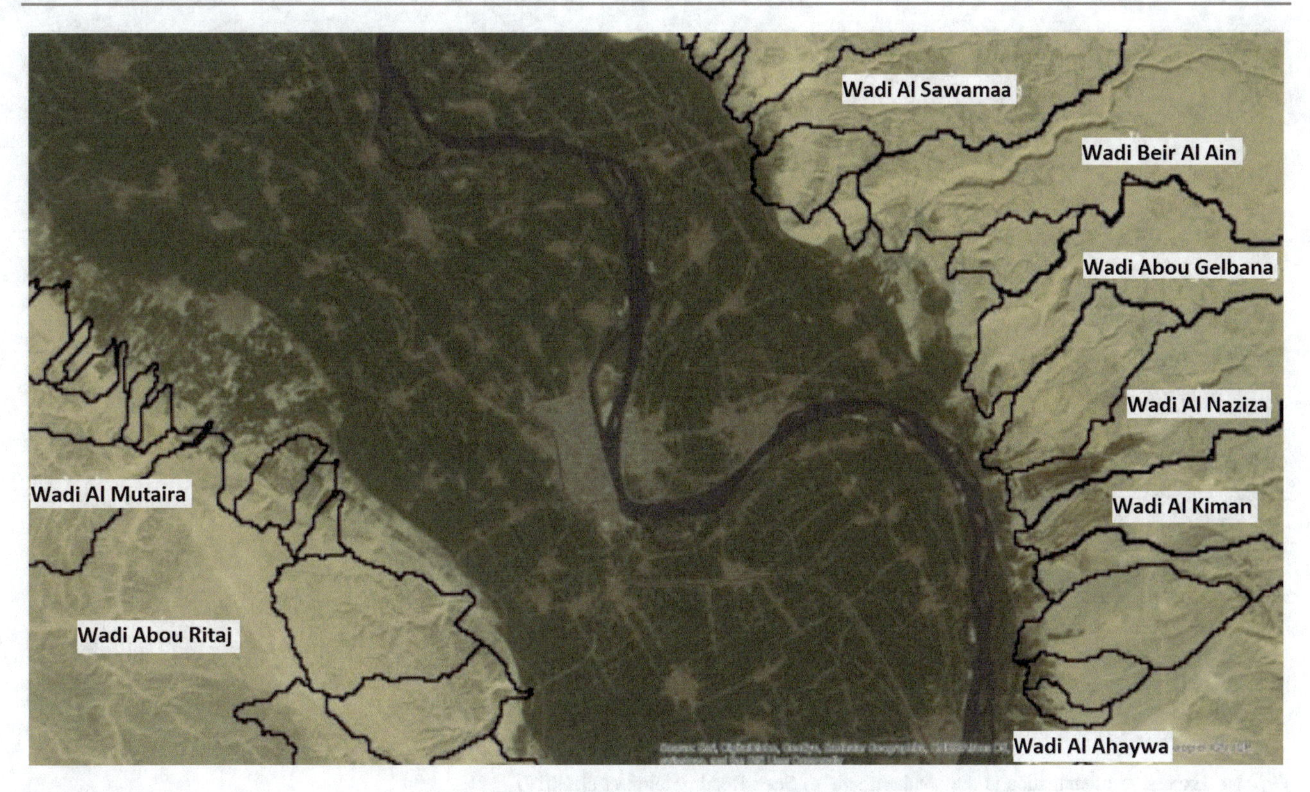

Fig. 12 Lost stream paths due to the expansion of urban areas and agricultural lands (MWRI 2016)

Fig. 13 Fifth settlement flooding (Helmi et al. 2019)

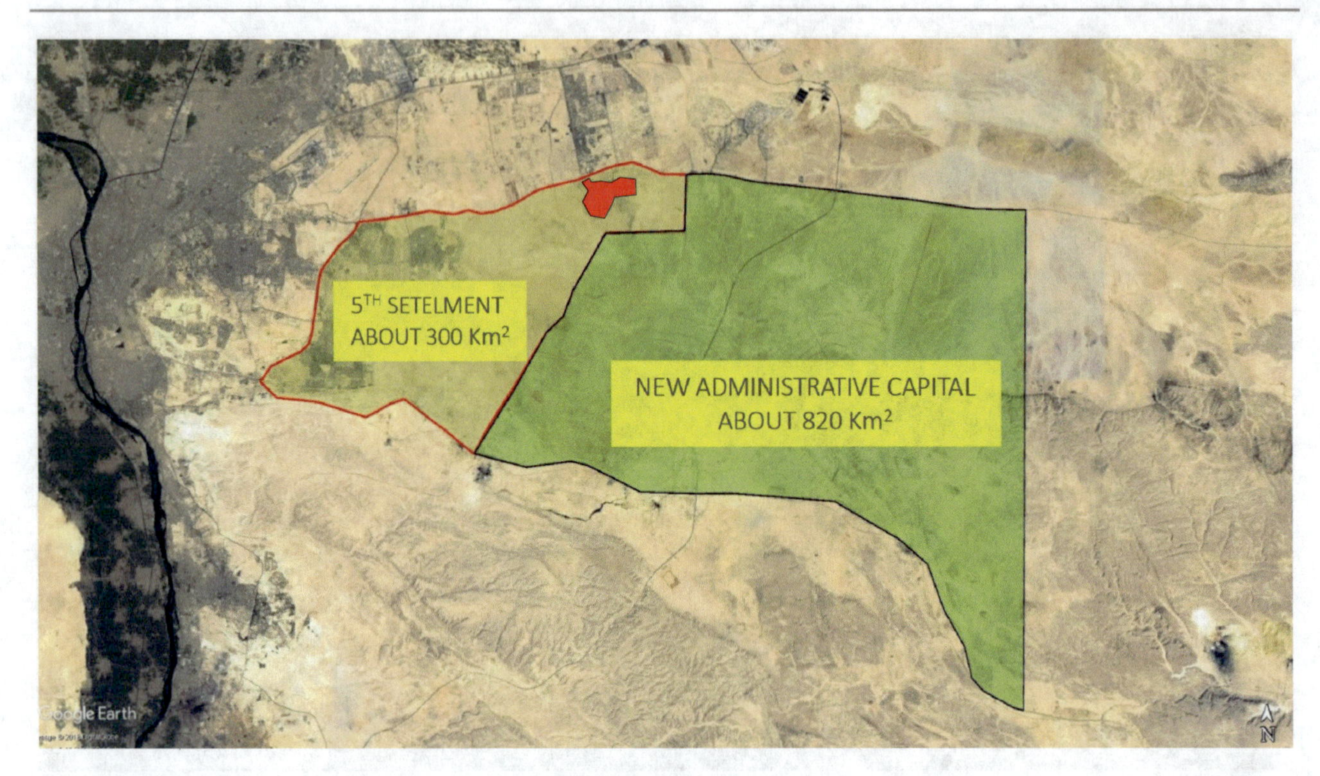

Fig. 14 Excessive urbanization at the southern side of Suez Road, (Helmi et al. 2019)

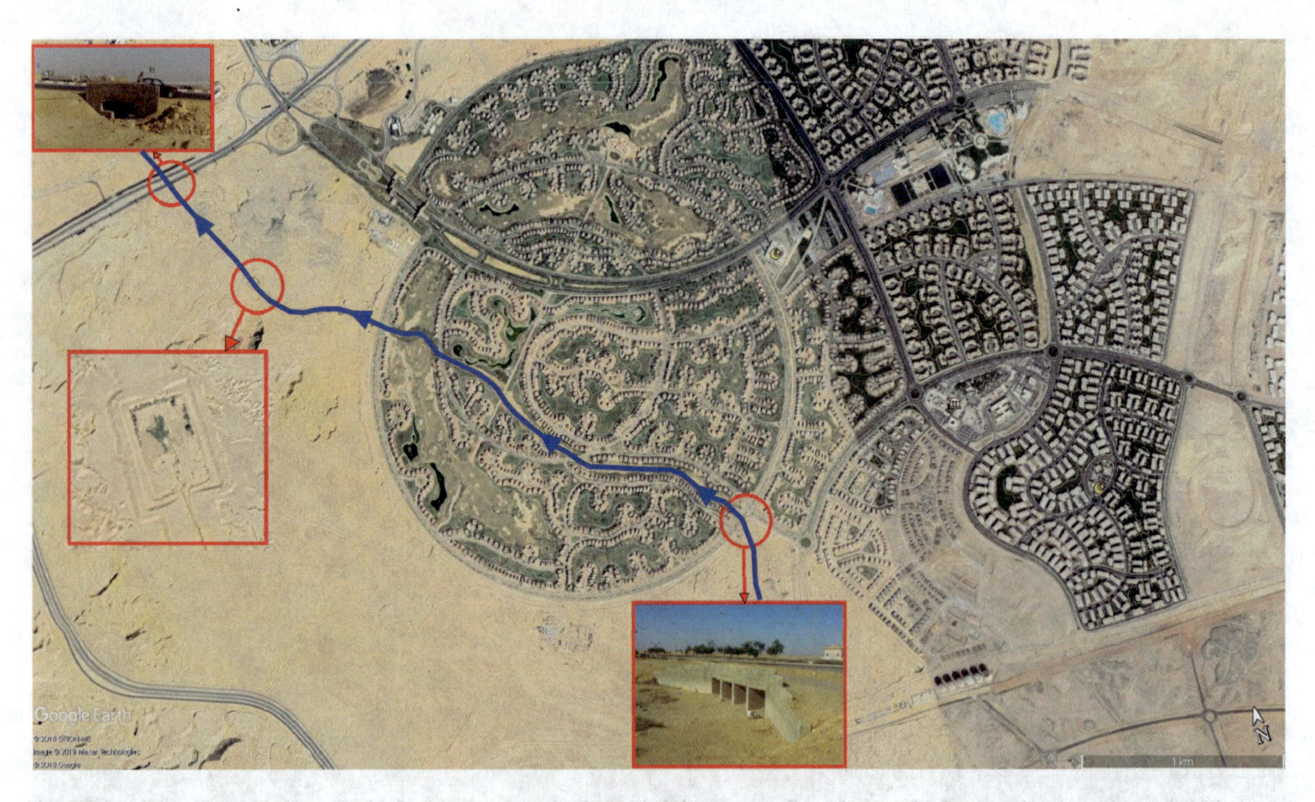

Fig. 15 Madinaty channelized wadi with four vents entrance culverts draining to a smaller culvert under Suez road and temporary detention pond

Fig. 16 Partial failure of Galawya dam due to the use of the dam top as an access road for quarries trucks (MWRI 2016)

7. The use of the top of the dams as access roads for the quarries to provide an easier path for heavy trucks as the case of (Galawiya Dam-Sohag Governorate) shown in Fig. 16.
8. There is no detailed storm drainage and flood mitigation design code to be used as a unified reference for all studies.
9. The high cost, and lengthy process of collecting rainfall data from the General Meteorological Authority even for universities and research institutes.
10. Many entities are responsible for projects affecting and/or affected by the wadies paths without any coordination with the main bodies responsible for these studies.

7 Risk Assessment of Flash Floods in Egypt

Many previous studies focused on the flash floods risk assessment in Egypt can be found in the literature. The majority of the articles relies on the catchments morphological characteristic and some studies considered the catchment discharge

(El-Shamy 1992) proposed a simple morphometric method to estimate the flash flood risk levels and the degree of hazardousness for each subbasin. Two different approaches were elaborated to determine hazardous sub-watersheds. The first is based on the relationship between bifurcation ration (R_b) and drainage density (D_d), whereas the second approach utilized the relationship between bifurcation ratio (R_b) and stream frequency (F_s) as given in Fig. 17. Drainage density (D_d) refers to topographic dissection, runoff potential, infiltration capacity of surface materials, climate, and land cover of the watershed. In this regard, low values of (D_d) indicate optimal conditions for infiltration, thus decreasing runoff potential, while high stream frequency (F_s) represents impermeable sub-surface materials, poor vegetation cover, high relief, and low infiltration capacity, hence increasing runoff potential. Applying this relationship separately to each subbasin will provide reasonable information on the estimation of flooding risk and recharge potential. In order to assess such relations, three morphometric parameters ($R_b, D_d,$ and F_s) must be calculated for each subbasin (Al-Saud 2010).

$$F_s = \frac{\mathrm{TN}_s}{A} \quad (1)$$

$$D_d = \frac{\mathrm{TL}_s}{A} \quad (2)$$

$$R_b = \frac{N_u}{N_{u+1}} \quad (3)$$

where

TN_s	Total number of streams.
TL_s	Total length of streams (m).
A	Catchment Area (km^2).
U	Stream orders according to (Horton 1945).
N_u	Number of streams of order (U).
N_{u+1}	Number of streams of order (U + 1).

Eman et al. (2002) evaluated the morphometric parameters of Wadi Um-Harika attacking Mersa Alam and the connecting road to and from DEM files. The data was used to define flash flood-vulnerable sites along the Idfo-Marsa Alam road.

El-rayes and Omran (2009) estimated the flood risks of Wadi Hagul basin. They concluded that three subbasins out of Wadi Hagul basin are recognized as risky flooding areas. The resulted runoff due to rainfall storm depth of 49.6 mm event occurred in October 1965 and reached about 7.2 million m^3.

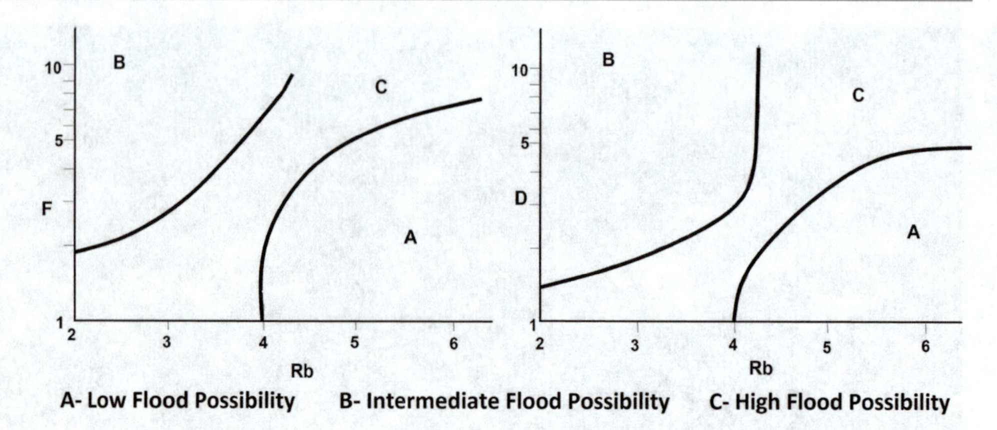

Fig. 17 Flooding possibilities based on (El-Shamy 1992) model

A multi-criteria standardized risk factors considering the (ASRF), (SSRF), (TCSRF), and Runoff Volume Risk Factor (RVSRF) was utilized by (El-moustafa 2012) to prioritize the flood protection work areas along Assiut Safaga Road in the eastern desert. Equal weight of each standardized risk factor was used to calculate (WSRF) to assess the risk of each catchment.

Mona Mohamed (2013) utilized multi-criteria analysis to assess the impact of morphological parameters on the risk categorization of some catchments located in the eastern desert and draining toward the Red Sea. The studied parameters were catchment Area Standardized Risk Factor (ASRF), Slope Standardized Risk Factor (SSRF), drainage Frequency Standardized Risk Factor (FSRF), drainage Density Standardized Risk Factor (DSRF), surface flow Length Standardized Risk Factor (LSRF), and Time of Concentration Standardized Risk Factor (TCSRF). The weight of each morphological factor was evaluated based on the impact on its peak flow discharges in order to obtain a Weighted Standardized Risk Factor (WSRF) that represents the equivalent catchment risk factor. As per the achieved weight of each risk factor, it was concluded that the (ASRF), (SSRF), and (TCSRF) are the most influential parameters affecting peak discharge.

Zaid et al. (2013) analyzed the flash flood of Wadi Abu-Hasah on Tall El-Amarna archeological area using GIS and Remote Sensing. They utilized (El-Shamy 1992) model to assess the flash flood hazard based on the morphological parameters, and concluded that the basin ranked as moderate to high hazard. They proposed a diversion channel to collect the surface water runoff away from the threatened, in addition to the construction of a dike around the tomb in order to minimize the water seeps into the tomb throughout the fractures and joints of the limestone section.

Abdel-fattah et al. (2017) investigated the relationship between variations in geomorphometric and rainfall characteristics and the responses of wadi flash floods. An integrated approach was developed based on geomorphometric analysis and hydrological modeling for Wadi Qina. Thirty-eight geomorphometric parameters representing the topographic, scale, shape, and drainage characteristics of the basins were considered and extracted using Geographic Information System (GIS) techniques. The results exhibited strong correlations between scale and topographic parameters and the hydrological indices of the wadi flash floods, while the shape and drainage network metrics have smaller impacts. The total rainfall amount and duration significantly impact the relationship between the hydrologic response of the Wadi and its geomorphometry.

Abdalla et al. (2014) utilized GIS-based morphometry and satellite imagery data to assess the flash floods occurrence and groundwater aquifers recharge relationship for Wadi El-Gemal, Wadi Umm El-Abas, Wadi Abu Ghuson, and Wadi Lahmi, along the southeastern Red Sea Coast in Egypt. The authors divided the studied areas into 45 subbasins and used (El-Shamy 1992) model to assess the potential for flash floods and groundwater recharge. Only two subbasins indicated the low potential of flooding and high potential of groundwater aquifers.

Elsadek et al. (2018) assessed the susceptibility of Wadi Qina watershed sub-catchments based on morphological parameters according to (El-Shamy 1992). The results illustrate that there are no subbasins with low risk of flooding. The subbasins with the highest hazard degree are concentrated in the middle of the watershed, although they have smaller areas compared with the surrounding subbasins. The subbasins located at the boundary of the watershed have an intermediate risk of flooding and moderate potential for groundwater recharge.

The Egyptian Ministry of Water Resources and Irrigation with cooperation with the National Water Research Center and Water Resources Research Institute published flood atlases for Aswan, Luxor, Qina, Assiut, Sohag, and North Sinai governorates while the Red Sea governorate is still under preparation. The risk assessment can be done by two approaches:

- Streamflow intensity: by evaluating the Intensity Factor (IF) obtained from multiplying the stream depth by the stream velocity at the catchments outlets. The results were classified into four ranges (low, medium, high, and high) for (IF) <1, 1–3, 3–5, and >5, respectively.
- Catchments Risk Assessment: by considering the catchment slope, Curve Number (CN), and 1-in-100 years' storm event as a governing factor in the determination of the catchment risk. The results were classified into four ranges (low risk, medium risk, high risk, and very high risk). No details were provided for the methodology used to achieve this classification.

In MWRI atlases the assessment is based on provincial boundaries and not a regional scale.

7.1 Available Data for the Study Area

Figure 18 shows the distribution of some of the Egyptian meteorological authority station, the 1-in-100 maximum daily precipitation in (mm) was collected from previous reports and studies for some meteorological stations located in the study area is given in Table 1. The 1-in-100 maximum daily precipitation is used to generate the isohyetal map for the study area as shown in Fig. 19. The 1:250000 topo map has been acquired, and georeferenced, to cover the study area as shown in Fig. 20. The regional SCS Curve Number (CN) is shown in Fig. 21 (Awadallah et al. 2016). The SRTM 90 × 90 DEM data has been collected and clipped to cover the study area as shown in Fig. 22.

7.2 Assessment of Previous Studies and Available Data

The available data in the study area can be utilized to obtain the catchment boundaries and streamlines from the DEM file for further verification versus the streamlines and wadies names available on topo maps. After the verification versus the topo maps, the morphological characteristics can be obtained for each catchment. The weighted average SCS CN for each catchment can be extracted using the GIS tool for each catchment. The GIS is utilized to produce an isohyetal map of the study area to obtain weighted 1-in-100 weighted average maximum daily precipitation for each catchment. Finally, each catchment peak 1-in-100 discharge and storm volume can be calculated. The proposed framework is summarized in Fig. 23 flow chart.

Most of the previous studies considered only the morphological parameters of the catchments in their weighted value or considered equal weigh morphological parameters and adding the volume of runoff as an additional parameter with the same weight. As well, all previous studies covered scattered parts of the eastern desert and no regional study was provided.

In this report, as long as the data collected are sufficient to calculate the runoff peak discharge, it will be used in its standardized value as an evaluating risk factor for the catchments. The Peak Flow Standardized Risk Factor (PFSRF) internally contains the morphological, geological, and meteorological parameters in their weighted form. PFSRF will be used to categorize the catchments based on their risk. The Runoff Volume Standardized Risk Factor (RVSRF) will be utilized in further catchments risk categorization, which will show the stormwater harvesting potential in addition to the risk generated if no proper flood mitigation measures are provided.

$$\text{Peak Flow Standardized Risk Factor (PFSRF)} = \frac{\mathrm{PF} - \mathrm{PF}_{\min}}{\mathrm{PF}_{\max} - \mathrm{PF}_{\min}} \tag{4}$$

$$\text{Runoff Volume Standardized Risk Factor (RVSRF)} = \frac{\mathrm{RV} - \mathrm{RV}_{\min}}{\mathrm{RV}_{\max} - \mathrm{RV}_{\min}} \tag{5}$$

The selected normalized ranges to classify the risk categories are given in Table 2.

7.3 Catchments Delineation

Many thresholds have been tested for catchments delineation in order to obtain a reasonable number of catchments for this regional-scale study. The selected threshold was set to 50 km^2.

7.4 Runoff Calculations

HEC-HMS software is used to calculate the catchment peak discharge and total volume of runoff. The selected calculation algorithm is the Soil Conservation Service SCS Curve Number method (USDA1986), where

$$R = \frac{(P - I_a)^2}{(P - I_a) + S} \tag{6}$$

$$S = \frac{25400}{\mathrm{CN}} - 254 \tag{7}$$

where

R Excess Runoff (mm).
P Rainfall in (mm).

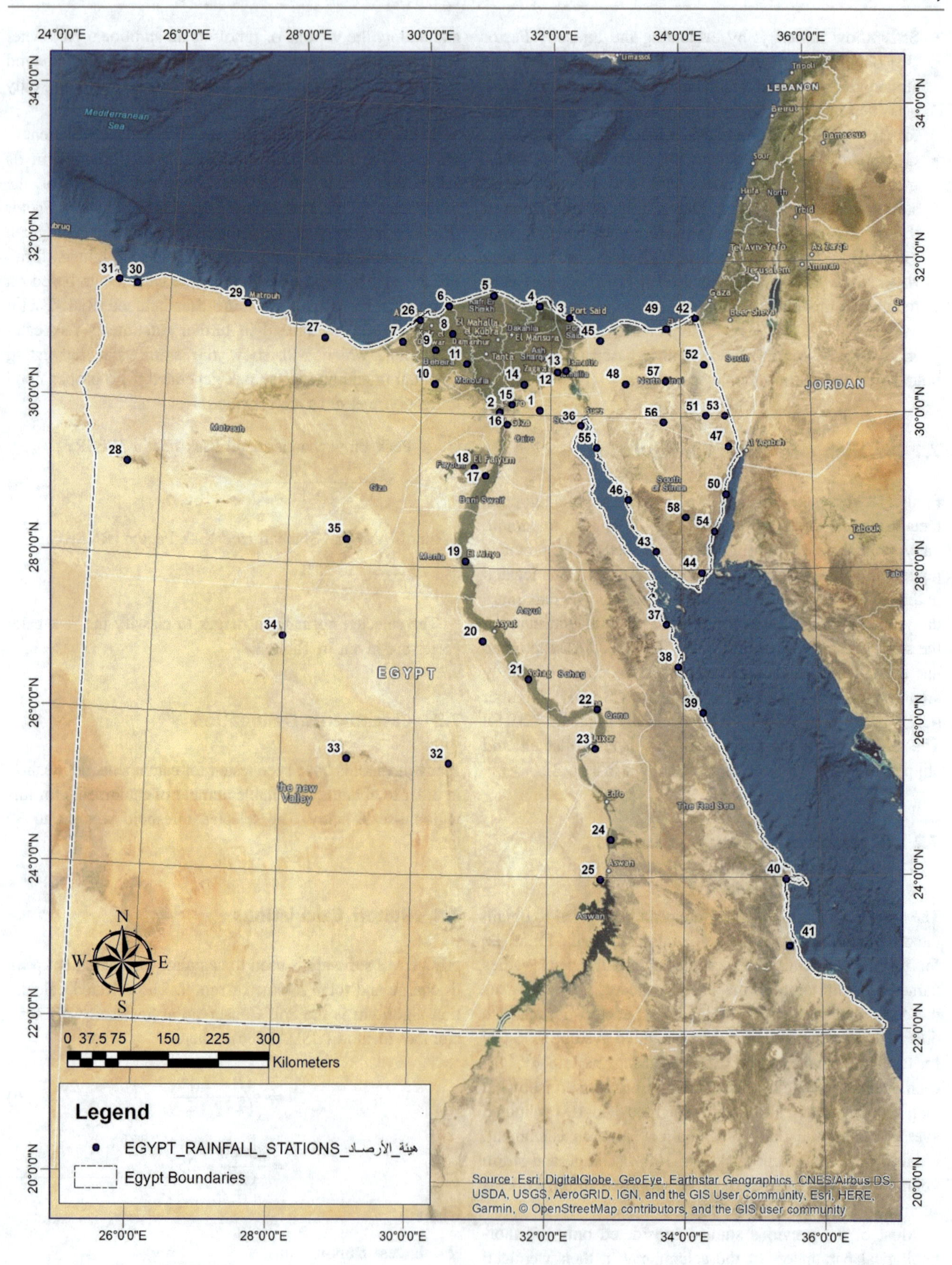

Fig. 18 Distribution of Egyptian meteorological authority stations

Table 1 Available rainfall stations frequency analysis results

Name	Lat	Long	1-in-100 Precipitation (mm)
Katamya	30.07°	31.83°	21.4
Sohag	26.57°	31.70°	34.8
Qina	26.18°	32.73°	61.7
Luxor	25.67°	32.70°	39.6
Komombo	24.48°	32.93°	43.7
Aswan	23.97°	32.78°	25.4
Helwan	29.87°	31.33°	57.6
Giza	30.03°	31.22°	23.3
Assyout	27.05°	31.02°	66.4
Suez	29.87°	32.47°	49.6
Hurghada	27.28°	33.77°	81.4
Quseir	26.13°	34.30°	76
Ras Banas	23.97°	35.50°	81.4
Sharm El-Sheikh	27.96°	34.30°	50
Abou Rdais	28.91°	33.19°	58
Nowebaa	28.98°	34.68°	47
Ras Sudr	29.58°	32.71°	46
Saint Catherin	28.68°	34.06°	87
Al Temed	29.40°	34.17°	107

S Potential maximum retention after runoff begins in (mm).
I_a Initial abstraction in (mm)

$$I_a = 0.2S \tag{8}$$

6-h storm duration with SCS type II distribution, given in Fig. 24, was selected for hydrologic modeling of the catchments.

The time of concentration is among the different parameters which are used to identify the response of any given watershed to a rainfall storm event. Time of concentration is defined as the time required by a water drop to travel from the hydraulically most distant point of any given watershed to the watershed outlet (Ramirez 2000; Durrans 2007), the hydraulically most distant point is the point with the longest travel time. There are various empirical formulas to estimate the time of concentration from topographic and/or rainfall characteristics. The well-known Kirpich equation was originally developed based on SCS data from seven rural watersheds on a farm in Tennessee, these watersheds were characterized by well-defined channels, and steep slopes and their catchment areas were ranging from 1.25 to 112 acres (0.005–0.45 km^2).

The Kirpich formula can be expressed as follows:

$$T_c = 0.019472 \frac{L^{0.77}}{S_l^{0.385}} \tag{9}$$

where

T_c Time of concentration in (min).
L Longest flow path in (m).
S_l Slope of the longest flow path.

The formula has been updated after (Rossmiller 1980) to consider the effect of CN

$$T_{c\text{-modified}} = T_c \times (1 + (80 - \text{CN}) * 0.04) \tag{10}$$

The Unit Hydrograph (UH) is the actual response of any given watershed (in terms of runoff volume and timing) to a unit input of excess rainfall. First proposed by (Sherman 1932), it can be defined as the Direct Runoff Hydrograph (DRH) resulting from one unit (e.g., one cm or one inch) of excess rainfall occurring uniformly over a watershed at a uniform rate over a unit time (Sherman 1941; Chow 1959; Ramirez 2000; Weaver 2003). The SCS dimensionless unit hydrograph given in Fig. 25 is used in the hydrologic simulation for the case in hand.

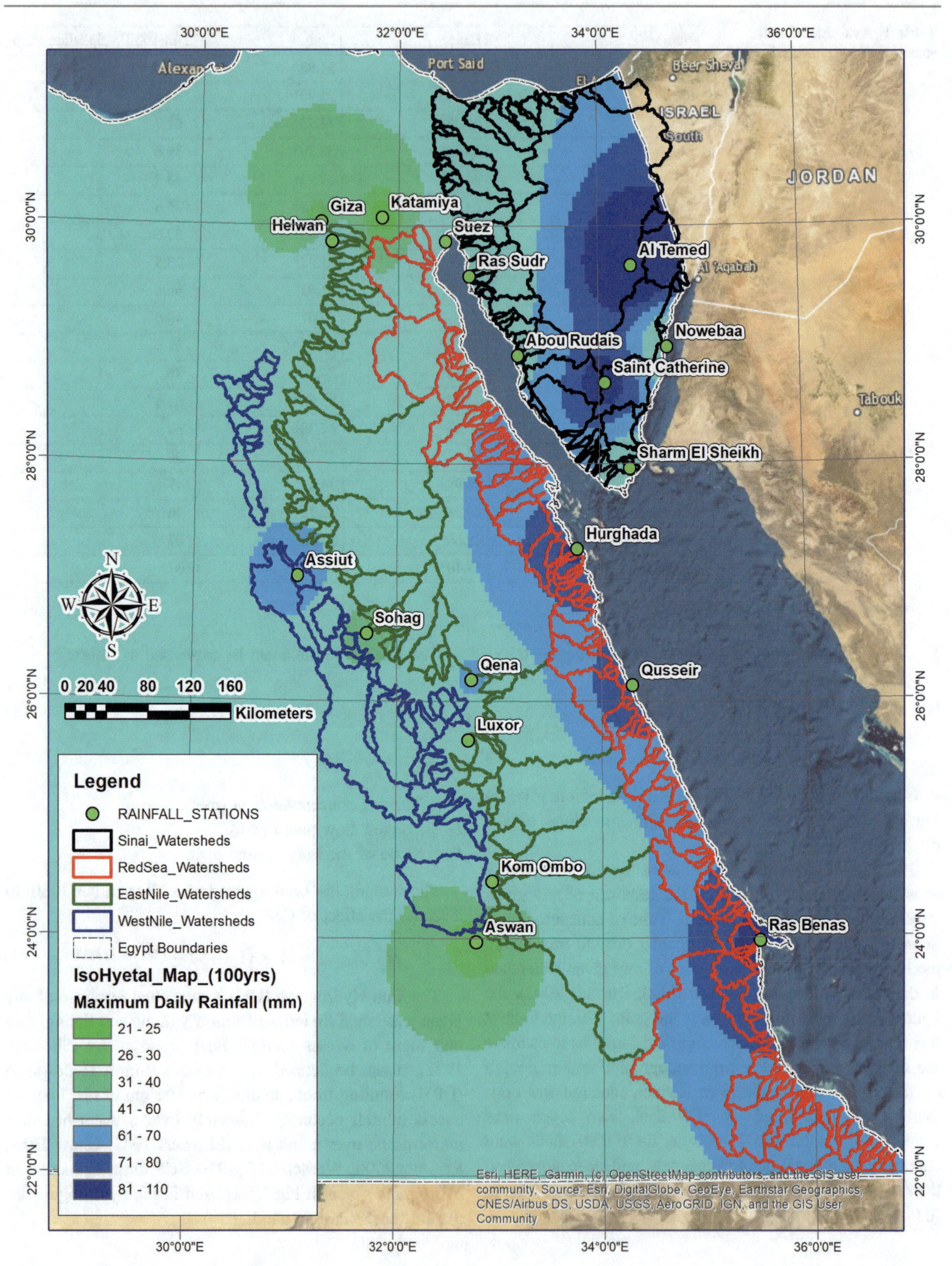

Fig. 19 Isohyetal map for the study area

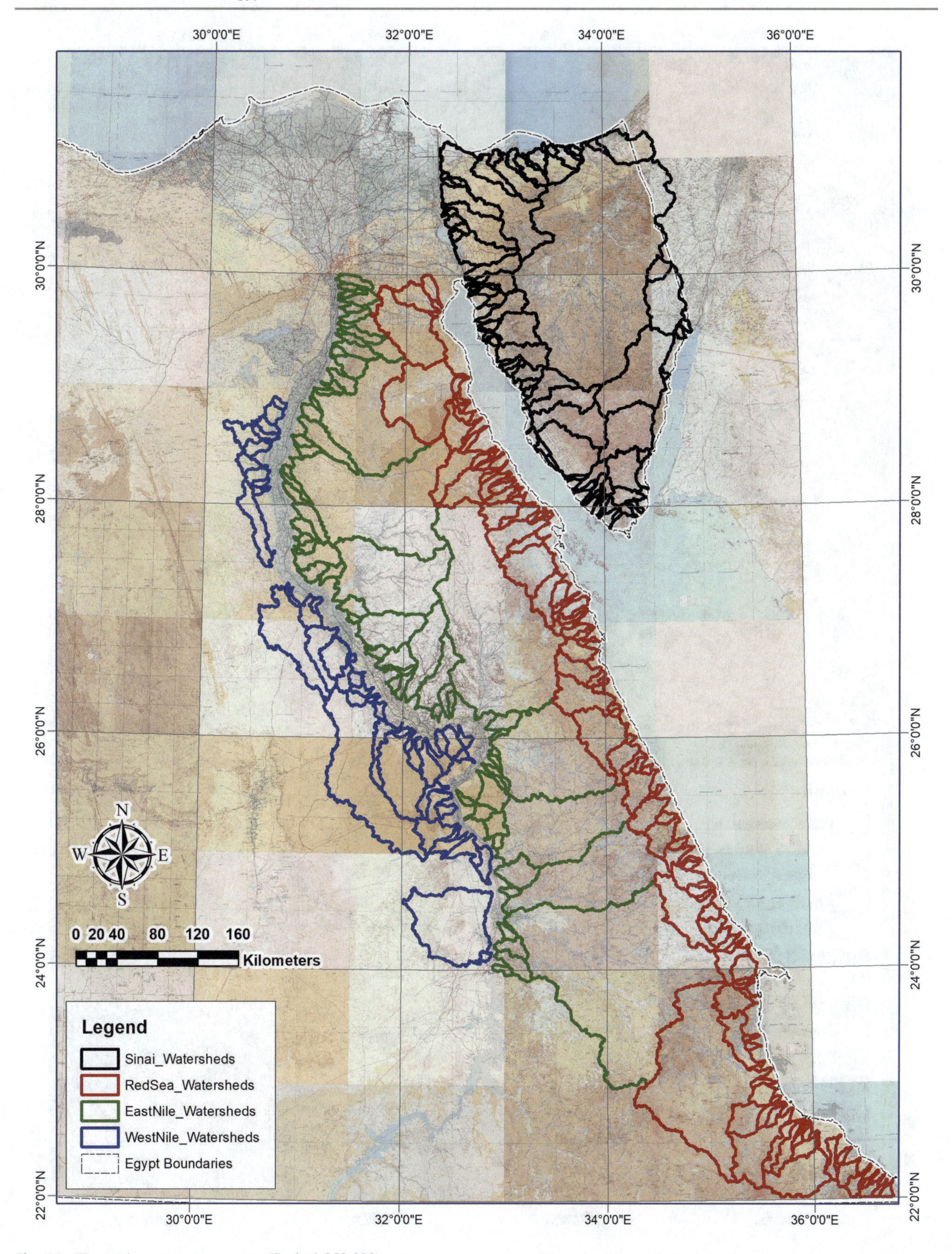

Fig. 20 The study area on topo maps (Scale 1:250,000)

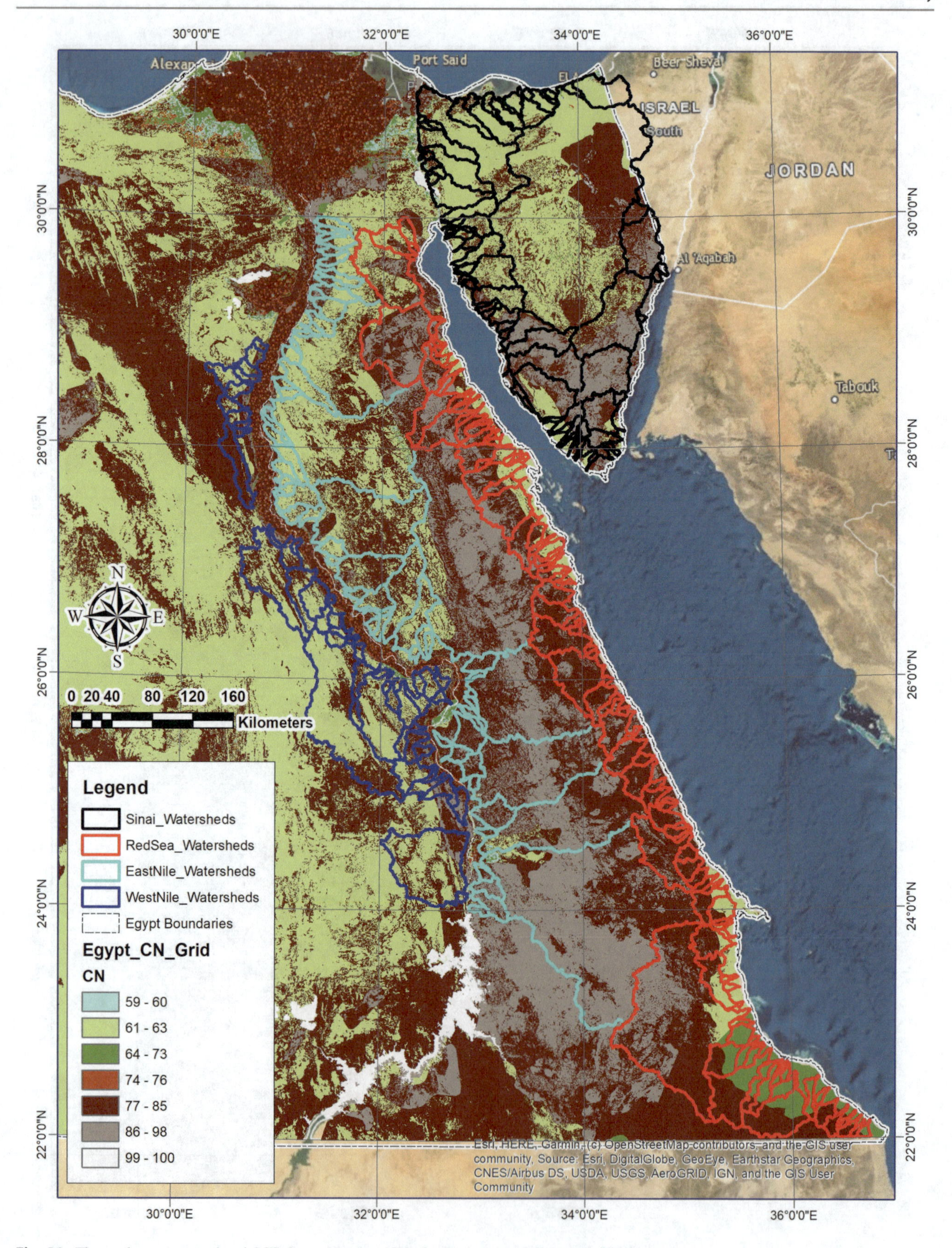

Fig. 21 The study area on regional SCS Curve Number (CN) for Egypt (Awadallah et al. 2016)

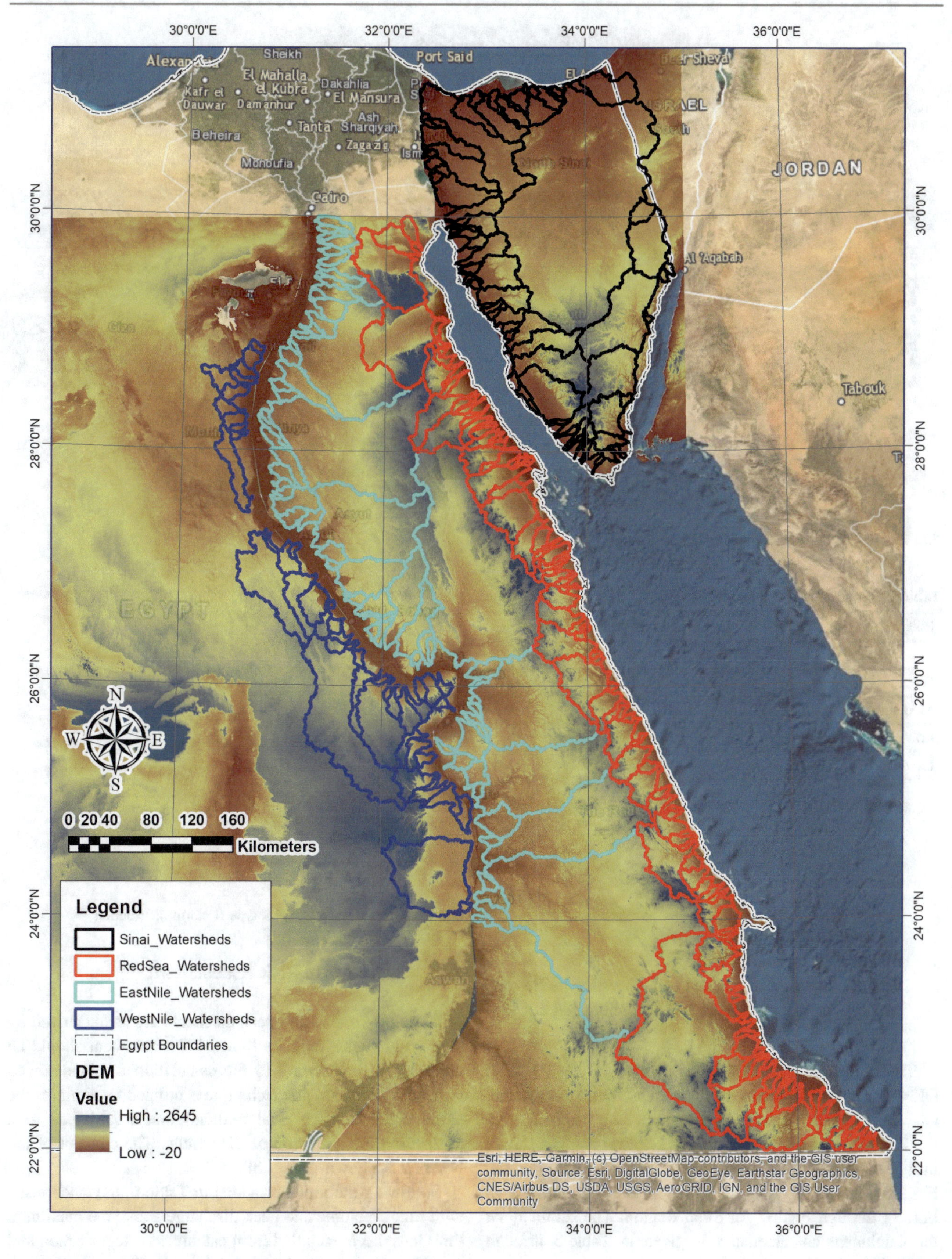

Fig. 22 Digital Elevation Model (DEM) for the study area [SRTM 90 × 90]

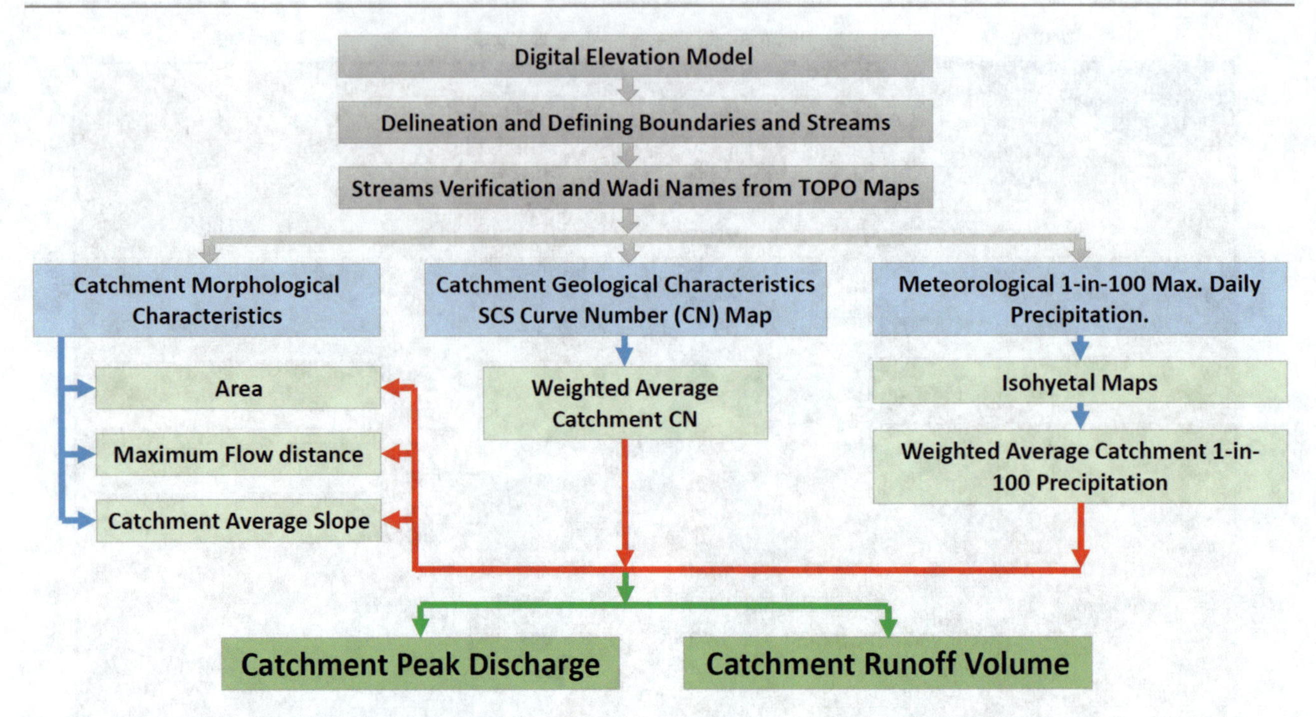

Fig. 23 Flow chart for data processing procedure

Table 2 Proposed risk categorization

Risk category	Normalized risk factor range
Very high	0.7–1.0
High	0.5–0.7
Moderate	0.3–0.5
Low to moderate	0.15–0.3
Low	<0.15

$$T_p = 0.5D + T_L \tag{11}$$

where

- T_p Time to peak discharge (min)
- T_L Lag time = $0.6 \times T_c$ (min)
- D Rainfall duration in minutes corresponding to the time step of calculations where it is recommended not to exceed 0.133–0.2 T_c.

The catchment boundaries projected on SRTM 90 × 90 DEM raster and projected on topo maps, weighted average CN, and weighted average maximum daily precipitation for the 1-in-100 return period for each catchment area are shown in Fig. 26 through Fig. 29, respectively, for the Nile Region, Fig. 30 through Fig. 33 for the Red Sea region, and in Fig. 34 through Fig. 37 for Sinai Region. The summary of the catchment characteristics is given in Table 3 through Table 5 for the three studied regions.

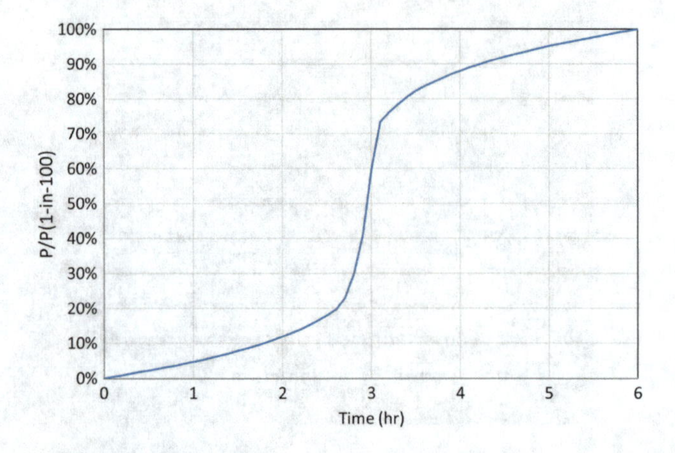

Fig. 24 6-h accumulated SCS type II storm distribution

7.5 Catchments Risk Assessment

Total runoff volume and peak discharge are characterized by their wide values range, a thoughtful assessment should be done before proceeding to the calculation of standardized risk factor. The box plot technique is utilized to eliminate the extreme values (Statistical outliers) that would lead to a misleading interpretation of the results. The concept of the box plot is shown in Fig. 38.

Following the limits provided in Table 6, the catchments with runoff volume and peak discharge above the maximum limit have been excluded from catchment categorization and considered as very high-risk catchments. Figure 39 through

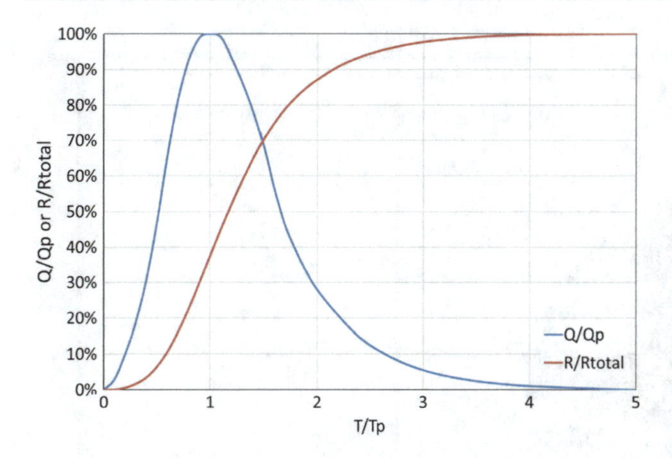

Fig. 25 SCS dimensionless unit hydrograph and mass curve

Fig. 44 shows the results of risk catchments risk categorization for the studied regions based on Peak Flow Standardized Risk Factor (PFSRF) and Runoff Volume Standardized Risk Factor (RVSRF).

8 Decision Support System Work Plan

Classifying the drainage catchments located in Egypt based on a standardized risk assessment basis is the first step in supporting decision-makers to achieve the prioritization of the detailed risk assessment studies. The detailed risk assessment can be expanded into two further steps:

- The risk of wadies and/or reaches which can be provided with proper drainage structures to its final discharge points for 1-in-100 years return period design storm, can be reduced to a low-risk category after the implementation and the assurance of proper maintenance for flood mitigation measures. The storm harvesting in the form of artificial lakes, and dams can be selected as the first option in flood mitigation measures for the catchments associated with very high, and High Runoff Volume Risk Factor (RVSRF) to increase the rate of investment return from both flood mitigation and the reduction of freshwater stress.
- The wadies and/or reaches which cannot be provided with proper drainage to its final discharge points for 1-in-100 years return period design storm, two-dimensional modeling is mandatory for proper assessment of hazard, exposure, and vulnerability of the people and assets located in the flood plain. Figure 45 illustrates the workflow chart for risk assessment.

9 Determination of Flood Hazard Using Two-Dimensional Modeling

Ras-Gharib city was selected as a case study for flood intensity calculations. The delineated catchment for wadi Abou-Had using DEM files of resolution 30 × 30 m showed the original flow direction of the wadi toward its outfall far to the north of Ras-Gharib city, as shown in Figs. 46 and 47. The DEM-based delineation may be sufficient for a preliminary assessment for new lands, but for any human intervention, it is necessary to conduct a topographical survey for the area under study with proper expansion to capture the wadies sections.

Accordingly, it should be highlighted that the use of the DEM files is not sufficient to capture the manmade variations to the topology but in order to capture the real variation of the topology the actual survey of Ras-Gharib–El-Sheikh-Fadl road was added to the DEM file.

The 1-in-100 rainfall weighted average precipitation (53.53 mm) with 6 h SCS type II distribution, weighted average curve number (CN = 81.75), and updated DEM files were used to build a rainfall-runoff two-dimensional HEC-RAS model to track the streams attacking Ras-Gharib city. The model boundaries and flow depths are given in Fig. 48. Figures 49, 50, and 51 show a close view for the flow depth, velocity distribution, and the Flood Intensity (FI) within Ras-Gharib and surrounding areas, respectively.

$$\mathrm{FI} = V \times d \tag{12}$$

where

FI Flood intensity (m^2/s)
V Flow Velocity (m/s)
d Flow depth (m).

Due to the lack of field measurements during the flooding event, and in order to verify the results of the model; some flood plain photos were used and showed consistency with the model output as shown in Fig. 52. Achieving accurate assessment of risk requires a topographical survey to capture the buildings and streets which cannot be captured in the DEM file and will significantly impact the flow depth and velocity. Additionally, a land-use map has to be utilized to define the vulnerability to achieve a final risk map.

Additional updated two-dimensional HEC-RAS rainfall-runoff model has been conducted in order to assess the effectiveness of the newly constructed flood mitigation measures (the new culvert with attached two dikes). The results are given in Figs. 53, 54, and 55 show a significant reduction in flood intensity in Ras-Gharib city.

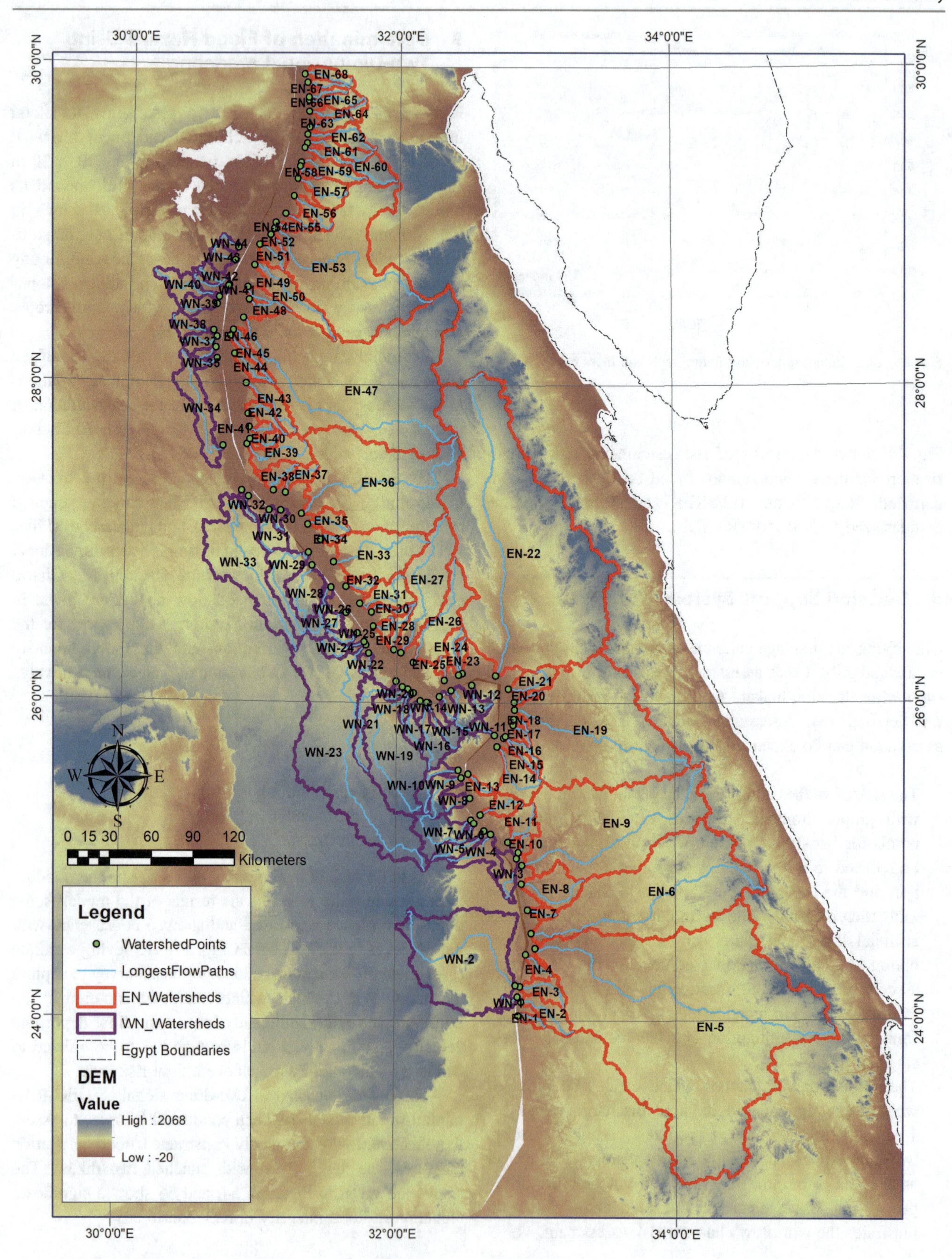

Fig. 26 Projected Nile catchments on SRTM 90 × 90 DEM

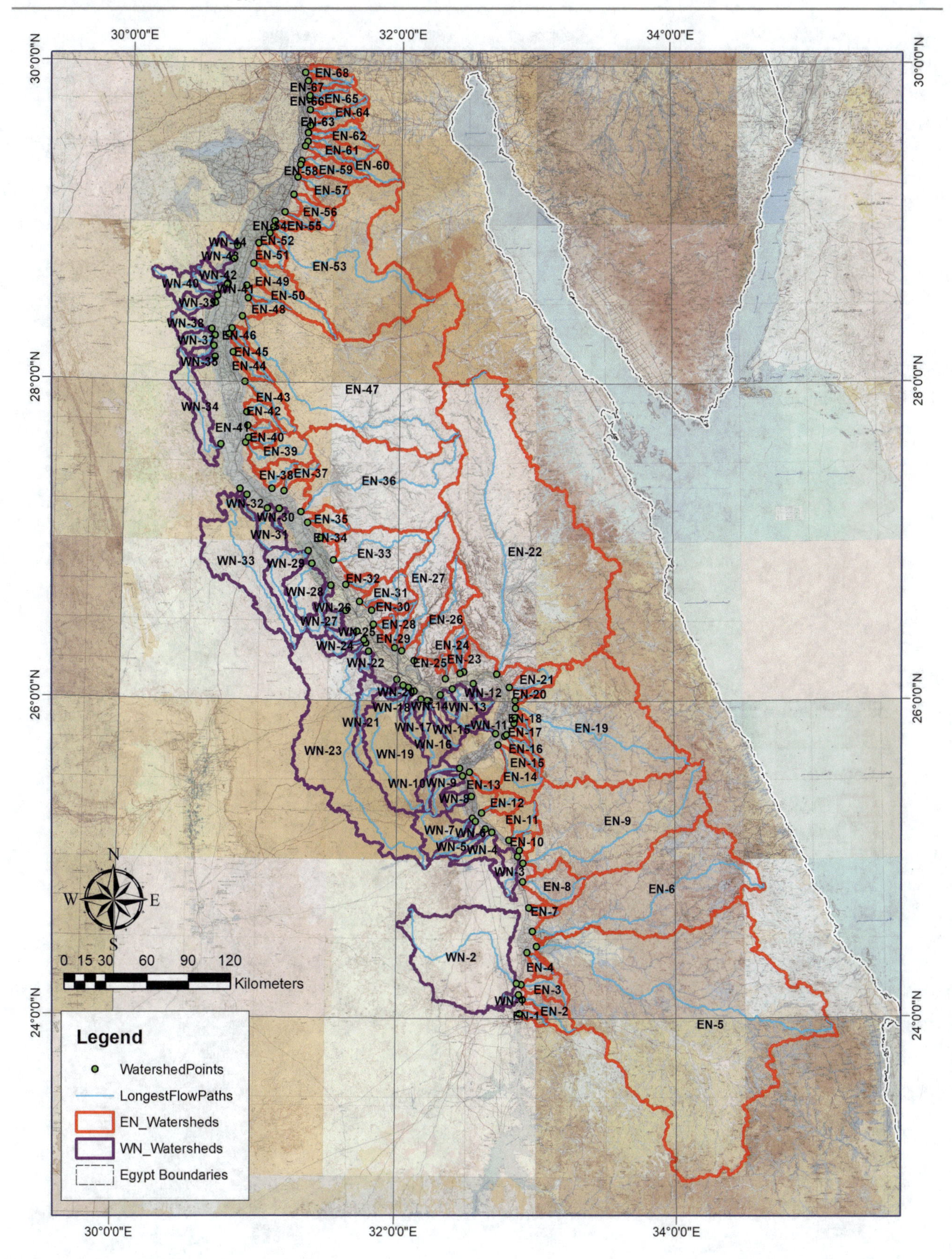

Fig. 27 Projected Nile catchments on topo maps

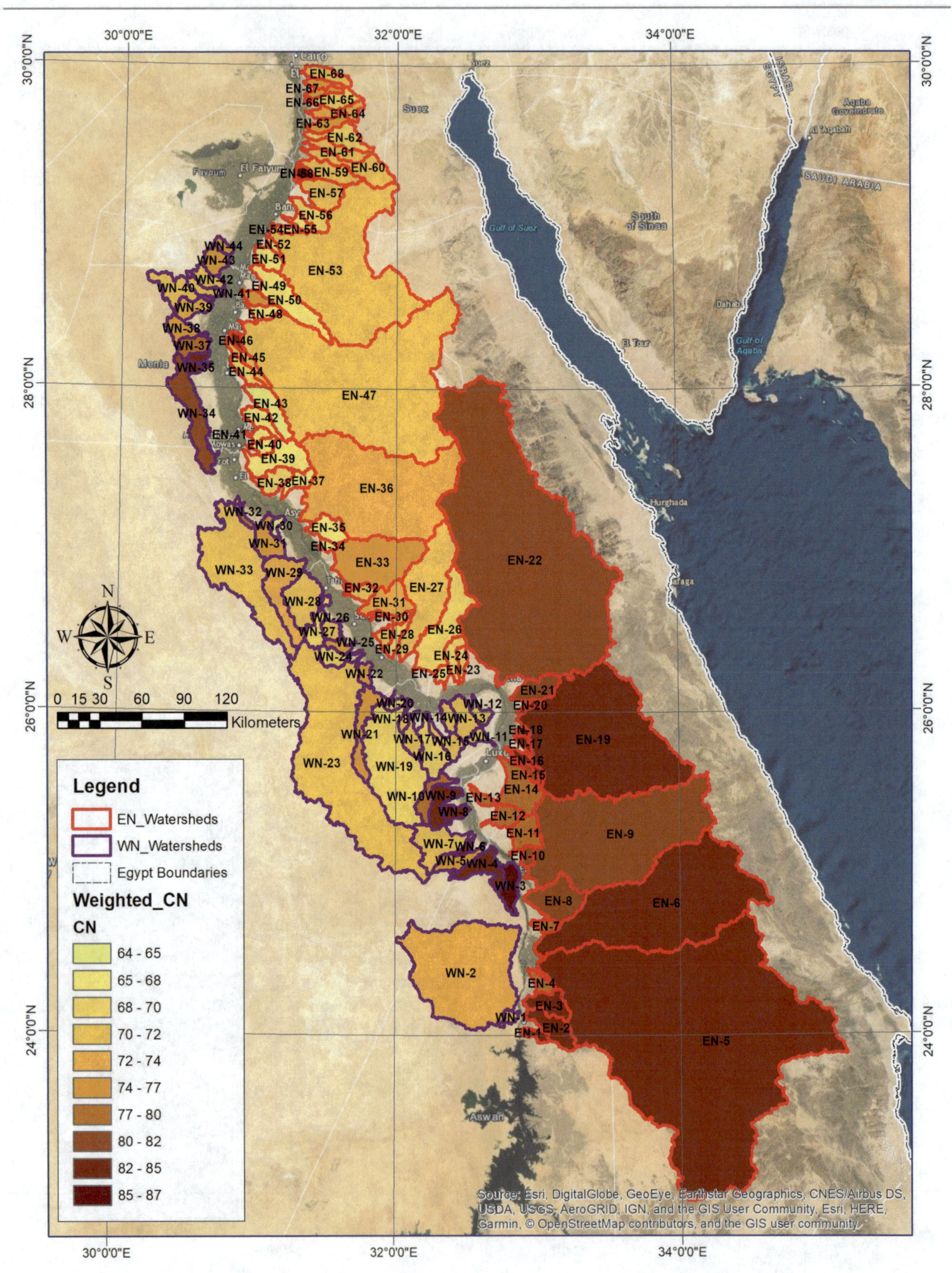

Fig. 28 Nile catchments weighted average CN

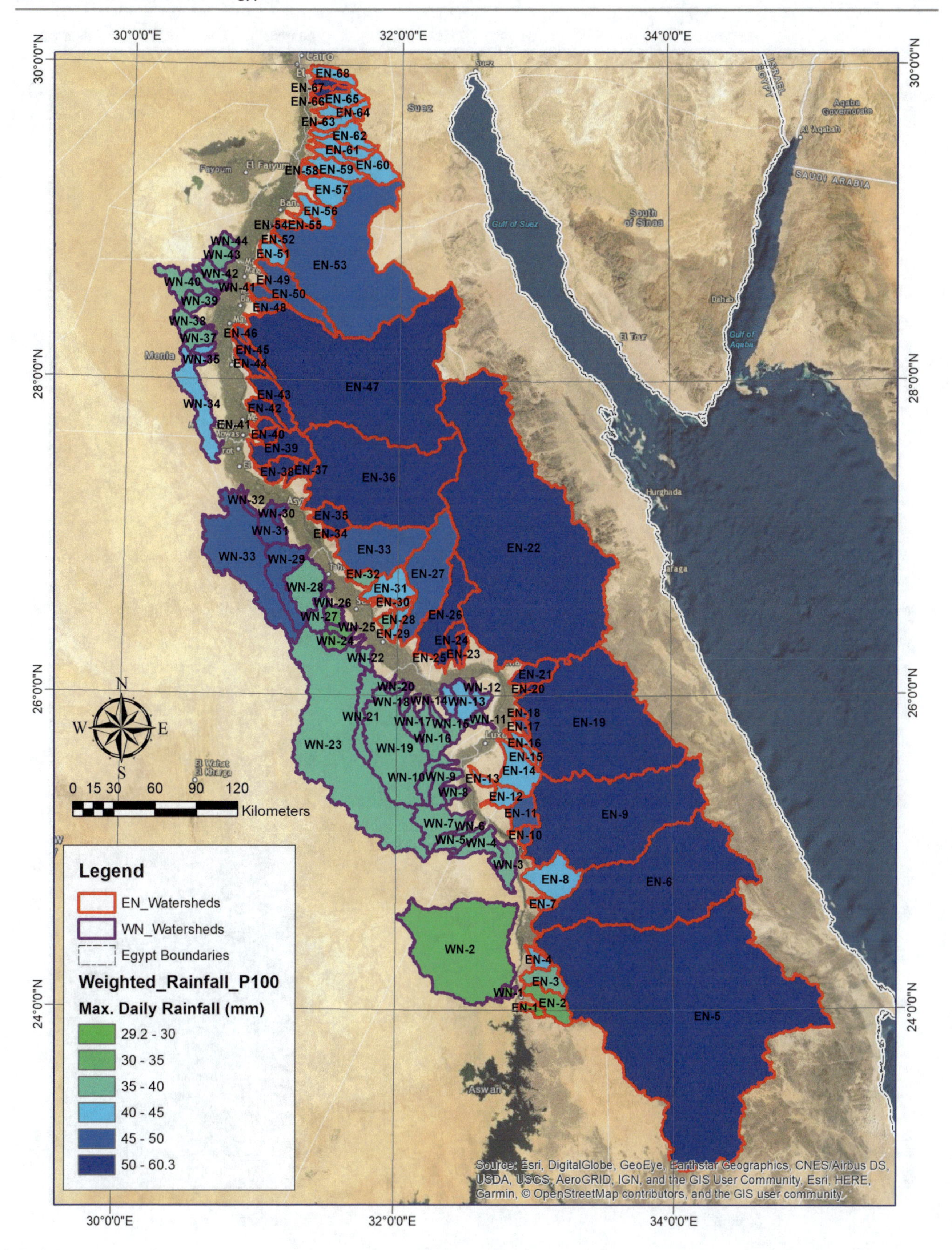

Fig. 29 Nile catchments weighted average 1-in-100 maximum daily precipitation

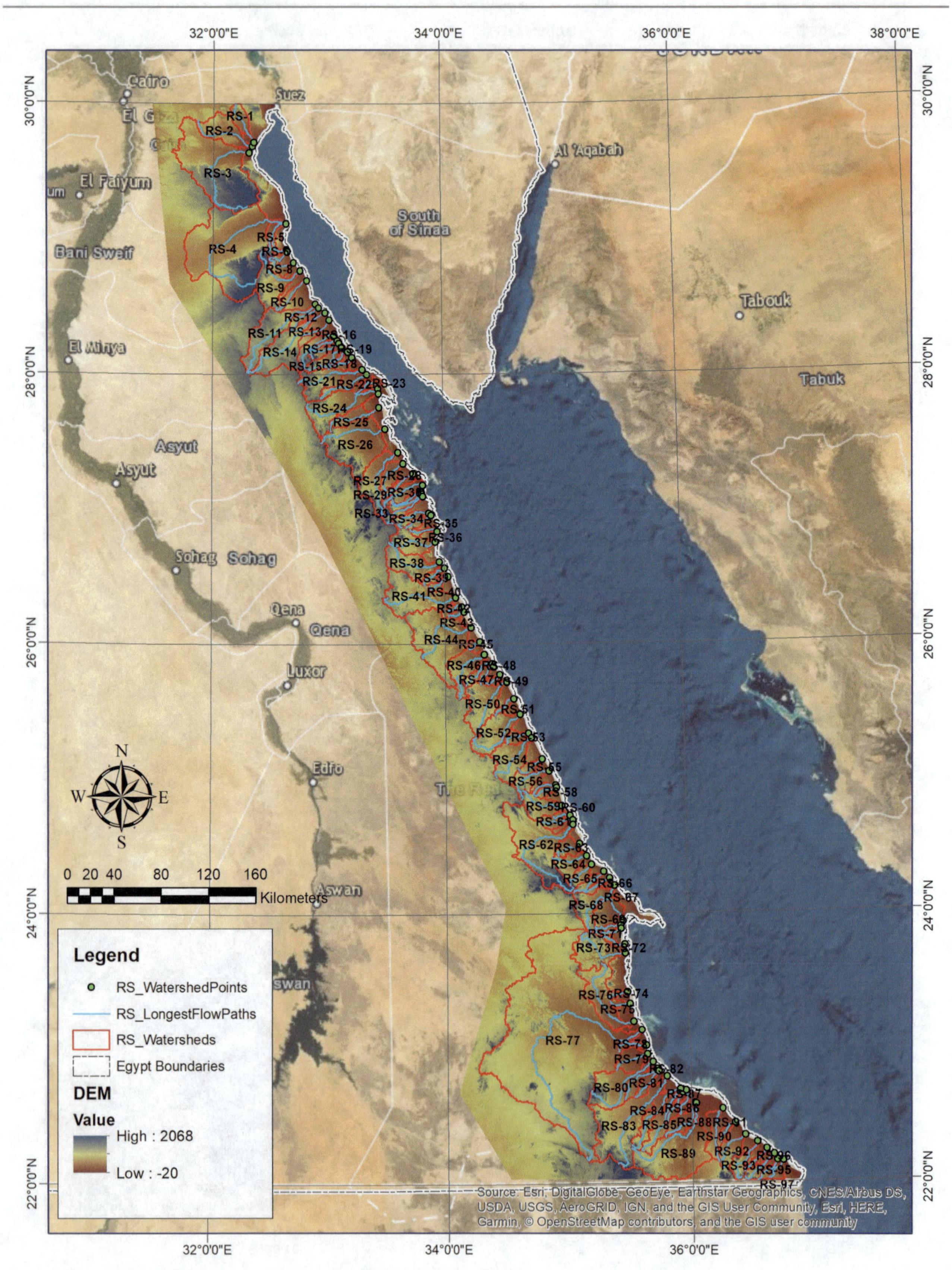

Fig. 30 Projected Red Sea catchments on SRTM 90 × 90 DEM

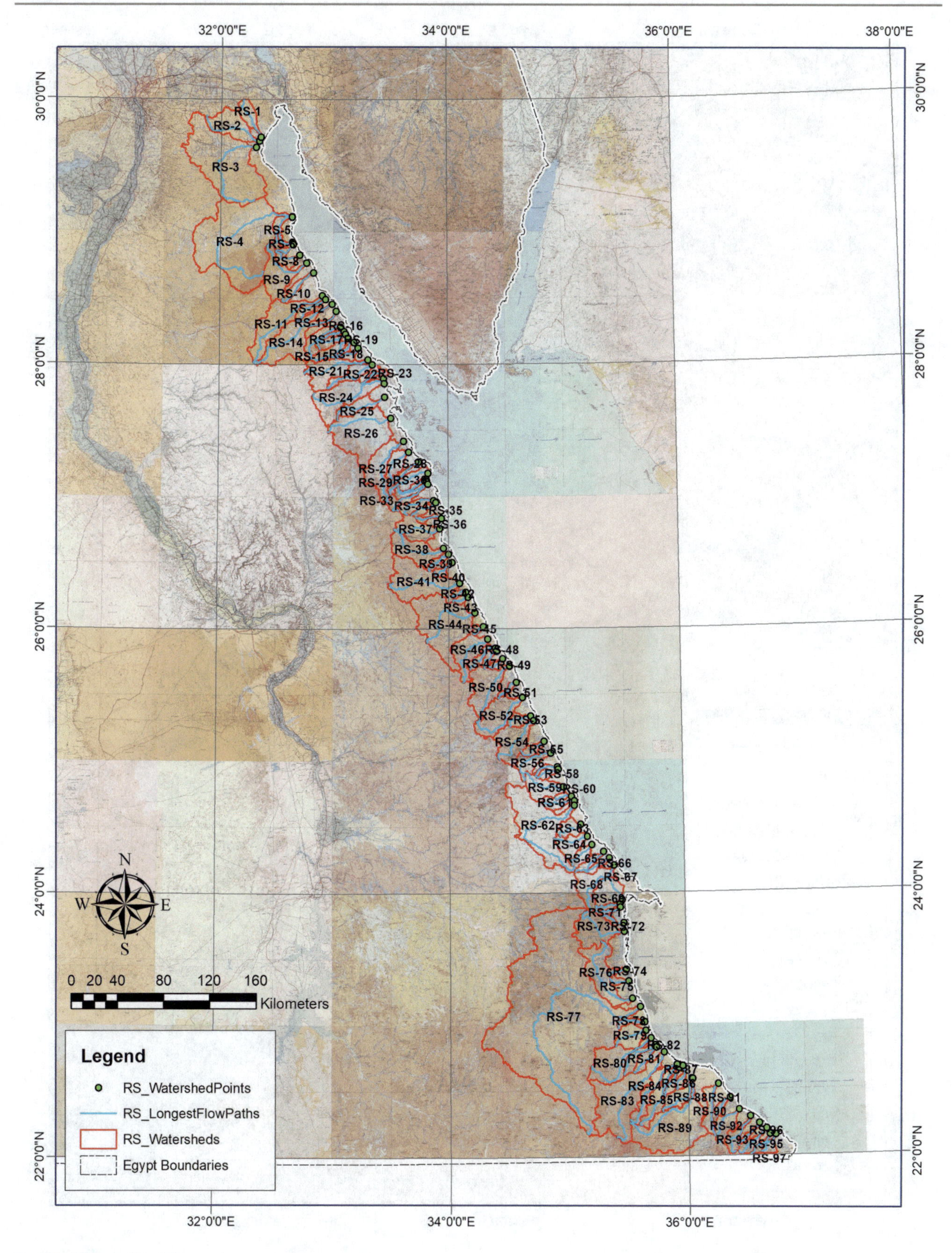

Fig. 31 Projected Red Sea catchments on topo maps

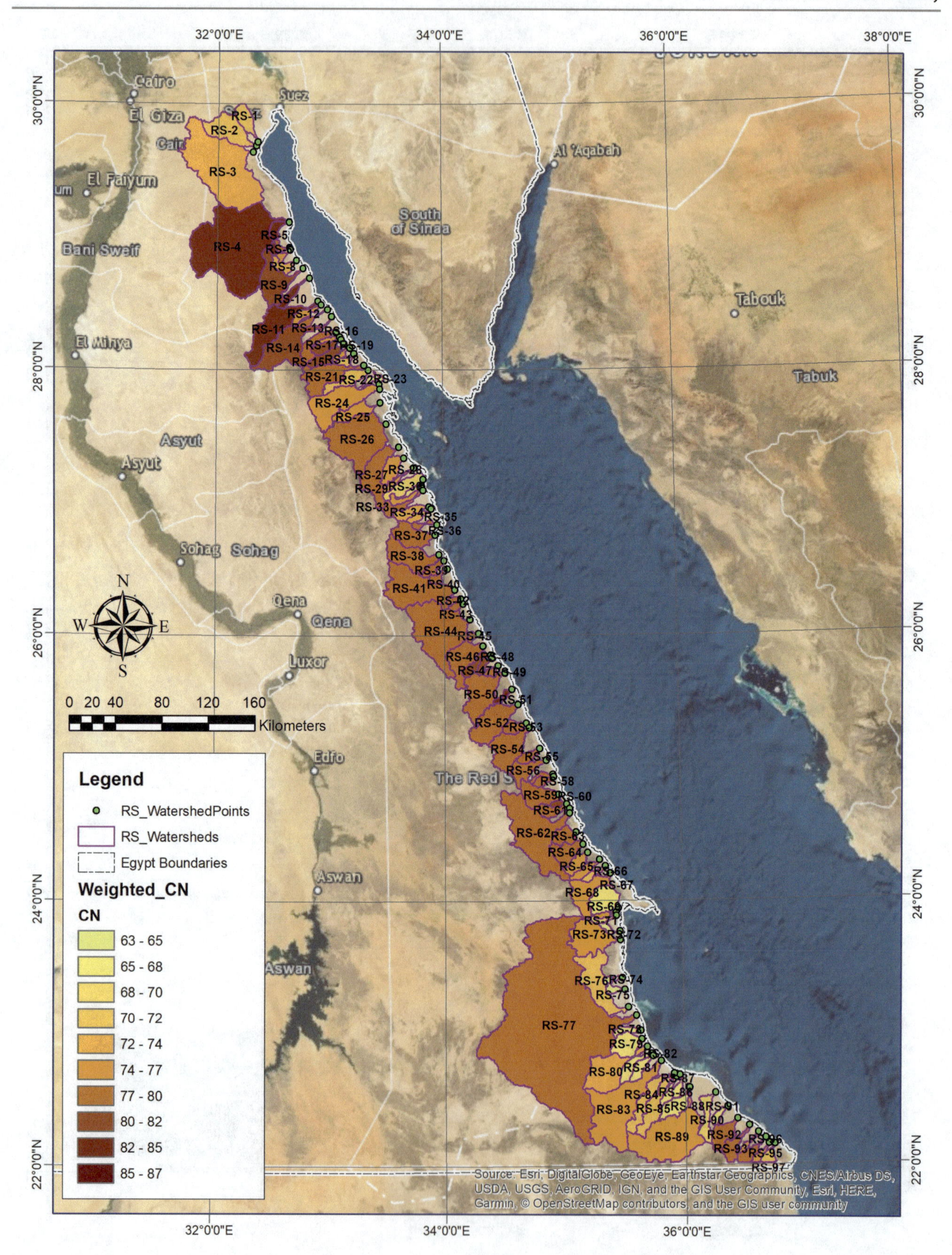

Fig. 32 Red Sea catchments weighted average CN

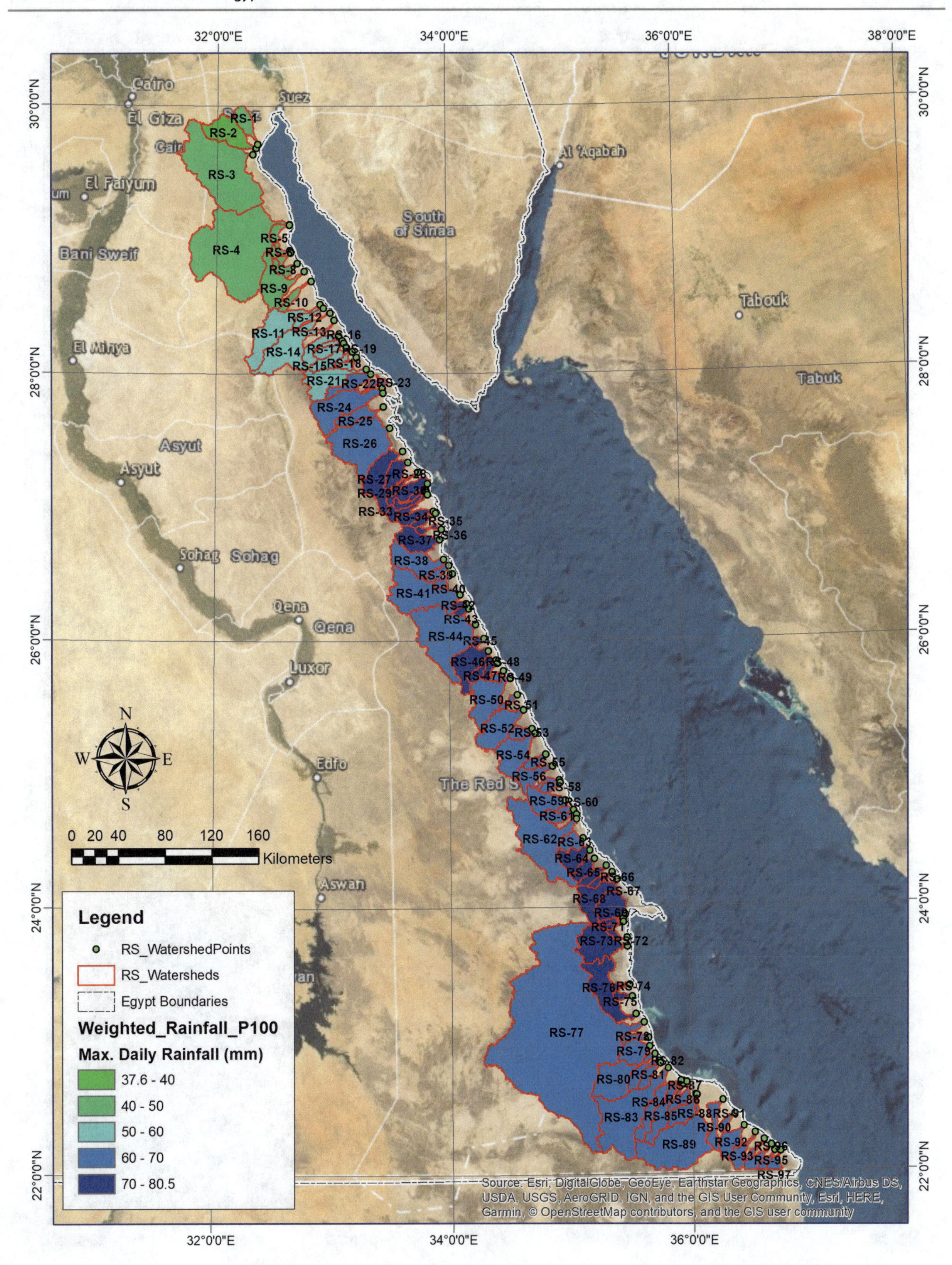

Fig. 33 Red Sea catchments weighted average 1-in-100 maximum daily precipitation

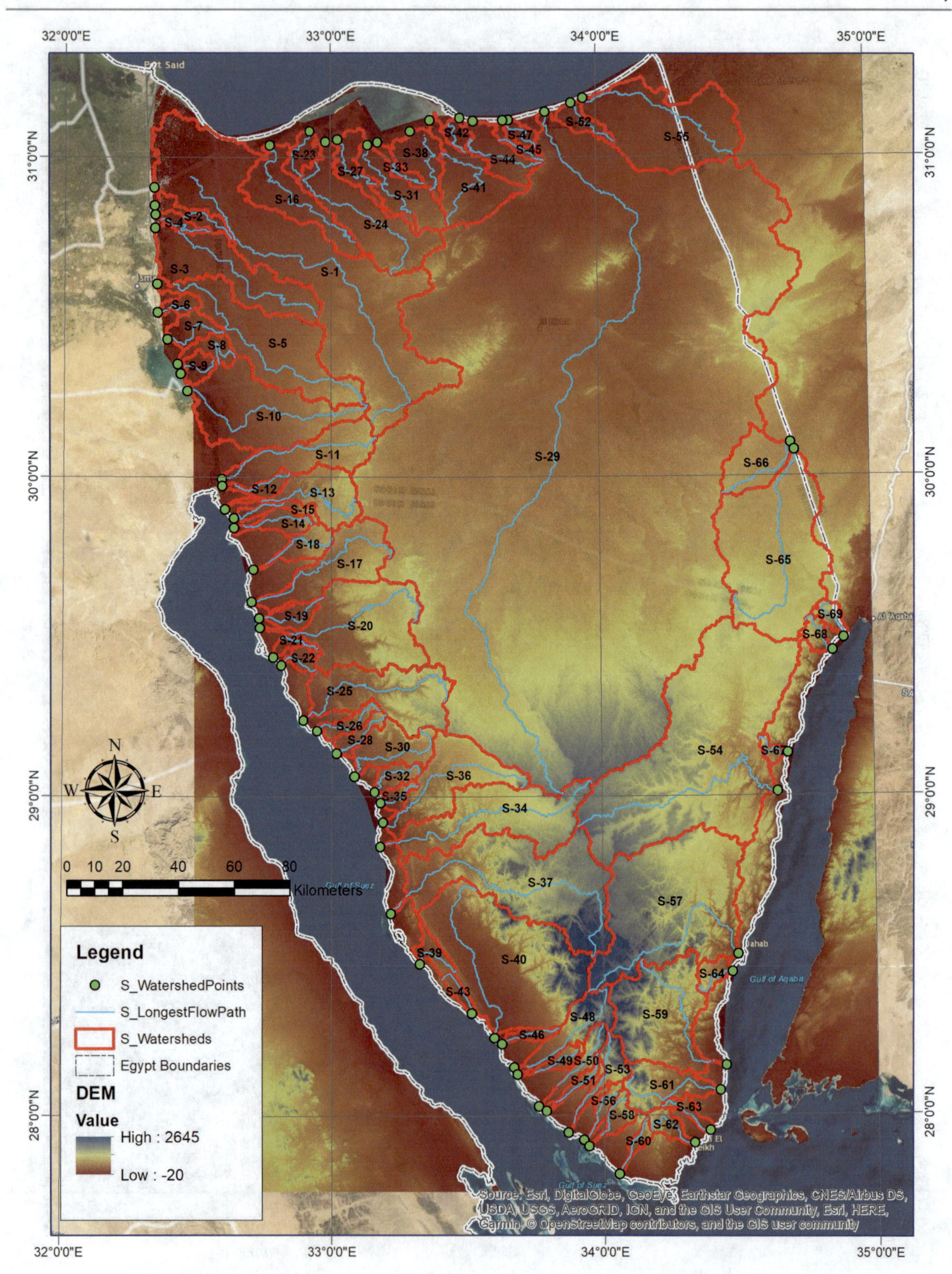

Fig. 34 Projected Sinai catchments on SRTM 90 × 90 DEM

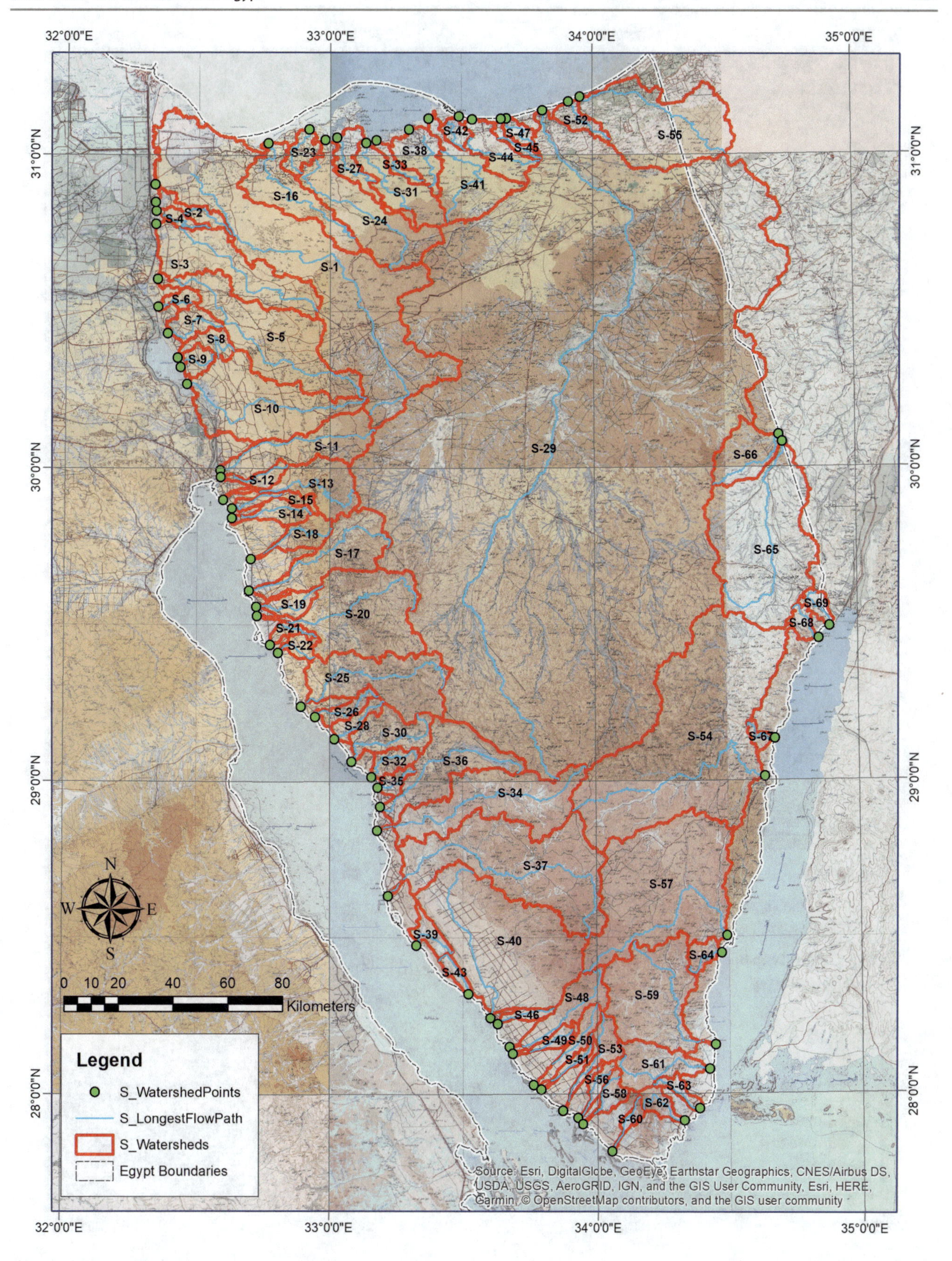

Fig. 35 Projected Sinai catchments on topo maps

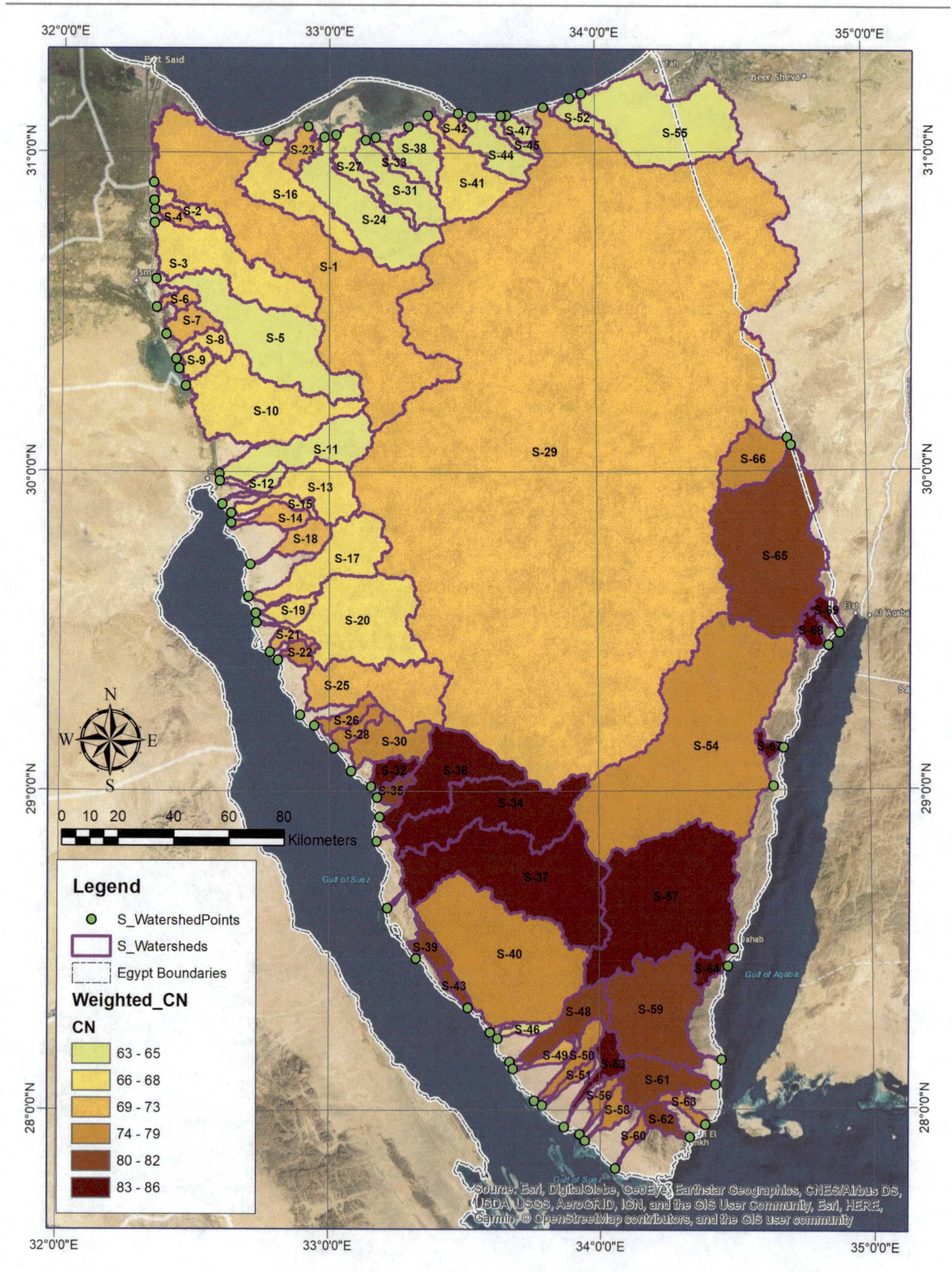

Fig. 36 Sinai catchments weighted average CN

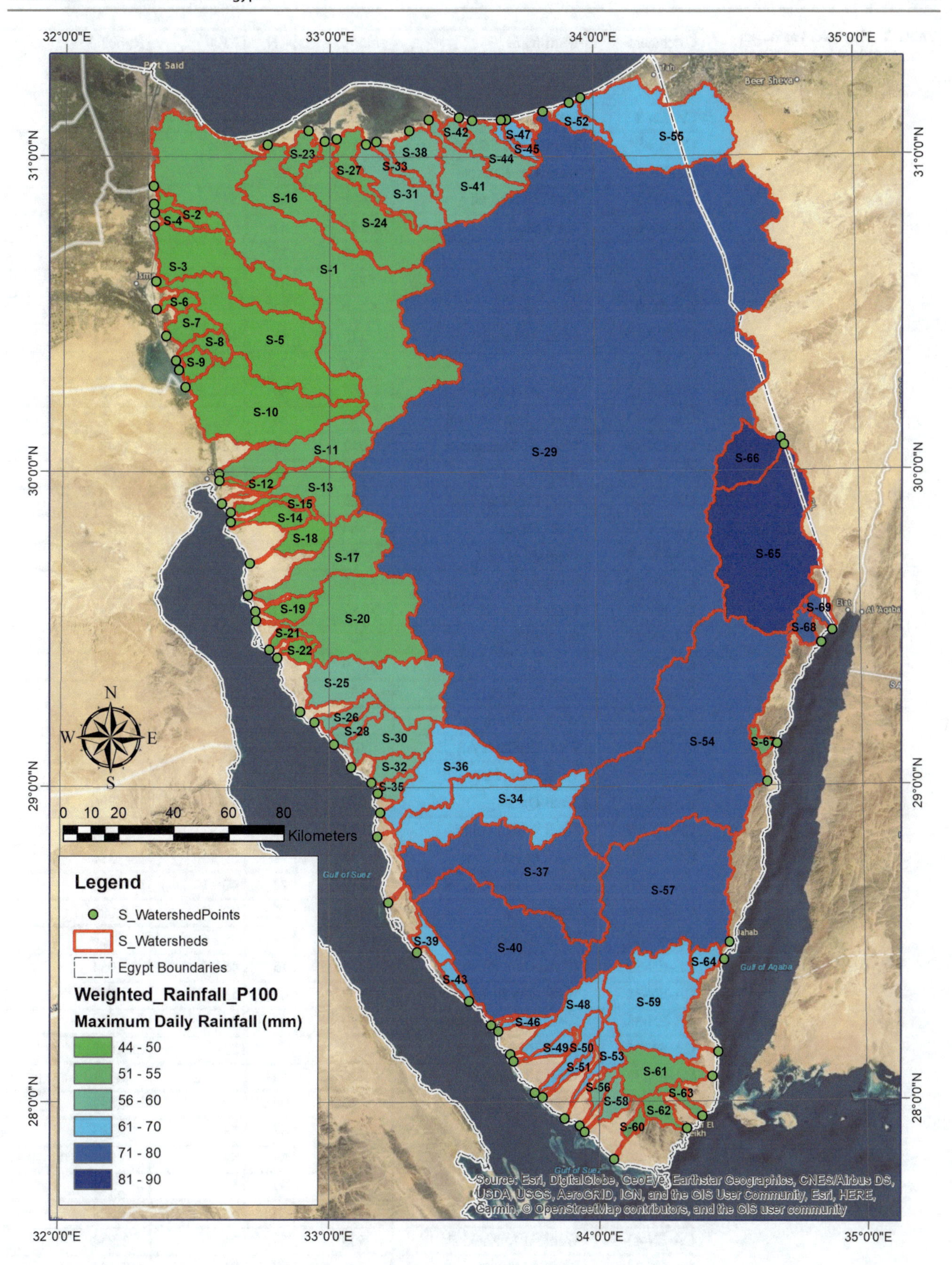

Fig. 37 Sinai catchments weighted average 1-in-100 maximum daily precipitation

Table 3 Nile river catchment characteristics

Catchment No.	Wadi name	Area (Km^2)	CN	P100 (mm)	Runoff Vol (1000 m^3)
CA-EN-1	Wadi Alakia	120	84	29	669
CA-EN-2	Wadi Abou Aggag	490	84	34	3,869
CA-EN-3	Wadi Abou Subaira	394	84	37	4,061
CA-EN-4	Wadi Um Roukba	163	79	41	1,284
CA-EN-5	Wadi Al Khurait	21,076	85	53	456,716
CA-EN-6	Wadi Sheit	7,156	82	52	126,771
CA-EN-7	Wadi AAied	129	78	43	1,119
CA-EN-8	Wadi Al Serag	822	81	45	9,224
CA-EN-9	Wadi Abbady	6,749	82	52	116,494
CA-EN-10	Wadi Al Domi	113	81	46	1,411
CA-EN-11	Wadi Hilal	407	79	45	4,242
CA-EN-12	Wadi Al Shouki	265	80	44	2,656
CA-EN-13	Wadi Abou Garawel	102	80	42	935
CA-EN-14	Wadi Al Madamoud	717	79	43	6,496
CA-EN-15	Wadi Banat Beri	101	81	43	1,085
CA-EN-16	Wadi Gabal Al Nazi	163	83	44	2,104
CA-EN-17	Wadi Hegaza	84	84	46	1,357
CA-EN-18	Wadi Al Hagiat	55	85	49	1,084
CA-EN-19	Wadi Al Qarn	7,210	83	56	151,972
CA-EN-20	Wadi El Sheikh Eida	81	83	56	1,734
CA-EN-21	Wadi El Seri	404	81	58	8,234
CA-EN-22	Wadi Qena	15,609	82	59	341,871
CA-EN-23	Wadi El Miat	131	69	57	1,060
CA-EN-24	Wadi El Shikh Omar	243	70	55	1,849
CA-EN-25	Wadi El Shikh Ali	106	66	54	507
CA-EN-26	Wadi Abou Nafoukh	1,238	70	51	7,730
CA-EN-27	Wadi Qasab	1,940	72	48	12,396
CA-EN-28	Wadi Awlad Amar	193	69	42	491
CA-EN-29	Wadi El Ahaywa	134	71	39	386
CA-EN-30	Wadi Abou Gelbana	117	72	40	397
CA-EN-31	Wadi Siflak	564	72	40	2,082
CA-EN-32	Wadi Al Galawya	173	71	38	466
CA-EN-33	Wadi Abou Shieh	1,766	75	45	13,054
CA-EN-34	Wadi emo El Kebly	78	67	53	395
CA-EN-35	Wadi eimo El Bahari	359	67	54	1,926
CA-EN-36	Wadi El-Asiouty	6,035	73	53	55,622
CA-EN-37	Wadi El Ibrahimi	234	68	59	1,801
CA-EN-38	Wadi El Gabarawy	322	69	60	2,897
CA-EN-39	Wadi Al Omrani	739	67	58	5,133
CA-EN-40	Wadi Abu Hasah	149	65	57	815
CA-EN-41	Wadi Alrashawy	110	65	56	574
CA-EN-42	Wadi El Maree	271	66	55	1,323
CA-EN-43	Wadi Al Mushakkak	500	64	54	1,932
CA-EN-44	Wadi Al Tahnawy	209	69	51	1,185

(continued)

Table 3 (continued)

Catchment No.	Wadi name	Area (Km^2)	CN	P100 (mm)	Runoff Vol (1000 m^3)
CA-EN-45	Wadi Garf El Deir	204	70	51	1,266
CA-EN-46	Wadi Sarirya	80	82	49	1,286
CA-EN-47	Wadi Al tourka	10,607	71	51	71,216
CA-EN-48	Wadi El Mohashm	199	65	47	532
CA-EN-49	Wadi Sharouna	152	75	47	1,153
CA-EN-50	Wadi El Sheikh	908	65	47	2,337
CA-EN-51	Wadi El Fakira	244	67	45	643
CA-EN-52	Wadi Ghayada	64	65	44	122
CA-EN-53	Wadi Sanour	6,241	70	46	27,987
CA-EN-54	Wadi Rood Ghorab	97	66	43	199
CA-EN-55	Wadi Bayad	121	66	43	258
CA-EN-56	Wadi Al Shoyab	327	69	43	1,066
CA-EN-57	Wadi Ramlya	498	70	43	1,710
CA-EN-58	Wadi Gabal Tarbool	107	87	43	1,847
CA-EN-59	Wadi El Atfihi	425	72	42	1,720
CA-EN-60	Wadi El Rashah	629	71	42	2,239
CA-EN-61	Wadi El Neomia	303	71	42	1,160
CA-EN-62	Wadi El Wadag	430	69	42	1,215
CA-EN-63	Wadi Um Ramath	210	77	45	1,732
CA-EN-64	Wadi Al Hira	267	72	44	1,277
CA-EN-65	Wadi Um Hassan	313	69	45	1,183
CA-EN-66	Wadi El Mahalawya	115	77	51	1,286
CA-EN-67	Wadi Houf	127	75	51	1,253
CA-EN-68	Wadi Degla	252	71	43	961
CA-WN-1	Wadi Al-Kotb	89	71	29	51
CA-WN-2	Wadi Al-Kobania	4,595	71	39	12,363
CA-WN-3	Wadi-Al-Kara	337	86	44	5,860
CA-WN-4	Wadi-Elhami	461	82	45	5,718
CA-WN-5	Wadi-Koum-Meir	290	72	45	1,558
CA-WN-6	Wadi-Abou-Aad	72	80	45	812
CA-WN-7	Wadi-Esna	722	69	45	2,781
CA-WN-8	Wadi-Al-Rokham	169	85	44	2,584
CA-WN-9	Wadi-El-Mahameed	369	83	45	5,168
CA-WN-10	–	304	75	46	2,249
CA-WN-11	–	110	68	46	366
CA-WN-12	–	138	70	54	1,055
CA-WN-13	–	372	69	51	2,093
CA-WN-14	–	181	68	51	902
CA-WN-15	–	268	68	47	972
CA-WN-16	–	521	67	47	1,795
CA-WN-17	–	247	70	48	1,200
CA-WN-18	–	92	69	48	427
CA-WN-19	–	2,497	67	47	8,499

(continued)

Table 3 (continued)

Catchment No.	Wadi name	Area (Km^2)	CN	P100 (mm)	Runoff Vol (1000 m^3)
CA-WN-20	–	80	66	48	243
CA-WN-21	Wadi-Samhoud	586	72	46	3,408
CA-WN-22	–	128	73	42	587
CA-WN-23	–	6,341	68	46	22,760
CA-WN-24	Wadi El-Yateem	221	67	40	353
CA-WN-25	Wadi-Tag-El-Deir	137	72	38	402
CA-WN-26	Wadi-Abou-Retag	167	70	37	331
CA-WN-27	Wadi-Juhaina	302	66	41	460
CA-WN-28	Wadi-Darb-El-Ghanayem	743	68	48	3,106
CA-WN-29	Wadi-Serga	913	70	56	7,705
CA-WN-30	–	89	64	65	651
CA-WN-31	–	405	73	66	6,328
CA-WN-32	–	71	77	65	1,355
CA-WN-33	–	2,696	70	60	26,814
CA-WN-34	–	947	78	54	12,901
CA-WN-35	–	92	81	51	1,377
CA-WN-36	–	88	82	50	1,387
CA-WN-37	–	262	77	49	2,781
CA-WN-38	–	268	73	49	1,976
CA-WN-39	–	268	71	47	1,426
CA-WN-40	–	486	68	46	1,658
CA-WN-41	–	166	69	46	700
CA-WN-42	–	212	69	46	832
CA-WN-43	–	262	72	45	1,366
CA-WN-44	–	96	73	44	529

Table 4 Red Sea catchment characteristics

Catchment no.	Wadi name	Area (Km^2)	CN	P100 (mm)	Runoff Vol (1000 m^3)
CA-RS-1	Wadi Hagoul	274	72	43	1,225
CA-RS-2	Wadi Bedaa	687	71	38	1,703
CA-RS-3	Wadi Ghoweba	2,882	73	40	11,505
CA-RS-4	Wadi Araba	3,910	84	45	57,134
CA-RS-5	Wadi Gabal Thalmat	183	84	46	2,773
CA-RS-6	Wadi El Garph	53	80	46	597
CA-RS-7	Wadi Al Beir	51	77	47	470
CA-RS-8	Wadi Abou Khalifi	150	74	48	1,102
CA-RS-9	Wadi Al Dahl	741	82	48	10,828
CA-RS-10	Wadi North Wadi Houshya	148	84	50	2,742
CA-RS-11	Wadi Hawashia North	1,020	82	51	17,604
CA-RS-12	Wadi Hawashia South	158	76	52	1,750
CA-RS-13	Wadi West Bakr Wells	99	79	53	1,479

(continued)

Table 4 (continued)

Catchment no.	Wadi name	Area (Km^2)	CN	P100 (mm)	Runoff Vol (1000 m^3)
CA-RS-14	Wadi Abou Had	1,048	82	54	18,862
CA-RS-15	Wadi EL Darb	279	79	55	4,470
CA-RS-16	Wadi Hareem	66	77	56	931
CA-RS-17	Wadi Um Yousr	170	81	56	3,161
CA-RS-18	Wadi Ghareb	247	80	57	4,602
CA-RS-19	Wadi Kharm El Oyoun	62	75	59	846
CA-RS-20	Wadi Garph	80	72	60	911
CA-RS-21	Wadi North Wadi Dara	723	80	59	14,026
CA-RS-22	Wadi Dara	313	71	62	3,738
CA-RS-23	Wadi Noth Wadi Dob	92	64	65	683
CA-RS-24	Wadi Dob	1,045	77	63	18,635
CA-RS-25	Wadi Abou Had (hurgada)	325	74	66	5,687
CA-RS-26	Wadi Malaha	1,659	78	70	40,612
CA-RS-27	Wadi Biali	735	79	74	21,159
CA-RS-28	Wadi Kharaza	282	74	78	7,014
CA-RS-29	Wadi Abou Malaka	104	74	78	2,549
CA-RS-30	Wadi Falek Al Sahl	318	69	79	5,863
CA-RS-31	Wadi Um Dalfa	65	65	79	922
CA-RS-32	Wadi Abou Eid	127	74	77	3,048
CA-RS-33	Wadi Um Gudari	196	78	74	5,426
CA-RS-34	Wadi Um Kbash	213	74	74	4,559
CA-RS-35	Wadi El Mamal	100	72	74	1,967
CA-RS-36	Wadi Al Mowasala	70	78	72	1,831
CA-RS-37	Wadi Al Baroud	506	80	70	13,449
CA-RS-38	Wadi Safaga	716	79	67	17,069
CA-RS-39	Wadi Gasous	137	80	69	3,586
CA-RS-40	Wadi Abou Shoukaili	100	78	70	2,468
CA-RS-41	Wadi Al Kareeh	1,390	78	67	31,089
CA-RS-42	Wadi Abou Oumra	62	77	73	1,600
CA-RS-43	Wadi El Hadadeen	60	78	74	1,635
CA-RS-44	Wadi El Nakheel	1,906	78	70	47,098
CA-RS-45	Wadi El Zarib	55	78	75	1,537
CA-RS-46	Wadi Esl	631	79	71	16,245
CA-RS-47	Wadi Sharm El Bahari	176	80	71	4,901
CA-RS-48	Wadi Sharm El Kebly	92	80	71	2,621
CA-RS-49	Wadi Wazar	67	79	70	1,688
CA-RS-50	Wadi Um Lasifa	841	80	66	20,330
CA-RS-51	Wadi Um Grifi	71	84	66	2,186
CA-RS-52	Wadi Mubarak	785	79	63	16,187
CA-RS-53	Wadi Abou Dabbab	165	77	63	3,082
CA-RS-54	Wadi El Nabe	742	78	61	13,631
CA-RS-55	Wadi Egla	134	79	62	2,736

(continued)

Table 4 (continued)

Catchment no.	Wadi name	Area (Km^2)	CN	P100 (mm)	Runoff Vol (1000 m^3)
CA-RS-56	Wadi Um Harika	315	78	62	5,856
CA-RS-57	Wadi Um Tandia	66	79	63	1,341
CA-RS-58	Wadi El ambaout	95	78	63	1,832
CA-RS-59	Wadi Ghadir	486	78	63	9,782
CA-RS-60	Wadi Sharm Al Foukairy	56	81	66	1,396
CA-RS-61	Wadi Arear	195	77	65	4,019
CA-RS-62	Wadi Al Gemal	1,951	79	65	44,347
CA-RS-63	Wadi Um Al Abs	236	78	70	5,596
CA-RS-64	Wadi Abou Ghousoun	368	78	72	9,177
CA-RS-65	Wadi Al Renga	210	77	75	5,444
CA-RS-66	Wadi El Rada	157	75	77	3,799
CA-RS-67	Wadi Al Khasheer	88	71	78	1,794
CA-RS-68	Wadi Lahmi	592	74	78	14,594
CA-RS-69	Wadi Naaeet	484	68	81	8,662
CA-RS-70	Wadi Um Selem	62	75	80	1,673
CA-RS-71	Wadi Kalalat	79	69	80	1,503
CA-RS-72	Wadi Kntroub	62	64	80	826
CA-RS-73	Wadi Khada	836	75	77	21,352
CA-RS-74	Wadi Marafai	63	65	75	742
CA-RS-75	Wadi Klibtab	77	63	74	773
CA-RS-76	Wadi El Rahba	950	71	73	16,804
CA-RS-77	Wadi Houdein	11,577	80	64	265,779
CA-RS-78	Wadi Al Wadah	111	66	69	1,194
CA-RS-79	Wadi Safira	391	68	68	4,587
CA-RS-80	Wadi Sab	906	73	65	13,971
CA-RS-81	Wadi Amrawy El Bahary	225	68	66	2,447
CA-RS-82	Wadi Amrawy El Kebly	180	68	66	1,973
CA-RS-83	Wadi Eib	1,969	75	63	32,262
CA-RS-84	Wadi Maysa-2	465	71	64	6,014
CA-RS-85	Wad Andri	479	71	64	5,765
CA-RS-86	Wadi Maysa-1	118	68	65	1,216
CA-RS-87	Wadi Ramram	92	68	65	965
CA-RS-88	Wadi Halal Rahandeeb	318	68	64	3,189
CA-RS-89	Wadi Deib	1,925	75	62	30,737
CA-RS-90	Wadi Daeit	301	70	63	3,372
CA-RS-91	Wadi Bashoya	95	68	64	940
CA-RS-92	Wadi Yowayder	352	78	63	6,768
CA-RS-93	Wadi Sarmatai	243	78	62	4,723
CA-RS-94	Wadi Merakwan	72	76	62	1,191
CA-RS-95	Wadi Shallal	180	77	62	3,263
CA-RS-96	Wadi Aklahok	82	76	62	1,323
CA-RS-97	Wadi Ay Kawan	111	70	62	1,231

Table 5 Sinai catchment characteristics

Catchment no.	Wadi name	Area (Km^2)	CN	P100 (mm)	Runoff Vol (1000 m^3)
CA-S-1	Wadi-Abou Amer	3,446	68	51	18,224
CA-S-2	Wadi-El Kantara Sharq-1	137	67	46	401
CA-S-3	Wadi-Mardoum	870	65	47	2,359
CA-S-4	Wadi-El Kantara Sharq-2	68	68	45	225
CA-S-5	Wadi-Elwa	1,199	65	49	3,282
CA-S-6	Wadi-Talia	71	73	44	384
CA-S-7	Wadi-Khoubaita	156	71	45	705
CA-S-8	Wadi-El-Habashi-North	122	66	46	321
CA-S-9	Wadi-El-Habashi-South	93	68	46	298
CA-S-10	Wadi-El Gadi	1,120	66	49	3,584
CA-S-11	Wadi-El Mor-1	607	64	51	1,805
CA-S-12	Wadi-El Mor-2	93	64	49	216
CA-S-13	Wadi-El Raha	442	68	51	2,099
CA-S-14	Wadi-El Rabina	150	68	49	630
CA-S-15	Wadi-Um Asagil	104	67	49	381
CA-S-16	Wadi-Um Okba	555	67	50	2,264
CA-S-17	Wadi-Sedr	631	67	51	2,664
CA-S-18	Wadi-Lhata	178	71	49	1,094
CA-S-19	Wadi-Roud El Raha	129	65	47	341
CA-S-20	Wadi-Werdan	1,148	68	54	6,468
CA-S-21	Wadi-Abou Hagar North	63	72	48	382
CA-S-22	Wadi-Abou Hagar South	81	78	49	931
CA-S-23	Wadi-El Kantara	134	71	51	897
CA-S-24	Wadi-El Beada	838	64	54	3,187
CA-S-25	Wadi Gharandal	861	69	57	7,002
CA-S-26	Wadi-Waset	113	76	56	1,499
CA-S-27	Wadi-El Sadat	221	65	54	915
CA-S-28	Wadi-Tal	108	74	57	1,248
CA-S-29	Wadi-El Arish	23,669	70	73	386,253
CA-S-30	Wadi-Taiba	357	77	58	5,788
CA-S-31	Wadi-El Abd	394	64	55	1,616
CA-S-32	Wadi-Nakhl	121	84	58	2,871
CA-S-33	Wadi-Musafak	75	64	55	298
CA-S-34	Wadi-Sedri	1,074	84	68	33,789
CA-S-35	Wadi-Defri	74	81	58	1,541
CA-S-36	Wadi-Bobo	718	83	62	18,613
CA-S-37	Wadi-Firan	1,780	83	73	62,523
CA-S-38	Wad-El Artah	262	63	56	1,092
CA-S-39	Wadi-Araba North	115	81	64	2,763
CA-S-40	Wadi-El Awag	1,918	76	70	41,726
CA-S-41	Wadi-El Nakhl	551	65	59	3,400

(continued)

Table 5 (continued)

Catchment no.	Wadi name	Area (Km^2)	CN	P100 (mm)	Runoff Vol (1000 m^3)
CA-S-42	Wadi-El Zaranique	76	65	58	435
CA-S-43	Wadi-Araba South'	65	80	66	1,551
CA-S-44	Wadi-El Gemal	284	63	60	1,497
CA-S-45	Wadi-El Medawara	67	63	61	353
CA-S-46	Wadi-El-Malaha North	66	64	68	572
CA-S-47	Wadi-Darb El Masaeed	75	63	61	394
CA-S-48	Wadi-Sulai	335	79	69	8,491
CA-S-49	Wadi-Abou Garph	74	68	66	839
CA-S-50	Wadi-Thaman	156	78	66	3,405
CA-S-51	Wadi-El Masein	62	73	64	906
CA-S-52	Wadi-Beir Abi Hani	123	63	63	730
CA-S-53	Wadi-El Raboud	166	82	62	4,092
CA-S-54	Wadi-Watir	3,513	78	73	91,291
CA-S-55	Wadi-Beir El Kharouba	1,238	64	65	8,677
CA-S-56	Wadi-Mukhairet 1	75	74	60	1,032
CA-S-57	Wadi-Dahab	2,066	85	76	83,634
CA-S-58	Wadi-Mukhairet 2	107	76	57	1,462
CA-S-59	Wadi-Keid	1,041	81	66	26,350
CA-S-60	Wadi-El Aat El Gharbi	57	78	54	824
CA-S-61	Wadi-Um Adawi	362	80	55	6,166
CA-S-62	Wadi-El Aat El Sharki	108	80	51	1,501
CA-S-63	Wadi-Um Tartir	79	73	51	644
CA-S-64	Wadi-El Atshan	98	86	67	3,496
CA-S-65	Wadi-Al-Garafi	1,824	81	90	82,129
CA-S-66	Wadi-Abou-Khadakhed	369	79	88	14,387
CA-S-67	Wadi-El Malha	52	85	53	1,141
CA-S-68	Wadi-El Abiad	84	84	76	3,148
CA-S-69	Wadi-El Harara	84	85	77	3,411

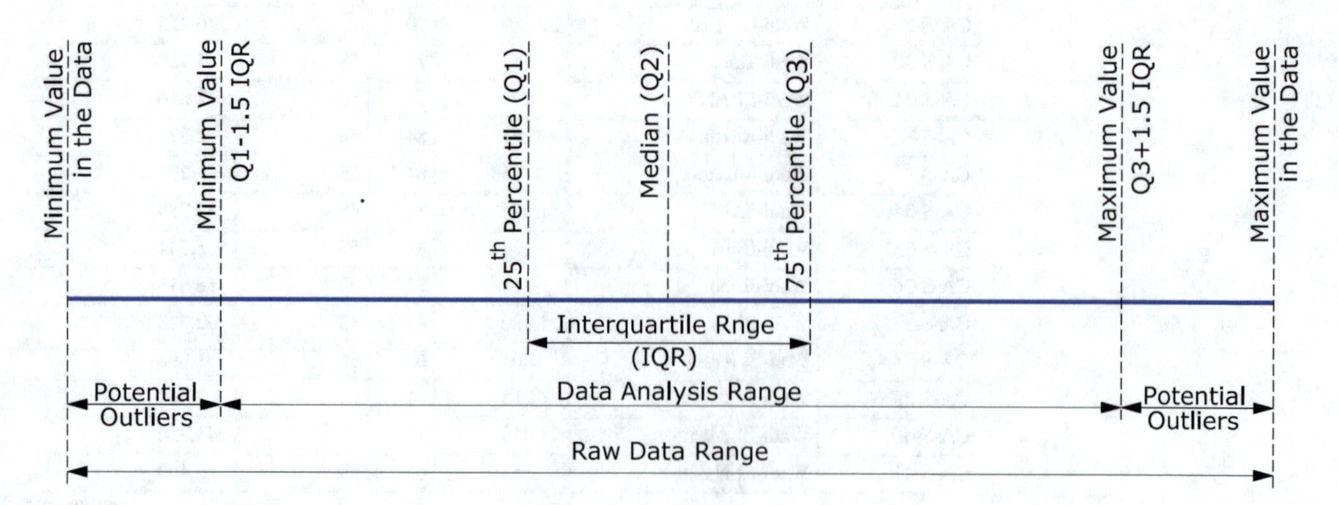

Fig. 38 Box plot concept

Table 6 Box plot limits

Limits	Runoff volume (1000 m^3)	Discharge (m^3/s)
Nile region		
Q1	885	15
Q2	1399	27
Q3	3523	61
IQR	2638	46
Max limit	7481	130
Red Sea region		
Q1	1600	41
Q2	3372	110
Q3	10828	220
IQR	9228	179
Max limit	24671	489
Sinai region		
Q1	730	15
Q2	1551	42
Q3	4092	150
IQR	3362	135
Max limit	9135	352

10 Conclusions

In Egypt, urbanized areas, and associated assets (e.g., agricultural lands, connecting roads, electrical transmission lines) are located at the catchments outfall points in addition to transversal highways. Some of the major and frequent incidents of flooding were recorded with a brief analysis of the causes.

Catchments were delineated using SRTM 90 × 90 DEM data and against versus 1:250,000 topo maps. The threshold for the delineation was set to 50 km^2. All morphological catchments' characteristics were extracted using ARC-GIS. Previously generated global Curve Number (CN) shapefile for Egypt was utilized to extract the weighted average CN each catchment. The 1-in-100 maximum daily precipitation from the sparse meteorological station was used to generate an isohyetal map of the study area to obtain the weighted average of 100 years' maximum daily precipitating for each catchment.

SCS method was chosen to calculate the peak discharge and runoff volume. The time of concentration modified from Kirpich equation that combines the effect of the CN was used to obtain the lag time. 6 h SCS type II storm distribution was selected to distribute the 1-in-100 maximum daily precipitation.

The peak discharge and total runoff volume for each catchment were calculated and standardized to calculate Peak Flow Standardized Risk Factor (PFSRF) and Runoff Volume Standardized Risk Factor (RVSRF). Due to the wide range of peak flow and runoff volume, the potential outliers obtained by the quartile technique were ignored during the calculations of the standardized risk factors.

Finally, the catchments were categorized and arranged based on five risk categories (very high, high, moderate, low to moderate, and low) for PFSRF and RVSRF, which prioritize the studies required for flood mitigation measures and show stormwater harvesting potentials.

Two-dimensional HEC-RAS rainfall-runoff model was conducted at Ras-Gharib area using 30 × 30 DEM. The DEM files could not capture the artificial manmade road of Ras-Gharib El-Sheikh-Fadl on the flow directions. The DEM file has been updated based on the available survey data of the road. The flood plain, flow depths, and velocities were obtained, and accordingly, the flood intensities were calculated. The model was verified against aerial photos of the 2016 incident for all stream affecting Ras-Gharib city. The effect of the newly constructed culvert (16 vent 3 m × 3 m) has been checked and the results showed a significant reduction in the flow intensities within the urban areas of Ras-Gharib city.

11 Recommendations

- The low-risk catchments cannot be considered as a safe catchment, but it has a lower priority for detailed assessment.
- The treatment of the locations with recorded incidents and providing proper flood mitigation measures are of top priority, even more than the high-risk catchments.
- Based on the runoff volume risk assessment, a priority should be given to stormwater harvesting projects in

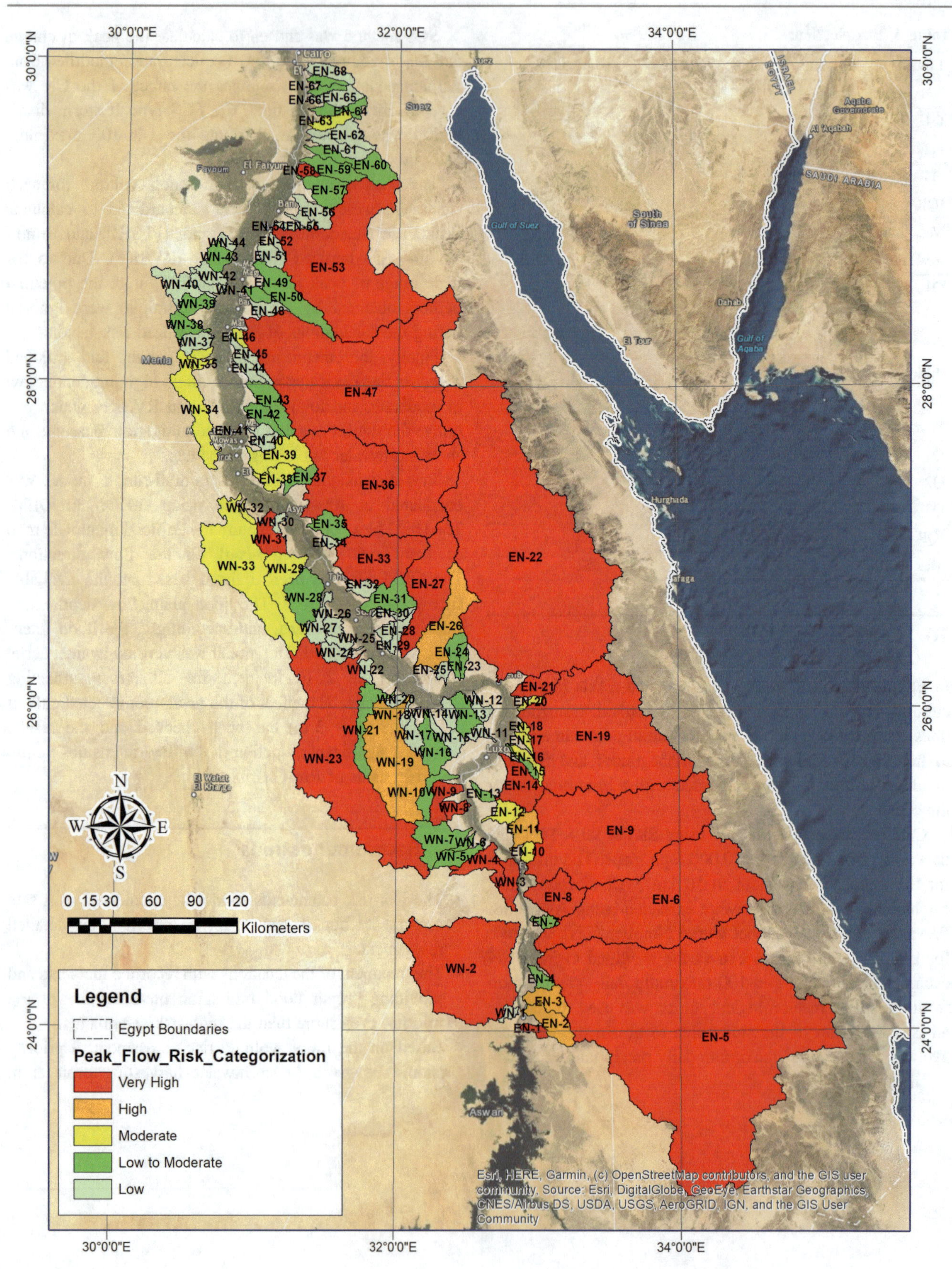

Fig. 39 Nile region catchments categorization as per PFSRF

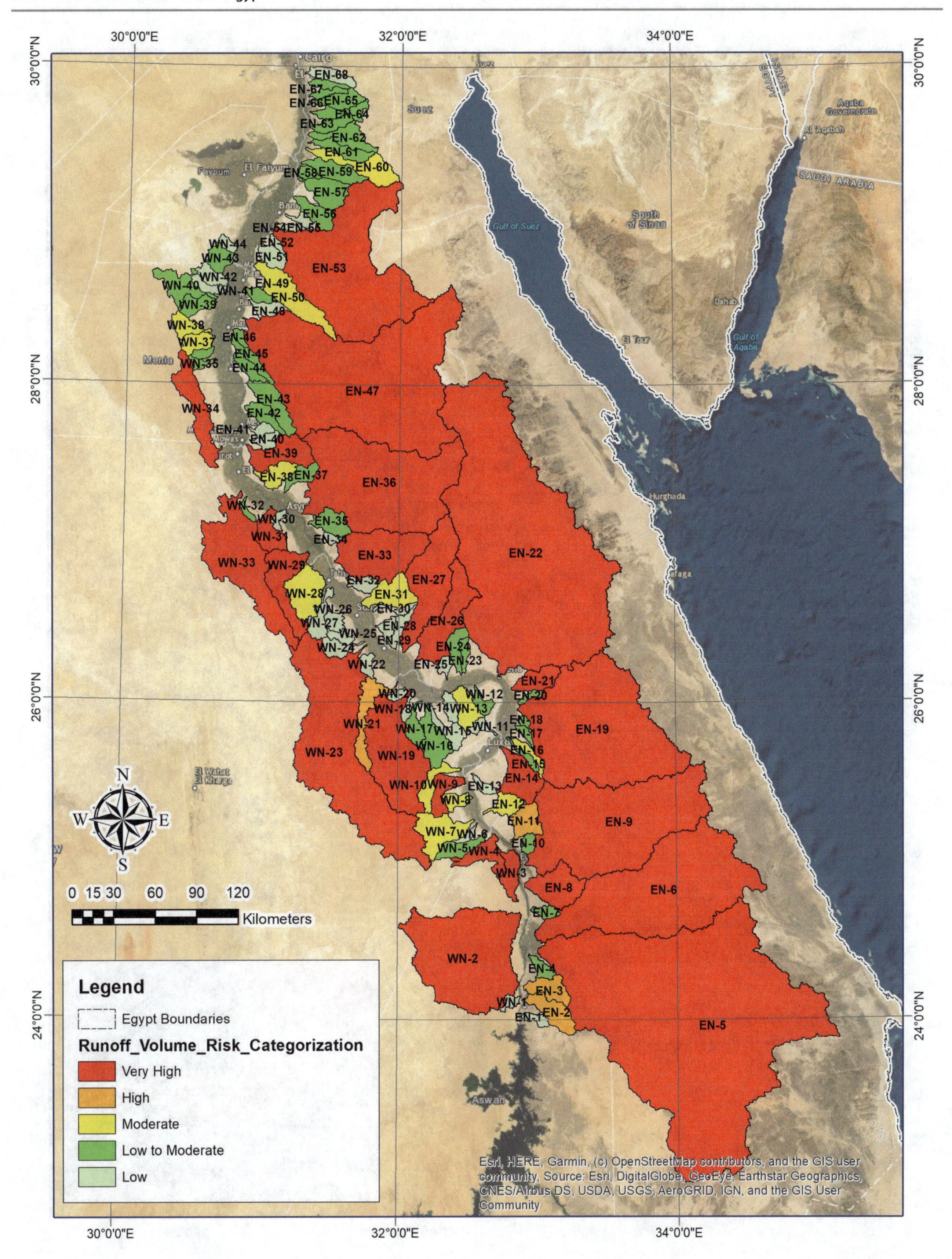

Fig. 40 Nile region catchments categorization as per RVSRF

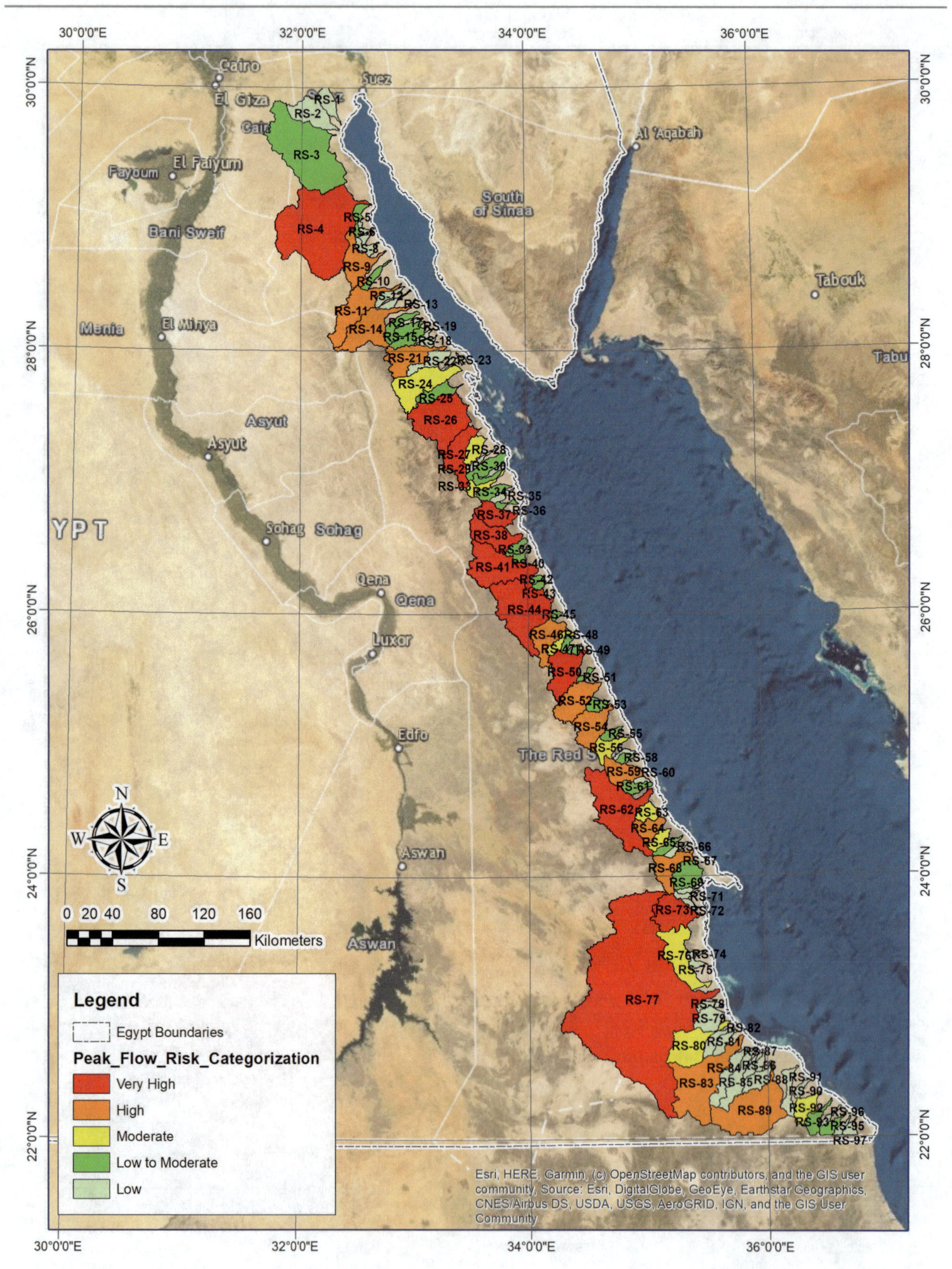

Fig. 41 Red Sea region catchments categorization as per PFSRF

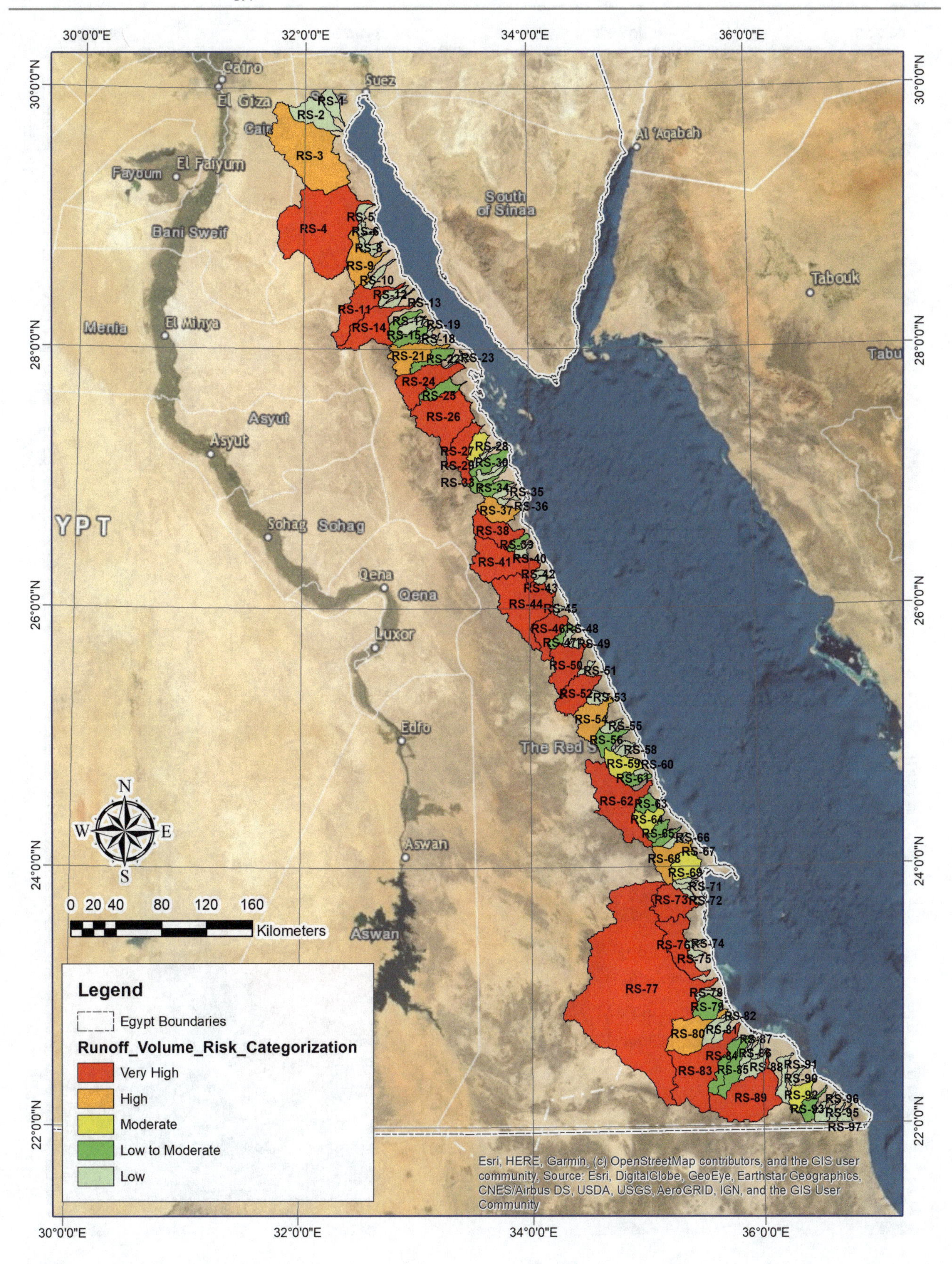

Fig. 42 Red Sea region catchments categorization as per RVSRF

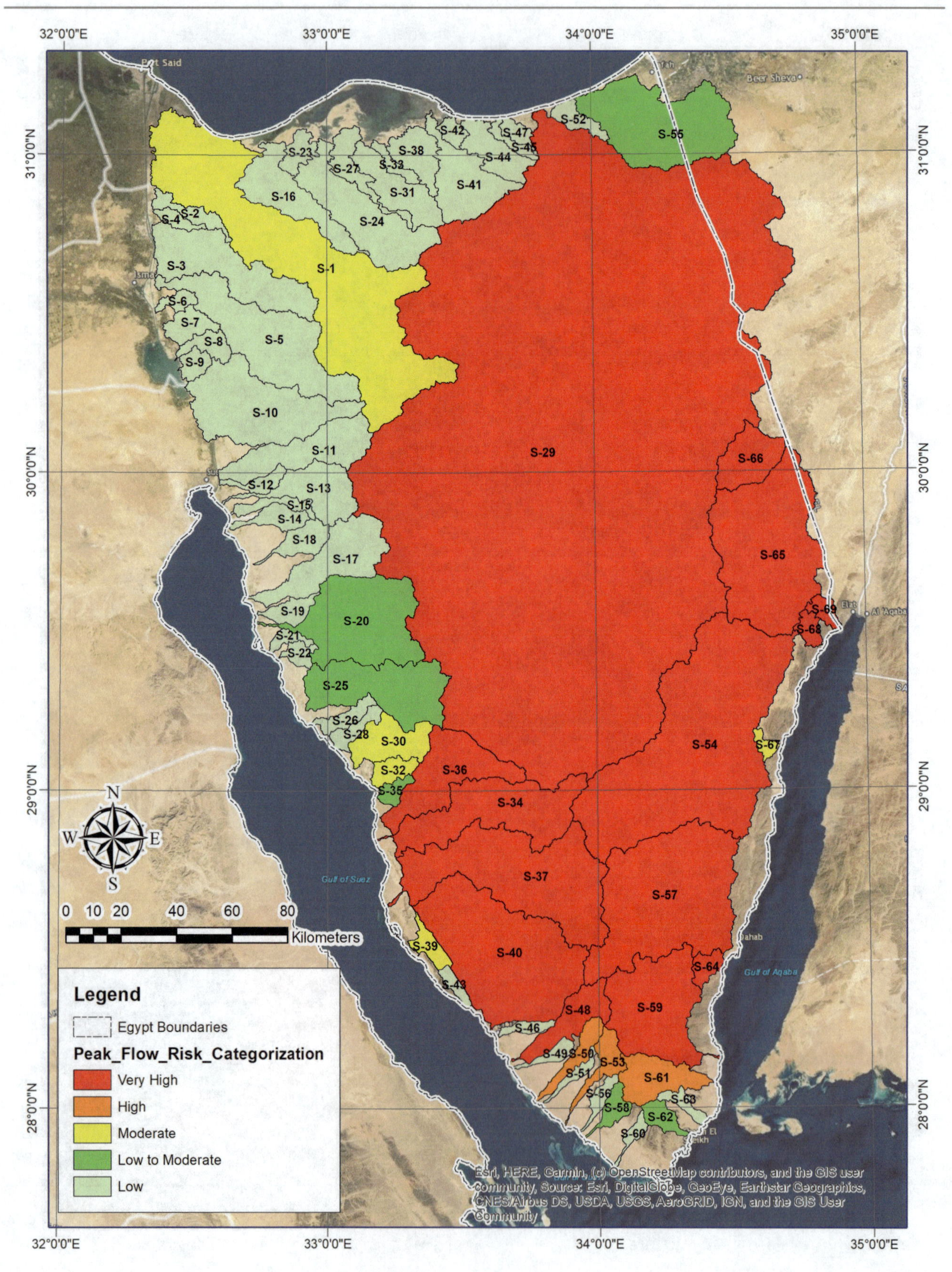

Fig. 43 Sinai region catchments categorization as per PFSRF

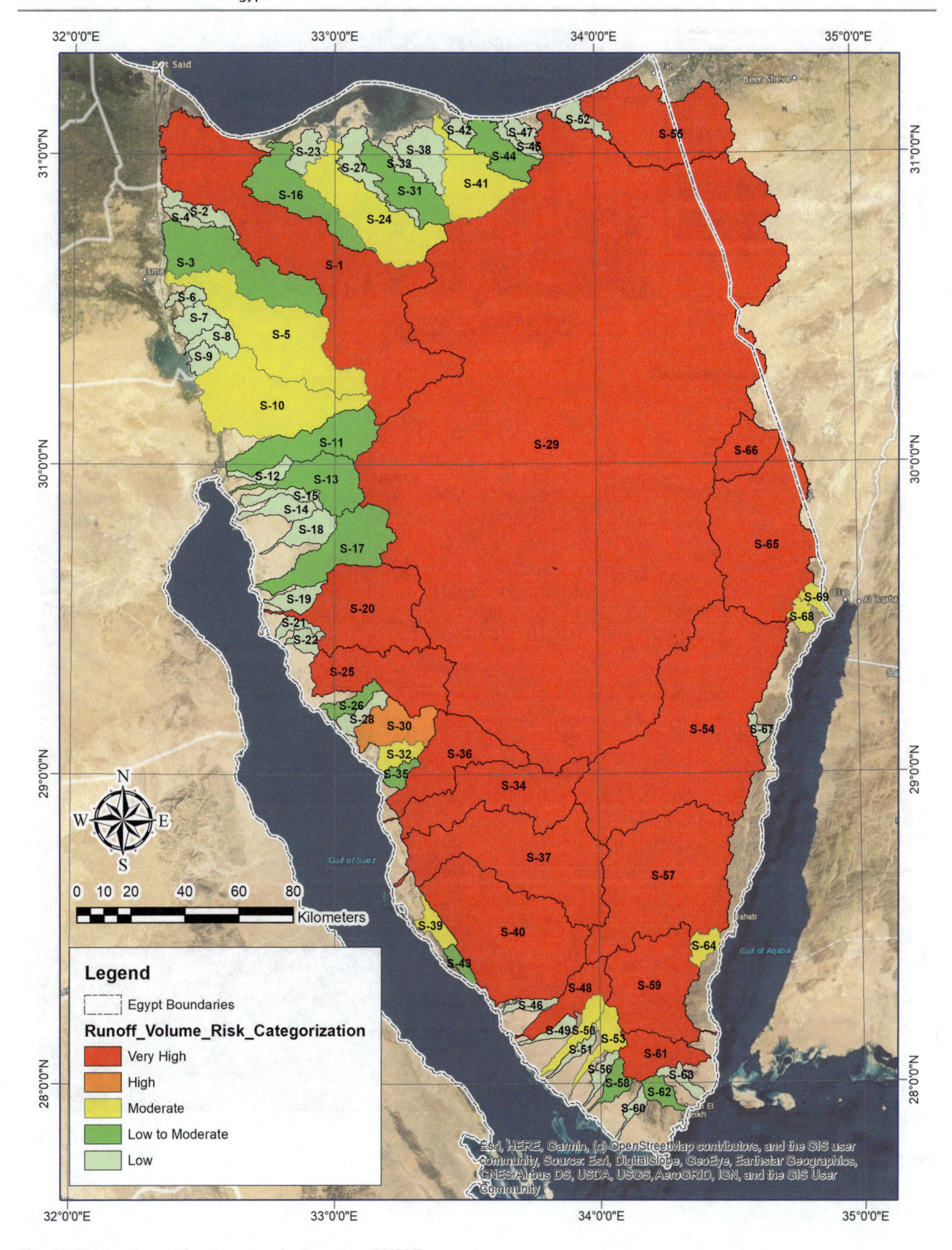

Fig. 44 Sinai region catchments categorization as per RVSRF

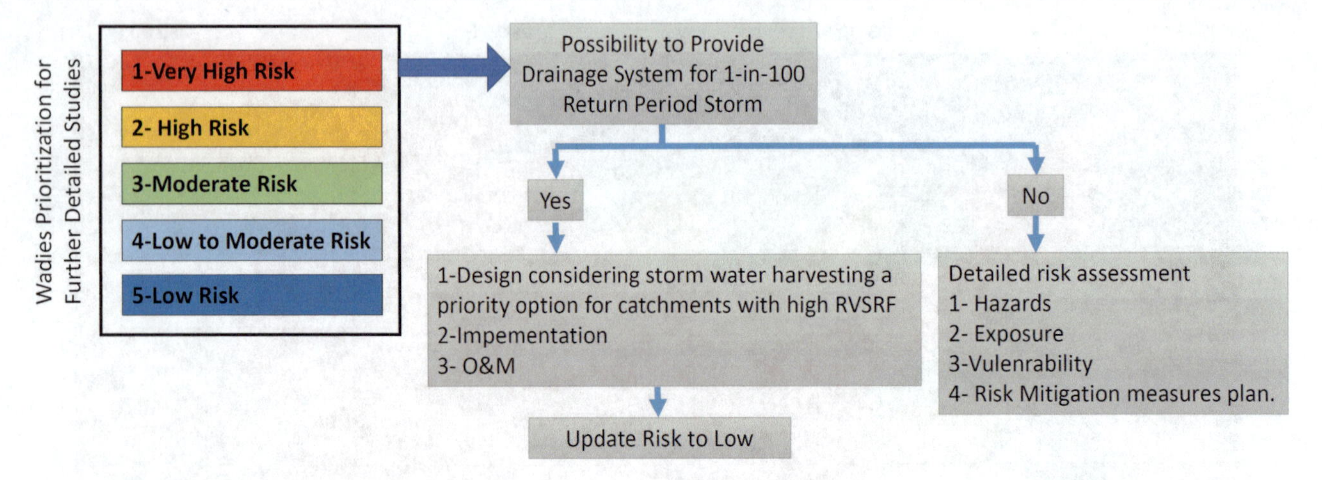

Fig. 45 Workflow chart for risk assessment

Fig. 46 Wadi Abou Had catchment on the satellite image

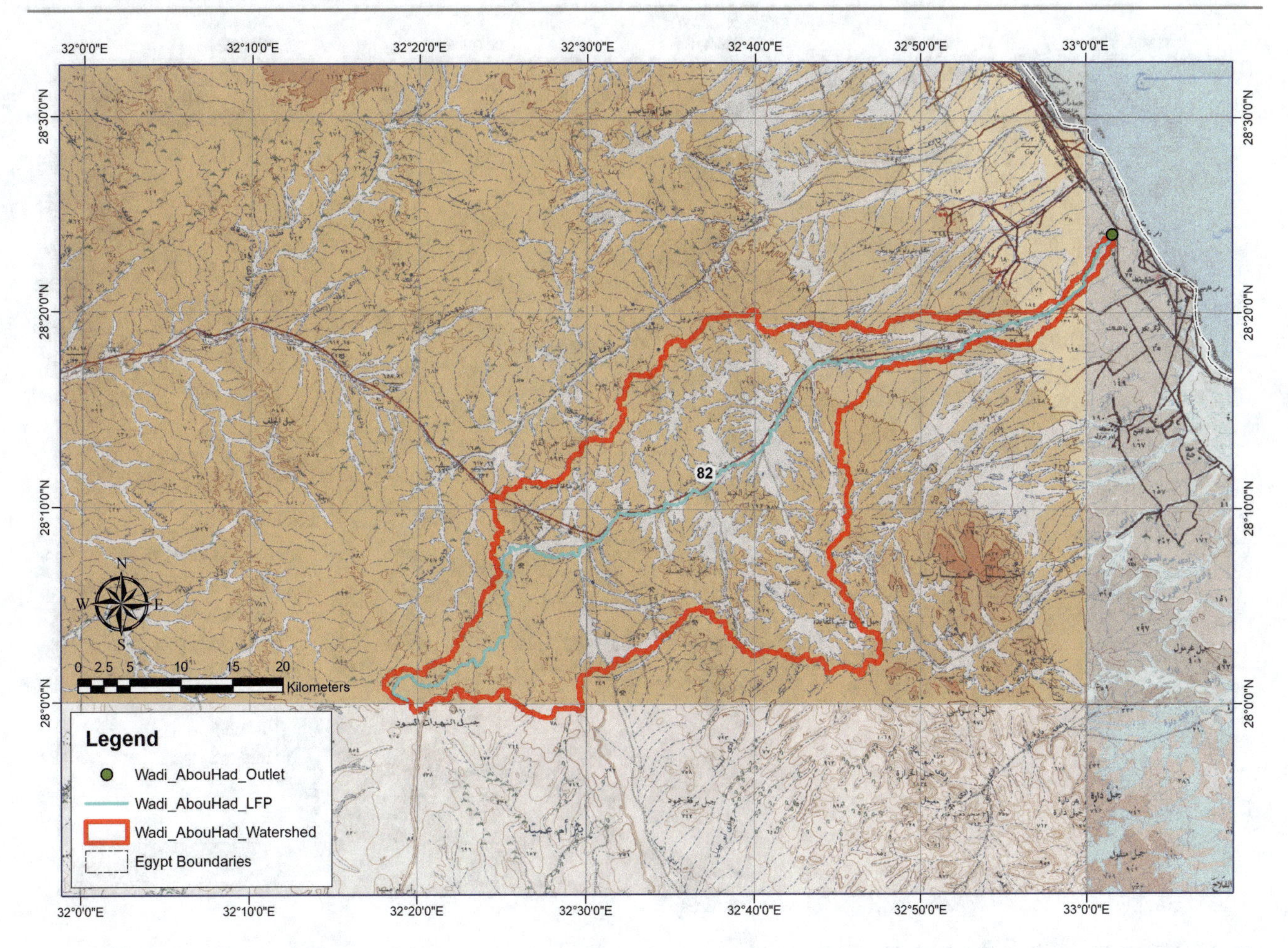

Fig. 47 Wadi Abou Had catchment on topo maps

order to maximize the return of the investment of flood mitigation measures as a step in solving the freshwater stress in Egypt. The risk of dams' construction should be considered, and the mitigation of providing the dams with spillways designed for a higher return period is mandatory. As a common practice and cost-benefit analysis, the dams' height can be designed to store the 1-in-10 years storm and the spillway design for passing 1-in-200 years' storm.

- Speeding the issuance of the Egyptian code for flood mitigation and storm drainage is a top priority task.
- A committee of the Ministry of Water Resources and Irrigation, Universities irrigation and Hydraulics departments, National Water Research Center, Egyptian Meteorological Authority, National Authority of remote sensing and space science, Water Resources Research Institute, and the National Research Center Department of Geological Science should be formulated. The task of this

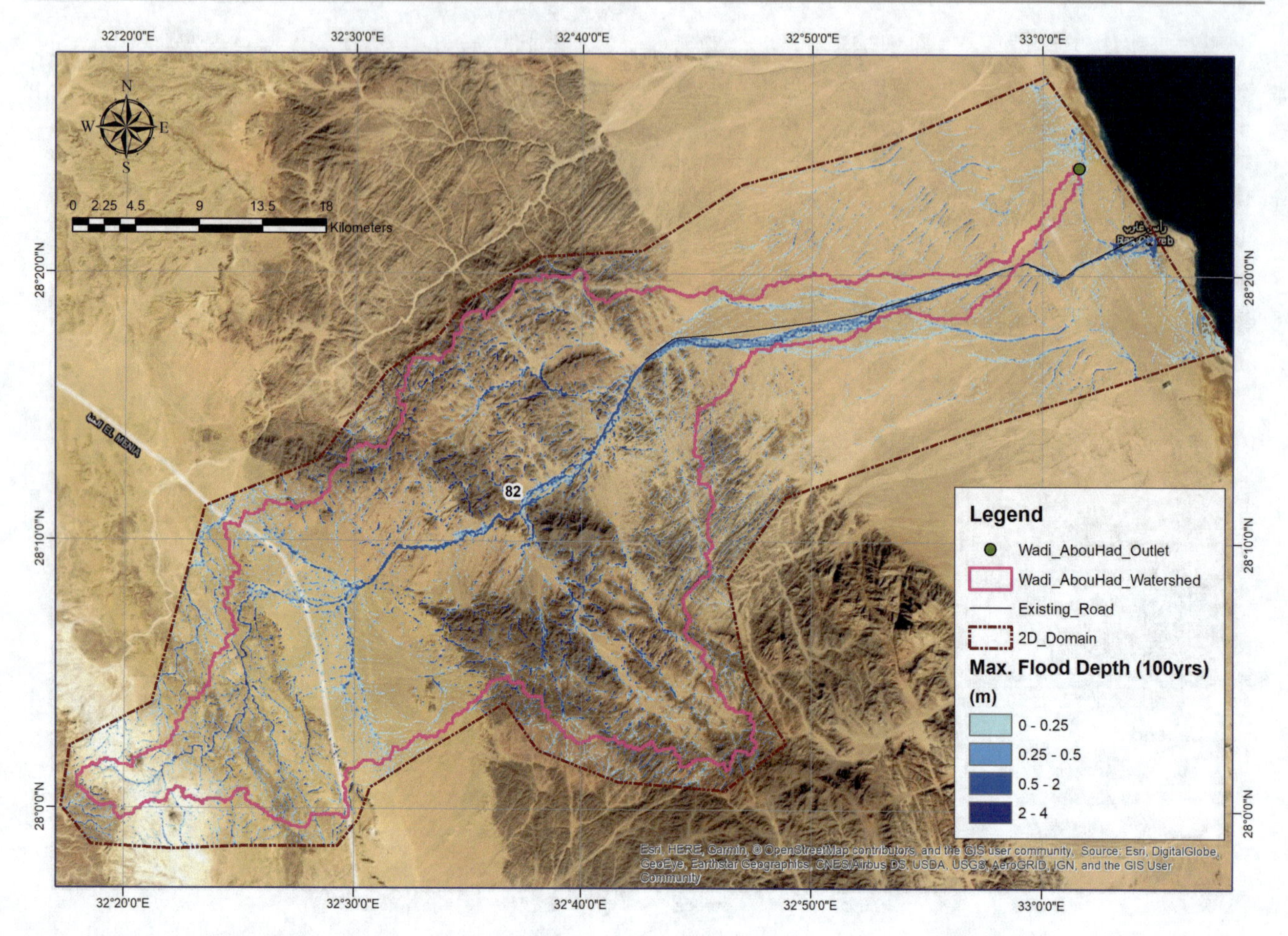

Fig. 48 HEC-RAS two-dimensional rainfall-runoff model boundaries

committee is to refine tune this study and ensure the latest data availability.

- Reassessment of previously designed flood mitigation measures as per the catchments priorities in light of the Egyptian Code.
- Updating the law and assigning one entity to approve any project hydrological study.
- Updating the law to assure that no permit for any rural road or urbanization extension will be provided without an approved hydrological study.

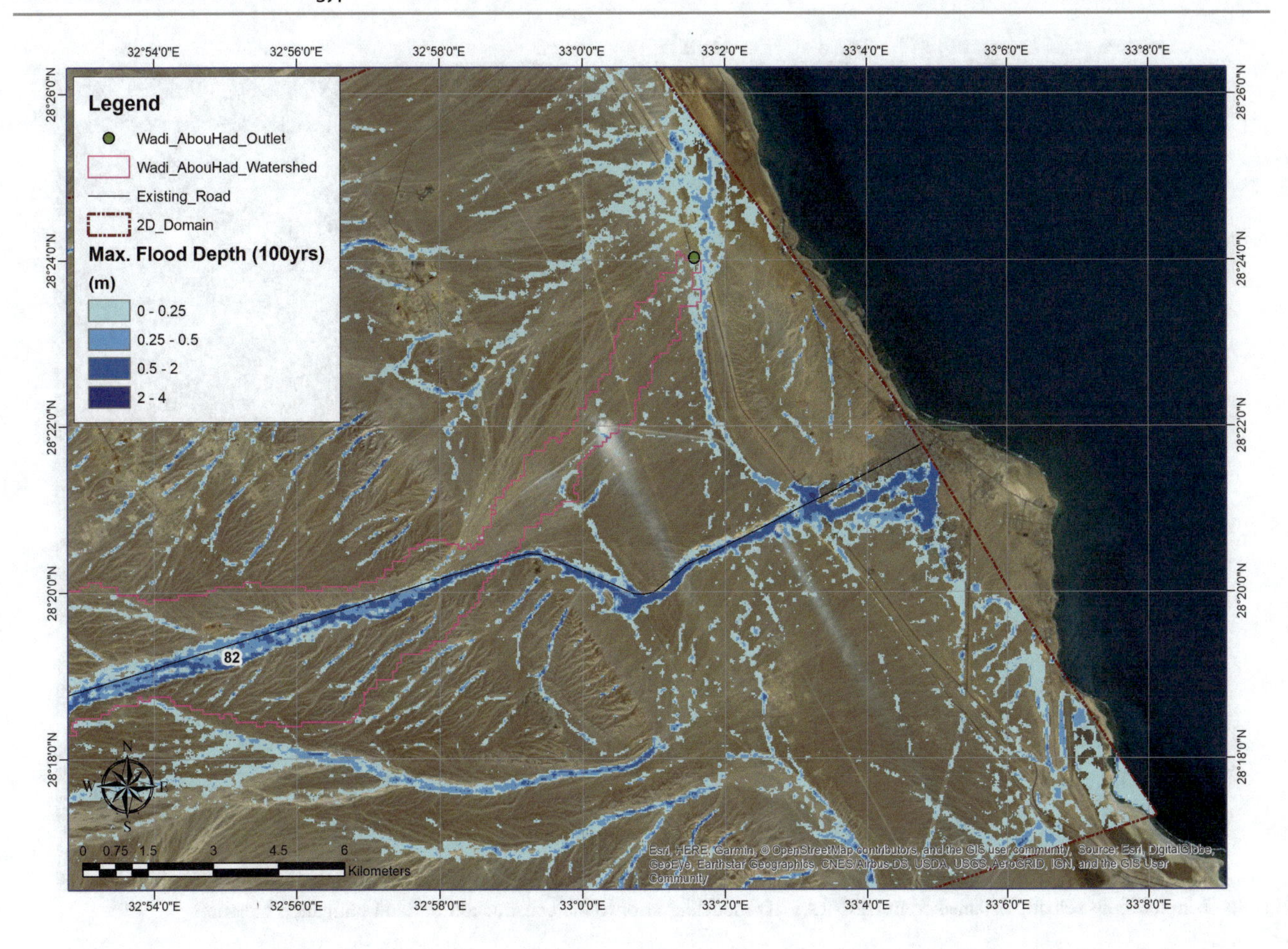

Fig. 49 1-in-100 flow depth obtained from HEC-RAS 2D-modeling prior to the construction of flood mitigation measures

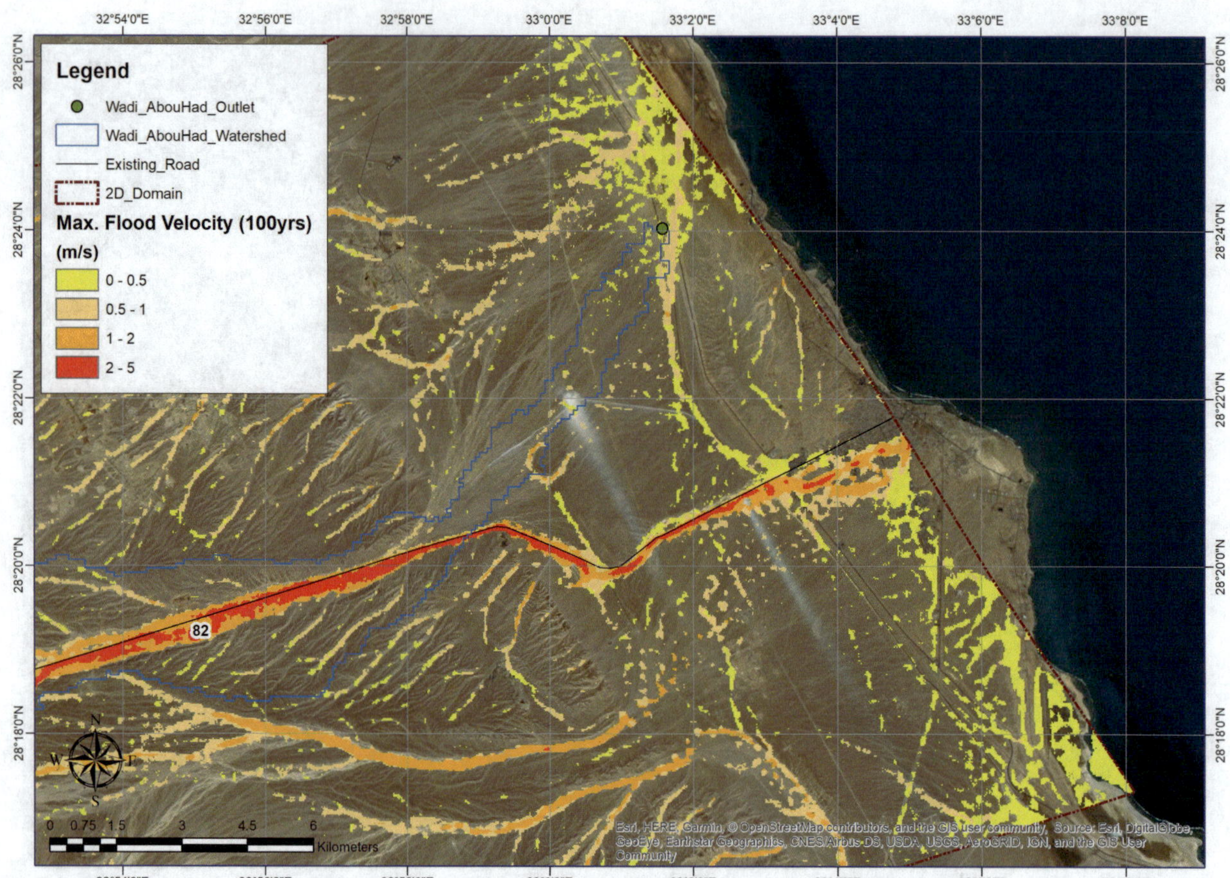

Fig. 50 1-in-100 flow velocity obtained from HEC-RAS 2D-modeling prior to the construction of flood mitigation measures

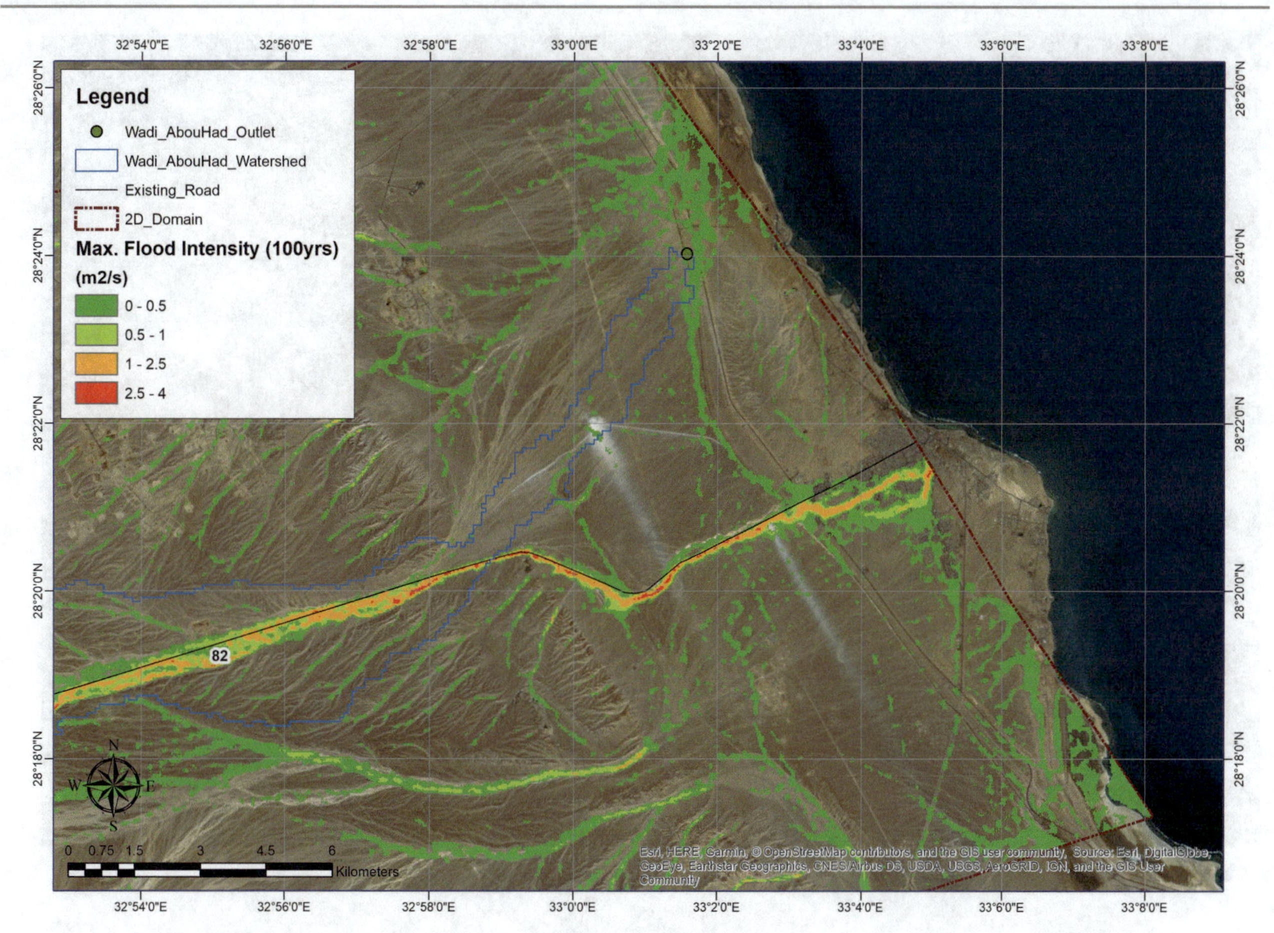

Fig. 51 1-in-100 flow Flood intensity from HEC-RAS 2D-modeling prior to the construction of flood mitigation measures

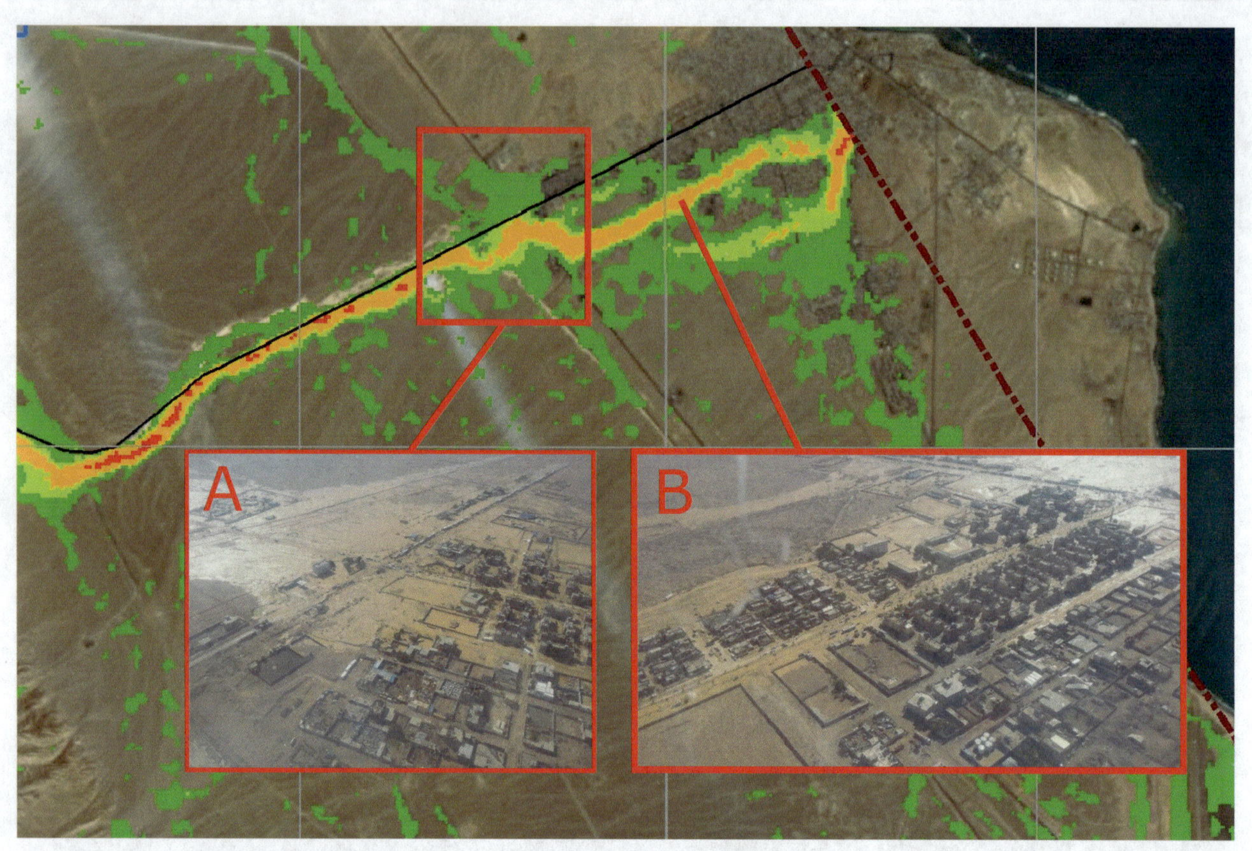

Fig. 52 **a** Matching water spread at the city entrance, **b** matching flow path inside the city

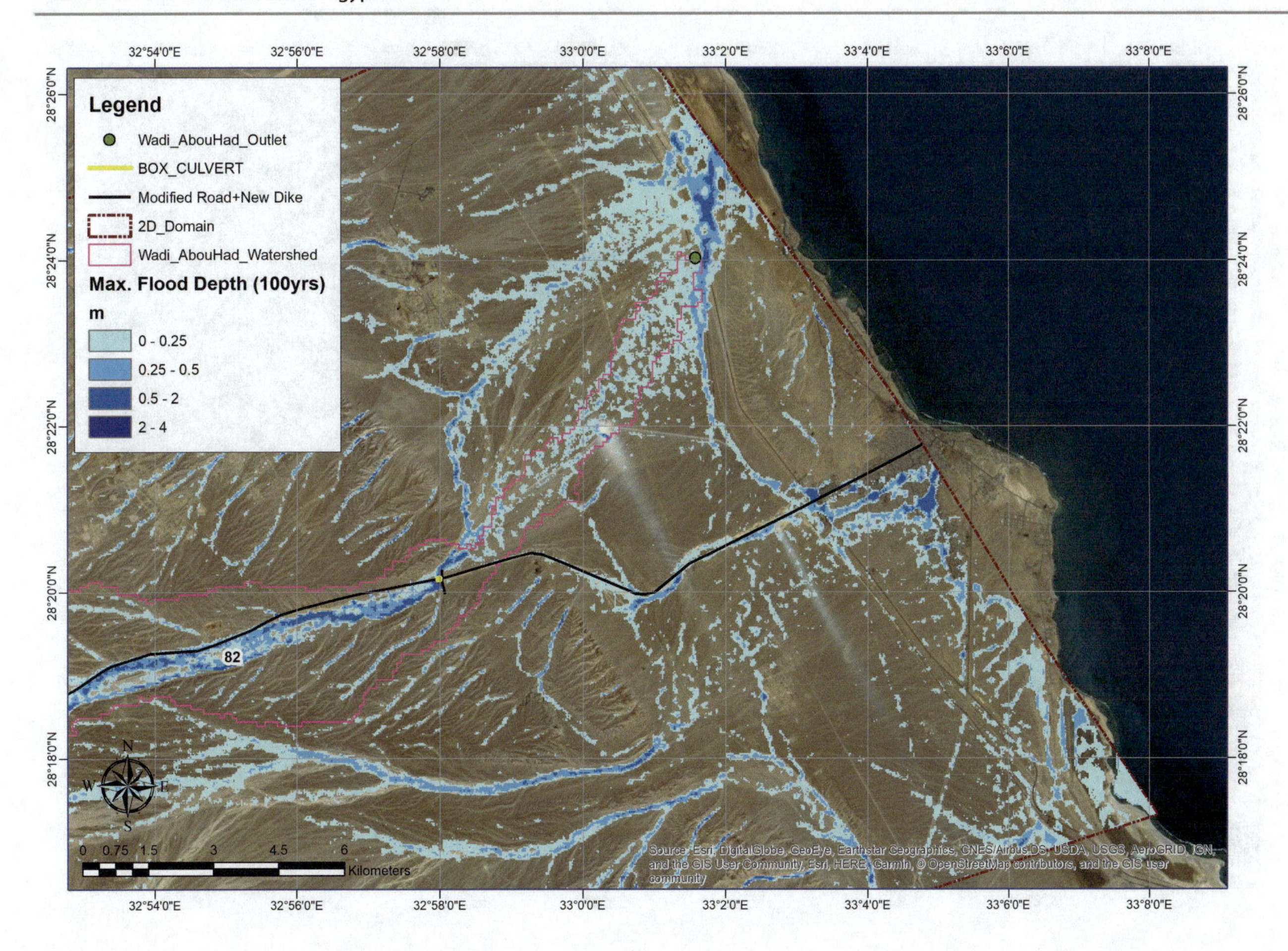

Fig. 53 1-in-100 flow depth obtained from HEC-RAS 2D-modeling after the construction of flood mitigation measures

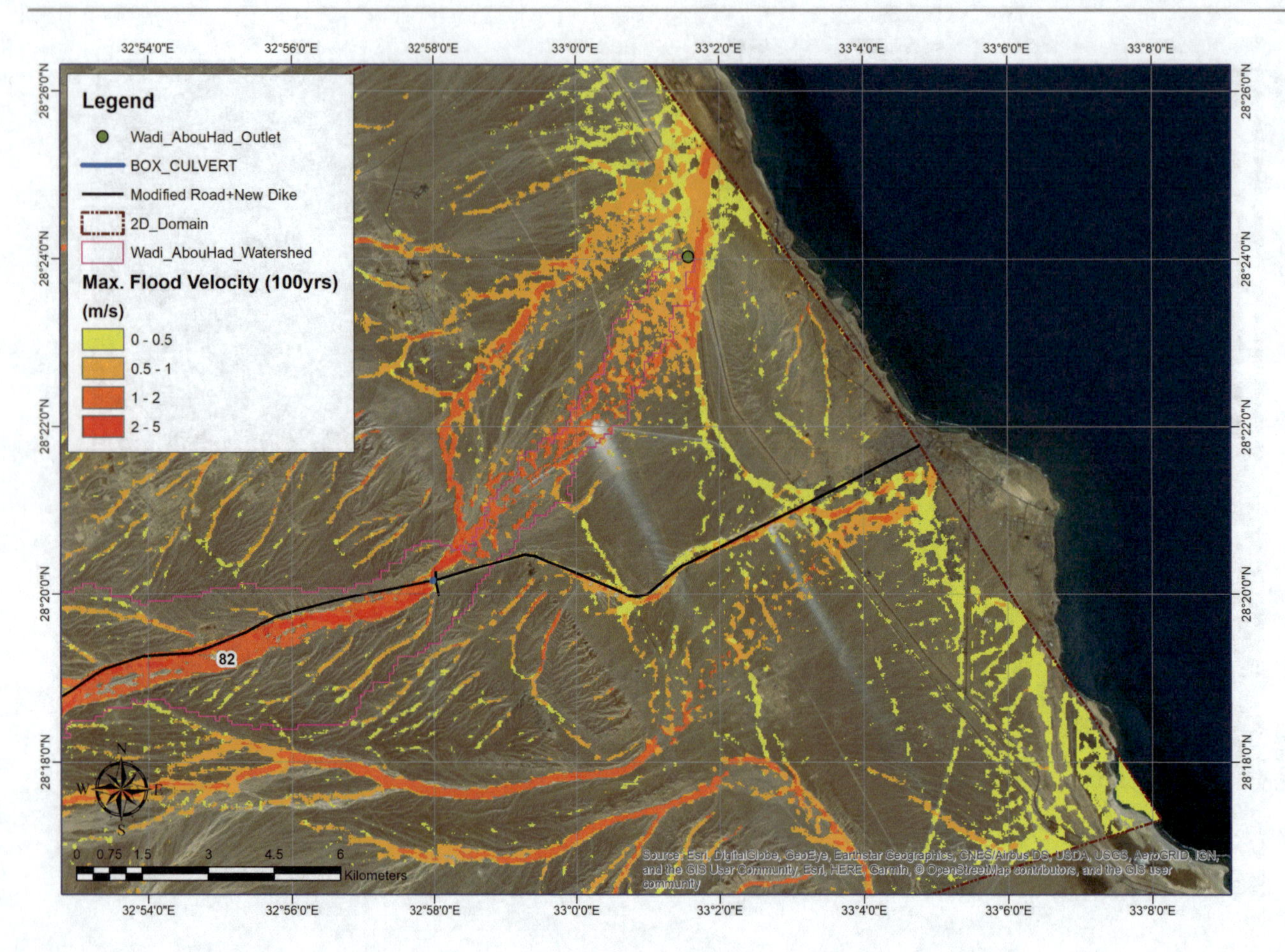

Fig. 54 1-in-100 flow velocity obtained from HEC-RAS 2D-modeling after the construction of flood mitigation measures

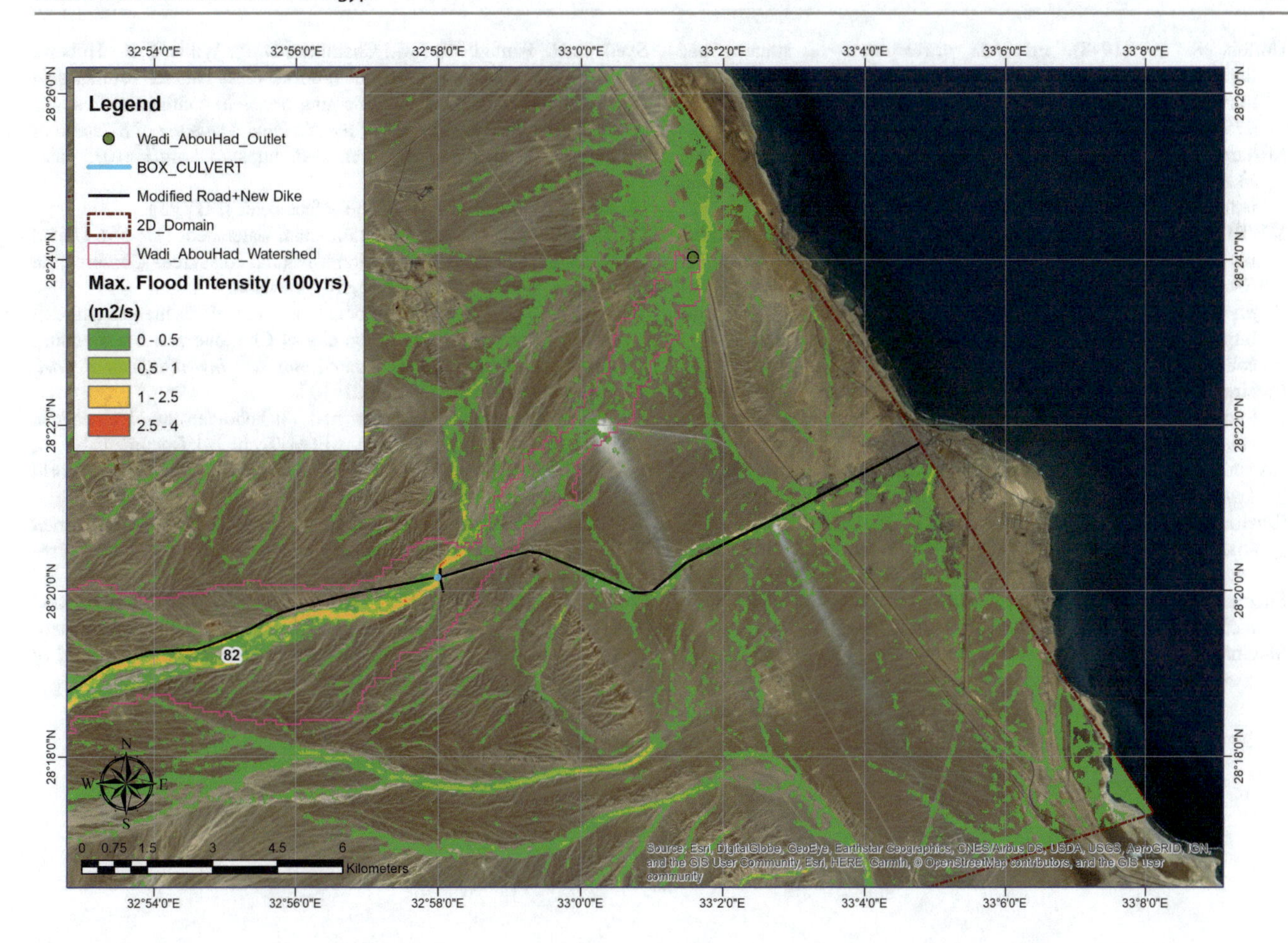

Fig. 55 1-in-100 flow Flood intensity from HEC-RAS 2D-modeling after the construction of flood mitigation measures

References

Abdalla, F., Shamy, I. El, Bamousa, A. O., Mansour, A., Mohamed, A., Tahoon, M. (2014). Flash floods and groundwater recharge potentials in arid land alluvial basins, Southern Red Sea Coast, Egypt. *International Journal of Geosciences*, *5*, 971–982. https://doi.org/10.4236/ijg.2014.59083.

Abdel-fattah, M., Saber, M., & Kantoush, S. A. (2017). A hydrological and geomorphometric approach to understanding the generation of Wadi flash floods. *Water, 9,* 1–27. https://doi.org/10.3390/w9070553.

Al-Saud, M. (2010). Assessment of flood Hazards of Jeddah Area 2009, Saudi Arabia. *Journal of Water Resource and Protection. 2*, 839–847. https://doi.org/10.4236/jwarp.2010.29099.

Awadallah, A. G., Saad, H., El-Moustafa, A., & Hassan, A. (2016). Reliability assessment of water structures subject to data scarcity using the SCS-CN model. *Hydrological Sciences Journal, 61,* 696–710. https://doi.org/10.1080/02626667.2015.1027709.

Chow, V. T. (1959). *Open-channel hydraulics*, McGraw-Hill Book Company.

Durrans, S. R. (2007). *Stormwater conveyance modeling and design*. Exton, PA: Bentley Institute Press.

El-Moustafa, A. M. (2012). Weighted normalized risk factor for floods risk assessment. *Ain Shams Engineering Journal, 3,* 327–332. https://doi.org/10.1016/j.asej.2012.04.001.

El-Rayes, A. E., Omran, A. (2009). Flood control and water management in arid environment : Case study on Wadi Hagul, Northwest Gulf of Suez region, Egypt. In *The international on water conservation in arid regions.*

Elsadek, W. M., Ibrahim, M. G., & Mahmod, W. E. (2018). Flash flood risk estimation of Wadi Qena watershed, Egypt using GIS based morphometric analysis. *Applied Environmental Research, 40*, 36–45.

El-Shamy, I. (1992). Recent recharge and flash flooding opportunities in the eastern Desert, Egypt. Annals of geological survey of Egypt. *Annals of the Geological Survey of Egypt, 18,* 323–334.

Eman, M. G., Nigel, W. A., Giles, M. F. (2002). Characterizing the flash flood Hazards potential along the Red Sea Coast of Egypt. In *The Extremes of the Extremes: Extraordinary Floods (IAHS Proceedings & Reports)* (pp. 211–216), Iahs Publication.

European-Commission. (2016). A communication on flood risk management; flood prevention, protection, and mitigation. Retrived June, 20, 19 from http://ec.europa.eu/environment/water/flood_risk/com.htm.

Helmi, A. M., Mahrous, A., Mustafa, A. E. (2019). Urbanization growth effect on hydrological parameters in mega-cities. In *Advances in sustainable and environmental hydrology, hydrogeology, hydrochemistry, and water resources. advances in science, technology & innovation (IEREK interdisciplinary series for sustainable development)*, Springer. https://doi.org/10.1007/978-3-030-01572-5_98.

Horton, R. E. (1945). Erosional development of streams and their Drainage Basins; hydrophysical approach to quantitative morphology. *Bulletin of the Geological Society of America, 56,* 275–370.

Mohamed, M. (2013). Flash flood risk assessment in the eastern desert. M.Sc. thesis, Irrigation, and Hydraulics Department. Ain Shams University.

MWRI. (2016). Ministry of Water Resources and Irrigation Procedures for Rainwater Management.

NASA. (2019). *Socioeconomic data and apllication center. gridded population world (GPW)* (Vol. 4), Retrieved June, 20, 19 from https://sedac.ciesin.columbia.edu/data/set/gpw-v4-population-count-rev11.

Ramirez, J. A. (2000). Prediction and modeling of flood hydrology and hydraulics. In *Inland flood Hazards: Human Riparian and Aquatic communities*, Cambridge University Press.

Rossmiller, R. L. (1980). The rational formula revised. In *International Symposium on Urban Storm Runoff*, University of Kentucky.

Rudari, R. (2017). Flood Hazard and risk assessment. Hazard-Specific Risk Assess. Modul. United Nations Off. Disaster Risk Reduct. 1–16.

Sherman, L. K. (1932). Stream flow from rainfall by the unit graph method. *Engineering News Records, 108,* 501–505.

Sherman, L. K. (1941). The unit hydrograph and its application. *Bull. Assoc. State Eng. Soc., 17,* 4–22.

Syed, T. H., Famiglietti, J. S., Chambers, D. P., Willis, J. K., Hilburn, K. (2010). Satellite-based global-ocean mass balance estimates of interannual variability and emerging trends in continental freshwater discharge. *Proceedings of the National Academy of Sciences of the United States of America*, 1–6. https://doi.org/10.1073/pnas.1003292107.

USAID. (2018). Climate risk profile fact sheet (EGYPT).

USDA. (1986). Urban hydrology for small watersheds (TR-55). United States Department of Agriculture. Natural Resources Conservation Service, Conservation Engineering Division.

Weaver, J. C. (2003). Methods for estimating peak discharges and unit hydrographs for streams in the city of Charlotte and Mecklenburg County. *North Carolina, Water-Resources Investigations Report*. https://doi.org/10.3133/wri20034108.

WMO. (2006). Social aspects and Stakeholder involvement in integrated flood management. APFM Technical Document No. 4, Flood Management Policy Series, Geneva, Switzerland: World Meteorological Organization.

WMO. (2012). Management of flash floods. In *A Tool for Integrated Flood Management, Version 1.0*, WMO/GWP Associated Programme on Flood Management.

Zaid, S. M., Zaghloul, E. S. A., & Ghanem, F. K. (2013). Flashflood impact analysis of Wadi Abu-Hasah on tell El-Amarna archaeological area using GIS and remote sensing. *Australian Journal of Basic and Applied Sciences, 7,* 865–881.

Determination of Potential Sites and Methods for Water Harvesting in Sinai Peninsula by the Application of RS, GIS, and WMS Techniques

Hossam H. Elewa, Ahmad M. Nosair, and Elsayed M. Ramadan

Abstract

This chapter focuses on determining optimum locations for runoff water harvesting in W. Dahab Watershed, southeastern Sinai, Egypt. A comprehensive approach involving the integration of geographic information systems (GIS), remote sensing (RS), and watershed modeling (WM) was applied through the present work to identify the potential areas for runoff water harvesting (RWH) in Wadi Dahab basin of southern Sinai, Egypt. These tools were effectively used in mapping, investigation, and modeling runoff processes. Eight thematic layers were used as a multi-decision support system (MDSS) for conducting a weighted spatial probability model (WSPM) to determine the potential areas for the RWH. These layers include the volume of the annual flood, basin area, basin length, maximum flow distance, drainage density, basin slope, overland flow distance, and basin infiltration number. The performed WSPM model was run through three different scenarios: (I) equal weights to criteria, (II) weights of criteria are proposed by authors' experience, and (III) weights are assigned by the sensitivity analysis. The resulted RWH potentiality maps classified the basin into five classes ranging from very low to very high class. According to the audited scenario (scenario 3), the major area of W. Dahab basin is categorized as of high and very high for the RWH potentiality (58.27 and 15.56% of the total watershed area, respectively). The WSPM's scenario III map gives the results in favor of the other scenarios, whether based on equal weights or those that were assumed by the authors. Four storage dams and five ground cisterns were proposed in the areas of moderate, high, and very high potentialities for the RWH to harvest runoff water and mitigate flash floods hazards.

Keywords

Sinai • Wadi dahab • Geographic information systems • Runoff water harvesting • Remote sensing • Spatial modeling

H. H. Elewa (✉)
Engineering Applications & Water Division (EAWD), National Authority for Remote Sensing & Space Sciences (NARSS), Cairo, Egypt
e-mail: elewa.hossam@gmail.com; hossam.eliewa@narss.sci.eg

A. M. Nosair
Geology Department, Faculty of Science, Zagazig University, Zagazig, Egypt

E. M. Ramadan
Water Structures Department, Faculty of Engineering, Zagazig University, Zagazig, Egypt

A. M. Negm (ed.), *Flash Floods in Egypt*, Advances in Science, Technology & Innovation,
https://doi.org/10.1007/978-3-030-29635-3_14

List of Acronyms and Abbreviations Listed in the Text

ANOVA	Analysis of Variance
ASTER	Advanced Spaceborne Thermal Emission & Reflection Radiometer
BA	Basin Area
BL	Basin Length
BS	Basin Slope
DD	Drainage Density
DEM	Digital Elevation Model
E	Degree of Effectiveness
EIA	Environmental Impact Assessment
ETM	Enhanced Thematic Mapper
GIS	Geographic Information System
GSA	Global Sensitivity Analysis
IF	Basin Infiltration Number
masl	Meter above sea level
MFD	Maximum Flow Distance
OFD	Overland Flow Distance
RS	Remote Sensing
RWH	RunoffWater Harvesting
SAM	Spatial Analysis Model
ST	Total effect, Sensitivity index
VAF	Volume of Annual Flood
V_i	A partial variance
W	Wadi (dry valley)
W_c	Criterion Weight
WMS	Watershed Modeling System (WMS Software) Water Management System
WSPM	Weighted Spatial Probability Model.

1 Introduction

The Sinai Peninsula is located in the northeastern portion of Egypt and is bounded by longitudes 32°20′–34°52′E and latitudes 27°45′–31°10′N (Fig. 1). It occupies a part of the arid belt of northern Africa and southwest Asia (UNESCO, 1977). It occupies an area of about 61,000 km^2 or about 6% of Egypt's total area with a population of about 400,000, which is mainly bedouins (60%), and the rest are located in small cities such as El-Arish and Sharm El-Sheikh. The Peninsula has a triangular shape; its apex is to the south at Ras Mohammed (south of latitude 28°), whereas its base is to the north extending along the Mediterranean Coast between Port Said and Rafah for about 210 km (Figs. 1 and 2). Sinai is bounded in the eastern side by the Gulf of Aqaba and the International Border and in the west by the Gulf of Suez and Suez Canal. The peninsula has coasts extending for 900 km, including about 155 km along the eastern bank of the Suez Canal. Sinai's history is intertwined with many societies. Previous settlers include the ancient Egyptians, Nabataean's, Romans, Byzantines, and bedouins. Their experience in water harvesting and its exploitation can teach basic lessons to the new settlers of the late twenty-first century (Dames and Moore 1985).

Sinai suffers from an overwhelming water crisis. Runoff water harvesting (RWH) may be the optimum solution for such a problem (Elewa et al. 2012). The present work is dedicated to help the decision-makers and environmental planners by proposing appropriate locations and controlling systems for the RWH and the implementing runoff farming and rain-fed agriculture. This main objective could be performed by determining the promising areas for the RWH.

Remote sensing (RS), geographic information systems (GIS), and watershed modeling (WM) tools are useful in determining the potential locations for the RWH (Elewa et al. 2013; Rahman et al. 2014). Drought management in dry areas of Sinai will heavily depend on the particular planning in exploitation, investigation, and utilization of water resources for favoring the human requirements. Planning should depend on the drainage basins that intercept intermittently the flash floods, where effective methodologies and systematic information of the resources base are selected to measure, analyze, and formulate objectives of water resources development by the RWH (Elewa et al.

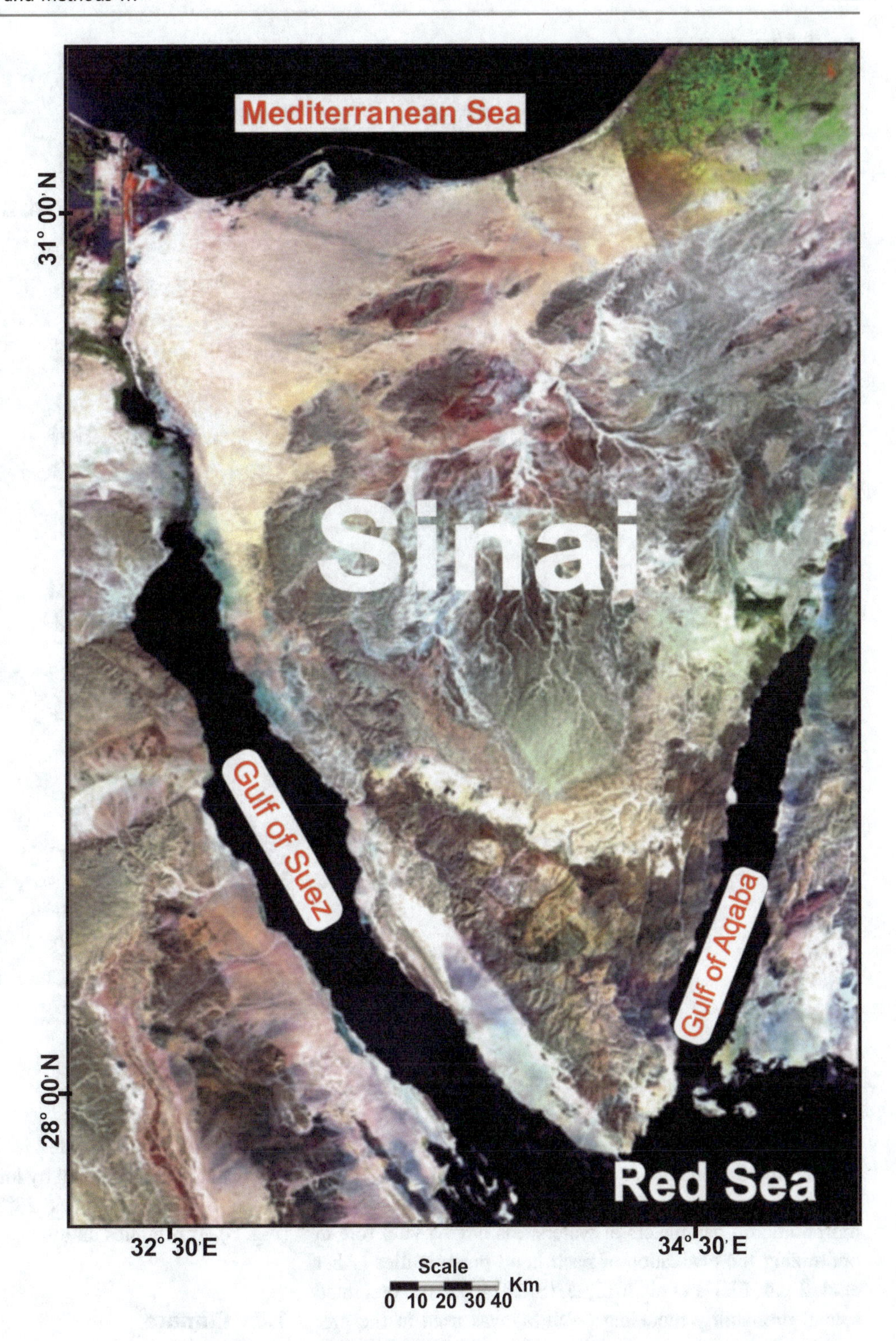

Fig. 1 ETM+ satellite image of Sinai Peninsula (Elewa et al. 2013)

2012). Accordingly, planning for maximizing the RWH, as an inherited historical technique, becomes an urgent need in order to cope with the human requirements and overcoming water scarcity problem. Maximizing the RWH will have its vital role in improving the groundwater recharging, raising water levels, and reducing its salinities to be suitable for the different uses. RWH could be efficient and may support the installment of new settlements in arid areas, with a direct impact on raising the quality of life of local inhabitants (Elewa and Qaddah 2011; Elewa et al. 2012, 2015, 2016; Yazar and Ali 2016). RWH is a well-known practice to improve water security and agricultural production (Paz-Kagan et al. 2017). Watershed management aims at enhancing the water availability in rain-fed areas through water conservation structures, which facilitate storage of water and recharge to groundwater by applying optimum

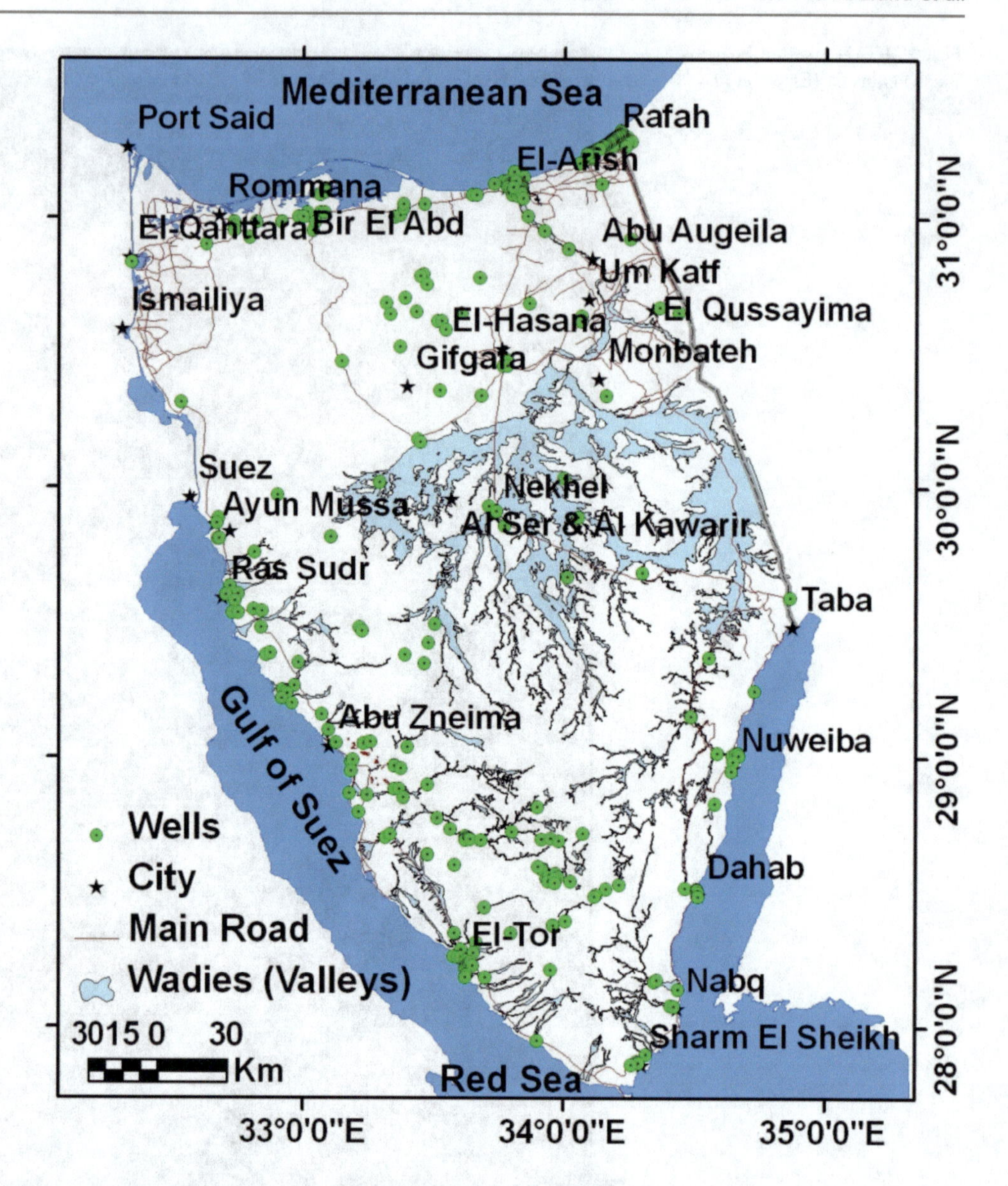

Fig. 2 The main physiographic features of Sinai Peninsula

RWH system (Vema et al. 2019). The present work involves an integrated framework of GIS, RS, and WM for determining optimum sites for RWH constructions in Dahab basin, south Sinai. Applications of RS, GIS, and statistical methods in extracting and analyzing the hydro-morphometric parameters of watersheds have a vital role in optimizing the evaluation of flash flood potentialities (Aher et al. 2014; Elewa et al. 2012, 2013, 2014, 2016). Weighted spatial probability modeling (WSPM) was used in the present context for determining the optimum sites for RWH in W. Dahab. The constructed model is based mainly on the most effective hydro-morphometric parameters of W. Dahab sub-watersheds. Wadi Dahab is considered a promising basin in southeastern Sinai, Egypt, where many tourist areas are located, which is very important for the Egyptian national income.

1.1 Site Description

Dahab watershed is located in the southeastern part of Sinai Peninsula and is bounded by longitudes 33°55′46.9″ and 34° 31′28.8″ E and latitudes 28°22′43.4″ and 28°52′18.5″ N (Fig. 3). It occupies an area of about 2071.26 km^2.

1.2 Climate

Wadi Dahab area generally has an arid climate with quite significant rainfall intensity. Severe floods from Wadi Dahab Watershed could cause significant damage and loss of lives at the City of Dahab and along the main streams. Due to the meteorological conditions of South Sinai, flood events are sporadic, but severe. The affected areas include tourist locations, as well as residences of local inhabitants. Flash

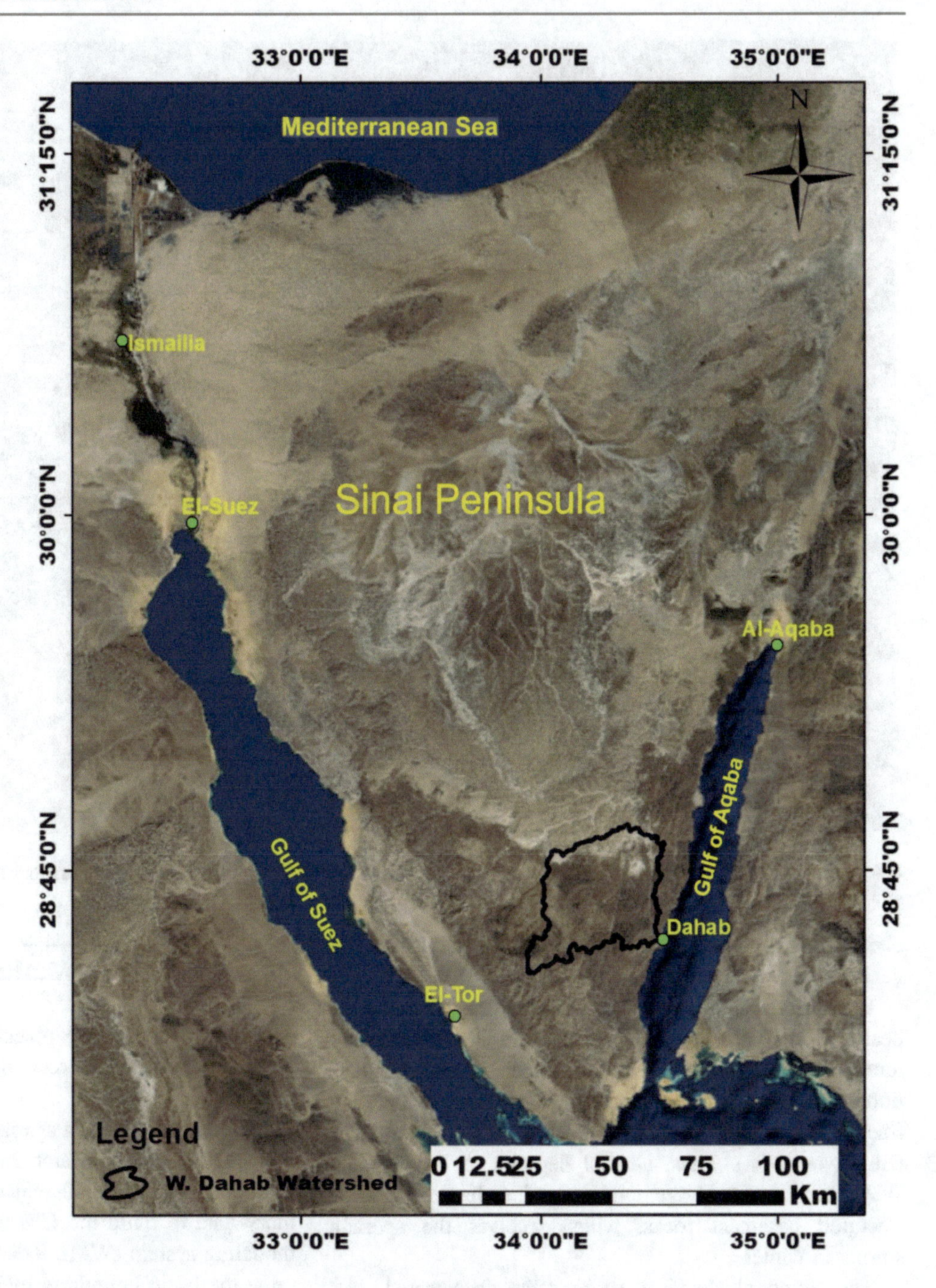

Fig. 3 Wadi Dahab study area

flood disaster is a major problem due to its diverse topography combined with meteorological and morphological conditions (Bisht et al. 2018). "The maximum intensity of rainfall, which has been recorded for a storm event of 24 h is 76 mm, which means that the area has received about 159 million m^3 with a return period of 100 years" (Omran 2013). Based on the climate conditions and the geology of the area, it could be concluded that about 50% of the precipitation is lost via evapotranspiration and infiltration, so that only half of the rainfall contributes to runoff. However, this is still enough to cause potential catastrophic effects on the infrastructure (Figs. 4 and 5).

1.3 Geological Conditions

Wadi Dahab is covered mainly with Cretaceous to Pre-Cambrian rocks. Sedimentary rocks of Cambrian to Cretaceous ages are located in the northern part of W. Dahab basin, while the rest of the basin is covered with pre-Cambrian rocks (igneous and metamorphic rocks), in addition to the recent wadi deposits, which cover the valley floors of the main streams (Said 1971) (Fig. 6). These deposits receive huge amounts of recharge from the fractured basement rocks, which receives the sporadic storms in winter. The topographic elevations range between 14 m and

Fig. 4 A photo showing the sabotage affecting the main road to Dahab City, which is caused by the rigorous flash floods of W. Dahab Watershed

about 2500 m (amsl), where the Wadi slope varies from gentle to very steep (Figs. 7 and 8).

Based on the lithological conditions, the groundwater-bearing formations are classified into the alluvial and fractured basement aquifer systems. Surface water from flash floods moves downward and laterally through the open fractures to recharge the alluvial deposits of the main valleys (El Rayes 1992). Thus, alluvial deposits of Wadi Dahab Watershed receive large amounts of recharge from the fractured basement rocks, which receives the sporadic storms in winter.

The structural elements, such as the deep-seated shear zones, tension fractures, and thrust faults act as conduits for groundwater recharge and movement (Shendi and Oada 1999). Rainwater, which is directly infiltrated to the highly jointed upper units is gradually transmitted into the fault zones and alluvial aquifers through the weakly jointed lower units. The intersections of fractures are good sites for the occurrence of groundwater and localization of water wells (El KiKi et al. 1992).

2 Materials and Methods

To accomplish the work objectives, the following tasks were performed which address the undertaken methods and techniques:

ASTER DEM (30-m spatial resolution), SPOT-4 satellite images (acquired on March 28, 2005) and topographic maps (1:50000 and 1:100000 scales) were used as essential primary data to build the GIS and performing the watershed modeling system (WMS 8.0©; AQUAVEO 2008) to determine the basin boundary, main sub-basins, and basin drainage net map. The stream orders were given for each stream according to Strahler method for stream ordering technique (Strahler 1964) using the ArcGIS 10.1©_ software platform.

The watershed hydrographic criteria derived from the WMS 8.0 software, which were used for the determination of the RWH optimum sites, include volume of annual flood (VAF), overland flow distance (OFD), maximum flow distance (MFD), basin infiltration number (IF), drainage density (DD), basin area (BA), basin slope (BS), and basin length (BL). These layers were used as multi-decision support criteria for conducting a weighted spatial probability model (WSPM) inside the GIS. The eight layers were used in raster

Fig. 5 A photo showing the destroyed main road to W. Dahab Watershed

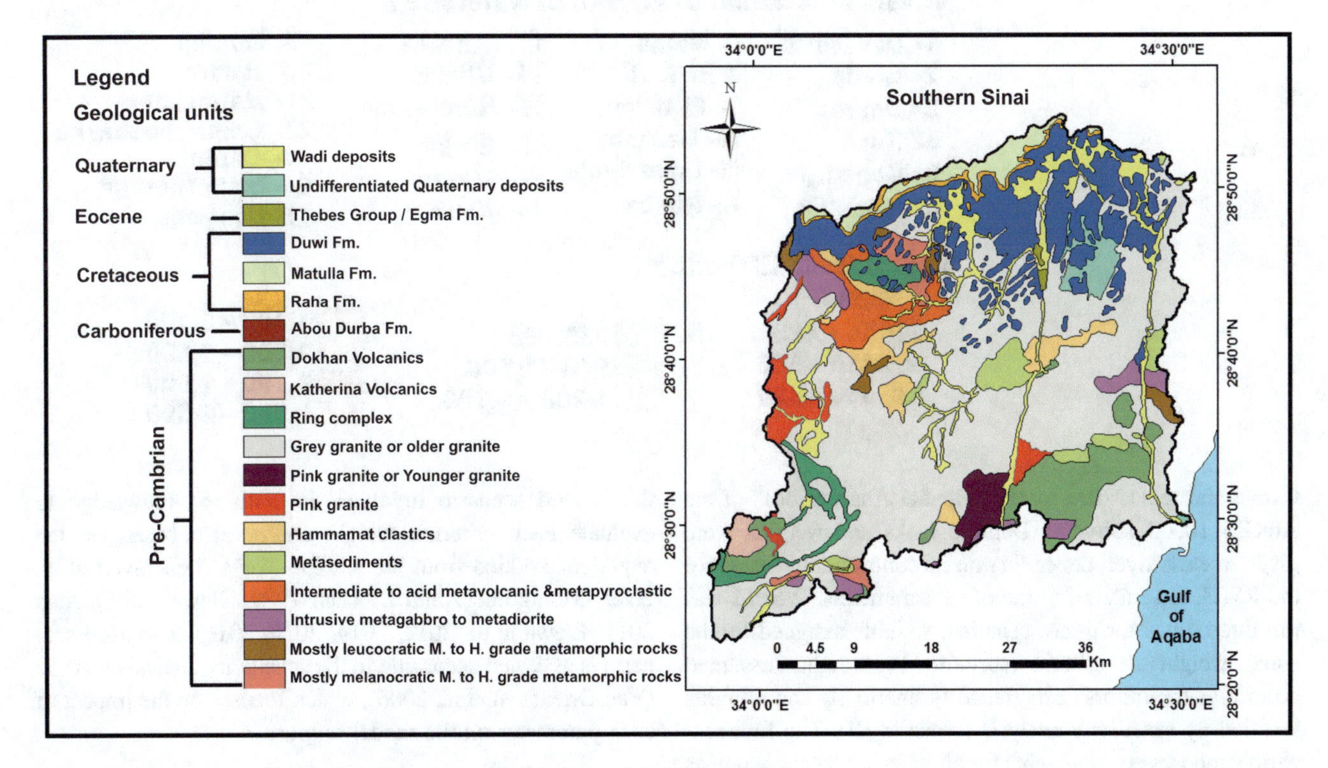

Fig. 6 Geological map of W. Dahab Watershed (simplified after CONOCO 1987)

Fig. 7 Orographic features and topographic elevations of W. Dahab Watershed (based on ASTER DEM of 30 m resolution)

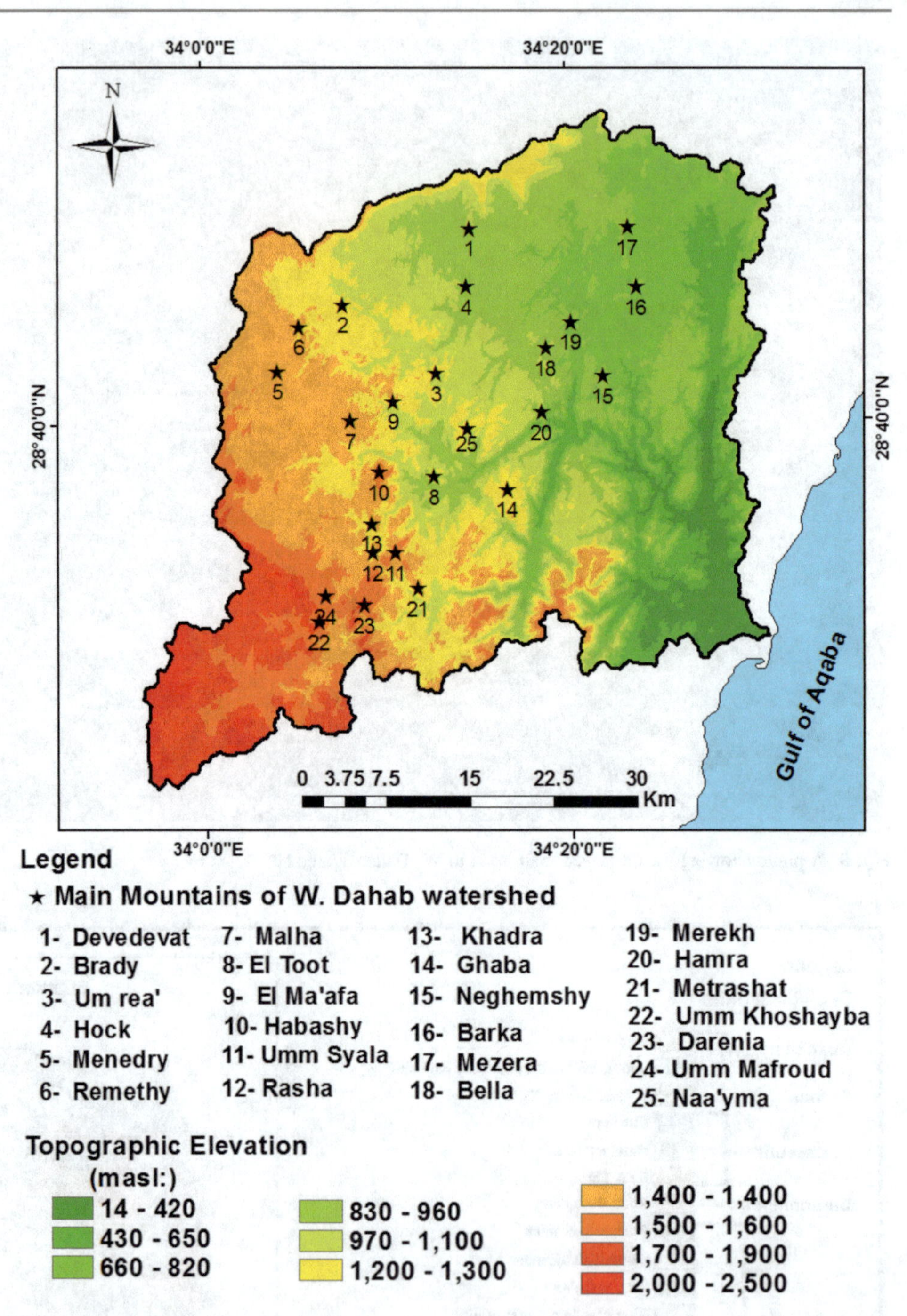

format and reclassified by the "Spatial Analyst tool" of the ArcGIS 10.1© software. Definite ranks and weights were given to each layer, depending on its contribution degree for the RWH and relation to the other parameters. WSPM was run three times for three scenarios; weights assigned by the equal weight to criteria (scenario I), weights assigned according to authors' experience (scenario II) and weights justified by sensitivity analysis (scenario III). The first scenario supposes equal weights for all thematic layers, whereas the second scenario involves the authors' knowledge to evaluate each criterion weight and/or also based on the experience added from the similar works (i.e., Javed et al. 2009; Montgomery and Dietrich 1989; Elewa and Qaddah 2011; Elewa et al. 2012, 2014, 2015, 2016). The third scenario is assigned according to the sensitivity analysis method (Van Griensven et al. 2006), which focuses on the impact of each parameter on the model output.

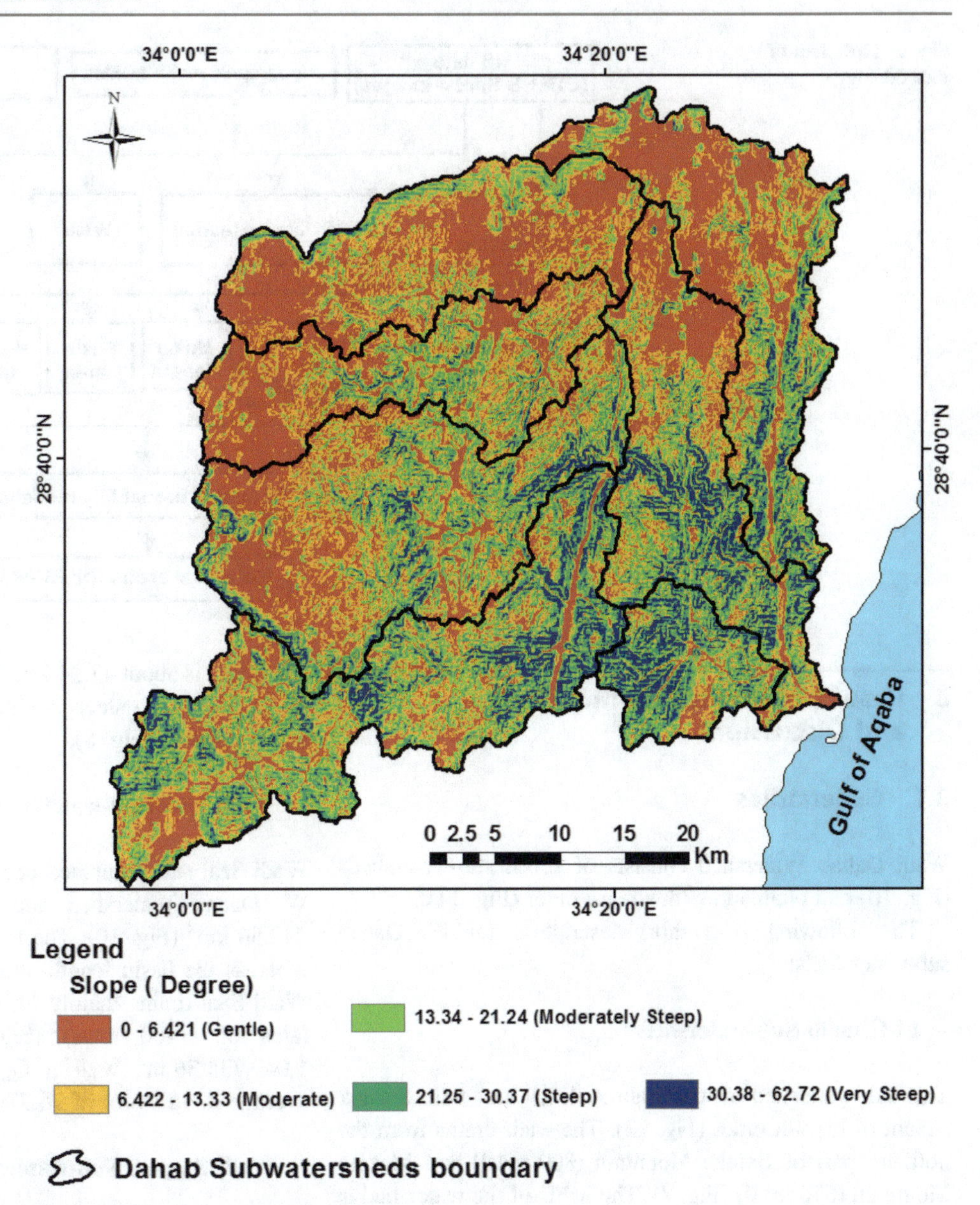

Fig. 8 Slope map of W. Dahab Watershed (based on 30 m resolution ASTER DEM)

The volume of the annual flood is calculated according to Finkel method (Finkel 1979), which was simulated by the WMS 8.0© software platform. Finkel (1979) used his method for the Araba Valley, which has similar climate conditions to Sinai Peninsula. It is a simple graphical method for determining the probability and frequency of occurrence of seasonal or yearly rainfall.

Finkel method (Finkel 1979) uses the following factors (Eqs. 1 and 2):

1. Peak flood flow (Q_{max})

$$Q_{\max} = K_1 A^{0.67} \tag{1}$$

where Q_{max} = Peak flood flows, in m^3/s.

2. Volume of the annual flood (v) in 1000 cubic meters

$$V = K_2 A^{0.67} \tag{2}$$

where A is basin area of the basin in km^2, and K_1 and K_2 are constants depending on the probability of occurrence

The overall flowchart of the methodology is given in Fig. 9.

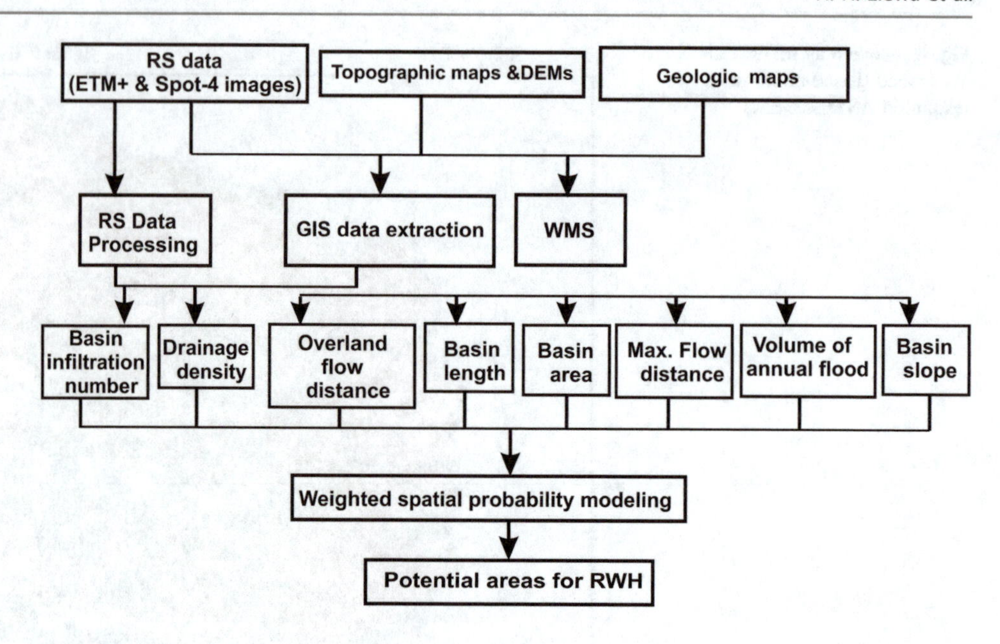

Fig. 9 Flowchart of methodology

3 Results of Watershed Modeling and Discussions

3.1 Generalities

Wadi Dahab Watershed consists of seven sub-watersheds (Fig. 10) and attains the 7th stream order (Fig. 11).

The following is a short description for W. Dahab sub-watersheds:

– **El-Ghaaib Sub-watershed**

This sub-watershed occupies about 297.9 km^2 with a trunk stream of the 4th order (Fig. 12). The wadi drains from the northern part of Baraka Mountain (800 masl) and Mezera Mountain (650 masl) (Fig. 7). The MFD of the watershed is 54.42 km with varying slopes from the gentle at the northern parts to the very steep in the southeastern side (Fig. 8). The average VAF resulting from W. El-Ghaaib is about 1,204,686.81 m^3 with a Q_{max} of 71.83 m^3/s occurring in an average flood duration of 16.77 h (Table 1).

– **Wadi Ganah Sub-watershed**

This sub-watershed is located at the northwestern part of W. Dahab Watershed and has the 5th order (Figs. 10 and 13). It occupies about 282.45 km^2. It drains from the Devedevat Mountain (420 m masl), Brady Mountain (1400 masl), and Remethy Mountain (1600 masl) (Fig. 7). The received VAF is about 1,162,453.75 m^3 with a Q_{max} of 69.31 m^3/s occurring in an average flood duration of 16.77 h (Table 1). The MFD is about 43.34 km. It is characterized by the gentle to moderately steep slopes in most parts of the sub-watershed (Fig. 8).

– **Wadi Saal Sub-watershed**

Wadi Saal sub-watershed occurs at the northwestern part of W. Dahab Watershed and occupies an area of about 242.56 km^2 (Fig. 10). The trunk channel has the 5th order, whereas the basin length attains about 34.19 km (Fig. 14). Wadi Saal drains mainly from Menedry and Malha Mountains of (1400 masl) (Fig. 7). The average VAF is 1,049,733.36 m^3 with a Q_{max} of 62.59 m^3/s within an average flood storm of 16.77 h (Table 1).

– **Wadi Zoghra Sub-watershed**

Wadi Zoghra is representing one of the largest sub-watersheds of W. Dahab among all W. Dahab sub-watersheds, where it encounters an area of about 415.99 km^2 (Fig. 10). Its trunk channel has the 5th order (Fig. 15). The MFD of the watershed is about 44.47 km. The VAF of W. Zoghra sub-watershed is about 1,506,716.76 m^3 with a Q_{max} of 89.83 within a flood storm of 16.77 h (Table 1).

– **Wadi Khoshaib Sub-watershed**

Wadi Khoshaib is the smallest sub-watershed of W. Dahab Watershed, where it encounters an area of about 108.57 Km^2 (Fig. 10). Its trunk channel has the 4th order (Fig. 16). It is characterized by a nearly circular shape and it is located in the southeastern side of W. Dahab Watershed close to W.

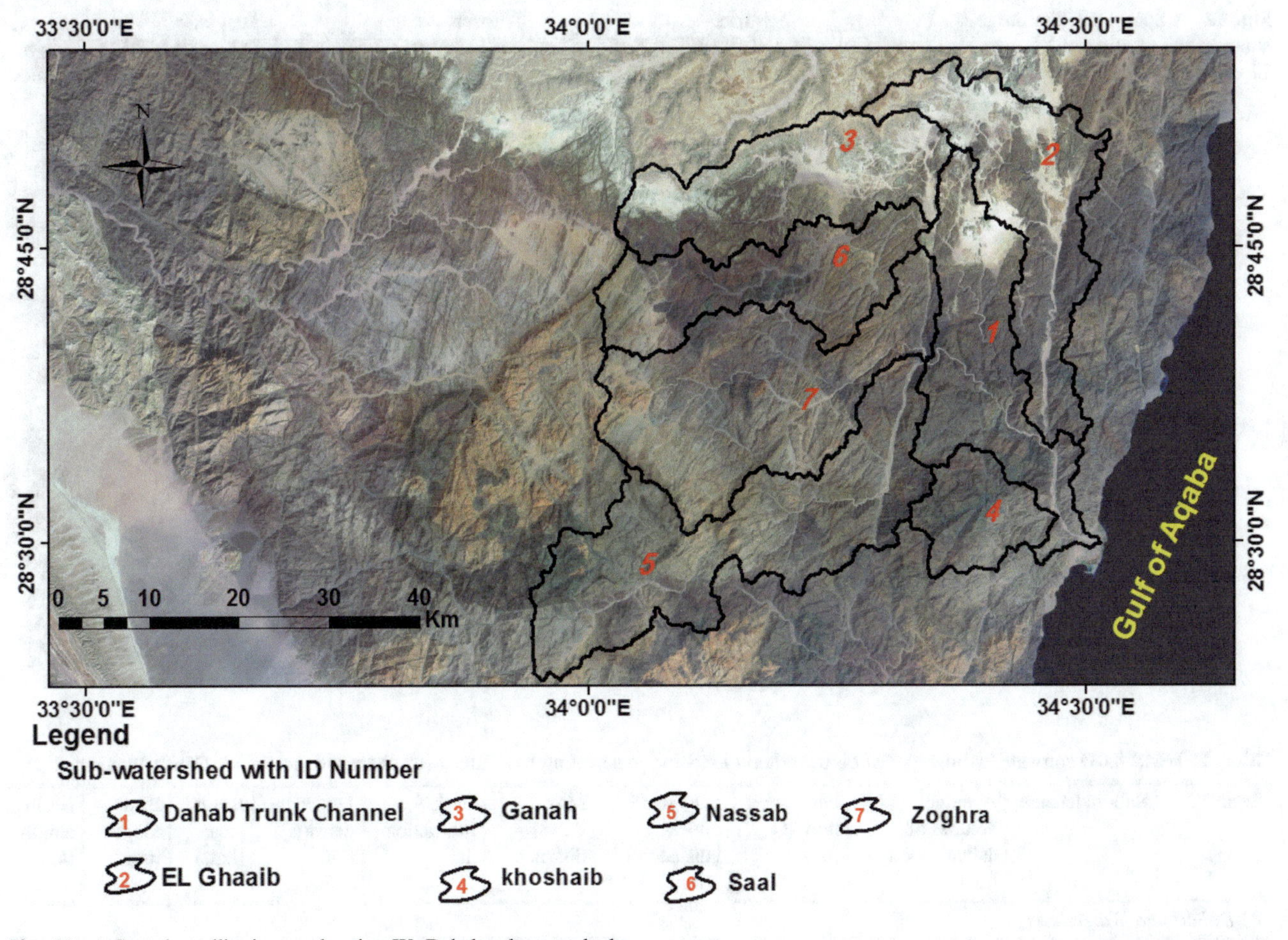

Fig. 10 A Spot-4 satellite image showing W. Dahab sub-watersheds

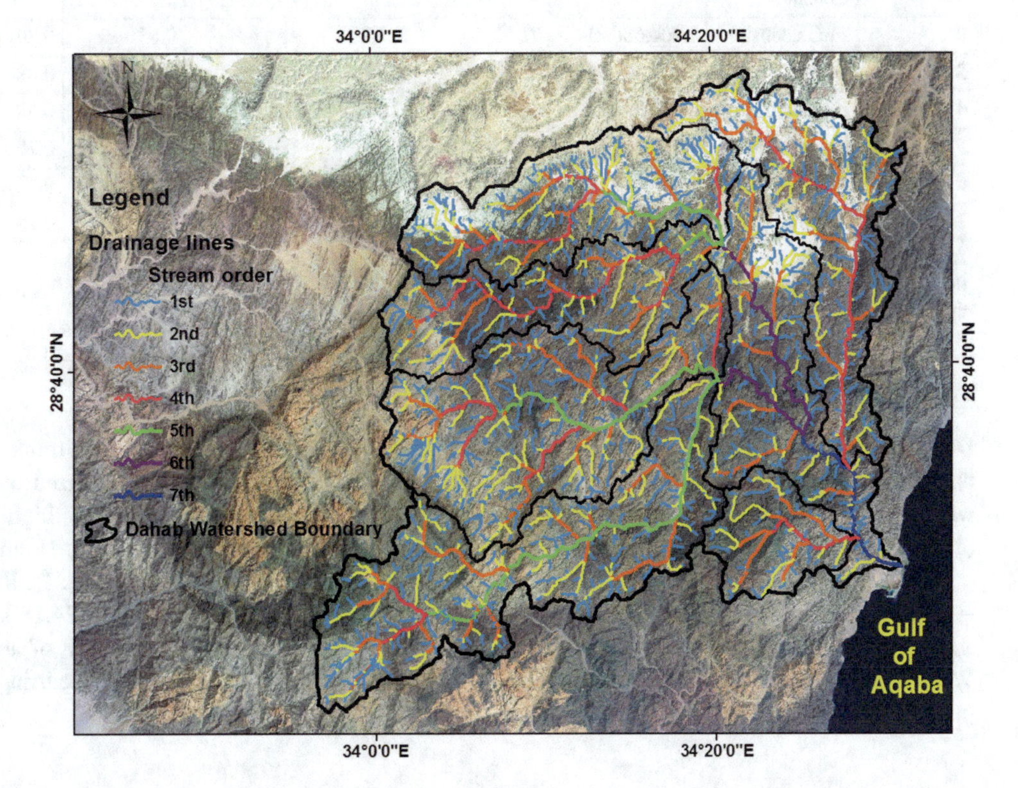

Fig. 11 A Spot-4 for W. Dahab watershed showing its stream orders

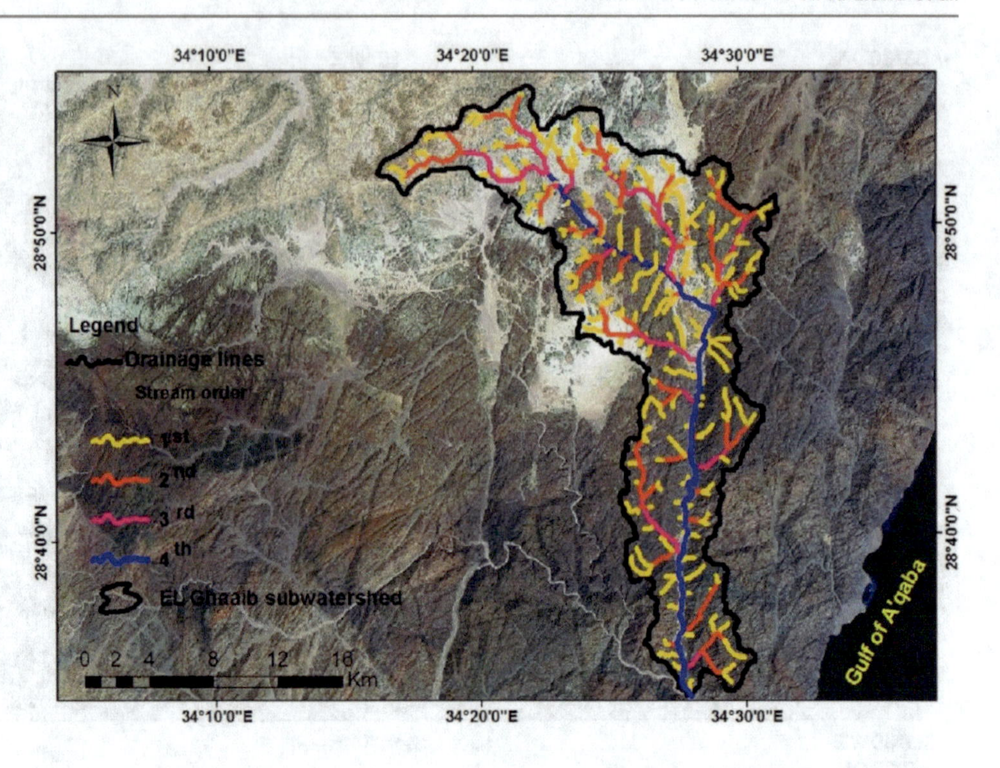

Fig. 12 A Spot-4 satellite image showing the location of W. El-Ghaaib sub-watershed

Table 1 WMS 8.0© software hydrographical output criteria used for demarcating the watersheds characteristics of W. Dahab Watershed

Basin ID	Sub-watershed	Average volume of annual flood (m^3)	Peak flood flow (Q_{max}) (m^3/s)	Overland flow distance (km)	Max. flow distance (km)	Basin infiltration no.	Drainage density (km^{-1})	Basin area (km^2)	Basin Slope (m/m)	Basin length (km)
W. Dahab Sub-Watersheds										
1	Dahab (Trunk Channel)	1,153,707.82	68.79	19.92	52.19	0.77	0.75	279.28	0.27	39.84
2	El-Ghaaib	1,204,686.81	71.83	18.92	54.42	0.83	0.79	297.90	0.20	37.84
3	Ganah	1,162,453.75	69.31	15.48	43.34	0.97	0.88	282.45	0.12	30.96
4	Khoshaib	612,592.39	36.52	7.37	19.63	0.59	0.58	108.57	0.37	14.74
5	Nassab	1,575,219.43	93.92	24.18	61.14	0.80	0.75	444.53	0.29	48.35
6	Saal	1,049,733.36	62.59	17.09	45.85	0.79	0.76	242.56	0.16	34.19
7	Zoghra	1,506,716.77	89.83	16.54	44.47	0.79	0.78	415.9	0.22	33.08
Average flood duration (h)	16.7									

Dahab Trunk Channel (Fig. 10). The MFD of the watershed is 19.631 km. It receives an average VAF of 612,592.39 m^3 with a Q_{max} of 36.52 m^3/s within a flood storm of 16.77 h (Table 1).

– **Wadi Nassab Sub-watershed**

Wadi Nassab sub-watershed is located in the southwestern part of W. Dahab Watershed with an area of 444.53 km^2 (Fig. 10) (Table 1). Its trunk channel has the 5th order (Fig. 17). It is characterized by an elongated shape, where torrents stem from the high mountains such as Umm Khoshayba (1800 masl), Umm Mafroud (1600 masl), and Darenia (1800 masl) (Fig. 7). It has a BL of about 48.35 km with an average OFD of 24.18 km, and MFD of 61.14 km. It receives an average VAF of about 1,575,219.43 m^3 with a Q_{max} of 93.92 m^3/s occurring within a flood storm of 16.77 h (Table 1).

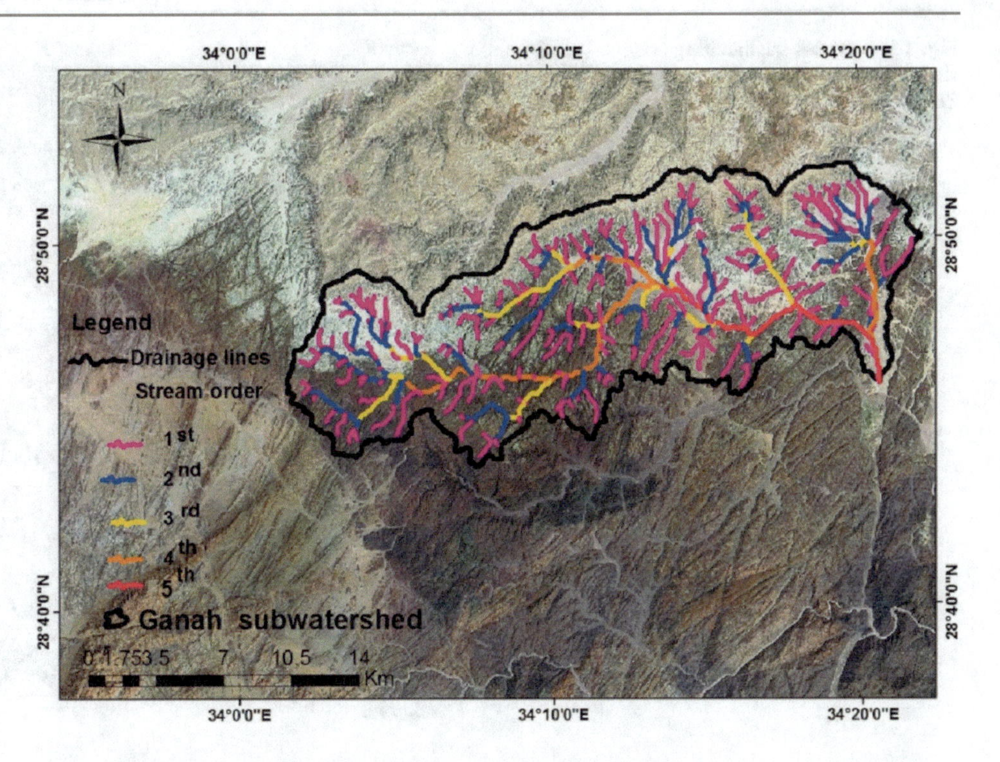

Fig. 13 A Spot 4-satellite image showing the location of W. Ganah sub-watershed

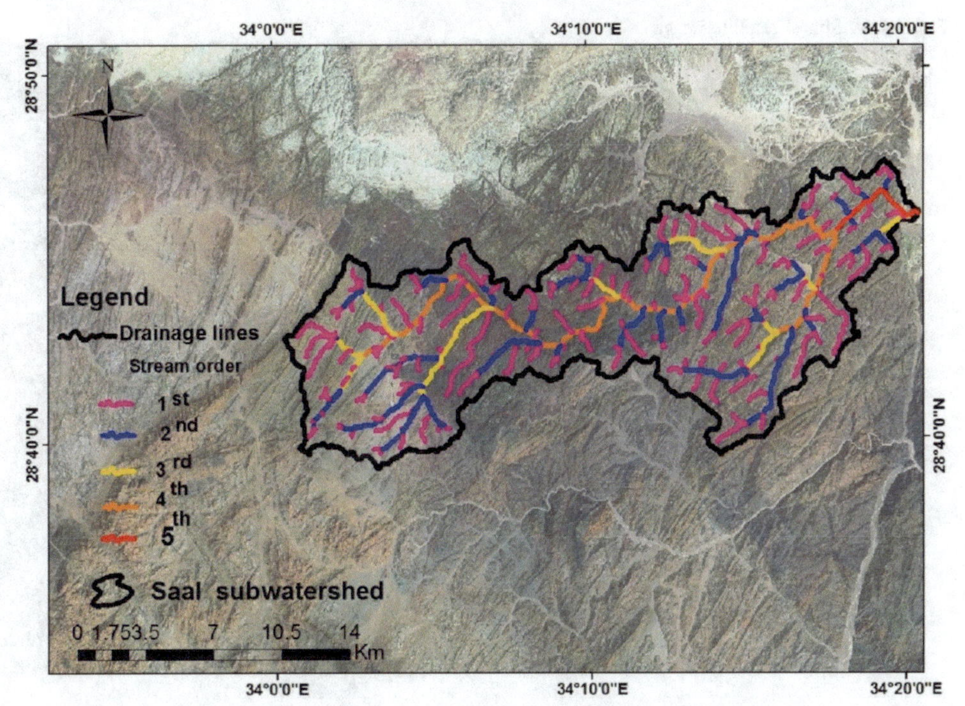

Fig. 14 A Spot-4 satellite image showing the location of W. Saal sub-watershed

– **Dahab Trunk Channel Sub-watershed**:

The Trunk Channel of W. Dahab Watershed debouches to the Gulf of Aqaba and occupies an area of about 279.28 km^2, with a trunk stream of the 5th order (Fig. 18) (Table 1). It originates from the northern parts of Dahab watershed to the outlet of the Wadi in the eastern side. It is characterized by varying slopes from gentle to very steep (Fig. 8). It has a BL of 39.8 km and a basin relief of 1.44 km. Average VAF is about 1,153,707.82 m^3 with a Q_{max} of 68.79 m^3/s occurring within a flood storm of 16.77 h. The MFD is 52.19 km and the average OFD is 19.92 km (Table 1).

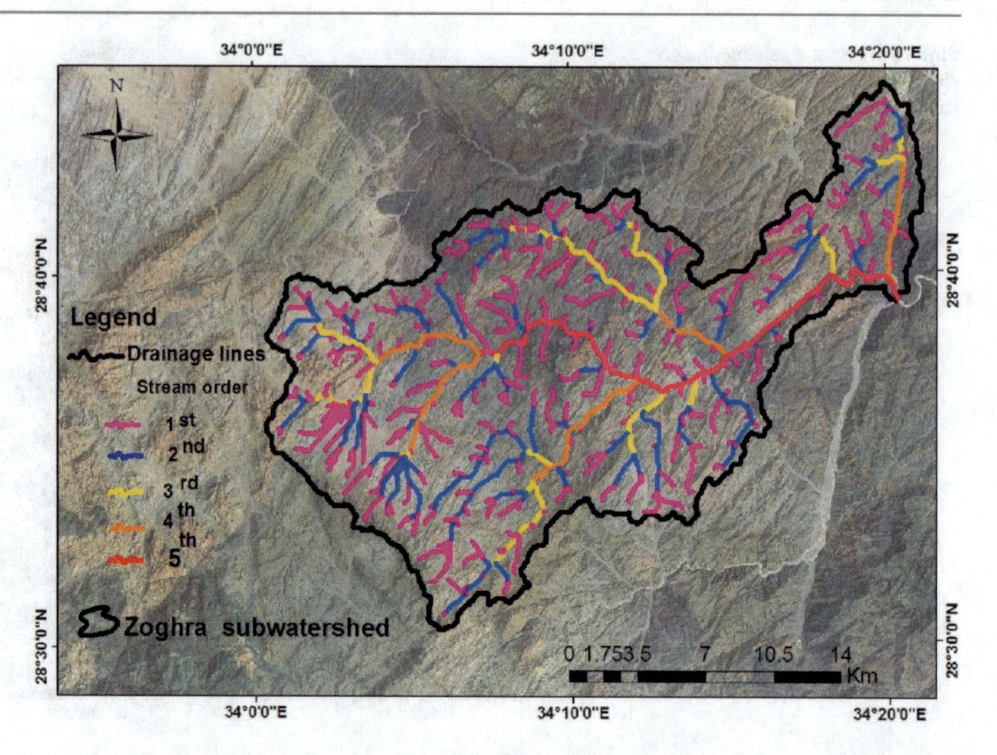

Fig. 15 A Spot-4 satellite image showing the location of W. Zoghra sub-watershed

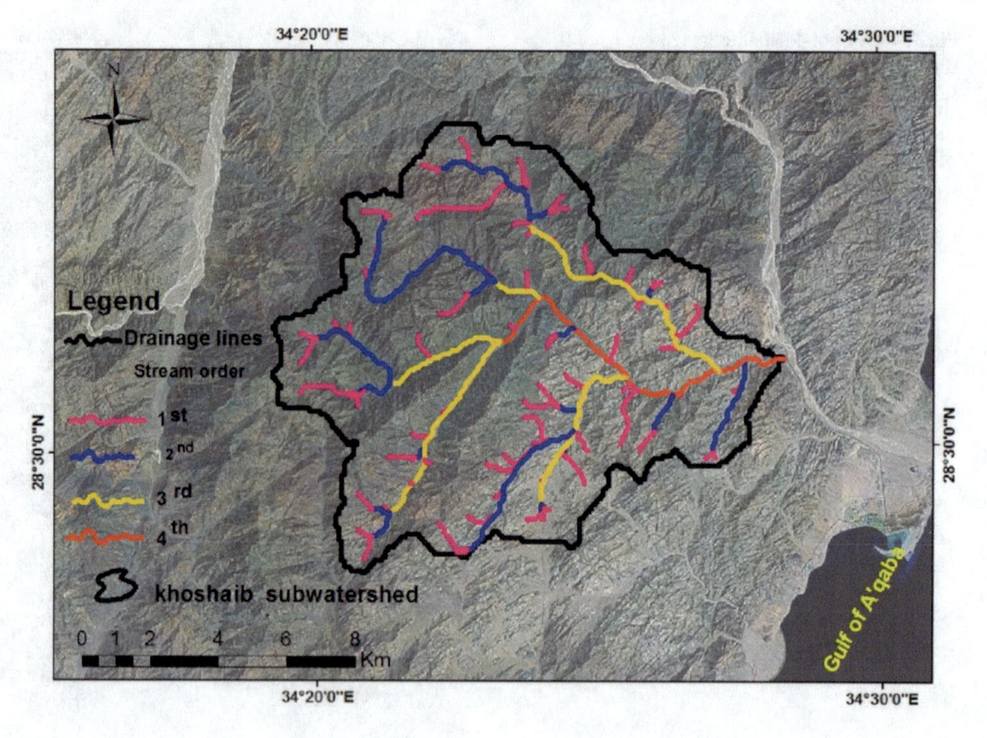

Fig. 16 A Spot-4 satellite image showing the location of W. Khoshaib sub-watershed

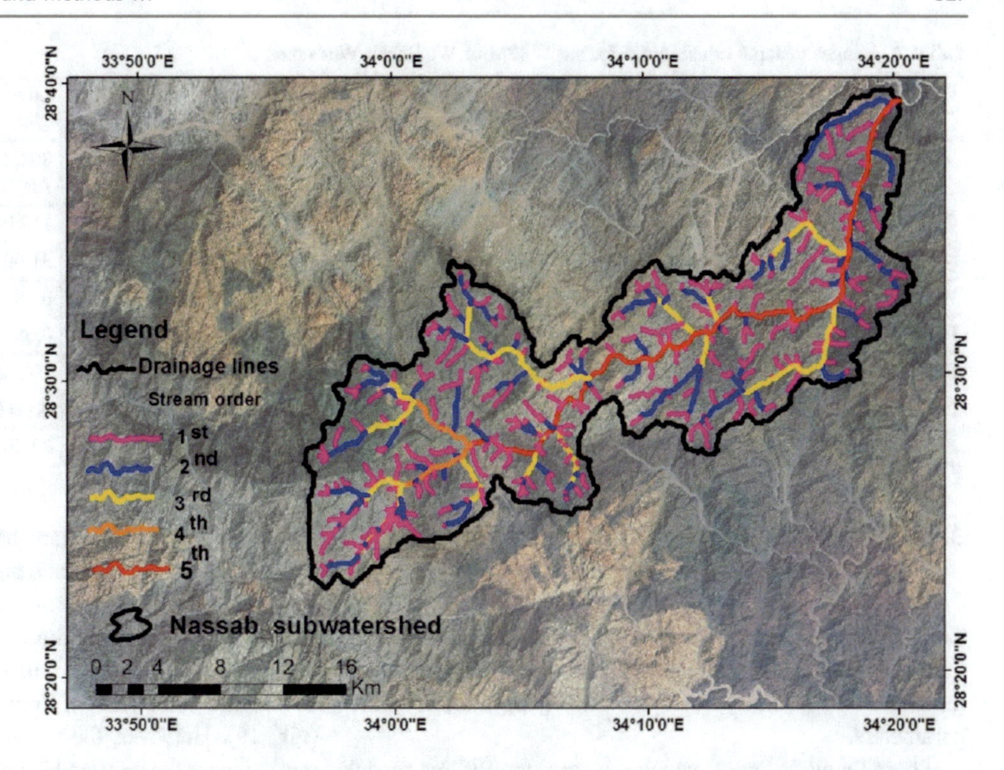

Fig. 17 A Spot-4 satellite image showing the location of W. Nassab sub-watershed

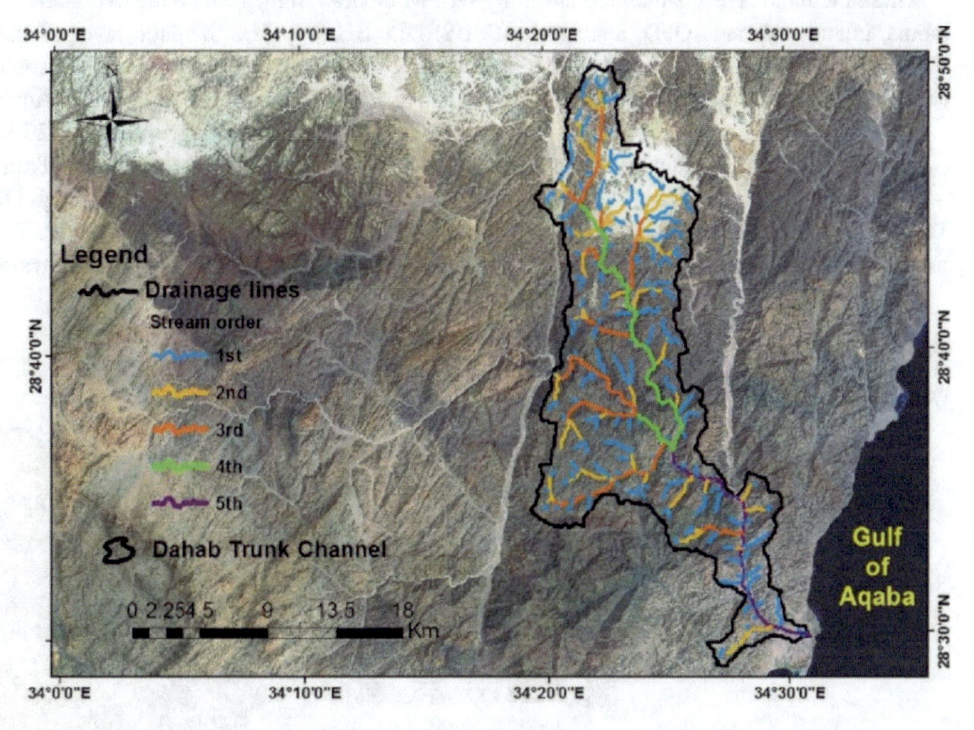

Fig. 18 A Spot-4 satellite image showing the location of W. Dahab Trunk channel sub-watershed

Table 2 Ranges of input criteria used for the WSPM of W. Dahab Watershed

Watershed RWH Criteria	Very high	High	Moderate	Low	Very Low
VAF (m^3)	>1,258,631.84	1,143,430.48–1,258,631.84	1,010,774.36–1,143,430.48	804,808.28–1,010,774.36	<804,808.28
OFD (km)	>20.80	18.00–20.70	15.10–17.90	11.30–15.00	<11.20
MFD (km)	>53.30	47.90–53.20	40.90–47.80	31.00–40.80	<30.90
IF	>0.89	0.83–0.89	0.76–0.83	0.68–0.76	<0.68
DD (km^{-1})	>0.81	0.76–0.81	0.71–0.76	0.65–0.71	<0.65
BA (km^2)	>385	321–384	263–320	192–262	<191
BS (m/m)	>0.32	0.27–0.32	0.22–0.27	0.17–0.22	< 0.17
BL (km)	>41.60	35.90–41.50	30.00–35.80	22.40–29.90	<22.40

3.2 Criteria of RWH Potentiality Determination in Wadi Dahab Watershed

The RWH potentiality of W. Dahab was determined by spatially integrating eight thematic layers, which represent the most decisive basin hydrographic and hydro-morphometric parameters.

These thematic layers, which represent the WSPM model inputs, include Average OFD, average VAF, BS, DD, BL, BA, IF, and the MFD (Table 1). The following is a short description of these themes for W. Dahab Watershed:

– **Average Overland Flow Distance (OFD)**

The OFD within the basin is computed by averaging the overland flow distance; traveled from the centroid of each triangle to the nearest stream. Most of W. Dahab Watershed is represented by the moderate and high classes for the average OFD (15.10–17.90 km and 18.00–20.70 km, respectively) (Table 2), which covers almost the areas of El-Ghaaib, Ganah, Saal, and W. Dahab Trunk Channel sub-watersheds, and covers also some parts of Zoghra and Nassab sub-watersheds (Fig. 19). However, the variation in relief and slopes determines where the overland is effective and generated.

The thematic layer of the OFD indicates a pronounced decrease in value on the southeastern side of W. Dahab near its outlet at the Gulf El-Aqaba, where low (11.3–15.00 km) and very low class (<11.20 km) occur and cover some areas of Nassab, Dahab Trunk Channel, and Khoshaib sub-watersheds. Very high OFD class (>20.80 km) occurs in southwestern parts of W. Dahab and covers some areas of Nassab and Zoghra sub-watersheds (Fig. 19).

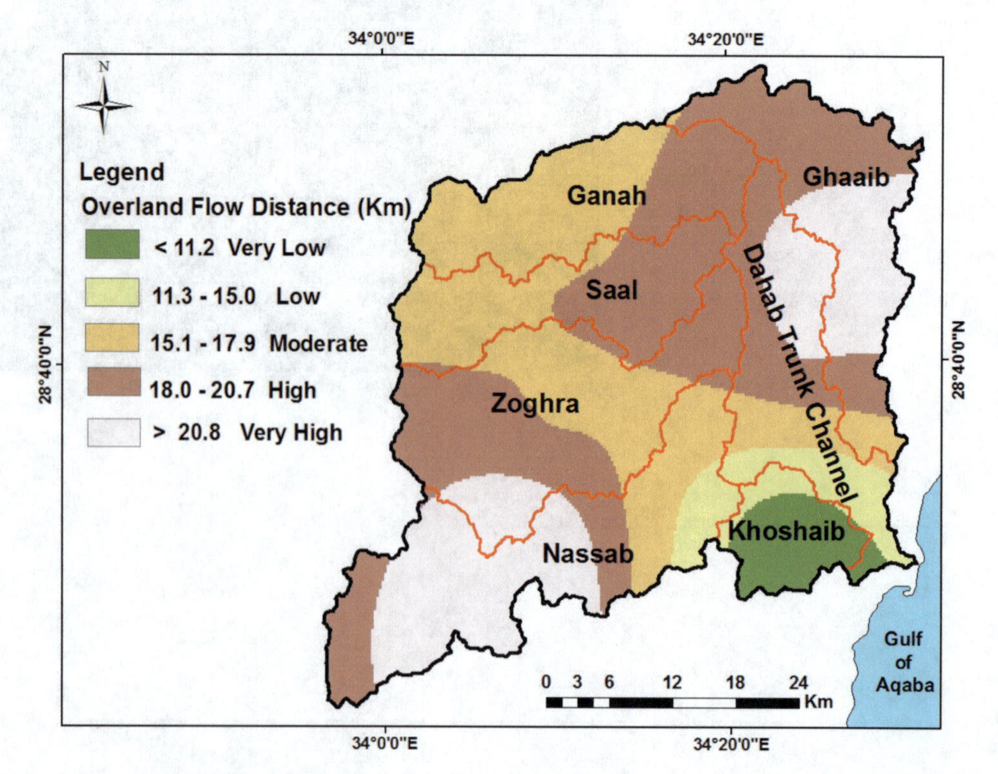

Fig. 19 The OFD thematic layer used in the WSPM of W. Dahab Watershed

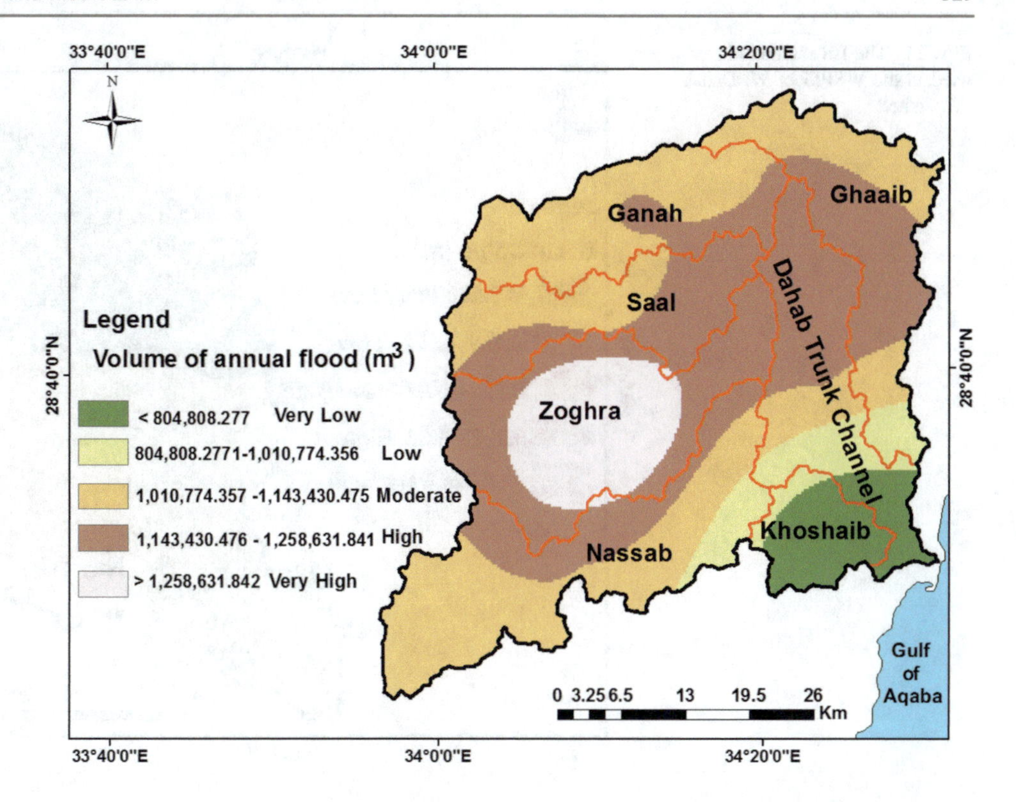

Fig. 20 VAF thematic layer used in the WSPM of W. Dahab Watershed

– **Average Volume of Annual Flood (VAF)**

The availability of a pronounced VAF in a drainage basin is one of the most significant parameters for the success of the RWH (Elewa and Qaddah 2011). W. Dahab was classified into five classes relative to its potential for the VAF generation, which was calculated according to Finkel's method. Figure 20 shows that almost the central parts of W. Dahab are classified as of high-very high classes for the VAF generation (>1,143,430.48 m^3), which represent the whole area of Zoghra and some areas of Nassab, Saal, Ganah, Ghaaib, and Dahab Trunk Channel sub-watersheds. Almost the remaining areas of these sub-watersheds were classified as of moderate class (1,010,774.36–1,143,430.48 m^3). Low-very low classes (<804,808.28 m^3) represent the lower reaches of Nassab, Khoshaib, and Dahab Trunk Channel sub-watersheds.

– **Basin Slope (BS)**

The BS map was generated from the ASTER DEM of 30-m resolution. The slope map was merged with the basin map to create the slope attributes of each drainage basin. Five BS classes were generated (Fig. 21). This map indicates a linear increase in slope values from the north, northwest, and southeast toward the eastern side at the outlet of W. Dahab Watershed (Fig. 21, Tables 1 and 2). The very steep BS is represented by Khoshaib and some areas of Dahab Trunk Channel sub-watersheds. However, this variation in BS is attributed to the general watershed slope and local geological and structural setting of W. Dahab Watershed.

– **Drainage Denisty (DD)**

W. Dahab Watershed was classified according to the DD into five classes (Fig. 22; Table 2). The thematic layer of the DD indicates a linear decrease in DD values from the upstream northern and western sides toward the southeastern side at the outlet of W. Dahab Watershed (Fig. 22). However, this variation in the DD values is mainly attributed to the lithological nature of exposed geological units, which are represented by the sedimentary rock units of Quaternary age (wadi and undifferentiated deposits), fractured limestones of Eocene Thebes Group, Duwi clay stone, siltstone with interbeds of phosphate-bearing units, Matulla Fm of fluvial and cross-bedded sandstone and varicolored glauconitic shale, Raha Fm of fine-grained glauconitic and pyritic sandstone with grey calcareous shale of Late Cretaceous (Said, 1962, 1990; CONOCO 1987). These rock units are characterized by soft surface soils with low vertical permeability that causes a noticeable increase in the DD values in these parts of the watershed (Figs. 6 and 22) (i.e., >0.814 km^{-1} and 0.764–0.813 km^{-1} for the very high and high DD classes, respectively). In contrary to the southeastern and southwestern parts of the watershed, which are covered by moderately fractured grey or older granites or

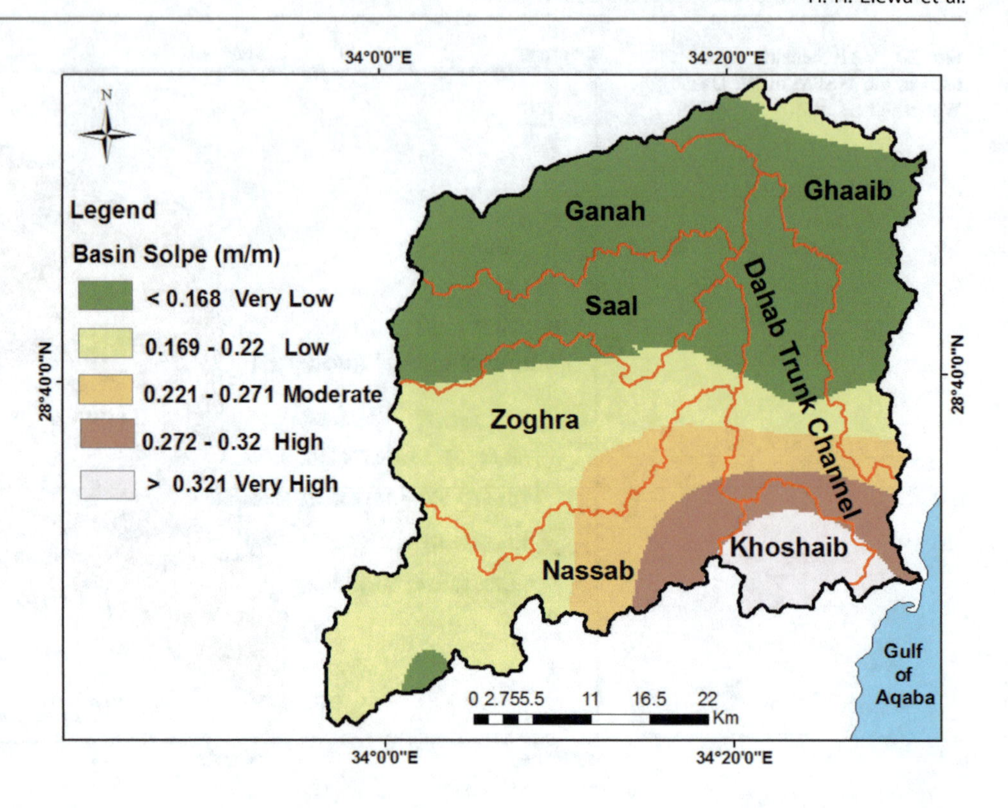

Fig. 21 The BS thematic layer used in the WSPM of W. Dahab Watershed

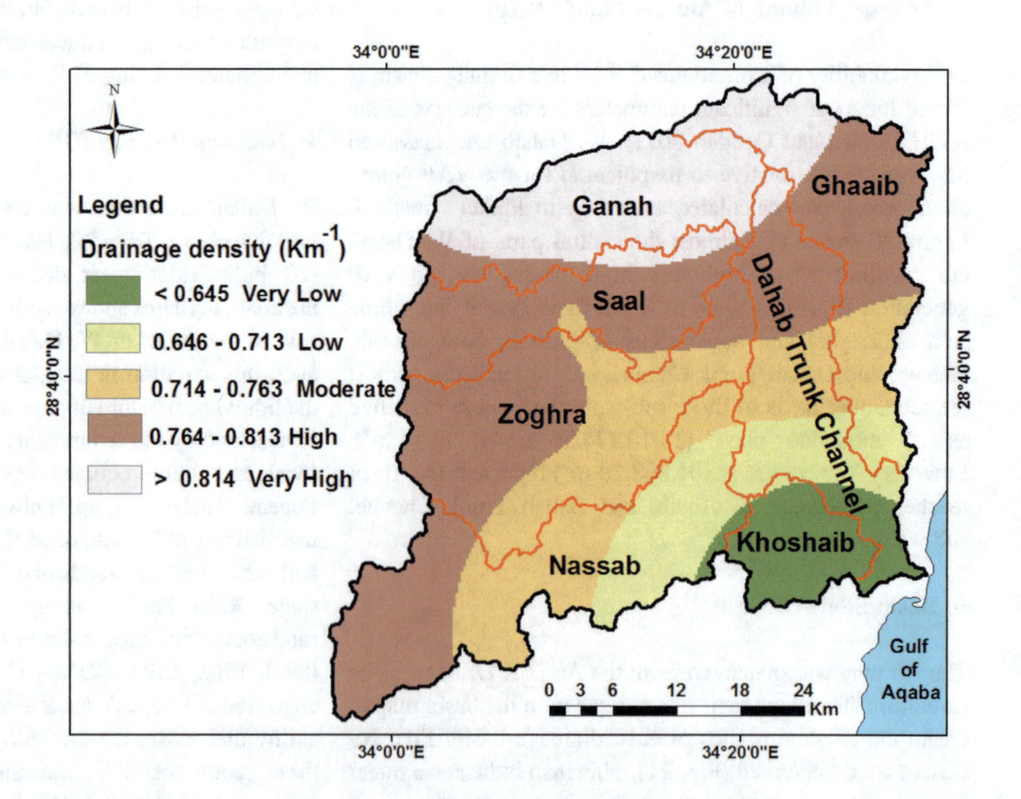

Fig. 22 The DD thematic layer used in the WSPM of W. Dahab Watershed

the highly metamorphosed rocks, which despite their very negligible rock porosity are having a reasonable vertical fracture permeability that causes the DD values to be low and very low (i.e., <0.65 km^{-1}). However, in consolidated sedimentary rocks and igneous and metamorphic rocks, fracture flow will almost invariably occur and hence a wide range of values of hydraulic conductivity can occur for any one lithology, depending on both the degree of fracturing and the size of the fractures (Lewis et al. 2006), which is clear in the southwestern areas of the watershed, where highly metamorphosed melanocratic rock of very low permeability occur (Figs. 6 and 22). The previously discussed facts that stand behind the reasons of spatial variation in the DD are also dramatically reinforced by the map of IF, where high-very high IF values are encountered in the sedimentary rock territories reflecting the low vertical rock permeability, hence their higher DD values (Fig. 22).

– **Basin Length (BL)**

The thematic layer of BL for W. Dahab Watershed was classified into five classes, where the more compacted basins, like those occurred in southeastern parts of W. Dahab Watershed represent the very low class (<22.30 km), which was occupied by W. Khoshaib sub-watershed (Fig. 23). This class was followed by the low class (22.4–29.9 km) that represents some areas in the downstream of Nassab and Dahab Trunk Channel sub-watersheds. The northeastern and central parts of W. Dahab Watershed were classified as of moderate class (30–35.8 km), which was represented by some parts of El-Ghaaib, Zoghra, Ganah and Dahab Trunk Channel and most areas of Saal sub-watersheds. The high (35.9–41.5 km) and very high (>41.6 km) classes of the BL were represented by some areas in the eastern and southwestern parts of W. Dahab Watershed, which were represented by the El-Ghaaib, Zoghra, Ganah, Saal, and Dahab Trunk Channel, Zoghra and Nassab sub-watersheds.

– **Basin Area (BA)**

Basin area has been identified as the most important of all "morphometric parameters controlling the catchment runoff pattern. This is because, the larger the size of the basin, the greater the amount of rain it intercepts and the higher the peak discharge that results" (Faniran and Ojo 1980). A thematic layer of BA with five classes was generated (Fig. 24). The very high (>385 km^2) and high (321–384 km^2). BA classes occur in the southwestern parts of the watershed, which are covering some areas of Zoghra and Nassab sub-watershed. Moderate class (263–320 km^2) (Table 2) represents the most area of W. Dahab in El-Ghaaib, Ganah, Saal, Zoghra, Nassab, and Dahab Trunk Channel sub-watersheds. Low class (192–262 km^2) is represented by the upstream parts of Saal and Ganah, and some parts are occurring at the southeast that is represented by El-Ghaaib, khoshaib, and Dahab Trunk Channel sub-watersheds. Very low BA class (<191 km^2) is represented by Khoshaib and outlet part of W. Dahab Trunk Channel sub-watersheds.

– **Basin Infiltration Number (IF)**

The IF map with five classes was produced to reveal the infiltration capabilities and their effect on the RWH capabilities of W. Dahab Watershed. The thematic IF layer indicates a linear decrease in IF values from the upstream northern–southwestern parts toward the southeastern downstream reaches of W. Dahab Watershed, indicating a

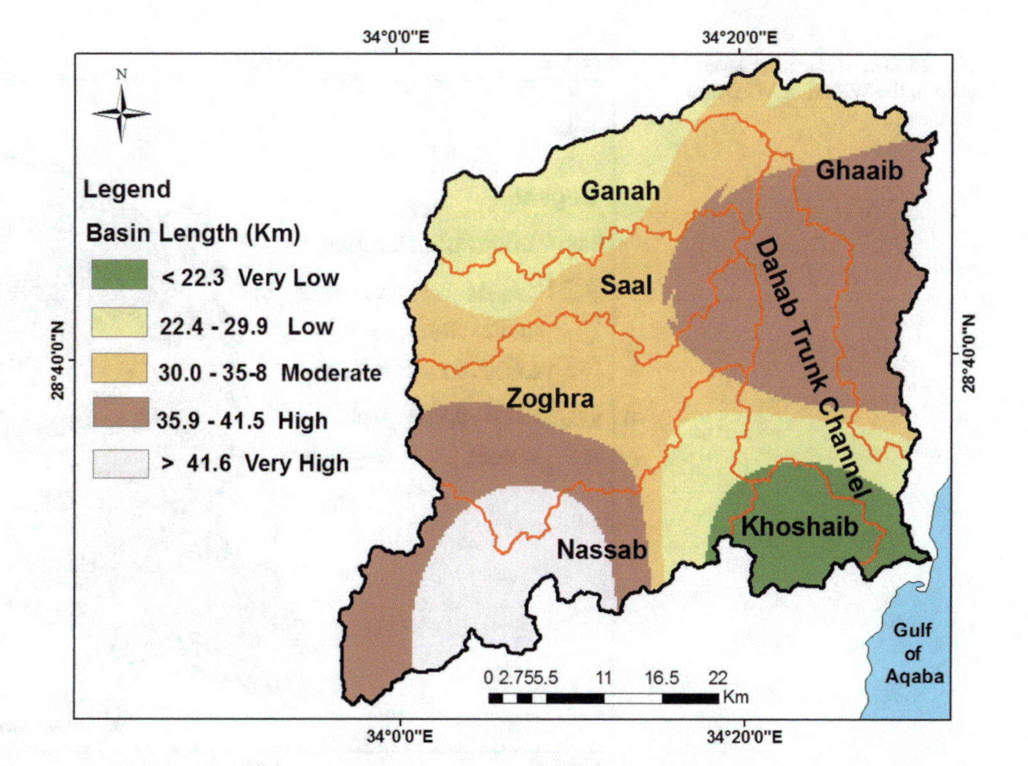

Fig. 23 The BL thematic layer used in the WSPM of W. Dahab Watershed

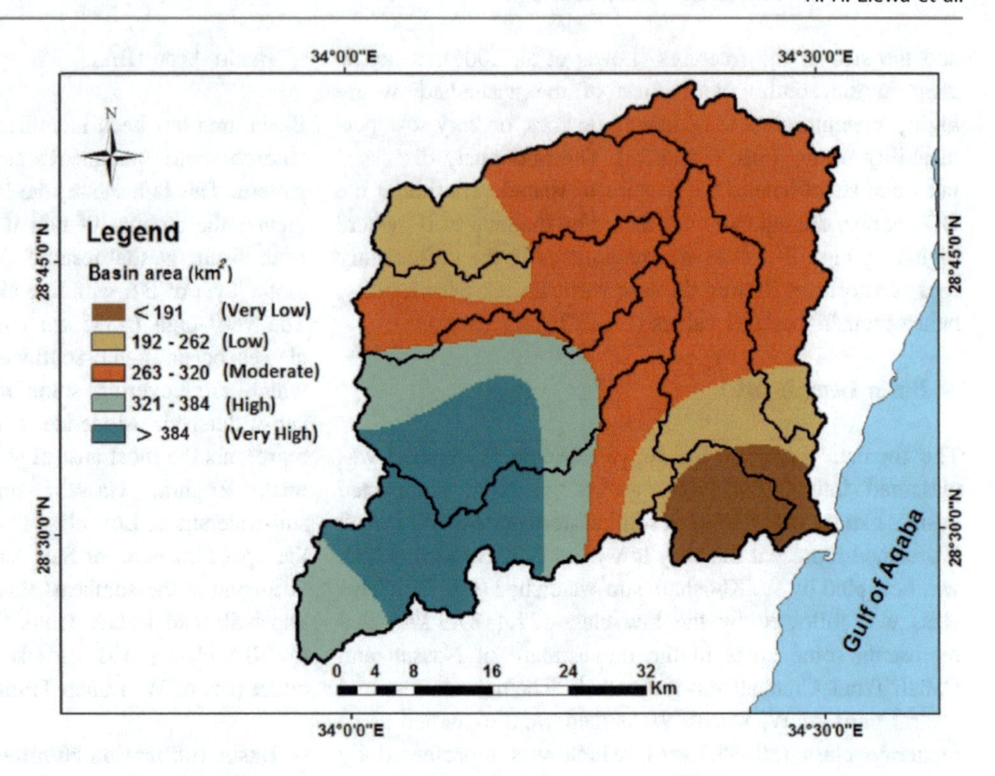

Fig. 24 The BA thematic layer used in the WSPM of W. Dahab Watershed

pronounced increase in infiltration capabilities in the same directions (Fig. 25). However, this variation in the IF has mainly attributed the facies change of different lithologic units exposed in the study area, as previously discussed. The IF classes are very high (>0.8939), high (0.8284–0.8938), moderate (0.763–0.8283), low (0.6792–0.7629), and very low (<0.6791) (Table 2; Fig. 25).

– **Maximum Flow Distance (MFD)**

The very low (<30.9 km) and low (31.0–40.8 km) MFD classes are represented by the southeastern parts of W. Dahab Watershed that cover the most areas of Khoshaib and some areas of Nassab, and Dahab Trunk Channel sub-watersheds (Fig. 26; Table 2). High (47.9–53.2 km) and

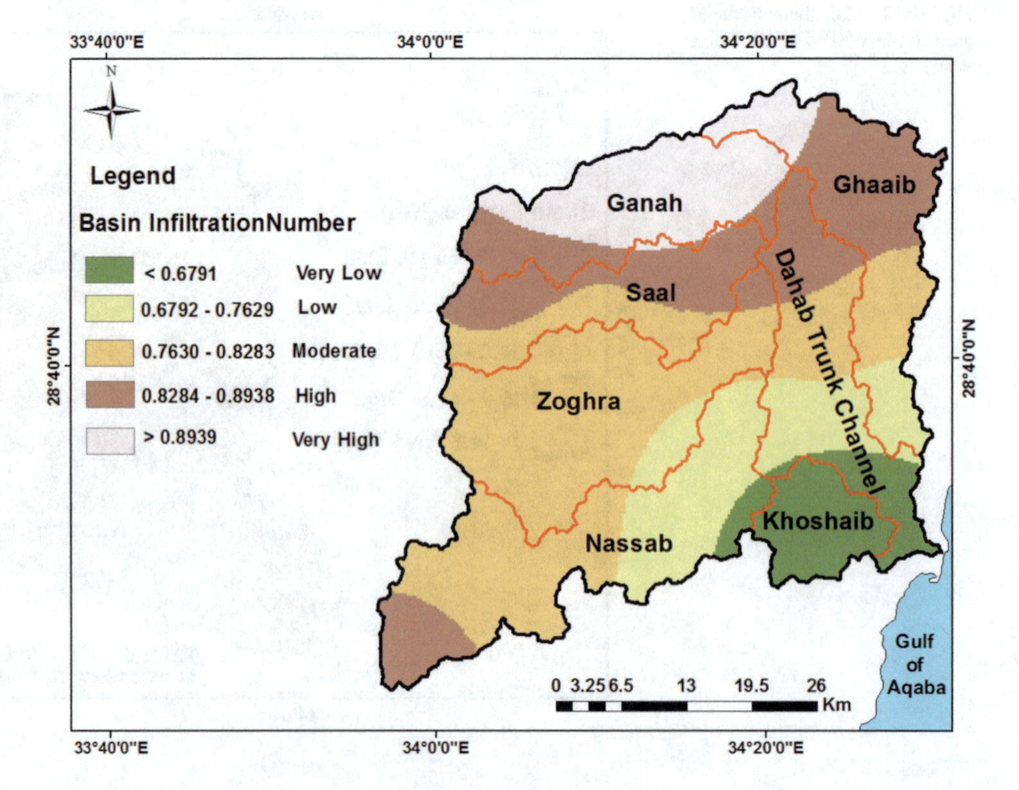

Fig. 25 The IF thematic layer used in the WSPM of W. Dahab Watershed

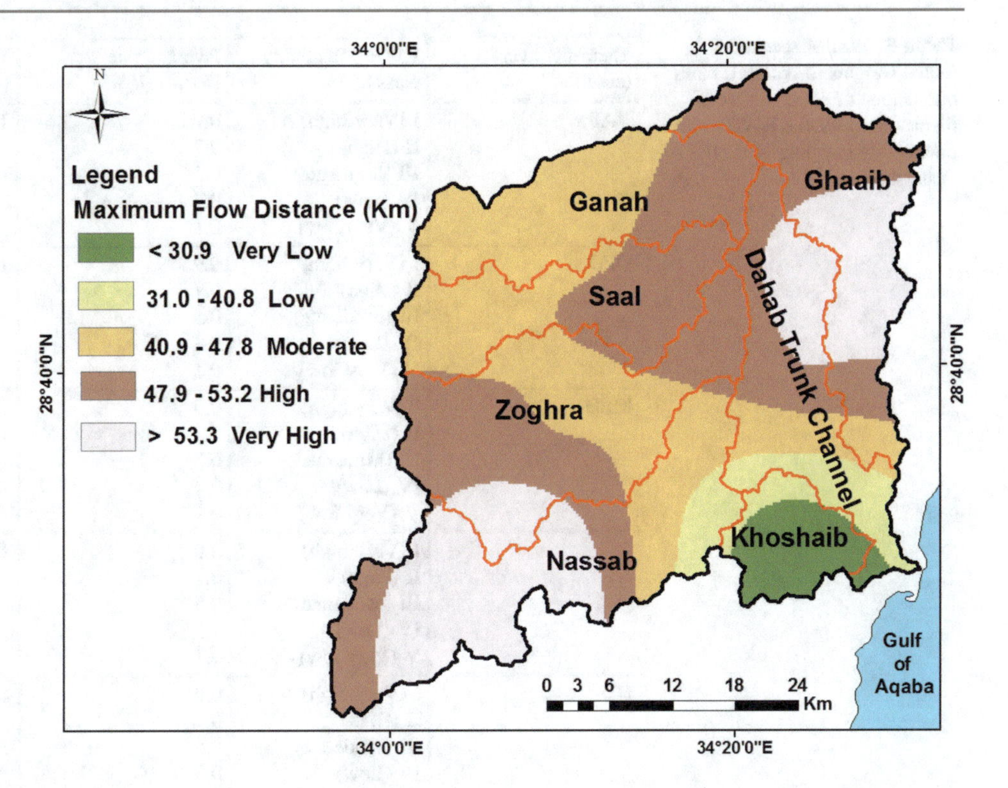

Fig. 26 The MFD thematic layer used in the WSPM of W. Dahab Watershed

very high (>53.3 km) MFD classes are represented in two localities; the first one occurs in the northeastern-eastern areas, which are occupied by some parts of El-Ghaaib, Ganah, Saal, Zoghra, and Dahab Trunk Channel sub-watersheds, whereas the second one is represented by the southwestern side, which is covered by some areas of Zoghra and Nassab sub-watersheds. The moderate MFD class (40.9–47.8 km) covers large areas of Ganah, Saal, Zoghra, Nassab, and Dahab Trunk Channel sub-watersheds.

3.3 Weighted Spatial Probability Modeling (WSPM) for Determining the RWH Potentialities of W. Dahab Watershed

The WSPM for W. Dahab Watershed was also run by the same previously discussed manner, where three scenarios were performed; i.e., equal weights to criteria, weights proposed by the authors, and weights justified by the sensitivity analysis.

– **WSPM's Scenario I (equal weights to criteria)**:

In this first WSPM's scenario, the previously discussed eight thematic criteria are proposed to have equal degrees of contribution in the potentiality mapping for RWH where the integrated criteria were given an equal weight of 12.5% with a summation of 100% for all data themes. The classes were categorized from I (very high potentiality) up to V (very low potentiality) in the RWH potentiality mapping (Table 3). Considering 100% for the maximum value of the rank, thus for the five classes; ranks will be classified as 100–80, 80–60, 60–40, 40–20, and 20–0%, respectively. Consequently, the average ranking for each class will be 0.90, 0.70, 0.50, 0.30, and 0.10% for classes from I to V, respectively (Table 3). The effectiveness degree (E) for each thematic layer was calculated by multiplying the criterion weight (W_c) with the criterion rank (R_c). For example, if the weight of VAF equals 12.5% and this is multiplied by the average rank of 90 (for a class I), the degree of effectiveness will be 11.25 (Eq. 3).

$$E = W_c \times R_f = 0.125 \times 90 = 11.25 \quad (3)$$

According to data manipulation method, the valuation of the effectiveness of each decision criterion provides a qualified analysis of the different thematic layers. A resulted WSPM map for RWH potentiality with five classes ranging from the very low to very high potentiality was obtained (Fig. 27).

The spatial distribution of these classes relative to the total study area is 3.82 (very low), 8.55 (low), 24.67 (moderate), 42.98 (high), and 19.96% (very high) for RWH potentiality mapping (Fig. 27; Table 4).

From this WSPM map, it could be concluded that the major area of W. Dahab is categorized as of high and very high potentiality for the RWH (i.e., 62.94% of total watershed area), especially in its northeastern, eastern, central, western, and southwestern parts of W. Dahab, the RWH is

Table 3 WSPM scenario I (equal weights to criteria), ranks and degree of effectiveness of themes used for the RWH potentiality mapping of W. Dahab Watershed

Thematic layer (criterion)	RWH potentiality class	Average rate (Rank) (R_c)	Weight (W_c)	Degree of effectiveness (E)
VAF	I (Very high)	0.9	12.5	11.25
	II (High)	0.7		8.75
	III (Moderate)	0.5		6.25
	IV (Low)	0.3		3.75
	V (Very low)	0.1		1.25
OFD	I (Very high)	0.9	12.5	11.25
	II (High)	0.7		8.75
	III (Moderate)	0.5		6.25
	IV (Low)	0.3		3.75
	V (Very low)	0.1		1.25
MFD	I (Very high)	0.9	12.5	11.25
	II (High)	0.7		8.75
	III (Moderate)	0.5		6.25
	IV (Low)	0.3		3.75
	V (Very low)	0.1		1.25
IF	I (Very high)	0.9	12.5	11.25
	II (High)	0.7		8.75
	III (Moderate)	0.5		6.25
	IV (Low)	0.3		3.75
	V (Very low)	0.1		1.25
DD	I (Very high)	0.9	12.5	11.25
	II (High)	0.7		8.75
	III (Moderate)	0.5		6.25
	IV (Low)	0.3		3.75
	V (Very low)	0.1		1.25
	V (Very low)			
BA	I (Very high)	0.9	12.5	11.25
	II (High)	0.7		8.75
	III (Moderate)	0.5		6.25
	IV (Low)	0.3		3.75
	V (Very low)	0.1		1.25
BS	I (Very high)	0.9	12.5	11.25
	II (High)	0.7		8.75
	III (Moderate)	0.5		6.25
	IV (Low)	0.3		3.75
	V (Very low)	0.1		1.25
BL	I (Very high)	0.9	12.5	11.25
	II (High)	0.7		8.75
	III (Moderate)	0.5		6.25
	IV (Low)	0.3		3.75
	V (Very low)	0.1		1.25

noticeably decreasing to low and very low potentiality for the RWH (8.55 and 3.82% of the total watershed area, respectively) at the downstream reaches and some areas in the southwestern part of the watershed (Fig. 27; Table 4).

– **WSPM's Scenario II (weights proposed by the authors):**

Based on the weights proposed by the authors, the eight thematic layers (WSPM criteria) were superposed by the ArcGIS 10.1© within the Spatial Analyst Model Builder to perform the WSPM by taking the new authors' proposed weights as OFD (17%), VAF (16%), BS (12%), DD (12%), BL (12%) MFD (10%), IF (10%), and BA (11%) (Table 5). The WSPM output map with five classes ranging from the very low to very high potentiality was obtained (Fig. 28). The spatial distributions of these classes relative to the total watershed area were 7.56 (very low), 14.13 (low), 21.35 (moderate), 42.53 (high), and 14.42% (very high) potentiality for the RWH (Fig. 28; Table 6). From the WSPM scenario II, it is clear that the high and very high potential classes for the RWH encompasses about 56.95% of the total watershed area, compared to 62.94% in scenario I. Furthermore, the same spatial distribution of RWH potentiality classes was established by the WSPM scenario II map (Fig. 28; Table 6).

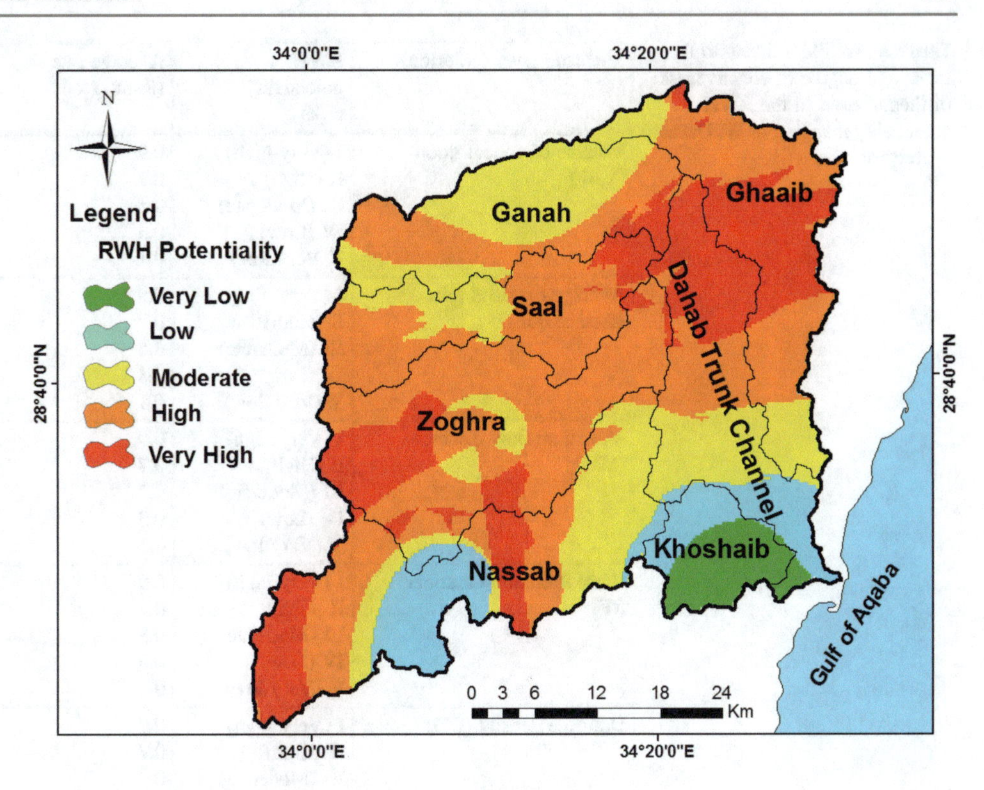

Fig. 27 WSPM's output map of Scenario I (equal weights to criteria) showing the RWH potentiality of W. Dahab Watershed

Table 4 Areas of RWH potentiality classes resulted from the WSPM Scenario I in W. Dahab Watershed

RWH potentiality map					
RWH potentiality class	Very low	Low	Moderate	High	Very high
Area (Km^2)	79.17	177.32	511.17	890.56	413.54
Area % relative to the total watershed area (Total watershed area: 2071.76 km^2) (%)	3.82	8.55	24.67	42.98	19.96

WSPM's Scenario III (Justified Weights by the Sensitivity Analysis) for the RWH Potentiality mapping of W. Dahab Watershed

In the third scenario, a sensitivity analysis (Van Griensven et al. 2006) was performed to justify the weights of the WSPM's criteria in order to attain a more justified or optimum RWH potential areas in W. Dahab.

However, to justify the WSPM's weights and results, we should take all the scenarios as different alternatives for assigning the criteria weights. The WSPM's sensitivity analysis for the determination of RWH potentiality was performed through the following steps:

- The first step (Scenario I) involves assuming that all the WSPM's eight thematic layers or criteria have the same magnitude of contribution or weights in the RWH potentiality mapping. In this scenario, all criteria were assigned an equal weight of 12.5% (equal effect) (Table 3).
- The second step was to determine the potentiality of RWH using the sensitivity analysis to justify the decisive criteria weights. The potentiality assessment of RWH sites involves the need to evaluate the consistency of the parameters used in the prioritization. A small perturbation in the decision weights may have an important influence on the rank ordering of the criteria, which afterward may ultimately modify the best choice and the model results. However, the uncertainties associated with MCDSS techniques are inevitable and the model outcomes are open to multiple types of uncertainty.

In each WSPM's running operation, seven parameters were kept with the same weights of 10%, while assigning only the remaining parameter with the residual 30% (Figs. 29 through 36) (Figs. 30, 31, 32, 33, 34, and 35).

The previously discussed WSPM's running practice is essential to apply the variance-based global sensitivity analysis (GSA) or what is called the analysis of variance (ANOVA), which subdivides the variability and apportions it to the uncertain inputs (Ha et al. 2012; Feizizadeh et al. 2004). GSA is depended on perturbations of the entire parameter space, where input factors are examined both

Table 5 WSPM's Scenario II; ranks and degree of effectiveness of themes used in the RWH potentiality mapping of W. Dahab Watershed

Thematic layer (criterion)	RWH potentiality class	Average rate (Rank) (R_c)	Weight (W_c)	Degree of effectiveness (E)
Volume of annual flood (VAF)	I (Very high)	0.9	16	14.4
	II (High)	0.7		11.2
	III (Moderate)	0.5		8
	IV (Low)	0.3		4.8
	V Very low)	0.1		1.6
Average overland flow distance (OFD)	I (Very high)	0.9	17	15.3
	II (High)	0.7		11.9
	III (Moderate)	0.5		8.5
	IV (Low)	0.3		5.1
	V (Very low)	0.1		1.7
Maximum flow distance (MFD)	I (Very high)	0.9	10	9
	II (High)	0.7		7
	III (Moderate)	0.5		5
	IV (Low)	0.3		3
	V (Very low)	0.1		1
Basin infiltration number (IF)	I (Very high)	0.9	10	9
	II (High)	0.7		7
	III (Moderate)	0.5		5
	IV (Low)	0.3		3
	V (Very low)	0.1		1
Drainage density (DD)	I (Very high)	0.9	12	10.8
	II (High)	0.7		8.4
	III (Moderate)	0.5		6
	IV (Low)	0.3		3.6
	V (Very low)	0.1		1.2
	V (Very low)			
Basin area (BA)	I (Very high)	0.9	11	9.9
	II (High)	0.7		7.7
	III (Moderate)	0.5		5.5
	IV (Low)	0.3		3.3
	V (Very low)	0.1		1.1
Basin slope (BS)	I (Very high)	0.9	12	10.8
	II (High)	0.7		8.4
	III (Moderate)	0.5		6
	IV (Low)	0.3		3.6
	V (Very low)	0.1		1.2
Basin length (BL)	I (Very high)	0.9	12	10.8
	II (High)	0.7		8.4
	III (Moderate)	0.5		6
	IV (Low)	0.3		3.6
	V (Very low)	0.1		1.2

individually and in combination (Ligmann-Zielinska 2013). Variance-based GSA has been used in this context, where this approach is recognized as one of the most proper procedures for qualifying the factors' weights of the WSPM (Saltelli et al. 2000; Saisana et al. 2005). The aim of variance-based GSA is to quantitatively determine the weights that have the most effect on model results, in this case the RWH categorization is computed for each cell of the basin area by the decisive parameter. With this method, we aim to create two sensitivity measures: first order (S) and total effect (ST) sensitivity index.

The importance of a given input factor (X_i) can be measured via the so-called sensitivity index, which is defined as the fractional contribution to the model output variance due to the uncertainty in (X_i). For (k) independent input factors, the sensitivity indices can be computed by using the following decomposition formula for the total output variance ($V(Y)$) of the output (Y) (Saisana et al. 2005) (Eqs. 4–6):

$$V(Y) = \sum_i V_i + \sum_i \sum_{j>i} V_{ij} + \ldots + V_{1,2,\ldots k} \quad (4)$$

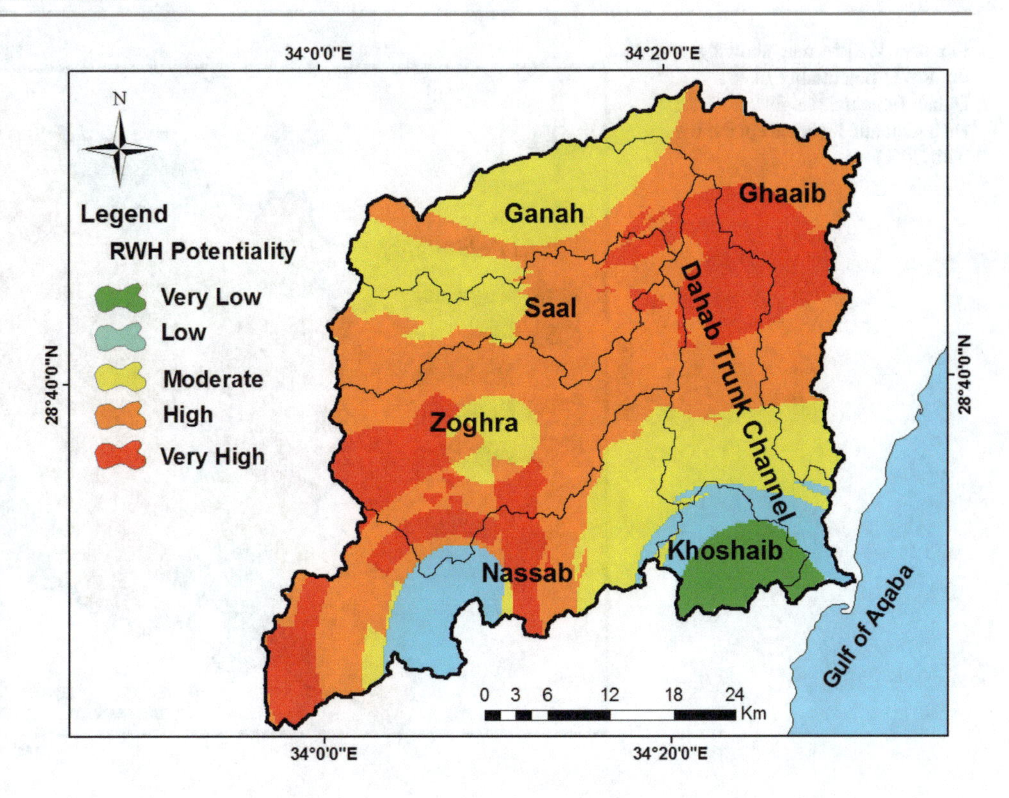

Fig. 28 WSPM map of Scenario II showing the RWH potentiality of W. Dahab Watershed

Table 6 RWH potential classes resulted from the WSPM's Scenario II map

RWH potentiality map					
RWH potentiality class	Very low	Low	Moderate	High	Very high
Area (Km^2)	156.64	292.66	442.45	881.19	298.85
Area % relative to the total watershed area (Total watershed area: 2071.75 Km^2) (%)	7.56	14.13	21.35	42.53	14.42

$$V_i = V_{xi}\{E_{x-i}(Y|X_i)\} \quad (5)$$

$$V_{ij} = V_{xixj}\{E_{x-ij}(Y|X_i, X_j)\} - V_{xi}\{E_{x-i}(Y|X_i)\} - V_{xj}\{E_{x-j}(Y|X_j)\} \quad (6)$$

A partial variance V_i represents the repeated variation of a single criterion (*i*) (i.e., one of the eight model criteria) that affects the other model criteria, which constitute the inputs of the WSPM. In other words, one of the WSPM eight parameters is changed while the rest remains constant. However, in Eq. 6, higher order effects ($V_{1,\ 2,\ \ldots\ p}$) are combined effect for two or more inputs. The partial effects can be calculated with special sampling schemes that are often computationally demanding (Saltelli et al. 2000).

Accordingly, the high-very high RWH potentiality classes and their total area (green columns), which are resulted from scenario I (equal weights to criteria) were compared to the Scenario III (unequal weighting), in which all criteria had been assigned an equal weight of 10% unless one criterion with 30%. Additionally, the percentage of variance in the total area of high and very high classes that resulted from scenario III was correlated with the scenario I (Figs. 37 and 38; Table 7).

Thus, it is observed that the BS has the highest percentage of variance value, followed by the MFD and Ba, whereas the least variance resulted from BL and VAF.

According to Fig. 38 and Table 7, the total of all variance ratios in the high-very high classes for the RWH potentiality in scenario III for each criterion with respect to their areas in scenario I is 60.64%. Accordingly, the justified weight of each criterion was determined by dividing the variance ratio of each criterion by the summation of all variance ratios (60.64%) (Fig. 38; Table 7), for example, the justified weight of the BS could be obtained by dividing its variance ratio (23.39%) by the summation of all variance ratios (60.64%), which is equal to 38.57% (Table 7).

Accordingly, the justified or optimum weights of each thematic layer shown in Table 7 were used to build a new arithmetic overlay approach within the ArcGIS 10.1© for performing the WSPM. The new justified weights are: BS

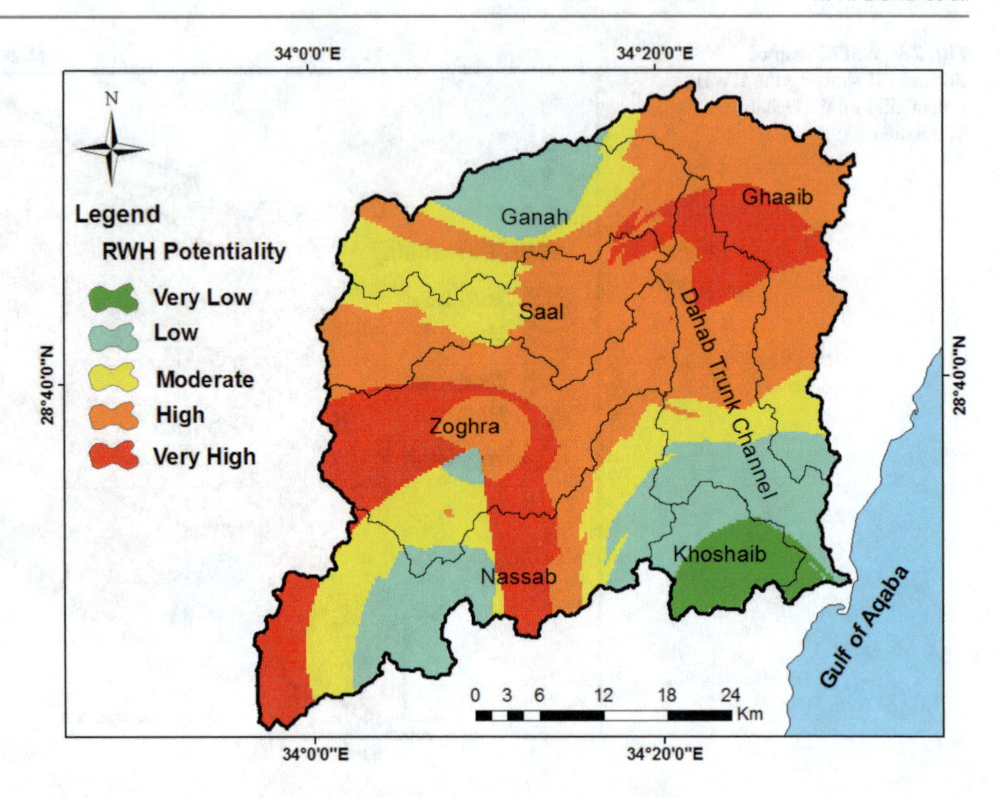

Fig. 29 WSPM map showing the RWH potentiality of W. Dahab (unequal weights: 10% for each thematic layer except the BA with 30%)

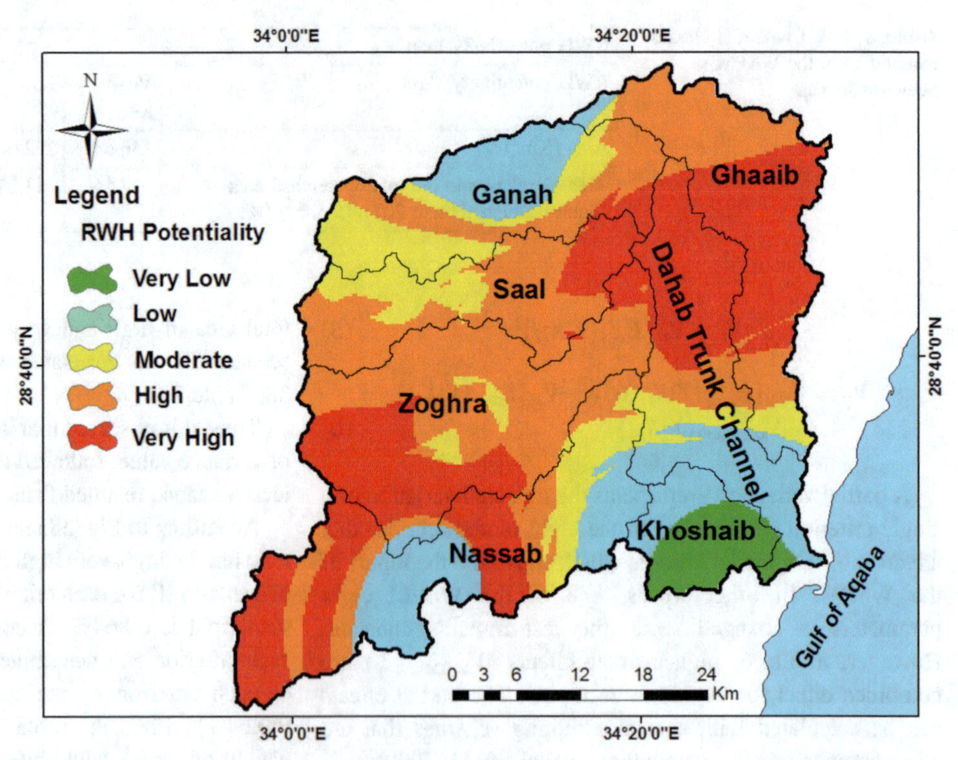

Fig. 30 WSPM map showing the RWH potentiality of W. Dahab (unequal weights: 10% for each thematic layer except the BL with 30%)

(38.57%), DD (8.71%), MFD (13.29%), VAF (1.86%), IF (8.67%), BL (7.44%), OFD (8.16%), and BA (13.29%). A WSPM output map with five classes ranging from the very low to the very high potentiality for the RWH was obtained. The spatial distribution of these classes relative to the total watershed area is 2.59% (very low), 8.85% (low), 14.7% (moderate), 58.27% (high), and 15.56% (very high) (Fig. 39; Table 8).

From the map produced by the audited weights (Table 8; Fig. 39), it could be concluded that the major area of W. Dahab is categorized as of high and very high for the RWH potentiality (58.27 and 15.56% of total watershed area,

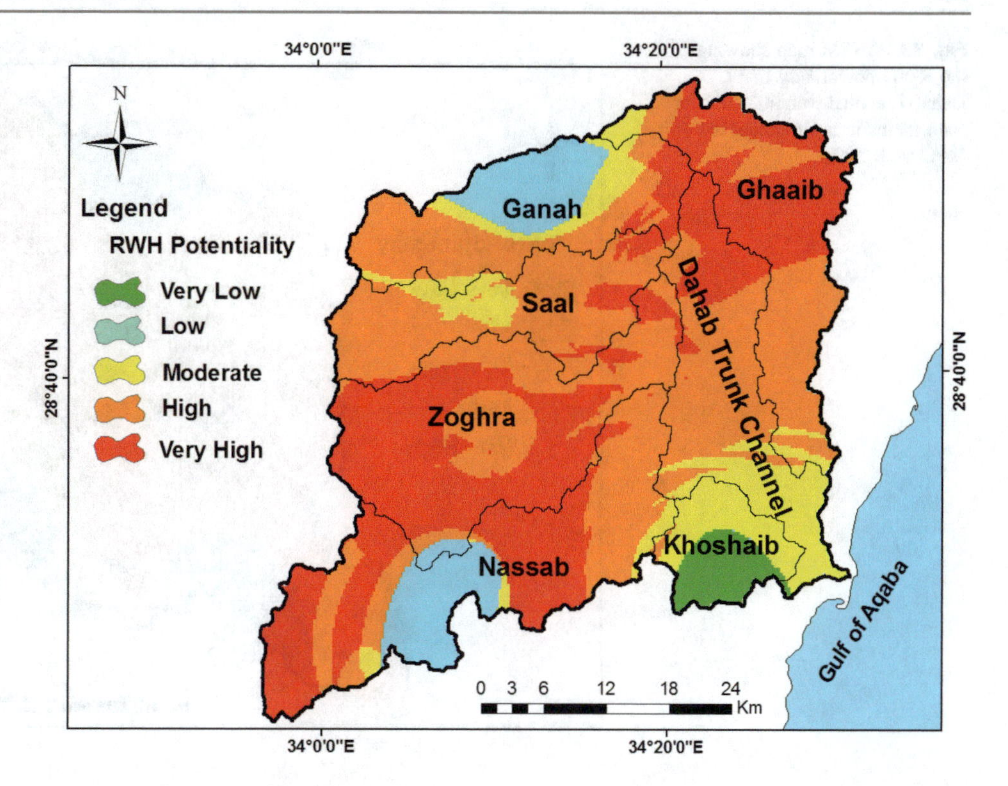

Fig. 31 WSPM map showing the RWH potentiality of W. Dahab (unequal weights: 10% for each thematic layer except the BS with 30%)

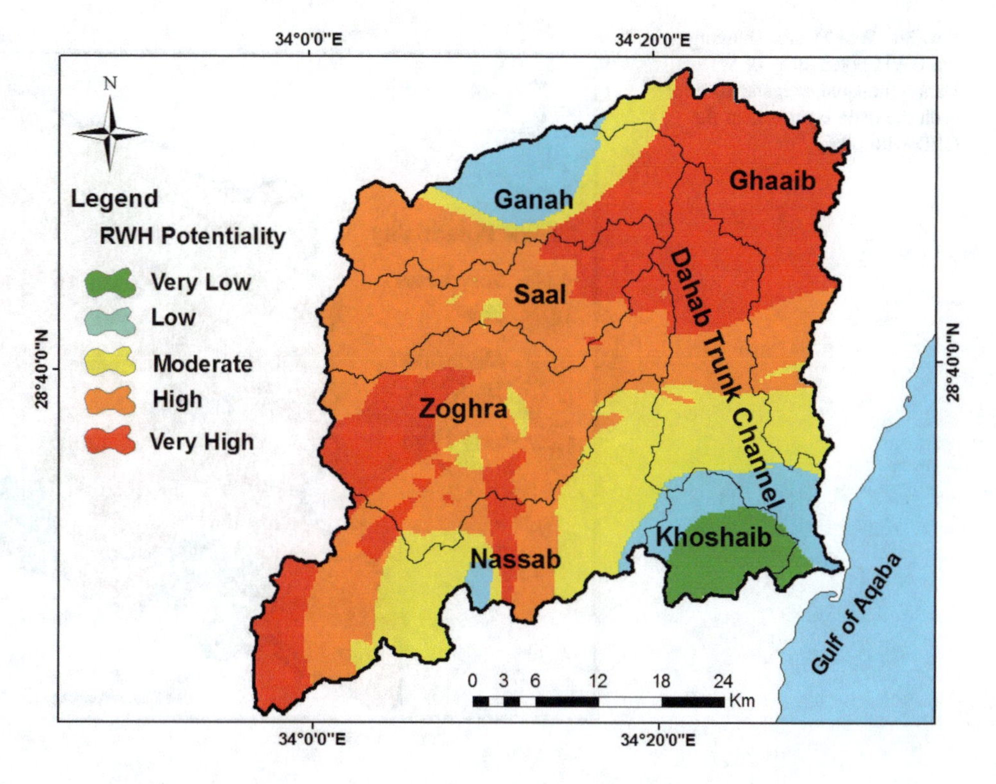

Fig. 32 WSPM map showing the RWH potentiality of W. Dahab (unequal weights: 10% for each thematic layer except the DD with 30%)

respectively), especially, in its northeastern, eastern, central, western, and southwestern parts, which are represented by El-Ghaaib, Dahab Trunk Channel, Zoghra, Nassab, Saal, Ganah, and Khoshaib sub-watersheds. The moderate class (14.7%), which is noticeably decreasing to low and very low (8.85 and 2.59%, respectively) are encountered in some areas in the upstream northwestern and southwestern parts of Saal, Ganah, Nassab, and Khoshaib sub-watersheds.

In general, the WSPM's scenario III map, which was audited by the sensitivity analysis, gives the results in favor of the other scenarios whether based on equal weights (scenario I) or those that were assumed by the authors

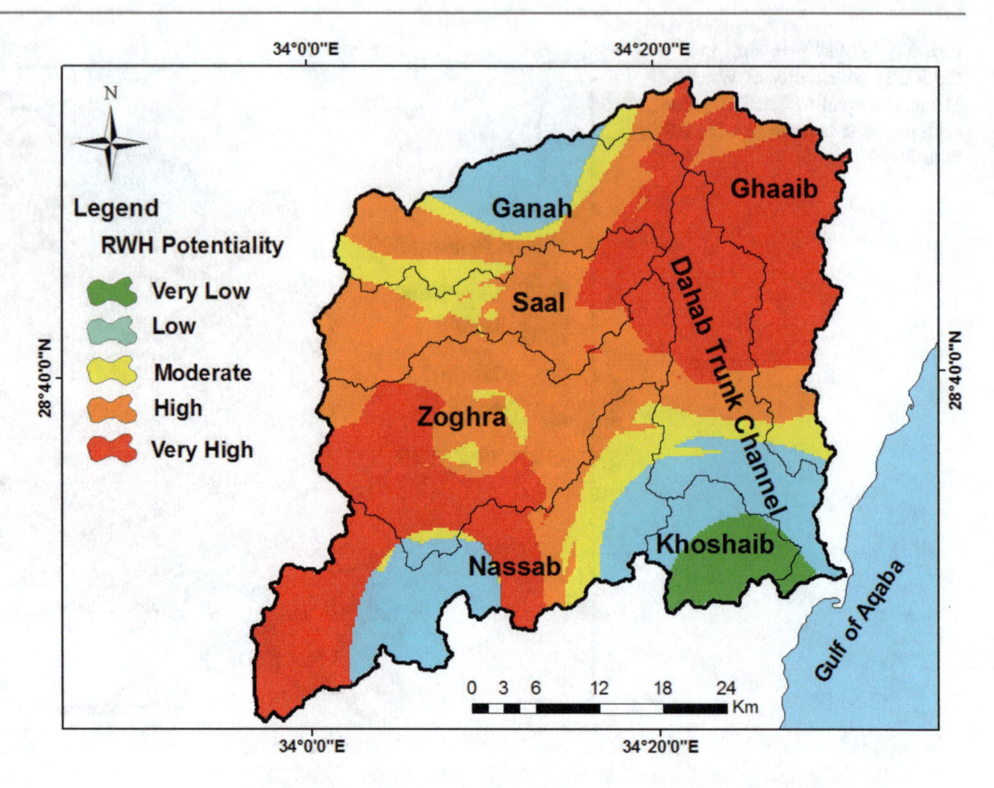

Fig. 33 WSPM map showing the RWH potentiality of W. Dahab (unequal weights: 10% for each thematic layer except the MFD with 30%)

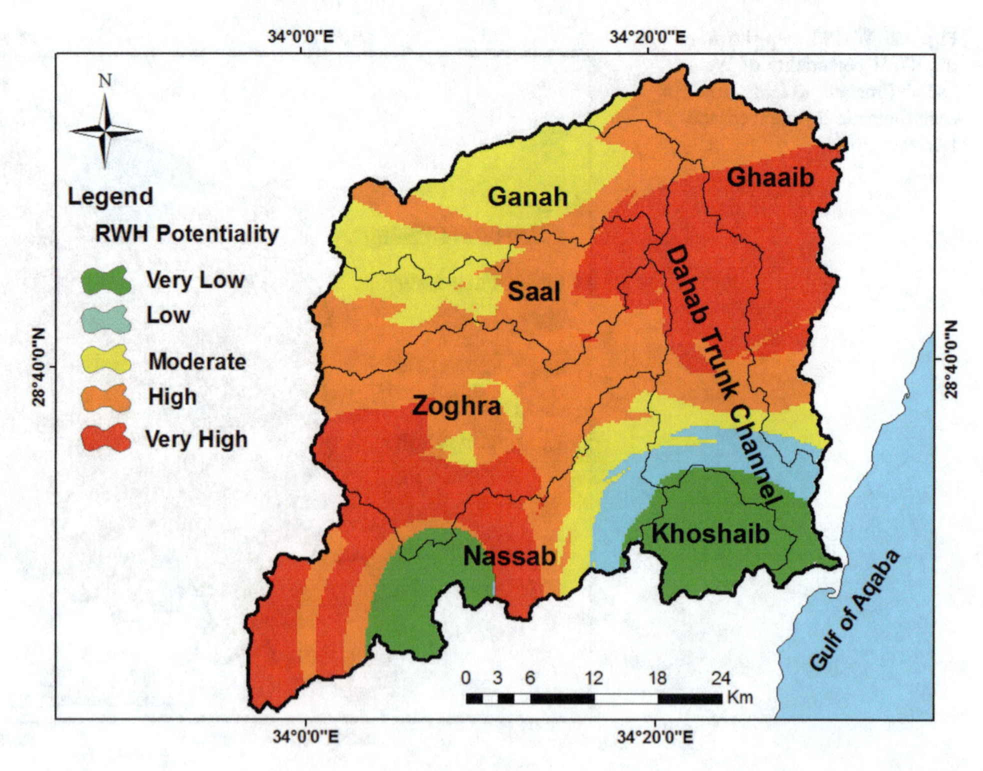

Fig. 34 WSPM map showing the RWH potentiality of W. Dahab (unequal weights: 10% for each thematic layer except the OFD with 30%)

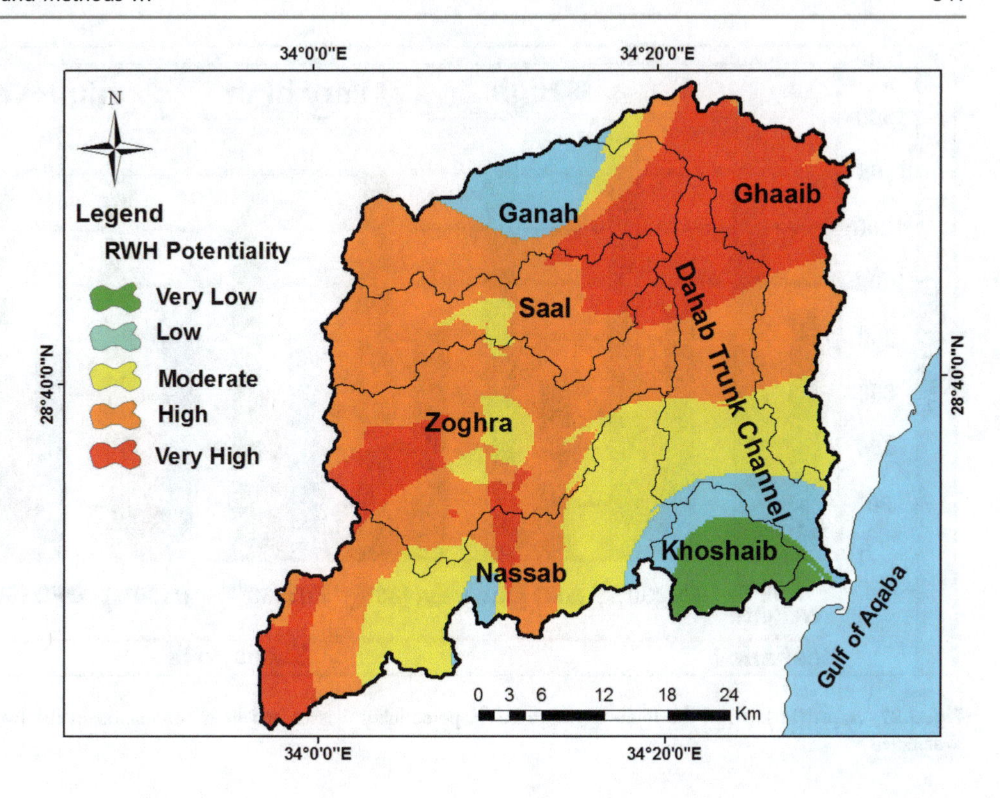

Fig. 35 WSPM map showing the RWH potentiality of W. Dahab (unequal weights: 10 % for each thematic layer except the IF with 30 %)

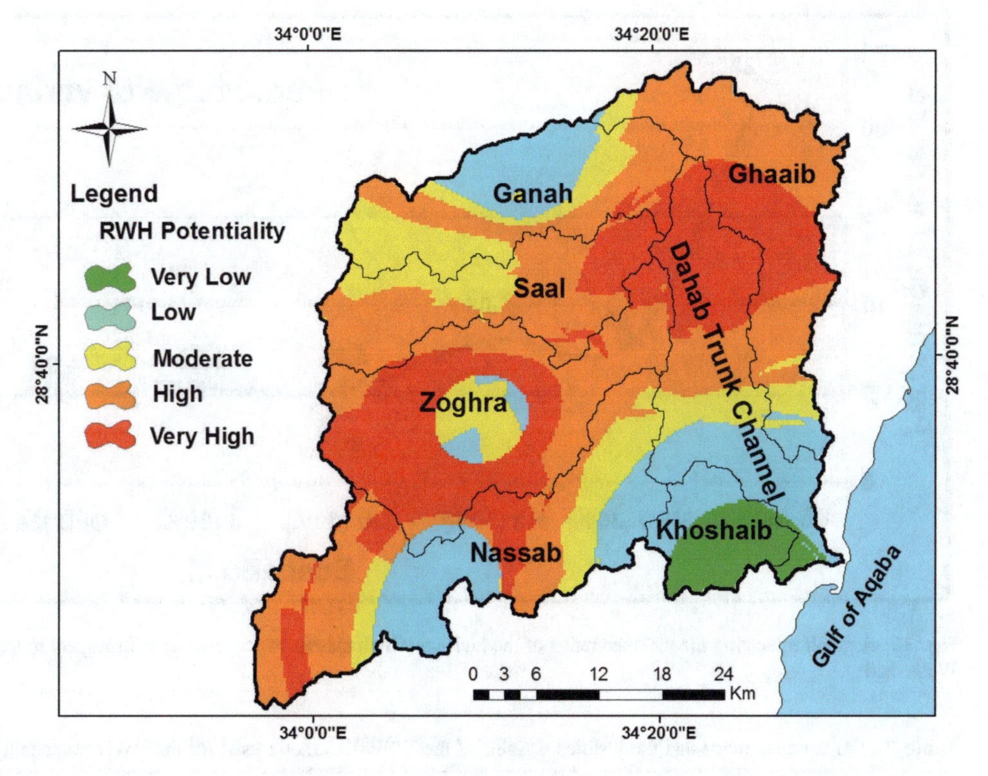

Fig. 36 WSPM map showing the RWH potentiality of W. Dahab (unequal weights: 10 % for each thematic layer except the VAF with 30 %)

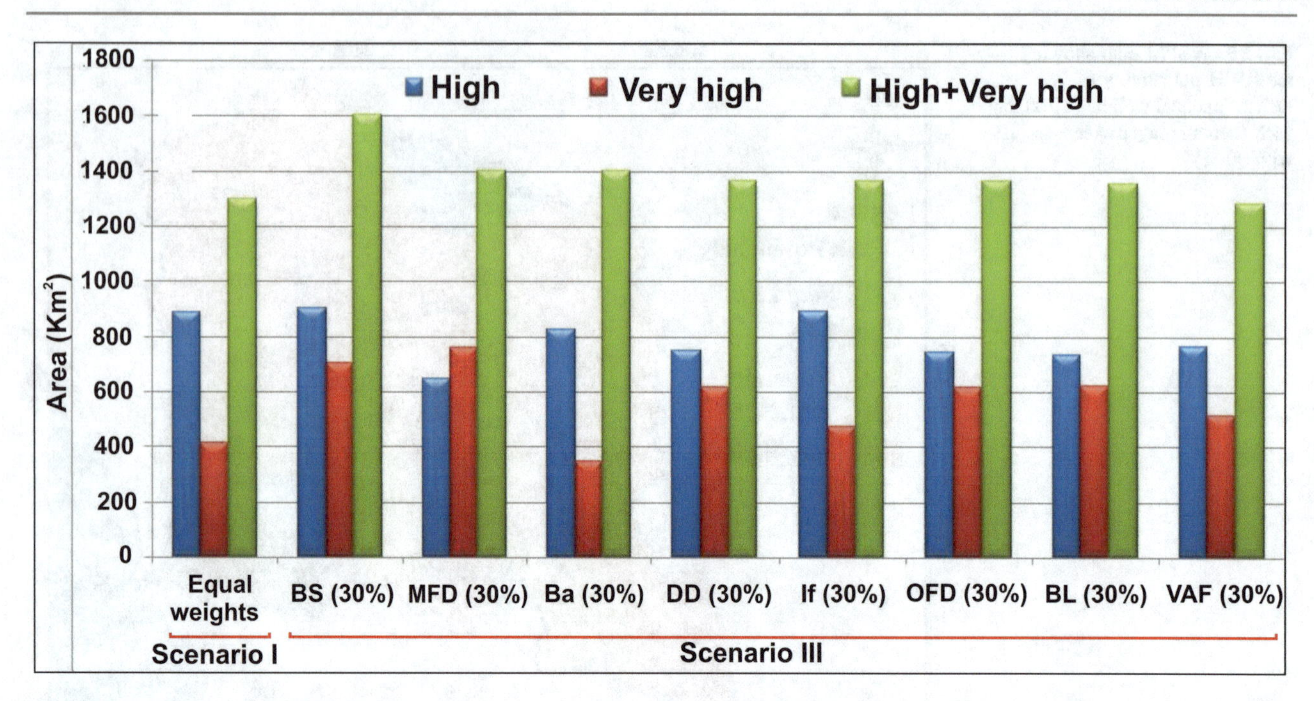

Fig. 3.37 Areas (in km^2) of the high-very high RWH potentiality classes and their summation in the two WSPM scenarios for W. Dahab Watershed

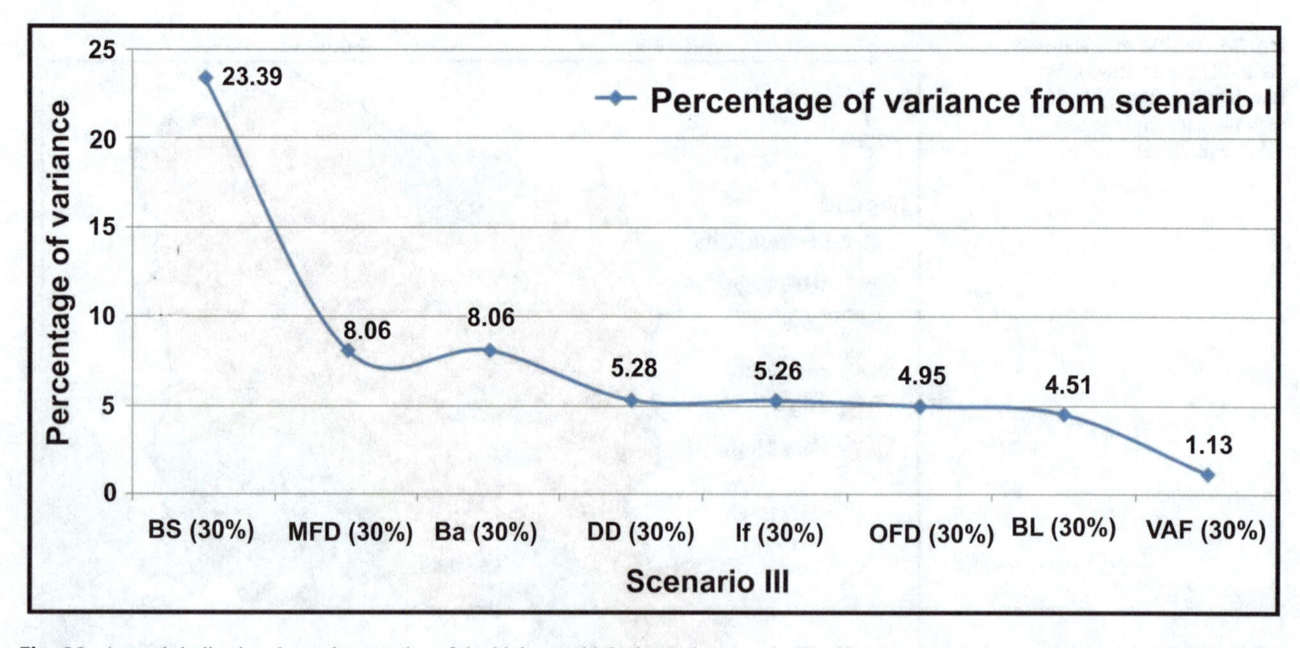

Fig. 38 A graph indicating the variance ratios of the high-very high classes in scenario III with respect to their areas in scenario I for W. Dahab Watershed

Table 7 The variance ratios and the justified weights of the WSPM's criteria used for the RWH potentiality mapping of W. Dahab

WSPM Criteria	BS	DD	MFD	VAF	IF	BL	OFD	BA
Variance ratio (%)	23.39	5.28	8.06	1.13	5.26	4.51	4.95	8.06
Justified weight (%)	38.57	8.71	13.29	1.86	8.67	7.44	8.16	13.29

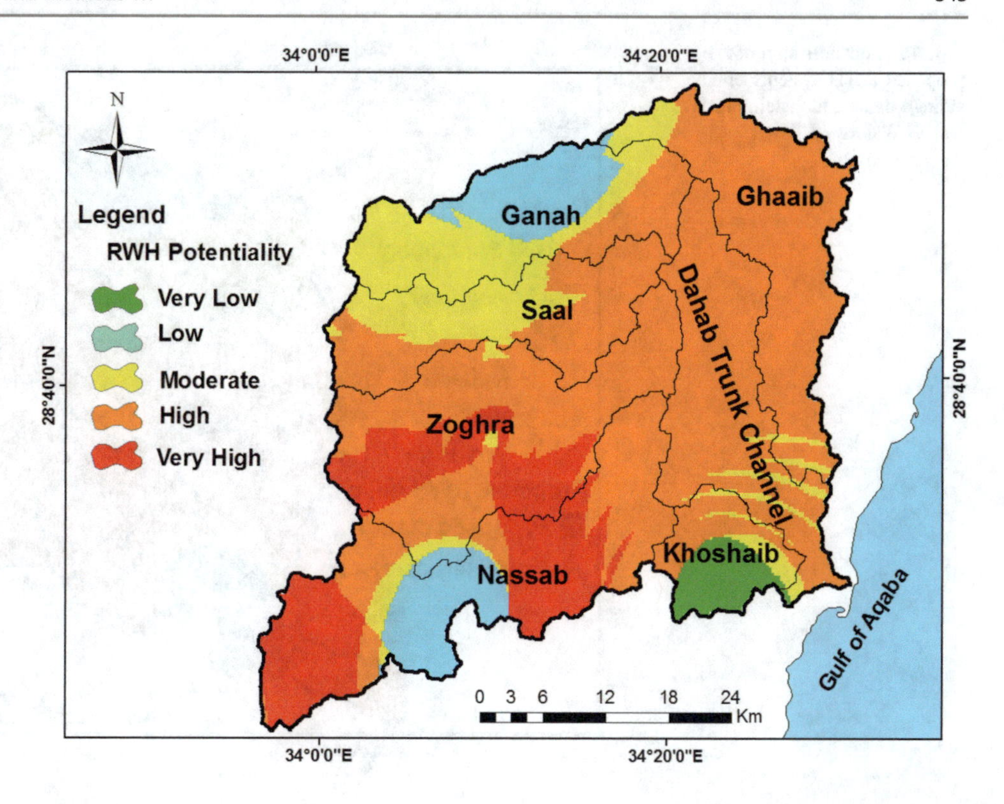

Fig. 39 WSPM map of scenario III showing the RWH potentiality of W. Dahab

Table 8 Areas of RWH potentiality classes resulted from the WSPM map of scenario III for W. Dahab

RWH Potentiality Map					
RWH Potentiality class	Very low	Low	Moderate	High	Very high
Area (Km^2)	53.75	183.54	304.71	1207.31	322.38
Area % relative to the total watershed area (Total watershed area: 2071.69 km^2) (%)	2.59	8.85	14.7	58.27	15.56

(scenario II). However, this gives very high credibility for the models used and, it confirms the performed field works and fulfills the local inhabitants' needs.

3.4 Proposed Action Plan for Applying RWH System

The present study proposed suitable locations for the construction of storage dams and ground cisterns to collect the runoff water. The moderate, high, and very high RWH classes are the most suitable areas for applying the RWH system to harvest any available quantities of runoff water and to mitigate the flash floods hazards. Four storage dams and five ground cisterns were proposed in optimum locations in the areas of moderate, high, and very high potentialities for the RWH (Fig. 40). The applied RWH system can help the decision-makers by proposing appropriate controlling systems for implementing runoff farming and rain-fed agriculture. Alluvial or wadi deposits in the main trunk channels of W. Dahab sub-watersheds will provide a good environment for the micro-catchment agriculture especially in the rainy seasons and for the establishment of new settlements in upstream and mid-stream areas of the studied watershed. Additionally, RWH has a direct impact on local Bedouin communities, which are the end users in these areas.

4 Summary and Conclusions

Wadi Dahab has very high importance in a new development in southeastern Sinai, for its touristic position and promising water resources. RS, WMS, and GIS techniques are modern research tools that proved to be highly effective in mapping, investigation, and modeling the runoff processes and optimization of the RWH. In the present work, these tools were

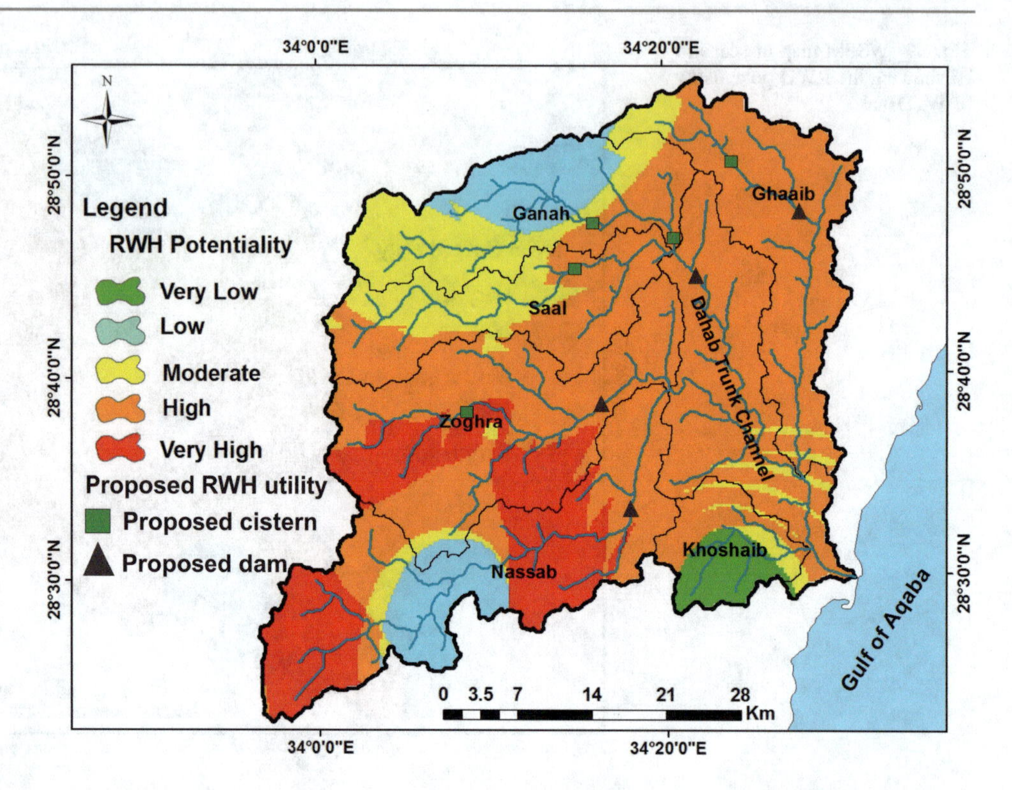

Fig. 40 Location map of proposed RWH systems of storage dams and cisterns in W. Dahab Watershed

used to determine the potential sites suitable for the RWH in W. Dahab. The performed WSPM for determining the potentiality areas for RWH depended on the hydro-morphometric parameters of drainage density, infiltration number, maximum flow distance, overland flow distance, basin slope, basin area, volume of the annual flood, and basin length. The WSPM model was accomplished through three scenarios: equal criteria weights (scenario I), authors' judgment (scenario II), and weights justified by the sensitivity analysis (scenario III). The obtained WSPM maps for defining the RWH potentiality areas classified W. Dahab basin into five RWH potentiality classes ranging from very low to very high. There are good matches between the three performed WSPMs' scenarios in results for the very high and high RWH potentiality classes, which are very suitable for RWH applications.

There are good matches between the three performed WSPMs' scenarios in results for the very high and high RWH potentiality classes, which are very suitable for RWH applications. These classes are frequently represented generally by El-Ghaaib, Dahab Trunk Channel, Zoghra, Nassab, Saal, and Ganah sub-watersheds, which represent about 62.94%, 56.95% and 73.83% of the total area of the basin for scenarios I, II, and III, respectively. RWH utility system was suggested in selected optimum locations for harvesting runoff water and mitigating flash floods hazards.

Acknowledgments The authors wish to express their great gratitude to the science & technology development fund (STDF) for kindly funding and supporting the project entitled "Determination of Potential Sites and Methods for Water Harvesting in Sinai Peninsula by Application of RS, GIS, and WMS Techniques" where the present work is derived. Deep gratitude is also dedicated to the national authority for remote sensing and space sciences (NARSS) for providing the facilities needed for conducting the present work.

References

Aher, P. D., Adinarayana, J., & Gorantiwar, S. D. (2014). Quantification of morphometric characterization and prioritization for management planning in semi-arid tropics of India: a remote sensing and GIS approach. *Journal of Hydrology, 511,* 850–860.

AQUAVEO. (2008). Water modeling solutions. Support forum for sub-water-shed modeling system software (WMS) www.aquaveo.com.

Bisht, S., Chaudhry, S., Sharma, S., & Soni, S. (2018). Assessment of flash flood vulnerability zonation through Geospatial technique in high altitude Himalayan watershed, Himachal Pradesh India. *Remote Sensing Applications: Society and Environment, 12,* 35–47.

CONOCO (Continental Oil Company). (1987). Geological Map of Egypt (Scale 1: 500,000). CONOCO Inc. in Collaboration with Freie Universitat Berlin, ISBN 3-927541-09-5.

Dames, & Moore. (1985). The vegetation of boyagin nature reserve. Internal report of department of conservation and land management, Perth(unpubl.).

El Kiki, M. F., Eweida, E. A., & El Refeai, A. A. (1992). Hydrogeology of the Aqaba rift border province. *Proceeding of the 3rd Conference on Geology Sinai Development* (pp. 91–100). Egypt: Ismailia.

El Rayes, A. E. (1992). Hydrogeological studies of St. Katherine area, South Sinai. M.Sc. Thesis, Suez Canal University Faculty of Science (95p).

Elewa, H. H., Abu El-Ella, E. M., Ramadan, E. M., El Feel, A. A. (2015). Determination of Potential sites and methods for water harvesting in Sinai Peninsula by application of RS, GIS, and WMS techniques. Unpublished report, 744p.

Elewa, H. H., & Qaddah, A. A. (2011). Groundwater potentiality mapping in the Sinai Peninsula, Egypt, using remote sensing and GIS watershed-based modeling. *Hydrogeology Journal, 19,* 613–628. https://doi.org/10.1007/s10040-011-0703-8.

Elewa, H. H., Qaddah, A. A., & El-Feel, A. A. (2012). Determining potential sites for runoff water harvesting using remote sensing and geographic information systems-based modeling in Sinai. *Journal of Environmental Sciences, 8,* 42–55.

Elewa, H. H., Ramadan, E. M., & Nosair, A. M. (2014). Water/land use planning of Wadi El-Arish Watershed, Central Sinai, Egypt using RS GIS and WMS techniques. *IJSER, 5*(9), 341–349.

Elewa, H. H., Ramadan, E. M., & Nosair, A. M. (2016). Spatial-based hydro-morphometric watershed modeling for the assessment of flooding potentialities. *Journal of Environmental Earth Sciences, 75,* 1–24.

Elewa, H. H., Ramadan, E. M., El Feel, A. A., Abu El Ella, E. A., & Nosair, A. M. (2013). Runoff water harvesting optimization by using RS, GIS and watershed modeling in Wadi El-Arish, Sinai. *The International Journal of Engineering Research and Technology, 2,* 1635–1648.

Faniran, A., & Ojo, O. (1980). *Man's physical environment* (p. 404). London: Heinemann Educational Books.

Feizizadeh, B., Jankowski, P., & Blaschke, T. (2004). A GIS based spatially explicate sensitivity analysis approach for multi-criteria decision analysis. *Computers & Geosciences, 64,* 81–95.

Finkel, H. H. (1979). Water resources in arid zone settlement. In G. Golany (Ed.), *A case study in arid zone settlement, the Israeli experience* (p. 567). New York: Pergamon Press.

Ha, W., Lu, Z., Wei, P., Feng, J., & Wang, B. (2012). A new method on ANN for variance based importance measure analysis of correlated input variables. *Structural Safety, 38,* 56–63.

Javed, A., Khanday, M. Y., & Ahmed, R. (2009). Prioritization of sub-watersheds based on morphometric and land use analysis using remote sensing and GIS techniques. *Journal of Indian Society of Remote Sensing, 37,* 261–274.

Lewis, M. A., Cheney, C. S., & Ódochartaigh, B. É. (2006). Guide to permeability indices. British Geological Survey Open Report, CR/06/160 N, 29p.

Ligmann-Zielinska, A. (2013). Spatially-explicit sensitivity analysis of an agent-based model of land use changes. *International Journal of Geographical Information Science, 27,* 1764–1781.

Montgomery, D. R., & Dietrich, W. E. (1989). Source areas, drainage density, and channel initiation. *Water Resources Research, 34,* 1907–1918.

Omran, A. F. (2013). Application of GIS and remote sensing for water resource management in arid area-Wadi Dahab basin–south Sinai-Egypt (case-study). Ph.D. thesis, Tübingen, University, Germany, 282p.

Paz-Kagan, T., Ohana-Levi, N., Shachak, M., Zaady, E., & Karnieli, A. (2017). Ecosystem effects of integrating human-made runoff-harvesting systems into natural dryland watersheds. *Journal of Arid Environments, 147,* 133–143.

Rahman, S., Khan, M. T. R., Akib, S., Che Din, N. B., Biswas, S. K., Shirazi, S. M. (2014). Sustainability of rainwater harvesting system in terms of water quality. *Hindawi Publishing Corporation e Scientific World Journal, 2014,* Article ID 721357, 10 pages. http://dx.doi.org/10.1155/2014/721357.

Said, R. (1962). *The geology of Egypt.* Amsterdam: Elsevier.

Said, R. (1971). Explanatory notes to accompany the geological map of Egypt. *Geological Survey of Egypt,* (special paper 56, Cairo).

Said, R. (1990). *Geology of Egypt* (p. 722p). Balkema Pub: Rotterdam.

Saisana, A., Saltelli, A., & Tarantola, S. (2005). Uncertainty and sensitivity analysis techniques as tools for the quality assessment of composite indicators. *Journal of the Royal Statistical Society, 168,* 307–323.

Saltelli, A., Tarantola, S., & Campolongo, F. (2000). Sensitivity analysis as an ingredient of modeling. *Statistical Science, 15,* 377–395.

Shendi, E. H., Oada K. (1999). Groundwater possibilities of wadi El-Nasb basin, southeastern Sinai, Egypt. *Annals of the Geological Survey* of *Egypt, VXXII,* 403–418.

Strahler, A. N. (1964). Quantitative geomorphology of drainage basin and channel network. In V. T. Chow (Ed.), *Handbook of applied hydrology.* New York: McGraw Hill.

Van Griensven, A., Meixner, T., Grunwald, S., Bishop, T., Diluzio, M., & Srinivasan, R. (2006). A global sensitivity analysis tool for the parameters of multi-variable catchment models. *Journal of Hydrology, 324,* 10–23.

Vema, V., Sudheer, K. P., & Chaubey, I. (2019). Fuzzy inference system for site suitability evaluation of water harvesting structures in rainfed regions. *Agricultural Water Management, 218,* 82–93.

Yazar, A., & Ali, A. (2016). Water harvesting in dry environments. In *Innovations in Dry Land Agriculture* (pp. 49–98). https://doi.org/10.1007/978-3-319-47928-63.

Mitigation of Flash Flood

Prediction and Mitigation of Flash Floods in Egypt

Ismail Fathy, Hany F. Abd-Elhamid, and Abdelazim M. Negm

Abstract

Egypt is located in a semi-arid region and the annual average precipitation is 12 mm/year which ranges from 0 mm/year in the desert areas to 200 mm/year in the coastal areas. Countries located in semi-arid regions such as Egypt have alluvial (wadi) systems that formed during the fluvial time of the Tertiary and Quaternary Periods. These wadis suffer from a flash flood, consequent to heavy precipitations. Two major factors are responsible for flood generation and impart specific features to it. The first is the physical process, which generates the change of position between the lithosphere, atmosphere, and hydrosphere. Second, the flooded area, depth of inundation, and its duration depend on the geographic situation of the region where flood takes place. Due to the great variety in operation of the natural processes and the endless variation in the condition of the geographic arena where they act, many different kinds and scales of floods can be distinguished. There are at least 18 major types of floods of natural origin. Growing human influence has become evident from the increasing frequency of floods of anthropogenic origin. The main purpose of this chapter is to identify the general system of floods in terms of types, causes, problems caused by flood, and how to determine the amount of floodwater and prevention with application to the areal case study. Wadi Sudr, Sinai, Egypt is used as a real case study. The description of this wadi is defined. Recorded rainfall depths are studied and analyzed, and the IDF curve is generated. Runoff hydrograph of the wadi is estimated, and locations of protection measures using a number of dams and open channels have been selected and designed to protect the area from flood hazards.

Keywords

Flood • Prediction • Mitigation • Hydrological models • Wadi Sudr • Egypt

1 Introduction

There are many definitions of a flood, but all meet in one direction is the relationship, or the boundary between water and land occurred by changing and overwhelmed the limits of water on the ground, and these definitions can be summarized as (Kingma 2014).

- Flood is any high streamflow which overtops the natural or artificial banks of a stream.
- Flood is a body of water that submerges land that is infrequently submerged and causes damage and loss of life.
- Flood is a waterway that submerges land that is inconsistently submerged and in doing so causes or takes steps to cause harm and death.
- Flood is a natural and recurring event for a river or stream.

Definitions of floods are useful for assessing the health effects, the damage to infrastructure, and the financial toll they can cause and to decide on a trigger for activation of emergency response. There is, however, no universal definition of what constitutes a flood. Examples of currently used definitions include

I. Fathy (✉) · H. F. Abd-Elhamid (✉) · A. M. Negm
Department of Water and Water Structures Engineering, Faculty of Engineering, Zagazig University, Zagazig, Egypt
e-mail: ismailfathy_eng@yahoo.com

H. F. Abd-Elhamid
e-mail: hany_farhat2003@yahoo.com

A. M. Negm
e-mail: amnegm@zu.edu.eg

H. F. Abd-Elhamid
Civil Engineering Department, College of Engineering, Shaqra University, Shaqraa, 11911, Dawadmi, Saudi Arabia

A. M. Negm (ed.), *Flash Floods in Egypt*, Advances in Science, Technology & Innovation,
https://doi.org/10.1007/978-3-030-29635-3_15

Fig. 1 Types of runoff

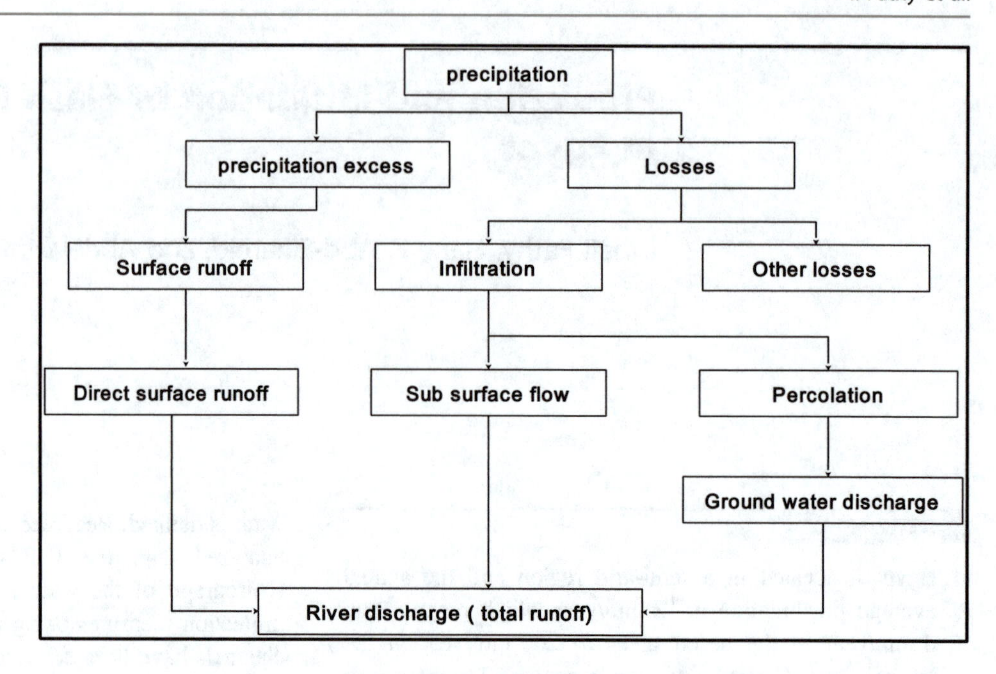

- The existence of water in areas that are generally dry.
- Flood disaster is a flood that significantly disrupts or interferes with human and social activity.
- An increase of water that has a significant impact on human life and well-being (Jonkman and Kelman 2005).
- A significant rise of water level in a stream, lake, reservoir, or coastal region.
- Any case where land not normally covered by water becomes covered by water (Vos 2009).

According to the above, if the amount of water is greater than the canal capacity, the flood will occur. In order to determine the amount of water, its components should be determined. Figure 1 shows the factors that affect flood quantity, and from Fig. 1 the relation between precipitation and runoff is defined by two stages: the first stage is losses and the second stage is how can water reach the stream. The intensity of flood is affected by many factors. These factors affect its peak and lag time. Figure 2 illustrates these factors. From this figure, we can see that there are three variables that affect flood propagation such as basin, network, and channel characteristics. The basin characteristics have the biggest impact because of many variables related to it, such as areas, slopes, and shape factor.

2 Causes of Flood

Insufficient capability of waterways to contain inside their banks the high discharges down from the upper catchment zones following substantial precipitation prompts flooding. The displacement of man around the waterways (flood plains) and the formation of societies have been from ancient times. Due to fluctuation of rainfall distribution, these areas experience severe inundation. Areas with poor protection or strong drainage systems are flooded by the accumulation of water from dense rainfall. Extra irrigation water applied to command areas and increase in groundwater levels due to leakage from canals and irrigated fields are also factors that emerge from the problem of water-logging (NDMG 2008). The main causes of flood are presented in Table 1.

3 Types of Flood

Flood can be classified into a number of types according to causes and intensity of flood as following. (Mandych 2015).

3.1 Flash Flood

Flash flood occurs as a result of several factors; the most important factor is higher rate of precipitation than the infiltration rate supported by increased land slopes and watershed drainage efficiency. Another cause of flash flood is the collapse of dams and barriers. The flash flood is characterized by a high water level in a short time, thus shortening the warning time. These floods have a short time that is less than an hour which destroy the facilities, roads, trees, and anything that hinders it. Although these floods have a short time and do not cover large areas, the short duration and major destruction distinguish them from other floods. Figure 3 shows a photo of flash flood that occurred in Detroit, USA (Tyra et al. 2016).

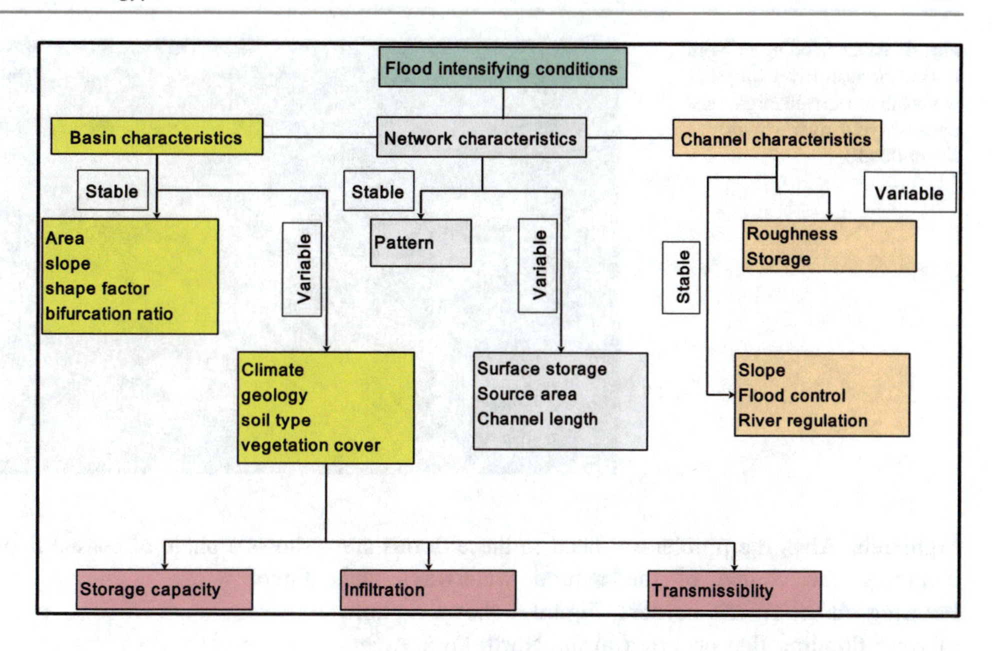

Fig. 2 Factors affecting floods

Table 1 Main causes of the flood (Bettina and Virginia 2013)

Cause	Examples
Extreme rainfall	Flash flood (rapid onset) Immersion of urban zones particularly its roads
Melting of snow	Glacial meltwater
Dam disaster	Dam failure Dam overtopping
Tsunami or extra wave propagation	Storm surge

Fig. 3 Flash flood in Detroit, USA (http://thevane.gawker.com/maps-which-parts-of-the-u-s-see-flash-floods-most-oft-1622076723)

3.2 River Flooding

The flood of rivers occurs when the runoff discharges of the rivers and waterways are higher than the capacity of the river and therefore the water level is higher than the safe level. These floods affect a large area, especially the areas behind the rivers. The causes of these floods are the high intensity of rainfall and the melting of the ice which raises the water level. Although river floods are predictable, their long-term impact is considerable due to the severe damage to the population of the surrounding area, especially the rivers that descend from mountains and

Fig. 4 River flooding in North Fork American River (https://www.americanrivers.org/rivers/discover-your-river/why-do-rivers-flood/)

highlands. Also, the problems related to these floods are changing the course of the natural waterways and building on the river's borders. Figure 4 shows a photo of river flooding that occurred in the North Fork American River.

3.3 Coastal Flood

This type of flood happens when seawater is pushed inlands. The main factors of these floods are Cyclones and tropical storms; Earthquakes can transfer huge amounts of water due to waves called tsunamis to flash inland, and incredible high tides associated with a full moon cause large waves and raise the sea level creating storm surge along beach. Figure 5 shows a photo of coastal flood that occurred in Alexandria, Egypt.

3.4 Urban Flood

Urban floods occurred by a high increase of rainfall intensity and urbanization which minimize infiltration rate and increase runoff flow as much as 2–6 times over what would occur on natural land. These floods have huge economic effects because they occur inside the city and works to disrupt traffic. In addition to high rainfall intensity, there are many factors that cause these floods such as flash river and coastal flooding. Figure 6 shows a photo of urban flood that occurred in Alexandria, Egypt.

Fig. 5 Coastal flood in Alexandria, Egypt (https://www.middleeastmonitor.com/20151107-egypt-blames-brotherhood-for-alexandria-floods/)

Fig. 6 Urban flood in Alexandria (https://www.middleeasteye.net/news/torrential-rain-heavy-flooding-hit-middle-east)

Fig. 7 Arial flood (https://www.weather.gov/bmx/outreach_flw)

3.5 Areal Flood

Areal floods occur by the heavy rainfall intensity on a short duration and collect runoff flow over the low-level lands and open areas. In addition, a long period of duration of rainfall can cause floods and damage agricultural lands. The pond and swales that occurred by these floods have stagnated water and serve as a breeding ground for pests and infection. Figure 7 shows a photo of the areal flood that occurred in Alabama, USA.

3.6 Dams and Levees Failure

Failure of flood control structures such as dams and levees due to poor design and construction or due to inadequate

Fig. 8 Dam failure in Columbia, USA (https://news.nationalgeographic.com/2015/10/151007-dam-failures-south-carolina-engineering-science/)

maintenance or operational mismanagement lead to sudden floods. These floods occur over short duration with huge amount of water that causes economic losses especially people and areas surrounding the river banks of these dams and levees. Also these floods occur due to Levee or dam overtopping or failure that typically occurs from floods beyond their capacity to handle, often with huge and catastrophic results. Figure 8 shows a photo of flash caused by dam failure that occurred in Columbia, USA (NWS 2013).

4 Effects of Flood

Harms and damages of floods can be summarized as the following items (IDNIE 2008) and are listed in Table 2:

Table 2 Potential effects of flooding, (Bettina and Virginia 2011)

Type of effect	Effect
Direct effects of flood	– Drowning and injuries from walking or driving through the flood, contact with debris in floodwater, falling into hidden manholes, injuries from submerged objects, injuries while trying to move possessions during floods – Building collapse and damage (injuries) – Electrocution – Diarrhoeal, vector- and rodent-borne diseases – Respiratory, skin, and eye infections – Chemical contamination, particularly carbon monoxide poisoning from generators used for pumping and dehumidifying – Water shortages and contamination due to loss of water treatment works and sewage treatment plants – Stress, short- and long-term mental health issues, including the impacts of displacement
Indirect effects of flood	– Loss of access to and failure to obtain continuing health care – Damage to health care infrastructure, and loss of access to essential care – Damage to or destruction of property, including hospitals and other vital – community facilities – Damage to water and sanitation infrastructure – Damage to crops, disruption of food supplies – Disruption of livelihoods and income – Population displacement – Mental health problems due to length of flood recovery and fear of recurrence; indirect effects of stress in dealing with insurance claims and refurbishing properties

Table 3 Multiplied factors of rational methods, TxDOT (2004)

Recurrence interval (years)	C_f
2, 5 and 10	1.0
25	1.1
50	1.2
100	1.25

- Death of humans and animals.
- Collapse of the infrastructure of the affected areas.
- Heavy losses in agricultural land and crops.
- Pollution of watercourses, especially those with drinking water purification plants.
- Formulation of marshes and ponds with stagnant water and thus the spread of epidemics and diseases.
- Shortage and lack of food.
- Erosion of watercourses due to high water speeds and the collapse of facilities erected on them.

5 Prediction of Flood

Flood studies require acquisition of data and the use of hydrological models for simulation and prediction. Flood discharge and lag time are important factors for studying flood that can be estimated using different methods as following.

5.1 Method of Maximum Discharge

One of the methods for calculating the maximum discharge is the rational methods

5.1.1 The Rational Method

The rational method is used to determine the rate of discharge using the following equation.

$$Q = C_f \times \text{CIA} \tag{1}$$

where

Q is the maximum rate of runoff (cfs),
C is dimensionless runoff coefficient, dependent upon land use,
I is the design rainfall intensity (inches/hour),
A is the drainage area (acres), and
C_f is the saturation factor. The values of saturation factors depend on the recurrence interval as presented in Table 3.

5.1.2 Synthetic Unit Hydrographs

The parameters for the synthetic unit hydrograph can be determined from gage data by applying the parameter optimization. Otherwise, these parameters can be determined from regional studies (Chow et al. 1988). There are several synthetic unit hydrograph methods that can be used as described below.

SCS Dimensionless Unit Hydrograph

SCS, dimensionless unit hydrograph method, (SCS 1972) consists of a single parameter, TL, which equals to the lag (hrs) between the center of mass of rainfall excess and the peak of the unit hydrograph as shown in Fig. 9. SCS suggests that the time of recession may be approximated as 1.67Tp. As the area under the unit hydrograph should equal to a direct runoff of 1 cm. The unit hydrograph can be drawn, as shown in Fig. 9.

$$T_p = \frac{T_r}{2} + T_L \tag{2}$$

$$T_L = 0.60\, T_c \tag{3}$$

$$q_p = \frac{CA}{T_p} \tag{4}$$

where

T_p is peak time (hr),
q_p is peak discharge, cms m, $c = 2.08$,
A is the drainage area (km^2).

Nash Unit Hydrograph

Nash developed an equation to calculate runoff hydrograph of gage watershed as represented below. The equation depends on two empirical parameters. These parameters depend on the geometric properties of the watershed.

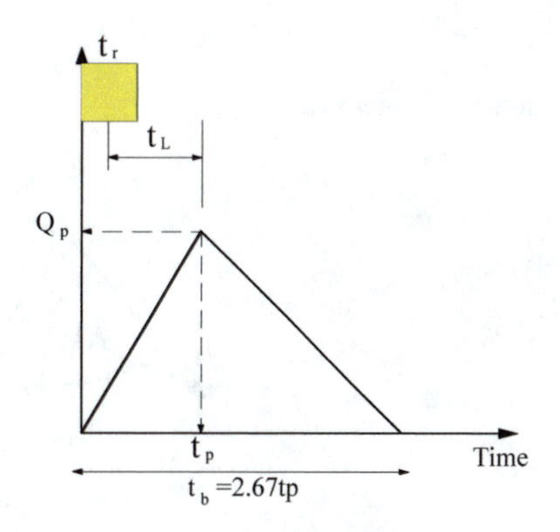

Fig. 9 SCS dimensionless unit hydrograph

$$u_{(t)} = \frac{1}{k(n-1)!} e^{-\frac{t}{k}} \left(\frac{t}{k}\right)^{n-1} \quad (5)$$

$$n = 2.4L^{0.10} \quad (6)$$

$$K = \frac{11A^{0.30}}{L^{0.10}S^{0.30}} \quad (7)$$

where

$u_{(t)}$ is instantaneous unit hydrograph(1/h),
n is a shape parameter,
K is delay time (h), which is a scale parameter,
L is the length of the mainstream (miles),
S is the slope of the basin, and
A is the area (sq. miles), (Raghunath 2006).

Clark's IUH (Time–Area Method)

On the concept of routing a time–area relationship through a linear reservoir, Clark developed a method to develop synthetic unit hydrograph, which is called Clark method. The basic concept used to develop the Clark method may be described as follows: water covering the catchment to a depth of 1 cm is released instantaneously and allowed to runoff the basin. A time–area relationship represents the translation hydrograph of this runoff as influenced by watershed characteristics, such as size, shape, and surface roughness, as shown in Fig. 10. The translation hydrograph is routed through a linear reservoir to capture additional storage effects of the watershed, Raghunath (2006).

$$\left(\frac{A_c}{A}\right) = 1.414T^{1.5} (\text{For } 0 \le T \le 0.50) \quad (8)$$

$$\left(\frac{A_c}{A}\right) = 1 - 1.414(1-T)^{1.5} (\text{For } 0.50 < T \le 1) \quad (9)$$

$$Q_2 = \left(\frac{2\Delta t}{2k+\Delta t}\right)I + \left(1 - \frac{2\Delta t}{2k+\Delta t}\right)Q_1 \quad (10)$$

where

A_c is the contributing area of total area A,
T is the fraction of concentration time t_c.

5.2 Hydrologic Models for Estimating Stream Flow

Modes that used to estimate the maximum discharge at the watershed outlet can be divided into three models: lumped, semi distribution, and distribution model. Many scientists and researchers studied these models and compared them and worked to show the advantages and disadvantages of each model. Figure 11 shows a comparison between the models and how they deal with watershed (Murari 2010).

5.2.1 Lumped Model

Sherman (1932) introduced the concept of the unit hydrograph. The unit hydrograph is the runoff that occurs at the downstream outlet of a drainage basin from a unit depth (i.e., 1 inch or 1 mm) of excess rainfall for a storm of uniform intensity for a specified duration over entire watershed drainage. Lumped model deals with the watershed as one identical unit. The parameters of watershed that effect on peak discharge averaged through model process (Refsgaard and Knudsen 1997). Through these models, the rainfall intensity averaged and has a uniform distribution through watershed; this case rarely happens in reality (Smith et al. 2004). These models do not depend on the watershed physics; the reliability of their results is valid only if the models are calibrated (Reed et al. 2004)

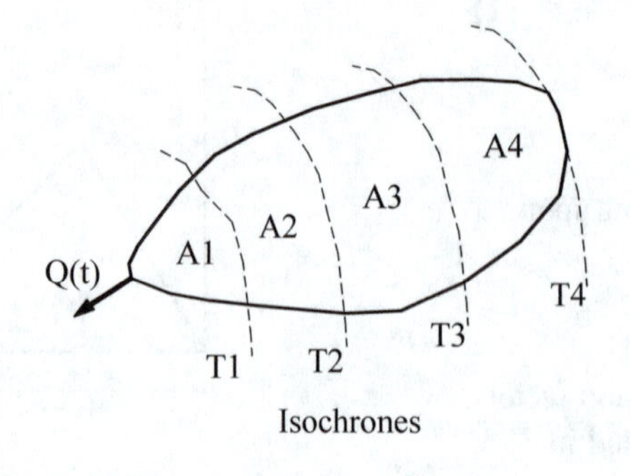

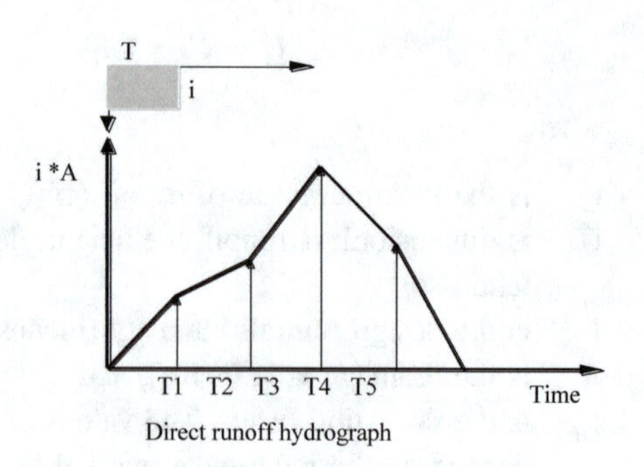

Fig. 10 Isochrones of watershed

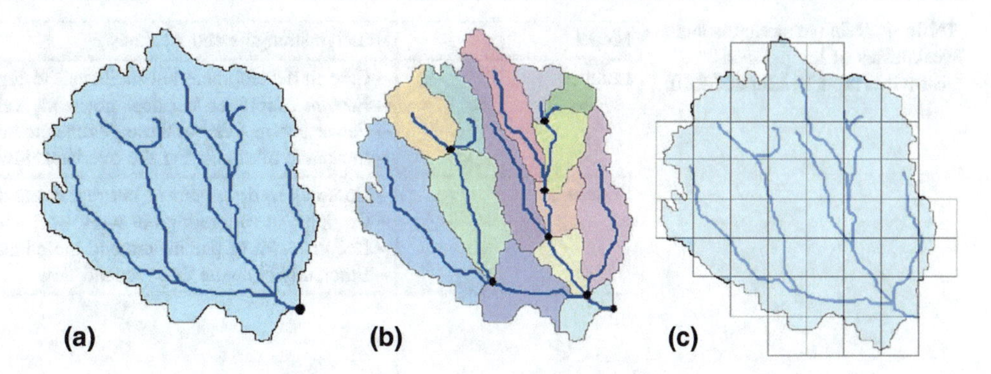

Fig. 11 Graphic representation of **a** Lumped, **b** semi-distributed and **c** Distributed models (Murari 2010)

Examples of lumped models available in the market:

- Rockwood et al. (1972) created SSARR model.
- Burnash et al. (1973) developed Sacramento model.
- Williams and Hann (1973) developed the Hydrologic model (HYMO).
- Sugawara et al. (1976) built Tank model.
- SCS (1986) Built the Technical Release TR-20, TR-55.
- HEC (1998) created The Hydrologic Engineering Center (HEC-1);
- Ries and Crouse (2002) created the National flood forecasting (NFF) now called National Streamflow Statistics (NSS) and
- HEC (2008) developed the HEC-HMS model.

5.2.2 Semi-distributed Model

In an attempt to consider the spatial variation of watershed characteristics, semi-distributed models were developed. In such models, the watershed is divided into smaller computational elements (or sub-basins), and hydrologic computation is carried out for each element. There is a wide variation on how these computational elements are formulated. Some of the models use natural watershed-divides as the criterion for dividing a watershed, whereas others use the hydrological response unit which is based upon the land use and/or soil characteristics (Bicknell et al. 1997).

The whole watershed was divided into sub-basins according to soil type and land use so the excess rainfall is estimated for each sub-basin and converted to discharge for sub-basins individually. Finally, the flood routing technique is used to calculate the total discharge at the outlet of the whole watershed (USACE 1994). These models take the spatial variation of watershed parameters as a sequence of sub-basins with constant appearances (Vieux et al. 2004).

5.2.3 Distributed Model

Distribution models are advanced models of hydrological simulations. where spatial variation was taken into account during the representation of watershed (Bobba et al. 2000). In these models, the watershed is divided into a group of cells, and each cell has a constant property to simulate the actual heterogeneous of the watershed. The flood routing methods are used to calculate the total discharge of the watershed. The cells' dimensions should be reasonable to give the accurate results; small cell dimensions take a lot of time during the solution on the other hand large cell dimension give inaccurate results (Vazquez et al. 2002). Due to the technological and information developments in the field of computer and software, it facilitated the use of distribution models as a result of easy data storage and processing. However, these advantages, the uncertainty of rainfall data as well as the difficulty of determining the watershed properties, have reduced the use of these models (Carpenter and Konstantine 2004).

The distributed models divide the basin into elementary unit areas such as grid cells and solve basic physical equations to simulate the watershed processes. While even at the finest grid resolution some information is still lumped into a grid cell, the distributed model can be used to account for the spatial variation of precipitation, land use, or soil type within a watershed (Murari et al. 2010).

Examples of disturbed models available in the market are as follows:

- DHI (1998) built MIKE-SHE model.
- Downer et al. (2002) created CASC2D (Predecessor of GSSHA) model;
- Vieux and Moreda (2003) created VfloTM model;
- Leavesley et al. (2004) developed Modular Modeling System (MMS);
- Koren et al. (2004) built Hydrology Lab's Research Modeling System (HLRMS);
- Carpenter and Konstantine (2004) built Hydrologic Research Center Distributed Hydrologic Model (HRCDHM) model.

Table 4 Relative strengths and weaknesses of lumped and distributed models, Murari (2010)

Model	Relative strengths and weaknesses
Lumped	– Ease in development but inefficient to represent the spatial variation – Ease in calibration but does not work well with other storms – Fewer parameters but these parameters are not related to watershed physics – Incapable of simulating the overland flow
Distributed	– Complex in development but represents the spatial variation efficiently – Complex in calibration but work well with other storms – Require a lot of parameters and the parameters are physically based – Efficiently simulate the overland flow

– USDA (2004) built Soil and Water Assessment Tool (SWAT) model; and
– Downer and Ogden (2006) created Gridded Surface/Subsurface Hydrologic Analysis (GSSHA).

5.2.4 Comparison Between Models

The most important and intuitive question is which models are the most accurate and the best representation of watershed. Many studies have worked on comparing the models Fitzpatrick et al. (2001). Reed et al. (2004). The comparison between distribution models and lumped models is inconclusive because each has its own advantages and disadvantages. Despite the superiority of distributive models, its complexity, and the frequent identification and measurement of variables, it reduces the overall efficiency— Smith et al. (2004). Comparison between lumped and distributed models is presented in Table 4.

6 Mitigation of Flood

Mitigation of flood requires a number of steps such as suitable warning systems, appropriate rainwater harvesting, and suitable protection techniques.

6.1 Warning Systems

The purpose of the flood warning is to alert people and warn them before the disaster, in order to preserve the lives and reduce the losses and serious damage. In order to complete the process, it is necessary to cooperate with local, government, and private institutions within the study area.

- ***Prediction***: detection of changes in the environment that lead to flooding and the prediction of future river levels during the flood.
- ***Interpretation***: identifying in advance the impacts of the predicted flood levels on the communities at risk.
- ***Message construction***: devising the content of the message in a way which will clearly warn people of impending flooding.
- ***Communication***: disseminating warning information in a timely way to people and organizations likely to be affected by the flood.
- ***Response***: getting the appropriate protective behavior from the threatened community and from the agencies involved.
- ***Review***: examining various aspects of the system with a view to improve its performance.

A number of warning systems were designed around the world. Cools et al. (2012) developed an early warning system for flash floods in Sinai Peninsula of Egypt and tested it based on the best available information. Also, a set of essential parameters has been identified to be estimated or measured under data-poor conditions. The results for flash flood studies showed that 90% of the total rainfall volume was lost to infiltration and transmission losses. The study showed that the effectiveness of an early warning system is only partially determined by technological performance. Daniele et al. (2008) evaluated the flash flood warning method, by considering a wide range of climatic and physiographic conditions.

Pierre et al. (2010) proposed a method to estimate antecedent soil moisture conditions to improve the accuracy of flash flood forecasts at ungauged locations. The method combines two indexes: a climatic temporal index calculated in each cell using an uncalibrated soil moisture accounting scheme and a spatial statistical index giving the average saturation state usually encountered before a flood. The proposed method was carried out on two different models. Simulation results were analyzed for 562 individual events, issued from 160 catchments located in South France. The presented method improved the efficiency of both models. Pierre (2012) applied Quantitative Precipitation Estimates and Forecasts (QPE/QPF) based on radar measurements on Gard region, France.

6.2 Water Harvesting

Rainwater harvesting (RWH) describes the methods of collection, storing, and spreading various forms of runoff from different sources for domestic, agricultural, and other uses. It can be described as all activities to collect available water resources, and temporarily storing excess water for use when required, especially in periods of drought.

Benefits of rainwater harvesting:

- Supply drinking water for animal and domestic water purposes.
- Make productive in food/forage/tree crops where it is normally not possible.
- Increase yield rainfall farming.
- Lessen erosion and the problem of siltation of reservoirs is minimized.
- Combating desertification by promoting tree cultivation.
- Improves farmer's standard of living and prevent migration.

Rainwater can be collected and stored in groundwater aquifers and can be used in dry period to overcome the scarcity of water. A number of methods can be used for storing rainwater into groundwater aquifers which are described as following.

- Sand dam
- Sub-surface dam
- Percolation ponds
- Vetiver contours
- Contour trenching
- Teras
- Tube recharge of groundwater.

A number of studies have been conducted for water harvesting. Al-Zayed (2010) used HEC-1 model to select water harvesting sites of Wadi Watier, Sinai, Egypt. The potential runoff was estimated using this model. The results showed a promising potential for water harvesting. In addition, a number of five different floodwater harvesting systems were sited and adopted. Elewa et al. (2013) integrated the geographic information systems, remote sensing, and watershed modeling to identify the suitable sites for implementing the runoff water harvesting constructions in Wadi El-Arish. The resultant map classified the area into three RWH potentiality classes ranging from low to high.

Consequently, the suitable sites for the construction of RWH dams were determined. The map suggested the collection of runoff water at the outlets of Wadi El-Arish upstream sub-watersheds with promising runoff potentialities. Dong and Moo (2014) used XP-SWMM model to calculate water harvesting and flash flood estimation in the flood-prone area in Suwon of South Korea.

6.3 Protection Measures

Protection measures and flood control are essential. Flood control refers to all methods used to reduce or prevent the detrimental effects of floodwater. A number of protection measures have been used all over the world.

1. Flood control on rivers (detention basins, levees, bunds, dams, reservoirs, and weirs)
2. Coastal flooding (coastal defenses such as sea walls, nourishment, and barrier)
3. Flood control on watershed.

Flood control on watershed can be done using a number of methods including

- Dams and Reservoirs: these hold back and regulate the flow of river water.
- Forestation: planting trees increases interception rates and reduces surface runoff.
- Diversion Canals: built canals which in turn divert the water to temporary holding ponds or other bodies of water where there is a lower risk or impact to flooding.
- Culverts: semicircular, smooth channels increase velocity and get water away from urban areas as quickly as possible.
- Water-Gate: water-gate flood barrier is a rapid response barrier.
- Self-closing flood barrier.
- Restricted use of flood plains: legislation, higher selective insurance premiums/refusal to ensure particular locations.
- Co-ordinated flood warning: emergence reaction procedures, e.g., Environment Agency Flood watch.

7 A Case Study (Wadi Sudr, Sinai, Egypt)

In this case, a hydrological study was conducted to predict and mitigate a selected area from flood hazards. The selected study area (Wadi Sudr) is situated at southwest Sinai, Egypt. It is located between latitudes 29° 35′ and 29° 55′, and longitudes 32° 40′ and 33° 20′. It drains rainwater directly into the Gulf of Suez at Sudr town as shown in Fig. 12. In the study area, there are some rainfall stations as shown in Fig. 13. This wadi was supported by Water Resources Research Institute (WRRI) for rainfall and runoff measurements since 1989, (El-Sayed 2006).

Digital elevation models (DEM) are used to derive the natural flow paths and the corresponding catchment boundaries for the study area. In the current study, 30 m resolution maps are used. The DEM, ASTER 30 for the study area, are derived and presented, as shown in Fig. 14.

The closest rainfall station with available daily rainfall data is Sudr station. It is located at a distance 1 km from the study area. The coordinates of this station are 29° 35′ 54.60″ N, 32° 50′ 37.24″ E. The design storm is generated based on measured data. The rainfall data are presented in time and in total rain record for 30 years, from 1980 to 2010. The period of the storm is presented too. A daily maximum time series of rainfall data is constructed for this station. Frequency analysis was performed, and a distribution curve was fitted to the data. Based on the fitted curve, 24-h rainfall depths for events of each return period 2, 5, 10, 25, 50, and 100 years were estimated using California method and are presented in Table 5 and Fig. 15. The hydrological estimation of peak flows is carried out for a storm of a given suitable frequency. A frequency analysis of the annual maximum rainfall time series is carried out to identify the rainfall depths for the corresponding return periods (Table 6).

Watershed Management System (WMS) was developed by Environmental Modeling Research Laboratory of Brigham Young University in cooperation with U.S. Army Corps of Engineering Waterways Experimental station is currently being developed by Aquaveo LLC. WMS offers

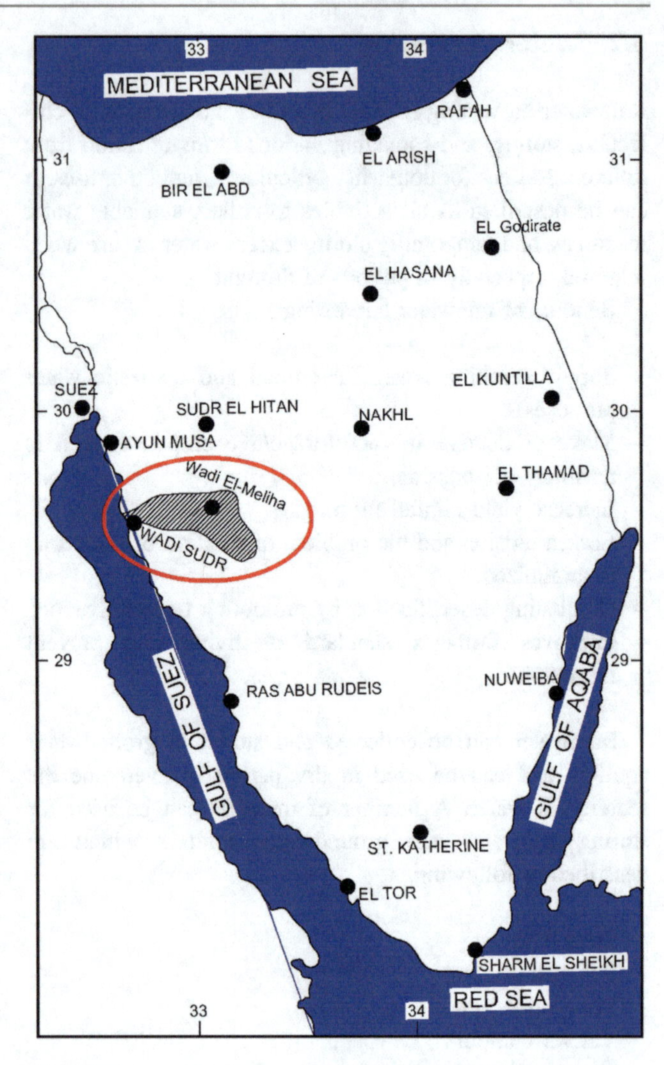

Fig. 12 Location map of the study area

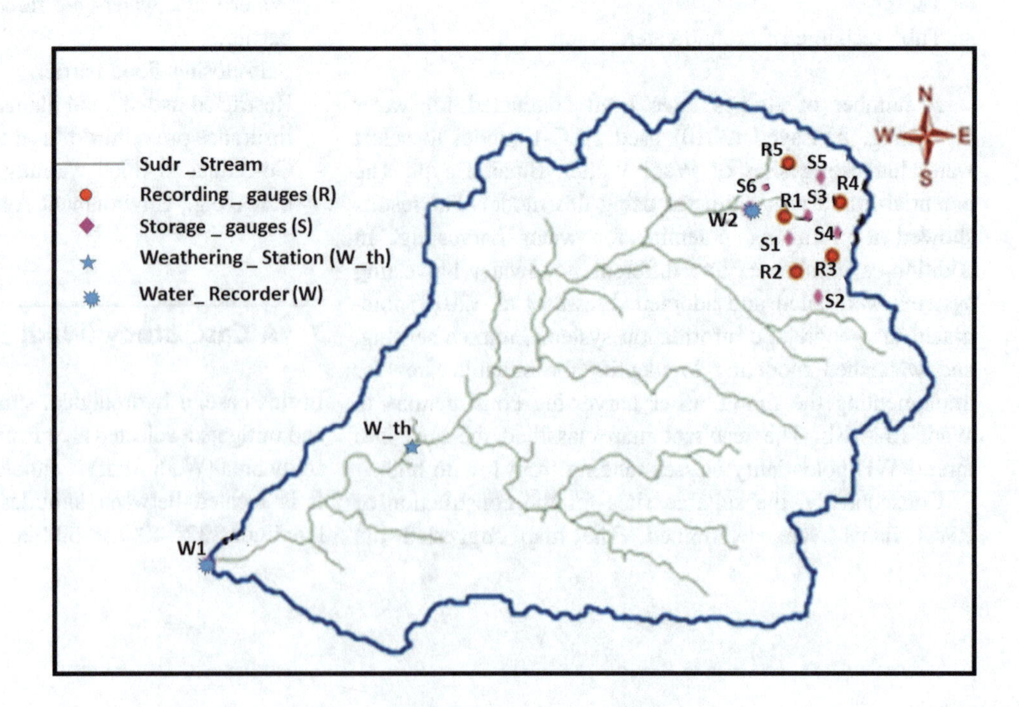

Fig. 13 Locations of rain gages in Wadi Sudr (El-Sayed and Habib 2008)

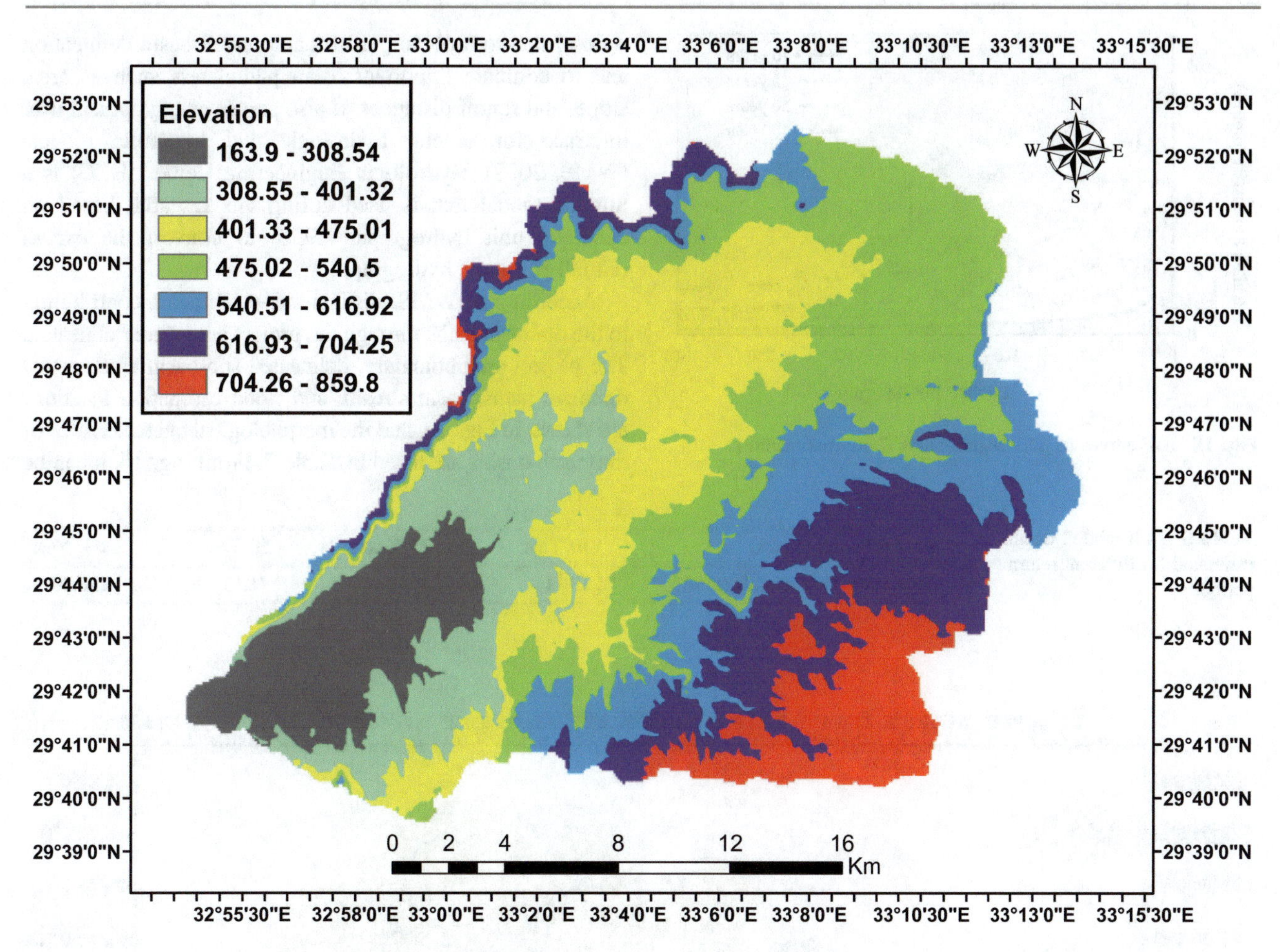

Fig. 14 Digital elevation model (DEM) of the study area, ASTER 30

Table 5 Derived rainfall intensity (mm/hr) with different storm durations at different frequency period

Storm duration (Min)	Different return periods (Years)			
	10	25	50	100
10	23.83	51.92	93.58	168.68
20	16.39	35.71	64.36	116.02
30	13.17	28.69	51.71	93.20
60	9.05	19.73	35.56	64.10
120	6.23	13.57	24.46	44.09
180	5.00	10.90	19.65	35.42
360	3.44	7.50	13.51	24.36
720	2.37	5.16	9.29	16.75

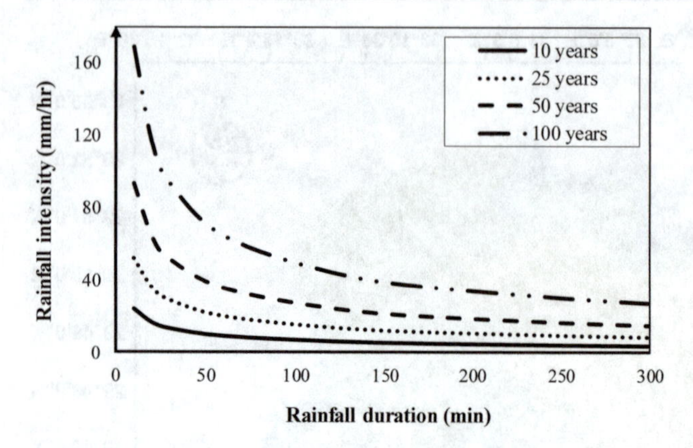

Fig. 15 IDF curves of Sudr region using California method

state-of-the-art tools to perform automated basin delineation and to compute important basin parameters such as area, slope, and runoff distances. It also serves as a graphical user interface for several hydrologic and hydraulic models (WMS, 2008). Hydrologic Engineering Center (HEC) is a lumped model that is used during this research based on synthetic unit hydrographs (SCS) to convert the excess rainfall to runoff hydrograph.

According to WMS, distinct sub-catchments contributing to the drainage paths through the project area were delineated. The project area boundary, delineated sub-basin catchments/ drainage paths, main stream, and flood estimation locations are shown in Fig. 16, and the morphological characteristics of these sub-basins are listed in Table 7. From Fig. 16, it can be

Table 6 24 h rainfall depths estimated for different return periods

Return Period (Years)	10 Year	25 Year	50 Year	100 Year
Maximum 24 h rainfall (mm)	18.11	39.46	71.13	128.20

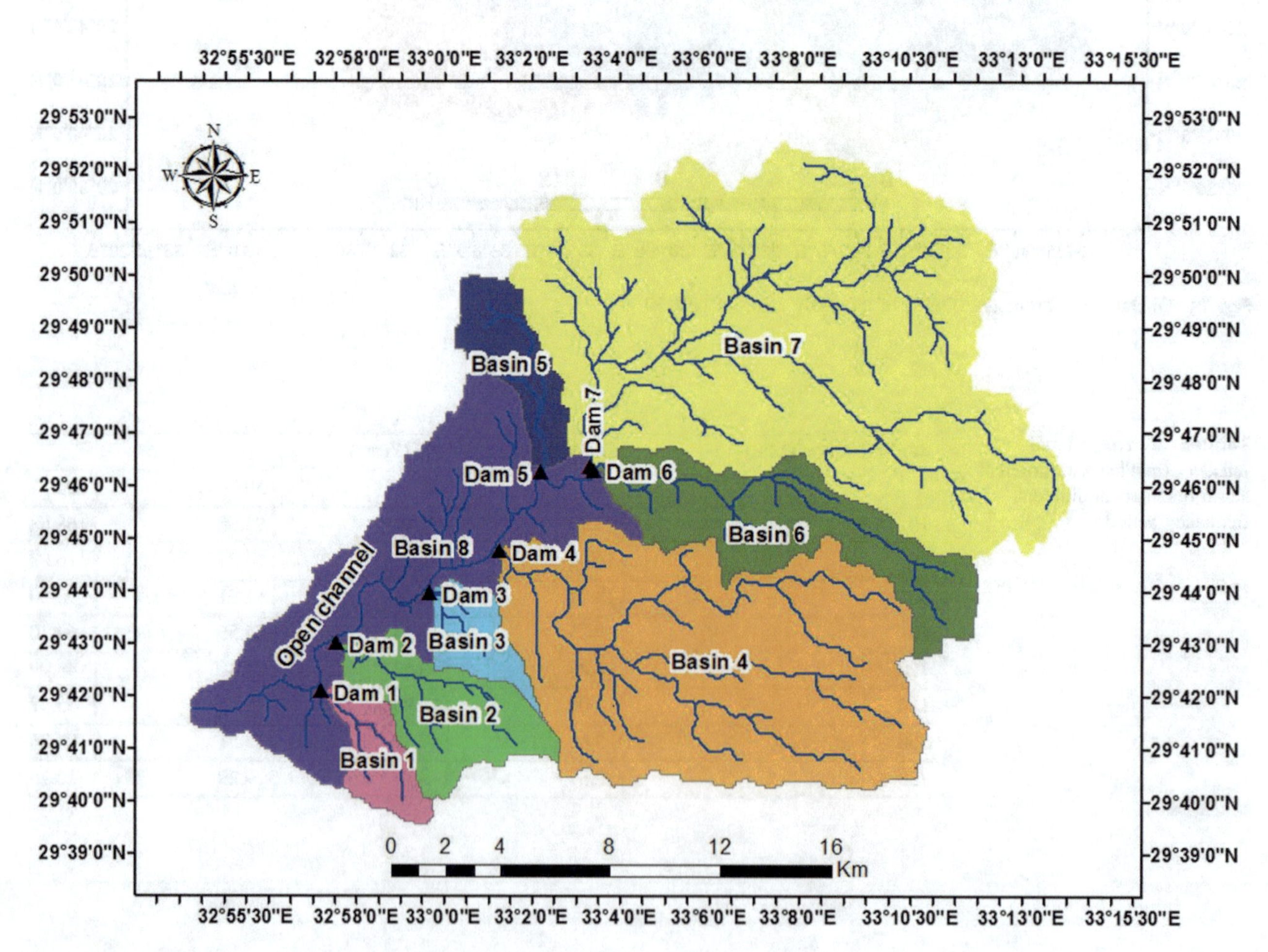

Fig. 16 Natural drainage network with sub-basin catchments affecting the study area

Table 7 Geomorphic properties of the target sub-basin

Name	Basin area (km^2)	Basin slop (m/m)	MFD (m)	MFD slope (m/m)	Shape fact	Sinuosity	Perimeter (m)	Mean elev (m)	CN
Basin 1	9.73	0.104	7187.26	0.042	3.753	0.936	19134.87	364.96	82
Basin 2	22.96	0.100	11159.33	0.037	3.606	1.015	30565.74	386.83	82
Basin 3	9.17	0.074	6934.58	0.046	4.172	0.577	19386.95	361.58	82
Basin 4	112.46	0.135	24345.54	0.021	2.585	1.311	64340.32	618.18	85
Basin 5	13.98	0.123	9038.50	0.042	3.875	1.033	23128.63	503.03	85
Basin 6	39.15	0.060	19221.82	0.017	5.640	1.190	52557.28	579.07	78
Basin 7	180.49	0.058	28487.93	0.009	1.669	1.529	87283.32	510.67	78
Basin 8	71.87	0.133	25774.36	0.013	4.302	1.343	77469.59	343.99	75

MFD basin length along the main channel from outlet to the upstream boundary

Table 8 Output of hydrology analysis for the study area

BASIN name	Q_{25} (m^3/s)	Q_{50} (m^3/s)	Q_{100} (m^3/s)	V25 (m^3)	V50 (m^3)	V100 (m^3)
1	10.3	40.0	105.4	92744.1	302410.8	771338.7
2	19.2	72.4	192.5	218832.3	713523.6	1819930
3	10.2	39.0	102.2	87466.5	285191.1	727416
4	75.3	239.2	586.1	1388543	4062155	9745588
5	18.8	59.5	144.0	172647	505080	1211742
6	12.4	58.8	173.8	255555	983563.2	2736820
7	39.8	178.7	528.5	1177938	4533608	12615008
8	12.3	69.6	227.1	340221.6	1520169	4541059

observed that the case study has eight sub-basin, which ranged from 9.17 to 180.49 km^2. According to this delineation, the case study area should be protected from a flash flood. Seven dams and one open channel are used to protect the case study, as shown in Fig. 16.

For the design of the seven dams that are used to protect the study area from flood hazard, embankment dams are used. The criteria should be met to ensure satisfactory earth and rock fill, which include embankment, foundation, abutments, freeboard, spillway, and outlet. Table 8 presents the storage volume upstream of each dam at different return periods that are used in the design of dams. The seven dams are capable of storing a large amount of water that can be used in the development of the area and estimated by 1.2 million m^3 from dam 1, 2.7 million m^3 from dam 2, 6 million m^3 from dam 3, 1.5 million m^3 from dam 4, 2.10 million m^3 from dam 5, 0.50 million m^3 from dam 6, and 20 million m^3 from dam 7, for return period 100 year as shown in Fig. 17.

To check the ultimate storage capacity of dams, the runoff hydrograph at seven dams for different return periods 10, 25, 50, and 100 years is calculated and listed in Table 8 and displayed in Fig. 18. The figure shows that the max runoff discharges ranged from 100 m^3/s at dam 1 to 600 m^3/s at dam 4. It is clear that the dams capture rainwater that attacked the study area.

Manning equation is used to design the main channel. The channel (CH) properties and design parameters are presented in Table 9, and the typical cross section of the channel is shown in Fig. 19. In this study, an investigation of the flood in a coastal tourism area is presented, and mitigation of flash flood hazards is also presented. The proposed protection measures are not costly as it can be built from available materials in the study area. One drainage channel and seven dams were designed to protect the area from flood risks. On the other hand, the stored water can be used as a source for both drinking and agriculture that can help in the development of this area. Protection of such areas can help in increasing the national income through expanding the tourism activities on the Red Sea coasts.

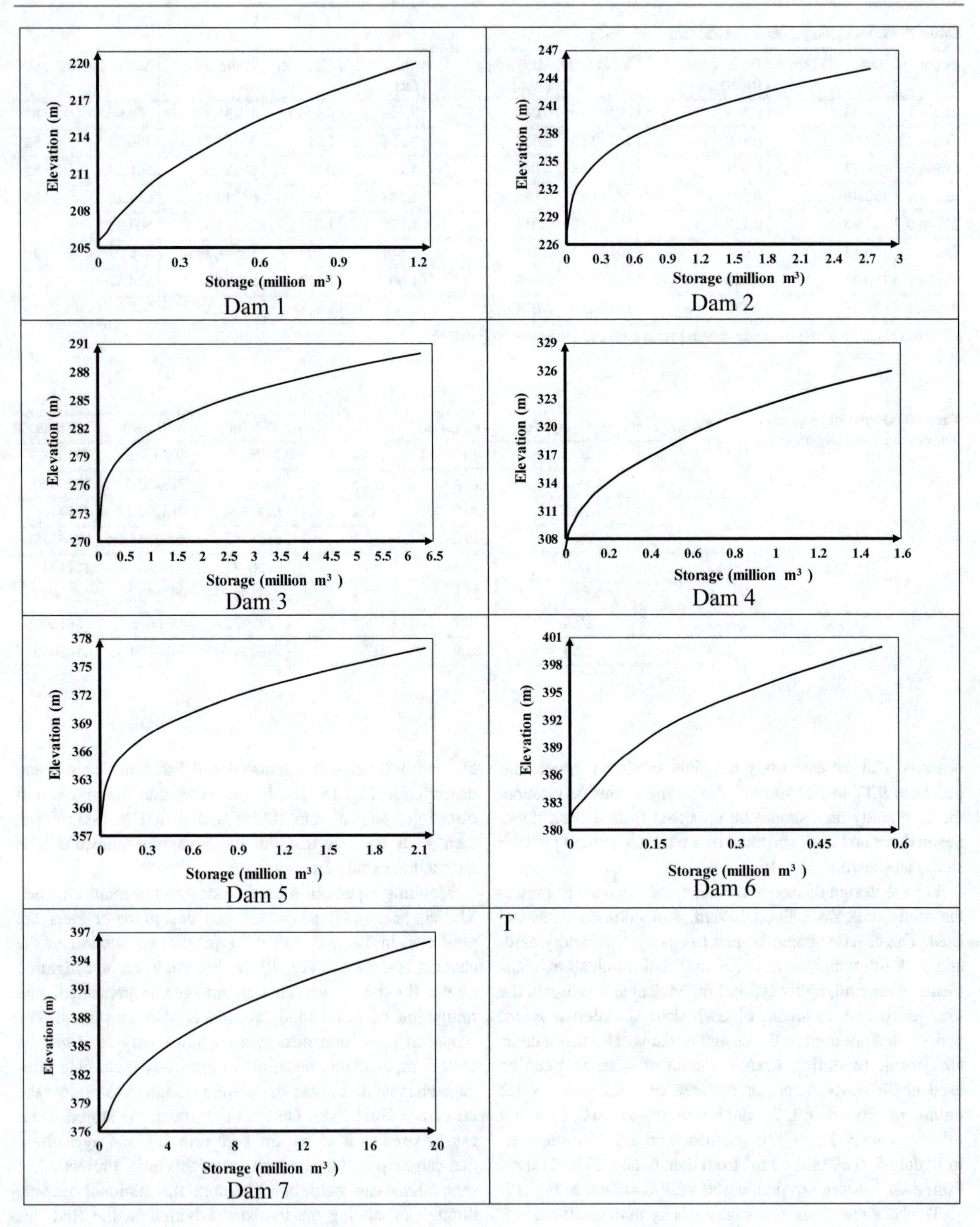

Fig. 17 Storage elevation curve for the seven dams

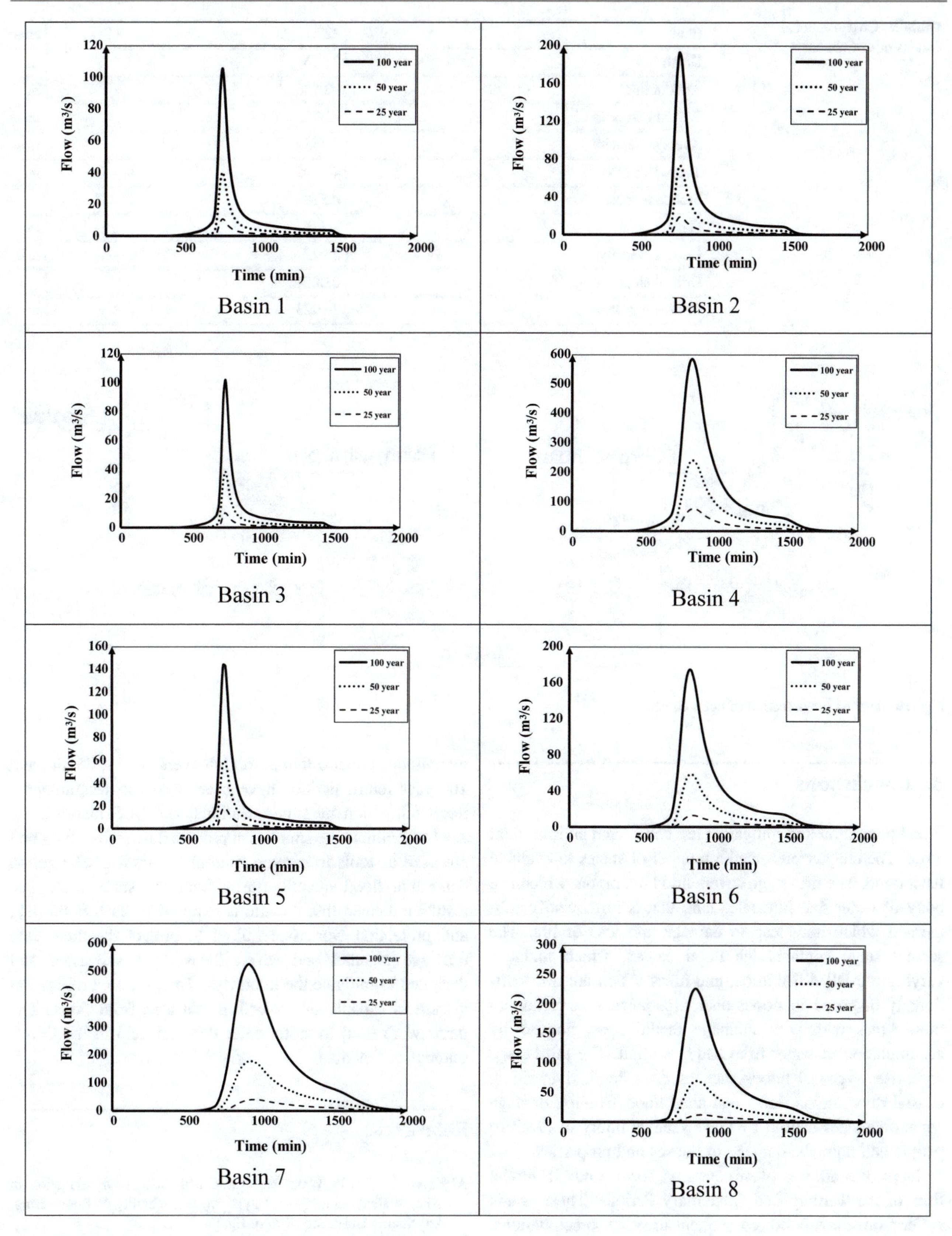

Fig. 18 Flow hydrograph of basins

Table 9 Cross-sectional dimensions of channels

Flow	227.1	cms
Depth	2.626	m
Area of flow	92.586	m^2
Wetted perimeter	41.745	m
Average velocity	2.453	m/s
Top width (T)	40.505	m
Froude number	0.518	
Critical depth	1.731	m
Critical velocity	3.922	m/s
Critical slope	0.00543	
Critical top width	36.923	m

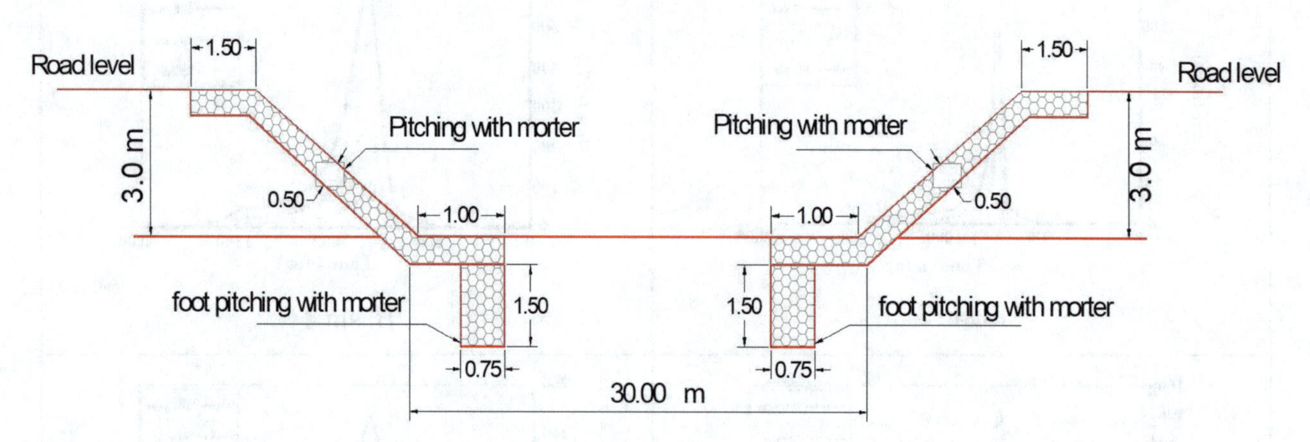

Fig. 19 Typical cross section of open channel

8 Conclusions

Flood prediction and mitigation are considered an important issue. The chapter presented a number of stages to evaluate flash flood. The first stage covers flood definition, which is a body of water that inundates land that is infrequently submerged which may lead to damage and loss of life. The second stage contains the flood causes, which includes varying rainfall distribution, and areas which are not traditionally disposed to floods also experience severe inundation. Areas with poor drainage facilities get flooded by accumulation of water from heavy rainfall. The third stage comprises types of floods such as flash flood, river flood, coastal flood, urban flood, and areal flood. The fourth stage covers damages caused by flood such as injury or death to people and animals, damage to houses and properties.

Egypt has alluvial (wadi) systems, formed during fluvial time of the Tertiary and Quaternary Periods. These wadis suffer from flash flood, consequent to heavy precipitations. Wadi Sudr, Sinai, is selected as case study to study the prediction and mitigation of flood in this area. The runoff flow paths are detected across the study area and their flow magnitudes under different rainfall events of 10, 25, 50, and 100-year return periods have been used for designing the flood mitigation measures. Rational and SCS methods are used to facilitate the simulation process during this study and are used as tools to convert rainfall to runoff discharges to determine flood quantity throughout the study area. The results indicated that the site is exposed to flash flood risk, and protection work is required to protect the area. One drainage channel and seven dams were suggested and designed to mitigate the flood risk. The proposed mitigation system is capable of protecting this area from flood. The dams were used to store water that can help in the development of this area.

References

Al-Zayed, I. (2010). Water harvesting and flash flood mitigation in Wadi Watier, south Sinai, Egypt. M. SC., Faculty of Engineering, Ain Shams University, Cairo, Egypt.

Bettina, M., & Virginia, M. (2011). Floods in the WHO European region. Periodic report, health effects and their prevention. World Health Organization.

Bettina, M., & Virginia, M. (2013). Floods in the WHO European region. Final report, health effects and their prevention. World Health Organization.

Bicknell, B., Imhoff, J., Kittle, J., Donigian, A., & Johanson, R. (1997). Hydrological simulation program-FORTRAN, user's manual for version 11. Environmental Protection Agency (EPA), USA.

Bobba, A., Singh, V., & Bengtsson, L. (2000). Application of environmental models to different hydrological systems. *Ecological Modelling*, *125*(1), 15–49.

Burnash, R., Ferral, L., & McGuire, R. (1973). A generalized streamflow simulation system—Conceptual modeling for digital computers. Technical Report NOAA, Depart of Water Resources Tech, California, USA.

Carpenter, T., & Konstantine, P. (2004). Continuous streamflow simulation with the HRCDHM distributed hydrologic. *Journal of Hydrology*, *298*, 61–79.

Chow, V. T., Maidment, D. R., & Mays, L. W. (1988). *Applied hydrology*. New York: McGraw-Hill Book Company.

Cools, J., Vanderkimpen, P., El Afandi, G., Abdelkhalek, A., Fockedey, S., El Sammany, M., Abdallah, G., El Bihery, M., Bauwens, W., & Huygens, M. (2012). An early warning system for flash floods in hyper-arid Egypt. *Natural Hazards and Earth System Sciences*, *12*, 443–457.

Daniele, N., Marco, B., Silvia E., Eric G., & Sandrine, A. (2008). Flash flood warning based on rainfall thresholds and soil moisture conditions: An assessment for gauged and ungauged basins. *Journal of Hydrology*, *362*, 274–290.

Department of Institutional Development National Institute of Education (IDNIE). (2008). Learning to live with FLOODS Natural Hazards and Disasters. Periodic report.

DHI. (1998). In MIKE-SHE v.5.3. User guide and technical reference manual, (No. 22). Technical University of Denmark, Danish Technical University.

Dong, G., & Moo, Y. (2014). The effect of flood reduction and water conservation of decentralized rainwater management system. In *11th International Conference on Hydroinformatics, HIC 2014*, New York City, USA.

Downer, C., & Ogden, F. (2006). GSSHA user's manual, gridded surface subsurface hydrologic analysis version 1.43 for WMS 6.1. ERDC Technical report. US Army Corps of Engineers.

Downer, C., Ogden, F., Martin, W., & Harmon, R. (2002). Theory, development, and applicability of the surface water hydrologic model CASC2D. *Hydrological Processes*, *16*(2), 255–275.

Elewa, H., Ramadan, E., El-Feel, A., Abu El Ella, E., & Nosair, A. (2013). Runoff water harvesting optimization by using RS, GIS and watershed modelling in Wadi El-Arish, Sinai. *International Journal of Engineering Research & Technology*, *2*(12), 1635–1648.

El-Sayed, E. (2006). Development and application of hydrologic models using geographic information systems. PHD in Civil Engineering, Faculty of Engineering, Ain Shams University, Cairo, Egypt.

El-Sayed, E. A., & Habib, E. (2008). Advanced technique for rainfall-runoff simulation in arid catchments Sinai, Egypt. In *The 3rd International Conference on Water Resources and Arid Environments*.

Fitzpatrick, A., Scudder, C., Lenz, N., & Sullivan, J. (2001). Effects of multi-scale environmental characteristics on agricultural stream biota in eastern Wisconsin. *American Water Resources Association*, *37*(1), 489–507.

HEC. (1998). HEC-1 flood hydrograph package user's manual. US Army Corps of Engineers Hydrologic Engineering Center.

HEC. (2008). HEC HMS, user's manual version 3.3. US Army Corps of Engineers Hydrologic Engineering Center.

Jonkman, N., & Kelman, I. (2005). An analysis of the causes and circumstances of flood disaster deaths. *Natural Disasters*, *1*(29), 75–97.

Kingma, C. (2014). Geomorphological aspects of flood hazard. Hydrology lectures, Department of Earth Systems Analysis, Faculty of Geo-Information Science and Earth Observation, university of TWENTE.

Koren, V., Reed, S., Smith, M., Zhang, Z., & Seo, D. (2004). Hydrology laboratory research modeling system (HL-RMS) of the US national weather service. *Journal of Hydrology*, *291*, 297–318.

Leavesley, H., Restrepo, J., Markstrom, L., Dixon, M., & Stannard, G. (2004). The modular modeling system (MMS)-a modeling framework for multidisciplinary research and operational applications. User's Manual, U.S. Geological Survey (USGS).

Mandych, A. F. (2015). Classification of floods. *Natural Disasters*, *II*, 215–250.

Murari, P. (2010). An examination of distributed hydrologic modeling methods as compared with traditional lumped parameter approaches. Ph.D., Department of Civil and Environmental Engineering, Brigham Young University.

Murari, P., Nelson, E. J., & Downer, W. (2010). Assessing the capability of a distributed and a lumped hydrologic model on analyzing the effects of land use change. *Journal of Hydroinformatics* (In Press), *18*, 65–80.

National Disaster Management Guidelines (NDMG). (2008). Management of floods. National Disaster Management Authority Government of India.

National Weather Service (NWS). (2013). Flood safety tips and resources. Periodical report, US Department of Commerce.

Pierre, V. (2012). Use of radar rainfall estimates and forecasts to prevent flash flood in real time by using a road inundation warning system. *Journal of Hydrology*, *416*, 157–170.

Pierre, J., Catherine, F., Patrick, A., & Jacques, L. (2010). Flash flood warning at ungauged locations using radar rainfall and antecedent soil moisture estimations. *Journal of Hydrology, 394,* 267–274.

Raghunath, H. (2006). Hydrology. Book, New Delhi: New Age International (P) Ltd., Publisher.

Refsgaard, J. C., & Knudsen, J. (1997). Operational validation and intercomparison of different types of hydrological models. *Water Resources Research, 32*(7), 2189–2202.

Ries III, K., & Crouse, M. (2002). The national flood frequency program, version 3: A computer program for estimating magnitude and frequency of floods for ungauged sites. Water-Resources Investigations Report 2–4168, U.S. Department of the interior. U.S Geological survey.

Reed, S., Koren, V., Smith, M., Zhang, Z., Moreda, F., Seo, J., & Participants, D., (2004). Overall distributed model intercomparison project results. *Journal of Hydrology*, *298*(4), 27–60.

Rockwood, D. M., Davis, E. D., & Anderson, J. A. (1972). User manual for COSSARR model. U.S. Engineer Division, North Pacific.

SCS, Soil Conservation Service. (1986). Urban hydrology for small watersheds, Technical Report 55. Technical Report tr55, USDA, Springfield, VA. Xi.

Sherman, L. K. (1932). Streamflow from rainfall by unit-graph method. *Engineering News-Record*, *108*, 501–505.

Smith, M. B., Seo, J., Koren, V., Reed, M., Zhang, Z., Duan, Y., Moreda, F., & Cong, S. (2004). The distributed model intercomparison project (DMIP): Motivation and experiment design. *Journal of Hydrology*, *298*, 4–26.

Soil Conservation Service (SCS). (1972). Snow suwey and water supply forecasting. In *National Engineering Handbook*, Section 22, Soil Conservation Service, Washington, DC.

Sugawara, M., Ozaki, E., Wantanabe, I., & Katsuyama, Y. (1976). Tank model and its application to bird Creek, Wollombi Brook, Bihin River, Sanaga River, and Nam Mune. National Center for Disaster Prevention, Tokyo, Japan.

Texas Department of Transportation (TxDOT). (2004). Hydraulic design manual. Design Division, Texas, U.S.A.

Tyra, B., Nicole, R., Eric, S., & Anna, O. (2016). Flood learning module. The University of Illinois, at Urbana-Champaign. Supported by the National Science Foundation CAREER Grant.

United States Army Corps of Engineers (USACE). (1994). Flood runoff analysis". Technical Report EM 1110-2-1417, US Army Corps of Engineers, Hydrologic Engineering Center.

U.S. Department of Agriculture (USDA). (2004). *National engineering handbook*. New York, U.S.A.

Vazquez, R. F., Feyen, L., Feyen, J., & Refsgaard, J. C. (2002). Effect of grid size on effective parameters and model performance of the MIKE-SHE code. *Hydrological Processes*, *16*(2), 4–18.

Vieux, B. E., & Moreda, F. G. (2003). Ordered Physics-based parameter adjustment of a distributed model. *Calibration of Watershed Models Water Science and Application, 6*(19), 267–281.

Vieux, B. E., Cui, Z., & Gaur, A. (2004). Evaluation of a physics-based distributed hydrologic model for flood forecasting. *Journal of Hydrology*, *298*(18), 155–177.

Vos, F. (2009). Annual disaster statistical review 2009: The numbers and trends. Final report of Brussels, Centre for Research on the Epidemiology of Disasters.

Williams, J. R., & Hann, R. W. (1973). Watershed HYMO: Problem-oriented language for hydrologic modeling. User's manual. 23.publ.ARS-S-9, U.S. Gov. Department of Agriculture, Washington, D.C.

Gray-to-Green Infrastructure for Stormwater Management: An Applicable Approach in Alexandria City, Egypt

Mahmoud Nasr and Ahmed N. Shmroukh

Abstract

Stormwater management is employed to provide adequate planning, analysis, and design of storm drainage networks. This trend would minimize the human and socio-economic losses caused by stormwater while maintaining benefits from flash floods. Although stormwater in residential areas creates flooding and pollution, it is currently considered a viable resource of water supply. The management of stormwater aims at achieving sustainable environmental, ecological, and social aspects. However, the management of heavy rainfall and flash floods involves a substantial economic investment. Recently, green infrastructure has been used for stormwater management, considering construction techniques, materials accessibility, transportation routes, site planning and restrictions, payments, and profits. Hence, this chapter represents the implementation of the green infrastructure concept in stormwater drainage projects to minimize the negative impacts of heavy rainfall and flash floods on the environment. The findings of this work are highlighted regarding previous case studies reported in the literature. Moreover, this work gives a future vision for the application of green infrastructure in Alexandria, Egypt.

Keywords

Best management practice • Egypt • Green infrastructure • Remediation • Stormwater runoff

M. Nasr (✉)
Sanitary Engineering Department, Faculty of Engineering, Alexandria University, Alexandria, 21544, Egypt
e-mail: mahmmoudsaid@gmail.com; mahmoud-nasr@alexu.edu.eg

A. N. Shmroukh
Department of Mechanical Engineering, Faculty of Engineering, South Valley University, Qena, 83521, Egypt
e-mail: eng_ahmednagah@yahoo.com

1 Introduction

Urban stormwater is runoff associated with rainfall or snow events, and it may carry various pollutants such as litter, nutrients, debris, toxic chemicals, and sediments to nearby residential areas (Grey et al. 2018). Moreover, stormwater can transfer animal waste, bird droppings, heavy metals, and various cancer-causing substances and anthropogenic pollutants from roofs and roads into downstream water bodies (Merriman et al. 2017). These contaminants can impair the hydrology, the aquatic environment, and the life of urban dwellers (Guertin et al. 2015). Moreover, stormwater runoff flows over the urban impervious surfaces and causes flooding, heavy casualties, and traffic paralysis (Vemula et al. 2019).

Stormwater infrastructure systems are designed to manage flash floods and to improve the water quality in rivers and streams (Granata et al. 2016). Currently, urban water infrastructure is facing growing pressure due to continuous fluctuations of climatic and socio-economic factors (Sörensen and Emilsson 2019). These impacts lead to flash flood problems, sewer pipe damage, and serious degradations in the quality of a citizen's life (Alves et al. 2019). In addition, the volume of water entering the sewerage systems is increasing due to urbanization and the wide application of water-resistant surfaces such as concrete and asphalt pavements (Shariat et al. 2019). Hence, upgrading, maintenance, expansion, and rehabilitation of the current water infrastructure systems are essentially needed as these projects are characterized by an extended long service life (e.g., over 50 years).

Recently, several regions in Egypt, especially the coastal strip, have been suffered from severe and unexpected flash floods (Mahmoud and Gan 2018). Although this stormwater runoff creates multiple problems, it can be harvested and used for non-potable applications such as road cleaning, toilet flushing, vehicle washing, and other non-irrigational purposes (Shehata 2018). Recently, several green infrastructural systems, which use tree plantation, grasslands,

A. M. Negm (ed.), *Flash Floods in Egypt*, Advances in Science, Technology & Innovation, https://doi.org/10.1007/978-3-030-29635-3_16

soils, and other ecological elements, have been designed to harvest and control stormwater runoff (Piro et al. 2019). However, the feasibility of this technology in Egyptian towns, cities, and governorates is still under investigation.

Hence, this chapter aims at reviewing the recent applications of the management of stormwater drainage projects. Moreover, this work would offer a better understanding of the barriers and facilitating factors that affect the management of reclaimed stormwater. The outputs of this work can assist water resource managers, government, professionals, and private and public sectors in maintaining flood risk management, especially in Egypt.

2 Stormwater Infrastructure

Recently, the exponential growth of population has resulted in increasing the civil and construction projects, including stormwater infrastructure (Gunnell et al. 2019). These stormwater drainage systems are considered crucial engineering strategies within the urban environment (Rabori and Ghazavi 2018). Stormwater drainage networks can be classified into gray and green infrastructure systems (McFarland et al. 2019). In the gray, also known as grey, stormwater drainage system, urban runoff is directly conveyed from impervious lands to downstream water bodies or wastewater treatment facilities (Irwin et al. 2018). This traditional stormwater management relies on curbs, gutters, drains, and other underground collection networks. However, gray sewer systems may suffer from various issues due to precipitation, gas formation, pipes corrosion, and odor problems (Fu et al. 2019).

Recently, the term "Green" has been used to manage stormwater by expanding the quantity and quality of green spaces (e.g., green roofs, woody plants, bio-retention basins, and porous pavements) that can capture rainwater at its source (Zhang and Muñoz Ramírez 2019). In green infrastructure, soil and vegetation undertake the abilities to infiltrate, reallocate, and store stormwater, providing valuable economic, environmental, and social advantages (Baker et al. 2019). For instance, Tirpak et al. (2019) found that suspended pavement systems (i.e., a tree-specific design approach) could be used to manage urban stormwater, providing adequate runoff volume reduction and pollutant removal capabilities. Alves et al. (2019) reported that gray infrastructure could provide a high reduction of flood risk, whereas green infrastructure could attain additional benefits such as water saving, air quality improvement, and carbon sequestration. Fu et al. (2019) reported that green infrastructure should be integrated with the traditional sewer structures, including pipes and storage tanks to provide an efficient stormwater management system. Wang et al. (2013) investigated the environmental benefits and economic opportunities associated with the application of green, gray, and integrated green/gray stormwater infrastructures. Their study (Wang et al. 2013) found that the bio-retention basins yielded the lowest greenhouse gas emissions and improved the water quality condition by filtering the runoff influents and removing the nonpoint source pollutants.

3 Green Infrastructure Types and Design Considerations

Green (or sustainable) infrastructure is a wide expression used to mimic the ecological processes of soils and plants instead of conventional stormwater drainage systems (Johnson et al. 2019). Green infrastructure is considered a stormwater management technology, which is designed to enhance the environmental quality and community welfare by offering various economic and human health benefits (Caplan et al. 2019). Some advantages of green infrastructure include water purification, erosion protection, air quality improvement (carbon sequestration), maintenance of soil structure and quality, hydraulic redistribution, and pollination (Gavrić et al. 2019). Furthermore, green stormwater infrastructure is able to mitigate the dramatic changes in land use and land cover in urbanized regions (Luan et al. 2019).

Various green infrastructures such as retention ponds, constructed wetlands, rain gardens, and bioswales have been used for groundwater recharge, water quality improvement, and erosion control. Some examples of green infrastructure can be represented as follows:

(a) Bio-retention basin is a cultivated and low-leveled landscape area installed from concrete, several engineered filter media, and a drainage pipe at the bottom. In this basin, stormwater runoff is filtered through vegetated species and different soil media layers (Wang et al. 2019a, b). The bio-retention basins should be shallow and vegetated with distinct species to collect and treat a sufficient amount of stormwater.

(b) Stormwater ponds are employed for the management of stormwater quantity and quality, and they can be classified into retention ponds and detention ponds. Retention (wet) ponds are designed for a permanent collection of stormwater runoff in large and open pools (Moura 2017). They can also offer landscape purposes and water quality advantages, such as sediments and pollutants reduction. Detention (dry) ponds are used to store rainwater for a limited period (e.g., 24–48 h) after storm episodes. They are also used to protect tributaries of bays, lakes, and rivers against flooding (Guzman et al. 2018).

(c) In constructed wetlands, stormwater runoff passes through a vegetated system in which soil, plant species, and the inherent microorganisms retain and remediate rainwater (Bousquin and Hychka 2019). Artificial wetlands can be classified into subsurface- and surface-flow schemes (Nasr and Ismail 2015).
(d) Rain gardens are shallow vegetated basins installed nearby impervious catchments such as roads and parking lots to store and infiltrate stormwater runoff. Malaviya et al. (2019) provided the physicochemical and biological characteristics of rain gardens that could capture runoff and avoid further groundwater pollution and surface water quality impairment.
(e) Bioswales use engineered soils that can capture, filter, and transport runoff volume via strips of vegetated, broad, and shallow channels (Everett et al. 2018).
(f) Urban street trees are essential constituents of green infrastructure as they improve stormwater interception regimes and enhance various ecosystem and ecohydrological services (e.g., habitat provision and pollination, crop improvement genes, and purification of water and air). Berland and Hopton (2014) emphasized the importance of expanding the tree planting density to reduce peak stormwater runoff via interception, evaporation, transpiration, and infiltration. Urban street trees in the USA can intercept rainfall in their crown with values ranging from 2.87 to 15.12 m^3/tree/year (Gonzalez-Sosa et al. 2017).
(g) Green roof, also known as a living roof, is an ecological roofing system covered by vegetation (impervious surfaces) to harvest runoff via infiltration and evapotranspiration (Kavehei et al. 2018).
(h) Permeable pavement is a pervious urban surface that can be used for the retention and detention of stormwater runoff. Jayakaran et al. (2019) found that porous asphalt pavements could be employed to manage stormwater in urbanized watersheds appropriately. Pavements should be provided with sufficient voids to infiltrate and store stormwater.

These stormwater green infrastructure projects are designed based on the following information:

(a) Future population and land use.
(b) Frequency and magnitude of stormwater, wastewater, and groundwater infiltration.
(c) Allowable slopes and diameters of sewers.
(d) Historical and future climate.
(e) Soil topography and contour lines.

4 Stormwater Management by Green Infrastructure

Heavy rainfall and flash floods have a significant impact on the national economic development and human settlements of various countries (Patel et al. 2019). Rainfall is often conveyed from buildings, roofs, sidewalks, and roads to the drainage system. When heavy rainfall exceeds the drainage capacity of sewer networks, the additional inflow may flood the streets and low-lying catchments (Jusić et al. 2020). The extreme rainfall episodes can cause severe flooding, resulting in several physical threats to people and failure to infrastructure and utilities. Hence, stormwater management is an essential approach that should be comprehensively investigated.

Stormwater management refers to the application of a standardized and organized problem-solving technique that balances between function, quality, and cost of stormwater infrastructure projects (Wang et al. 2019a, b). The application of stormwater management is based on the concept that the value of each infrastructure element can be improved by increasing its function (or performance). For this purpose, the essential project components are defined, systematized, and segregated, and then the lowest possible cost of each function is determined (Meng and Hsu 2019). The management procedures can be considered during any stage of the project (e.g., planning, design, drawings, construction, operation, and maintenance) to minimize risks, create new concepts, and solve problems. Accordingly, stormwater management attempts to complete the project's tasks using limited resources, adequate period, and feasible cost.

The application of the management concept in the field of stormwater infrastructure provides various advantages such as

(a) Maximize customer satisfaction, organization performance, and environmental safety and reliability (Xu et al. 2019).
(b) Constitute ecosystem services for both present and future generations while reducing the life-cycle cost.
(c) Analyze and audit the performance of stormwater under various climatic conditions and unpredicted quantities of rainfall. This trend will positively save a large portion of the construction and maintenance costs without limiting the functions and performances of stormwater collection systems (Diogo and do Carmo 2019).

However, stormwater management can be negatively influenced by

(a) Lack of knowledge about historical and future extreme rainfall data.
(b) Insufficient practices and experiences.
(c) Inadequate standards and specifications.

The components of stormwater infrastructure that can be considered by the management technique include (Delgrosso et al. 2019)

(a) Pump station: Pump station is used in gravity sewer systems to lift the flow (e.g., sewage) from a certain location to another place at a specific height.
(b) Manhole: Manhole is a hole constructed of brick or concrete masonry, and it used to monitor, test, clean, and maintain the sewers.
(c) Ventilation: The municipal pipe system may be equipped with ventilation supplies to prevent the formation of harmful gases under the anaerobic condition.
(d) Sewers: sewer pipes are installed underground to collect and carry wastewater from buildings and industries and transport it to the final destination. These pipes are manufactured from plastic, concrete, and steel materials.

The value of each item can be expressed as a ratio between function and cost, in which the management approach tends to eliminate or prevent unnecessary expenses.

5 Green Infrastructure Vision for Alexandria

5.1 Alexandria City Description

Alexandria is a Mediterranean port and the second-largest city in Egypt. It is located 70 km northwest of the Nile Delta (31.20° N; 29.92° E) and 179 km north of Cairo (the capital of Egypt) (Soliman et al. 2008). The governorate has a population of about 5.2 million inhabitants in 2018, and it occupies an area of about 2818 km^2.

5.2 Infrastructure Description

The city infrastructure is mainly composed of a combined sewer system that collects both sewage and stormwater from domestic, commercial, and industrial areas (Brudler et al. 2019). These wastewater sources are conveyed to centralized wastewater treatment plants leading to hydraulic and organic shock load conditions. The infrastructure has a limited storage capacity, and it is vulnerable to overflowing during stormwater episodes. However, some suburban areas of the city are composed of separate sewer systems, in which stormwater and sanitary sewage are conveyed via two separate pipe networks (Abbas et al. 2019).

5.3 Flood Conditions

The average annual rainfall of Alexandria is almost 169 mm, which distributes heavily along the north coastal zone (Soliman and Reeve 2010). Unfortunately, this coastal city suffers from periodic flash floods due to rapid urbanization and various infrastructure-related problems (Zevenbergen et al. 2017). Moreover, the performance of existing drainage systems fails to properly flood-proof the city against foreseeable stormwater events. Other factors such as the shortage of flood susceptibility maps (e.g., historical floods), and insufficiency of information related to the impact of climate change on surface runoff have negatively affected the ability of the city to manage the stormwater episodes.

During flash flood events, raw stormwater is directly discharged into receiving water bodies such as canals, drains, and the Mediterranean Sea. As a result of high runoff depth, which amounts to 180 mm/yr, dramatic flood risks have occurred, and several private properties have been paralyzed. Moreover, dense rainfall regimes increase the soil moisture content, leading to the deterioration of groundwater resources. A study by Zevenbergen et al. (2017) has described the October 2015 unexpected severe storm in Alexandria (i.e., rainfall over 100 mm in 2 h), which was the worst flooding over the previous decades. The runoff depth increases at the northeastern and northwestern locations due to the influence of impervious surfaces, and it slightly reduces at the southern and western parts of Alexandria due to the existence of agriculture lands. About 15.9, 33.5, and 41.0% of the city area are classified as "Very high", "High", and "Moderate" flood susceptibility zones, respectively.

Elboshy et al. (2019) reported the negative impacts of extreme rainfall events (e.g., December 1991; January 2004; November 2010; December 2010; November 2011; September 2015; October 2015; November 2015) on the public infrastructure in Alexandria. Their study (Elboshy et al. 2019) developed a framework that could minimize the pluvial flood risk at some locations in Alexandria. The pluvial floods in November 2015 have resulted in several economic and human losses, which could be due to the lack of proper sewer system cleaning and maintenance, high population density, and insufficient vegetation areas. Ibrahim et al. (2019) used remote sensing and a geographic information system (GIS) to determine the appropriate urban green space systems that could be used to avoid the extreme rainfall and pluvial flood hazard in Alexandria. Their work (Ibrahim et al. 2019) also suggested that more effort should be conducted to upgrade the sewage and drainage systems to mitigate potential flood damages.

5.4 Management Practice

Based on the management concept, it is recommended to employ decentralized stormwater management practices including green roofs (partially or completely covered with vegetation), rainwater harvesting structures, and infiltration schemes. Moreover, impermeable surfaces such as asphalt or concrete pavements that limit infiltration should be reduced. Furthermore, urban street trees capable of interacting with the hydrologic cycle and eliminating water from the soil via transpiration should be widely vegetated. For instance, Shehata (2018) used a survey study and questionnaire to represent a future vision of stormwater management in Alexandria via the application of green infrastructure. The study (Shehata 2018) suggested the implementation of permeable pavements, green roofs, public gardens, and planted areas to harvest stormwater, improve air quality, enhance community liveability, and reduce energy consumption.

5.5 Vision of Green Infrastructure

Figure 1 shows the decision tree used to select the appropriate green infrastructure type that can be used for stormwater management in Alexandria. The green infrastructures can be classified into two groups:

(a) Elongated green infrastructures:

Some shapes, such as vegetated filter strip, swale, and infiltration trench, have a length far larger than their width dimensions (Guertin et al. 2015). The low-lying vegetated filter strips lack a visual level difference from the surrounding field, and hence, they can be classified as "No" regarding "Visual elevation difference with the road nearby". Vegetated filter strips are composed of deeply rooted plant species used to manage stormwater runoff and to deal with nonpoint source pollutants such as sediments, nutrients, and particles (Reichenberger et al. 2019). However, swale and infiltration trench are shallow as they are characterized by low elevation differences (<0.5 m) from the surrounding area (Shafique et al. 2018). Further, the degree of vegetation is classified as "Trees", "Grass", and "None" for vegetated filter strips, swale, and infiltration trench, respectively. In grass swales, stormwater is collected in an open channel and then percolates into the ground. Infiltration trench is filled with rubble, stone, and other porous material, which temporarily stores stormwater runoff (Naghedifar et al. 2019).

(b) Non-elongated green infrastructures:

Ponds and gardens are characterized by approximately square or circular shapes. The wet pond is a green infrastructure that can permanently carry water (Ryan et al. 2010). This pond is used for flood management and stormwater runoff treatment. Bio-retention cell and rain garden are composed of tree features including bushes, flowers, shrubs, and native vegetation to retain and remediate stormwater before being discharged into a nearby stream (Johnson and Hunt 2019). Detention basin, also known as infiltration basin, is located on permeable and coarse media to allow stormwater to percolate into underlying soils and groundwater. Dry ponds are stormwater-control ponds designed to capture and temporarily hold water and prevent flash flooding (Guzman et al. 2018).

5.6 Stormwater Management Procedures

The management of stormwater engineering projects includes consecutive procedures, which can be explained as follows:

(a) Project selection and defining objectives.
(b) Information phase: in this stage, the relevant information is collected from various sources such as people, documentation, standards and regulations, and test results. This step identifies primary key constraints, e.g., expected cost, implementation schedule, and energy consumption.
(c) Analysis phase: in this step, the project functions are identified, classified, and organized via a function hierarchy diagram. This diagram defines the functions of a particular system in the hierarchical model. Each criterion such as initial expenses, performance, maintenance and operating costs, and land saving is given a weight and score value. Scores may vary from 1 (poor) to 5 (very good).
(d) Creative phase: this step aims at generating plenty of ideas and suitable alternatives via lateral thinking and brainstorming.
(e) Evaluation phase: the selected ideas are assessed, ranked, and rated to identify viable alternative ideas, and each choice is reviewed regarding its advantages and disadvantages.
(f) Development phase: in this stage, cost–benefit analysis, technical data package, implementation plan, and final

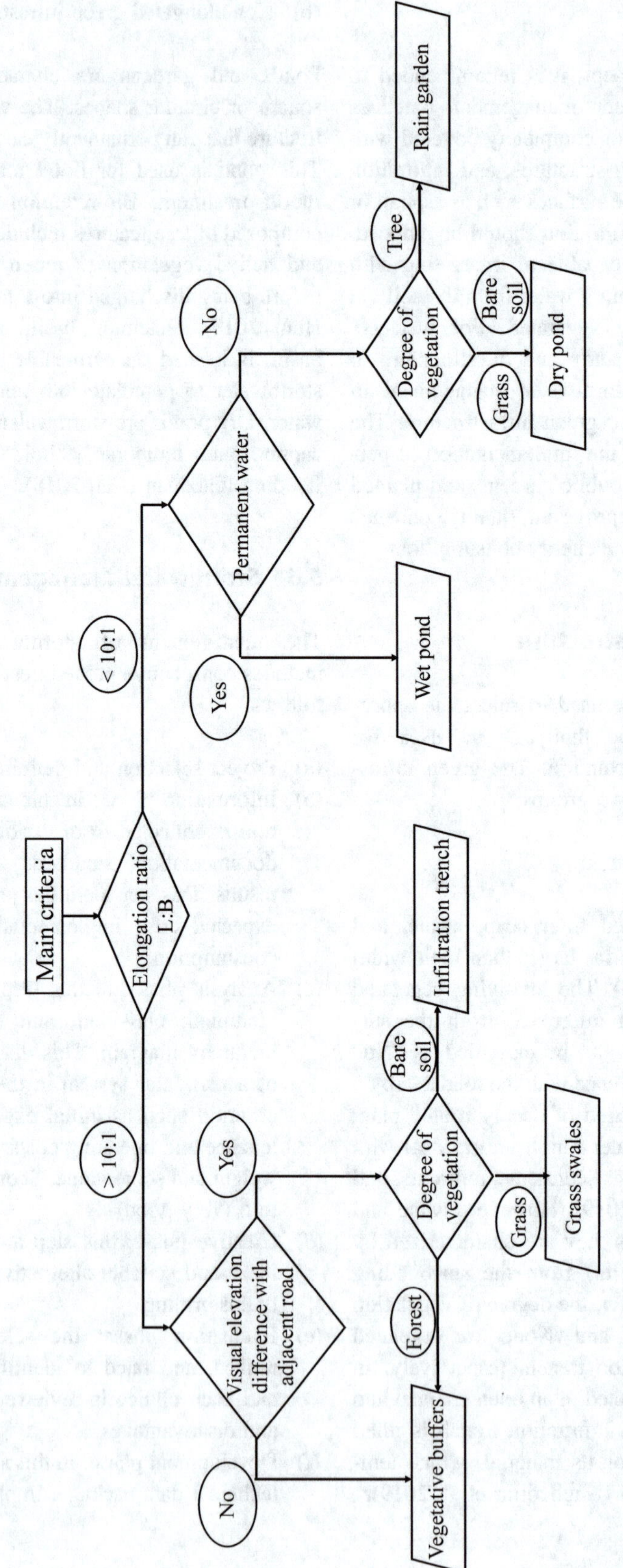

Fig. 1 Decision-making flowchart of green infrastructure for stormwater management in Alexandria, Egypt

into value engineering proposals regarding expenses and performances.

(g) Presentation phase for delivering oral and written reports to the client, owner, or project manager.

(h) The final stage includes implementation, completion, and monitoring phases.

6 Case Studies for Stormwater Management Using Green Infrastructure

Recently, various studies have investigated the possibility of applying the management concept in stormwater drainage networks. The literature survey for flood management in the Scopus database reveals an exponential increase in the number of documents during 2000–2018 (Fig. 2a–d). This finding suggests that more effort has been recently exerted to minimize the negative impact of flooding on the environment.

Gao et al. (2015) used green infrastructure systems (e.g., wet ponds, permeable pavements, bio-retention basins, and vegetated swales) for the management of flooding and stormwater pollution. Their study (Gao et al. 2015) demonstrated that the management practices reduced the total runoff volume by 41%, and the pollution load decreased by 57%, 55%, and 60% for total nitrogen, total phosphorus, and chemical oxygen demand, respectively.

Lu et al. (2019) synthesized a permeable pavement material using recycled ceramic aggregate and bio-based polyurethane binder. The porous pavement showed high ability to infiltrate and percolate stormwater compared to those of conventional porous asphalt. Accordingly, the obtained material could mitigate the increasing risk of urban flooding and reduce greenhouse gas emissions.

Yazdi (2019) presented a reliable and efficient optimization tool, known as EPA's Storm Water Management Model, to improve the operation of urban detention ponds during various external (rainfalls) loads. The proposed model was also appropriate for simulating stormwater quantity and quality in the urban environment.

Bryant et al. (2019) presented an overview of the design strategies of green stormwater infrastructure to manage stormwater runoff from drainage areas, reduce pollutants in separate and combined municipal sewer systems, and mitigate localized flooding.

Finewood et al. (2019) applied the green infrastructure approach in a post-industrial environmental city, and reported that this system could capture and slow down stormwater before it reached the municipal sewer systems.

Tsegaye et al. (2019) investigated the conversion of drainage systems into green infrastructure to address the stormwater management tasks. This transformation was carried out using the following procedures:

(a) Mapping the current site hydrology and identifying the green infrastructure components,
(b) Siting and locating areas suitable for green infrastructure,
(c) Estimating and predicting pollution loads and stormwater runoff volumes,
(d) Selecting appropriate green infrastructure for the chosen site, and
(e) Evaluating alternatives and scenarios to expand and implement the infrastructure system.

7 Recommendations

The green infrastructure systems have recently found successful applications for stormwater management and flood control. Based on the chapter outputs, some recommendations that are helpful for the case study of Alexandria can be derived:

(a) Flood insurance programs should be developed to reduce the negative impacts of flooding on private and public properties.
(b) A framework for implementing the stormwater management approach should be proposed to reduce cost, damages, and human and socio-economic losses.
(c) Stormwater mitigation and soil management should be significantly considered in the urban stormwater policy.
(d) Urban planners and architects should apply environmental and ecological methods in designing green infrastructures.
(e) Integration of engineering, social, and environmental criteria is needed to identify the most appropriate and effective stormwater infrastructure for a particular area.
(f) Dynamic computer simulation models should be developed to predict the performance of drainage systems under various extreme meteorological and climatic conditions.

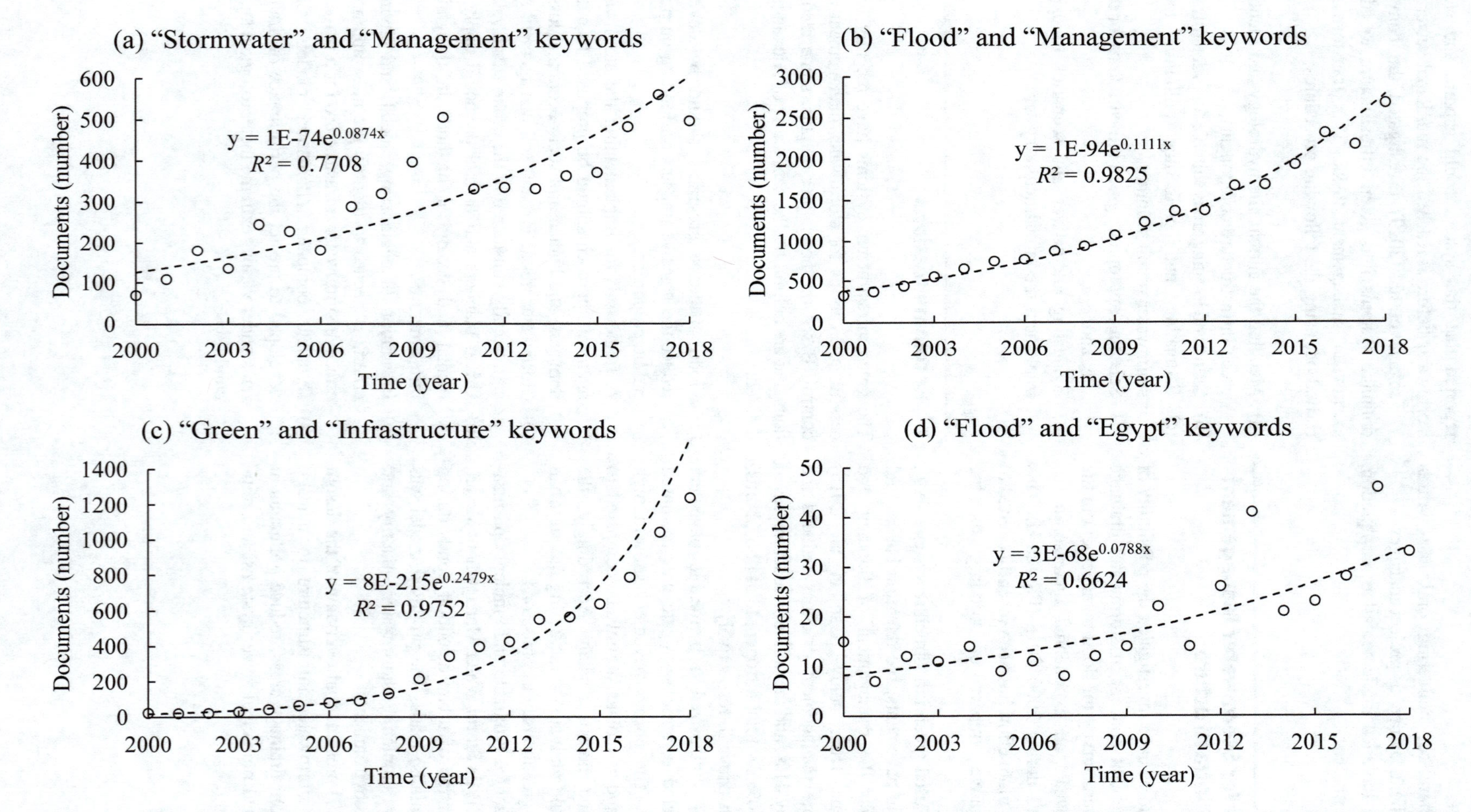

Fig. 2 Number of documents in Scopus database during (2000–2018) including **a** "Stormwater" and "Management" terms, **b** "Flood" and "Management" terms, **c** "Green" and "Infrastructure" terms, and **d** "Flood" and "Egypt" terms

8 Conclusions

In this chapter, a literature survey and case studies are presented to draw sufficient conclusions about the implementation of stormwater management in residential areas. The discussed green infrastructures include bioswales, retention basins, ponds, wetlands, rain gardens, permeable pavements, and urban green spaces. These stormwater infrastructures tend to control runoff volume and timing and promote ecosystem services. Stormwater management can be applied to a case study of Alexandria Governorate to exemplify the impact of climate change and urbanization on the drainage system performance of a city subjecting to successive periods of heavy rain, flash flood, and strong winds. The combination of traditional drainage systems with green infrastructure could be a viable solution to mitigate the stormwater runoff in Alexandria city, Egypt. In addition, urban trees should have a vital role in stormwater management, as trees can be connected to urban hydrology and increase the infiltration of stormwater.

Acknowledgements The first author would like to acknowledge Nasr Academy for Sustainable Environment (NASE).

References

Abbas, A., Ruddock, F., Alkhaddar, R., Rothwell, G., & Andoh, R. (2019). Improving the geometry of manholes designed for separate sewer systems. *Canadian Journal of Civil Engineering, 46*(1), 13–25.

Alves, A., Gersonius, B., Kapelan, Z., Vojinovic, Z., & Sanchez, A. (2019). Assessing the co-benefits of green-blue-grey infrastructure for sustainable urban flood risk management. *Journal of Environmental Management, 239,* 244–254.

Baker, A., Brenneman, E., Chang, H., McPhillips, L., & Matsler, M. (2019). Spatial analysis of landscape and sociodemographic factors associated with green stormwater infrastructure distribution in Baltimore, Maryland and Portland, Oregon. *Science of the Total Environment, 664,* 461–473.

Berland, A., & Hopton, M. (2014). Comparing street tree assemblages and associated stormwater benefits among communities in metropolitan Cincinnati, Ohio, USA. *Urban Forestry & Urban Greening, 13*(4), 734–741.

Bousquin, J., & Hychka, K. (2019). A geospatial assessment of flood vulnerability reduction by freshwater wetlands—A benefit indicators approach. *Frontiers in Environmental Science, 7*(APR), 54.

Brudler, S., Arnbjerg-Nielsen, K., Hauschild, M. Z., Ammitsøe, C., Hénonin, J., & Rygaard, M. (2019). Life cycle assessment of point source emissions and infrastructure impacts of four types of urban stormwater systems. *Water Research, 156,* 383–394.

Bryant, R., Stein, J., & Tonning, D. (2019). Reimagining parks as stormwater infrastructure-stormwater parks of all sizes, designs, and funding sources. *Journal of New England Water Environment Association, 53*(1), 32–42.

Caplan, J., Galanti, R., Olshevski, S., & Eisenman, S. (2019). Water relations of street trees in green infrastructure tree trench systems. *Urban Forestry and Urban Greening, 41,* 170–178.

Delgrosso, Z., Hodges, C., & Dymond, R. (2019). Identifying key factors for implementation and maintenance of green stormwater infrastructure. *Journal of Sustainable Water in the Built Environment, 5*(3), 05019002.

Diogo, A., & do Carmo, J. (2019). Peak flows and stormwater networks design-current and future management of urban surface watersheds. *Water* (Switzerland), *11*(4), 759.

Elboshy, B., Kanae, S., Gamaleldin, M., Ayad, H., Osaragi, T., & Elbarki, W. (2019). A framework for pluvial flood risk assessment in Alexandria considering the coping capacity. *Environment Systems and Decisions, 39*(1), 77–94.

Everett, G., Lamond, J. E., Morzillo, A. T., Matsler, A. M., & Chan, F. K. S. (2018). Delivering green streets: An exploration of changing perceptions and behaviours over time around bioswales in Portland, Oregon. *Journal of Flood Risk Management, 11,* S973–S985.

Finewood, M., Matsler, A., & Zivkovich, J. (2019). Green infrastructure and the hidden politics of Urban stormwater Governance in a Postindustrial City. *Annals of the American Association of Geographers, 109*(3), 909–925.

Fu, X., Goddard, H., Wang, X., & Hopton, M. (2019). Development of a scenario-based stormwater management planning support system for reducing combined sewer overflows (CSOs). *Journal of Environmental Management, 236,* 571–580.

Gao, J., Wang, R., Huang, J., & Liu, M. (2015). Application of BMP to urban runoff control using SUSTAIN model: Case study in an industrial area. *Ecological Modelling, 318,* 177–183.

Gavrić, S., Leonhardt, G., Marsalek, J., & Viklander, M. (2019). Processes improving urban stormwater quality in grass swales and filter strips: A review of research findings. *Science of the Total Environment, 669,* 431–447.

Gonzalez-Sosa, E., Braud, I., Becerril Piña, R., Mastachi Loza, C., Ramos Salinas, N., & Chavez, C. (2017). A methodology to quantify ecohydrological services of street trees. *Ecohydrology & Hydrobiology, 17*(3), 190–206.

Granata, F., Gargano, R., & de Marinis, G. (2016). Support vector regression for rainfall-runoff modeling in Urban Drainage: A comparison with the EPA's stormwater management model. *Water, 8,* 69. https://doi.org/10.3390/w8030069.

Grey, V., Livesley, S., Fletcher, T., & Szota, C. (2018). Establishing street trees in stormwater control measures can double tree growth when extended waterlogging is avoided. *Landscape and Urban Planning, 178,* 122–129.

Guertin, D. P., Korgaonkar, Y., Burns, I. S., Barlow, J., Unkrich, C., Goodrich, D. C., & Kepner, W. (2015). Evaluation of green infrastructure designs using the automated geospatial watershed assessment tool. In *11th Watershed Management Symposium 2015: Power of the Watershed.* ASCE National HeadquartersReston; United States.

Gunnell, K., Mulligan, M., Francis, R., & Hole, D. (2019). Evaluating natural infrastructure for flood management within the watersheds of selected global cities. *Science of the Total Environment, 670,* 411–424.

Guzman, C. B., Cohen, S., Xavier, M., Swingle, T., Qiu, W., & Nepf, H. (2018). Island topographies to reduce short-circuiting in stormwater detention ponds and treatment wetlands. *Ecological Engineering, 117,* 182–193.

Ibrahim, M. G., Elboshy, B., & Mahmod, W. E. (2019). Integrated approach to assess the Urban green infrastructure priorities

(Alexandria, Egypt). In: H. Chaminé, M. Barbieri, O. Kisi, M. Chen, B. Merkel (Eds.), *Advances in sustainable and environmental hydrology, hydrogeology, hydrochemistry and water resources. advances in science, technology & innovation (IEREK interdisciplinary series for sustainable development)*, Springer, Cham.

Irwin, S., Howlett, C., Binns, A., & Sandink, D. (2018). Mitigation of basement flooding due to sewer backup: Overview and experimental investigation of backwater valve performance. *Natural Hazards Review, 19*(4), 04018020.

Jayakaran, A., Knappenberger, T., Stark, J., & Hinman, C. (2019). Remediation of stormwater pollutants by porous asphalt pavement. *Water (Switzerland), 11*(3), 520.

Johnson, C., Tilt, J., Ries, P., & Shindler, B. (2019). Continuing professional education for green infrastructure: Fostering collaboration through interdisciplinary trainings. *Urban Forestry and Urban Greening, 41,* 283–291.

Johnson, J., & Hunt, W. (2019). A retrospective comparison of water quality treatment in a bioretention cell 16 years following initial analysis. *Sustainability (Switzerland), 11*(7), 1945.

Jusić, S., Hadžić, E., & Milišić, H. (2020). Urban stormwater management—New technologies. *Lecture Notes in Networks and Systems, 76,* 790–797.

Kavehei, E., Jenkins, G., Adame, M., & Lemckert, C. (2018). Carbon sequestration potential for mitigating the carbon footprint of green stormwater infrastructure. *Renewable and Sustainable Energy Reviews, 94,* 1179–1191.

Luan, B., Yin, R., Xu, P., Wang, X., Yang, X., Zhang, L., et al. (2019). Evaluating green stormwater infrastructure strategies efficiencies in a rapidly Urbanizing catchment using SWMM-based TOPSIS. *Journal of Cleaner Production, 223,* 680–691.

Lu, G., Liu, P., Wang, Y., Faßbender, S., Wang, D., & Oeser, M. (2019). Development of a sustainable pervious pavement material using recycled ceramic aggregate and bio-based polyurethane binder. *Journal of Cleaner Production, 220,* 1052–1060.

Mahmoud, S., & Gan, T. (2018). Urbanization and climate change implications in flood risk management: Developing an efficient decision support system for flood susceptibility mapping. *Science of the Total Environment, 636,* 152–167.

Malaviya, P., Sharma, R., & Sharma, P. (2019). Rain gardens as stormwater management tool. In S. Shah, V. Venkatramanan, & R. Prasad (Eds.), *Sustainable green technologies for environmental management*. Singapore: Springer.

McFarland, A., Larsen, L., Yeshitela, K., & Engida, A. L. N. (2019). Guide for using green infrastructure in urban environments for stormwater management. *Environmental Science: Water Research & Technology, 5*(4), 643–659.

Meng, T., & Hsu, D. (2019). Stated preferences for smart green infrastructure in stormwater management. *Landscape and Urban Planning, 187,* 1–10.

Merriman, L. S., Moore, T. L. C., Wang, J. W., Osmond, D. L., Al-Rubaei, A. M., Smolek, A. P., et al. (2017). Evaluation of factors affecting soil carbon sequestration services of stormwater wet retention ponds in varying climate zones. *Science of the Total Environment, 583,* 133–141.

Moura, N. (2017). The Jaguaré Creek revitalization project: transforming São Paulo through a green stormwater infrastructure. *Procedia Engineering, 198,* 894–906.

Naghedifar, S., Ziaei, A., & Naghedifar, S. (2019). Optimization of quadrilateral infiltration trench using numerical modeling and Taguchi approach. *Journal of Hydrologic Engineering, 24*(3), 04018069.

Nasr, M., & Ismail, S. (2015). Performance evaluation of sedimentation followed by constructed wetlands for drainage water treatment. *Sustainable Environment Research, 25*(3), 141–150.

Patel, P., Ghosh, S., Kaginalkar, A., Islam, S., & Karmakar, S. (2019). Performance evaluation of WRF for extreme flood forecasts in a coastal urban environment. *Atmospheric Research, 223,* 39–48.

Piro, P., Turco, M., Palermo, S.A., Principato, F., & Brunetti, G. (2019). A comprehensive approach to stormwater management problems in the next generation drainage networks. In F. Cicirelli, A. Guerrieri, C. Mastroianni, G. Spezzano, A. Vinci (Eds.), *The internet of things for smart Urban ecosystems. internet of things (Technology, communications and computing)* (pp. 275–304), Springer, Cham, 275–304.

Rabori, A., & Ghazavi, R. (2018). Urban flood estimation and evaluation of the performance of an urban drainage system in a semi-arid urban area using SWMM. *Water Environment Research, 90*(12), 2075–2082.

Reichenberger, S., Sur, R., Kley, C., Sittig, S., & Multsch, S. (2019). Recalibration and cross-validation of pesticide trapping equations for vegetative filter strips (VFS) using additional experimental data. *Science of the Total Environment, 647,* 534–550.

Ryan, P., Wanielista, M., & Chang, N.-B. (2010). Nutrient reduction in stormwater pond discharge using a chamber upflow filter and skimmer (CUFS). *Water, Air, and Soil pollution, 208*(1–4), 385–399.

Shafique, M., Kim, R., & Kyung-Ho, K. (2018). Evaluating the capability of grass swale for the rainfall runoff reduction from an Urban parking lot, Seoul, Korea. *International Journal of Environmental Research and Public Health, 15*(3), 537.

Shariat, R., Roozbahani, A., & Ebrahimian, A. (2019). Risk analysis of urban stormwater infrastructure systems using fuzzy spatial multi-criteria decision making. *Science of the Total Environment, 647,* 1468–1477.

Shehata, S. (2018). The impact of green infrastructure rainwater harvesting on sustainable Urbanism: Case study in Alexandria, Egypt. *WIT Transactions on Ecology and the Environment, 217,* 797–805.

Soliman, A., Moghazy, H., & El-Tahan, A. (2008). Sea level rise impacts on Egyptian Mediterranean coast (case study of Alexandria). *AEJ—Alexandria Engineering Journal, 47*(1), 75–87.

Soliman, A., & Reeve, D. (2010). Applying the artificial submerged reefs techniques to reduce the flooding problems along the Alexandria Coastline. In *Coasts, Marine Structures and Breakwaters: Adapting to Change—Proceedings of the 9th International Conference* (Vol. 2, pp. 188–199).

Sörensen, J., & Emilsson, T. (2019). Evaluating flood risk reduction by urban blue-green infrastructure using insurance data. *Journal of Water Resources Planning and Management, 145*(2), 04018099.

Tirpak, R., Hathaway, J., Franklin, J., & Kuehler, E. (2019). Suspended pavement systems as opportunities for subsurface bioretention. *Ecological Engineering, 134,* 39–46.

Tsegaye, S., Singleton, T. L., Koeser, A. K., Lamb, D. S., Landry, S. M., Lu, S., et al. (2019). Transitioning from gray to green (G2G)—A green infrastructure planning tool for the urban forest. *Urban Forestry and Urban Greening, 40,* 204–214.

Vemula, S., Raju, K., Veena, S., & Kumar, A. (2019). Urban floods in Hyderabad, India, under present and future rainfall scenarios: A case study. *Natural Hazards, 95*(3), 637–655.

Wang, J., Chua, L., & Shanahan, P. (2019a). Hydrological modeling and field validation of a bioretention basin. *Journal of Environmental Management, 240,* 149–159.

Wang, M., Zhang, D., Cheng, Y., & Tan, S. (2019b). Assessing performance of porous pavements and bioretention cells for stormwater management in response to probable climatic changes. *Journal of Environmental Management, 243,* 157–167.

Wang, R., Eckelman, M., & Zimmerman, J. (2013). Consequential environmental and economic life cycle assessment of green and gray stormwater infrastructures for combined sewer systems. *Environmental Science and Technology, 47*(19), 11189–11198.

Xu, Z., Xiong, L., Li, H., Chen, K., & Wu, J. (2019). Runoff simulation of two typical urban green land types with the stormwater management model (SWMM): Sensitivity analysis and calibration of runoff parameters. *Environmental Monitoring and Assessment, 191*(6), 343.

Yazdi, J. (2019). Optimal operation of Urban storm detention ponds for flood management. *Water Resources Management, 33*(6), 2109–2121.

Zevenbergen, C., Bhattacharya, B., Wahaab, R. A., Elbarki, W. A. I., Busker, T., & Salinas Rodriguez, C. N. A. (2017). In the aftermath of the October 2015 Alexandria flood challenges of an Arab city to deal with extreme rainfall storms. *Natural Hazards, 86*(2), 901–917.

Zhang, S., & Muñoz Ramírez, F. (2019). Assessing and mapping ecosystem services to support urban green infrastructure: The case of Barcelona, Spain. *Cities, 92,* 59–70.

Conclusions

Update, Conclusions, and Recommendations for "Flash Floods in Egypt"

Abdelazim M. Negm and El-Sayed E. Omran

Abstract

The present situation in Egypt is framed by the scarcity of land and water that is under serious pressure. Soil and water resources, on the one hand, are at the core of sustainable development and critical to socio-economic growth. On the other hand, water utilized from flash floods is considered an important source of water supply in Egypt after the River Nile and groundwater. This chapter captures the key Flash Floods utilization (in terms of findings and suggestions) and provides ideas extracted from the volume cases. In addition, some (update) findings from a few recently published research work related to the Flash Floods covered themes. This chapter provides the present problems faced by the Flash Floods with a set of recommendations to safeguard the water to supply the populations and farmers with water.

Keywords

Assessment • Sustainability • Flash floods • Environment • Hazards • Mitigation • Egypt

1 Introduction

In reaction to the loss of twenty-six life across three governorates in 2016 and seven lives in Alexandria in 2015, meanwhile, the issue of our preparedness to deal with the implications of climate change continues unanswered. Is Egypt actually ready?

On the one hand, while past floods have resulted in a loss of life, damage to personal and public property, power cuts, highway closures, road accidents, and public outrage, perhaps the amount of rainfall in this autumn could be the greatest quantity of rainfall in 2019. On the other hand, however, water from flash floods in arid areas can be a significant source of water supply. Particularly affected are desert regions like the Egyptian Sinai Peninsula. Wadis are especially susceptible to such a case.

Therefore, the wise use of floodwater to enable the viable management of water resources is a significant challenge in these fields. A barrier is that observation information is usually scarce and the findings of the model are too coarse to allow precise predictions. The significance of the challenge will only improve as it is anticipated that the frequency and effect of flash floods will increase as a consequence of climate change. So, the general goal of this book was to accomplish sustainable water resource management in Egypt.

Therefore, the intention of the book is to improve and address the following main theme.

- Analysis and Design
- Recent Technologies For Investigating Flash Flood
- Environmental Approaches In Flash Flood
- Hazards, Risk, and Utilization of Flash Floods in Sinai
- Prediction, Mitigation, and Management of Flash Flood.

The next section presents a brief of the important findings of some of the recent (updated) published studies on the Flash Flood, then the main conclusions of the book chapters

A. M. Negm (✉)
Water and Water Structures Engineering Department, Faculty of Engineering, Zagazig University, Zagazig, 44519, Egypt
e-mail: amnegm85@yahoo.com; ; amnegm@zu.edu.eg

E.-S. E. Omran
Soil and Water Department, Faculty of Agriculture, Suez Canal University, Ismailia, 41522, Egypt
e-mail: ee.omran@gmail.com

E.-S. E. Omran
Institute of African Research and Studies, Nile Basin Countries. Aswan University, Aswan, Egypt

A. M. Negm (ed.), *Flash Floods in Egypt*, Advances in Science, Technology & Innovation,
https://doi.org/10.1007/978-3-030-29635-3_17

in addition to the main recommendations for researchers and decision-makers. The update, conclusions, and recommendations presented in this chapter come from the data presented in this book.

2 Update

Under various circumstances, flash floods can happen. Flash floods most often happen in usually dry regions where precipitation has lately occurred, but they can be seen anywhere downstream from the precipitation source, even many miles from the source.

The following are the major updates for the book project based on the main book theme:

1. Analysis and Design. An overview of the Flash Floods analysis and design is discussed in two chapters. Chapter 2 deals with the statistical behavior of rainfall in Egypt. A thorough study of the statistical characteristics of extreme rainfall events in Egypt is discussed in this chapter. The extreme rainfall events have critical impacts on hydrologic systems and society, especially in arid countries. Recently, Egypt has been subject to some flash floods due to extreme rainfall events, in particular regions (e.g., Sinai, North Coast, and Upper Egypt) that caused severe damages in lives and vital infrastructure and buildings. This chapter investigates, therefore, the statistical characteristics of rainfall over Egypt based on historical daily rainfall records at 30 stations throughout the country. Six types of rainfall data were extracted from daily records: monthly rainfall, annual rainfall, the monthly number of rainy days, the annual number of rainy days, monthly maximum daily rainfall, and annual maximum daily rainfall. To estimate rainfall intensity of different durations and return periods, intensity–duration–frequency (IDF) relationships were developed by using different distribution models in different neighborhoods (see, e.g., (Jiang and Tung 2013). Many studies evaluated the performance of different probability models to identify the most appropriate distribution that could provide accurate estimates of extreme rainfall in different regions of the world, (see, e.g., Wdowikowski et al. 2016). A recent study (Salama et al. 2018) assessed the performance of some of the most popular probability distributions to identify the most suitable model that could accurately estimate extreme rainfall events in different regions in Egypt. The results indicated that the Log-Normal and Log-Pearson Type III distributions are the best models for most studied stations in Egypt. The impact of extreme precipitation events on both water resources quality and water supplies was investigated in the Alexandria region and Upper Egypt (Yehia et al. 2017). In Qena, the flash flood that occurred on January 28, 2013 was studied, and the surface runoff was estimated (Moawad et al. 2016).

The third chapter discusses the critical analysis and design aspects of flash floods in arid regions: Cases from Egypt and Oman. Most of the world can be classified as arid or semi-arid. Such a climate is well known for flash flood development due to convective rainfall and poor soil. Such a combination creates a hazard, noticeably different from other climates. Due to the remoteness of such areas and sparsely population, the flash flood information and data required for simulation are lacking. This chapter addresses several cases of peculiar flash flood events that took place in the arid region. The addressed cases covered extreme events generated from Cyclonic storms and convective storms. All cases caused damages to the properties and loss of lives in some events. The mountainous arid areas in Egypt and the gulf (Sultanate of Oman) are suffering from flash flood attacks. The two countries have been selected for detailed investigation due to the availability of enough data to carry out detailed simulations. The available data covers temporal and spatial long rainfall and flow records. Moreover, the Sultanate of Oman has experienced severe cyclones over recent decades.

2. Recent Technologies for Investigating Flash Flood. Two chapters are identified in the book related to recent technologies for investigating flash floods. The first technology is developing an Early Warning System for Flash flood in Egypt: Case Study Sinai Peninsula. The good rainfall forecasting is very much necessary for providing an early warning to avoid or minimize flash flood disasters. Weather research and forecasting (WRF) model is an important and good tool in developing early warning systems of flash flood (Dhote et al. 2018) over the Sinai Peninsula. These systems will help and support the related authorities and decision-makers to avoid many disasters accompanied by a flash flood. In addition, this research could apply and spread over the whole of Egypt as it gives warnings several days in advance.

A hydrodynamic flash flood model for the region of El Gouna (Northern Red Sea Governorate) is the second technology, which is presented to simulate the spreading of water and flooding areas generated by heavy rainfall inside the domain as well as by the runoff from the large catchment of Wadi Bili. Prediction of flash flood events is, first of all, a challenge in terms of meteorology. Long-term knowledge of changing weather conditions in the future is almost impossible, and therefore, reliable forecasts are only available for a few days. But besides the amount of precipitation as the source of the flooding, the morphology and composition of the hydrologic basin affect the impact of a flash flood as well. Rainfall events that cause major flash floods are very rare in the Eastern Desert and their occurrence is infrequent (Hadidi 2016). These differences in reporting flood occurrences may be explained because of two main reasons: first, rainfall events are usually of small-scale convective systems (5–10 km) (Hadidi 2016); second, the Eastern Desert is a remote area, and there is no official observation for all wadis. This chapter describes the specific situation of flash floods around El Gouna, Red Sea, and is focused on general morphological, geological, and hydrological settings of the catchment and detailed estimation of flooding effects close to the city using numerical modeling. Flash flood models represent an important tool to better understand the flow processes during and after heavy rainfall events. They can be applied to investigate different scenarios regarding increased rainfall intensities on the one hand and structural measures such as dams, canals, and retention basins on the other hand to find suitable solutions to mitigate flood damages. Flash flood models should also be part of early warning systems and be used to create hazard and risk maps raising the awareness regarding zones which are particularly prone to flash floods so that the communities can be better prepared. Also, methods to effectively store and use the freshwater amounts supplied from flash floods can be investigated by combining a flash flood model and a geological model.

3. Environmental Approaches in Flash Flood. Four environmental approaches are used to delineate and discuss the Flash Flood, which are presented in the following five chapters. The first approach is related to Environmental Flash Flood Management (Osti 2018) in Egypt, which presents the tools for implementing disaster risk reduction in environmental hazards zones. It explains the strategic environmental assessment (SEA): Flood risk management strategies, SEA objectives and assessment method, environmental assessment of the flood risk management, and its mitigation and monitoring. The national environmental action plan (NEAP) is Egypt's plan for environmental actions for the coming fifteen years. This complements and incorporates economic growth and social development business strategies. The NEAP is the foundation for local environmental policies, actions, and activities to be established. It is designed to be the framework that coordinates for future environmental activities in support of sustainable development of Egypt.

The second approach is Flash flood management and harvesting via groundwater recharging in wadi systems: An integrative approach of remote sensing and direct current resistivity techniques. An integrated approach is carried out to study the hydrogeological conditions in wadi systems. In this chapter, the use of remote sensing and direct current (DC) resistivity techniques were considered to manage the flashflood and explore groundwater in desert lands. It is possible to recognize sites for successful dam construction and groundwater bearing zones exploration. The electrical geophysical methods are not routinely integrated with RS datasets. Direct current (DC) resistivity method is one of the most popular electrical methods for groundwater exploration and geoenvironmental investigations (e.g., Al-Manmi and Rauf 2016). In high conductive soils, penetration depth and resolution of the DC resistivity method are less than that of resistive medium and resistivity imaging under conductive soils. They can be quite challenging if not impractically used (Attwa and Günther 2013). Consequently, the DC resistivity method, especially when combined with RS and GIS can be capable of categorizing and characterizing areas prone to geohazards.

The third approach is environmental monitoring and evaluation of flash floods risks using remote sensing and GIS techniques. It presents a model for the application of advanced technologies capable of identifying and analyzing climatic factors and various ground and that contribute to the understanding of the nature of the floods in Egypt, which is witnessing dynamic changes fail with old maps and traditional methods in the work of actual usefulness in responding to the dynamics of the floods. Remote sensing technology along with GIS has become the key tool for load hazard and risk maps for the vulnerable areas. El Bastawesy et al. (2012) and Yousif et al. (2013) have applied the techniques of GIS and remote sensing for watershed delineation.

Sustainable development of mega drainage basins of the Eastern Desert of Egypt; Halaib-Shalatin as a case study area is the fourth approach in this update. It focuses on using effective tools of monitoring and management of natural resources, based on the integration of remote sensing (RS) and geographic information systems (GIS) techniques with a field survey in surface and groundwater resources evaluation. The overall goal is to construct a developmental and management system for water resources in this important area to increase and improve agricultural activity, food production and improve the standard of living for natives. Additionally, it includes the construction of the rainwater harvesting (RWH) system which helps in the mitigation of flood hazards that are frequently threatening different developmental activities. Rainwater harvesting (RWH), as a historical and worldwide trend, could fulfill the gap of water scarcity in arid or semi-arid regions. This proposed work is to use the modern techniques of remote sensing (RS), geographic information systems (GIS), and watershed modeling systems (WMS) to provide a plan for the RWH. RWH is the accumulation and storage of rainwater for reuse before it reaches the aquifer system (Groundwater). Multi-spectral remote sensing (MSRS) and geographic information systems (GIS) are vital tools to optimize the surface water usage of episodic rainfalls, where the concept of runoff water harvesting (RWH) in promising watersheds should be applied

(Elewa et al. 2016). GIS and digital elevation models (DEM) enable the development of hydrological models to investigate every ancient terraced field in a non-invasive manner, without disturbing the archeological remains (Bruins et al. 2019).

4. Hazards, Risk, and Utilization of Flash Floods in Sinai. Three areas are used to utilize Flash Floods in Sinai. First is Egypt's Sinai Desert Cries: Flash Flood Hazard, Vulnerability, and Mitigation. Sinai is identified as a flood-prone area where flash floods were recorded and resulted in significant infrastructure damages, population displacement, and sometimes loss of lives (Abuzied and Mansour 2018; Mohamed and El-Raey 2019). The establishment of different dams leads to the presence of communities around these dams to work in agriculture and grazing. In addition to reducing water losses, reducing the speed of floods is the main factor in protecting the soil from water erosion. Floods such as the floods in El-Arish (Moawad, 2013) may not be due to flooding alone, but due to the nature of the randomness of the construction of buildings in the corridors of floods and without the work of a previous geological study of the area on which the various facilities will be built. One effective way to reduce the risk of flash floods lies in the implementation of an early warning system. The early warning information system (EWIS) consists of a number of components, linked and activated through an automatic platform. Remote sensing (RS) and geographic information system (GIS) are utilized to provide improved spatial considerate of basin response to storm rain events (Moawad, 2013) and flood monitoring. It is important to understand that factors (Ţîncu et al. 2018) and risks are determined not only by the climate and weather events, i.e., the hazards but also by the exposure and vulnerability to hazards, which have been induced by human activity (IPCC 2012). Flash flood hazard mapping is a supporting component of non-structural measures for flash flood prevention. Therefore, effective adaptation and disaster risk management strategies and practices also depend on a rigorous understanding of the dimensions of exposure and vulnerability (Azmeri and Isa 2018), as well as a proper assessment of changes in those dimensions.

The second area is Egypt's Sinai Desert Cries: Utilization of Flash Flood for Sustainable Water Management. The eastern desert and the Sinai Peninsula are subjected to flash floods, where floods from the mountains of the Red Sea and Sinai are causing heavy damage to man-made features (Mohamed and El-Raey 2019). The central focus of this chapter was to achieve sustainable water resources management in the Sinai Peninsula. The objective of the current chapter is to mitigate and utilize the flood water as a new supply for water harvesting in Sinai. Applying water harvesting of the flash flood will reduce the flood risk at the outlet. In addition, it could be used for recharging the groundwater aquifers, which are the basis for sustainable development in Sinai. Furthermore, Bedouins usually move from place to place searching for fresh grazing for their animals and water for their families. These locations sometimes are hazardous areas as flash floods occur there. This chapter helps to develop sustainable planning for the Bedouins, as hazard areas were defined. Proved locations to get use of the flooded water and store it will encourage the Bedouins to resettle the area. Flash floods are furthermore not well understood from a hydrological, meteorological as well as from a socio-economical point of view because they come unexpected and often at a very localized scale (WRRI 2010). It is, therefore, extremely difficult to forecast flash floods with sufficient lead-time to take emergency actions. In arid regions, flash floods (Abdelkarim et al. 2019) are the main cause of loss in infrastructure, property, and human life. Despite this, floodwater could be used to create a great amount of Sinai water resources, and be utilized to fulfill part of the increasing water demand in areas prone to flooding that are currently experiencing high population growth and economic development (Shokhrukh-Mirzo 2018). This is especially true in the wadi system of Southern Sinai, where flash floods have become an unlucky annual occurrence.

The third area is the determination of potential sites and methods for water harvesting in Sinai Peninsula by the application of RS, GIS, and WMS techniques. A comprehensive approach involving the integration of geographic information systems (GIS), remote sensing (RS), and watershed modeling (WM) was applied through the present work to identify the potential areas for runoff water harvesting (RWH) in Wadi Dahab basin of southern Sinai, Egypt. These tools were effectively used in mapping, investigating, and modeling runoff processes. Accordingly, planning for maximizing the RWH, as an inherited historical technique, becomes an urgent need in order to cope with the human requirements and overcoming the water scarcity problem. Maximizing the RWH will have its vital role in improving the groundwater recharging, raising water levels, and reducing its salinities to be suitable for the different uses. RWH could be efficient and may support the installment of new settlements in arid areas, with a direct impact on raising the quality of life of local inhabitants (Elewa et al. 2016; Yazar and Ali 2016). RWH is a well-known practice to improve water security and agricultural production (Paz-Kagan et al. 2017). Watershed management aims at enhancing the water availability in rain-fed areas through water conservation structures, which facilitate the storage of water and recharge to groundwater by applying an optimum RWH system (Vema et al. 2019).

The last chapter in this section is titled "Torrents Risk in Aswan Governorate, Egypt." Identification of areas vulnerable to torrent hazard is the basis for decision-making to take the necessary management plan and mitigate the risk to minimize the effects of flood risk. Historical records show that the governorate of Aswan has been and continues to be exposed to the hazards of geomorphological torrents. Aswan's torrents arise as a result of precipitation on the mountains East of the government where water heads west to the river Nile in a group of valleys. The rivers flowing into Lake Nasser reflect the minimum risk of torrents, as urban communities do not exist. Eastern Nile basins in the Edfu-Aswan region are very risky, particularly in the Kom Ombo area and east of Aswan city.

5. Prediction, Mitigation, and Management of Flash Flood. Two Mitigations of Flash. Flood issues are updated. The first issue related to the prediction and mitigation of flash floods in Egypt. Two key drivers are essential for the generation of floods and giving it specific features. The first is the physical process that results in a change of position between the lithosphere, atmosphere, and hydrosphere. Second, the flooded area, depth of inundation, and its duration depend on the geographic situation of the region where flood takes place. Several different types and sizes of floods can be differentiated due to the great variety in the activity of the natural processes and the infinite variability in the state of the geographic arena in which they operate. There are at least 18 major types of floods of natural origin.

The second issue is gray-to-green infrastructure for Stormwater management: an applicable approach in Alexandria City, Egypt. The management of stormwater aims at achieving sustainable environmental, ecological, and social aspects. However, the management of heavy rainfall and flash floods involves a substantial economic investment. Recently, green infrastructure has been used for stormwater management, considering construction techniques, materials accessibility, transportation routes, site planning and restrictions, payments, and profits. Urban stormwater is runoff associated with rainfall or snow events, and it may carry various pollutants such as litter, nutrients, debris, toxic chemicals, and sediments to nearby residential areas (Grey et al. 2018). Moreover, stormwater runoff flows over the urban impervious surfaces and causes flooding, heavy casualties, and traffic paralysis (Vemula et al. 2019). Currently, urban water infrastructure is facing growing pressure due to continuous fluctuations of climatic and socio-economic factors (Sörensen and Emilsson 2019). These impacts lead to flash flood problems, sewer pipe damage, and serious degradations in the quality of a citizen's life (Alves et al. 2019). In addition, the volume of water entering the sewerage systems is increasing due to urbanization and the wide application of water-resistant surfaces such as concrete and asphalt pavements (Shariat et al. 2019). Recently, several green infrastructural systems, which use tree plantation, grasslands, soils, and other ecological elements, have been designed to harvest and control stormwater runoff (Piro et al. 2019).

3 Conclusions

Throughout the course of the present book project, several conclusions drawn from this book were reached by the editorial teams. In addition to methodological ideas, the chapter draws important lessons from the bookcases, in specific the promising aspects of both the present and historical local flood hazard. In order to improve sustainable food supply in Egypt, these findings are crucial. Based on the materials described in all chapters of this volume, the following findings could be indicated:

1. When urban growth tended to torrents risks, torrents become obstacles to development and must be properly exploited and need continuous studies to mitigate the risk. Results presented here demonstrate the potential for Geospatial techniques to determine the vulnerable areas of torrents risk. The lowest elevation is more risk than the highest, where Eastern Nile basins in the area between Edfu and Aswan cities are very risky, particularly in the area of Kom Ombo and East of Aswan city. Streams that flow into Lake Nasser represent the minimum risk of torrents as there are no urban communities. Avoiding the risks of torrents and benefit from them.
2. In general, there is a great variation over the whole country in all different aspects of rainfall. The rainfall indices have higher values in the north of the country than those in the middle and the south. Six types of rainfall data were extracted from daily records for 30 stations throughout the country: monthly rainfall, annual rainfall, the monthly number of rainy days, the annual number of rainy days, monthly maximum daily rainfall, and annual maximum daily rainfall. The results conclude that the rainy season in Egypt extends from October to March when some stations receive a significant amount of rainfall. The dry season, on the other hand, extends from April to September. Furthermore, this research seeks to derive the design rainfalls through the GEV distribution along with the L-moments using annual maximum daily rainfalls in these stations. In the case study, the GEV distribution fits well the annual maximum daily rainfall data with a shape parameter having negative values at all stations except only Ras

Sedr which implies an upper bound. Thus, the Gumbel distribution may be better than the GEV in this station. Although all stations not located on the north have low values of rainfall characteristics, some locations (e.g., Hurghada) have high values of design extreme rainfall.

3. Critical analysis and design aspects of flash floods in arid regions: Cases from Egypt to Oman are concluded for the major cyclone Gonue that took place in June 2007 in the following aspect: The maximum rainfall was 904 mm/d, recorded on the mountainous rain gauge. Such value corresponded to 10,000-years return period. Such storm generated a huge volume of water. All dams were full, and the spillways reached the maximum design values, the risk of dam failure was very near. The damage that took place was major. The storm distribution as a function of time was nearly uniform over 24 h, and the generated flows were comparable with the generated design storm of Oman. Oman developed a new (HDS) highway design standards (2010) to accommodate such events. The HDS was updated in 2017. It was observed in all study cases that the right of way (ROW) of all Wadis is not respected, with massive constructions by the locals.
4. Evaluation of the performance of WRF model to predict heavy rainfall, thunderstorms, and flash floods in different study cases that occurred over the Sinai Peninsula is investigated in this study via comparisons of its predictions with ground stations' measurements. The results revealed that the weather research and forecasting (WRF) model could be used to predict the synoptic situations that cause thunderstorms and heavy rainfall. So it may be concluded that WRF model can be used as a pre-warning system for flash floods forecasting. The predicted total rainfall by WRF model is in similar accordance and in harmony with observed rainfall in all rain gauges, which show very significant consistency between WRF and measurements; the WRF is presumably reliable in short-term forecasting. WRF has success in representing small or mesoscale hazards such as severe thunderstorms, squall line,s and flash floods so it can be used as a good tool in the assessment of the threat and rapid dissemination of warnings to the population and appropriate authorities. Useful forecasts of the behavior of the larger scale weather systems such as tropical storms and cyclones, intense depressions, and large flash floods can be prepared several days in advance using WRF model.
5. Recent construction activities in the area of El Gouna covering the prevention of damages caused by flash floods were undertaken. After the experiences of the 2014 flooding, a protection dam was constructed in the southern part of El Gouna where most of the destruction took place. Flash flood models can be applied to study the effectiveness of structural mitigation measures. They also enhance the understanding of flow processes during and after heavy rainfalls and should be used to support the risk analysis. The represented results of the flash flood model of El Gouna include fields of water depths and flow velocities in the whole simulated domain at different times. The representation of the model results facilitates the understanding of effects from infiltration and retention and their interactions in the overall considered domain. It was shown that the retention basins combined with infiltration processes could completely prevent critical water depths in and around the city area, while the effects of retention and infiltration were significantly less distinct if they were considered separately. The Eastern Desert has even more impacts with flash floods which are analyzed more and more recently as the density of inhabitants rises. The flash flood event of Ras Gareib 2016 shows dramatically the results of unprotected areas and comes more and more into consideration of urban planners and governmental institutions along the Red Sea coast. Prediction of flash floods and their impact, therefore, is a substantial concern of urban planning and can avoid casualties and in the second place material losses.
6. Far from the Nile, especially in the Eastern Desert and Sinai, renewable water resources are limited to the very low rainfall that sometimes causes rapid flooding. If floods are properly managed, they will serve the needs of the fragile communities in the desert. Moreover, social stability and economic development are strongly linked to the sustainable existence of water resources. Sudden flooding in the Sinai in terms of infrastructure (roads, natural gas, oil pipelines, airports, ports, resorts, settlements, etc.), causes severe damage to human life and socio-economic development. The Egyptian experience with the floods has shown that much remains to be done to improve the use of water resources and to protect our infrastructure that crosses dangerous areas. The Ministry of water resources and irrigation (MWRI) has taken some positive measures to manage flood risk. These include: Developing an Atlas of floods throughout Egypt—Early warning system registration as a tool for daily management of potential flooding—Development of engineering code for construction in flood-prone areas, Expanding our water harvesting strategy, Meeting demand from existing communities away from the Nile.
7. In desert lands, flashflood management and groundwater resources assessment are urgently required. In this book chapter, an integrative approach of RS and DC resistivity data was discussed. As case studies, the efficiency of this approach was presented at two wadis from Egypt. This approach illustrated and proved how

efficient was the integration between the RS, GIS, and DC resistivity techniques for scarce water resources management in wadi systems. The proposed approach was useful not only for the mitigation of flashflood hazards but also for replenishing aquifers in an attempt to urban development even in small scales. RS and GIS techniques can be applied to delineate the basin of a high chance for runoff water harvesting and groundwater recharge possibilities.

8. This chapter demonstrates how remote sensing techniques and geographic information system technologies can be used effectively in watershed extraction and flash flood monitoring. Drainage flow directions, (upslope) areas, and catchments were extracted from digital elevation models in flash flood areas. Risk flood areas have been delineated. The methodology was applied in Firan catchment, Sinai, and wadi Hashim2 in the northern coast of Egypt as case studies. In this study, GIS-based SCS-Curve Number method was used to estimate the runoff from the watershed. It may conclude that the land-use planning and watershed management can be done effectively and efficiently using the SCS-CN number method with GIS. The study demonstrates that SCS-Curve Number with GIS is a powerful tool for estimating runoff of ungauged watersheds for better watershed management and conservation purposes. The study reveals that the methodology followed in this study can be applied in all other wadis watersheds, but the runoff/rainfall percentage will depend on its watershed area and its physical characteristics.
9. Halaib-Shalatin area is considered to be a promising region for different developmental activities such as tourism and mining. The present work presented an overview of the optimum use for the natural resources in HS area. Remote sensing, watershed modeling, and GIS techniques are modern research tools that proved to be highly effective in mapping, investigating, and modeling the runoff processes and optimizing the runoff water harvesting (RWH). In the present work, these tools were used to determine the potential sites or areas suitable for the RWH in Halaib-Shalatin area (Southeastern Desert). The runoff water harvesting potential areas were determined by spatially integrating thematic layers, which represent the most important hydrographic and hydro-morphometric parameters or criteria for determining the RWH potentiality. The performed WSPMs segregated these watersheds into five potential classes for the RWH potentiality, which are graded from the very low to very high. WSPM is a very effective method for determining the potential sites for RWH, especially that based on justified weights by sensitivity analysis. A land-use master plan (LMP) was developed by the present work to foster the sustainable development of Halaib-Shalatin area.

10. Sinai is progressively suffering from an overwhelming water crisis. Flash flood and runoff water could be an answer to this issue. The produced risk map is useful to know the locations that have a high flood risk in order to avoid loss of life and reduce damages to property. The main watersheds flowing through Sinai are classified into four categories where 4% of watersheds have very high risk, 10% have high risk, 38% have moderate risk, and 48% have moderate to low risk. The aim of any risk study is to minimize the harm caused by flooding. This involves reducing the likelihood of flooding and reducing the impacts when flooding occurs. At the same time, there are underlying pressures that are increasing risks, such as climate change, housing development, or changes in land use. Flood risk assessment and flood mapping will help to show which places are most at risk and in what circumstances. After that, governments can take the correct strategy for flood risk reduction or mitigation. In addition, they can select suitable locations of weirs and dams for two reasons. The first is to prevent the risk and the second is to collect water in this arid area for Bedouins. The establishment of different dams leads to the presence of communities around these dams to work in agriculture and grazing. In addition to reducing water losses, reducing the speed of floods is the main factor in protecting the soil from water erosion. Floods may not be due to flooding alone, but due to the nature of the randomness of the construction of buildings in the corridors of floods. So, local knowledge of flash floods of Bedouins is needed for the development of the early warning system. Because of the rapidity of flash flood occurrence and its power, flash flood experts recommend the use of early warning systems for reducing vulnerability. Flood risk assessment helps to create flood vulnerable map, and from the historical rainfall data, we can make an early warning system. The early warning system is very important to protect the city by reducing the losses and victims in the region.

11. Sinai is progressively suffering from an overwhelming water crisis. Flash flood and runoff water could be an answer to this issue. The central focus of this chapter rotates around the utilization of the floodwater as a new supply for water harvesting in Sinai. Applying water harvesting of the flash flood will reduce the flood risk at the outlet. In addition, it could be used for recharging the groundwater aquifers, which are the basis for sustainable development in Sinai. Furthermore, Bedouins

usually move from place to place searching for fresh grazing for their animals and water for their families. These locations sometimes are hazardous areas as flash floods occur there. This chapter helps to do developed sustainable planning for the Bedouins, as hazard areas will be defined. Proved locations to get used of the flooded water and store it will encourage the Bedouins to resettle the area. The total amounts of rainfall affecting the different drainage systems in Sinai are of the order of 2000 million m^3/year, and most of the possible runoff water is about 150 million m^3/year. Structural dams and artificial lakes are highlighted in this chapter to utilize the flash flood.

12. Wadi Dahab has very high importance in a new development in southeastern Sinai, for its touristic position and promising water resources. RS, WMS, and GIS techniques are modern research tools that proved to be highly effective in mapping, investigating, and modeling the runoff processes and optimizing the RWH. In the present work, these tools were used to determine the potential sites suitable for the RWH in W. Dahab. The performed WSPM for determining the potentiality areas for RWH depended on the hydro-morphometric parameters of drainage density, infiltration number, maximum flow distance, overland flow distance, basin slope, basin area, the volume of the annual flood, and basin length. The WSPM model was accomplished through three scenarios: equal criteria weights (scenario I), authors' judgment (scenario II), and weights justified by the sensitivity analysis (scenario III). The obtained WSPM maps for defining the RWH potentiality areas classified W. Dahab basin into five RWH potentiality classes ranging from very low to very high. There are good matches between the three performed WSPMs' scenarios in results for the very high and high RWH potentiality classes, which are very suitable for RWH applications. There are good matches between the three performed WSPMs' scenarios in results for the very high and high RWH potentiality classes, which are very suitable for RWH applications.

13. Flood prediction and mitigation are considered an important issue. The chapter presented a number of stages to evaluate flash floods. The first stage covers flood definition, which is a body of water that inundates land that is infrequently submerged that may lead to damage and loss of life. The second stage contains the flood causes, which include varying rainfall distribution; areas which are not traditionally disposed to floods also experience severe inundation. Areas with poor drainage facilities get flooded by the accumulation of water from heavy rainfall. The third stage comprises types of floods such as flash flood, river flood, coastal flood, urban flood, and areal flood. The fourth stage covers damages caused by floods such as injury or death to people and animals and damage to houses and properties. Egypt has alluvial (wadi) systems, formed during fluvial time of the Tertiary and Quaternary periods. These wadis suffer from flash floods, consequent to heavy precipitations. Wadi Sudr, Sinai is selected as a case study to study the prediction and mitigation of floods in this area. The results indicated that the site is exposed to flash flood risk, and protection work is required to protect the area. One drainage channel and seven dams were suggested and designed to mitigate the flood risk. The proposed mitigation system is capable of protecting this area from flood. The dams were used to store water that can help in the development of this area.

14. The discussed green infrastructures include bioswales, retention basins, ponds, wetlands, rain gardens, permeable pavements, and urban green spaces. These stormwater infrastructures tend to control runoff volume and timing and promote ecosystem services. Stormwater management can be applied to a case study of Alexandria Governorate to exemplify the impact of climate change and urbanization on the drainage system performance of a city subjecting to successive periods of heavy rain, flash flood, and strong winds. The combination of traditional drainage systems with green infrastructure could be a viable solution to mitigate the stormwater runoff in Alexandria city, Egypt. In addition, urban trees should have a vital role in stormwater management, as trees can be connected to urban hydrology and increase the infiltration of stormwater.

4 Recommendations

The capacity to adapt to future problems is the main element of water sustainability. We contend that to accomplish this objective, sustainable flash flood systems need integrated flexibility. The editor observed certain aspects that could be explored for further enhancement throughout the course of this book project. Based on the results and conclusions of the contributors, this chapter provides a number of recommendations that provide suggestions for future scientists to exceed this book's scope.

1. Flood risk recommendations and how to address them can be categorized into two parts; the first dealing with floods in their streams, the second deals with the precautions to be taken in establishing urban facilities, agricultural land, and others. Taking into account that most villages in Aswan are actually built adjacent to

hell slopes and without pre-planning. It is important to apply several complementary torrents plan management techniques including (1) establishing an advanced early warning system. (2) Construction of Obstruction dams. (3) Conducting geotechnical studies to select the most suitable places for building houses and avoid building in areas threatened by serious torrents.

2. The inferences made in this chapter represent a starting point in regard to the analysis of extreme rainfall events in Egypt. As the case study is based on rainfall data from only 30 stations, a recommendation for future studies is directly related to the use of more stations to have a better understanding of the statistical characteristics of extreme rainfall events in Egypt. Furthermore, different probability distributions along with different parameter estimation methods can be used to identify the best method that could provide the most accurate extreme rainfall estimation in the country. Also, regional rainfall frequency analysis should be used to estimate extreme rainfall quantiles at ungauged sites in Egypt. Lastly, a comprehensive study of the effect of climate change on rainfall characteristics in Egypt should also be conducted to develop robust methods for estimating extreme rainfalls considering climate change.
3. Related to critical analysis and design aspects of flash floods in arid regions, some of the protection works constructed over the last few decades were not adequately designed/constructed. There is a clear lack of maintenance in the hydraulic and drainage structures, which caused blocking of the drainage area, causing overtopping, flooding, and failure of the structures. These areas may become one direction for future research.
4. Future investigations related to Flash flood investigations in El Gouna can foster the knowledge about the principles of flash floods in Wadi Bili and surrounding catchments. Numerical modeling of surface flow in the whole catchment can help to predict the impact of flash floods but also can calculate the amount of water and its location during the flood. In combination with an underground flow model, a holistic water balance can be designed, and the usage of water at surface and underground can be quantified. Construction of dams for protection and artificial recharge can be built on the basis of the model results. Nevertheless, the quality of a numerical model strongly depends on the amount and quality of data. In the catchment of Wadi Bili, many investigations are ongoing concentrating on modeling surface flow and the quality and origin of groundwater. Most of the research concentrates on the coastal plain of El Gouna. However, the distal parts of the catchment are not very well recognized by measurements yet. Future plans are focusing on basic investigations of the catchment starting with the installation of weather stations and measurements of infiltration rates on the Abu Shar plateau.
5. In future work, the impact of spatially variable infiltration rates based on measurements as well as other scenarios of structural measures in terms of canals, dams, or combinations of different structures can be investigated with the model. Also, scenarios with higher rain intensities and the resulting flooding areas can be studied. In the future, setting up an early warning system would be desirable by coupling weather observations, a hydrological catchment model, and a flash flood model with a sufficiently fast computational effort to get real-time flood predictions providing information to the community in time to take necessary steps before the flood reaches the city. To use flash flood models for more detailed planning of structural protection measures, more accurate data of topography as well as a higher mesh resolution of the model would be necessary. Additionally, the construction of protection and infiltration dams in the Wadi Bili itself seems to be a reasonable action to face both challenges, prevention of destruction as well as enhancement of the water recourses. The porous deposits of the Wadis' soils can catch relevant amounts of water. Those underground storage may help to face the increasing amount of water needed by the city of El Gouna.
6. Conclusions related to environmental Flash Flood management in Egypt are summarized as follows. There is the capacity to manage rapid flooding in Egypt and the Arab region, such as hydrological data analysis; use of hydrological models in the design of the water collection structure; development of flood atlas flood risk maps; evidence for flood management and flood protection; Simple hydrological methods sufficient to manage floods; positive response by officials to warning signals. Egypt flood management is still facing some constraints. These include Data are sporadic and insufficient—Lack of awareness of flood risk—Bedouins and local people are not fully involved in flood mitigation planning—No networks exist with neighboring countries. However, there are major challenges in the region, such as lack of continuous data for both flow and precipitation; lack of real-time data transmission; soil erosion problems in mountain areas; maximization of flood use rather than flood risk management; the absence of a flash flood early warning system; the absence of a disaster risk management plan; there is little interest in the users of the flood; the availability of a model to describe the hydrological conditions of the Arab region; social gaps in terms of minimal awareness of frequent catastrophic events;

communication technology is very unreachable and expensive to upgrade; the absence of an integrated approach to rapid flood management, not only the warning system but the full cycle and continuity; the lack of methodology for risk assessment; the lack of design standards for precipitation network; inter-state technical and administrative conflict, no flood insurance policy, lack of access to data collected by previous studies through international partners; conflict and disputes among institutions responsible for flood management, the absence of a legal framework for preventive construction; the view of rapid flood as a source and blessing rather than risk; a knowledge gap that begins with data, information, and data analysis, human skills; attitude and the value of data., there is a need to train trainers for all aspects of rapid flood aspects. More and better quantitative data are needed on health impacts associated with all flood categories. This data is required for a better understanding of vulnerability. There is a need for government-supported research on the health effects of floods. There is a need to assess the health effects of adaptation options. Case studies should be identified where health impact assessments can be included in flood mitigation strategies.

7. The DC resistivity method can be used to delineate/image the subsurface layer distributions and aquifers extension. Accordingly, the successful applications of the present integrative approach open the way for sustainable development in wadi systems by decision-makers to (i) overcome water scarcity problems and (ii) reduce the flash flood hazards.
8. According to the results of the present study, the following recommendations must be taken into consideration. Installing meteorological stations in all watersheds to have spatial and temporal rainfall data that can be used for hydrological modeling. Carrying out field measuring of runoff volume at many positions and comparing the results with the estimated values (SCS-CN method calibration). The results indicated that accurate watershed analysis can be developed successfully using remotely sensed data and open source available software. This tool saves time, effort, and cost. Finally, validation of the current technologies for watershed modeling and flash food mapping is required through further researches and studies.
9. However, for future work, relatively simple calculation methods for flash flood hazard mapping are necessary. With the development of computational technologies, the hydrodynamic model can be used to model flash floods to provide more detailed information. With the rapid development of LiDAR and sonar measurement technology, it is becoming easier and more practical to obtain high-resolution DEM data, which can be used to develop more accurate flash flood hazard maps especially in an area with limited data. Moreover, except for the danger of the life of the population, floods cause also considerable material and financial losses, which need to be explored. Measures for the reducing losses are undertaken depending on the capital value of the threatened objects and the expected consequences from their damage or destruction. Disasters have a negative impact on economic activity, and the associated economic uncertainties hamper investment in long-term commercial relationships. Conversely, particular types of economic activity and a truncated policy focus can increase a country's economic vulnerability to natural disasters.

10. Because of their features, flash floods are hard to handle through traditional approaches to flood management. The following must be regarded during the development of a management plan for flash floods. Flash flood prediction and warning play a key role in flash flood management. However, providing users with precise and timely forecasting and warning data is still a challenge at times. Appropriate technical methods and suitable legal and institutional frameworks are needed to address these issues. Appropriate spatial planning can assist decrease exposure and decrease flash flood risks.

11. The green infrastructure systems have recently found successful applications for stormwater management and flood control. Based on the chapter outputs, some recommendations that are helpful for the case study of Alexandria can be derived: Flood insurance programs should be developed to reduce the negative impacts of flooding on private and public properties. A framework for implementing the stormwater management approach should be proposed to reduce cost, damages, and human and socio-economic losses. Stormwater mitigation and soil management should be significantly considered in the urban stormwater policy. Urban planners and architects should apply environmental and ecological methods in designing green infrastructures. The integration of engineering, social, and environmental criteria is needed to identify the most appropriate and effective stormwater infrastructure for a particular area. Dynamic computer simulation models should be developed to predict the performance of drainage systems under various extreme meteorological and climatic conditions.

References

Abdelkarim, A., Gaber, A. F. D., Youssef, A. M., & Pradhan, B. (2019). Flood Hazard assessment of the Urban area of Tabuk City, Kingdom of Saudi Arabia by integrating spatial-based hydrologic and hydrodynamic modeling. *Sensors, 19,* 1024.

Abuzied, S. M., & Mansour, B. M. H. (2018). Geospatial hazard modeling for the delineation of flash flood-prone zones in Wadi Dahab basin, Egypt. *Journal of Hydroinformatics, 21,* 180–206.

Al-Manmi, D. A. M., & Rauf, L. F. (2016). Groundwater potential mapping using remote sensing and GIS-based, in Halabja City, Kurdistan, Iraq. *Arabian Journal of Geosciences, 9*(5), 357.

Alves, A., Gersonius, B., Kapelan, Z., Vojinovic, Z., & Sanchez, A. (2019). Assessing the co-benefits of green-blue-grey infrastructure for sustainable urban flood risk management. *Journal of Environmental Management, 239,* 244–254.

Attwa, M., & Günther, T. (2013). Spectral induced polarization measurements for environmental purposes and predicting the hydraulic conductivity in sandy aquifers. *Hydrology and Earth System Sciences (HESS), 17,* 4079–4094.

Azmeri, A., & Isa, A. H. (2018). An analysis of physical vulnerability to flash floods in the small mountainous watershed of Aceh Besar Regency, Aceh province, Indonesia. Jàmbá. *Journal of Disaster Risk Studies*, *10*, a550.

Bruins, J. H., Guedj, H. B., & Svoray, T. (2019). GIS-based hydrological modelling to assess runoff yields in ancient-agricultural terraced wadi fields (central Negev desert). *Journal of Arid Environments, 166,* 91–107.

Dhote, P. R., Thakur, P. K., Aggarwal, S. P., Sharma, V. C., Garg, V., Nikam, B. R., Chouksey, A. (2018). Experimental flood early warning system in parts of Beas Basin using integration of weather forecasting, hydrological and hydrodynamic models. In *The International Archives of the Photogrammetry, Remote Sensing and Spatial Information Sciences, Volume XLII-5, 2018 ISPRS TC V Mid-term Symposium "Geospatial Technology—Pixel to People*", Dehradun, India.

El Bastawesy, M., El Harby, K., & Habeebullah, T. (2012). The hydrology of Wadi Ibrahim Catchment in Makkah City, the Kingdom of Saudi Arabia: The interplay of urban development and flash flood hazards. Life science journal, in an arid environment. *Journal of Hydrology, 292,* 48–58.

Elewa, H. H., Ramadan, E. M., & Nosair, A. M. (2016). Spatial-based hydro-morphometric watershed modeling for the assessment of flooding potentialities. *Journal of Environment and Earth Sciences, 75,* 1–24.

Grey, V., Livesley, S., Fletcher, T., & Szota, C. (2018). Establishing street trees in stormwater control measures can double tree growth when extended waterlogging is avoided. *Landscape and Urban Planning, 178,* 122–129.

Hadidi, A. (2016). *Wadi Bili catchment in the Eastern Desert—flash floods, geological model and hydrogeology*. Dissertation. Berlin: Fakultät VI—Planen Bauen Umwelt der Technischen Universität Berlin.

IPCC. (2012). Managing the risks of extreme events and disasters to advance climate change adaptation. In C. B. Field, V. Barros, T. F. Stocker, D. Qin, D. J. Dokken, K. L. Ebi, M. D. Mastrandrea, K. J. Mach, G.-K. Plattner, S. K. Allen, M. Tignor, & P. M. Midgley (Eds.), *A special report of working groups I and II of the intergovernmental panel on climate change* (p. 582). Cambridge: Cambridge University Press.

Jalilov, S.,M., Kefi,M., Kumar, P., Masago, Y., Mishra, B., K. (2018). "Sustainable Urban Water Management: Application for Integrated Assessment in Southeast Asia,'. Sustainability, MDPI, Open Access Journal, 10(1), 1–22, January.

Jiang, P., & Tung, Y.-K. (2013). Establishing rainfall depth–duration–frequency relationships at daily rain gauge stations in Hong Kong. *Journal of Hydrology, 504,* 80–93.

Moawad, M. (2013). Analysis of the flash flood occurred on 18 January 2010 in wadi Al-Arish, Egypt. *Geomatics, Natural Hazards and Risk Journal, 4,* https://doi.org/10.1080/19475705.2012.731657.

Moawad, M. B., Abdelaziz, A. O., & Mamtimin, B. (2016). Flash floods in the Sahara: A case study for the 28 January 2013 flood in Qena, Egypt. *Geomatics, Natural Hazards and Risk, 7*(1), 215–236. https://doi.org/10.1080/19475705.2014.885467.

Mohammed, S. A. & El Raey, M. (2019). Vulnerability assessment for flash floods using GIS spatial modeling and remotely sensed data in El-Arish City, North Sinai, Egypt. Natural Hazards.

Osti, R. P. (2018). Integrating flood and environmental risk management principles and practices. ADB east Asia working paper series. NO. 15. Rabindra Osti is a senior water resources specialist at the Asian Development Bank.

Paz-Kagan, T., Ohana-Levi, N., Shachak, M., Zaady, E., & Karnieli, A. (2017). Ecosystem effects of integrating human-made runoff-harvesting systems into natural dryland watersheds. *Journal of Arid Environments, 147,* 133–143.

Piro, P., Turco, M., Palermo, S.A., Principato, F., & Brunetti, G. (2019). A comprehensive approach to stormwater management problems in the next generation drainage networks. In F. Cicirelli, A. Guerrieri, C. Mastroianni, G. Spezzano, A. Vinci (Eds.), *The internet of things for smart Urban ecosystems*. (*Technology, communications and computing*) (pp. 275–304), Springer, Cham.

Salama, A. M., Gado, T. A., & Zeidan, B. A. (2018). On selection of probability distributions for annual extreme rainfall series in Egypt. In *Twenty-first International Water Technology Conference, IWTC21* (pp. 383–394). Ismailia.

Shariat, R., Roozbahani, A., & Ebrahimian, A. (2019). Risk analysis of urban stormwater infrastructure systems using fuzzy spatial multi-criteria decision making. *Science of the Total Environment, 647,* 1468–1477.

Sörensen, J., & Emilsson, T. (2019). Evaluating flood risk reduction by urban blue-green infrastructure using insurance data. *Journal of Water Resources Planning and Management, 145*(2), 04018099.

Ţîncu, R., José Luis, Z., & Gabriel, L. (2018). Identification of elements exposed to flood hazard in a section of Trotus River, Romania, Geomatics. *Natural Hazards and Risk, 9,* 950–969.

Vema, V., Sudheer, K. P., & Chaubey, I. (2019). Fuzzy inference system for site suitability evaluation of water harvesting structures in rainfed regions. *Agricultural Water Management, 218,* 82–93.

Vemula, S., Raju, K., Veena, S., & Kumar, A. (2019). Urban floods in Hyderabad, India, under present and future rainfall scenarios: A case study. *Natural Hazards, 95*(3), 637–655.

Wdowikowski, M., Kaźmierczak, B., & Ledvinka, O. (2016). Maximum daily rainfall analysis at selected meteorological stations in the upper Lusatian Neisse River basin. *Meteorology Hydrology and Water Management*, *4*(1).

WRRI. (2010). Flash floods in Egypt protection and management, final report. National Water Research Center. Ministry of Water Resources & Irrigation Egypt LIFE06 TCY/ET/000232.

Yazar, A., & Ali, A. (2016). Water harvesting in dry environments. In *Innovations in dry land agriculture* (pp. 49–98). https://doi.org/10.1007/978-3-319-47928-63.

Yehia, A. G., Fahmy, K. M., Mehany, M. A. S., & Mohamed, G. G. (2017). Impact of extreme climate events on water supply sustainability in Egypt: Case studies in Alexandria region and upper Egypt. *Journal of Water and Climate Change*. https://doi.org/10.2166/wcc.2017.111.

Yousif, M., Abd, E. S. E., & Baraka, A. (2013). Assessment of water resources in some drainage basins, northwestern coast, Egypt. *35TApplied Water Science35T*, *3*, 35TI. 235T, 439–452.